“第十届中国钢铁年会”暨“第六届宝钢学术年会”

The 10th CSM Steel Congress & The 6th Baosteel Biennial Academic Conference

Proceedings of the 10th CSM Steel Congress & the 6th Baosteel Biennial Academic Conference

论文集 摘要

中国金属学会
宝钢集团有限公司 编

Organized by
The Chinese Society For Metals
Baosteel Group Coporation

北京
冶金工业出版社
2015

内 容 简 介

本书共收录997篇论文摘要。全部论文约1200万字，内容包括炼铁与原料、炼钢与连铸、轧制与热处理、表面与涂镀、钢材深加工、钢铁材料、能源与环保、冶金设备与工程、冶金自动化与智能管控、冶金分析、信息情报等方面，全面反映了近两年来我国及世界钢铁行业科研、生产、管理等方面的最新成果，内容全面、新颖，具有较高学术水平。本书可供钢铁行业的科研人员、管理人员、工程技术人员等学习参考。

本论文集摘要以纸质图书出版（不含大会报告），另以电子版方式同时出版所有论文的全文。

图书在版编目(CIP)数据

第十届中国钢铁年会暨第六届宝钢学术年会论文集摘要／中国金属学会，宝钢集团有限公司编. —北京：冶金工业出版社, 2015.10

ISBN 978-7-5024-7006-7

Ⅰ.①第…　Ⅱ.①中…　②宝…　Ⅲ.①钢铁工业—学术会议—文集摘要　Ⅳ.①TF4-53

中国版本图书馆 CIP 数据核字 (2015) 第 226575 号

出 版 人　谭学余

地　　址　北京市东城区嵩祝院北巷39号　邮编　100009　电话　(010)64027926

网　　址　www.cnmip.com.cn　电子信箱　yjcbs@cnmip.com.cn

责任编辑　李培禄 等　美术编辑　彭子赫　版式设计　孙跃红

责任校对　王永欣　责任印制　牛晓波

ISBN 978-7-5024-7006-7

冶金工业出版社出版发行；各地新华书店经销；北京印刷一厂印刷

2015年10月第1版, 2015年10月第1次印刷

210mm×297mm；48印张；1彩页；1592千字；698页

600.00 元(含盘)

冶金工业出版社　投稿电话　(010)64027932　投稿信箱　tougao@cnmip.com.cn

冶金工业出版社营销中心　电话　(010)64044283　传真　(010)64027893

冶金书店　地址　北京市东四西大街46号(100010)　电话　(010)65289081（兼传真）

冶金工业出版社天猫旗舰店　yjgycbs.tmall.com

“第十届中国钢铁年会”暨
“第六届宝钢学术年会”
组　委　会

大　会　主　席　徐匡迪

大会执行主席　干　勇　徐乐江

大会秘书长　王天义

大会副秘书长　张丕军

学术工作委员会主任　赵　沛

学术工作委员会副主任　张丕军（兼）

顾问委员会（按姓氏字母排序）

Edwin BASSON　Harry BHADESHIA　才　让　陈德荣

Alan W. CRAMB　Anthony J. DeArdo　洪及鄙　黄伯云

靳　伟　John J. JONAS　Eiki KASAI　Young-Kook LEE

李静海　李新创　李文秀　林忠钦　刘　玠　逯高清　罗宏杰

John W. MORRIS　Kotobu NAGAI　John ROWE　沈文荣

John G. SPEER　Marcos STUART　Brian G. THOMAS　邹若齐

Fumitaka TSUKIHASHI　王国栋　王锡钦　王一德

翁宇庆　徐滨士　杨　锐　殷瑞钰　余艾冰　于　勇　张寿荣

张晓刚　张欣欣　赵　继　仲增墉　左　良　左铁镛

《第十届中国钢铁年会暨第六届宝钢学术年会论文集》编 委 会

前　言

中国钢铁年会自1997年首次召开以来，每两年召开一次，通过交流钢铁科技最新成果，研讨和展望中国与世界钢铁科技发展前景，受到了越来越多的国内外钢铁同行的关注，已成为国内最重要的综合性冶金学术会议。

宝钢学术年会第一届会议于2004年召开，至今已连续举办了5次年会（包括2006年、2008年、2010年、2013年年会），吸引了全球众多著名钢铁企业、高等院校和科研机构的专家学者参与，作为高层次、跨学科、开放式的国际型大型冶金学术交流会，已负盛名。

今年，为提升行业学术交流水平，实现会企学术交流的强强联合，中国金属学会主办的“第十届中国钢铁年会”与宝钢集团有限公司主办的“第六届宝钢学术年会”于2015年10月21～23日在上海联合举办（以下简称年会），同期还举办“第十八届上海国际冶金工业展览会”。本次年会以“更好的钢铁、更好的生活”为主题，将聚焦产业结构调整和转型、冶金工业技术进步与创新、冶金生态文明等行业热点展开交流研讨。

自2014年初开始筹备以来，本届年会的论文征集工作得到了国内外钢铁企业、科研院所及广大科技人员的广泛响应和积极支持，共收到国内外投稿1275篇，其中来自美国、英国、日本、德国、法国、韩国、澳大利亚、瑞典、比利时、波兰、奥地利、意大利、加拿大、荷兰、西班牙、巴西、印度、中国台湾等20多个国家和地区的非中国大陆投稿有121篇。经年会学术委员会专家评审，录用997篇结集出版，内容包括：炼铁与原料、炼钢与连铸、轧制与热处理、表面与涂镀、钢材深加工、钢铁材料、能源与环保、冶金设备与工程、冶金自动化与智能管控、冶金分析、信息情报等方面。

本书是本次年会论文集的摘要集（不含大会报告），全文以电子版方式出版。

由于编辑、出版时间较紧，疏漏与错误之处，恳请读者批评指正。

中国金属学会
宝钢集团有限公司
2015年10月

Preface

The CSM Steel Congress has been biennially convened since it was first held in 1997. Through the exchange of new scientific results in iron & steel technology and the discussion on the future developing tendency and prospects of steel industry in China and in the world, the CSM Steel Congress has received more attention from the domestic and international steel industry, and has become the most important domestic conference for synthetic metallurgical science.

The First Baosteel Biennial Academic Conference (BAC) was held in 2004, and then convened in 2006, 2008, 2010 and 2013, up to now BAC has been conducted for 5 times. BAC has attracted many famous experts and scholars from noted steel enterprises, universities and research institutes in the world to attend. As the high level and inter-discipline international symposium on metallurgical science, BAC has become famous for years.

To elevate the academic exchanging level and to realize the strength alliance between academy and enterprise, this year the 10th CSM Steel Congress sponsored by Chinese Society for Metals and the 6th Baosteel Biennial Academic Conference sponsored by Baosteel Group are jointly convened on Oct.21-23, 2015 in Shanghai (following abbreviated as the Biennial Meetings). During the same period, the 18th Shanghai International Metallurgical Industry Exhibition is also opened. Taking "Better Steel, Better Life" as the motif, the Biennial Meetings will focus on the hot topics for metallurgical industry, including the adjustment and transformation of enterprises, metallurgical technology progress and innovation, metallurgical ecology.

Since the preparation started in the beginning of 2014, the Biennial Meetings attained the extensive response and active support in the paper collection from the domestic and international steel enterprises, research institutes and scientific workers, and 1275 papers have been received, among them 121 papers are from more than 20 foreign countries and areas outside the mainland China, including the United States, England, Germany, Japan, France, Korea, Australia, Sweden, Belgium, Poland, Austria, Italy, Canada, Netherlands, Spain, Brazil, India and China Taiwan. After the evaluation by the scientific committee, 997 papers are collected in the proceedings for publication, the contents dealing with iron making and raw materials, steel making and continuous casting, steel rolling and heat treatment, metal coating technology, steel products deep-working (further processing), iron & steel materials, energy saving and environment protection, metallurgical equipment and engineering, metallurgy intelligent management and control, metallurgical analysis, information etc.

This proceedings is the abstracts collection for all the published papers (without plenary reports), the full papers are published with the electronic version in a disk.

Due to the tight time-table for compile and publication, some oversights and mistakes could be inevitable, please point out so that they can be corrected.

Chinese Society for Metals
Baosteel Group Corporation
October 2015

目 录

1. 炼铁与原料
Iron Making and Raw Materials

1.1 焦化
Coking

1.2 矿与烧结
Mining and Sintering

1.3 高炉炼铁
Iron Making

1.4 炼铁新技术与资源综合利用

New Technology of Iron Making and Resource Comprehensive Utilization

2. 炼钢与连铸
Steel Making and Continuous Casting

2.1 铁水预处理、转炉电炉
Hot Metal Pretreatment, Converter, Electric Arc Furnace

2.2 炉外精炼及钢水质量控制
Second Refining and Liquid Steel Quality Control

2.3 连铸工艺
Continuous Casting Technology

2.4 品种及铸坯质量控制
Steel Grade Development and Quality Control

2.5 钢铁近终形连铸
Near Net-Shape Continuous Casting

3. 轧制与热处理
Steel Rolling and Heat Treatment

4. 表面与涂镀
Metal Coating Technology

5. 钢材深加工
Further Processing for Rolled Products

6. 钢铁材料
Steel Products

6.1 汽车用钢
Automotive Steel

6.2　电工钢
Electrical Steel

6.3　管线钢和管线管
Pipeline Steel and Linepipe

6.4 不锈钢
Stainless Steel

6.4.1 铁素体及马氏体不锈钢 Ferritic and Martensitic Stainless Steel

6.4.2 奥氏体及双相不锈钢 Austenitic and Duplex Stainless Steel

6.4.3　不锈钢的加工及应用 Processing and Application

6.5 特殊钢 Special Steel

7. 能源与环保
Energy Saving and Environmental Protection

8. 冶金设备与工程
Metallurgical Equipment and Engineering Technology

9. 冶金自动化与智能管控

Metallurgy Intelligent Management and Control

10. 冶金分析
Metallurgical Analysis

11. 信息情报

Information and Intelligence

1 炼铁与原料

Iron Making and Raw Materials

1.1 焦化/Coking

转型升级中炼铁与焦化的协调与发展
——焦炭质量研究与改善焦炭质量的措施

孟庆波

（中钢集团鞍山热能研究院有限公司炼焦技术国家工程研究中心）

摘　要　综述了焦炭在高炉冶炼过程的变化及焦炭质量研究进展。焦炭在高炉中主要发生碳溶反应、石墨化和溶解以及受到碱金属侵蚀、灰中矿物质还原、风口高速鼓风等作用而降解粉化。焦炭在高炉不同区域发生的不同变化与焦炭不同尺度的结构/组成（孔隙结构、显微结构、碳微晶结构、矿物组成及晶相等）相关。提出根据高炉不同区域焦炭劣化机制来研究控制和/或改善焦炭本质质量的措施。建议从煤种选择、煤岩组成、煤中矿物质组成等入手进行精细优化配煤并与煤粉碎、炼焦工艺及制度相配合改善焦炭质量；干熄焦可以提高焦炭质量应进一步推广；捣固炼焦应合理配煤保障焦炭本质质量。钢铁企业应关注独立焦化企业的转型升级，应与焦化企业建立稳定的供应关系，实现二者转型升级的协调和共同发展。

关键词　焦炭质量，优化配煤，反应性，反应后强度，矿物质，石墨化，高炉炼铁

Coordination and Development of the Ironmaking and Cokemaking Industries during Their Transformation and Upgrading ——Coke Quality Research and Measures to Improve Coke Quality

Meng Qingbo

(Sinosteel Anshan Research Institute of Thermo-Energy Co., Ltd. National Engineering Research Center for Cokemaking Technology)

Abstract It is summarized that the previous research work about the deterioration of coke in blast furnace and the progression of coke quality research. Coke encounters carbon loss reaction, graphitization and dissolution, as well as alkali attack, reduction of minerals, high speed blast in BF. All these factors cause coke deterioration and coke fines generation. The degradation of coke in the different zones of the blast furnace related to the coke structure, texture and composition in different size scale, such as the porosity and cracks, micro-texture, carbon crystal, minerals and crystalline phases etc. It is pointed out that the coke intrinsic quality can be controlled and/or improved according to the degradation mechanisms of coke in the different zones of the blast furnace. It is suggested that (a) the selection of coals, the coal petrographic composition, the minerals in coals and so on should be concerned and match the coal crushing, coking process and coking regulations when optimizing coal blending to improve coke quality; (b) CDQ should be further promoted as it can improve the coke quality; (c) rational coal blends should be adopted to guarantee the coke intrinsic quality in stamp charging cokemaking process; (d) Iron and steel enterprises should focus on the transformation and upgrading of independent coking enterprises and establish stable supply relationship with the coking enterprises for the coordination and common development of both during the transformation and upgrading.

Key words coke quality, optimizing coal blending, CRI, CSR, mineral, graphitization, BF operation

内蒙古煤的性质研究及其应用

黄世平 刘 需 李立业 田京雷 刘宏强

（河北钢铁集团钢铁技术研究总院，河北石家庄 052165）

摘 要 对内蒙古煤炭的性质进行全面研究，结果表明内蒙古煤的全硫含量较高，灰分偏高，胶质体塑性温度期间宽，奥亚膨胀度很大，胶质体最大流动度高，适合配煤炼焦的使用；内蒙古煤炭在邯钢生产中的应用表明，用 6%的内蒙古煤替代 6%的 NG 肥煤作为焦炉工业化生产的配煤比，不但保证了焦炭质量的稳定和提高，还能有效降低焦炭成本。

关键词 内蒙古煤炭，配煤炼焦，肥煤，焦炉，焦炭质量

The Properties Study of Inner Mongolia Coal and Its Application

Huang Shiping, Liu Xu, Li Liye, Tian Jinglei, Liu Hongqiang

(Hebei Iron and Steel Technology Research Institute, Hesteel Group, Shijiazhuang 052165, China)

Abstract According to the properties study of Inner Mongolia coal, we got the results that Inner Mongolia coal, with high sulfur content, a little high ash content, big plastic temperature range, high maximal flowing of plastic and high maximal expanding of plastic, is fit for coal blending and coking. We use 6% Inner Mongolia coal instead of 6% NG fat coal in coordinate coal of big coke-oven, and the results show that not only it can keep the quality of coke steady-going, but also the production cost of coking reduces significantly.

Key words Inner Mongolia coal, coal blending and coking, fat coal, coke-oven, the quality of coke

弱黏结煤使用技术研究与应用

彭永根[1] 陈 鹏[2]

（1. 华菱湘钢技术中心，湖南湘潭 411101；2. 武钢研究院，湖北武汉 430080）

摘 要 本文介绍了弱黏结煤种类和质量特点，对其成焦显微结构和结焦过程行为开展了研究，开发了高变质程度弱黏结煤的预破碎控制技术和低变质程度弱黏结煤的粒度控制技术，通过减少低 *G* 值焦煤配比和配用高膨胀度、高流变性煤等优化配煤结构方法，实现了弱黏结煤的生产配用。开发配用多个矿点弱黏结煤，提高了低价弱黏结煤配比，在稳定焦炭质量的基础上降低了配煤成本。

关键词 弱黏结煤，贫瘦煤，气煤，成本，质量

Usage of Weakly Caking Coal

Peng Yonggen[1], Chen Peng[2]

(1. Technology Center of Valin Xiangtan Steel，Xiangtan 411101, China；
2．Research Institute of Wuhan Steel，Wuhan 430080, China)

Abstract This thesis introduced category and quality characteristics of weakly caking coal, studied its coke micro-structure and coking process, developed pre crushing control technology for highly deteriorated weakly caking coal and granularity control technology for slightly deteriorated weakly caking coal. And by reducing proportion of low *G* value coking coal and increasing the use of high quality highly expanded, rheological coal, the application of weakly caking coal in production was achieved. With these efforts, the use of weakly caking coal from various mines has been increased on the basis of stable coking coal quality, which decreased the cost of blending.

Key words weakly caking coal, meager lean coal, gas coal, cost, quality

“焦–化”一体配煤炼焦技术

杜 屏[1] 吕青青[1] 白新革[2] 周俊兰[1] 钱如刚[2]

（1. 江苏省（沙钢）钢铁研究院炼铁与环境研究室，江苏张家港 215625；
2. 江苏沙钢集团有限公司宏发焦化厂，江苏张家港 215625）

摘 要 研究了煤岩指标、流动度指标、焦炉操作条件等因素影响焦炭质量的规律，开发了综合考虑煤质条件与工艺条件的焦炭质量控制方法；研究了焦化副产品收率规律，建立能量预测的方法，实现对化学副产品收率的控制；开发了既考虑焦炭质量，又能兼顾化学副产品收益的焦炭成本控制方法。

关键词 焦炭质量，化产预测，焦炭质量预测，成本控制

Considering the Quality of Coke and Chemical Products Yield Coking Coal Blending Technology

Du Ping[1], Lv Qingqing[1], Bai Xinge[2], Zhou Junlan[1], Qian Rugang[2]

(1. Iron Making and Environment Research Group of Iron and Steel Research Institute of Shasteel, Zhangjiagang Jiangsu 215625, China; 2. Hongfa Coking Plant of Jiangsu Shagang Group Co., Ltd., Zhangjiagang Jiangsu 215625, China)

Abstract The factors affecting the quality of the coke, such as coal and rock, the plastic index, coke oven operation, and so on, are studied. The coke quality control method has been developed for the comprehensive consideration of the quality of coke and process conditions. The yield rule of coking chemical by-product was studied, the energy prediction method was established, and the yield of chemical by-products was controlled. The cost control method of the coke is developed, which considers the coke quality and the yield of chemical byproducts.

Key words coke quality, chemical product forecast, coke quality prediction, cost control

焦炭原始灰成分对焦炭热性能影响规律的研究

余国普 曹银平 程乐意 许永跃 姚国友

（宝山钢铁股份有限公司炼铁厂，上海 200941）

摘 要 根据炼焦煤种类的不同，对比了常规灼烧法制灰和CO_2还原法两者的试验结果，研究了不同制灰方法对焦炭热性质的影响。结果表明，二氧化碳还原法制灰过程气体流通性更好，反应气体分散更为均匀，更有利于避免灰化过程中的干扰影响，灰化温度更高，所得灰分结果更佳。对焦炭反应性的预测值与实测值的对比表明CO_2还原法预测较好。

关键词 焦炭，制灰，灰成分，反应性

Study on the Influence of the Raw Ash Composition on the Thermal Properties of Coke

Yu Guopu, Cao Yinping, Cheng Leyi, Xu Yongyue, Yao Guoyou

(Iron-making Plant, Baoshan Iron & Steel Co., Ltd., Shanghai 200941, China)

Abstract According to the different types of coking coal,compared to the results of the conventional burn method grey and CO_2 reduction method,and then effects of different grey method on coke thermal properties was studied.The results show that the carbon dioxide reduction method of the grey process gas flow better, reaction gas disperse more evenly, more conducive to avoid interference effects of ashing processes, higher ashing temperature, income ash results better.The comparison between the predicted value and the measured value of the coke reactivity shows that the CO_2 reduction method is better.

焦炭灰分控制技术应用研究

张启锋[1] 曹银平[2] 刘 全[1] 程乐意[2]

（1. 宝山钢铁股份有限公司制造部，上海 200941；
2. 宝山钢铁股份有限公司炼铁厂，上海 200941）

摘 要 评价焦炭质量有冷热强度、灰、硫及粒度共8项指标，近年来国内炼焦煤煤质劣化，特别是灰分的上升，给焦炭质量的控制带来了严重影响。焦炭灰分的逐步上升，带来高炉燃料使用的增加，同时对焦炭冷强度有影响。本文通过对焦炭灰分上升原因的分析，将石油焦、长焰煤、无烟煤用于炼焦配煤进行研究，在稳定焦炭冷强度的基础上，找到了有效控制焦炭灰分的方法，为高炉提高喷煤比、降低燃料比创造了有利条件。

关键词 石油焦, 长焰煤, 无烟煤, 炼焦, 配煤, 焦炭

Study on Controlling Coke Ash

Zhang Qifeng[1], Cao Yinping[2], Liu Quan[1], Chen Leyi[2]

(1. Manufacturing Management Department, Baoshan Iron & Steel Co., Ltd., Shanghai 200941, China;
2. Iron-making Department, Baoshan Iron & Steel Co., Ltd., Shanghai 200941, China)

Abstract There are eight coal property indexes to evaluate the coal quality.The degradation of the quality of coking coal,especially the rising of the ash content, leads to the rising of the ash content of the coke.Accordingly ash content of coke effects the coke cold strength and the usage of blast furnace fuel.In this paper,the causes of the rising of the coking ash content are analysed and the coal blending with petroleum coke ,meta-lignitous coal and anthracite is studied. A method to decrease coke ash content and stabilize the coke cold strengthen has been obtained,which has provided the favorable conditions for the high coal injection ratio and the lower fuel ratio.

Key words petroleum coke, meta-lignitous coal, anthracite, coal blending, coking, coke

鞍钢鲅鱼圈焦化稳定焦炉集气管压力的新方法

赵恒波 武 斌

（鞍钢股份有限公司鲅鱼圈钢铁分公司炼焦部，辽宁营口 115007）

摘 要 焦炉集气管压力稳定与否，直接关系到焦炉炉体和护炉设备的寿命，集气管压力波动过大还会造成荒煤气放散污染环境，且造成煤气资源和副产品的损失。我们通过对稳定焦炉集气管压力的研究，经过技术攻关和实践应用，确定了新的调节方法，取得了良好的运行效果。稳定焦炉集气管压力依靠两级调节来实现，首先初冷器前吸力要保持稳定，属于粗调，一般波动范围在几百帕以内，在此基础上再由集气管压力调节机进行微调，保持集气管压力稳定在设定值附近，一般要求波动范围在几十帕以内。初冷器前吸力稳定是基础条件，如果初冷器前吸力波动大，则集气管调节机是无法保持集气管压力稳定的。2014 年 7 月 16 日开始，初冷器前吸力调节方式改为翻板调节，配合新的操作方法，取得良好的效果，初步解决了焦炉集气管压力波动过大这一行业难题。

关键词 鲅鱼圈，焦炉，集气管，压力

A New Adjustment Method of the Collector Pressure of Coke Oven in Bayuquan Ansteel

Zhao Hengbo, Wu Bin

(Ansteel Bayuquan Iron & Steel Subsidiary Department of Coking,
Yingkou 115007, China)

Abstract Whether the collector pressure of coke oven is stable or notdirectly relates to the life of the furnaces and coke furnace protection equipments. Too largecollector pressure fluctuation cancausethe emission of the unterted coke oven gas, which will pollute the environment and result in the loss of gas resources and by-products. Throughresearching the stability of coke oven collector pressure,technical research and practical applications, we determined the new adjustment method, and achieved good operating results. The stable pressure of collecting pope is achieved by two adjustments, firstly, the suction should be kept stable before the early cold, which belongs to coarse adjustment and usually fluctuates within the range of a few hundred Pa, and on this basis, it's adjusted by the collector pressure adjustment machine a little to keep the collecting pope pressurestable at around the set value and fluctuate within the range of tens of Pa generally. Keeping the suction stable is the basis condition before the early cold and the collecting pipe conditioner is unable to maintain a stable level if the suction fluctuates seriously. The suction adjustment mode was changed as flap type in July 16, 2014 and achieved good results with the help of new methods and solved the large problem in this industry of the serious fluctuation of the oven's collector pressure.

Key words　Bayuquan, coke oven, collector, pressure

New Energy Recovery Pilot Coke Oven Concept and Design

Al-dojayli Maher[1], Wittich Kirby[2], Quancl John[3], Kruse Richard[4], Brown Rick[1], Cameron Ian[1]

(1. Hatch Ltd., Canada; 2. CanmetENERGY/Natural Resources Canada, Canada; 3. SunCoke Energy Inc., USA; 4. ArcelorMittal, USA)

Abstract　Interest in Heat Recovery Coke Oven (HRCO) usage and processes has increased due to their improved environmental performance and ability to use a wide range of metallurgical and non-coking coals.　To better understand the HRCO process, the Canadian Carbonization Research Association (CCRA) embarked on a program to construct a pilot-scale HRCO. Hatch completed the design of this unique pilot oven needed to reproduce the industrial scale HRCO heating profile and coal charge, and to minimize coal heating edge effect. Different aspects of the pilot coke oven design concept are presented, including oven geometry and wall insulation design, heating strategy using electric heaters, structural buckstay and oven door designs. Several 2D and 3D thermal transient finite element analyses were carried out to calculate the temperature profile of the coal/coke bed and oven walls during coking, capturing the thermal nonlinear behavior of the coal and oven wall bricks and to size the coal bed insulation for minimum edge heating effect. The structural stability of the oven arches was achieved using a spring loaded buckstay for lateral support and expansion allowance. The pilot oven, referred to as the Energy Recovery Coke Oven (ERCO), is being constructed at the CanmetENERGY Laboratory in Ottawa, Canada.

Key words　energy recovery coke oven, cokemaking, pilot scale design, buckstay

炼焦节能新技术在宝钢的应用

程乐意　曹银平　许永跃　余国普　姚国友

（宝钢股份公司炼铁厂，上海　201900）

摘　要　炼焦作为高能耗、环保问题突出的产业，具有巨大的节能减排潜力。结合宝钢炼焦现状，介绍了宝钢焦炉在提高传热效率、余热回收以及减少焦炉炉体散热等方面的节能技术。

关键词　炼焦，焦炉，节能

High-Temperature Characteristics of Coke in Blast Furnaces

GUPTA Sushil, KOSHY Pramod, SORRELL Charles Christopher

(School of Materials Science and Engineering, UNSW Australia, Sydney, NSW, 2052, Australia)

Abstract High-temperature reactivity and strength of cokes are critical considerations for the stable operation of high-productivity blast furnace (BFs). High coke strength after reaction (CSR) generally is desired owing to its having an empirical association with high process efficiency. However, coke is subjected to reaction conditions as it descends towards the tuyere zones, with these conditions' being significantly harsher than those used in the CSR test. The present work discusses the critical high-temperature coke attributes determined by examination of tuyere-level cokes from a European blast furnace. The graphitisation behaviour, phase mineralogy, and gasification reactivity of the cokes were analysed using X-ray diffraction (XRD) and reactivity testing in a fixed bed reactor. In tuyere-level cokes, the amounts of quartz and mullite were lowered significantly while silicon carbide and gupeiite phases were present in the raceway regions. The apparent reaction rates of the tuyere-level cokes were ten times greater than those observed for the feed cokes; this effect is attributed to the presence of adsorbed recirculating potassium in the BF. The present work highlights the strong influences of the raceway temperature, alkali loading, and presence of hot metal on the modification of the mineralogy and reactivity of tuyere-level cokes. The results have implications to the reliability of bench-scale test criteria, which are not representative of actual BF conditions.

Key words blast furnace, coke, graphitisation, gasification reactivity, tuyere-drilling, X-ray diffraction

不同单种煤焦炭光学显微结构的反应性研究

孙维周 胡德生

（宝山钢铁股份有限公司研究院，上海 201900）

摘 要 根据单种煤变质程度，焦炭灰成分催化指数模型（MBI,MMCI）选取了 9 种单种煤焦炭，缩取焦炭破碎制成 0.2mm 以下的焦粉，将每种焦分别在同步热分析仪中进行程序升温的高温反应试验并对结果进行分析，再对 9 种单种煤焦炭的光学显微结构进行定性分析。试验结果对比发现，各向同性与细粒镶嵌结构的碳溶反应性最高，中粒与粗粒镶嵌结构次之，其他各向异性组织的碳溶反应性差别不大，纤维状结构多的焦炭碳溶反应性相对最低。

关键词 焦炭显微光学结构，反应性，单种煤焦炭

The Reaction Test for Coke's Optical Microstructure by Different Individual Coal

Sun Weizhou, Hu Desheng

(Research Institute, Baoshan Iron & Steel Co., Ltd., Shanghai 201900,China)

Abstract According to the individual coal's rank and coke's ash catalytic index model(MBI,MMCI),select 9 kinds of individual coke and take broken powder below 0.2mm,which bring to STA(Simultaneous thermal analyzer)for high

temperature reaction test. The result found that as the coal rank rising the coke reactivity has a obvious decrease trend,which mainly relate to optical microstructure of coke. The calculation of reaction activation energy indicate that show a "reverse V" with the rank rising,and Rr =1.35% reach the highest value,but the difference is little overall.
Key words　optical microstructure of coke, coke's reactivity, individual coal's coke

对焦炭高温冶金性能及检测方法的思考

郝月君　郝素菊　蒋武锋　张玉柱

（华北理工大学冶金与能源学院，河北唐山　063009）

摘　要　随着高炉的大型化，对焦炭的质量要求越来越高，优质炼焦资源的短缺直接影响到我国钢铁业的可持续发展。对焦炭的高温冶金性能进行了重新认识，分析了高炉内有害元素对焦炭高温冶金性能的影响，对焦炭的反应性及反应后强度的检测方法进行再思考。建议焦炭高温冶金性能的检测应具有针对性，模拟高炉内焦炭粒度降级最为严重区域的温度和有害元素的量，计算含铁炉料的还原度与CO_2气体含量的关系，针对不同的生产条件，设计不同的焦炭高温冶金性能的检测方法。这将对高炉现场生产更加具有指导意义，将更有效合理地利用焦炭资源。
关键词　焦炭，反应性，反应后强度，检测方法，高炉

Thinking on the High Temperature Metallurgical Properties and Testing Method of Coke

Hao Yuejun, Hao Suju, Jiang Wufeng, Zhang Yuzhu

(College of Metallurgy & Energy, North China University of Science and Technology, Tangshan 063009, China)

Abstract　With the large volume of blast furnace, the higher quality of coke is required, the shortage of high quality coking resources will affect directly on the sustainable development of the iron and steel enterprise. The high temperature metallurgical properties of coke were re-recognition, the influence of the harmful elements in the blast furnace on the high temperature metallurgical properties of coke was analyzed, with a further consideration on the reactivity and post-reaction strength of coke. The testing of high temperature metallurgical properties of coke should be targeted, The amount of harmful elements and temperature in the most serious area of the coke particle degradation in the blast furnace should be simulated, The relationship between the reduction degree of blast furnace burden and CO_2 gas content should be calculated, we should design the different methods for high temperature metallurgical properties of coke with different production conditions. This will have a guiding significance to the blast furnace production, and to utilization of coke resources is more effective and reasonable.
Key words　coke, reactivity, post-reaction strength, testing method, blast furnace

捣固焦的高温性能及抗碱性研究

王　炜[1]　林文康[2]　张志强[2]　王　杰[1]　朱航宇[1]　章东海[1]

（1. 武汉科技大学钢铁冶金及资源利用省部共建教育部重点实验室，湖北武汉　430081；

2. 攀钢集团西昌钢钒有限公司，四川西昌　615000）

摘　要　为了研究捣固焦和顶装焦的冶金性能，本文实验对比研究了捣固焦和顶装焦的热反应性和反应后强度，并且通过“浸泡法”研究了富碱条件对捣固焦和顶装焦反应性和反应后强度的影响，并结合碳的失重曲线分析了捣固焦和顶装焦在高温下的规律。研究结果表明，碱金属对焦炭与CO_2反应具有强烈的催化作用，它使焦炭反应性增加，反应后强度降低，焦炭质量劣化；并且随着富碱量的增多，焦炭的强度先是缓慢降低，而后急剧降低。工业分析数据相近的捣固焦和顶装焦在相同实验条件下，顶装焦比捣固焦的反应性高，反应后强度低，碳损反应速率大，即捣固焦的高温性能及抗碱性优于顶装焦。

关键词　捣固焦，抗碱性，碳损反应，焦炭反应性

Anti-alkaline and Thermal Reactivity of Stamp Charged Coke

Wang Wei[1], Lin Wenkang[2], Zhang Zhiqiang[2], Wang Jie[1],
Zhu Hangyu[1], Zhang Donghai[1]

(1. Key Laboratory for Ferrous Metallurgy and Resources Utilization of Ministry of Education, Wuhan 430081, China;
2. Pangang Group Xichang Iron & Steel Co., Ltd., Xichang 615000, China)

Abstract　To study the metallurgical properties of stamp charged coke and top charged coke, the thermal reactivity and reaction strength were studied by experiment. And the effect of alkali-rich condition on coke reactivity and reaction strength were studied by “immersion”. Combined with carbon weight loss curve analysis of stamp charged coke and top charged coke at high temperature. The results show that the alkali metal has a strong catalytic effect on reaction of coke with CO_2, which increases the reactivity of coke, decreases strength, deteriorates the quality of coke. The increase of the amount of alkali, coke strength decreased slowly at first, then drastically reduced. And under the same experimental conditions, the reactivity of top charged coke is higher than stamp charged coke, the strength is lower, the rate of carbon loss reaction is higher, the anti-alkaline of stamp charged coke is better than top charged coke.

Key words　stamp charged coke, anti-alkaline, carbon loss reaction, coke reactivity

冶金焦炭热态性能试验研究

张文成[1,2]　郑明东[1]

（1. 安徽工业大学化工学院，安徽马鞍山　243002；
2. 上海梅山钢铁股份有限公司技术中心，江苏南京　210039）

摘　要　通过对国内十几家冶金焦炭的试验研究，探讨焦炭热性能的主要影响因素，并通过等反应性强度试验探讨了焦炭热态性能新评价方法。试验结果显示，从炼焦工艺上，顶装焦与捣固焦之间焦炭热态性能有差异，顶装焦炭CSR比捣固焦炭高5%~10%；从熄焦方式上，干熄焦比湿熄焦焦炭CSR高2%~3%。研究表明影响焦炭CSR的关键因素是焦炭的内在质量特性，焦炭催化指数MCI低则焦炭CSR高，焦炭各向异性指数OTI高则焦炭CSR高。研究还表明焦炭等反应后强度TCSR与焦炭CSR有联系也有差别，TCSR主要受到反应温度、反应速率的影响，TCSR及焦炭粉化率为焦炭热态性能评价提供了新的思路。

关键词　焦炭，反应后强度，强度

Experimental Study on Thermal Performance of Metallurgical Coke

Zhang Wencheng[1,2], Zheng Mingdong[1]

(1. School of Chemistry and Chemical Engineering, Anhui University of Technology, Maanshan 243002, China;
2. Meishan Iron and Steel Corporation Technology Center, Nanjing 210039, China)

Abstract Through the experimental study of domestic dozen metallurgical coke, the main influence factors of the thermal performance of coke were discussed, and the new evaluation method for the thermal performance of coke was discussed by the test of the strength of the coke. The test show that the index CSR of top loding coke is higher than that of tamping coke. The index CSR of top loading coke is 5%~10% higher than tamping coke, the Coke dry quenching coke is 2%~3% higher than wet quenching coke. The research also shows that the key factor effecting coke is the immanent properties of coke, namely the index MCI and the index OTI. The lower MCI and the higher OTI is contribute to improve the CSR. Studies also shows that coke strength after fixed reaction 30% (TCSR) and CSR have the difference meaning. TCSR is mainly affected by the reaction temperature, reaction rate and pulverization rate. The index TCSR and pulverization rate provides a new idea for the coke thermal performance evaluation.

Key words coke, CSR, strength

Effect of the Addition of Steel Slag on the Properties of Coke: Preparation of High Strength and Highly Reactive Coke

Sun Zhang, Li Peng, Guo Rui, Liu Pengfei, Liang Yinghua

(College of Chemical Engineering, North China University of Science and Technology, Tangshan 063009, China)

Abstract The steel slag (0.1~0.2mm and <0.1mm) was added to a coking coal (Liuling coal) with different blending ratios and cokes were then prepared from the blended coal samples. The CO_2-gasification reactivity of the resulting cokes (*PRI*) increases with the increase of blending ratio of the steel slag. And the smaller the size of the steel slag, the higher the improvement of the reactivity of cokes. The specific surface area of the cokes after reaction first increases then decreases at *PRI* > 25%, which shows that the much more carbon loss ratio enhanced the degradation trend ($\Delta PSR/\Delta PRI$) of coke. In addition, the highly reactive cokes were prepared from coal blends by addition of a steel slag with 0.5 wt.% and 1 wt.% blending ratios. The mechanical strength of these highly reactive cokes is a little different from the one of the base coke (BC). The addition of steel slag increases the CO_2-gasification reactivity of these cokes, and lowers the reaction starting temperature of the cokes. However, the difference of the reaction rate (CRR_{25}) and strength after reaction (CSR_{25}) between these cokes is small. The average pore diameter of these cokes after reaction is similar to the one of BC. In conclusion, the steel slag is a suitable additive for preparation of the highly reactive cokes.

Key words coke, steel slag, highly reactive, reaction starting temperature, CSR_{25}

焦化厂小品种物料无尘装车装置的应用

甘秀石 王 健 梁 波

（鞍钢股份公司炼焦总厂，辽宁鞍山 114021）

摘 要 通过伸缩套筒、通风夹套、槽边吸尘罩等设备，有效处理小品种物料装车过程中，由于粉尘真密度小、落差大、吸尘口距产尘点远等原因导致的二次扬尘十分严重的问题。

关键词 小品种物料装车，伸缩套筒，通风夹套，槽边吸尘罩，真密度，二次扬尘

Application of No-Dust Carloader about the Little Variety Material in the Coke-Making Plants

Gan Xiushi, Wang Jian, Liang Bo

(General Coke-Making Plant of Angang Steel Co., Ltd., Anshan Liaoning 114021, China)

Abstract By the telescopic sleeve, the ventilate flue in the telescopic sleeve, the dust hood at the groove edge, the secondary dust can be effectively caught during the course of loading the little variety material in the chemical plants. The reason of the generating secondary dust is the dust-true-density is little, the falling-distance of the material is long, and the dust-collection orifice is far away from the source of the generating dust.

Key words the loading the little variety material, the telescopic sleeve, the ventilate flue in the telescopic sleeve, the dust hood at the groove edge, the dust-true-density, the secondary dust

焦化废水剩余污泥的特性及处置研究

刘峰材 谭 啸 龙朝阳 陈艳伟 高 薇

（鞍钢股份有限公司炼焦总厂，辽宁鞍山 114000）

摘 要 焦化废水剩余污泥，是生物脱氮处理工艺中产生的过盛活性污泥。由于其属于危险废物，一直以来国内焦化厂大多配入洗精煤中回焦炉处置。在当前环保、成本压力越来越大的形势下，如何更加合理、环保、低成本的处置这一危险废物值得研究。本文详细分析了焦化废水剩余污泥的组分特性及回配煤处置对焦炭质量、焦炉炉窑及环境保护等方面的影响，寻找处理焦化污泥的最佳方式，为企业绿色发展、降本增效提供支撑。

关键词 焦化废水剩余污泥，组分特性，合理处置，绿色发展

Abstract Coking wastewater sludge，Is the biological denitrification process of excess activated sludge.Due to its belong to hazardous waste, most of domestic coking plant have been into the washed coal to coke oven disposal. In the current under the situation of environmental protection, increasing cost pressure, how to more reasonable, environmental protection, low cost of the disposal of the hazardous waste is worth studying. This paper analyzes the composition characteristics of coking wastewater sludge and coal blending back disposal quality of coke, coke furnaces and environment protection, etc, To find the best way to deal with coking sewage sludge, Green development, authors efficiency to provide support for the enterprise.

Key words coking wastewater sludge, component characteristics, reasonable disposal, green development

1.2　矿与烧结/Mining and Sintering

新疆帕尔岗铁矿尾矿铁资源回收利用研究

王振刚　刘　明

(宝钢集团八钢公司矿山管理(事业)部，新疆　830022)

摘　要　当前，八钢公司自产铁原料严重紧缺，自有矿山生产的铁原料只能满足需求的35%。依靠技术进步，提高八钢公司自有铁矿山资源利用效率，提升选矿厂技术经济指标已迫在眉睫。本文通过对新建帕尔岗铁矿选矿厂综合尾矿进行工艺矿物学分析及铁元素回收试验研究，探索在现有选矿技术条件下，进一步提高帕尔岗铁矿资源回收利用效率的有效方式。

关键词　帕尔岗，尾矿铁资源，回收利用

Carry on a Recall Research to the Resources in the Tail Mineral in the Paergang Iron Mine in Xinjiang

Wang Zhengang, Liu Ming

(Mine Management Department, Bayi Iron & Steel Go, Baosteel Group, Xinjiang 830022, China)

Abstract　Eight steel companies self product iron raw material to seriously and tightly lack, from have iron raw material that the mineral mountain produces 35% that can meet the demands.Depend on a technique progress, raise eight steel companies from have the iron mine mountain the using of resource efficiency, promoting to choose mineral factory a technique the economic index sign already of the utmost urgency.This text passes to lately set up the Paergang iron mine to choose a comprehensive tail mineral of mineral factory to carry on craft mineralogy analysis and iron chemical element recall to experiment a research, the quest raises the effective way that the Paergang iron mine resources recall makes use of an efficiency further under the sistuation that is existing to choose a mineral technique.

Key words　Paergang, resources in the tail mineral, carry on a recall research

新疆某粗精矿制取超级铁精矿的试验研究

任新年　李付龙　关　翔

（新疆钢铁雅满苏矿业有限责任公司，　新疆哈密　839000）

摘　要　针对新疆某铁矿石嵌布粒度细的特点，在生产普通铁精矿的基础上，进行了制取超级铁精矿的试验研究。当再磨细度–320 目 94%的极细条件下，采用磁选、反浮选工艺，可使二氧化硅含量降低到 0.324%，超级铁精矿品

位达到 71.92%，回收率接近 70%。试验取得理想的技术指标，使该铁矿资源得到合理、更有效的利用。

关键词 超级铁精矿，磁选，反浮选，联合流程

Experimental Study on Producing Super Concentrate Iron with Iron Ore in Xinjiang

Ren Xinnian, Li Fulong, Guan Xiang

(Xinjiang Steel Ya Su Mining Co., Ltd., Hami 839000, China)

Abstract In view of the characteristics of the fine granularity of iron ore in Xinjiang, the experimental study on the iron concentrate was carried out on the basis of the ordinary iron concentrate. When the fine conditions of the fineness 320 mesh 94%, the magnetic separation, reverse flotation process, can make the silica content decreased to 0.324%, the super iron concentrate grade reached 71.92%, the recovery rate is close to 70%. The ideal technical index of the test is obtained, which makes the iron ore resources reasonably and more effective.

Key words super iron concentrate, magnetic separation, reverse flotation, combined process

雅矿尾砂可选性实验研究

任新年 王振刚 李付龙

（新疆钢铁雅满苏矿业有限责任公司， 新疆哈密 839000）

摘 要 通过对雅矿尾砂矿石性质的物相分析，确定了一磁—磨矿—分级—二磁的选矿工艺，通过实验室流程试验，确定了流程工艺参数。当磨矿粒度为–0.074mm 占 86%，磁场强度为 105kA/m 时，通过单一磁选实验室流程试验取得了最终品位 51.84%，回收率 62.93%，产率 62.93%的良好指标。

关键词 物相分析，工艺参数，产率，品位，回收率

Test Research on Xinjiang Desulfurization of Iron Concentrate with High Sulfur Content by Flotation

Ren Xinnian, Wang Zhengang, Li Fulong

(Xinjiang Steel Ya Su Mining Co., Ltd., Hami 839000, China)

Abstract Through the JAS mine tailings ore properties phase analysis, determine the magnetic - grinding ore - grade - two magnetic dressing process, the process experiments in laboratory to determine the process parameters. When grinding the ore granularity is –0.074mm 86%, the magnetic field strength is 105kA Podesta M-1, through a single magnetic separation laboratory flow tests made the final grade 51.84%, recovery rate of 62.93% ,yield 62.93% good indicators.

Key words phase analysis, technological parameters, yield, grade, rate of recovery

新疆某高硫磁铁精矿脱硫试验研究

李付龙　关　翔　李　田

（新疆钢铁雅满苏矿业有限责任公司，新疆哈密　839000）

摘　要　由于磁黄铁矿易氧化、被抑制，而且不同地区磁黄铁矿性质差异较大，因此磁黄铁矿的可浮性较差，使得磁铁矿和磁黄铁矿的分离成为一个世界难题[1~3]。文中介绍了磁铁矿精矿的性质，通过运用以稀硫酸为 pH 介质调整剂，硫酸铜+硫化钠为活化剂以及复合捕收剂丁黄药+Y89，采用浮选方法成功将铁精矿中铁品位提高至 68.72%，硫含量降至 0.055%，为矿山的提铁降硫提供了新的途径。

关键词　磁铁矿，磁黄铁矿，调整剂，活化剂，组合捕收剂

Test Research on Xinjiang Desulfurization of Iron Concentrate with High Sulfur Content by Flotation

Li Fulong, Guan Xiang, Li Tian

(Xinjiang Steel Ya Su Mining Co., Ltd., Hami 839000, China)

Abstract　Due to the magnetic pyrite oxidation, inhibition, and in different parts of the properties of pyrrhotite differences larger, and therefore of pyrrhotite floatability is poor, making the separation of magnetite and pyrrhotite has become a difficult problem in the world. The ore properties are described, by using in dilute sulfuric acid medium adjustment for pH, copper sulfate + sodium sulfide as activating agent and compound collector butyl xanthate Y89, the iron grade of the magnetite can be increased to 67.72%, and the sulfur content can be decreased to 0.055%, which provides a new effective way for raising the iron and reducing the sulfur.

Key words　magnetite, pyrrohotite, adjustment agent, activator, compound collector

雅满苏铁矿选矿厂节能降耗途径浅析

纪永东　任新年　关　翔　王振刚

（新疆钢铁雅满苏矿业有限责任公司，新疆哈密　839000）

摘　要　“节能降耗”是现代化管理的大型工厂企业降低成本、促进本企业发展的根本措施。也是当前经济模式下企业提高竞争力的有效手段。节能降耗，简言之就是节约能源、降低消耗，用最少的投入去获取最大的经济效益。多年来雅满苏铁矿选矿厂为谋求发展，一直注重节能降耗工作，从自身实际出发。只有通过对选矿厂节能降耗的总结与分析，从技术改造、引进消化节能新设备、新工艺和强化现场管理入手，加大了改造降耗和管理降耗，从而使设备的经济运行水平得到了进一步提高，才能获得经济效益最大化。

关键词　选矿厂，节能降耗，技术改造，新工艺新设备

Jamin Cycad Ore Dressing Plant Energy Saving Way

Ji Yongdong, Ren Xinnian, Guan Xiang, Wang Zhengang

(Xinjiang Steel Jamin Sue Mining Co., Ltd., Hami 839000, China)

Abstract "Energy saving" is the modern management of large-scale factory enterprises to reduce costs, the fundamental measures to promote the development of the enterprise. Is the current mode is the effective means to enhance competitiveness of enterprises. Saving energy and reducing consumption, in short is to save energy, reduce consumption, with the least amount of inputs to obtain the biggest economic benefits. Years jamin cycad ore concentrator to seek development, always pay attention to energy saving work, from their own actual conditions. Only through the basis of the summary and analysis of saving energy and reducing consumption, from the technical innovation, the introduction of digestion and energy saving new equipment, new technology, and to strengthen field management, increased the transformation and reducing consumption, consumption and management so that the economic operation level of the equipment has been further improved, in order to get maximum economic benefits.

Key words New technology of energy saving technical reformation of concentrator new equipment

大型带式焙烧机球团技术创新与应用

张福明[1] 王渠生[1] 韩志国[1] 黄文斌[2]

(1.北京首钢国际工程技术有限公司，北京 100043；
2.首钢京唐钢铁联合有限责任公司，河北唐山 063200)

摘 要 本文介绍了首钢京唐 504m^2 带式焙烧机球团工艺和技术装备。为提高球团矿质量和性能，满足 2 座 5500m^3 巨型高炉的高效低耗生产的需求，采用合理的炉料结构，配置了年产 400 万 t/a 带式焙烧机球团生产线。该项目采用了一系列先进技术与创新，如球团工程的流程设计、工艺布局、设备开发、节能减排、自动化控制系统等。开发与应用了一系列先进技术和设备，如大型干燥窑、直径为 7.5m 的造球盘、梭式布料器、大型带式焙烧机、高效燃烧器等。工程投产后，建立了基于配料研究、造球、布料、焙烧温度控制等多工序的综合控制管理技术体系，球团生产运行和球团技术指标达到先进水平，取得了显著的经济效益和环境效益。

关键词 球团，带式焙烧机，原料，节能，工程设计

Innovation and Application on Pelletizing Technology of Large Travelling Grate Machine

Zhang Fuming[1], Wang Qusheng[1], Han Zhiguo[1], Huang Wenbin[2]

(1. Beijing Shougang International Engineering Technology Co., Ltd., Beijing 100043,China;
2. Shougang Jingtang Iron & Steel Union Co., Ltd., Tangshan 063200, China)

Abstract The pelletizing process and technical equipment of 504m^2 travelling grate machine at Shougang Jingtang steel plant are introduced in this paper. In order to improve the pellet quality and performance, meet the requirement of two

5500m³ huge blast furnaces high efficiency - low consumption operation, adopt reasonable burden composition, therefore a large scale induration travelling grate machine pelletizing plant with annual output 4.0 million tons is configured at Shougang Jingtang. In this project, a number of advanced technologies and innovations are implemented, such as the pelletizing process design, general layout, equipment development, energy saving and emission control, automatic operation system, etc. A series of advanced technologies and large scale equipment are developed and applied, such as the large drying rotary kiln, the balling disc with diameter 7.5 meter, the shuttle type green pellet distributor, the large travelling grate machine and the seal facility, the high efficiency burner and so on. After commissioning, the comprehensive operation system is established based on charging mixture investigation, balling, green pellet distribution, and baking temperature control. As a result, the operational parameters and pellet quality have achieved advanced technical level; huge operational benefit and magnificent environmental benefit have been obtained.

Key words pellet, induration travelling grate machine, raw material, energy saving，engineering design

Recent Advances in Evaluation of Iron Ore for the Sintering Process

Lu Liming[1], Higuchi Takahide[2]

(1. CSIRO Mineral Resources P. O. Box 883, Kenmore, QLD 4069, Australia;
2. Steel Research Laboratory, JFE Steel Corporation 1 Kokan-cho, Fukuyama, 721-8510 Japan)

Abstract With the rapid change in iron ore supplies and composition, steel mills now have to adjust the composition of the ore mixture to the sinter plant more frequently than ever before. Hence, different scales of test methods have been developed to help sinter plants evaluate new ore types and ore blends in order to optimise their ore mixtures. The reactions in the combustion-melting zone of a sintering bed have a great effect not only on the structural formation of sinter cake but also the permeability of the sintering bed. Understanding of the sintering reactions occurring in this zone is therefore particularly important to improve coke combustion and control melting. Progress has been recently made to carry out in-situ studies of sintering reactions. This paper provides an overview of various techniques which have been developed recently in order to evaluate iron ore and improve the understanding of the iron ore sintering process.

Recent Advances in Iron Ore Sintering in NSSMC

Kenichi Higuchi, Masaru Matsumura, Osamu Ishiyama, Seiji Nomura

(Process Research Laboratories, Nippon Steel & Sumitomo Metal Corporation, Chiba, Japan)

Abstract Sintering is the most economic and widely used agglomeration process to prepare iron ore fines for blast furnace use. Due to the depleting reserves of traditional high grade iron ore, there have been considerable changes in iron ore resources available throughout the world, especially in steel mills in East Asia. Corresponding to the changes in the availability of iron ore resources, the amount of impurities in iron ore has been slowly increasing. Some of these impurities have been found to have deleterious impacts on sinter quality and sintering performance. In the mean time, an increasing number of large blast furnaces with inner volumes of more than 5000 m³ have been built, which require more sinter and are often more demanding in terms of the quality requirements of the ferrous materials. Finally, sinter plants are facing increasing pressures and more stringent regulations regarding their environmental impact. This paper gives an overall review of a variety of technologies developed worldwide to tackle the changing raw material characteristics and mitigate emissions from sintering operations.

Key words agglomeration, granulation, blast furnace, ironmaking

铁矿粉烧结优化配矿技术的研究进展

吴胜利　苏　博　宋天凯　翟晓波

(北京科技大学冶金与生态工程学院，北京　100083)

摘　要　本文首先分阶段回顾了铁矿粉烧结优化配矿技术的发展历程，归纳了国内外炼铁工作者的研究工作；其次，介绍了铁矿粉的烧结高温特性及其研究现状，重点阐述了铁矿粉的熔融性、吸液性；再者，举例叙述了基于铁矿粉同化性、液相流动性的互补配矿技术在实际生产中的应用，以及基于抑制褐铁矿吸液性的大颗粒分加技术；在此基础上，针对铁矿粉烧结优化配矿技术的未来发展趋势，从兼顾铁矿粉高温特性的烧结优化配矿专家系统、应对铁矿粉劣质化的烧结工艺优化等两大方面进行了展望。

关键词　铁矿粉，烧结，高温特性，优化配矿，技术进展

Research Progress of Optimization Technology of Ore-Blending for Sintering

Wu Shengli, Su Bo, Song Tiankai, Zhai Xiaobo

(School of Metallurgical and Ecological Engineering, University of Science and Technology Beijing, Beijing 100083, China)

Abstract　In the present study, the development procedure of optimization technology of fine iron ore blending for sintering was firstly discussed, and the research works in this field were world widely analyzed. Then the contents and research status of high temperature characteristics in sintering of fine iron ores were introduced, and the fusion characteristics, liquid absorbability of fine iron ores were emphatically discussed. Furthermore, the industrial practice of complementary optimization technology based on the assimilability and liquid phase fluidity of iron ores were also discussed, as well as large particles divided adding technology based on the control of liquid absorbability of limonite. Finally, the future prospects of optimization technology of iron ore blending for sintering was discussed from two sides, one is the sintering optimization expert system based on the high temperature characteristics of fine iron ores, another is the sintering process optimization based the deterioration of fine iron ores.

Key words　Iron ores, sintering, high temperature characteristics, optimization blending, technology development

秘鲁原矿矿相结构及组成研究

潘　文[1]　裴元东[1,2]　程峥明[3]　马怀营[1]　张晓晨[1]

（1. 首钢技术研究院，北京　100043；2. 北京科技大学，北京　100083；
3. 首钢京唐钢铁联合有限责任公司，河北曹妃甸　064000）

摘　要　本文利用光学显微镜矿相分析、扫面电镜分析，以及 X 衍射物相分析等方法对首钢秘鲁原矿的矿物组成

和显微结构进行了分析和鉴定。结果显示，秘鲁原矿中主要矿物为磁铁矿、赤铁矿、脉石和硫化物，并含有少量的黄铜矿和磷酸钙，其中脉石以石英矿物为主，硫化物为 FeS、Fe_2S_3 和 FeS_2（黄铁矿）。硫化物主要以条带状、片状或斑状结构嵌布于铁氧化物基体内，并多与脉石矿物共生存在。硫化物和脉石矿物与铁氧化物具有较高的解离度，颗粒直径在 200~500μm 之间，少数颗粒直径超过 1000μm，破碎后具有较好的选矿脱硫条件。秘鲁原矿在首钢京唐选矿厂经先浮后磁的联合选矿后，品位由 56.68%提高至 65.75%，S 含量由 3.44%降至 0.48%。

关键词 秘鲁原矿，矿物组成，显微结构，选矿

Research on Mineral Structure and Compositions of Peru Raw Ore

Pan Wen[1], Pei Yuandong[1,2], Cheng Zhengming[3], Ma Huaiying[1], Zhang Xiaochen[1]

(1. Shougang Research Institute of Technology, Beijing 100043, China;
2. University of Science and Technology Beijing, Beijing 100083, China;
3. Jingtang Iron and Steel Union Company Limited, Capital Iron and Steel Company, Caofeidian Hebei 064000, China)

Abstract In current study, mineral structures and compositions of Peru raw ore was investigated by using optical microscope, scanning electron microscope, and X-ray diffraction. The results show that Peru raw ore is composed primarily of magnetite, hematite, gangue, and sulfide. A small amount of chalcopyrite and calcium phosphate also exist in Peru raw ore. Gangues are mainly quartz mineral. Sulfides exist as FeS, Fe_2S_3, and FeS_2 respectively. Sulfides, most of which accrete with gangues, are embedded among ferrous bodies in the form of banding, schistose, or porphyritic structures. The particle size of sulfides and gangues is between 200μm and 500μm. Some of them are over 1000μm. It can be concluded that high liberation degree of sulfides and gangues will be obtained after crushing. After flotation and magnetic separation in Shougang Jingtang dressing plant, the Fe content of Peru raw ore was improved from 56.68% to 65.75%. The S content dropped to 0.48% from 3.44%.

Key words peru raw ore, mineral compositions, micro structure, beneficiation

一种含钛精粉试验研究

赵 勇[1,2] 张晓晨[2] 马怀营[2] 吴 铿[1]
潘 文[2] 裴元东[2] 申 威[1]

（1. 北京科技大学，北京 100083；2.首钢技术研究院，北京 100043）

摘 要 本文对一种澳洲含钛精矿粉毅星精粉进行了实验室研究，包括化学成分全分析、烧结基础特性检测、扫描电镜分析、X 衍射物相分析、成球性检测、烧结杯试验等。该矿粉粒度偏细（<200 目超过 65%），钛含量高（TiO_2 含量 3.69%），并且在受热后易产生刺激性气味。研究发现，毅星精粉中的主要矿物为磁铁矿（Fe_3O_4）和氧化铁黄（FeOOH），该矿粉成球性差，烧结配加后料层透气性恶化，根据烧结杯试验结果，认为秘鲁球团粉与毅星精粉在小比例配比条件下可进行互替，而随着 TiO_2 含量的增加，应适当地提高燃料配比，从成本上需要分析可行性。

关键词 含钛精粉，烧结，显微形貌，矿物组成

Experimental Study on a Ti-bearing Concentrate

Zhao Yong[1,2], Zhang Xiaochen[2], Ma Huaiying[2], Wu Keng[1],
Pan Wen[2], Pei Yuandong[2], Shen Wei[1]

(1. University of Science and Technology Beijing, Beijing 100083, China;
2. Shougang Research Institute of Technology, Beijing 100043, China)

Abstract In this paper, a Ti-bearing Australia iron ore of Yi-xing concentrate was studied with a fully analysis of the chemical compositions, basic sintering characteristics test, SEM analysis, X-ray diffraction phase analysis, Balling test, sintering pot test and so on. The particle size of over 65% Yi-xing concentrate is under 300 meshes. The result of chemical compositions analysis shows the TiO_2 content in Yi-xing concentrate of 3.69%. According to the XRD analysis, magnetite (Fe_3O_4) and ferric hydroxide (FeOOH) are the major mineral components in Yi-xing concentrate. With the sintering pot test, the permeability of the sinter bed was deteriorated after Yi-xing concentrate added to the iron ore blends due to its poor balling performance. The fuel consumption will be increased with higher TiO_2 content in sinter mixer. It can be concluded from the sintering pot test that the proportion of Yi-xing concentration in the sinter mixer should be controlled within a proper range so as to maintain the quality of the iron ore sinter.

Key words Ti-bearing concentrate, sinter, micro morphology, mineral compositions

IOC 精粉烧结试验及生产实践

杨冬峰　易　平

（宝钢不锈钢有限公司炼铁厂，上海　200431）

摘　要　IOC 精粉是加拿大低铝精粉，进行基础研究、烧结杯试验并用于生产实践。IOC 精粉是一种粒度较细精粉，研究表明，在烧结过程中采用提高生石灰配比、增加燃料等措施，可以在烧结配矿替代巴西矿，在试验当期具有经济效益。

关键词　烧结矿，精粉，生产实践

Study and Practice on IOC Iron Concentrates

Yang Dongfeng, Yi Ping

(Iron Making Plant, Baosteel Stainless Steel Co., Shanghai 200431, China)

Abstract Through the study of IOC iron concentrates，Physico-chemical and sintering properties has been studied.This iron concentrates been use in production, with optimizing the ration of the iron mines , increasing the proportion of lime and coal. Finally, with max 14% replacement of Brail iron and other iron ,high qualified sinter is obtain ,as well as economic benefits.

Key words sinter, high iron and low silica, practice

宝钢不锈烧结使用三种澳粉新品种的实践和评估

易 平 杨冬峰 李 辉

（宝钢不锈钢有限公司炼铁厂，上海 200431）

摘 要 近几年，进入投产期的铁矿石矿山不少，陆续有新的铁矿石矿种投放市场，例如 BHP 公司的金布巴粉、FMG 公司的矿种、南非昆巴矿、乌克兰矿、几内亚矿等。宝钢不锈钢有限公司本部炼铁厂自 2014 年 3 月至 8 月初进行了三种澳粉新品种在烧结生产中使用的实践，包括 FMG 公司火箭粉、BHP 公司金布巴粉、FMG 公司超特粉。本文介绍了宝钢不锈炼铁厂使用这些矿种的情况、结果和经验，包括烧结杯、烧结大生产、高炉使用这些烧结矿的数据，并对其作出了初步的评估。

关键词 铁矿石，澳粉，生产实践

The Practice and Evaluation on Using Three New Australia Ores in the Sinter of Baosteel Stainless

Yi Ping, Yang Dongfeng, Li Hui

(The Iron-making Plant of Baosteel Stainless Steel Co., Ltd., Shanghai 200431, China)

Abstract Recent years, some new kinds of iron ore have put into markets one after another. For example, the Jimblebar fine in BHP Company、some ores in FMG Company、the Kumba ore in South Africa、Ukraine ores and so on. The iron-making plant of Baosteel stainless steel Co., Ltd. carried out the practice on three new Australia fine used in sinter production during March to August 2014. The three fines have Rocket fine、Jimblebar fine in BHP and Chaoter fine in FMG. This article introduces the case and experience that these iron ore fines used in the sinter of iron-making plant of Baosteel stainless. A preliminary evaluation was made about the three fines.

Key words iron ore, Australia fine, production practice

Minas Rio 赤铁精粉球团制备特性研究

朱德庆 师本敬 潘 建 钟 洋

（中南大学资源加工与生物工程学院，湖南长沙 410083）

摘 要 针对赤铁精粉成球性较差、焙烧温度高，导致在链箅机—回转窑球团原料中配比低的问题，对赤铁精粉配加磁铁精矿及采用高压辊磨预处理造球原料的方法改善 Minas Rio 赤铁精粉的成球性及焙烧性能进行了研究。结果表明：Minas Rio 赤铁精粉的成球性及焙烧性能相对磁铁精矿较差，膨润土用量大；按 6:4 的比例与不同种类的国内磁铁精矿混合经高压辊磨处理后造球，膨润土添加量由 1.5%以上最低可降至 0.6%，球团焙烧温度可降低 50~100℃，生球和焙烧球团强度均可满足生产要求。

关键词 赤铁精粉，球团特性，磁铁精矿，高压辊磨

Studies on the Pelletization of Minas Rio Hematite Concentrates

Zhu Deqing, Shi Benjing, Pan Jian , Zhong Yang

(School of Minerals Processing and Bioengineering, Central South University, Changsha 410083,China)

Abstract Considering its lower ball-ability and higher firing temperature required during pelletizing, hematite concentrates ratio in pellet feed is restricted in rotary kiln process. Serials experimental studies on Minas Rio hematite concentrates pelletizing performance were conducted through varying Minas Rio blending ratio in hematite - magnetite concentrates mix and pretreating of the materials by high pressure roller (HPGR). The results show that the ball-ability index and firing performance of Minas Rio hematite concentrates are inferior to domestic magnetite concentrates. Without pretreatment of the pellet feeds higher bentonite dosage is required and lower strength of fired pellets is expected although at higher firing temperature. In contrast, after pretreatment by HPGR, the pellet feeds consisting of 40% domestic magnetite concentrates and 60% Minas Rio hematite concentrates reduce bentonite usage from 1.5% to 0.6%, and decrease firing temperature by 50 ~100℃. The strength of green pellets meets pelletizing demands and the fired pellets satisfy Blast Furnace operation.

Key words hematite concentrates, pelletizing characteristics, magnetite concentrate, HPRG

TiO_2对高钛钒钛磁铁烧结矿性能的影响

何木光

（攀枝花钢钒有限公司炼铁厂，四川攀枝花 617024）

摘 要 为了探索在现有原料条件下不同TiO_2对高钛型磁铁矿烧结性能的影响，通过实验得出，在烧结矿中TiO_2含量6%～10%内，随着TiO_2含量增加，混合料初熔和熔化温度均上升，烧结矿强度和成品率下降，低温还原粉化率上升，还原性降低，熔滴性能变差，贮存性能变化不大。当TiO_2含量从6%提高到10%时，烧结矿转鼓强度下降4.69个百分点，成品率下降4.43个百分点，低温还原粉化率上升6.02个百分点，还原度下降3.9个百分点，烧结矿的平均粒径由22.10mm下降到19.28mm；随着TiO_2含量增加，钛赤铁矿、铁酸钙含量减少，钛磁铁矿以及含钛矿物钙钛矿、钛辉石、钛榴石含量增加；烧结矿的孔洞增加，矿物形态和结构变差。

关键词 TiO_2，高钛磁铁矿，烧结，性能

Effect of TiO_2 on High Titanium Vanadium-titanium Magnetite Sinter Properties

He Muguang

(Panzhihua Steel Vanadium Ironmaking Plant, Panzhihua 617024, China)

Abstract In order to explore the effects of high titanium magnetite sintered materials performance on TiO_2 under different conditions existing，obtained by experiment，in the sinter TiO_2 content of 6% to 10%，with the increase of TiO_2 content，early melt and mix melting temperatures wed rised，sinter strength and yield a decreased，RDI rate reducing reduce droplet poor performance, storage performance changed little. When TiO_2 increased from 6% to 10%, the sinter tumbler strength decreased 4.69 percentage points, the yield fell 4.43 percentage points, RDI rate rose 6.02 percentage points, reduction decreased 3.9 percentage points, the average particle sinter diameter decreased from 22.10mm to 19.28mm; with increasing TiO_2 content, hematite titanium iron, calcium content decreased, titanium and titanium magnetite mineral perovskite、titanium pyroxene、garnet titanium content increases; sinter hole increases, mineral morphology and structure deteriorated.

Key words TiO_2, high titanomagnetite, sintering, performance

SiO_2/TiO_2 比对高镁型钒钛矿烧结过程和烧结矿性能的影响

唐昭辉 魏 国 董 越 张 强 丁学勇

（东北大学材料与冶金学院，辽宁沈阳 110819）

摘 要 相关研究表明，SiO_2/TiO_2 比对钒钛矿烧结料原始透气性，烧结阻力，烧结液相量以及成品率都有一定影响。本文以高镁型钒钛矿为主，通过配加普通钒钛矿、澳矿进行了不同 SiO_2/TiO_2 比、二元碱度 R_2 对烧结过程和烧结矿性能影响试验研究。

关键词 SiO_2/TiO_2 比，高镁型钒钛矿，二元碱度 R_2，烧结过程，烧结矿性能

Effect of SiO_2/TiO_2 Ratio on Sintering Process and Sinter Property of High Magnesium of V-Ti-magnetite

Tang Zhaohui, Wei Guo, Dong Yue, Zhang Qiang, Ding Xueyong

(School of Materials & Metallurgy, Northeastern University, Shenyang 110819, China)

Abstract Research shows that SiO_2/TiO_2 ratio of vanadium titanium sinter raw material permeability, resistance to sintering, sintering liquid phase amount and rate of finished products have certain effect. In this paper, we studied the effect of different SiO_2/TiO_2 ratio, binary basicity R_2 on the sintering process and the properties of sinter, basing on the high magnesium of V-Ti-magnetite, ordinary V-Ti-magnetite and Australian ore.

Key words SiO_2/TiO_2 ratio, high magnesium of V-Ti-magnetite, binary basicity, sintering process, sinter properties

含铬型钒钛磁铁矿配矿正交优化研究

杨松陶 周 密 姜 涛 张立恒 薛向欣

（东北大学材料与冶金学院，辽宁沈阳 110819）

摘 要 在实验室条件下，以国内某钢铁企业生产实际为参考，以含铬型钒钛磁铁矿为原料，利用烧结杯进行了 3 因素 4 水平的正交试验研究，检测了烧结矿冶金性能；然后运用综合加权评分法对试验结果进行分析评价，获得

了含铬型钒钛烧结矿配矿的适宜配比：大阪通运 20%，恒伟矿业 15%，远通矿业 45%，建龙矿业 20%。对烧结矿综合指标影响从大到小依次为：远通矿业，恒伟矿业，大阪通运。

关键词 含铬型钒钛铁矿，烧结，配矿，正交研究

Orthogonal Study on Ore Blend for Sintering Process Using Cr-bearing Vanadium -Titanium Magnetite

Yang Songtao, Zhou Mi, Jiang Tao, Zhang Liheng, Xue Xiangxin

(School of Materials and Metallurgy, Northeastern University, Shenyang Liaoning 110819, China)

Abstract The sintering pot tests , designed through orthogonal method in form of 3 factors and 4 levels , were conducted using Cr-bearing vanadium - titanium magnetite (V-Ti-Cr) ores under laboratory conditions,based on the actual production of a iron and steel company .In addition, the sinter properties were determined and the method of comprehensive evaluation was applied to obtain the optimal matching proportions for V-Ti-Cr mixture .The results showed that 20% DB,15%HW, 45%YT,and 20% JL was the optimal proportions . The sequences of factoring from high to low were YT, HW, DB.

Key words Cr-bearing vanadium and titanium iron ores, sinter, ore blend, orthogonal study

B_2O_3对高铬型钒钛磁铁矿球团抗压强度的影响

程功金 薛向欣 姜 涛 段培宁

（东北大学材料与冶金学院，辽宁沈阳 110819）

摘 要 鉴于高铬型钒钛磁铁矿的综合利用技术和有价组元的利用率仍需提高的现状，目前国内的高铬型钒钛磁铁矿仍未大规模开采和利用。本文研究了以B_2O_3为添加剂生产高铬型钒钛磁铁矿氧化球团，考察了外配B_2O_3质量分数为 0%、0.3%、0.5%、1.0%和 1.5%对模拟链箅机—回转窑工艺生产的高铬型钒钛磁铁矿氧化球团抗压强度的影响。研究表明：随着B_2O_3含量的增加，高铬型钒钛磁铁矿球团的抗压强度显著增强，并且抗压强度指标均满足高炉生产要求。外配质量分数为 0.5%、1.0%、1.5%B_2O_3的高铬型钒钛磁铁矿氧化球团的抗压强度分别达到 3112N/个、4447N/个、5510N/个，其值比未配入B_2O_3的抗压强度值分别提高 23.10%、75.91%、117.96%。配入B_2O_3的高铬型钒钛磁铁矿球团发现有含硼物相$(Mg,Mn)_2Mn(BO_3)_2$生成，并且脉石相和孔洞明显减少，并且有相当量的镁铝尖晶石相生成，这也很可能是配入B_2O_3后焙烧所得氧化球团抗压强度显著增强的重要原因。

关键词 B_2O_3，高铬型钒钛磁铁矿，球团，抗压强度，矿相

Effect of B_2O_3 on the Crushing Strength of High-Chromium Vanadium-titanium Magnetite Pellets

Cheng Gongjin, Xue Xiangxin, Jiang Tao, Duan Peining

(School of Materials & Metallurgy, Northeastern University, Shenyang Liaoning 110819, China)

Abstract Domestic high-chromium vanadium-titanium magnetite has not been exploited and made full use of on a large scale so far because of the immature utilization technology and the fact that the utilization efficiency of valuable metals still urgently needs to be improved.Effect of B_2O_3 on the crushing strength of high-chromium vanadium-titanium magnetite pellets prepared with B_2O_3 additive agent of 0, 0.3, 0.5, 1.0 and 1.5 mass-% under simulating the conditions of grate rotary kilnwas studied in the present paper. It is obtained that the crushing strength values clearly increase with increasing the B_2O_3 contents, and the crushing strength meets the requirements of blast furnace production. The crushing strength values of high-chromium vanadium-titanium magnetite pellets with B_2O_3 of 0.5, 1.0 and 1.5 mass-% are 3112N/a、4447N/a、5510N/a, and the increasing rates are 23.10%, 75.91%, 117.96% compared with the pellets without B_2O_3. Boron-bearing phases $(Mg, Mn)_2 Mn(BO_3)_2$ is found in the high-chromium vanadium-titanium magnetite pellets, also less gangue phase and pores and considerable quantities of magnesia-alumina spinel are observed, which is probably the reason that the crushing strength is enhanced obviously.

Key words B_2O_3, high-chromium vanadium-titanium magnetite, pellet, crushing strength, mineral phases

Al_2O_3、SiO_2在铁酸钙系熔体中的溶解动力学

喻 彬 吕学伟 向升林 徐 健

（重庆大学，重庆 400044）

摘 要 铁矿石烧结是一个依靠液相产生和再结晶来实现颗粒粘结的高温物理化学过程。复合铁酸钙是高碱度烧结矿的主要粘结相。Al_2O_3、SiO_2 等脉石成分在铁酸钙中的溶解过程对烧结过程中复合铁酸钙的形成有着重要影响。本研究采用固体试样旋转法探索了 Al_2O_3、SiO_2 在铁酸钙系熔体中的溶解动力学。结果表明：Al_2O_3 在边界层的扩散是 Al_2O_3 溶解反应的限制性环节。SiO_2 的溶解速率主要由浓度差和渣的初始化学成分决定。计算得到的 Al_2O_3、SiO_2 的扩散系数的范围分别是（0.43～2.45）$\times10^{-6}$ cm^2/s 和（2.09～6.40）$\times10^{-6}$ cm^2/s。对于 Al_2O_3 的溶解过程，体系 AI–1 的溶解活化能是 245.35 kJ/mol。对于 SiO_2 的溶解过程，体系 SI–3 的溶解活化能是 174.70 kJ/mol。

关键词 溶解速率，扩散，铁酸钙系熔体，浓度差

Dissolution Kinetics of Al_2O_3, SiO_2 into Calcium Ferrite Based Slag

Yu Bin, Lv Xuewei, Xiang Shenglin, Xu Jian

(Chongqing University, Chongqing 400044, China)

Abstract Iron ore sinter is a high-temperature physicochemical process that depends on the formation and crystallization of liquid phase to bond the ore particles. SFCA is the main binding phase of high-basicity sinter. Solid Al_2O_3, SiO_2, the main composition of gangue, dissolving into molten calcium ferrite (CF) based slag has a significant effect on the formation of SFCA in sinter. In this study, a rotating cylinder method was used to explore the dissolution kinetics of Al_2O_3, SiO_2 into CF-based slag. It shows that the diffusion of Al_2O_3 at boundary layer was the rate limiting step at the dissolution process. The dissolution rate of SiO_2 was primary controlled by concentration difference and initial composition of the slag. The calculated diffusion coefficient of Al_2O_3, SiO_2 during the dissolution process respectively ranged in $(0.43\sim2.45)\times10^{-6}$ cm^2/s and $(2.09\sim6.40)\times10^{-6}$ cm^2/s. For Al_2O_3 dissolution process, the activation energy of samples AI-1 was 245.35 kJ/mol. For SiO_2 dissolution process, the activation energy of samples SI-3 was 174.70 kJ/mol.

Key words dissolution rate, diffusion, calcium ferrite based slag, concentration difference

铁酸一钙等温还原动力学研究

丁成义　宣森炜　吕学伟

（重庆大学材料科学与工程学院，重庆　400044）

摘　要　铁酸钙体系是熔剂型烧结矿主要的液相成分，铁酸钙还原性能是评价烧结矿品质的重要角度，热分析动力学可为还原性研究提供更好的定量化描述手段。铁酸一钙作为最基础且重要的铁酸钙体系成员，研究其还原动力学可为更为复杂的复合铁酸钙体系提供重要的铺垫。此次研究采用热重分析法探究铁酸一钙在 30%CO+70%Ar 气氛中以 850℃、900℃和 950℃恒温还原的动力学问题。为保证求解过程的严谨性和准确性，分别以两种方法验证求解描述反应过程机理的模式函数和反应活化能。实验结果表明：以 lnln 分析法和约化时间法相互验证得到模式函数为 $G(\alpha)=1-(1-\alpha)^{1/2}$，通过等转化率法和模式函数法求得还原反应活化能分别为 46.89 kJ/mol 和 43.74kJ/mol，对应的指前因子分别为 2.41min^{-1} 和 1.75 min^{-1}，两种方法所得结果的一致性证明了动力学参数求解的准确性。

关键词　铁酸一钙，等温还原，动力学，活化能，模式函数

Isothermal Reduction Kinetics for $CaO·Fe_2O_3$

Ding Chengyi, Xuan Senwei, Lv Xuewei

(School of Materials Science and Engineering, Chongqing University, Chongqing 400044, China)

Abstract　The $CaO-Fe_2O_3$ system is the most significant liquid phase for fluxed sinter, of which the reducibility becomes a vital index having great significance on the evaluation of sinter qualities. Thermal kinetics providing a powerful tool to quantificationally describe the reducibility of $CaO-Fe_2O_3$ system was selected to this research. As the most fundamental and important member of $CaO-Fe_2O_3$ system, investigation on the reducibility of $CaO·Fe_2O_3$ will open the way to explore more complex $CaO-Fe_2O_3$ system. The isothermal reduction of $CaO·Fe_2O_3$ in a continuous stream of 30% CO and 70% Ar at 850、900 and 950℃ was carried out by thermogravimetric analysis(TGA). In order to guarantee the rigour and the accuracy of the results, two different methods were applied to calculate the model function and activation energy respectively. The results indicate that the model function are both shown as $G(\alpha)=1-(1-\alpha)^{1/2}$ by lnln-Analysis and Sharp method. The activation energy lies at 46.89 kJ/mol and 43.74 kJ/mol based on iso-conversional method and model-limited method, and the pre-exponential factor was calculated as 2.41 min^{-1} and 1.75 min^{-1} corresponding to the two methods respectively.The results based upon the two methods are very consistent with each other, which ensure the accuracy of the results.

Key words　$CaO·Fe_2O_3$, isothermal reduction, kinetics, activation energy, model function

Effect of MgO on Melting and Crystallization of $CaO-Fe_2O_3$ System

Guo Xingmin, Yang Nan

(School of Metallurgical and Ecological Engineering, University of Science and Technology Beijing, Beijing 100083, China)

Abstract To solve some problems of sinter properties and slag performance existed on use of iron ores with high-Al_2O_3 content in blast furnace, a favour method is to add amount of MgO into sinter. In this work, effect of MgO on melting and crystalization of binding phase in high-basicity siner has been investigated by using TG-DSC and crystalizing test of CaO-Fe_2O_3 system (molar ratio=1:1) combining X-ray diffraction and Scanning electron microscopy. The result showed that MgO added into CaO-Fe_2O_3 system increases the initial melting temperature from the eutectic temperature of $CaO·Fe_2O_3$ and $CaO·2Fe_2O_3$ to the peritectic temperature of $CaO·Fe_2O_3$ and $2CaO·Fe_2O_3$ in the heating process, simultaneously the beginning temperature of crystallization of calcium ferrite increases with increase of MgO content in the cooling process. It was found that magnetite formed in CaO-Fe_2O_3 system was mainly in the heating process, especially over 1338℃, and the amount of magnetite formed increased with increase of MgO added. It was also observed in the experiments that MgO can get into magnetite, and exist in (Ca, Mg, Fe)O·Fe_2O_3 phase, while the MgO could not be in phases of $CaO·Fe_2O_3$ and $2CaO·Fe_2O_3$.

Key words magnesia, calcium ferrite, binding phase, melting temperature, crystallization

New Insights into Alumina Types in Iron Ore and Their Effect on Sintering

O'dea Damien, Ellis Ben

(BHP Billiton, 1 Technology Court, Pullenvale, Qld 4069, Australia)

Abstract Most of the research published on the effect of alumina on iron ore sintering is based on work with Indian or African ores. These ores tend to contain alumina in the mineralogical form of gibbsite. Most Australian iron ores contain alumina in the form of kaolinite or aluminous goethite. Pilot-scale sintering studies reported here found that when Brockman – Marra Mamba fines from BHP Billiton containing kaolinite and aluminous goethite were added to the blend, sinter quality (strength, size, RDI, RI) could be maintained with small increases in the coke addition as alumina was increased from 1.7% to 2.3%. As alumina level swere further increased to 2.7%, which is significantly higher than most sinter plant operations, productivity could be maintained but sinter strength decreased. Even at these high alumina levels the low temperature reduction degradation of the sinter did not deteriorate. This is very different to what has been reported for ores with gibbsite alumina. Optical microscopy and SEM EPMA investigations found that for the higher alumina cases the extra fuel and higher sintering temperatures led to more replacement of hematite by magnetite and SFCA phases in the sinter. Also the alumina concentrated more into these phases. This resulted in a sinter composition and morphology that maintained its resistance to reduction degradation. Newer techniques such as X-ray micro-tomography are being applied to investigate the 3D internal structure of the sinter and its effect on quality. This forms part of future work that is required to further improve understanding of the influence of alumina type (especially aluminous goethite) and help optimize sintering with Australian iron ores.

Key words iron ore, sintering, alumina, kaolinite, aluminous goethite, gibbsite

鞍钢炼铁总厂烧结提高精矿配比试验研究

徐礼兵 周明顺 翟立委 刘 杰 张 辉

（鞍钢集团钢铁研究院，辽宁鞍山 114009）

摘 要 为了探究鞍钢炼铁总厂烧结生产提高铁精矿配比的可行性，在实验室条件下，进行了提高铁精矿配比的烧结试验研究，同时对烧结矿进行了冶金性能及矿相结构的分析研究。结果表明：当使用粉矿 A 时，随着铁精矿 C 配比的提高，烧结矿转鼓强度、燃耗、利用系数均降低，成品率升高，烧结矿品位提高，SiO_2 含量降低，烧结矿 $RDI_{+3.15mm}$ 值升高，还原度降低，建议铁精矿 C 配比为 65%左右；当使用粉矿 B 时，烧结矿转鼓强度、成品率、利用系数均降低，燃耗升高，烧结矿品位提高，SiO_2含量降低，烧结矿 $RDI_{+3.15mm}$ 值升高，还原度降低，铁精矿 C 配比在 50%至 75%范围内均可。

关键词 铁精矿，烧结，转鼓强度，低温还原粉化

Experimental Study on Improving the Ratio of Concentrate in Sintering of Ansteel Iron-making Plant

Xu Libing, Zhou Mingshun, Zhai Liwei, Liu Jie, Zhang Hui

(Ansteel Group Iron and Steel Research Institute, Anshan 114009, China)

Abstract In order to explore the feasibility of improving the ratio of concentrate in sintering of Ansteel iron-making plant, on the condition of laboratory, carrying on the study of sintering test that improving the ratio of concentrate, and researching the metallurgical properties and mineral phase structure of sinter. The results show that: when fine A is used, with the improvement of concentrate, the sinter tumbler strength, fuel consumption, utilization coefficient are reducing, yield is improving, sinter grade is improving, content of SiO_2 is reducing, $RDI_{+3.15mm}$ is improving, reducibility is reducing, proposing the ratio of concentrate is about 65%. When using fine B, the sinter tumbler strength, yield, utilization coefficient are reducing, fuel consumption is improving, sinter grade is improving, content of SiO_2 is reducing, $RDI_{+3.15mm}$ is improving, reducibility is reducing, the ratio of concentrate range among 50% to 75%.

Key words concentrate, sintering, tumbler strength, low temperature reduction degradation

高比例微细精矿超高料层烧结技术研究

贺淑珍

（山西太钢不锈钢股份有限公司，山西太原 030003）

摘 要 随着烧结技术的发展，从低料层到高料层，取得了很好的应用效果，现在已发展到超高料层，如韩国浦项的烧结料层厚度提高到 1000mm，国内马钢烧结料层厚度提高到 900mm，京唐提高到 800mm 以上，不仅使得烧结矿质量提高，而且固体燃料消耗大大减少，对于减少 C 的排放也有好处。目前采用超高料层烧结的企业都是粉矿烧结，精粉率不足 10%，但也会带来透气性变差，产量降低的情况。针对太钢烧结原料特点，为了探求高精粉率超高料层烧结技术，进行了实验室试验研究。研究结果表明，在高精粉原料条件下，在强化制粒效果、优化燃料粒度、配用复合粘结剂等有效措施后，将料层厚度提高到 800~900mm，烧结矿的矿物结构改善，产、质量指标得到了保证，固体燃料消耗大大降低，达到了预期效果。

关键词 高精粉率，超高料层，透气性，产、质量指标，固体燃料消耗

Research of Sinter Technology with High Fine Ore Ratio and Super-High Bed Height

He Shuzhen

(Taigang Stainless Steel Co., Ltd., Taiyuan 030003, China)

Abstract The bed height has been increased rapidly as the development of sinter technology. The bed height of raw mix has reached 1000mm in PoHang Iron & Steel company, Korea, as well as 900mm in Maanshan Iron & Steel company, China. They all have got good affection on sinter quality, additionally they also decreased the specific fuel consumption remarkably and had benefit to reduce the carbon emission to the circumstance. The companies adopted super high bed height sintering technology are mostly sintering with fine ore. Based on the raw material conditions within TISCO, high fine ore ratio in the blended ore, lab test and research has been carried out for super high bed height sintering technology. With serial of measurements such as optimized fuel grain size and addition of intensive complex additive, the trial has proved that the sinter productivity and quality can be guaranteed with high fine ore ratio sintering and more than 900mm bed height. Meanwhile, the mineral structure of the product sinter has been improved, the specific fuel consumption has been reduced enormously.

Key words high fine ore ratio, super high bed height, permeability, productivity and quality index, specific fuel consumption

烧结混合料制粒过程研究

熊　林　毛晓明　李　建

（宝山钢铁股份有限公司研究院，上海　201999）

摘　要　为改善制粒过程优化制粒效果，通过对比不同制粒阶段的粒度组成、化学成分标准差、粘附层强度、原始颗粒分布和颗粒形貌研究了两段圆筒混合制粒工艺的制粒过程。研究结果表明：随着水分含量在制粒过程中的逐渐增加，可作为粘附粉的颗粒或准颗粒越来越大，各粒级粘附粉成分也逐渐趋同，进而降低了各粒级之间化学成分偏差。在制粒过程中，低强度粘附层厚度始终大于高强度粘附层厚度，增幅却小于高强度粘附层。对最终制粒结果而言，原始–0.5mm 颗粒基本为粘附粉，0.5~1mm 颗粒大部分为核粒子，少部分为粘附粉，+1mm 颗粒几乎完全为核粒子。准颗粒在一混阶段主要遵循滚雪球式长大机理，在二混阶段则同时遵循滚雪球式和粘合式长大机理。

关键词　烧结，制粒，粘附粉，长大机理

The Study of the Sinter Mixture Granulation Process

Xiong Lin, Mao Xiaoming, Li Jian

(Research Institute, Baoshan Iron & Steel Co., Ltd., Shanghai 201999,China)

Abstract To improve granulation process and optimize granulation performance, the two cylinder mixing granulation process was studied by comparing the size distribution, chemical composition standard deviation, strength of the adhesive layer, original particle size distribution and particle morphology at different granulation stages. The result shows that along with the moisture content gradually increasing in the granulation process, the layering particles size grows, the chemical composition of layering particles in each size fractions gradually become the same, and this reduces the chemical composition deviation between each size fraction. During granulation process, the thickness of low-strength adhesive layer is always larger than the high-strength adhesive layer, but the rate of increase is less than the high-strength adhesive layer. In terms of the final granulation results, basically –0.5mm are layering particles, most of 0.5~1mm particles are seed particles, small part of them are the layering particles, +1mm particles are almost seed particles. Quasi-particles in first mixing stage mainly follow snowballing growth mechanism, in the second mixing stage, while follow snowballing and adhesive-type growth mechanism.

Key words sintering, granulation, layering particles, growth mechanism

Enhancing the Sintering Performance of Blends Comprising High Ratio of Brazilian Hematite Concentrate by SMBGS Process

Zhu Deqing, Shi Benjing, Pan Jian

(School of Minerals Processing and Bioengineering, Central South University, Changsha 410083,China)

Abstract Considering that only about 10% Brazilian hematite concentrate could be proportioned in the sinter blends because of its fine size in granulation, enhancing its sintering performance was studied by separated microballing granulating sintering (SMBGS) processes to further elevate its ratio up to 30% in the sinter blends. Compared with traditional granulation-sintering process, much better results were achieved by the SMSP as follows: elevating the unit productivity from 1.44 $t \cdot m^{-2} \cdot h^{-1}$ to 1.71 $t \cdot m^{-2} \cdot h^{-1}$, augmenting the tumble index from 57.60% to 63.80% , decreasing the solid fuel rate from 76.46 $kg \cdot t^{-1}$ to 65.24$kg \cdot t^{-1}$. Therefore, the SMBS process can significantly enhance the sintering performance of blends comprising high ratio of Brazilian hematite concentrate. The mechanism of the SMBS processes is demonstrated that the permeability of sinter mixture is increased markedly by separately microballing the fine hematite concentrate, and the coke level is cut appropriately, leading to a higher oxidation potential in sintering bed and more calcium ferrite bonding phase being formed finally. In the meantime, minerals in the product sinter crystalyse fully and compactly, and more middle or small sized round pores form compared to loose microstructure and irregular big pores in the traditional sinter. Good quality sinter with higher strength and better reducibility can be manufactured by the SMBS process.

Key words brazilian hematite concentrate, separated microballing-sintering process, metallurgical performance

基于离散单元法的烧结偏析布料过程数值仿真

庄佳才 闫红杰 周 萍 石朋雨

（中南大学能源科学与工程学院，湖南长沙 410083）

摘 要 烧结工艺是高炉炼铁的一个重要环节，布料状况影响着烧结质量及产量。料层高度方向上的混合料粒径分布会显著影响烧结层的透气性，合理的粒径分布对于烧结机的高效生产至关重要。本文以某钢铁企业烧结机的偏析布料过程为研究对象，采用离散单元法建立了对应的数学模型，借助 EDEM 软件实现了反射板布料器及多辊布料器布料过程的数值仿真。结果表明：反射板布料器及多辊布料器布料所获得的料层在水平方向的粒度分布均较平稳，最大无量纲化粒度差不超 10%；且多辊布料方式所对应的料层的质量平均直径 d_m 沿水平方向变化较反射板式平缓，水平方向上的粒度偏析不明显；采用多辊布料器后，料层上部区域混合料的平均粒径明显变小，粒径自上而下增大，最大无量纲化粒度差由原先的 0.236 增至 0.321，各层料层中的混合粒度组成更为集中，可以提高料层的透气性。

关键词 多辊布料器，布料偏析，离散单元法

A Numerical Simulation of Segregation Distribution of Sintering Process Based on the Discrete Element Method

Zhuang Jiacai, Yan Hongjie, Zhou Ping, Shi Pengyu

(School of Energy Science and Engineering, Central South University, Changsha 410083, China)

Abstract Sintering process plays a crucial role in iron-making process. The distribution of raw material after feeding process has significant influence on sintering quality and economic benefit. The mechanism is that natural size segregation occurs while the feed flows along the chute and settles on the slope of feed bed. Due to differences in permeability and density of the charged materials, the size segregation of the sinter bed should be controlled exquisitely. A mathematical model was developed to analyze the feeding process by discrete element method. Numerical simulation of traditional and multi-roller charging was achieved by EDEM software. Simulation results showed that particle size segregation along horizontal direction in both charging pattern are slight as well as the maximum dimensionless size difference, which are smaller than 10%. Besides, the mean diameter of layer obtained by multi-roller changes gently, which is more conducive to balance the heat load. In addition, mean particle size in the upper part of the feeding bed obtained by multi roller devices is smaller than that of traditional charging equipment. As a result, heavier particle size segregation is gained in the vertical direction, with the maximum dimensionless size difference increases from precious value 0.236 to 0.321, and size distribution is more concentrated, which can enhance permeability of feed bed.

Key words multi-roller charger, segregation, discrete element method

烧结生产进一步提质节能的途径
——均热高料层烧结

姜 涛 李光辉 许 斌 张元波

（中南大学烧结球团与直接还原工程研究所，湖南长沙 410083）

摘 要 根据高料层烧结存在的热量不均和由此带来的问题与影响，提出了均热高料层烧结的概念。以宝钢生产原料为对象，研究了沿料层高度方向热量分布规律。在分析料层厚度对料层高度和宽度方向热量分布影响的基础上，提出了均热高料层烧结的模式与要求。简述国内外均热烧结技术现状，研究了气流偏析布料实现均热高料层烧结的原理、技术与效果。

关键词 烧结，节能，均热，高料层

An Approach of Quality-improving and Energy-saving for Sintering Process: Heat-Homogenizing and Deep-Bed Sintering

Jiang Tao Li Guanghui Xu Bin Zhang Yuanbo

(Institute of Agglomeration and Direct Reduction, Central South University, Changsha Hunan 410083, China)

Abstract The heat inhomogeneity during the deep-bed sintering has an adverse effect on the sintering process and reduces the quality of the final sinters. Hence, the technology of heat homogenizing for deep-bed sintering has been developed to avoid the shortage above. Taking the iron raw material from Baosteel as research object, the heat distribution law in the height direction of the sinter bed was investigated in this study. The mode and requirements for heat-homogenizing and deep-bed sintering were put forward based on the analysis of the effect of sinter bed depth on the heat distribution. In addition, the research status and progress of heat-homogenizing sintering at home and abroad were reviewed. And the specific technical, principles and performance of heat-homogenizing sintering by airflow segregation feeding were also studied in this paper.

Key words sintering, energy saving, heat homogenizing sintering, deep-bed sintering

鞍钢烧结工艺参数优化试验研究

翟立委 周明顺 刘 杰 张 辉 徐礼兵

（鞍钢集团钢铁研究院，辽宁鞍山 114009）

摘 要 为探求鞍钢烧结厂合理烧结工艺参数，采用三元二次正交回归组合设计方法进行了烧结杯试验。结果表明，混合料水在7.0%~7.6%时，对烧结矿转鼓强度、利用系数和垂直烧结速度影响较大；混合料碳在3.3%~4.0%时，对烧结矿FeO含量和成品率指标影响最大；抽风负压在1300~1450mmH_2O时，对燃料消耗，烧结速度，成品率影响最大。故在碱度为2.05时，经过优化后，鞍钢烧结最佳参数组合为混合料水为7.0%，抽风负压为1450mmH_2O，混合料碳为3.86%。

关键词 正交试验，烧结杯，烧结指标，工艺参数，优化

An Experimental Study on Reasonable Processing Factors for Ansteel Sintering Plant

Zhai Liwei, Zhou Mingshun, Liu Jie, Zhang Hui, Xu Libing

(Iron & Steel Research Institute of Angang Group, Anshan 114009, China)

Abstract In order to get the suitable processing factors for Ansteel sintering plant, the sinter pot test has been done with three-element quadratic regression orthogonal composite method. It was shown that, water content within 7.0%~7.6% has a greater impact on drum index, utilization coefficient and vertical sintering speed. Carbon content within 3.3%~4.0% plays a most important role for FeO content and productity in sinter. The negative pressure alkalinity within 1300~1450mmH_2O has a greater impact on carbon consumption, vertical sintering speed and productity. Aftering optimization and sinter basicity is 2.05, the reasonable control strategy is like this: water conternt is 7.0%, negative pressure alkalinity is 1450mmH_2O, carbon content is 3.86%.

Key words three-elemem quadratic regression orthogonal composite method，sinter pot, sintering index, processing parameters, optimization

唐钢烧结矿 FeO 含量的试验研究

胡启晨[1] 王 娜[2] 胡宾生[3] 赵 军[4]

（1. 河北钢铁技术研究总院，河北石家庄 052000；
2. 北京工业大学材料科学与工程学院，北京 100024；
3. 华北理工大学冶金与能源学院，河北唐山 063000；
4. 河北钢铁集团唐山分公司，河北唐山 063000）

摘 要 烧结矿中 FeO 含量对烧结矿质量、固体燃耗和生产成本影响至关重要，对高炉生产的顺行状况、燃料比水平和产量影响也比较大。基于降低成本的需要，国内各钢铁厂都在降低烧结矿 FeO 含量，由于企业之间装备水平的差异，原燃料条件限制等因素限制，各企业之间烧结矿 FeO 含量控制存在很大差异。本研究通过实验室烧结杯模拟试验、烧结生产和高炉生产的工业试验过程，对烧结矿粒度、转鼓强度、还原性和低温还原粉化等指标的实验分析，结合对高炉生产的影响程度，确定出唐钢目前烧结矿生产状态下适宜 FeO 含量在 7.5%~8.5%范围内。

关键词 烧结矿, FeO, 高炉

The Study of FeO Content of Sinter in Tangshan Iron and Steel Co., Ltd.

Hu Qichen[1], Wang Na[2], Hu Binsheng[3], Zhao Jun[4]

(1. Hebei Iron and Steel Technology Research Institute, Shijiazhuang 052000, China;
2. School of Materials Science and Engineering, Beijing University of Technology,
Beijing 100024, China; 3. School of Metallurgical Energy Engineering,
North China University of Science and Technology, Tangshan 063000, China;
4. Tangshan Iron and Steel Co., Ltd., Tangshan 063000, China)

Abstract The content of FeO in the sinter is very important to the quality of sinter, solid fuel consumption and production cost. Based on the need to reduce costs, the domestic steel plants are in the reduction of FeO content in sinter, due to differences in the level of equipment enterprises, raw fuel conditions and other factors limit, among enterprises in the FeO content of the sintered ore control there is a big difference. In this study, the industrial test laboratory sintering pot process simulation, sintering and blast furnace, sinter experimental analysis of particle size, drum strength, restore, and RDI and other indicators, combined with the impact of the blast furnace, determine the Tangshan Iron and Steel sinter ore production currently appropriate state FeO content in the range of 7.5 to 8.5 percent range.

Key words sinter, FeO, blast furnace

宝钢烧结工序技术升级改造实践

周茂军 张代华

（宝山钢铁股份有限公司炼铁厂，上海 200941）

摘 要 在实施烧结工序技术升级改造过程中，宝钢股份通过明确烧结工序功能定位和立足点，积极采用新工艺新装备新技术，形成了烧结生产典型工艺流程，夯实了“高效、低耗、节能、环保”的烧结工艺定位，将促进宝钢打造示范性样板烧结工厂。文章围绕对烧结工艺流程的再解析，总结了宝钢烧结技术改造实践。

关键词 烧结，技术升级，混合制粒，布料，烟气净化，余热回收

Practice of Technological Upgrading and Revamping of Baosteel Sintering Process

Zhou Maojun, Zhang Daihua

(Iron-making Plant, Baoshan Iron & Steel Co., Ltd., Shanghai 200941, China)

Abstract With the implementation of sintering process upgrading and revamping, Baosteel has adopted new process, new technology and new equipment positively and formed the typical sintering process by defining the sintering process functions and basic principles, which can strengthen the concept of “high efficiency, low consumption, energy saving, environmental protection”, and promote the building of a demonstration sintering plant. Focused on the the reanalysis of the sintering process, the reform practice of Baosteel sintering technique was summarized.

Key words sintering process, technology upgrading, mixing and granulating, feeding equipment, gas purification, waste heat recovery

The Practice of No. 4 Sintering Machine Putting into Production in Baosteel

Ma Luowen, Wang Xuming, Wu Wangping

(Ironmaking Plant, Baoshan Iron & Steel Co., Ltd., Shanghai 200941, China)

Abstract In order to meet the blast furnace’s need of the sinter and protect the environment in shanghai, Baosteel newly built the No.4 sintering machine that had an area of 600 square meters. The new sintering machine follows the principle of “advanced technology, mature process, reasonable economy” in technological process and equipment selection. It's built and put into operation that will ease the bottle-neck problem of energy saving and environmental protection in sintering process, but also make the material structure of blast furnace tend to be reasonable. After the No.4 sintering machine put into production the original sintering machine will be gradually eliminated. No.4 sintering machine in Baosteel adopted advanced, mature, stable and reliable technological process. The equipment and automatic control of sintering process have achieved domestic first-class level and international advanced level contrast to the similar type. In the aspect of energy-saving, it integrated with the mainstream and independent R&D equipment technology which can meet the “iron and steel industry clean production standards (sintering)” requirements. In the aspect of environmental protection, the use of efficient 4 electric field of electric dust catcher in the head of sintering machine and the the bag pulse dust collector in tail which can meet the stringent environmental protection requirements about dust discharge. In the aspect of flue gas treatment, using semi-dry desulfurization process which operated and maintained more economically and also set aside the denitration interface.In the aspect of raw material adaptability, the uses of three section of mixer include a strong mixing machine which greatly enhanced the adaptability of raw materials. The No.4 sintering machine will improve the competitiveness of Baosteel sintering unit and achieve the target of cleaner, low consumption, low cost and high efficiency iron making.

Key words snitering machine, energy saving, environmental protection, device configuration

提高宝钢原料场石灰石粉产量的生产实践

仇晓磊　李　刚　马洛文　冯军利　姜伟忠

（宝山钢铁股份有限公司炼铁厂，上海　200941）

摘　要　受破碎系统整合改造以及＜10mm 碳酸钙使用的影响，宝钢原料场石灰石粉产量无法满足烧结需求。本文介绍了为提高石灰石粉产量，宝钢原料场从降低＜10mm 碳酸钙的水分和提高粉碎系统作业效率两方面入手，采用天然料场翻晒的方式减少＜10mm 碳酸钙水分；使用指定原料槽和按比例输送＜10mm 碳酸钙和未选石灰石的方式提高输送台时；应用筛网粘附物料自行清理装置来减少清料时间，提高粉碎作业时间；通过上述技术进步和加强管理，宝钢原料场石灰石粉月产量提升了约 50%，有力地保证了烧结的使用量。

关键词　石灰石粉，粉碎，产量，水分，筛网

Practice of Increasing the Limestone Powder Production in Baosteel Raw Material Yard

Qiu Xiaolei, Li Gang, Ma Luowen, Feng Junli, Jiang Weizhong

(Ironmaking Plant, Baoshan Iron & Steel Co., Ltd., Shanghai 200941,China)

Abstract　Influced by the integrated renovation project of the crushing system and the use of the fine limestone,the limestone powder production in Baosteel raw material yard could not meet the sintering requirement. This paper introduces the measures using to increase the yield of limestone powder. Based on lowing the fine limestone moisture and enhancing the operating efficiency of the crushing system, Baosteel raw material yard aireded the fine limestone on the yard to reduce the water, used the specialized bin and the mixed transportation way to improving the delivery capability, applied a self-cleaning device of screen mesh blockage to reducing the cleaning time which is helpful for improving the working time of the crushing system. Through the technological progress and management strengthening, the limestone powder yield of Baosteel raw materials yard increased by about 50%, which effectively guarantees the usage of sintering.

Key words　limestone powder, crush, production, moisture, screen

近几年首钢京唐烧结技术进步

裴元东[1,3]　安　钢[2]　熊　军[2]　赵志星[1]　史凤奎[2]

康海军[2]　石江山[2]　吴胜利[3]

（1.首钢技术研究院，北京　100043；2. 首钢京唐公司，河北唐山　063200；
3.北京科技大学，北京　100083）

摘　要　首钢京唐烧结 2×550m^2 烧结机，承担为 2×5500m^3 高炉提供优质原料的任务。近年来，随着资源的逐渐

劣化，烧结矿质量受到考验。技术进步是应对原料劣化、提高烧结矿质量的重要手段。京唐烧结开展了优化配矿和熔剂结构、自然镁烧结、物料有害元素控制、料层分区测定评价、防箅条黏结、单辊导料槽加耐材保护、优化环冷配风等研究和攻关，使烧结矿的质量持续改进，烧结矿 TFe 含量逐步升高到57%以上，MgO 含量降到1.3%水平，返矿率（–6.3mm）从30%降到27%，转鼓指数保持在82.5%以上。

关键词 烧结，技术进步，配风，导料槽，MgO

解决烧结机箅条糊堵问题的生产实践

张 辉[1] 周明顺[1] 夏铁玉[2] 谢永清[2] 刘 帅[2]
翟立委[1] 刘 杰[1] 徐礼兵[1]

（1．鞍钢集团钢铁研究院，辽宁鞍山 114009；2. 鞍钢股份有限公司炼铁总厂，辽宁鞍山 114000）

摘 要 在对烧结机箅条糊堵原因进行分析的基础上，通过优化烧结原料结构，改进混合料预热系统、提高烧结终点温度和改善铺底料结构等措施，新烧车间烧结机箅条糊堵问题基本解决，烧结混合料温度升高；烧结主管负压降低 0.8kPa，利用系数提高 3.57%；烧结固体燃耗降低 0.8 kg/t，烧结矿转鼓强度提高 2.12 个百分点。

关键词 烧结机，箅条糊堵，黏结物，过湿层

The Production Practice to Solve the Problem of Pasting and Blocking Grate of Sintering Machine

Zhang Hui[1], Zhou Mingshun[1], Xia Tieyu[2], Xie Yongqing[2], Liu Shuai[2],
Zhai Liwei[1], Liu Jie[1], Xu Libing[1]

(1. Iron & Steel Research Institute of Angang Group, Anshan 114009, China;
2. Iron-making Plant of Ansteel Company Limited , Anshan 114000, China)

Abstract Based on the analysis of the reason of pasting and blocking grate of sintering machine, through of the optimization of sintering raw material structure, improving preheating system of sinter mixture, raising the temperature of BTP, improving the structure of hearth layer for sinter, the problem of pasting and blocking grate of sintering machine in new workshop was basically solved. The temperature of sinter mixture is increased, sintering utilization coefficient is improved by 3.57%, the solid fuel consumption is reduced by 0.8 kg/t, the sinter drum strength is improved by 2.12 percentage points.

Key words sintering machine, blocking grate, adhesion, wet layer

方大特钢烧结固体燃料及熔剂经济运行评价

胡小清 肖炸和 王 毅 钟国英

（方大特钢科技股份有限公司炼铁厂，江西南昌 330012）

摘 要 为适应原料结构的变化，方大特钢炼铁厂对烧结固体燃料及熔剂结构进行了优化及调整，生产实践结果表明，调整后，在保证烧结矿产质量无明显变化的情况下，配加无烟煤代替部分焦粉，石粉代替部分生石灰，燃料及熔剂成本大幅降低，实现了经济的低成本生产运行，为公司降本增效探索出一条新途径。

关键词 烧结，燃料，熔剂，经济运行

The Big Special Steel Sintering Solid Fuel and Flux Economy Evaluation

Hu Xiaoqing, Xiao Zhahe, Wang Yi, Zhong Guoying

(Fangda Special Steel Technology Co., Ltd., Nanchang 330012, China)

Abstract In order to adapt to the change of raw material structure, the major steel mills of sintering solid fuel and flux structure has been optimized and adjusted, production practice results show that, under the condition of ensure the quality of sintering mineral has no obvious change, with addition of anthracite coal in place of partial coke powder, replacing parts of the lime stone powder, fuel and flux greatly reduced cost, achieve the low cost of economic operation, reduce the efficiency for the company to explore a new way.

Key words sintering, fuel, flux, economic operation

环冷机配风技术在京唐烧结的应用

石江山[1] 樊统云[1] 裴元东[2] 张效鹏[2] 曹刚永[1]
史凤奎[1] 康海军[1] 许树生[1] 赵景军[1]

（1. 首钢京唐公司炼铁部，河北唐山 063200；2. 首钢技术研究院，北京 100043）

摘 要 环冷机冷却作为烧结工艺中最主要的冷却流程，其冷却效果的好坏直接影响了烧结矿的质量。针对京唐烧结环冷存在的冷却不均问题，通过仿真技术诊断出原因，并通过配风技术的研究及应用，提高了环冷冷却的效果，改善了烧结矿的质量，降低了烧结的返矿率 1.5%，并降低环冷鼓风机电量消耗至少 0.66kW·h/t，具有很大的应用与推广价值。

关键词 环冷配风，环冷机，烧结

Application of Ring Cooler Air Distribution Technology in Jingtang Sinter

Shi Jiangshan[1], Fan Tongyun[1], Pei Yuandong[2], Zhang Xiaopeng[2], Cao Gangyong[1],
Shi Fengkui[1], Kang Haijun[1], Xu Shusheng[1], Zhao Jingjun[1]

Abstract Ring cooler cooling cooling process as the main sintering process, the cooling effect will have a direct impact on the quality of sinter. Aiming at Jingtang sintering circulating cooling the existence of uneven cooling problems, through the simulation technique in the diagnosis of reason, and through the research and application of air distribution technology, improves the loop cooling effect, improve sinter quality, reduce the sinter return rate 1.5% sintering ring cold air blower, and reduce the power consumption of at least 0.66kW·h/t, with great application and popularization value.

Key words cold air distribution ring, ring cooler, sinter

高炉炼铁系统铁矿粉采购评价方式研究

张韶栋[1] 高 斌[1] 李洪玮[2] 国宏伟[2]

（1. 北京科技大学冶金与生态工程学院，北京 100083；
2. 苏州大学沙钢钢铁学院，江苏苏州 215000）

摘 要 针对钢铁行业发展面临严峻的成本挑战的现状，选择高性价比的铁矿粉对于降低炼铁成本至关重要。传统的铁矿粉性价评价方法主要有铁矿粉品位吨度价法、综合品位吨度价法、特殊成分价格折算法等，这些方法难以全面、准确地表征铁矿粉对于整个高炉炼铁系统有效的适应性；基于此，本文提出一种基于“铁矿粉-烧结-高炉炼铁”工艺流程模拟计算的铁矿粉性价评定方法，该方法综合考虑铁矿粉的价格、品位、脉石成分及其他特殊成分组成等因素，通过模拟计算，以炉渣性能为控制目标，得到利用该种铁矿粉进行高炉炼铁系统生产所对应的铁水成本，并以此作为铁矿粉性价评定的统一指标，从而为钢铁企业选购高性价比的铁矿粉提供指导。

关键词 铁矿粉，烧结矿，高炉炼铁，评价

Study on the Evaluation Methods of Iron Ores Purchased for Blast Furnace Ironmaking System

Zhang Shaodong[1], Gao Bin[1], Li Hongwei[2], Guo Hongwei[2]

(1. School of Metallurgical and Ecological Engineering, University of Science and Technology Beijing, Beijing 100083, China;
2.School of Iron and Steel, Soochow University, Suzhou 215000, China)

Abstract Nowadays, confronted with the increasing pressure from the production cost, it is of vital importance for the worldwide steel companies to use the type of cost-effective iron ore fines. Conventional evaluation methods regarding the cost performance of iron ore fines usually include applying the price unit of dry metric per ton about either the average Fe content or the comprehensive grade of iron ore fines, cost conversion method based on the price of specific compositions, etc. The application of these evaluation methods, however, often fails to fully characterize the adaptability of iron ore fines to the whole system of blast furnace ironmaking with a satisfactory accuracy. Therefore, in this paper, a novel evaluation system concerning the cost performance of iron ore fines was established on the basis of the calculation results of engineering simulation about the process involving iron ore fines, sintered ore, and blast furnace technology. It is noteworthy that, several factors including the price of iron ore fines, Fe content, and the chemical composition of gangue and other special components, etc. were taken into account to conduct the simulation calculation. Besides, for the purpose of forming the blast furnace slag with favorable properties, the corresponding hot metal cost was derived under the assumption that the iron-bearing material consisted of only sintered ore. The procedure described briefly above yields the evaluation index capable of determining the cost performance of iron ore fines, thus providing a valuable guide for steel companies in the course of buying iron ore.

Key words iron ore fines, sintered ore, blast furnace ironmaking, evaluation

含铁原料评价系统的建立及采购决策软件的开发

史志文　张建良　李新宇

（北京科技大学冶金与生态工程学院，北京　100083）

摘　要　国内外含铁原料性能差异较大，为了更合理有效地评价和利用各种铁矿资源，建立单种铁矿综合性能评价体系，对含铁原料采购提出建议。针对某钢厂实际情况，本文提出含铁原料评价标准及计算公式，建立了矿粉评级制度和采购建议，采用 VC++开发软件模型，对各种铁矿石做出快速、定量的综合评价，为钢铁企业采购性价比高的含铁原料提供决策建议。

关键词　含铁原料，性能评价，采购决策

Establishment of Iron-containing Material Evaluation System and Development of Software Purchasing Decision-making

Shi Zhiwen, Zhang Jianliang, Li Xinyu

(School of Metallurgical and Ecological Engineering, University of Science and Technology Beijing, Beijing 100083, China)

Abstract　Iron-containing materials are quite different performance at home and abroad, in order to evaluate and use iron ore resources more reasonably and effectively , we establish a single iron ore comprehensive performance evaluation system, which can offer proposals for the iron-containing raw material procurement . For the actual situation of a steel company, this article present the evaluation criteria of containing material and formulas, establish fine ore rating system and purchase proposals, use VC ++ to develop software model, it is aimed to make rapid, comprehensive quantitative evaluation to the various iron ore and to take advice for iron and steel enterprise about procuring cost-effective iron-containing materials.

Key words　iron-containing material, performance evaluation, purchase decision-making

1.3　高炉炼铁/Iron Making

“新常态”下高炉炼铁技术转型升级和创新之路

杨天钧　张建良　刘征建　李克江

（北京科技大学冶金与生态工程学院，北京　100083）

摘　要　面对新的经济形势，本文试图阐述中国炼铁工业的“新常态”：产量接近饱和、产能严重过剩、先进与落后指标并存、现代与传统理念并存。作者分析了高炉炼铁行业所面临的新挑战：在国家政策的约束条件下，炼铁工序肩负着整个钢铁行业在资源、能源和减少污染排放方面的艰巨责任，与此同时，加快高炉大型化步伐，完善高效、安全长寿高炉技术已经成为对炼铁行业的迫切要求。面对“新常态”的挑战，炼铁工作者需要转变技术创新的观念和理念，以实现知识、技术和管理创新；大力开发和推广炼铁新技术，诸如：烧结—高炉配料一体化技术，拓展兰炭等辅助燃料在高炉、烧结中应用技术，高炉可视化及控制技术，高炉粉尘处理技术，镁质球团技术，超级烧结技术等；与此同时，不断提高高炉操作水平，在高风温、富氧喷煤、安全长寿等方面取得新进展，从而实现高炉炼铁技术的转型升级和创新。

关键词　新常态，炼铁，高炉

The Transformation, Upgrade and Innovation of China's Blast Furnace Ironmaking Technologies in the "New Normal" Economy

Yang Tianjun, Zhang Jianliang, Liu Zhengjian, Li Kejiang

(School of Metallurgical and Ecological Engineering, University of Science and Technology Beijing, Beijing 100083, China)

Abstract　Facing the updated economical circumstance, the "New Normal" of Chinese ironmaking technologies was elaborated in detailed in this paper: the production is close to saturation with excess production capacity; the advanced and the backward technical-economical indicators and concept coexist. The new challenges of blast furnace ironmaking were also analyzed: under the high pressure from the national strict constraint policy, ironmaking industry shoulders the responsibilities in terms of resources, energy and pollution emission in the whole iron and steel industry; Meanwhile, the development of large-scale, long campaign life and safe blast furnace becomes the new demand in ironmaking field. Facing those new challenges, Chinese ironmaking workers should transform the traditional concept to achieve innovation in knowledge, technologies as well as management. To ultimately achieve transformation and upgrade, new technologies such as comprehensive optimization of sintering-blast furnace process, extension of supplementary fuels for blast furnace and sintering process, develop of visualization and controlling system for blast furnace, dust treatment technology, production of magnesium containing pellet, super sintering, etc. should be developed and adopted. Meanwhile, the operation techniques should also be improved to achieve high blast temperature, economical coal injection rate and oxygen enrichment rate, and long campaign life blast furnace.

Key words　new normal, ironmaking, blast furnace

国外炼铁生产及技术进展

沙永志　曹　军

（中国钢研科技集团，北京　100081）

摘　要　2014 年，世界高炉炼铁产量为 11.795 亿吨，其中国外全部产量为 4.679 亿吨。直接还原和熔融还原产量在增长，分别为 7522 万吨和 730 万吨（2013 年），新工艺技术的研发在继续。欧盟的生铁产量有所下滑，但单炉的生产率提高，炉料结构得到改进，烧结烟气得到有效治理，应用数学模型和专家控制系统等技术使高炉燃料比得到进一步降低。北美生铁产量下降到 4000 万吨，普遍采用球团为主要原料以及高炉煤和气的混合喷吹，推广应用高炉操作远程监控诊断及标准化系统。日本开发应用 LCC，烧结喷天然气，RCA 技术，铁焦技术，高炉可视化技术等，取得良好效果。Course50 项目完成喷吹 COG 和 RCOG 试验和模拟计算，降低高炉碳耗 3%。韩国炼铁的技术动向是高炉大型化和 Finex 工艺的开发应用。南美的炼铁主要解决铁矿质量下降带来的烧结和高炉生产问题。

关键词 国外炼铁，高炉，烧结，炼铁技术

Ironmaking Production and Technology Progress Overseas

Sha Yongzhi, Cao Jun

(China Iron & Steel Research Institute, Beijing 100081, China)

Abstract Total iron production from BF around world was 1.1795 billion tons in 2014, 467.9 million tons from overseas. New DRI process and smelting process are under development. Iron production in EU decreased, with increase of single BF productivity, optimal burden constitutes, and effective treatment of sinter flue gas. BF efficiency was improved with application of computer simulation and expert system. Iron production in North America decreased to 42 million tons in 2014. Most of BF use pellet as only or major ferrous burden. Co-injection of coal and natural gas is common practice. RMDS is being applied to BF operation. Japan applied several techniques to commercial operations with good results, including LCC, super-sinter ROXY, RCA, ferrous coke, and 3D-Venus. In Course 50 project, 3% of carbon consumption reduction was realized from injection of COG or RCOG, confirmed by both experiment BF and simulation calculation. Two trends of ironmaking in South Korea are BF enlarging and application of Finex process. Major problems faced and solved for ironmaking in South America are sintering productivity and BF operation, caused by lowering quality of iron ores.

Key words ironmaking oversea, BF, sinter, ironmaking technology

落实高炉低碳炼铁生产方针的探讨

项钟庸[1] 王筱留[2] 刘云彩[3] 邹忠平[1] 欧阳标[1]

（1. 中冶赛迪工程技术股份有限公司，重庆 401122；2. 北京科技大学冶金与生态工程学院，北京 100083；3. 北京首都钢铁公司，北京 100083）

摘　要 以精料为基础，高效、低耗、优质、长寿、环保的炼铁技术方针，符合低碳炼铁、节能降耗及降低成本的要求。低碳炼铁是当前高炉炼铁发展的必然趋势，符合国家的长远目标。通过评价高炉炼铁生产的新方法来具体落实低碳炼铁的方针。

关键词 炼铁方针，低碳炼铁，评价方法

Study on Implementation of Blast Furnace Low-carbon Production Principle

Xiang Zhongyong[1], Wang Xiaoliu[2], Liu Yuncai[3], Zou Zhongping[1], Ouyang Biao[1]

(1. CISDI Engineering Co., Ltd., Chongqing 401122, China;
2. School of Metallurgical and Ecological Engineering, University of Science & Technology Beijing, Beijing 100083, China; 3. Shoudu Steel Work, Beijing 100083, China)

Abstract Based on beneficiated burden material, iron-making technical principles, including high efficiency, low consumption, high quality, long campaign life and environment friendly, meet the requirements of low-carbon production, energy-saving and cost reducing. Low-carbon iron-making production, which is the inevitable trend of current iron-making development, meets the country's long-term objectives. This study evaluated new methods of blast furnace iron-making production to specify implementation of low-carbon iron-making principles.

Key words iron-making principles, low-carbon iron-making, evaluation method

邯宝炼铁厂低成本炼铁实践

刘志朝

（邯郸钢铁股份有限公司）

摘　要 在原燃料条件日益劣化的不利条件下，邯宝炼铁厂通过优化原燃料结构、提高技术操作水平以及控制合理的操作炉型等举措，在保证炉况基本稳定的前提下，实现了低成本炼铁，收到了很好的经济效益。

关键词 低耗炼铁，炉料结构，操作制度，操作炉型

Low Cost Ironmaking Practice in Hanbao Ironmaking Plant

Liu Zhichao

(Handan Iron & Steel Co., Ltd.)

Abstract Under the adverse conditions of raw material and fuel condition deteriorates increasingly, Hanbao ironmaking plant optimized the raw material and fuel structure, raised the level of technical operation and came into being a reasonable operation furnace profile. In the premise of ensuring the basic stability of the furnace.We realized the way of low cost of ironmaking and received the good of economic benefits.

Key words low consume ironmaking, charge structure, operating system, furnace profile of operation

鞍钢鲅鱼圈降低炼铁成本的生产实践

孙俊波　杨金山　马贤国

（鞍钢股份鲅鱼圈钢铁分公司炼铁部，辽宁营口　115007）

摘　要 为了应对当前钢铁市场形势，降低企业工序成本，鞍钢鲅鱼圈高炉实施稳定产量低成本生产的运行模式，进行低成本配矿，混匀矿造堆过程配加混杂料，烧结配料使用廉价固体燃料，高炉喷吹炼焦除尘灰和提高烟煤配比，使用粒铁和炉头焦替代烧结矿和焦炭，炼铁成本大幅度降低。

关键词 低成本配矿，经济料，经济效益

Practice to Reduce the Iron Making Produced Costing in Ansteel Bayuquan Iron & Steel Subsidiary

Sun Junbo, Yang Jinshan, Ma Xianguo

(Ironmaking Department of Ansteel Bayuquan Iron & Steel Subsidiary, Yingkou 115007, China)

Abstract　In order to reply present steel market condition, and to reduce process cost. The running model of controlling productivity and decreasing produced cost is put into use in Ansteel Bayuquan Iron & Steel Subsidiary, with inexpensive ore blend, blend ore with mix material in piling process, inexpensive solid fuel used in sintering blend ores, coking dust are injected in blast furnace and enhance the proportion of bituminous coal, a part of sintering ore and cokes are replaced by granular iron and coke of oven end, iron making cost is reduced remarkable.

Key words　inexpensive ore blend, inexpensive material, economic benefit

鞍钢 10 号高炉降低燃料比操作实践

韩淑峰　赵正洪　曾　宇

（鞍钢股份有限公司炼铁总厂，辽宁鞍山　114001）

摘　要　对鞍钢 10 号高炉降低燃料比的生产操作实践进行总结。通过采取改善入炉原燃料质量，调整高炉风口面积、布料角度、布料矩阵、放料顺序等技术手段，提高了高炉的煤气利用率，改善了高炉的煤气流分布，降低了高炉的消耗水平，实现高炉稳定、顺行、低燃料比运行，提升了高炉煤气流控制技术和应对技术。

关键词　燃料比，布料制度，放料顺序，煤气利用率

Production Practice of No.10 BF on Reducing Fuel Ratio in Ansteel

Han Shufeng, Zhao Zhenghong, Zeng Yu

(Ansteel Ironmaking Plant, Anshan 114001, China)

Abstract　The paper summarizes production practice of No.10 BF on reducing fuel ratio in Ansteel. In this furnace a series of advanced technologies are taken，such as Adjusting tuyere area, the distribution schedule, distribution angle, discharge sequence. That increases utilization rate of blast furnace gas to improve the gas flow distribution and decrease fuel consumption. The blast furnace gas control and handling techniques were improved, which helped the BF run at low fuel ratio smoothly and steadily.

Key words　fuel ratio, distributing schedule, discharge sequence, utilization rate of gas

CO-H_2还原浮氏体的竞争行为研究

王冬东　胡招文　严　明　黄雅萍　徐　健　吕学伟　白晨光

（重庆大学材料科学与工程学院，重庆　400044）

摘　要　本研究以浮氏体还原为对象，采用热重法实验研究在850℃下，三种不同还原气氛下的还原机理，由此初步探索CO-H_2间的竞争行为。通过计算分析不同气体组成下，浮氏体还原转化率在0.1~0.5范围内的经验机理函数幂指数n和类化学速率常数的均值k^*，结合两种气体混合还原反应的速率与两种气体分别进行还原的速率的关系，获得以下结论：本实验条件下浮氏体的还原遵守三维随机成核随后成长机理；在转化率0.1~0.5之间，随着转化率的升高，三种还原气体组成下浮氏体的还原速率均呈增加的趋势，并且在各个还原度水平下，CO气体的还原速率最慢，CO-H_2混合气体的还原速率最快但要小于CO与H_2单一成分还原速率的线性加和，H_2气体的还原速率则介于两者之间。

关键词　CO，H_2，浮氏体还原，热重分析

Competitive Behavior of Wustite Reduction by CO-H_2 Mixture Gas

Wang Dongdong, Hu Zhaowen, Yan Ming, Huang Yaping,
Xu Jian, Lv Xuewei, Bai Chenguang

(College of Materials Science and Engineering, Chongqing University, Chongqing 400044, China)

Abstract　To explore competitive behavior of wustite reduction by CO-H_2 initially, the reduction reaction of the wustitebased on TGAunder the temperature at 850℃ and three different gas compositions.Based on exponent n of mechanism function and the average value of based chemical rate constant k^*, the relationship between the rate of using two gas mixed reductions and the rate of reduction of the two gases used respectively are analyzed. The results show that: Under the experimental conditions, the reaction is subject to random nucleation followed by growth of three-dimensional.When the conversion rate increases from 0.1 to 0.5, the reduction rate of all are increasing.While under each of conversion the reduction of CO is the slowest, the mixed gas is the fastest and the reduction rate of H_2 is between them.

Key words　carbon monoxide, hydrogen, wustite reduction, TGA

秘鲁氧化块矿在首钢高炉上的冶炼实践与分析

李荣昇[1]　孙　健[1]　徐　萌[1]　陈　辉[1]　朱　利[2]　梁海龙[1]

（1. 首钢技术研究院，北京　100043；
2. 秦皇岛首秦金属材料有限公司，河北秦皇岛　066300）

摘　要　为了验证秘鲁氧化块矿对高炉实际冶炼生产的影响，首先在实验室对其理化性能和冶金性能进行了测试，之后进行了为期一个月的工业入炉实验。实验室理化性能和冶金性能检测结果表明：秘鲁氧化块矿铁品位比较低，所含有害元素较多；铁相多为磁铁矿，质地不致密、空隙较多，脉石含量多；低温还原性能和还原性一般，长期大量使用可能会对高炉透气性产生不利影响；热爆裂性能好，几乎不发生爆裂，明显优于其他入炉生矿；综合炉料熔滴性能检测结果显示秘鲁氧化块矿可基本满足高炉生产需要，建议实际氧化块矿入炉比例不超过5%。工业入炉实验结果表明：在配比低于3%的情况下，秘鲁氧化块矿的配加对高炉实际稳定顺行生产不会造成大的影响；通过铅平衡计算，配加秘鲁氧化块矿会大幅增加入炉铅负荷，最高达到0.0822kg/t，但是低于0.15kg/t的行业标准。

关键词　秘鲁氧化块矿，理化性能，冶金性能，工业试验

Practice and Analysis of Smelting on the Peru Oxidation Ore in the Blast Furnace of Shougang

Li Rongsheng[1], Sun Jian[1], Xu Meng[1], Chen Hui[1], Zhu Li[2], Liang Hailong[1]

(1. Shougang Research Institute of Technology,Beijing 100043, China;
2. Qinhuangdao Shouqin Metal Materials Co.,Ltd.,Qinhuangdao 066300, China)

Abstract To validate the influence of oxidation ore of Peru in the progress of blast furnace smelting, physicochemical properties and metallurgy performance were carried out in the lab and the industrial test was carried on for a month. The laboratory test results show that: the grade of Peru oxidation ore was lower , there were more harmful elements in the Peru oxidation ore; magnetite was the main phase and the amount of gangue was larger; the poor performance of RDI and RI might cause the air permeability problem in blast furnace; the DI was better than the other raw ore according to the laboratory test results; the droplet test results show that Peru oxidation ore could meet the blast furnace production needs; the proposed ratio was no more than five percent. The industrial test indicated that the adding of Peru oxidation ore would not affect the normal production when its proportioning was less than three percent; the lead load of furnace would sharply elevated with the adding of Peru oxidation ore, which might top out at 0.0822kg/t by the balance calculation of lead, while the industry standard of the lead load in blast furnace is 0.15kg/t.

Key words the oxidation ore of Peru, physicochemical properties, metallurgy performance, industrial test

宣钢高炉合理炉料结构的试验研究

赵晓杰[1]　刘　然[1]　李豪杰[2]

（1. 华北理工大学，河北唐山　063009；2. 河北钢铁集团宣钢公司，河北宣化　075100）

摘　要　本文根据宣钢目前高炉炉料结构进行了熔滴性能试验研究。单矿的熔滴试验结果表明：宣钢常用的 2 种烧结矿软化性能较好，但滴落性能较差，1520℃仍未滴落；3 种球团矿中自产球总体熔滴指标最好但软化区间最长（313℃）；2 种块矿中 PB 矿开始软化温度最低软熔区间最长，试验中未能滴落。蒙古矿软熔性能较好，但滴落温度高达 1518℃。通过对 15 种配矿方案进行的熔滴试验，结果表明：降低烧结矿配比，提高 PB 矿比例可改善炉料的熔滴性能。在试验条件下，配矿方案为“烧结矿（68%）+球团矿（16%）+PB 矿（16%）”的炉料结构熔滴性能最佳。

关键词　高炉，炉料结构，配矿，熔滴试验

Experimental Study on Xuanhua Steel BF Reasonable Burden Structure

Zhao Xiaojie[1]，Liu Ran[1]，Li Haojie[2]

(1. North China University of Science and Technology, Tangshan 063009，China;
2. The Xuanhua Steel Company of HBIS, Xuanhua 075100，China)

Abstract According to Xuanhua Steel blast furnace structure currently, we carried out the softening-dripping experiment. Single mine drop test results show that the commonly used two kinds of plant of Xuanhua Steel sinter softening with good performance, but drop performance is poorer, 1520℃ still dripping; The soft melting of zichan pellet is the best in three kinds of pellets,but its softening range is longest(313℃).The start softening temperature of PB ore is minimum and its soft melting ranges longest. In addition, PB ore failed to drip in the test. Menggu ore is good of soft melting performance,but drop temperature as high as 1518℃. Through the 15 kinds of ore programs for drop test, the results show that reducing sinter ratio and increasing the proportion of ore PB can improve the softening-dripping performance. Under the experiment condition, ore matching scheme as "sinter(68%) + ore pellets(16%)+PB ore(16%)", the burden structure of molten drops of best performance.

Key words blast furnace, burden structure, burden design, melting and dripping experiment

炉料结构对高炉冶炼的影响

林成城[1] 沈红标[2]

（1. 宝钢股份炼铁厂，上海 201900；2. 宝钢股份研究院，上海 201900）

摘 要 高炉炉料结构是精料技术的重要组成部分，精料技术不仅仅局限于建立完善的原燃料质量管理标准，还要关注各种炉料合理搭配，以及炉料高温软熔性能。高炉的软熔带与炉料高温软熔性能密切相关，不同炉料结构对高炉炉况稳定顺行、煤气流分布以及透气性等高炉冶炼操作有至关重要的影响。本文根据宝钢不同炉料结构生产实绩，结合各种炉料高温性能分析，解析不同炉料结构对高炉煤气流分布和透气性的影响规律，以及不同炉料结构下高炉操作技术对策，探讨优化炉料结构技术方向。

关键词 高炉，炉料结构，软熔性能，透气性

The Influence of Burden Structure on Blast Furnace Smelting

Lin Chengcheng[1], Shen Hongbiao[2]

(1. Ironmaking Plant, Baoshan Iron & Steel Co., Ltd., Shanghai 201900, China;
2. Research Institute, Baoshan Iron & Steel Co., Ltd., Shanghai 201900, China)

Abstract Blast furnace burden structure is an important part of quality-charge technique. The quality-charge technique are not limited to establish perfect the raw material quality management standards, but also focus on a variety of burden reasonable collocation, and the high temperature softening-melting property of burden. The cohesive zone of blast furnace is closely related to the high temperature softening-melting property of burden. The different burden structure on the stability of blast furnace, the gas flow distribution and permeability of blast furnace smelting operation has a crucial impact. According to different burden structure of baosteel production performance, combining with a variety of burden high temperature performance analysis, the paper parses the influence law on gas flow distribution and permeability of blast furnace under the different burden structure, and the technology countermeasures of blast furnace operation under different burden structure, and explores technology direction of optimizing the burden structure.

Key words blast furnace, burden structure, softening-melting property, permeability

Mechanisms Determining Softening/Melting and Reduction Properties of Lumps and Pellets

Loo Chin Eng[1], Penny Gareth[1], Matthews Leanne[1], Ellis Benjamin[2]

(1. Centre for Ironmaking Materials Research, Discipline of Chemical Engineering School of Engineering, Faculty of Engineering & Built Environment, The University of Newcastle, NSW 2308, Australia; 2. BHP Billiton, 1 Technology Court, Pullenvale, Qld 4069, Australia)

Abstract Although softening/melting tests are widely used to assess ferrous material properties in the cohesive zone of a blast furnace, there is little understanding of the mechanisms causing the softened bed to densify. Quenched beds of three acid feed materials–a porous hematite goethite lump (PHGL), a dense high silica lump (DHSL) and a high silica pellet (HSP)-were used to provide information on the role of silica and porosity on softening and melting behavior. Reduction of all three materials is topochemical, resulting in the formation of a distinct metallic iron shell and a wustite core. The efficiency of the wustite to iron reduction process is dependent on the penetration of reducing gas through the shell and the structure of the core. This is enhanced for samples with higher porosity and low silica content, which minimizes the formation of fayalite, a low melting compound formed from the reaction of silica with iron oxide. With higher levels of fayalite, the core can become sealed off from the reducing gases, leading to greater levels of wustite that must undergo direct reduction at temperatures above 1000℃. Furthermore, the presence of fayalite in significant quantities can act as a lubricant between particles, facilitating sliding and bed densification further reducing bed permeability and reduction efficiency. For the three materials tested, the higher porosity and lower silica content of the PHGL improved the low temperature reducibility of the sample relative to DHSL and HSP, leaving much less wustite for direct reduction. This study also shows that increasing the reduction rate of wustite to iron also has two other benefits. Firstly, with higher reducibility, fayalite levels decrease because iron oxide, not metallic iron, is required in its formation. Secondly, the level of exuded liquid wustite in inter-particle pores decreases which will improve cohesive zone permeability. These results suggest that as a complementary feed to basic sinter, a porous low-silica lump such as PHGL should yield better performance in a blast furnace.

Key words blast furnace, cohesive zone, softening and melting, bed properties

Softening Behaviors of High Al_2O_3 Iron Blast Furnace Feeds

Wang Dongqing[1,2], Chen Mao[2], Zhang Weidong[1], Zhao Zhixing[1], Zhao Baojun[2]

(1. Shougang Research Institute of Technology, Beijing, China; 2. The University of Queensland, Brisbane, Australia)

Abstract High temperature softening experiments of industrial blast furnace feeds – sinter, pellet and lump – with two different levels of Al_2O_3 content have been carried out to simulate the blast furnace condition. The iron feeds were packed in sandwich structure (coke/Fe-material/coke) with load on top and 30 CO/70 Argas was blowing from bottom of crucible through the beds.The temperatures and displacements were recorded simultaneously to construct the softening curves. The softening starting temperatures (T_s), softening ending temperatures (T_e) and meltdown temperatures (T_m) are obtained from the softening curves and compared for different feeds. The effects of Al_2O_3 to the softening behaviorshave been revealed. Separated quenching experiments were carried out under the same atmosphere and temperature profile. The quenching temperatures were selected from the critical temperatures based on the softening curves. The phase assemblages and compositionswereanalyzed by Electron Probe X-ray Microanalysis (EPMA). FactSage calculations are also used to assist the prediction of phase proportions at different temperatures.

Key words softening, Al_2O_3, FactSage prediction, blast furnace

高炉软熔带内铁矿石初渣生成行为研究

吴胜利 王 哲 刘新亮 寇明银

（北京科技大学冶金与生态工程学院，北京 100083）

摘 要 本研究通过试验模拟高炉冶炼条件，采用“历程中断法”试验研究了五种典型铁矿石的初渣生成行为。试验结果表明，酸性铁矿石的初渣在较低温度下即开始生成，且疏松块矿的初渣开始生成温度最低；碱性铁矿石的初渣在较高温度下才开始生成，且烧结矿的初渣开始生成温度显著高于自熔性球团矿；酸性铁矿石的初渣完全生成温度则介于烧结矿与自熔性球团矿之间。酸性铁矿石软熔带下沿初渣主要以 FeO、SiO_2、Al_2O_3 为主，矿物主要是高熔点的莫来石，因而使得其初渣黏度较高，在试验过程中发生了明显的液泛现象；碱性铁矿石软熔带下沿初渣则以 CaO、SiO_2 为主，自熔性球团矿初渣矿物以黄长石为主，而烧结矿初渣含有较多的硅酸二钙，碱性铁矿石软熔带下沿初渣黏度均低于酸性铁矿石，其初渣生成过程未发生明显的液泛现象。

关键词 初渣，软熔带，高炉

Primary-slags Formation Behaviors of Iron Ores in Cohesive Zone of Blast Furnace

Wu Shengli, Wang Zhe, Liu Xinliang, Kou Mingyin

（School of Metallurgical and Ecological Engineering, University of Science and Technology Beijing, Beijing 100083, China）

Abstract By simulating the blast furnace smelting condition, a process-interruption experimental method was used to study the primary-slags formation behaviors of five different respective iron ores. The results shows that primary-slags of acidic iron ore are generating in low temperature, and the primary-slags generation temperature of loose lump is the lowest. Alkaline iron ores primary-slags are generating in high temperature, the generation beginning temperature of sinters is higher than fluxed pellets obviously. The generation completing temperature of loose iron ores are between sinters and fluxed pellets. The loose iron ores primary-slags composition is mainly contain FeO, SiO_2 and Al_2O_3, and the main part of mineral composition is mullite with high melting point, causing higher primary-slags viscosity, flooding phenomena can be found easily during experiment. The alkaline iron ores primary composition is mainly contain CaO and SiO_2, the mineral composition is melilite in fluxed pellets primary-slags and dicalcium silicate in sinters primary-slags. Flooding phenomena was not obviously during experiment.

Key words primary-slags, cohesive zone, blast furnace

Phase Equilibria in the "FeO"-CaO-SiO_2-Al_2O_3-MgO System for Blast Furnace Primary Slags

Jang Kyoung-oh[1], Ma Xiaodong[1], Zhu Jinming[2], Xu Haifa[2], Wang Geoff[1], Zhao Baojun[1]

(1. School of Chemical Engineering, The University of Queensland, Brisbane, Australia;

2. Baoshan Iron and Steel Co., Ltd., Shanghai, China)

Abstract In modern iron blast furnace operations, utilising low grade iron ores and poor quality fuels is essential for low cost ironmaking. The accurate information of phase equilibria for complex slag system is necessary for better understanding the changes of slag chemistry in blast furnace.The compositions and liquidus temperatures of primary slag in cohesive zone was determined with the reduction of industrial sinters. The phase equilibria in the "FeO"-CaO-SiO_2-Al_2O_3-MgO system with metallic iron saturation have been comprehensively investigated by quenching technique and EPMA analysis.The pseudo-ternary section (CaO+SiO_2)-Al_2O_3-MgO at 50 wt% "FeO" was constructed and a series of pseudo-binaries were derived to be easily used in industrial applications.The major primary phase was found to be oxide[(Fe, Mg)O] and the liquidus temperatures increase significantly with increasing MgO in this primary phase field. Al_2O_3 has little effect on liquidus temperature in this primary phase field.Liquidus temperatures increase in melilite primary phase field and decrease in Ca_2SiO_4 primary phase field with increasing Al_2O_3 concentration.Liquidus temperatures are not sensitive to MgO concentration in both melilite and Ca_2SiO_4 primary phase fields.

Key words ironmaking blast furnace, liquidus temperature, primary slag, sinter

Viscosity Model for Blast Furnace Slags Including Minor Elements

Chen Han[1], Mao Chen[1], Weidong Zhang[2], Zhixing Zhao[2], Tim Evans[3], Baojun Zhao[1]

(1. The University of Queensland, Brisbane, Australia;

2. Shougang Research Institute of Technology, Beijing, China;

3. Rio Tinto Iron Ore, Australia)

Abstract Viscosity is one of the most important metallurgical properties of ironmaking slags. In the present study, the viscosities of SiO_2-CaO-Al_2O_3-MgO based slags containing minor elements, including Na_2O, K_2O, FeO, CaF_2, B_2O_3, CaS, TiO_2 and MnO have been systematically studied. The viscosities of synthetic slags with the additions of minor elements have been measured using the rotational rheometer at high temperature. After viscosity measurement, the sample was quenched, mounted, polished and analysed by electron probe micro-analysis. It was found that all of minor element additives will decrease the slag viscosity. The ability to reduce the viscosity can be ranked as CaF_2 > B_2O_3 > CaS > Na_2O >K_2O > TiO_2 > MnO > FeO. Based on the present measurements and available viscosity data from literatures, a viscosity model was proposed to describe the viscosity of SiO_2-CaO-Al_2O_3-MgO based slags containing minor elements. The model predictions were found to have an outstanding agreement with the experimental data at blast furnace (BF) slag composition range. The present work will improve the understanding of the behaviours of BF slags with inclusion of minor elements. The viscosity model developed will be used for industrial process optimization.

Key words blast furnace, ironmaking slag, minor element, viscosity

Sulphide Capacity and Sulphur Partition between Blast Furnace Slags and Hot Metal

Ma Xiaodong[1], Xu Haifa[2], Zhu Jinming[2], Wang Geoff[1], Zhao Baojun[1]

(1. School of Chemical Engineering, The University of Queensland, Brisbane, Australia;
2. Baoshan Iron and Steel Co., Ltd., Shanghai 201900, China)

Abstract Desulphurization has been of more importance to control the quality of hot metal with increasing utilization of low-grade iron ores and poor-quality fuels and reducing the slag volume by decreasing MgO concentration. To improve performance with better understanding the desulphurization in blast furnace, sulphide capacities of CaO-SiO_2-Al_2O_3-MgO system relevant to blast furnace slags were experimentally determined using gas-slag equilibration technique. The results show that sulphide capacity increases with the increase in temperature, CaO/SiO_2 weight ratio and MgO at fixed Al_2O_3 concentration. The experimental results were further used to compare with reported models of sulphide capacity. In addition, the equilibrium sulphur partition ratio between slag and hot metal was calculated from sulphide capacity with oxygen activity in the melt, which is also compared with industrial data and the laboratory experiments of kinetic desulphurization of hot metal.

Key words blast furnace slag, sulphide capacity, sulphur partition

高炉渣熔点计算模型

干 磊

（江西理工大学冶金与化学工程学院，江西赣州 341000）

摘 要 熔点是高炉渣最基本的热物理性质之一，但其计算模型几乎未有报道。本文在高炉渣完整成分范围内，建立了其 CaO-SiO_2-Al_2O_3-MgO 体系合成高炉渣和工业高炉渣的熔点计算模型。该模型采用两步法建立，即首先采用多项式确定初晶相种类，然后在初晶相内采用多项式计算熔点。模型的计算值与实测值的误差仅为 10℃，计算得到的等熔点线与广泛使用的平衡相图非常一致。对于工业高炉渣，每 1%的少量成分（非 CaO、SiO_2、Al_2O_3 、MgO）会降低对应的合成高炉渣熔点 15℃。本文提出的高炉渣熔点计算模型具有准确、简便的优点，并已编制成 Excel 计算程序。

关键词 高炉渣，熔点，工业渣，初晶相

Liquidus Temperature Models for Synthetic and Industrial Blast Furnace Slags

Gan Lei

(School of Metallurgical and Chemical Engineering, Jiangxi University of Science and Technology, Ganzhou 341000, China)

Abstract Liquidus temperature (T_{liq}) is one of the most fundamental properties of blast furnace (BF) slags. However, calculation model for T_{liq} is extremely scarce. In this work, liquidus temperature models CaO-SiO_2-Al_2O_3-MgO system for synthetic and industrial BF slags were proposed. The model covers the full composition ranges of BF slags. It was established according to a two-stage strategy, which involves firstly identifying the primary phase by polynomials, and then describing the T_{liq} within identified primary phase by another one. The average error between calculated and measured T_{liq} is only 10℃. The calculated iso-T_{liq} lines agree very well with the classic equilibrium phase diagram. For industrial slags, it is found that 1% of minor constituents (except for CaO, SiO_2, Al_2O_3 and MgO) reduce the T_{liq} by 15℃. Current model is highly accurate and convenient. It is available as an Excel calculator.

Key words blast furnace slag, liquidus temperature, industrial slag, primary phase

Cr_2O_3对高炉渣黏度及熔化性温度的影响

钱 毅 张建良 王志宇 王 朋

（北京科技大学冶金与生态工程学院，北京 100083）

摘 要 利用RTW-10熔体综合物性测定仪，采用内柱体旋转法测量Cr_2O_3含量对CaO-SiO_2-MgO-Al_2O_3-Cr_2O_3渣系的黏度及熔化性温度的影响，并利用FactSage热力学计算软件分析CaO-SiO_2-MgO-Al_2O_3-Cr_2O_3渣系的相图。实验结果表明，固定高炉渣碱度为1.11时，随着炉渣中Cr_2O_3含量的增加，炉渣黏度下降，炉渣熔化性温度升高。当Cr_2O_3含量为1%~3%时，炉渣熔化性温度增加趋势较慢，当Cr_2O_3含量为4%时，炉渣熔化性温度急剧升高。结合相图表明，该渣系随着温度的降低、Cr_2O_3含量的增加，析出高熔点初晶相为Cr_2O_3，其熔点约为2435℃，其会使炉渣熔化性温度升高。计算条件为1500℃时，从相图中可以看出，随着渣系中Cr_2O_3含量的增加，相图中液相区在不断缩小，表明炉渣的流动性变差。综上所述，Cr_2O_3可降低炉渣黏度但熔化性温度偏高会使炉渣稳定性变差，不利于高炉顺行。

关键词 黏度，Cr_2O_3，熔化性温度

Influence of Cr_2O_3 Content on Viscosity and the Running Temperature of CaO-SiO_2-MgO-Al_2O_3-Cr_2O_3 Slags

Qian Yi, Zhang Jianliang, Wang Zhiyu, Wang Peng

（School of Metallurgical and Ecological Engineering, University of Science and Technology Beijing, Beijing 100083, China）

Abstract The effect of Cr_2O_3 on the viscosity and the running temperature of CaO-SiO_2-MgO-Al_2O_3-Cr_2O_3 were investigated by rotating cylinder method.The phase diagrams of CaO-SiO_2-MgO-Al_2O_3-Cr_2O_3 slag system were analyzed using FactSage. The experimental results revealed that viscosity slightlydecreased as the content of Cr_2O_3 gradually increasing at CaO/SiO_2 ratio of 1.11. The free running temperature of slag is raised with the Cr_2O_3 addition. When the content of Cr_2O_3 less than 3%, the running temperature raises slowly, while it raises sharply with 4% Cr_2O_3.It can be concluded with the phase diagrams that with the decrease of temperature and the increase of Cr_2O_3 content, the precipitated primary crystal phase with high melting temperature is Cr_2O_3, whose melting temperature is about 2435℃. This may increase the melting temperature of the slag system. The liquid phase region shrinks at the temperature of 1500℃ with the increase of Cr_2O_3 content in slag system, which demonstrates the decreasing slag fluidity. The slag viscosity decreases with the increase of Cr_2O_3 content at the basicity of 1.11. In conclusion, Cr_2O_3 can reduce the viscosity and the stability of the slag, which is not favorable to the operation of blast furnace.

Key words viscosity, Cr_2O_3, the free running temperature

梅钢 5 高炉高铝炉渣优化技术研究

韩宏松[1,2] 张正好[2] 毕传光[2] 沈峰满[1] 姜 鑫[1]

（1. 东北大学，辽宁沈阳 110004；
2. 梅钢公司技术中心，江苏南京 210039）

摘 要 梅钢高炉在生产中炉渣 Al_2O_3 含量一般在 15.5%左右，长期以来在高炉炉料中加入蛇纹石熔剂，以获得合适的炉渣黏度等冶金性能，但产生了渣量增加问题。为了进一步研究本问题，对不同 Al_2O_3 含量的高炉渣进行了温度-黏度试研究，研究分析了 MgO、R_2 对高铝炉渣的熔化性温度及高温区黏度（1400~1500℃）的影响，研究结果表明：Al_2O_3 在 15%~16.5%，MgO 在 7%~7.5%，R_2 在 1.2~1.3，渣系熔化性温度及黏度能够满足高炉生产要求。为此，在梅钢 5 高炉进行了生产试验，试验期间，高炉炉况维持了稳定顺行，渣比降低 10kg/t，利用系数维持在 2.2t/(m^3·d)，煤比与基准期相比提高 8kg/t，达到 145kg/t，焦比由 357kg/t 降低到 349kg/t，技术经济指标得到明显改善。

关键词 高炉，炉渣，高铝，冶炼

Optimization Research on No.5 Blast Furnace Operation with High Alumina Slag in Meishan Steel

Han Hongsong[1,2], Zhang Zhenhao[2], Bi Chuanguang[2], Shen Fengman[1], Jiang Xin[1]

(1. Northeastern University, Shenyang 110004, China;
2. Meishan Steel, Nanjing 210039, China)

Abstract Many experiments related to investigate the properties of higher alumina slag had been carried out through adjustments of slag composition and test temperature. The influence of MgO and basicity (CaO/SiO_2) on melting temperature and high temperature(1400~1500℃) viscosity was specially paid attention. The experimental results show that when Al_2O_3 range in 15%~16.5%, MgO range in 7%~7.5% and basicity range in 1.2~1.3, the BF slag can perfectly satisfy the requirements of blast furnace operations. Thus the industrial trial of using higher alumina slag (>16%) had been implemented at No.5 BF (inner volume 4000m^3) in Meishan steel based on the experimental results.The blast furnace maintained smooth and steady performances, the productivity maintained 2.2t/(m^3·d), the slag rate was reduced about 10kg/tHM, the PCI increased 8kg/tHM (up to 145kg/tHM), and the coke rate was decreased from 357kg/tHM to 349 kg/tHM.

Key words blast furnace, slag, high aluminum, ironmaking

碱度对钒钛磁铁矿高炉渣冶金性能的影响

储满生 冯 聪 唐 珏 柳政根 李 峰

（东北大学材料与冶金学院，辽宁沈阳 110819）

摘　要　基于钒钛磁铁矿显著的综合利用价值以及造渣制度在高炉炼铁生产过程中所占的重要地位，本研究以国内某钢铁企业的钒钛磁铁矿现场高炉渣为基准，采用纯化学试剂配制渣样，在氩气气氛条件下运用 RTW-10 熔体物性综合测定仪并结合 Factsage 6.4 热力学软件分析探索了炉渣二元碱度对钒钛磁铁矿高炉渣熔化性温度、黏度及稳定性的影响规律。最后，对实验后的渣样进行了 XRD 物相组成分析。研究结果表明，随着钒钛磁铁矿高炉渣系二元碱度的升高，炉渣熔化性温度呈升高的趋势，初始黏度和高温黏度呈降低的趋势，渣系热稳定性和化学稳定性变好。各不同二元碱度钒钛磁铁矿高炉实验渣系基体物相均为黄长石相，且随炉渣二元碱度升高，渣中低熔点物相数量相对减少，而高熔点物相数量相对增多。

关键词　钒钛磁铁矿，高炉渣，二元碱度，冶金性能

Effect of Binary Basicity on Metallurgical Properties of Blast Furnace Slag of Vanadium-titanium Magnetite

Chu Mansheng, Feng Cong, Tang Jue, Liu Zhenggen, Li Feng

(School of Materials & Metallurgy, Northeastern University, Shenyang 110819, China)

Abstract　With the high comprehensive utilization value of vanadium-titanium magnetite and the important position of slag system in ironmaking production process, the effect of binary basicity on melting temperature, viscosity and stability of blast furnace slag of vanadium-titanium magnetite was researched by RTW-10 melt property tester and Factsage 6.4 thermodynamic software under Ar atmosphere, the phases of slag samples were analyzed by XRD when the experiments were end at the same time. The slag samples were confected by using pure chemical reagents in this study, based on the components of practical blast furnace slag of vanadium-titanium magnetite provided by a domestic enterprise. The conclusions were as follows: with increase of binary basicity, melting temperature of blast furnace slag of vanadium-titanium magnetite was increased, initial viscosity and high-temperature viscosity of slag were decreased, and thermostability and chemical stability of slag became better. The main phase of each slags with different binary basicity was melilitite. Low melting point phases in slag were decreased, but high melting point phases in slag were increased, with the binary basicity of slag increasing.

Key words　vanadium-titanium magnetite, blast furnace slag, binary basicity, metallurgical properties

Value-in-use Analysis of Using Olivine Flux for Agglomerate-and Iron-making Processes

Wang Chuan[1], He Zhijun[2], Mousa Elsayed[1], Sundqvist Lena[1], Wikström Jan-Olov[1]

(1. Swerea MEFOS, Box 812, SE-971 25 Luleå, Sweden;
2. University of Science & Technology Liaoning, Anshan, Liaoning 114051, China)

Abstract　Olivine, as one MgO containing material, can be used as one type of fluxing additives for sinter, pellet and blast furnace. In this work, the benefits of using olivine flux for agglomerate- and iron-making processes were analyzed. The results show that olivine mainly contributes on (1) an increased productivity at the sinter bed; (2) a stable and better blast furnace operation by improving the permeability inside the blast furnace and an effective sulfur and alkalis removal; and (3) lower energy consumption. In addition, a theoretical low CO_2 emission can be achieved for olivine compared to other fluxes of dolomite and serpentine. Perspectives and challenges of using olivine were further addressed.

Key words olivine, sinter, pellet, blast furnace, value-in-use

高炉喷吹单煤种经济性评价

彭 觉

（湖南华菱湘潭钢铁有限公司技术中心，湖南湘潭，411101）

摘 要 文章通过对煤的各项检测指标分析入手，指出高炉喷吹用单种煤的评价可用单位价格可购买的有效发热量来衡量，无烟煤与烟煤分别计算排队比较。对指导喷吹煤的采购和降低喷吹煤的成本有一定的指导意义。

关键词 喷吹煤，无烟煤，烟煤，有效热量，评价

Economic Evaluation of Simple-type Coal for Injection into Blast Furnace

Peng Jue

(XISC Technological Center, Xiangtan 411101, China)

Abstract In this paper, based on the analysis of various tested indexes, the evaluation of single-type coal for injection into blast furnace may be done with the real calorific value calculated based on a coal buying price, and the anthracitic coal and bituminous coal are calculated and compared and analyzed separately. This is useful to guide both the purchase of coal and the lowering of injection costs.

Key words injection coal, anthracitic coal, bituminous coal, real calorific power, evaluation

迁钢 2 号高炉煤粉有效热及碳的燃烧率研究

卫广运[1] 马金芳[1,2] 王宇哲[2] 张建良[1] 杨天钧[1] 柴轶凡[1] 陈宇廷[1]

(1.北京科技大学冶金与生态工程学院，北京 100083；
2.首钢股份公司迁安钢铁公司，河北迁安 064404)

摘 要 根据迁安 2 号高炉的实际生产条件和基础数据，通过计算煤粉在风口回旋区的不完全燃烧放热、煤粉的分解吸热、煤粉的成渣耗热以及煤粉带入硫分的脱硫耗热，计算出单位质量的煤粉的有效热量，通过高炉鼓风参数和炉顶的煤气成分计算高炉内碳的燃烧率，得出以下结论：煤粉能够用于高炉的有效热为 5613.14~5614.49kJ/kg，碳在风口前的燃烧率为 53.04%。

关键词 大型高炉，喷煤，有效热燃烧率

Research on Qian'an Steel No.2 Blast Furnace's PCI Effective Calorific Value and Combustion Rate of Carbon

Wei Guangyun[1], Ma Jinfang[1,2], Wang Yuzhe[2], Zhang Jianliang[1], Yang Tianjun[1], Chai Yifan[1], Chen Yuting[1]

(1. School of Metallurgical and Ecological Engineering,University of Science and Technology Beijing, Beijing 100083, China; 2. Shougang Qian'an Iron and Steel Co., Ltd., Qian'an 064404, China)

Abstract According to the Qian'an No.2 blast furnace of the actual production conditions and basic date, By means of calculate the calorific value of pulverized coal in theraceway ofincomplete combustion, the decomposition heat of pulverized coal, the heat consumption of slag forming and desulfurization, we work outthe effective calorific value of per unitquality of pulverized coal in theraceway. Through the parameters of blower and the content of top gas of blast furnace, we obtain the combustion rate of carbon,conclusions are drawn as follows: the effective calorific value of pulverized coal is 5613.74~5614.49 kJ/kg,the combustion rate of pulverized coal in raceway is 53.04 percent.

Key words large sized blast furnace, pulverized coal injection, effective calorific value combustion rate

Evolution of Chars from Anthracite and Bituminous Pyrolysis Catalyzed by CaO and Its influence on Char Combustion Reactivity

Wang Peng, Zhang Jianliang, Xu Runsheng, Qian Yi

(School of Metallurgical and Ecological Engineering, University of Science and Technology Beijing, Beijing 100083, China)

Abstract Coal combustion is divided into two stages. One is the emission and combustion of volatiles and the second is the combustion of coal char.In this paper, effects of CaO on structure evolution and combustion reactivity of chars obtained from anthracite and bituminous coals were investigated by a thermo-gravimetric analyzer, SEM and XRD analysis. The chars were obtained from pyrolysis of coals under a N_2 atmosphere. Combustion parameters such as ignition, burnout, peak rate, ignition index and combustibility index were studied. The results show that the ignition temperature was decreased and the TG and DTG curves removed to the low temperature region. The ignition index and combustibility index of catalytic chars increased, which means the combustion characteristics of chars were improved by CaO addition. The SEM images of chars indicated that catalytic pyrolysis made the chars featuring a relatively loose structure, high porosity and big specific surface area with plenty of active sites which are favorable for the absorption of O_2. The XRD analysis of chars indicated that the order and graphitization degree were decreased by CaO, with d_{002} increased and L_c and L_a decreased.

Key words catalyst, XRD, coal char, pyrolysis

兰炭燃烧性能研究

刘思远　张建良　王广伟　徐润生　宋腾飞　王海洋

（北京科技大学冶金与生态工程学院，北京　100083）

摘　要　本实验主要探究兰炭与不同变质程度煤粉燃烧特性的差别，并单独比较了兰炭与几种无烟煤燃烧性的差异，采用热重分析仪探究不同变质程度煤粉（兰炭、烟煤、无烟煤、改质煤、焦粉）燃烧特性的规律。实验研究结果表明：不同变质程度煤粉燃烧特性不同，随着变质程度的增加，煤粉燃烧的着火点、燃尽温度、表观活化能均呈增大趋势，综合燃烧特性指数呈变小趋势，其中，两种兰炭的燃烧性要优于焦粉的燃烧性，但小于烟煤的燃烧性。且两种兰炭的燃烧性与几种无烟煤燃烧性差别不是很大，但要优于无烟煤。

关键词　兰炭，变质程度，燃烧特性

The Study of Combustion Characteristics of Semi-coke

Liu Siyuan, Zhang Jianliang, Wang Guangwei, Xu Runsheng,
Song Tengfei, Wang Haiyang

(School of Metallurgical and Ecological Engineering, University of Science and Technology Beijing, Beijing 100083, China)

Abstract　This study has mainly exploredthe differences between the semi-coke and different metamorphic degree of coal in the characteristics of combustion, and compare the differenceof combustion between thesemi-coke and several anthracite coal. The combustion characteristics of different metamorphic grade coal (semi-coke, bituminous coal, anthracite, modified coal, coke) were studied by thermo-gravimetric analyzer. The experimental results show that with metamorphic grade increases, coal combustion ignition point, burnout temperature, the apparent activation energy showed an increasing trend, the comprehensive combustion characteristic index showed a decrease. Wherein the two semi-coke combustion is better to coke nut, but less than the bituminous coal. And the combustion between thesemi-coke and several anthracite coalshowed several little difference, But better than the anthracite.

Key words　semi-coke, metamorphic grade, combustion characteristic

Preliminary Evaluation of Injecting Brown Coalsin Ironmaking Blast Furnace

Shen Y S[1], Liao J H[1], Yu A B[1], Li Y T[2], Zhu J M[2]

(1.Department of Chemical Engineering, Monash University, Clayton, VIC 3800, Australia;
2.Ironmaking, Baosteel, Shanghai 201900, China)

Abstract　Pulverized coal injection (PCI) iswidely practised in ironmaking industry due to many benefits in particular the reduction of production cost, where black coal was largely used at present. Victorian brown coal with aneven lower cost has a great potential to be used in PCI technology and further reduce the cost by replacing the black coal. Victorian brown coal needs upgrading before burning in the furnace aiming for removing the massive moisture ~ 60%, usually by briquetting or pyrolysis. It is unclear whether the brown coal products can be used in ironmaking blast furnace (BF) and replicate the performance of black coal in aspects of aerodynamic and physicochemical behaviours. In this study, an industry-scale three-dimensional CFD PCI model is used to preliminarily study the feasibility of using brown coals in PCI operation by comparing the coal combustion efficiency of brown coalsupgraded by briquetting andpyrolysis with one given black coal under typical PCI conditions, in terms of flow, temperature, gas composition and coal combusts characteristics. The simulation results indicate that the brown coal upgraded by briquettingshowsquite different coal burnoutat the raceway end while the brown coal upgraded by pyrolysis shows similar burnout at the raceway end, compared to the given black coal.

Numerical modelling offers a cost-effective way to optimize and control the injection of different brown coals upgraded by different methods under PCI conditions.

Key words　blast furnace, brown coal, PCI, CFD

外场处理对煤粉燃烧性能影响的实验研究

闫海祥　何志军　庞清海　高立华　康从鹏　杨高旸

（辽宁科技大学材料与冶金工程学院，辽宁省化学冶金重点实验室，鞍山　114051）

摘　要　煤粉燃烧性能难以提高是限制高炉提高煤粉喷吹量的主要因素之一，本文首先采用微波-超声波外场协同处理高炉喷吹用焦作烟煤煤粉，然后利用卧式煤粉燃烧器对处理煤粉的燃烧性能进行分析。研究表明：微波处理和微波-超声波外场协同处理后的焦作烟煤燃烧率比未处理的原始煤粉的燃烧率分别提高 10.46%和 15.91%，外场处理对煤粉能够明显提高煤粉的燃烧性能。

关键词　煤粉，超声波，微波，燃烧率

Research on Pulverized Coal Combustion Properties after Microwave and Ultrasonic Wave Coupled Treatment

Yan Haixiang，He Zhijun, Pang Qinghai, Gao Lihua, Kang Congpeng, Yang Gaoyang

(School of Materials and Metallurgy，University of Science and Technology Liaoning，
Key Laboratory of Chemical Metallurgy Liaoning Province，Anshan 114051, China)

Abstract　Difficulty in improving combustion performance of pulverized coal is considered a major factor in limiting injection amount of pulverized coal in blast furnace. In this paper, microwave-ultrasonic coupling co-processing in the external field is utilized in modification of pulverized coal injection in blastfurnace, a comprehensive study has been conducted on combustion performance of modified pulverized coal by making use of lab-made burner for analysis and testing.Studies have shown that, Compared to the original pulverized coal, the combustion rate of Jiaozuo bituminous coal was increased by 10.46% and 15.91% respectively after microwave and coupled microwave -ultrasonic wave field treatment.

Key words　pulverized coal injection, ultrasonic wave, microwave, combustion rate

鼓风温度对煤粉利用率的影响

刘　奇　程树森

（北京科技大学冶金与生态工程学院，北京　100083）

摘　要　采用岩相显微分析方法测定了某钢厂 1 号高炉不同喷煤比下炉尘中未消耗煤粉和焦炭的比例，结合炉尘的碳含量分析得到不同喷煤比下煤粉利用率，分析了鼓风温度对煤粉利用率影响。结果表明，高喷煤比下，鼓风温度

对煤粉利用率影响较大；高炉鼓风温度由 1172℃升高到 1185℃时，煤粉利用率由 98.11%增加到 98.90%。提高鼓风温度可以促进风口回旋区内煤粉燃烧，增加煤粉利用率。

关键词 高炉，喷煤比，鼓风温度，利用率

Effect of Blast Temperature on Utilization Ratio of Pulverized Coal at Large Pulverized Coal Injection Rates

Liu Qi, Cheng Shusen

(Metallurgy and Ecology Engineering School, University of Science and Technology of Beijing, Beijing 100083, China)

Abstract The ratio of unconsumed pulverized coal and coke under different pulverized coal injection (PCI) rates in No.1 blast furnace dust was determined by petrographic microscopic analysis, the utilization ratio of pulverized coal under different pulverized coal injection was carried out based on petrographic microscopic analysis results and the carbon amounts of dust. The effect of blast temperature on utilization ratio of pulverized coal was investigated. The results show that blast temperature has a significant influence on the utilization ratio of pulverized coal in blast furnace dust at large PCI rates; The utilization ratio of pulverized coal increases from 98.11% to 98.90% as the blast temperature increases from 1172℃ to 1185℃. Increasing the blast temperature can improve the combustion of pulverized coal in the raceway and increasetheutilization ratio of pulverized coal.

Key words blast furnace, pulverized coal injection rate, blast temperature, utilization ratio

鞍钢西区高炉优化配煤研究与应用

张立国 赵长城 赵立军 赵正洪 范振夫 张海明 刘宝奎 张飞宇

（鞍钢股份有限公司，辽宁鞍山 114009）

摘 要 在分析所用喷煤煤粉理化性能指标优劣的基础上，以煤种作为影响因素，配入比例作为水平参考，采用混合煤粉的挥发分、灰分、固定碳、可磨性、发热值和灰熔融性能等作为配煤依据，通过数学优化方法，确定了不同煤种、不同配入水平下分公司配煤方案。在此基础上，开展采用新配煤标准下的喷吹煤粉工业试验，炼铁总厂西区两座高炉取得了提高喷煤比 3.82kg/t（1 号高炉）和 4.45kg/t（2 号高炉），降低燃料比 5.52kg/t 和 6.71kg/t 的良好效果。

关键词 数学优化，配煤，喷煤比，燃料比

A Reasonable Injecting Coal Proportion Study for Blast Furnace and Its Application in Angang West Section

Zhang Liguo, Zhao Changcheng, Zhao Lijun, Zhao Zhenghong, Fan Zhenfu, Zhang Haiming, Liu Baokui, Zhang Feiyu

(Angang Steel Co., Ltd., Anshan 114009, China)

Abstract A reasonable injecting coal proportion study has been conducted by mathematic optimization, and the relative physical and chemical properties have been tested and analysised, and choose coal proportion as factor and coal performance in BF as target feature. Finally an industrial practice has been carried out with the new coal blend, and the coal injection rate has been improved 3.82 kg/t Fe and 4.45 kg/t Fe relatively, the fuel ration has been decreased 11.84 and 10.03 kg/t Fe.

Key words mathematical optimization, injecting coal proportion, coal ratio, fuel ratio

对韶钢 3200m^3 高炉风温的研究

杨国新　陈国忠　钟树周

（宝钢集团韶关钢铁炼铁厂，广东韶关　512123）

摘　要 通过对比韶钢 8 号高炉与宝钢高炉的热风炉结构、烧炉烟气理论燃烧温度包括煤气发热值、预热器预热效果的差异，富氧烧炉等其他影响风温因素的查找，得出热风炉结构的差异是影响风温水平的决定性因素，宝钢新日铁外燃式热风炉对比 8 号高炉的内燃式热风炉在燃烧能力、风温水平上具有先天性优势。在假设同样掺烧 4.5%的焦炉煤气的条件下，由于宝钢高炉煤气热值高，预热效果好，助燃空气和煤气带入的物理热量高，经计算其理论燃烧温度、拱顶温度和风温均高出 8 号高炉约 70℃。根据对韶钢 8 号高炉开炉以来风温的研判和不同烧炉条件下风温的推算，得出可通过提高焦炉煤气的掺烧比例至 6%，提高助燃空气、煤气预热温度至 160℃和进行富氧烧炉工艺等可将 8 号高炉的风温水平提高到 1190℃水平。

关键词 高炉，热风炉，风温

3200m^3 of Shaoguan Steel Blast Furnace Blast Temperature Research

Yang Guoxin, Chen Guozhong, Zhong Shuzhou

(Baosteel Shaoguan Steel Ironworks Shaoguan City, Shaoguan 512123, China)

Abstract Stove in contrast Shaoguan Steel blast furnace No.8 with Baosteel blast furnace, burning furnace flue theoretical combustion temperature, including locating gas calorific value, the difference preheater effect, oxygen-enriched air burning furnace temperature impact and other factors, differences draw stove structure is the decisive factor affecting the level of air temperature, BNA external combustion stove contrast on the 8th of internal combustion type hot blast furnace with a capacity of innate advantages in combustion air temperature level. Under the assumption that the same blending 4.5% of coke oven gas conditions, due to BFG calorific value, good warm-up effect, the physical heat combustion air and gas into the high combustion temperature calculated theoretical temperature, vault temperature and air temperature were higher than the 8th blast about 70℃. According to the Shaoguan Steel since No.8 blast furnace in air temperature of judgments, and under different conditions burning furnace air temperature projections, obtained by increasing the blending ratio of coke oven gas to 6% increase combustion air preheat temperature to gas 160 ℃ and be enriched combustion furnace process, etc. can raise the level of the 8th blast furnace blast temperature to 1190 ℃ level.

Key words BF, hot blast, temperature

顶燃式热风炉送风制度优化试验研究

刘德军
（鞍钢股份有限公司技术中心，辽宁鞍山　114021）

摘　要　针对高炉热风炉固定周期的风温波动大、平均风温低等缺陷不足，在高炉热风炉进行了非固定周期的工业试验，试验结果表明试验期间送风持续时间缩短，拱顶温度持续稳定在较高水平，热风温度均维持在工艺要求的最低温度值以上且波动较小，取得了成功。
关键词　热风炉，非固定周期，拱顶温度，热风温度

Optimization Test Research on Blast System for Top-Combustion Hot-Blast Stove

Liu Dejun
(Technology Center of Angang Steel(Group) Co., Ltd., Anshan 114021, China)

Abstract　Considering hot blast the average temperature low and temperature fluctuation range big,etc, A industrial test of variable cycle for hot stoves was done. The industrial test results showed that the on-blast duration in the variable operation mode is shorten, the arch temperature is steady at a high level, the hot blast temperature is above the minimum temperature value which is meet the process requirements and fluctuates in a narrow range.
Key words　hot blast stove, variable cycle, arch temperature, hot blast temperature

Investigation of Burden Distribution in the Blast Furnace

Tamoghna Mitra, Henrik Saxén
(Thermal and Flow Engineering Laboratory, Faculty of Science and Engineering, Åbo Akademi University, Biskopsgatan 8, FI-20500 Åbo, Finland)

Abstract　In the ironmaking blast furnace a proper control of the burden distribution is important as this is the primary means of affecting the conditions in the upper part, the shaft, of the furnace, where the burden is heated and the indirect reduction of the iron oxides takes place. The present paper studies the charging of pellets and coke in the furnace both by experiments and mathematical modeling. A small-scale bell-less charging device is used to study the burden distribution experimentally for different charging programs. Corresponding programs are simulated using the discrete element modeling (DEM) approach. The results are compared and conclusions concerning the findings are drawn. The coke push effect, where lighter coke on the burden surface is shifted by charging of the heavier pellets, is also discussed.
Key words　blast furnace, burden distribution, discrete element method, bell-less top, ironmaking, coke push

Development of a Cold Model for Burden Distribution at Bell-less Top of Blast Furnace

Zhao Huatao[1], Du Ping[1], Ren Liqun[1], Liu Jianbo[1], Tang Manfa[2]

(1. Ironmaking & Environment Research Group, Institute of Research of Iron & Steel, Shasteel, Jinfeng, Zhangjiagang 215625, China; 2. Hongfa Ironmaking Plant, Shasteel, Jinfeng, Zhangjiagang, 215625, China)

Abstract　The burden profile distribution in the radial direction of the blast furnace determines the gas flow passage, thus it is of vital importance for stable and economical operation of a blast furnace. In this study, an experimental bell-less top charging model was built on a 1:15 scale of 5800m^3 blast furnace in Shasteel to investigate how burden layer behaves inside the large scale blast furnace. To keep the same similarity condition between the cold charging model and the actual blast furnace, the particle size distribution, burden falling trajectory, and the U/Umf were kept the same. Automatic charging of materials into the cold model is done by a rotating distributor at preset charging and blowing parameters which are determined by similarity condition. By making use of this model furnace, the burden layer distribution in the radial direction of 5800m^3 at four stable operation periods was reproduced. The result shows that with increase of center coke amount, the slope angle decreases and less coke is pushed to the center part of blast furnace, while large amount of center coke leads to a large center coke column and result in a low gas utilization ratio in actual blast furnace operation. Furthermore it is found that great non-uniformity of the descent velocity in the radial direction exists, 40% in difference between the wall and center part.

Key words　charging model, burden profile distribution, center coke column, gas utilization ratio, non-uniformity, burden descent

无钟炉顶高炉多环布料过程料流轨迹数学模型

赵国磊　程树森　徐文轩　李　超

（北京科技大学冶金与生态工程学院，北京　100083）

摘　要　多环布料方式是无钟炉顶高炉生产中最主要的布料方式。在跨档位布料时，布料溜槽同时发生旋转和倾动，使得料流轨迹更加复杂，而长期以来前人建立的众多布料过程中炉料运动数学模型主要针对溜槽倾角固定仅做旋转运动时的布料过程。本文基于炉料颗粒运动过程机理，首次建立了多环布料时溜槽内颗粒综合运动的三维数学模型，考虑了溜槽同时旋转和倾动对溜槽内颗粒轴向运动和切向偏转的影响。同时，考虑了炉顶煤气流对布料过程的影响，建立了炉料颗粒在空区下落过程的数学模型。本文所建立模型对于实际布料过程中准确预测料流轨迹及落点分布有着重要意义。

关键词　无钟炉顶高炉，多环布料，料流轨迹，数学模型

Mathematical Model of Burden Trajectory in Multi-ring Burden Distribution Process of Bell-less Top Blast Furnace

Zhao Guolei, Cheng Shusen, Xu Wenxuan, Li Chao

(School of Metallurgical and Ecological Engineering, University of Science and Technology Beijing, Beijing 100083, China)

Abstract Multi-ring burden distribution is the most important charging pattern in the production of bell-less top blast furnace. During crossing rings, the distributing chute rotates and tilts simultaneously, making burden flow trajectory more complex, nevertheless for a long time numerous established mathematical models for burden flow in charging process by predecessors mainly aim at the charging process with chute tilting angle fixed and rotating only. Based on the motion mechanism of burden particles, the three dimensional mathematical model for particles comprehensive motion within chute during multi-ring burden distribution process is firstly developed in the paper, with the consideration of the influence of chute rotation and tilt on axial motion and tangential deflection of particles. Meanwhile, the model of particles falling in the freeboard is also established, considering the effects of top gas. The developed models have great significance in accurately predicting burden trajectory and falling point in actual charging process.

Key words bell-less top blast furnace, multi-ring burden distribution, burden trajectory, mathematical model

京唐高炉精准布料实践

张贺顺　刘长江　王牧麒　陈　建

（首钢京唐公司炼铁作业部，河北唐山　063000）

摘　要　布料是高炉操作中重要的一环，尤其对首钢京唐两座 5500m^3 超大型高炉而言布料的精准性更为重要。为此，京唐高炉在开炉之前专门按照无料钟炉顶设备的实际尺寸开展了 1:1 冷态布料实验，采用激光网格和图像采集测定对排料流量与料流调节阀开度关系、焦炭和烧结矿的料流轨迹等进行了研究，并将研究结果运用到实际开炉生产中，取得了良好效果。正常投产后，通过在线取样机实时获得原燃料粒度变化，根据粒度变化探究最佳的筛分控制法，实现了筛分调整与原燃料粒度协动变化，并通过设备控制方面不断优化实现了大型高炉精确化布料。通过开炉前 1:1 布料实验研究及正常投产后多项措施并举，京唐高炉布料精准度大幅提高，取得了长时间生产顺稳，燃料比低于 500kg/t 的良好效果。

关键词　布料精度，重量，煤气利用率

Practice of Improving Distribution Precision for Jingtang Iron and Steel Co., Ltd. BF

Zhang Heshun, Liu Changjiang, Wang Muqi, Chen Jian

(Shougang Jingtang Iron and Steel Co., Ltd. Iron Works Department, Tangshan 063000, China)

Abstract Charging operation was one of the most important operation of blast furnace. The volume of Jingtang BF is 5500m^3, as a oversize blast furnace, has more strict requirement to the precision of charging system. Therefore taking great care of charging operation, charging experiment was carried out and experimental results were applied to BF production. The blow-in of BF was smooth and rapid advancement its design index. After normal production, the granularity change of BF raw material was obtained by using of online sampling machine. The reasonable sieving control method was explored and sieving control could be taken at any time. Continuous optimization methods were carried out at equipment controlling. By adopting diversified measures, the charging precision was increased greatly, the BF operating was smooth for a long time and the fuel ratio was below 500kg/t.

Key words charging precision, weight, gas utilization

高炉入炉焦炭均匀性评价

梁海龙 陈 辉 孙 健 武建龙 刘文运 李荣昇

（首钢技术研究院，北京 100043）

摘 要 目前，对于高炉入炉焦炭的技术要求是强度高，灰分低，有害杂质少，冶金性能好。上述要求可用“高、低、少、好”四个字概括，然而并未对焦炭的粒度分布提出要求，本文在上述四字方针的基础上提出了“匀”的概念，并对高炉入炉焦炭均匀度进行了定义，找到了适合评价焦炭均匀性的焦炭均匀度计算公式，该评价方法对高炉顺行和提高煤气利用率具有十分重要的指导意义。

关键词 高炉，精料，焦炭，均匀性

The Evaluating Homogeneity of Coke into Blast Furnace

Liang Hailong, Chen Hui, Sun Jian,
Wu Jianlong, Liu Wenyun, Li Rongsheng

(Shougang Research Institute of Technology, Beijing 100043, China)

Abstract Currently, the technical requirements for coke into blast furnace is high-strength, low ash, less harmful impurities, good metallurgical properties. These requirements can be summed up with “high, low, little, good”, but the particle size distribution of coke has not been request, the concept of homogeneity is developed on the basis of the four words principle mentioned above, and the homogeneity of coke into blast furnace is defined, finding a formulation for the evaluation of the homogeneity of coke, this evaluation method has very important significance for the blast furnace and increasing gas utilization.

Key words blast furnace, burden beneficiation, coke, homogeneity

烧结矿 3mm 以上分级入炉技术

范小刚[1] 王小伟[1] 周宗彦[2]

（1. 中冶南方工程技术有限公司，湖北武汉 430223；
2. Monash 大学化学工程系，澳大利亚，墨尔本）

摘 要 本文提出了烧结矿 3mm 以上分级装入高炉技术，基于 CFD-DEM 数值模拟研究，对分级效果进行了分析，并探讨了分级点的选取。与烧结矿 5mm 以上不分级相比，该技术可改善料层透气性，并能有效回收 3～5mm 小粒度烧结矿，提高烧结矿成品利用率。烧结矿 3mm 以上分级可促进高炉稳定、顺行、低耗，具有良好的经济效益。

关键词 高炉，烧结矿分级，料层透气性，回收 3~5mm 小矿

Classification Technique of Sinter above 3 Millimeter

Fan Xiaogang[1], Wang Xiaowei[1], Zhou Zongyan[2]

(1. WISDRI Engineering & Research Incorporation Limited, Wuhan 430223, China;
2. Monash University, Melbourne, Australia)

Abstract Classification technique of sinter above 3 millimeter was proposed in this paper. With CFD-DEM numerical simulation technique， analyzed the classification effort and discussed the classification point selection principle. Compared sinter above 5 millimeter without classification, this technique can not only improve burden permeability, but also recovery 3~5mm little scale sinter which increase sinter utilization. Furthermore, it can assure blast furnace stability, low energy consumption and considerable economic benefit.

Key words blast furnace, sinter classification, burden permeability, 3~5mm sinter recovery

高炉中心加焦技术初论

车玉满 郭天永 孙 鹏 姚 硕 姜 喆

（鞍钢股份有限公司技术中心，辽宁鞍山 114009）

摘 要 中心加焦有利于高炉顺行，但如果选择不当，则燃料比上升。利用溜槽结构参数，提出中心加焦溜槽倾角计算方法，模拟计算炉料在炉内二次分布，认为球团更容易滚入中心，因此，需要根据炉料结构中球团所占比例，准确取消中心加焦或减少中心加焦量。球团配比<10%，可以取消中心加焦，当球团配比>15%时，则需要实施中心加焦，中心加焦量与球团比例相关，有关结论得到高炉实验验证。

关键词 高炉，溜槽倾角，中心加焦量，球团配比

Discussing on Adding Coke in the Central of Blast Furnace

Che Yuman, Guo Tianyong, Sun Peng, Yao Shuo, Jiang Zhe

(Technology Center of Ansteel Co., Ltd., Anshan 114009, China)

Abstract Technology of adding coke in blast furnace central can efficiency make blast furnace work smoothly. But the gas usage will decreased and the fuel consuming will rose in some time if the technology is put into use in inaccuracy way. A method to calculate the chut angle of adding coke is work out by using the chut some parameters to help the blast furnace to realize technology of adding coke . The pellet can run easy toward the center by calculating the charge to distribute second inner blast furnace. Both the way how to using accuracy and how to unusing accuracy the technology of adding coke keeps relationship with the pellet proportion in a charge. It can discarded the technology of adding coke if the pellet proportion less than 10 percentage in a charge. In other way, the technology of adding coke muse be used if the pellet proportion greater than 15 percentage in a charge, and the adding coke proportion will increase with the pellet proportion. The result was tested by a blast furnace(2600m^3) experiment.

Key words blast furnace, chut angle, adding coke proportion, pellet proportion

本钢七号高炉布料矩阵的探索与优化

董 悦 丁洪海

（本钢集团炼铁厂，辽宁本溪 117000）

摘 要 本钢集团七号高炉有效容积 2850m^3，年产铁水 200 万吨。为应对国际钢铁市场的严峻形势，本钢炼铁厂积极挖潜降耗，在保证焦比和产量不变的前提下，采取平台漏斗的布料矩阵使燃料比由 528kg/t 降低至 510kg/t，为企业带来巨大的经济效益。

关键词 布料矩阵，平台漏斗，燃料比，经济效益

The Explore and Optimize of the Distribution Mode in Benxi Steel No.7 Blast Furnace

Dong Yue, Ding Honghai

(The Ironworks of Benxi Steel and Iron Group, Liaoning Benxi 117000, China)

Abstract The volume of Benxi steel group No. 7 blast furnace is 2850m^3, Produce hot metal 2 million tons every year. In order to cope with the serious situation of the international steel market,The ironworks of Benxi Steel and Iron Group positive to reduce costs,On the premise that coke rate and production is not changed,take the distribution mode of platform-funnel to make fuel rate reduce from 528kg/t to 510kg/t,Taking huge economic benefits for the enterprise.

Key words cloth matrix，platform funnel，fuel rate，economic benefits

湘钢 3 号高炉炸瘤恢复炉况生产实践

但家云

（湖南华菱湘潭钢铁有限公司技术中心，湖南湘潭 411101）

摘 要 湘钢 3 号高炉 2005 年开炉，快速达产达效，之后由于原燃料质量劣化及操作制度不合理等，炉墙结瘤，炉况顺行破坏，高炉指标大幅下滑。在 2014 年 7 月，通过炸瘤，瘤体 70%脱落，复风后，通过合理的上下部制度调整，炉况快速稳定，指标恢复正常水平。

关键词 结瘤，操作炉型，操作制度，炉况顺行

Production Practice of XISC No.3 Furnace to Recover Stable and Smooth Operation by Blasting Lumps

Dan Jiayun

(XISC Technological Center, Xiangtan 411101, China)

Abstract After successful blowing-in in 2005, the No.3 blast furnace realized its capacity and profit quickly. However, the quality degradation of the raw material/fuel and the unreasonable operating system caused scaffolding in the wall and the bad furnace condition, so the blast furnace index fell sharply. In July 2014, 70% of the limps fell off by blasting. After reblowing, the furnace condition was stabilized rapidly by adjusting the top and bottom systems reasonably, and the index returned to the normal levels.

Key words scaffolding, operating furnace type, operation system, smooth operation

鞍钢3高炉冷却壁破损后稳产操作实践

赵正洪　曾　宇　谢明辉　刘　建

（鞍钢股份有限公司炼铁总厂，辽宁鞍山　114001）

摘　要　对鞍钢3高炉冷却壁大量破损情况下的生产操作实践进行总结。通过采取加强焦矿筛监管、提高高炉操作水平与管理水平，创造性地在破损铜冷却壁上加装柱状冷却器等措施，实现了高炉在冷却壁水管大量破损的不利情况下，高效、稳定运行。

关键词　高炉冷却壁，布料制度，稳定生产

Production Practice in the Presence of Ansteel No.3 BF Run Steadily in Many Cooling Pipes Leaking Environment

Zhao Zhenghong, Zeng Yu, Xie Minghui, Liu Jian

(General Iornmaking Plant of Angang Steel Co., Ltd., Anshan 114001, China)

Abstract The paper summarizes production practice in the presence of Ansteel No.3 BF many cooling pipes leaking. In this furnace a series of advanced technologies are taken, such as reinforced censorship of coke sieve and oil sieve, improved operation level and management level, creative design of columnar stave installed in copper stave, which helped the BF run steadily at cooling pipes of Blast Furnace degrade condition.

Key words blast furnace stave, distributing schedule, stable production

鞍钢3号高炉布料溜槽掉恢复炉况实践

赵正洪　肇德胜　李永胜　曾　宇

（鞍钢股份有限公司炼铁总厂，辽宁鞍山　114021）

摘　要　近几年来，鞍钢炼铁厂已经有多次高炉布料溜槽掉事故。一旦发现不及时，送风后炉况把握不到位，势必会影响到高炉的稳定顺行，造成炉况长时间失常。2014年5月18日三高炉布料溜槽掉，处理的整个过程存在很多不足，拿出来供大家借鉴。

关键词　布料溜槽，管道，热量

Practice on Distribution Chute off Recovery Furnace Condition of No.3 BF in Ansteel

Zhao Zhenghong, Zhao Desheng, Li Yongsheng, Zeng Yu

(Ansteel Ironmaking Plant, Anshan 114021, China)

Abstract　In recent years, There have been many times about the falling accident of distribution chute in Ansteel Ironmaking Plant. Once it is not timely found, the master of furnace condition does not reach the designated position after sending wind, it is bound to affect the stability of the blast furnace, causing the furnace condition is in disorder for a long time. On May 18,2014, the distribution chute fell off in No.3 BF, there were many deficiencies during handing the whole process, and take it out for your reference.

Key words　distribution chute, piping, heat

梅钢4号高炉减少小套损坏的生产实践

张银鹤　王本伟　卢开成

（上海梅山钢铁股份有限公司炼铁厂，江苏南京　210039）

摘　要　对梅钢4号高炉减少风口小套的生产实践进行分析总结，生产实践表明：改善原燃料质量、合理的上下部制度、优化造渣制度、休风后的快速恢复、调整喷煤角度和煤枪直径、增加小套冷却水流量以及使用内腔有耐磨涂层的小套等措施，使更换小套的数量逐年递减。

关键词　风口小套，上部制度，造渣制度

The Practice of Reducing the Number of Tuyere Cooler Damaged in Meisteel's No.4 Blast Furnace

Zhang Yinhe, Wang Benwei, Lu Kaicheng

(Ironmaking Plant of Meishan Iron & Steel Co., Nanjing 210039, China)

Abstract This paper analyzes and summarizes the practice of reducing the number of tuyere cooler damaged in Meisteel's No.4 blast furnace. It shows that the appropriate burden conditioning and blast conditioning, improving the quality of raw material, optimizing slagging regime, quick recovery after delay, adjusting the angle and diameter of coal lance, increasing the cooling water flow of tuyere cooler, using the tuyere cooler with wear-resistant coating can reduce the damage of tuyere cooler.

Key words tuyere cooler, burden conditioning, slagging regime

宝钢股份高炉休风减矿的优化及效果

华建明

（上海宝山钢铁股份有限公司炼铁厂生产技术室，上海 200941）

摘 要 总结了近年来宝钢股份高炉历史休风减矿基本情况及特点，提出简洁、高效的“一次方程式”休风减矿标准公式对原来休风减矿标准作了优化和改进，分析、对比了采用新的减矿标准前后在送风恢复时间、炉温水平等方面的成效，得出结论：宝钢在大高炉休风减矿方面积累了成功的技术和经验，休风减矿标准不断改进、完善，为高炉顺利休风、复风奠定了坚实的技术基础。高炉休风减矿的核心关键在于：根据休风时间、基础矿焦比、炉况状态、休风工事情况确定合适的总减矿率；根据炉料下达时间及顺行、补热需要，分段制定合理减矿率。按照最新优化的“一次方程式”休风减矿标准，高炉总体复风情况明显改善，说明该休风减矿标准总体是较为科学、有效的；实际操作若偏离标准过大，则休风或者复风过程将不够稳定、高效。为使宝钢休风、复风过程尽可能稳定、高效化，最大程度降低冶炼成本，执行统一的标准是必要和必须的。

关键词 高炉，休风，定修，减矿率

Optimization and Effect of BF Blowing-down and Ore Reduction in Baosteel

Hua Jianming

(Baoshan Iron & Steel Co., Ltd. Production Technical Department,Shanghai 200941, China)

Abstract General conditions and characteristics of BF blowing-down and ore reduction in Baosteel over recent years are summarized in the present paper, where a more concise and efficient linear formula is proposed for optimizing and improving the blowing-down and ore reduction operations. A new standard of ore reduction based on the formula is employed and results are analyzed and compared in terms of blowing-in recovery time and furnace temperature. The conclusions are as follows. Baosteel has accumulated successful techniques and practical experience in blowing-down and ore reduction of large-scale BFs. The standard of blowing-down and ore reduction is continuously improved, playing a fundamental role in BF blowing-down and recovery. The key of BF blowing-down and ore reduction lies in the following aspects. A rational total ore reduction ratio must be determined by comprehensively considering blowing-down period, basic ore to coke ratio, in-furnace conditions and blowing-down operation requirement. In addition, sequential ore reduction ratio is determined by considering burden descending time and the requirement of smooth-running and heat supplement. According to the new standard, recoveries of BF are obviously improved, showing the standard is scientific and reasonable. The process of blowing-down or recovery seems to be less stable and efficient if practical operation deviates from the standard. In order to make the process of blowing-down or recovery more stable and efficient in Baosteel and thus lower production costs to the largest extent, it is necessary and compulsory to adopt a uniform standard.

Key words blast furnace, blowing-down, scheduled repair, ore reduction ratio

高炉风口前理论燃烧温度的在线计算研究

陈令坤[1] 李向伟[2]

（1. 武汉钢铁集团公司研究院工艺所，武汉 430081；
2. 武汉钢铁集团公司炼铁室，武汉 430081）

摘 要 风口前理论燃烧温度是一个重要的操作参数，本文为了评估大喷煤状况下风口区理论燃烧温度的变化情况，根据基本的热平衡理论，提出了一个可以在线计算的理论燃烧温度计算公式，并利用该公式计算了武钢 5 号高炉在 2013 年的理论燃烧温度变化状况，在求得理论燃烧温度的基础上，利用一次多项式进行理论燃烧温度的计算拟合，得到与理论分析相对应的经验多项式，经验多项式的建立有利于生产过程的直接使用，该回归公式明显优于目前 5 号高炉使用的理论燃烧温度计算公式，更符合高炉的实际状况，更为实用。

关键词 高炉，理论燃烧温度，煤质

Analysis of the Theoretical Flame Temperature(T_f) in the Front of BF Tuyeres

Chen Lingkun[1], Li Xiangwei[2]

(1. Research and Development Center of Wuhan Iron & Steel (Group) Corporation, Wuhan 430081, China;
2. Ironmaking Plant of Wuhan Iron & Steel (Group) Corporation, Wuhan 430081, China)

Abstract It is an important parameter for the operation of blast furnace, a new methods used for calculating the theoretical flame temperature in the front of BF tuyere on-line has been modified based on the heat balance theory in the condition of big pulverized coal, the variation of theoretical flame temperature in the front of tuyere for blast furnace No.5 in WISCO has been calculated by using such formula according to the operational data in 2011, a multiitem formula of experience which is associated with the theoretical analysis has been set up based on the fitting principle of the least square method. The setting up of the equation is useful to the direct application of it in industrial production. The formula is more useful than that of used before, the equation is more practical for the performance of blast furnace No.5 in WISCO.

Key words blast furnace, flame temperature(T_f), coal quality

A Numerical Study of Raceway Formation in an Ironmaking Blast Furnace

Miao Zhen, Zhou Zongyan, Yu Aibing

(Laboratory of Simulation and Modelling of Particulate Systems, Department of Chemical Engineering, Monash University, VIC 3800, Australia)

Abstract This work presents a numerical study of gas-solid flow in a blast furnace raceway region by CFD-DEM coupled approach. In the simulation, 120000 spherical particles with particle diameter of 80mm are packed into a real blast furnace geometry. Blast is injected at high velocities to generate a void space in the front of tuyere. The results show that three types of raceways are formed, including anti-clockwise circulating motions of particles, clockwise circulating motions of particles, and plume-like raceway. The results are analysed mainly in terms of particle flow patterns, particle velocity, porosity,and gas flow field around the raceway region. The observation of two opposite circulating gas vortexes locating upon and below tuyere during raceway formation has a good agreement with experiments. Further, the effects of different variables such as tuyere length and angles, and particle properties on raceway size and shape are examined.

Key words blast furnace, raceway, CFD-DEM, solid flow

基于料层分布的高炉软熔带形状和位置的数值模拟

周 萍[1] 伍东玲[1] 李昊岚[1] 闫红杰[1] 周 谦[1,2]

（1.中南大学能源科学与工程学院，湖南长沙 410083;

2. Center for Innovation through Visualization and Simulation, Purdue University Calumet, 2200 169th Street, Hammond, IN 46323, USA）

摘 要 本文基于 Fluent 软件，自编程开发高炉炉身内部分层布料时气固两相流动、传热和传质的二维数学模型。利用所建立的数学模型对高炉软熔带的形状和位置进行数值计算，讨论不同料层分布和下部回旋区计算结果作为入口条件时对高炉软熔带形状和位置的影响。研究结果表明：随着料层中心区域焦炭量的增加，中心气流柱的流速增大，边缘气流逐渐被抑制，软熔带的形状从“W”形变化为倒“V”形。当中心区域全为焦炭时，软熔带的顶部远离中心轴，气流直接冲出炉顶造成热量利用损失。采用高炉下部的计算结果作为边界时，软熔带仍然呈现倒“V”形，但更为扁平。

关键词 高炉，软熔带，料层，数值模拟

Simulation of the Cohesive Zone in the Blast Furnace Shaft with Layered Burden

Zhou Ping[1], Wu Dongling[1], Li Haolan[1], Yan Hongjie[1], Zhou Chenn Q.[1,2]

(1. School of Energy Science and Engineering, Central South University, Changsha 410083, China;

2. Center for Innovation through Visualization and Simulation, Purdue University Calumet, 2200 169th Street, Hammond, IN 46323, USA)

Abstract A set of symmetric, two dimensional, steady models of momentum, heat and mass transfer of gas-solid based on CFD software Fluent were adopted to simulate the shape and location of cohesive zone in the blast furnace with a layered burden. The cohesive zone was got in different burden distribution and inlet conditions. The results indicate that charge bulk coke at center can lead to central gas column, and then the gas escape from the center of blast furnace and the gas utilization rate decrease. The shape of cohesive zone varies from a "W" to an inverse "V" with increasing the charged coke in the center of the blast furnace. The inlet condition including velocity, temperature and gas species distribution calculated by coal combustion in the raceway has impact on the shape of cohesive zone, the shape of cohesive zone becomes more flat but maintain a reverse "V" shape.
Key words blast furnace, cohesive zone, burden, numerical simulation

焦炭在高炉炉身料柱内的滞留时间及其影响因素分析

李洋龙　程树森

（北京科技大学冶金与生态工程学院，北京　100083）

摘　要　高炉内焦炭运动规律较为复杂，随着高炉炉况和操作条件的变化，焦炭在高炉内的滞留时间会发生改变，如何定量评价高炉焦炭的滞留时间一直是高炉操作者关注的问题。本文从高炉炉料和铁水中含铁量的收支平衡角度出发，建立了焦炭在高炉炉身料柱内滞留时间的计算公式。基于该公式，分析了某 3200 m^3 高炉在产铁量、燃料比、焦比、矿石品位、铁水含铁量等参数影响下，高炉炉身料柱内焦炭滞留时间和平均运动速度的变化。计算发现高炉产铁量增加、燃料比不变时焦比增加、煤比不变时焦比增加、铁水含铁量增加、矿石品位降低等条件都会减小焦炭在高炉炉身料柱内的滞留时间，增加焦炭的平均运动速度。该公式可用于定量化分析在不同操作条件下的高炉炉身料柱内焦炭的滞留时间。

关键词　高炉，产铁量，焦比，品位，滞留时间

Analysis of Residence Time of Coke in Blast Furnace Stock Column and Influence Factors

Li Yanglong, Cheng Shusen

(School of Metallurgical and Ecological Engineering, University of Science and Technology Beijing, Beijing 100083, China)

Abstract The movement of coke is very complex in a blast furnace (BF). With the variation of BF working and operating conditions, the residence time of coke in blast furnace will change. How to quantitatively evaluate the residence time of coke in blast furnace stock has always been a concern for blast furnace operators. Based on the relationship of iron content between charged burden and productivity, a calculation equation of residence time of coke in blast furnace stock column were established. Based on this equation, the influences of productivity, fuel ratio, coke ratio, ore grade and iron content in hot metal on the residence time and average velocity of coke in a 3200 m^3 BF stock column were analyzed. It was found that increasing productivity, increasing coke ratio at constant fuel ratio, increasing coke ratio at constant coal ratio, increasing iron content in hot metal or decreasing ore grade would lead to decrease the residence time and increase average velocity of coke in blast furnace shaft. The equation can be used to quantitatively analyze residence time of coke in blast furnace stock column under different operating conditions.
Key words blast furnace (BF), productivity, coke ratio, ore grade, residence time

A Multi-fluid Model for Simulation of Iron-making Blast Furnace Processes

Huang Jianbo, Mao Xiaoming, Xu Wanren, Hu Desheng, Li Yuntao

(Central Research Institute, Baosteel Group, Shanghai, China)

Abstract This paper presents a computational model for simulating the thermal-physical and chemical processes , taking place in iron-making blast furnaces. The model arranges the different material types in the furnace into two phase-groups: the gas phase group and the solid-liquid phase group. This treatment significantly simplifies the model. It has the advantage that in each group the transport properties vary smoothly. Mass, heat and momentum exchanges taking place among and between the two groups. Unlike the usual multi-phase treatments which assumes that all phases share the same pressure, this model drops the single pressure field assumption, therefore the two fluid groups have their own pressure fields. The model is implemented by applying the popular computational libraries, and using the so-called "PIMPLE" algorithm, which combines the advantages of PISO (Pressure-Implicit Splitting Operator) and SIMPLE (Semi-Implicit Method for Pressure-Linked Equations) algorithms. The model presented in the paper has been shown to be efficient and effective, thus can be used as a basis for developing a systematic numerical modeling platform for simulating processes taking place in blast furnaces.

Key words blast furnace, computational fluid dynamics, multi-fluid model, multiphase

Numerical Simulation and Visualization of Blast Furnaces

Chenn Q. Zhou

(Steel Manufacturing Simulation and Visualization Consortium (SMSVC), Center for Innovation through Visualization and Simulation (CIVS), Purdue University Calumet, Hammond, IN, 46323)

Abstract A blast furnace involves various capital and energy intensive processes. Due to complex phenomena and the difficulties in taking measurements, the knowledge needed for optimization of blast furnace operation can be most readily obtained through the development of high fidelity computational fluid dynamics (CFD) simulations. However, with increasingly complex CFD capable of simulating and analyzing ever-larger amounts of data, interpolating and presenting the numerical data in a meaningful fashion is a key for effective communication between CFD experts and plant engineers. Recently, virtual reality (VR) visualization technology made it possible for people to analyze huge amounts of CFD data in a virtual environment. This paper will present the methodology and results of integrating CFD simulation and VR visualization of blast furnaces. Such integration allows the creation of virtual blast furnaces which can be used for optimization, troubleshooting, design and training to improve the process performance and productivity.

Key words blast furnace, CFD simulation, virtual reality visualization

大型高炉热风炉工作强度等若干问题的讨论

武建龙 陈 辉 孙 健 梁海龙 刘文运

（首钢技术研究院，北京 100043）

摘　要　对比分析了不同立级高炉的热风炉的设备参数和工艺参数，并对可能给热风炉所带来的影响进行了分析。分析表明，高炉的大型化可能给热风炉带来诸多风险，如：（1）热风炉内氮氧化物的生成量升高；（2）热风炉内结露温度提高；（3）热风炉炉壳所受到的膨胀力和冲击力加大；（4）热风炉工作强度提高。

关键词　大高炉，热风炉，氮氧化物，温度，受力

延长高炉炉龄的有效技术

胡俊鸽　郭艳玲

（鞍钢股份有限公司技术中心，辽宁鞍山　114009）

摘　要　介绍了高炉铜冷却壁损坏后的维护措施，包括中国台湾中钢开发和应用的安装金属软管技术，中国台湾中钢和宝钢开发应用的安装雪茄式冷却器技术，以及新日铁住金和宝钢成功应用的冷却壁更换技术。阐述了美钢联对利于形成炉缸渣皮的操作技术的研究结果，以及欧洲开发的风口喷吹含钛物料的护炉技术及其应用效果。也介绍了韩国浦项科技大学通过控制炉渣成分延长炉缸寿命的技术研究现状。

关键词　高炉长寿，铜冷却壁，含钛物料，渣皮，高炉

Available Technology for Prolonging Blast Furnace Life

Hu Junge, Guo Yanling

(Technology Center of Angang Steel Company Ltd., Anshan 114009, China)

Abstract　Maintenance countermeasures for defective copper stave of blast furnace are introduced, which include installing flexible pipe technology developed by China Steel and Bao Steel, cigar cooler technology developed by Nippon Steel and Sumitomo Metal, and replacing stave technology applied successfully in Nippon Steel and Sumitomo Metal and Bao Steel. Some operational technologies available for skull formation researched by US Steel, and new technologies and application effect of using titanium bearing material for prolonging hearth life are presented. The technology by controlling slag composition for prolonging BF life by Pohang University of Science and Technology is introduced.

Key words　blast furnace, copper stave, titanium bearing material, skull, long life of blast furnace

高炉长寿低成本冷却系统的选用与维护新技术

胡源申

（安徽工业大学冶金工程学院，安徽马鞍山　243002）

摘　要　对使用开路循环水、预处理循环水、软水密闭、纯水密闭四种冷却水质和循环方式，使用灰铸铁、球墨铸铁、耐热铸铁、铸钢和铜五种材质的一代服役后高炉冷却壁进行了系统解剖研究。分析比较了不同材质冷却壁的力学性能、导热性能、抗结垢性能、易加工性能和经济性能;揭示了铁基材质冷却壁的微观破损机理和壁内水管的成垢机制;提出了合理选用高炉冷却水系统和冷却设备使高炉进一步长寿并降低成本的维护新技术。

关键词 长寿，低成本，高炉，冷却系统，选用

Advanced Technology of Selection and Maintenance for the Cooling System of BF with Long Life and Lowcost

Hu Yuanshen

(Anhui University of Technology, Maanshan 243002, China)

Abstract The BF cooling stave after a campaign life made by the materials of grey cast iron, nodular cast iron, heat-resistant cast iron，cast steel and copper and used of open loop water，pretreatment circulating water，soft water closed circulation and pure water closed circulation have been systematically dissectional studied respectively. The comparative analysis of the mechanical property，conducting property，scab retardation property，easy processing and the economical of different materials of cooling stave have revealed the micro damage mechanism of iron-based BF cooling stave and the scaling mechanism of the internal pipe.

Key words BF longevity, lowcost, cooling system, selection, maintenance technology

碳复合砖抗渣侵蚀性能及挂渣能力研究

王志宇[1] 张建良[1] 焦克新[1] 王 聪[1] 赵永安[2]

（1. 北京科技大学冶金与生态工程学院，北京 100083；

2. 河南五耐集团实业有限公司，河南巩义 451250）

摘 要 本文对碳复合砖的抗渣侵蚀性能及挂渣能力进行研究。在高温实验条件下探究不同高炉渣对碳复合砖的侵蚀机理，通过 SEM-EDS 及 XRD 等分析手段分析侵蚀界面的微观组织结构和物相组成，并分析了碳复合砖的挂渣机理。实验结果表明，高炉渣与碳复合砖在侵蚀界面发生反应，反应生成的 CA_2、CA_6、MA 尖晶石及碳复合砖中的 Al_2O_3 和 SiC 等高熔点物质促进高炉渣黏附在碳复合砖表面，促使渣皮的形成，阻碍高炉渣对碳复合砖的进一步侵蚀。

关键词 耐火材料，侵蚀，高炉渣，碳复合砖

Carbon Composite Brick Slag Erosion Resistance and Hang Research

Wang Zhiyu[1], Zhang Jianliang[1], Jiao Kexin[1], Wang Cong[1], Zhao Yong'an[2]

(1. School of Metallurgical and Ecological Engineering,University of Science and Technology Beijing, Beijing 100083, China; 2. Henan Winna Industrial Group Co., Ltd., Gongyi Henan 451250, China)

Abstract In this paper, the slag corrosion resistance and the adherent dross behavior of carbon composite brick are investigated. The corrosion mechanisms of carbon composite brick by different blast furnace slags are studied at high temperature. The microstructure and the phase composition of corrosion interface are analyzed by XRD and SEM-EDS, and

the corrosion mechanism is also discussed. The results show that blast furnace slags react with carbon composite brick on the corrosion interface generating CA_2, CA_6 and spinel (MA). These productions and some high melting point compounds in carbon composite brick such as Al_2O_3 and SiC promote blast furnace slags adhering to carbon composite brick surface, leading the formation of slag skin which can hinder the corrosion process of carbon composite brick by blast furnace slag.
Key words refractory, erosion, blast furnace slag, carbon composite brick

微孔刚玉砖在武钢高炉上的应用研究

邹祖桥　卢正东　刘栋梁

（武钢研究院，湖北武汉　430080）

摘　要　研制出一种性能优良的用于高炉陶瓷杯的微孔刚玉砖。该砖具有优良的抗碱性、抗渣铁侵蚀性，并且透气度低，平均孔径小,<1μm 孔容积大于 70%，可以有效防止铁水、K、Na、Zn、Pb 有害物质的渗透侵蚀。该砖在武钢 6 号、4 号、7 号、5 号、8 号高炉用作陶瓷杯壁砖，使用效果均很好。并且在武钢 5 号和 8 号高炉风口带中应用，减少了风口上翘的问题。
关键词　微孔，刚玉砖，陶瓷杯，组合砖

Research and Application of Micropore Corundum Brick in WISCO Blast Furnace

Zou Zuqiao, Lu Zhengdong, Liu Dongliang

(Research and Development Center of WISCO, Wuhan 430080, China)

Abstract A kind of good performance micropore corundum brick for blast furnace ceramic cup is developed. The brick has the advantages of alkali resistance, slag iron erosion resistance, low permeability and small average pore diameter,and less than 1μm pore volume of the brick is more than 70%, which can also effectively prevent penetrated erosion of the harmful substances, such as molten iron, K, Na, Zn and Pb. The brick is used in ceramic cup wall of No.6, No.4, No.7, No.5, No.8 blast furnace in WISCO, and application effect is very good. The brick is also used in the tuyeres zone of No.5 and No.8 blast furnace in WISCO, which has reduced the phenomena of tuyeres tilting upwards.
Key words micropore, corundum brick, ceramic cup, combined brick

大型高炉用 TD-Ⅰ型陶瓷杯的抗渣铁、抗碱性能研究

王治峰[1]　高长贺[1]　张积礼[2]　马淑龙[1]　夏文斌[1]　马　飞[1]　程树森[3]

（1. 通达耐火技术股份有限公司，北京　100085；2. 巩义通达中原耐火技术有限公司，河南郑州　451261；3. 北京科技大学冶金与生态工程学院，北京　100083）

摘　要　大型高炉用 TD-Ⅰ型陶瓷杯是通达新一代高炉炉缸用耐火材料，在原有金属塑性相结合的基础上，引入有

机结合剂形成碳结合，以两种结合方式使刚玉和碳化硅深度复合，制备出氧化物和非氧化物复合材料，此产品具有低气孔、高致密、适度的导热性能及极好的热震稳定性。通过感应炉法动态模拟对比了TD-Ⅰ型陶瓷杯与市场同类产品的抗渣铁侵蚀冲刷性能，采用碱蒸气法对比评价了TD-Ⅰ型陶瓷杯与市场同类产品的抗碱性能，发现TD-Ⅰ型陶瓷杯抗渣铁侵蚀冲刷性能及抗碱侵蚀性能要优于市场同类刚玉质产品，可适用于大型高炉，特别是经济料冶炼的高炉陶瓷杯中。

关键词 TD-Ⅰ型陶瓷杯，抗渣铁侵蚀，抗碱侵蚀

The Slag Iron Erosion and Alkaline Resistance of TD-Ⅰ Ceramic Cup for Large Blast Furnace

Wang Zhifeng[1], Gao Changhe[1], Zhang Jili[2], Ma Shulong[1], Xia Wenbin[1], Ma Fei[1], Cheng Shusen[3]

(1. Tongda Refractory Technologies Co., Ltd., Beijing 100085, China;
2. Gongyi Tongda Zhongyuan Refractory Technologies Co., Ltd., Zhengzhou 451261, China;
3. School of Metallurgical and Ecological Engineering, University of Science and Technology Beijing, Beijing 100083, China)

Abstract TD-I ceramic cup is a new generation hearth lining for blast furnace of Tongda, which is a kind of composite of oxides and non-oxides. On the basis of original combination of plastic metal, it is introduced organic binder to form carbon combine. These two combinations make the composite characteristics of oxides and non-oxides. This new ceramic cup has low porosity, high density, moderate thermal conductivity and excellent thermal shock resistance. The slag iron erosion resistance of this product and similar products on the market are compared through the induction furnace method dynamic simulation, the alkaline resistance of these two products is compared through alkali vapor method, the study found that it has higher slag iron erosion resistance and alkaline resistance than similar products on the market. Therefore, it can be applied as ceramic cup in large blast furnace, especially using economic burden.

Key words TD-I ceramic cup, slag iron erosion, alkaline- resistance

高炉炭砖抗铁水侵蚀测定标准探讨

夏欣鹏[1] 余仲达[2]

（1. 宝山钢铁股份有限公司，上海 201900；2. 上海大学，上海 200444）

摘 要 随着高炉往大型化和长寿化发展，炉缸炭砖的重要性越发引人注目，但目前大高炉用炭砖还主要依赖进口。对于炉缸炭砖的质量，特别是左右炭砖寿命的抗铁水侵蚀等的重要指标方面，目前还没有特定用于大高炉炭砖的检测方法和标准。本文通过分析大高炉的工作特征，优化了高炉炭块抗铁水侵熔蚀性试验方法（GB/T 24201—2009）的测试条件，并对目前常用的多种炉缸用炭砖作了该项目的测试；通过将其测试结果和相对应的炭砖溶解于铁水的理论模型进行了多因子干涉的理论分析，验证了本研究提倡的高炉炭砖抗铁水侵蚀试验方法的合理性。另外，对长寿高炉所必须具备的炭砖抗铁水侵蚀的绝对条件进行了有益的探讨。

关键词 高炉炭砖，铁水侵蚀，铁水渗透，试验方法

耐蚀低耗型炮泥在武钢 8 号高炉的应用

卢正东[1]　王齐武[2]　李友武[2]　孙　戎[3]　李勇波[1]

（1. 武钢研究院，湖北武汉　430080；2. 武钢炼铁厂，湖北武汉　430083；
3. 武钢耐材公司，湖北武汉　430083）

摘　要　针对 4000m^3 以上特大型高炉炮泥的使用要求，从耐火原料的选择、粒级搭配、添加剂和结合剂的选取以及配方优化等方面进行了研究。研制的炮泥具有良好的作业性能，抗渣铁侵蚀性能好，出铁流速稳定。在武钢 8 号高炉（4117m^3）上进行了工业试验，使用效果良好，可以满足特大型高炉出铁要求。

关键词　高炉，炮泥，耐蚀，低耗

Taphole Clay of High Corrosion Resistance and Low Consumption Applied in No.8 BF of WISCO

Lu Zhengdong[1]，Wang Qiwu[2]，Li Youwu[2]，Sun Rong[3]，Li Yongbo[1]

(1. R & D Center of WISCO，Wuhan 430080，China；2. Iron Making Plant of WISCO, Wuhan 430083, China；
3. Refractory Company of WISCO，Wuhan 430083, China)

Abstract　Aiming at the use requirements of taphole clay on huge blast furnaces(over 4000m^3), the research was carried out considering the selection of refractory raw materials, the match of raw material sizes, the selection of addictive and anchoring agent, the formula optimization , and so on. The taphole clay has good performance of working, corrosion resistance to slag-iron and stable flow rate of tapping. Based on industrial experiment on WISCO's No.8 blast furnace(4117m^3), it has good effect for use and can meet the requirements of taphole clay for the huge blast furnaces.

Key words　blast furnace, taphole clay, corrosion resistance, low consumption

Al_2O_3 含量 60%的矾土基莫来石均质料的性能和结构研究

叶　航[1]　安建成[2]　王林俊[1,2]　李　平[1]　冯运生[1]

（1. 通达耐火技术股份有限公司，北京　100085；
2. 阳泉金隅通达高温材料有限公司，北京　100085）

摘　要　随着我国矾土资源利用率的提高，合成矾土基均质料作为重要耐火原料之一，以其显微结构分布均匀和优异的高温性能已得到广泛应用。本文以 Al_2O_3 含量为 60%的均质莫来石原料（M60）为对象，借助 XRD，SEM，OM 等仪器，研究了 M60 均质料的结构与性能，并通过显微观察分析 M60 均质料的气孔和分布、晶体结构以及相组成（XRD），阐明了材料结构与高温性能之间的关系。结果表明：M60均质料莫来石晶体发育良好，呈网络结构均匀分布，玻璃相量少，荷重软化温度高，抗蠕变性能好。材料中含有较多的封闭气孔，有利于提高其热震稳定性，更适合高温窑炉或冶炼容器使用条件苛刻的部位使用。

关键词 均质料，莫来石，显微结构，气孔，高温性能

The Property and Structure Investigation of Bauxite-based Homogenized Mullite Grogs with the Al_2O_3 Content of 60%

Ye Hang[1], An Jiancheng[2], Wang Linjun[1], Li Ping[1], Feng Yunsheng[1]

(1. Tongda Refractory Technologies Co., Ltd., Beijing 100085, China;
2. Yangquan Jinyu Tongda High Temperature Materials Co., Ltd., Beijing 100085, China)

Abstract With the increase of resource utilization rate of China Bauxite, as one of the important refractory raw material, the synthetic bauxite-based homogenized grogs has been widely used with its uniformly microstructure distribution and excellent high temperature performance. In this paper, a homogenized mullite raw material with the Al_2O_3 content of 60% (M60) has been investigated. The microstructure and the property of M60 have been studied by XRD, SEM and OM. The porosity, pore distribution, crystal structure and phase composition (XRD) of M60 has been analyzed by microscopic observation. The study shows: The M60 homogenized grogs has the following characteristics: well-developed mullite crystal, uniform crystal distribution of network structure, less amount of glass phase, high temperature of refractoriness under load; there are a lot of closed-pores in the material, it's helpful to improve the thermal shock resistance, and suitable to use at the harsh part of the high temperature furnace and smelting vessel.

Key words homogenized grogs, mullite, microstructure, pore, high temperature performance

采用西格里炭砖的大型长寿命设计高炉砌筑技术要点

朱相国 游 卫

（鞍钢建设集团有限公司工业炉分公司，辽宁鞍山 114021）

摘 要 介绍了鞍钢11号高炉大修改造后有效容积2580m³的长寿命设计，采用西格里炭砖的高炉内衬砌筑技术。西格里炭砖砌筑工艺要求严格，施工关键点难于把握，砖缝为≤0.5mm，铁口区域为≤0.3mm，为了达到较高的高炉内衬砌筑质量，阐述了从砌筑材料由炉外到炉内运输设备布置，炉底找平层施工，炉底满铺炭砖下砖、砌筑及满铺炭砖铲平技术，陶瓷杯底砌筑，环炭砌筑技术，铁口区域砌筑，陶瓷杯壁砌筑，风口组合砖砌筑，以及施工过程中各处缝隙的处理。提出了炉底和炉缸砌筑质量对长寿命设计高炉至关重要，对在砌筑工艺过程的质量控制关键点提供了解决方法。本文介绍的西格里炭砖砌筑技术已成功应用于鞍钢10号高炉大修改造，炉况稳定，内衬质量高，实现了长寿命设计理念。

关键词 大型高炉，西格里炭砖，长寿命高炉，高炉砌筑

The SGL Carbon Brick in Large Blast Furnace Building Techniques of Long Life Design

Zhu Xiangguo, You Wei

(Industrial Furnace Branch，Ansteel Construction Consortium Co., Ltd., Anshan 114021,China)

Abstract　Anshan Iron and Steel No. 11 blast furnace overhaul transformation after the effective volume 2580m^3 long life design, uses the SGL carbon brick of blast furnace lining technology. SGL carbon brick strict technical requirements, construction key points difficult to grasp, the brickwork is less than or equal to 0.5mm, iron mouth area is less than 0.3mm. In order to achieve high quality of blast furnace lining, The layout of transportation equipment from the furnace to the furnace is described, Bottom leveling layer construction, The bottom and covered with brick, carbon brick under covered carbon brick leveling technology, Ceramic cup bottom masonry, Ring carbon masonry Technology, Iron mouth area masonry, Ceramic cup wall masonry, Air outlet combination brick masonry, And the processing of the cracks in the construction process. The bottom and hearth lining quality is critical for the long-life design of blast furnace, The key points of quality control in masonry process are provided.

Key words　large blast furnace, SGL carbon brick, long life blast furnace, blast furnace masonry

鞍钢新 1 号高炉炉缸内衬破损调查

郭天永[1]　车玉满[1]　王宝海[2]　谢明辉[2]　孙　鹏[1]

张立国[1]　姚　硕[1]　姜　喆[1]

（1. 鞍钢股份有限公司技术中心，辽宁鞍山　114009；

2. 鞍钢股份有限公司炼铁总厂，辽宁鞍山　114021）

摘　要　利用新 1 号高炉炉缸换衬机会，对新 1 号高炉炉缸内衬的侵蚀状况进行测绘、照相，取不同部位的残存炭砖、渣皮以及黏结物进行理化性能检测，分析碱金属及锌在炉缸内衬的分布状况以及钒钛矿护炉效果。根据炭砖理化检测结果结合炉缸冷却制度，总结分析新 1 号高炉炉缸炭砖侵蚀特征及异常侵蚀原因，并提出减轻炉缸炭砖发生异常侵蚀措施。

关键词　炉缸，炭砖，异常侵蚀，破损调查

Investigation on Hearth Lining Erosion of New No.1 Blast Furnace in Ansteel

Guo Tianyong[1], Che Yuman[1], Wang Baohai[2], Xie Minghui[2], Sun Peng[1],

Zhang Liguo[1], Yao Shuo[1], Jiang Zhe[1]

(1. Technology Center of Angang Steel Co., Ltd., Anshan 114009, China;

2. General Ironmaking Plant of Angang Steel Co., Ltd., Anshan 114021, China)

Abstract Using the opportunity of changing hearth lining in new No.1 Blast Furnace in Ansteel, this BF hearth erosion was mapped and photographed,and the physical-chemical properties of remaining carbon brick , slag skin and adhesive materials in different parts of the BF hearth was studied, the author analyzed furnace protection effect with vanadium titanium ore and alkali metal and Zn distribution. According to physical-chemical properties test of carbon brick and cooling system in hearth, the paper summarized abnormal erosion reasons of the BF hearth and provide measures on how to reduce abnormal erosion.
Key words hearth, carbon brick, abnormal erosion, investigation on erosion

高炉炉缸侧壁温度升高与控制等相关问题的探讨

孙 健[1,2]，武建龙[2]，储满生[1]，陈 辉[2]，梁海龙[2]，马泽军[2]

（1. 东北大学材料与冶金学院，沈阳 110004; 2. 首钢技术研究院，北京 100043）

摘 要 近年来，随着高炉冶炼强度的增加，炉缸发生烧穿、侧壁温度急剧升高等问题的高炉明显增多。本文对近年来国内高炉炉缸烧穿现象、高温点现象进行了统计分析，并以国内某大型高炉为例进一步阐述了高温点分布区域集中性、铁口使用频率与高温点爆发的规律性，提出了合理的出铁制度有利于减少炉缸侧壁高温点的出现。随后，对比分析了目前常用护炉措施，得出目前此大型高炉所采用的常态化加钛护炉可长期有效控制炉缸温度，为其他高炉预防炉缸高温点出现及合理选择护炉措施提供了理论和技术支持。
关键词 炉缸烧穿，炉缸侧壁温度高，铁口使用频率，加钛护炉

Discussion on the High Temperature of the BF Hearth Sidewall and the Control Measures

Sun Jian[1,2], Wu Jianlong[2], Chu Mansheng[1], Chen Hui[2], Liang Hailong[2], Ma Zejun[2]

(1. School of Materials & Metallurgy, Northeastorn University, Shenyang 110004, China;
2. Shougang Research Institute of Technology, Beijing 100043， China)

Abstract In recent years, with the increase of the intensity of the blast furnace, the issues of the hearth burn occurs and blast furnace sidewall temperature rapid rise increased significantly. In this paper, the BF hearth burn and the hot spots were statistically analyzed, and further elaborated concentration, iron mouth outbreak frequency and regularity hot spots hot spots distribution areas proposed sound system helps to reduce the iron appears hearth Sidewall high temperature points. Subsequently, the comparative analysis of the current common furnace protection measures come now normalized Titanium protect this large blast furnace can be used in long-term effective control of hearth temperature.
Key words hearth burn, high temperature of the hearth sidewall, taphole frequency, titanium furnace protection

高炉炉缸侧壁侵蚀过程监测和分析

徐 萌 刘洪松 陈 辉 孙 健 武建龙

（首钢技术研究院，北京 100043）

摘　要　论文以两种典型炉缸结构为例，通过双点法计算了热流强度、炉衬剩余厚度、自保护渣皮厚度，分析了炉缸侧壁的侵蚀过程、陶瓷杯和炭砖的侵蚀速率、自保护渣皮的生长与剥蚀速率。结果表明 A 炉 NMA 炭砖的平均侵蚀速率为 1.5mm/h，最大侵蚀速率接近 3mm/h，浇注层的平均侵蚀速率为 1mm/h，自保护渣皮的平均剥蚀速率为 1.5~1.7mm/h，平均生长速率为 1.3~1.6mm/h，但剥蚀速率波动幅度大；B 炉国内大块炭砖的侵蚀速率为 0.8~2mm/h，陶瓷杯的平均侵蚀速率为 0.3~0.6mm/h，渣皮的剥蚀和生长速率均为 0.3~0.4mm/h，波动幅度相对小。建议建立炉缸侧壁自保护层厚度日常管理模型，通过自保护层的当前厚度、发展趋势和变化速率直观判断炉缸状况，采取有效高炉操作和护炉措施控制炭砖不直接暴露于铁水，提高炉缸寿命。实践证明，控制冶炼强度和加钛护炉是抑制渣皮剥蚀或促进渣皮重新形成、稳定和生长的主要手段。

关键词　高炉炉缸侧壁，炭砖，陶瓷杯，自保护渣皮，侵蚀，剥蚀

Analysis and Monitoring on the Corrosion Course of BF Hearth Sidewall

Xu Meng, Liu Hongsong, Chen Hui, Sun Jian, Wu Jianlong

(Shougang Research Institute of Technology, Beijing 100043, China)

Abstract　For the two cases of typical BF hearth structure, the heat flux, the thickness of refractory lining and the thickness of self protecting slag skull were calculated by the two point temperature. And the corrosion course of hearth sidewall, the corrosion rate of carbon brick or ceramic cup and the erosion rate of slag skull were analyzed. For A BF, it was concluded that the average and maximum corrosion rate of NMA brick was 1.5mm/h and 3mm/h respectively, the average corrosion rate of pouring layer 1mm/h, the average erosion rate of slag skull 1.5~1.7mm/h but with large fluctuation range, and its average growth rate 1.3~1.6mm/h. For B BF, the average corrosion rate of domestic bulk carbon brick and ceramic cup was 0.8~2mm/h and 0.3~0.6mm/h respectively, and either the erosion rate or the growth rate of slag skull 0.3~0.4mm/h with small fluctuation range. It had been suggested that the hearth situation should be visually justified by the current thickness, developing trend and rate throughout daily management, and some effective measures be applied for protecting carbon brick from hot metal. It had been practically verified that reducing smelting intensity and [Ti] were major methods for controlling the erosion of slag skull or improving its formation, growth and stabilization.

Key words　BF hearth sidewall, carbon brick, ceramic cup, self protecting slag skull, corrosion, erosion

高炉炉缸炉底温度场控制技术

张福明[1]　赵宏博[2]　钱世崇[1]　程树森[2]

（1. 北京首钢国际工程技术有限公司，北京　100043；
2. 北京科技大学冶金与生态工程学院，北京　100083）

摘　要　通过对高炉炉缸炉底内衬侵蚀和温度过热现象的解析，阐述了当代高炉炉缸炉底温度场控制的理论。解析了高炉炉缸炉底温度过热的现象和形成机制，论述了高炉炉缸炉底工作过程中，内衬侵蚀破损的过程和机理。提出了“无过热-自保护”的炉缸炉底内衬设计理念，强调了通过设计合理的炉缸内衬结构、采用优质耐火材料和高效冷却系统，在线监测炉缸炉底温度和热流强度变化，控制炉缸炉底温度场合理分布，从而有效抑制炉缸炉底内衬侵蚀速率、延长高炉寿命。

关键词　高炉，炉缸，长寿，温度场，耐火材料

Control Technology on Temperature Field of Blast Furnace Hearth and Bottom

Zhang Fuming[1], Zhao Hongbo[2], Qian Shichong[1], Cheng Shusen[2]

(1. Beijing Shougang International Engineering Technology Co., Ltd., Beijing 100043, China;
2. School of Metallurgical and Ecological Engineering, University of Science and Technology Beijing, Beijing 100083, China)

Abstract The blast furnace (BF) hearth lining corrosion and the temperature overheat phenomenon are analyzed, the temperature field control concept of contemporary BF hearth is expounded. The phenomenon and formation mechanism of BF hearth and bottom temperature overheat are analyzed. The mechanism of hearth lining corrosion is analyzed under the BF working condition. The temperature distribution of hearth lining, and the hot metal flowing field distribution in the hearth are researched by CFD model. The design concept of "no overheat - self protection" of hearth lining is proposed. Reasonable measures such as structure design of hearth lining, high quality refractory and the high efficiency cooling system, hearth and temperature and heat flux intensity on-line monitoring, reasonable control of hearth bottom temperature field distribution, will be adopted to avoid the abnormal wear of hearth lining and prolong the BF campaign life. As a result, satisfied performance has been achieved in application.

Key words blast furnace, hearth, long campaign life, temperature field, refractory

高炉炉缸活性的基础研究

代 兵 梁 科 王学军 惠国东 董 辉 孙洪军

（本钢板材股份有限公司炼铁厂，辽宁本溪 117000）

摘 要 通过对高炉炉缸活性的基础研究，重点分析了炉缸活性的概念及其影响因素，认为焦炭所提供的“透气-透液通道”数量、熔体（渣铁）流动性能和风口回旋区的位置是影响炉缸活性的三个重要因素。为了加强对炉缸活性的监测，提出了炉缸活性量化计算模型，通过采集与处理现场高炉数据以及编程技术完成了模型的在线和离线计算，并成功应用于本钢新一号高炉的生产中。最后，以本钢新一号高炉曾经发生的炉缸堆积事故为例，分别从热制度、造渣制度、送风制度、装料制度四大操作制度阐述了恢复炉缸活性的具体方法和过程。实践证明，这些研究可以为维护炉缸活性的长期良好稳定提供参考。

关键词 高炉，炉缸活性，量化计算模型，操作制度

The Basic Research of Blast Furnace Hearth Activity

Dai Bing, Liang Ke, Wang Xuejun, Hui Guodong, Dong Hui, Sun Hongjun

(Ironmaking Works of Benxi Steel Plate Co., Ltd., Benxi 117000, China)

Abstract Through the basic research of blast furnace hearth activity, the concept of hearth activity and its influencing factors were emphatically analyzed, which indicated that the coke quality, the slag and iron flow performance and tuyere raceway location are the three most important factors affecting hearth activity. In order to strengthen the monitoring on the hearth activity, two hearth activity quantitative calculation models were put forward, whose on-line and off-line calculations were realized and practiced in the production of BX steel new No.1 blast furnace by the technologies of data acquisition, processing and programming. At last, taking BX steel new No.1 blast furnace hearth accumulation accident for instance, the specific methods and procedures to recovery hearth activity by four major operating systems, such as thermal system, slag-making system, blasting system and charging system, were illustrated in detail. Practice has proved that these studies could provide the reference to maintain good and long-term stability of the hearth activity.

Key words blast furnace, hearth activity, quantitative calculation models, operating systems

鞍钢4高炉炉缸钒钛粉自由式喷吹保护技术研究

刘德军

（鞍钢股份有限公司技术中心，辽宁鞍山　114021）

摘　要　鉴于目前多种炉缸钒钛矿护炉方法普遍存在对现场场地的适应性差、对炉缸具体需要补护部位的针对性差、方法的灵活性差和护炉料的加入量可控性差等缺陷，借鉴高炉喷吹煤粉的相关技术，开发了可移动罐式高炉炉缸钒钛粉自由式喷吹护炉技术及装置，并在鞍钢4高炉取得了良好的使用效果。

关键词　高炉，炉缸，钒钛粉，自由式喷吹

Technology Research to Protect Hearth by Vanadium Titanium Powder Injection in the Freestyle in 4# BF of Ansteel

Liu Dejun

(Angang Steel (Group) Co., Ltd., Anshan 114021, China)

Abstract Given the variety of vanadium titanium ore protecting hearth methods poor adaptability to the site and specific need to fill for hearth care of poor pertinence and poor flexibility of those methods and methods poor controllability for the amount of those methods, etc shortcomings, draw lessons from the relevant technology of the blast furnace coal injection, developed a technology and a tank device to protect hearth by vanadium titanium powder injection in the freestyle, in 4# blast furnace of Ansteel, has achieved good using effect.

Key words blast furnace, hearth, vanadium titanium powder, injection in the freestyle

Transient Simulation of Titanium Compounds in a Blast Furnace Hearth during Titania Injection

Guo Baoyu[1,2], Komiyama K. Matthew[1], Yu Aibing[2], Zughbi Habib[3], Zulli Paul[3],

(1. School of Materials Science and Engineering, University of New South Wales Sydney, NSW 2052, Australia;

2. Department of Chemical Engineering, Monash University, VIC 3800, Australia;

3. BlueScope Steel Research, P.O. Box 202, Port Kembla NSW 2505, Australia)

Abstract A transient 3-D CFD model is developed to simulate the TiC particle behaviour in the blast furnace hearth during titania addition via tuyere injection for a single tapping cycle. Analyses are conducted with respect to the distribution of TiC mass fraction along the inner hearth surface. The effects of key operational parameters are investigated. The following recommendations are made. To protect the side walls, titania injecting tuyere located upstream from the hot spot location, by an offset angle as proposed, should be selected. To protect the hearth bottom corner, titania injecting tuyere should be selected such that the hot spot location is between the active taphole and tuyere are around 60° apart.

Key words blast furnace hearth, CFD modelling, Ti solidification

宝钢高炉凉炉操作

赵思杰　王 俊　桂 林

（宝山钢铁股份有限公司炼铁厂，上海　200941）

摘　要　三十年来，宝钢炼铁深入持久地推进炼铁系统的技术创新，高炉炼铁技术得到大幅度发展，产能规模逐步扩大，技术水平不断提高，高炉高利用系数、高煤比、长寿等技术取得了长足进展。近年来，宝钢股份对现有的四座高炉陆续进行了停炉改造，其中，高炉停炉之后的凉炉工作至关重要。宝钢在各高炉凉炉过程中分别采用了放残铁储水凉炉、不放残铁储水凉炉的方式，形成了一套全面、系统的凉炉技术。凉炉技术涉及到安全技术、凉炉效率核算、凉炉设备研发等诸多方面。凉炉过程中，凉炉速度、冷却介质的利用效率受到炉内残铁量、排水速度等诸多因素的影响。本文就宝钢四座高炉的凉炉实际操作及改进进行探讨，以期为今后高炉凉炉操作提供一定的参考依据。

关键词　高炉，凉炉

Cooling Practice of Blast Furnace in Baosteel

Zhao Sijie, Wang Jun, Gui Lin

(Iron-making Plant Baoshan Iron & Steel Co., Ltd., Shanghai 200941, China)

Abstract The iron making technology in Baosteel has been greatly improved during the last 30 years, the production capacity has been expanded and the smelting technology has been continuously improved. The technology of high utilization coefficient, high coal ratio and longevity of blast furnace has made considerable progress. In recent years, the four blast furnace of Baosteel have been overhauled one after another, and it is a very important step to cool down the blast furnace after blow-down. In the process of cooling, the technology of discharge of remaining iron and direct-cooling technology have both been used. The safety technology, cooling efficiency calculation, research and development of cooling equipment, and many other aspects have been well considered. The cooling operation and improvement were discussed in this paper, so as to provide reference and optimization suggestion for cooling operation in the future.

Key words blast furnace, cooling

修编《高炉炼铁工程设计规范》

邹忠平　项钟庸　王　刚

（中冶赛迪工程技术股份有限公司，重庆　401122）

摘 要 低碳炼铁、节能降耗及降低成本都要求降低燃料比,这是当前高炉炼铁发展的必然趋势，也是长远的目标。本次修编仍然落实以精料为基础，全面贯彻高效、低耗、优质、长寿、环保的炼铁技术方针，重点是进一步解决中国高炉炼铁高燃料比的问题。采用了按照低燃料比操作的高炉炼铁新指标，同时解决了主要设备选型不合理的问题，从设计上为高炉最终实现低燃料比操作奠定了基础。

关键词 高炉炼铁，设计规范，燃料比

Revision for Design Specification of Blast Furnace Engineering

Zou Zhongping, Xiang Zhongyong, Wang Gang

(CISDI Engineering Co., Ltd, Chongqing 401122, China)

Abstract low carbon iron making, energy saving and cost reduction are required to reduce fuel ratio, which is the inevitable trend of the development of blast furnace iron making, but also a long-term goal. In this revision, iron making technology policy of high efficiency, low cost, good quality, long life and environmental protection is still implemented based on the theory of fine material, which is to solve problem of high fuel ratio of Chinese blast furnaces. Adopting the new index of low fuel ratio blast furnace iron making, problem of unreasonable selection of the main equipment is also solved, and the basis for the design of low fuel ratio operation is established.

Key words blast furnace iron making, design specification, fuel ratio

1.4 炼铁新技术与资源综合利用/ New Technology of Iron Making and Resource Comprehensive Utilization

Non-Blast Furnace Ironmaking and COREX

Zou Zongshu

(School of Metallurgy, Northeastern University, Shenyang 110819, China)

Abstract Non-blast-furnace ironmaking, or non-coking ironmaking, is one of the most concerned focuses in iron and steel industry, particularly in China. Although it covers a great number of various processes for both direct reduced iron (DRI) and hot metal production, those for hot metal production, particularly the COREX process are discussed in comparison with blast-furnace process in this presentation.

Smelting for melting and separation of metal and slag is an indispensable, and mostly the final step for non-blast-furnace hot metal production, and this also leads to the topic of smelting reduction. In the original concept of smelting reduction, there is, however the incompatibility issue of oxidation heat supply and reduction reaction. Considering that the thermodynamic states are independent of the process route, the process sequence, *i.e.* melting and then reduction or reduction and then smelting, is analyzed, leading to the conclusion that reduction and then smelting is the reasonable sequence for hot metal production. This is also the reason of the succeeding of the so-called two-step smelting reduction process, like COREX, as well as blast furnace process. In view of this, various smelting reduction processes are analyzed and summarized.

Essentially, COREX is the same as blast furnace, or more precisely a two-segmented oxygen blast furnace. Mass and heat balances of COREX and oxygen blast furnace, as well as their varieties with different top gas recyclings, are studied. The inherent feature of top gas recycling of blast furnace though hot-blast stove is also analyzed. The results show that, being born from blast furnace and similar as oxygen blast furnace but without top gas recycling, it is not fair to expect the current COREX process to achieve a better performance of fuel consumption than hot blast furnace and oxygen blast furnace with top gas recycling. With top gas recycling, however, COREX can show its advantages in fuel consumption, as well as in the utilization of low grade fuels and raw materials. In view of this, the further development of COREX process is analyzed and proposed.

Hot metal production, or general ironmaking is not the only function of various ironmaking reactors and processes with coal as fuel and reductant. Energy transformation (particularly coal gasification) and solid waste treatment are other two major functions which have drawn great attentions in the worlds of energy, metallurgy and environment or geology. The former is of greater significance particularly in China where the major energy resource is coal. The integrations of coal gasification and hot metal production with various ironmaking reactors and processes including COREX are analyzed and discussed. The results show that, together with hot metal production, coal gasification with ironmaking reactors has its advantage over the current coal gasification reactors in the post-processing of generated raw gas. Iron bath smelting reduction for coal gasification and hot metal production is finally proposed.

Key words non-blast-furnace ironmaking, COREX, smelting reduction, coal gasification

COREX® - An Answer for Hot Metal Production in a Changing Environment

Shibu John[1], Wolfgang Sterrer[2], Johann Wurm[3], Barbara Rammer[4]

(all: PRIMETALS Technologies Austria GmbH, Linz, Austria; 1.Head Direct and Smelting Reduction; 2. Head of Technology Department Smelting Reduction; 3. Senior Expert Direct and Smelting Reduction; 4. Process Technology Smelting Reduction)

Abstract The COREX® Process is a well-established route for the production of hot metal and beside the FINEX® Technology the only industrially realized alternative to the blast furnace route.

Changes in the technology itself subsequently improved operation of the established COREX® plants and process economics.Changes of external factors e.g.general decrease in raw material quality, increase in energy cost and availability of energy (especially natural gas), strict environmental regulations, pressure to lower the investment costs and lower operation cost,make it worth to carefully re-evaluate the COREX® technology.

Current COREX® pre-projects clearly show an increasing demand for the "by-product" COREX® export gas for power generation and for the production of DRI. Based on the economic evaluation of different COREX® gas based plant concepts, the comparison with the traditional blast furnace route shows that the COREX® technology is highly competitive in the challenging iron and steel producing environment.

Key words COREX®, MIDREX™, economics, export gas utilization, hot metal, DRI

米德雷克思煤制气直接还原技术

钱良丰

（米德雷克思，上海　200040）

摘　要　本文介绍了米德雷克思煤制气直接还原技术的发展历史，工艺流程，取得的技术突破及商业化应用范例说明。提出了 MXCOL®煤制气直接还原技术的主要经济技术指标，突出说明煤制气直接还原技术的优点，并对煤制气直接还原技术的污染排放量与传统高炉流程进行了比较。分析了直接还原技术在中国的发展前景，直接还原铁的市场前景。在我国发展高端制造过程中，直接还原技术可以助推特钢技术的发展，确保关键钢铁材料的供应。

关键词　煤制气直接还原，直接还原铁，海绵铁，热压铁块, Midrex®，MXCOL®

Midrex Syngas Based Direct Reduction Technology

Qian Liangfeng

(Midrex, Shanghai 200040, China)

Abstract　This article makes simply introduction about the development of Midrex syngas based direct reduction technology in the view of history, technology and process. The significance of syngas based direct reduction technology is described and the commercial reference plant is disclosed. Also the detail economic and technical performance index are described and the environment emission data is compared with the traditional blast furnace process. The development future of syngas direct reduction process is discussed, and the market trend of direct reduction iron is evaluated. The direct reduction product is able to help the development of high grade manufacture, which is guided by the center government.

Key words　syngas direct reduction, DRI, direct reduction iron, HBI, Midrex®, MXCOL®

适用于气基竖炉还原工艺的煤气化工艺评价及合理选择

储满生　李　峰　唐　珏　冯　聪　汤雅婷　柳政根

（东北大学材料与冶金学院，辽宁沈阳　110819）

摘　要　基于现有煤气化工艺特征和竖炉还原工艺需求，并同时考虑投资成本、氧耗、煤耗、冷煤气转化效率、煤气中 CO/H_2、煤气氧化度、煤气中有效还原气含量、净热效率、碳转化率、单炉产能等指标，本文运用多指标综合加权评分法对 Lurgi、Ende、Texaco 和 Shell 四种主要煤气化技术进行定量化评价，并给出了选择适宜煤气化技术的相关建议。结果表明，Shell 气流床干粉煤加压气化法更适于作为气基竖炉直接还原用煤气的成产技术，Ende 流化床粉煤常压气化法次之。

关键词　煤气化，竖炉，还原性气体，多指标，综合加权评分

Evaluation and Rational Choice of Coal Gasification Process Applied to Gas-based Shaft Furnace Reduction Process

Chu Mansheng, Li Feng, Tang Jue, Feng Cong, Tang Yating, Liu Zhenggen

(School of Materials & Metallurgy, Northeastern University, Shenyang 110819, China)

Abstract Based on the characteristics of the existing coal gasification process and shaft furnace reduction process demand, this paper quantitatively evaluated the four main technologies of coal gasification through multi index synthetic weighted mark method in which many indexes considered include the cost of the investment, oxygen consumption, coal consumption, efficiency of cold gas transformation, the CO/H_2 in the gas, the gas oxidation degree, the content of effective reducing gas in the coal gas, the net thermal efficiency, the carbon conversion rate, single furnace capacity etc. At the same time, relevant proposals for coal gasification technology were given in this paper. Above all, Shell flow bed powder coal gasification was more suitable for the gas production used for gas-based shaft furnace production, followed with the Ende fluidized bed pulverized coal gasification.

Key words coal gasification, shaft furnace, reducing gas, multi index, synthetic weighted mark method

含硼铁精矿回转窑直接还原-电炉熔分工艺研究

赵嘉琦　付小佼　王　峥　柳政根　储满生

（东北大学材料与冶金学院，辽宁沈阳　110819）

摘　要 硼铁矿是我国重要的硼资源，但该矿结构、成分复杂，综合利用难度大，至今未实现合理的综合开发利用。硼铁矿经选矿完成硼铁的第一级分离之后，含硼铁精矿中的硼约占原矿硼总量的25%～30%，含硼铁精矿的高效利用是硼铁矿综合利用的重要环节。实验室条件下对含硼铁精矿煤基回转窑直接还原-电热熔分工艺进行了研究。实验结果表明：含硼铁精矿是优良的造球原料，在1190℃焙烧20min后，成品球团抗压强度达到2000N以上，可满足回转窑直接还原的要求；以非焦煤为还原剂，在950℃下还原150min即可得到金属化率超过90%的直接还原铁，满足电炉熔分的要求；在高温下进行电热熔分，渣铁分离效果较好，实现硼铁高效分离，其中富硼渣中B_2O_3含量达到22.29%，活性达到86.62%，是一步法生产硼酸的优良原料。含硼铁精矿回转窑直接还原-电热熔分可以实现硼铁高效分离。

关键词 含硼铁精矿，直接还原，回转窑，电炉熔分

Study on the Process Utilizing Boron-bearing Iron Concentrate on Rotary Kiln Direct Reduction and Electric Furnace Smelting Separation

Zhao Jiaqi, Fu Xiaojiao, Wang Zheng, Liu Zhenggen, Chu Mansheng

(School of Materials & Metallurgy, Northeastern University, Shenyang 110819, China)

Abstract In this work, oxide pellet was prepared from boron-bearing iron concentrate. The process that the boron-bearing iron concentrate was directly reduced and separated by rotary kiln and electric furnace smelting, respectively, was studied. The results showed that boron-bearing iron concentrate pelletization can fully meet the requirements of the rotary kiln direct reduction process. After boasting at 1190℃ for 20min, the compressive strength of the finished pellet is higher than 2000N. Using non coking coal as reducing agent and selecting 950℃ as the reduction temperature, the time needs to reduce 90% of the boron-bearing iron concentrate was below 150min, which completely meets the requirements of the electric furnace smelting separation. Boron and iron can be efficiently separated by electric furnace at high temperature. The activity and the mass fraction of B_2O_3 in the slag can reach 86.62% and 22.29%, respectively, suggesting that this kind of slag can be used as high quality raw material for the one-step boric acid production. The new process can efficiently separate boron-bearing iron concentrate.

Key words boron-bearing iron concentrate, direct reduction, rotary kiln, electric furnace smelting separation

高铁铝土矿热压块还原-电炉熔分工艺探索研究

王 峥 柳政根 储满生 赵嘉琦 王宏涛 赵 伟

（东北大学材料与冶金学院，辽宁沈阳 110819）

摘 要 高铁三水铝土矿是一种富含铁铝的复合矿产资源，由于其铁铝质量比大于 1、铝硅质量比小于 3，其铁铝品位不满足各自钢铁工业和铝工业的要求。为了实现高铁铝土矿中铁、铝的高效分离，在实验室条件下进行了高铁铝土矿热压块还原-电炉熔分工艺探索研究。研究表明，高铁铝土矿热压块还原-熔分工艺可实现高铁铝土矿铁铝的高效分离，得到的铁水质量类似于高炉水平，得到的铝酸钙自粉渣适用于碳酸钠溶液浸出提取 Al_2O_3；高铁铝土矿热压块具有较高的冷态抗压强度和还原冷却后强度，其还原装置的选取更为广泛，可选择反应器包括转底炉、车底炉，或煤基竖炉；随着配碳比的增加，冷态抗压强度呈上升的趋势，还原冷却后强度呈降低的趋势，还原速率加快，但熔分效果变差，适宜的配碳比为 1.0。本研究为高铁铝土矿的综合利用提供了新途径。

关键词 高铁铝土矿，含碳热压块，直接还原，熔分，铁铝分离

Exploratory Research on the Reduction and Melting Process of High Iron Gibbsite Ore Based on Hot Briquette

Wang Zheng, Liu Zhenggen, Chu Mansheng, Zhao Jiaqi, Wang Hongtao, Zhao Wei

(School of Materials & Metallurgy, Northeastern University, Shenyang 110819, China)

Abstract High iron gibbsite ore is a kind of composite mineral resources and rich in iron and aluminum. As the $w(Fe_2O_3)/w(Al_2O_3)$ is no less than 1.0 and the $w(Al_2O_3)/w(SiO_2)$ is no more than 3.0, the grade of iron and aluminum do not meet the demands of steel industry and alumina industry, respectively. In order to achieve the effective separation iron and alumina from high iron gibbsite ore, the exploratory research on the reduction and melting process of high iron gibbsite ore based on hot briquette is carried out under laboratory conditions. The results show that, the reduction and melting process of high iron gibbsite ore based on hot briquette could separate iron and alumina from high iron gibbsite ore effectively. The hot metal which is similar to blast furnace pig iron and calcium aluminate slag are obtained. The Al_2O_3 could be leached from calcium aluminate slag by adding odium carbonate solution. As the high iron gibbsite ore hot briquette have enough compressive strength before and after reduction, the reactor could have many choices, such as rotary hearth furnace, car-bottom furnace, coal agent shaft furnace, and so on. With increasing carbon ratio, the compressive strength and reduction rate of high iron gibbsite hot briquette tend to increase, while the compressive strength after reduction and melting separation effect tend to decrease. The proper carbon ratio of high iron gibbsite hot briquette is 1.0. This paper could provide a new technological approach for the comprehensive utilization of high iron gibbsite ore.

Key words high iron gibbsite ore, carbon composite hot briquette, direct reduction, melting separation, Fe-Al separation

水冷处理在提高高铬型钒钛磁铁矿固相还原效果中的作用

姜 涛 周 密 关山飞 杨松陶 卢 乐 薛向欣
（东北大学材料与冶金学院，辽宁沈阳 110819）

摘 要 将高铬型钒钛磁铁精矿、还原煤粉和黏结剂按一定比例配料、混匀、模压成型后进行直接还原。通过化学分析、XRD、SEM 等研究了水冷处理对不同温度下的高铬型钒钛磁铁矿固相还原产物金属化率以及物相的影响,并对不同温度下水冷处理的高铬型钒钛磁铁矿固相还原产物的显微结构、有价组元铁、钒、钛、铬的分布进行了分析。实验结果表明:水冷处理可以显著提高高铬型钒钛磁铁矿固相还原的金属化率，避免二次氧化的发生，在 1300℃下，通过水冷处理，高铬型钒钛磁铁矿固相还原金属化率可达到 96.79%。同时，通过水冷处理可以发现，随着温度的提高，高铬型钒钛磁铁矿固相还原产物中金属铁晶粒长大，V、Cr 逐步富集于铁相中，Ti 则进一步富集于脉石相中，从而有利于后续铁、钒、钛、铬的分离提取。

关键词 高铬型钒钛铁矿，直接还原，水冷，金属化率

Effect of Water Cooling on Improving Solid-phase Reduction for High Chromium Vanadium - titanium Magnetite

Jiang Tao, Zhou Mi, Guan Shanfei, Yang Songtao, Lu Le, Xue Xiangxin
(School of Materials and Metallurgy, Northeastern University, Shenyang 110819, China)

Abstract The mixture was mixed by the high chromium vanadium-titanium magnetite(V-Ti-Cr), reducing coal and binder, and molded by the tableting machine, then on the direct reduction.The effect of water cooling on metallization and phase of V-Ti-Cr after solid-phase reduction was studied by chemical analysis, XRD. In addition, the micro-structures, the distribution of Fe, V, Ti, Cr of the solid-phase reduced V-Ti-Cr followed by water cooling at different temperatures were observed via SEM. The results showed that water cooling could prevent reoxidation and improve the metallization significant. Moreover, the maximum metallization degree could get 96.79% at 1300℃ by water cooling. Meanwhile, through the method of water cooling , it is found that, with the temperature increasing, the grain of metallic iron grew gradually and the elements V and Cr enriched in iron-bearing phases while the Ti enriched in slag phases.And this is beneficial to the separation of Fe,V,Ti,Cr.

Key words high-chromium vanadium-titanium magnetite, direct reduction, water cooling, metallization

富铁铜渣碳热还原过程的热力学分析

宫晓然 赵 凯 邢宏伟
（华北理工大学冶金与能源学院，河北唐山 063000）

摘　要　通过 FactSage 热力学软件中的 Equilib 模块对铜渣中含铁、铜物相碳热还原过程进行理论分析，热力学计算结果显示：添加碳及碱性氧化物 CaO 会降低还原反应的初始温度，约 200℃。并且通过实验研究了温度、碱度、配碳量等工艺参数对富铁铜渣碳热还原过程及其收得率的影响，研究表明：Fe 的收得率主要受温度的影响较大，碱度的影响不大，配碳量有一定影响，其中最适宜工艺参数为：温度为 1200℃，碱度为 0.6，C/O=1.4，铁的收得率最高可以达到 91%以上。

关键词　铜渣，FactSage，金属收得率，碳热还原

Iron Rich Copper Slag Thermodynamic Analysis of Carbon Thermal Reduction Process

Gong Xiaoran, Zhao Kai, Xing Hongwei

(North China University of Science and Technology, College of Metallurgy and Energy, Tangshan 063000, China)

Abstract　Through Equilib modules of the FactSage thermodynamics software. Theory analysis to the Iron，copper phase carbon thermal reduction process. Thermodynamic calculation results display：Adding carbon and alkaline oxide CaO will reduce the initial temperature of the reduction reaction，about 200℃. And studied the effects of temperature，alkalinity，carbon content to the carbon thermal reduction process of the rich iron copper slag and its yield by the experiment.The research showed that the yield of Fe is mainly affected by temperature，the influence of basicity is small，the carbon content had a certain effect.One of the most suitable process parameters：the temperature is 1200℃，the alkalinity is 0.6，C/O is 1.4.The iron yield can reach more than 91%.

Key words　copper slag, FactSage, metal yield, carbothermic reduction

Unconventional Raw Materials for Ironmaking

Babich Alexander, Senk Dieter

(Dept. of Ferrous Metallurgy (IEHK), RWTH Aachen University, Aachen, Germany)

Abstract　Keynote deals with use of biomass and waste plastics by means of injection and embedding in pellets, composites, coke and sinter, with re-use of top gas and COG, with efficiency of nut coke in the BF, coal briquettes in Corex, mini-pellets in sintering and, finally, with use of DRI/LRI in the BF. All these topics are discussed considering our research results at IEHK/RWTH Aachen University.

Key words　biomass, waste plastics, off gas, nut coke and coal briquettes, DRI / LRI

基于低碳要求的国际钢铁工业突破性项目现状分析

刘清梅　杨学梅　王铭君

（首钢技术研究院，北京　100043）

摘　要　随着全球构建碳排放权交易经济体系，钢铁企业对碳成本的重视程度逐渐增强。本文对全球二氧化碳排放现状进行调查，阐述了基于当前二氧化碳排放基准上的欧盟、日本和中国碳交易建设情况。为了适应未来减少碳排放的

要求，国际上先进的钢铁生产国家对钢铁行业低碳减排突破性项目的推进进入了实质性阶段。本文着重分析了日本环境友好型炼铁项目 COURSE50、欧盟超低二氧化碳排放项目 ULCOS 和韩国浦项低碳 FINEX 及全氢高炉炼铁项目的技术进展和未来推进计划，并对项目的前景进行探讨，为我国钢铁行业低碳减排突破性技术的开发提供指导。

关键词 国际，钢铁，低碳排放，交易，突破技术

Current Status Analysis of the International Steel Industry Breakthrough Projects Based on Low Carbon Requirement

Liu Qingmei, Yang Xuemei, Wang Mingjun

(Shougang Research Institute of Technology，Beijing 100043, China)

Abstract With the construction of the global carbon emissions trading system, iron and steel enterprises pay more attention to the carbon cost gradually. This paper investigates the status of global carbon dioxide emissions, and expounds the carbon trading construction in the European Union, Japan and China on the basis of the current carbon dioxide emission standards. In order to meet the requirements of reducing carbon emission, the low-carbon emission reduction projects in the international advanced steel production countries are into the substantive stages. This paper focuses on the technological process analysis and future plans of Japan environmental friendly iron-making project COURSE50, the European Union ultra-low carbon dioxide emissions project ULCOS and Korea Posco low carbon FINEX and full hydrogen blast furnace iron-making project, and has a discussion on the prospect of these projects, which provide a guidance for the development of Chinese low carbon emission reduction technologies.

Key words international, iron and steel, low-carbon emission, trading, breakthrough technology

Properties of Ferro-coke and Its Effect on Blast Furnace Reactions

Li Peng[1], Bi Xuegong[1], Zhang Huixuan[1], Shi Shizhuang[2], Cheng Xiangming[1], Ma Yirui[1], Wu Qiong[2]

(1. The State Key Laboratory of Refractories and Metallurgy, Wuhan University of Science and Technology, Wuhan 430081，China; 2. Hubei Coal Conversion and New Carbon Materials Key Laboratory, Wuhan University of Science and Technology, Wuhan 430081, China)

Abstract The reactivity of ferro-coke is much higher than the conventional coke and its application to the blast furnace is expected to improve the blast furnace reactions efficiency and to reduce fuel consumption and CO_2 emission. Ferro-coke could thus be a new kind of BF burden materials with broad development prospects. In order to under the effect of using ferro-coke on BF reactions, a series of laboratory tests has been conducted in this paper. Firstly, ferro-cokes were made in a laboratory cokemaking oven of chamber type by adding three kinds of iron ores into the cokemaking coal blend and the ferro-cokes with higher reactivity as well as comparatively high strength after reaction were chosen as experimental samples for the further study. Secondly, the carbon solution loss reaction tests, the coupled reactions tests between iron ore reduction and ferro-coke gasification, and the primary slag formation tests under the condition of ferro-coke and iron ores mixed charging and simulating blast furnace conditions were carried out. The experimental results showed the gasification reaction beginning temperature of ferro-coke is obviously lower than that of the conventional coke, that ferro-coke mixed charging is more effective than nut coke mixed charging in terms of accelerating iron ore reduction, and that ferro-coke mixed charging can decrease the melt-down temperature as well as the pressure drop of stock column and narrow down the softening-melting range, which are beneficial to the improvement of blast furnace performance.

Key words ferro-coke, high reactivity, iron ores reduction, primary slag formation

新型炼铁炉料热压铁焦制备及冶金性能

王宏涛　王　峥　储满生　赵　伟　柳政根

（东北大学材料与冶金学院，辽宁沈阳　110819）

摘　要　高炉炼铁是钢铁工业节能减排的关键环节。铁焦是一种新型碳铁复合炉料。高炉使用铁焦后，焦比降低、冶炼效率提高、CO_2排放减少，从而实现节能减排。本文总结了国内外铁焦的研究现状，并基于我国原燃料条件，提出了热压铁焦制备新工艺。在实验室条件下进行了热压铁焦的制备研究，并研究了热压铁焦的抗压强度、反应性、反应后强度和对焦炭热态性能的影响等冶金性能。结果表明，热压铁焦是一种新型炼铁炉料，可以用于高炉冶炼。当铁矿粉配比为15%、烟煤配比为65%、无烟煤配比为20%、热压温度为300℃、炭化温度为1000℃、炭化时间为4 h时，热压铁焦的抗压强度达到5048.50 N，反应性为62.41%，反应后强度为10.65%。热压铁焦对焦炭具有保护作用，且当铁焦添加比例不超过10%时，保护作用更为明显。

关键词　新型炼铁炉料，热压铁焦，高炉炼铁，二氧化碳减排

Preparation and Metallurgy Properties of New Ironmaking Material-Iron Coke Hot Briquette

Wang Hongtao, Wang Zheng, Chu Mansheng, Zhao Wei, Liu Zhenggen

(School of Materials & Metallurgy, Northeastern University, Shenyang 110819, China)

Abstract Blast furnace process is the critical link of steel industry in energy conservation and emissions reduction. Ferro coke is one of new carbon iron composite materials. By using ferro coke in blast furnace, coke ratio is decreased and smelting efficiency is improved and CO_2 emission is reduced, namely realizing energy saving and emission reduction of blast furnace. In this paper, present research status of ferro coke domestic and overseas is summarized. Simultaneously, based on the conditions of raw materials in our country, the new process of Iron Coke Hot Briquette (ICHB for short) preparation is put forward. Under laboratory conditions, preparation, compressive strength, CRI and CSR of ICHB were investigated. And the effects of ICHB on coke hot metallurgy properties (CRI and CSR) were studied. The results showed that, ICHB is a new typical ironmaking material, which can be used to blast furnace process. With 15% iron ore, 65% bituminous coal, 20% blind coal, 300℃ hot briquette temperature, 1000℃ carbonization temperature and 4 h carbonization time, ICHB was prepared, the compressive strength, CRI and CSR of which are 5048.50 N, 62.41%, 10.65%, respectively. In addition, ICHB has certain protective effect for coke, which is more obvious with ICHB ratio no more than 10%.

Key words new ironmaking material, iron coke hot briquette, blast furnace ironmaking, carbon dioxide emission reduction

含焦油渣铁焦的制备实验研究

任　伟　李金莲　张立国　胡俊鸽　王　亮　韩子文

（鞍钢股份有限公司技术中心，辽宁鞍山　114021）

摘　要　为探索焦油渣用于铁焦制备的可行性，以铁盒实验方式进行了1/3焦煤和铁矿粉配加8%~15%焦油渣的制备实验。研究表明：在1/3焦煤和铁矿粉配比为6∶4的基础上，添加8%~15%的焦油渣，铁焦生球的强度可达到110N/球以上，且随着焦油渣配比的提高，生球强度也得到改善，铁焦成品球团的抗压强度高于3000N/球，最高可达6900N/球，铁焦CRI为32%，生产中配加8%以上的焦油渣可生产出强度满足高炉要求的铁焦。

关键词　铁焦，焦油渣，铁盒实验，抗压强度

The Iron Coke Preparation with Tar Residue

Ren Wei, Li Jinlian, Zhang Liguo, Hu Junge, Wang Liang, Han Ziwen

(Angang Steel Company Limited, Technological Centre, Anshan 114021, China)

Abstract　An investigation on iron coke preparation with tar residue was carried out by box coking way, and 8%~15% tar residue was add to the mixture of 1/3 coking coal and iron ore concentrates. It was shown that 8~15 percent tar residue benefited the iron coke with 60% 1/3 coking coal and 40% iron ore concentrates, and the green pellet reached more than 110N per ball, moreover, the more proportion of tar residue, the higher compressive strength. After coking for green pellet, the compressive strength reached more than 3000N per ball and the max value would get 6900N per ball, and CRI index can reach 32% above. Finally it was feasible the preparation of iron coke with tar residue.

Key words　iron coke, tar residue, box coking, compressive strength

钒钛磁铁矿含碳热压块新型炉料的制备及优化

赵伟　王峥　储满生　王宏涛　柳政根　唐珏

（东北大学材料与冶金学院，辽宁沈阳　110819）

摘　要　钒钛磁铁矿在我国储量巨大，如何高效利用这一特色资源对我国钢铁工业的健康发展意义重大，而含碳热压块作为一种具备良好冶金性能的新型炼铁炉料备受关注。基于此，本文提出了钒钛磁铁矿含碳热压块这一新型炉料的制备，考察了其制备过程中影响其抗压强度的重要因子，进一步采用田口法计算各因子对抗压强度的贡献率，最终给出钒钛磁铁矿热压块的最佳制备参数。实验结果表明，在热压温度、配碳比、煤粉粒度和矿粉粒度四个影响因素中，热压温度对抗压强度的影响程度最大，其贡献率达到了79.11%，配碳比、煤粉粒度和矿粉粒度的贡献率分别为1.42%、16.58%和1.64%。优化后钒钛磁铁矿热压块的制备参数为热压温度300℃、配碳比1.6、煤粉和矿粉粒度均为–200目。在优化后的参数下进行验证实验，得到的钒钛矿热压块的平均抗压强度达到1306.5N。

关键词　钒钛磁铁矿，含碳热压块，抗压强度，田口法

Preparation and Optimization of Vanadium-titanium Magnetite Carbon Composite Hot Briquette

Zhao Wei, Wang Zheng, Chu Mansheng, Wang Hongtao, Liu Zhenggen, Tang Jue

(School of Materials & Metallurgy, Northeastern University, Shenyang 110819, China)

Abstract With the huge reverse in our country of vanadium-titanium magnetite, how to ultilize this special resource efficiently is significant to the healthy development of China's iron and steel industry. As a new type burden for ironmaking, carbon iron composite hot briquette has been payed great attention. Based on these, the preparation of vanadium-titanium magnetite carbon composite hot briquette (VTM-CCB) was put forward in this paper, and the influencing factors to the crushing strength of VTM-CCB was explored, and then the contribution rates of these factors to crushing strength were calculated based on Taguchi methods. Finally, the best preparation conditions of VTM-CCB were established. The results showed that in the processing parameters including hot-briquetting temperature, carbon ratio, coal particle size and iron particle size, hot-briquetting temperature was the most important factor to the crushing strength, and its contribution rate was up to 79.11%, the contribution rates of carbon ratio, coal size and iron size were 1.42%, 16.58% and 1.64% respectively. The optimized parameters of the preparation process of VTM-CCB were as follow: hot-briquetting temperature 300℃, carbon ratio 1.6, coal particle size -200 mesh, iron particle size -200 mesh. Verification experiments were conducted under the optimized parameters, and the average crushing strength of VTM-CCB was up to 1306.5N.

Key words vanadium-titanium, carbon composite hot briquette, crushing strength, Taguchi methods

高炉煤气提纯技术在低碳炼铁中应用和创新

刘文权[1] 胡鸿频[2]

（1. 冶金工业规划研究院，北京 100711；2. 杭州东安科技有限公司，浙江杭州 310018）

摘 要 本文论述了变压吸附（VPSA）对煤气提纯的应用，论述了煤气提纯技术在普通高炉、氧气高炉、UCLOS、COURSE50 和 FINEX 等工艺中的应用，并对高炉煤气提纯采用变压吸附进行高炉喷吹的可行性和经济效益分析，指出高炉煤气提纯采用变压吸附技术可实现低碳炼铁的技术创新。

关键词 高炉煤气提纯，变压吸附，低碳炼铁

生物质燃料热解过程动力学研究

左海滨[1] 章鹏程[1] 王广伟[2] 张建良[2]

（1. 北京科技大学钢铁冶金新技术国家重点实验室，北京 100083；
2. 北京科技大学冶金与生态工程学院，北京 100083）

摘 要 本文利用非等温热重法研究了四种生物质原料，分别为木头、小麦壳、玉米芯以及竹子在 N_2 气氛下的热解特性，研究了升温速率等因素对生物质原料热解过程的影响，同时利用随机孔模型对生物质热解过程进行了模拟并计算了其动力学参数。结果表明：提高升温速率，生物质热解开始阶段以及最大失重速率对应的热解温度均向高温区移动，热解行为变差。动力学分析可知，四种生物质热解过程均符合随机孔模型，在热解过程主要受到升温速率以及孔结构等因素的影响。采用随机孔模型计算木头、小麦壳、玉米芯以及竹子的热解活化能分别为 91.5kJ/mol, 83.0kJ/mol,101.7 kJ/mol 和 89.2kJ/mol。

关键词 生物质，热解，非等温，动力学模型

Kinetic Study of Pyrolysis Process of Biomass

Zuo Haibin[1], Zhang Pengcheng[1], Wang Guangwei[2], Zhang Jianliang[2]
(1. State Key Laboratory of Advanced Metallurgy, University of Science and Technology Beijing, Beijing 100083, China; 2. School of Metallurgical and Ecological Engineering, University of Science and Technology Beijing, Beijing 100083, China)

Abstract Pyrolysis process of various biomass under different heating rates were investigated by the TG-DTG thermal analysis device. The experiments were carried out with a HCT-3 thermo-balance analyzer under high purify N_2, and random pore model (RPM) was used for analyzing kinetic parameters of biomass. The experimental results stated that with increase of heating rate, both initial temperature of heat decomposition and the maximum decompose rate and its corresponding temperatures of biomass moved into high temperature zone, pyrolysis behavior deteriorated. It was concluded from kinetics analysis that RPM model could apply to describe pyrolysis process of biomass , the activation energy of pyrolysis process was 91.5kJ/mol，83.0kJ/mol,101.7 kJ/mol and 89.2kJ/mol for biomass of wood, wheat shells, corn cobs and bamboo respectively.

Key words biomass, pyrolysis, non-isothermal, kinetic model

新型抑尘环保技术在宝钢原料场的应用

吴旺平　马洛文　谢学荣　燕双武
（宝山钢铁股份有限公司炼铁厂，上海　201900）

摘　要　随着2015年新环保法的推出，宝钢原料场针对自身物料露天堆放易产生可视扬尘的特点，从减少料场贮存扬尘、卡车作业扬尘及皮带机洒料、冒灰等实际情况出发，积极建设B/C/E型封闭式料场，做到煤进仓、矿进棚物料封闭储存；同时对胶带机通廊及转运站进行整体封闭，减少物料在输送过程中的扬尘扩散至周边居民区，避免胶带机故障跑偏时造成物料直接洒落路面污染环境；此外原料场积极引入喷雾炮和微雾抑尘等环保抑尘技术，通过使水雾化喷洒抑制粉尘，大大降低了物料带入水分同时返程皮带的二次落料污染明显改善，从根本上解决了物料输送过程中的环境落料问题。随着上述多种新型抑尘环保技术在宝钢的开发应用，宝钢原料场降尘量大为降低，为宝钢原料场实现绿色环保可持续型发展打下良好基础。

关键词　封闭式料场，微雾抑尘，胶带机通廊封闭，原料场

Application of the New Dust Prevention Technology for Raw Material Yard in Baosteel

Wu Wangping, Ma Luowen, Xie Xuerong, Yan Shuangwu
(Ironmaking Plant, Baoshan Iron & Steel Co., Ltd., Shanghai 201900,China)

Abstract With the new environmental protetion laws issues in 2015 ,Baosteel is put at a serious disadvantage for its open-air yard that could easily cause visual dust pollution.Aiming at decreasing the dust pollution from raw material storage、truck

transportation、conveyor spilling、 dust emission and so on, Baosteel activily begins to construct type B/C/E enclosed stockyards to make all kinds of ores in the sheds and coals in the storehouses, also using the covering blocks out the walkway of coveyors and transferring spots completely to reduce the dust pollution spreading into residential areas and raw material polluting the roads directly when conveyors are off tracking.In addition, raw material yard has introduced spray guns and micro-flog dust suppression device which can make water atomization to depress dust,and when the device is put into production ,they can cut down the moisture content of raw material then reduce spilled material from the return belts,thus Baosteel can fundamentally solve the environment pollution that produces from belts conveying process in the yard. With the development of the above new kinds of dust prevention technologies in Baosteel,the dustfall content of raw material yard has been declining, ultimately lay a good foundation for raw material yard to achive environmental-friendly sustainable development.
Key words　enclosed stockyard, micro-atomization spayer, coveyors blocked, raw material yard

宝钢煤场改建筒仓探讨

张世斌　马洛文　李　刚　燕双武　仇晓磊

（宝山钢铁股份有限公司炼铁厂，上海　201941）

摘　要　通过对宝钢原料场煤场、煤输出系统现状和存在的安全生产运行、环保及物料损失问题进行分析，阐述了煤场改建筒仓的必要性，并结合原料场目前运行工艺，设定了初步改造方案，改造后将有效改善环境，提高贮煤量，为原料场大修逐步实施创造条件，保障安全生产。
关键词　煤场，筒仓，环境

Discussiong on Rebuilding Coal Yard of Baosteel to Soli

Zhang Shibin, Ma Luowen, Li Gang, Yan Shuangwu, Qiu Xiaolei

(Ironmaking Plant, Baoshan Iron & Steel Co., Ltd., Shanghai 201941,China)

Abstract　It is necessary to rebuild coal yard to silo based on the anlaysis of coal yard of Baosteel raw material yard,the current status of output system of coal, exisitng issue of production safety ,enviroment and the loss of raw material.The initial rebuiding program is desgined with the combination of current process in raw material yard. The rebuild will improve enviroment efficiently, and stock of coal volume.It will help the implementation of repairment of raw material yard and ensure the safety production.
Key words　coal yard, soli, enviroment

Contribution of Green Ore on Environment Protect

Marcos Cantarino[1],　Hamilton Pimenta[1],　Rogerio Carneiro[1],
Alex Castro[2],　Macus Emrich[1], Dai Yuming[2]

(1. Vale, Brazil; 2. Vale, China)

Abstract　In recent years, smog issue attracts more and more attention in China. A lot of studies aim to find the reason to cause smog.　Steel and iron industry is considered as one of the reasons. How to reduce emission in steel and iron industry becomes an urgent problem to be solved. In China, steel and iron industry normally includes sintering, pelletizing, coking, ironmaking, steelmaking, refining, continues casting and rolling process. Sintering is one of the main processes to cause emission. In this paper, emission in sintering process is researched. In addition, green ore solution is studied and discussed to save energy and reduce emission. According to the numerical simulation result, high grade ore like Brazilian ore IOCJ is helpful to reduce fuel consumption and PM emission during sintering.

Key words iron ore, sintering, green ore, environment, PM emission

Successful Integration of Partial FGD and SCR at Sinter Plant in China Steel Corporation

Yu-Bin Hong, Po-Yi Yeh

(Sinter Session of Iron –making Department in China Steel Corporation)

Abstract End-of-pipe technology for flue gas treatment to remove NO_x, DXN, and SO_x is successfully exercised at sinter plant in China Steel Corporation (CSC). As for removal of NO_x and DXN, Selective Catalytic Reduction (SCR) was adopted in 1990s. Furthermore, the SCR process combined FGD (Flue Gas Desulphurization) at sinter plant was developed for reduction of SO_x emission. CSC has evaluated the investment, limited space, residual disposal, saving energy, removal efficiency and operation flexibility. The corporation of design, construction and commissioning are executed by design and engineering department, research department and operation department in CSC group. Finally, partial FGD process including wet-method absorber, Gas/Gas heater, electrostatic precipitator , booster fan and switched damper of wind-box was established. 40 percent of flue gas with high SO_x concentration are desulphurized thought the FGD, the rest of the untreated flue gas are mixed together to SCR. Since 2013, friendly, environmental sintering process is fulfill successfully to minimize the NO_x, DXN, and SO_x emission by means of integration of existing SCR and partial FGD at No.4 sinter plant in CSC.

Key words end-of-pipe technology, SO_x, NO_x, SCR, pFGD

高炉锌富集的新认识

李肇毅 姜伟忠

（宝山钢铁股份有限公司炼铁厂，上海 201900）

摘 要 高炉炼铁技术的提升，对一些过程现象认识也在不断深化。锌是作为微含量元素参与了高炉冶炼过程，其危害过去已有一定认识。本文试图定量化讨论锌的富集程度，以及对升高燃料比的不良作用。

关键词 高炉，锌富集，燃料比

The New Understanding of Blast Furnace Zinc Enrichment

Li Zhaoyi, Jiang Weizhong

(Ironmaking Branch, Baoshan Iron & Steel Co., Ltd., Shanghai 201900, China)

Abstract Along with progress of blast furnace ironmaking technology, the deepening understanding are also in the

process of some phenomenon. Zinc is a trace elements participate in blast furnace smelting, the harm existing must know in the past. Discussed in this paper, quantitative zinc enrichment degree in the furnace, and the effect of fuel rate analysis.
Key words blast furnace, zinc enrichment, fuel rate

含锌粉尘的综合处理研究进展

郝月君 蒋武锋 郝素菊 张玉柱

（华北理工大学冶金与能源学院现代冶金技术教育部重点实验室，河北唐山 063009）

摘 要 随着我国含锌粉尘排放量的不断增加，对环境及工人健康的危害也越来越严重，因此对钢铁厂含锌粉尘的处理及锌的回收利用提出了更高的要求。针对我国含锌粉尘处理和再利用率低的问题，研究了国内外对含锌粉尘处理工艺的发展过程、现状及其特点，并进行了系统深入的理论分析和对比。介绍了几种含锌粉尘处理的新方法，提出了利用熔融钢渣显热处理含锌粉尘的新思路。

关键词 含锌粉尘，回收利用，熔融钢渣

Research of Disposal on Zinc-containing Dust

Hao Yuejun, Jiang Wufeng, Hao Suju, Zhang Yuzhu

(College of Metallurgy & Energy, Ministry of Education Key Laboratory of Modern Metallurgy Technology, North China University of Science and Technology, Tangshan 063009, China)

Abstract As the displacement of Zinc-containing increasing continuously, the hazards has become increasingly serious for environment and worker's health, so steel enterprise has put forward higher requirements for processing and recycling of Zinc-containing dust. In view of the problems on the zinc-containing dust treatment and reuse utilization, studied the treatment process and characteristics at home and abroad on zinc-containing dust. Moreover, combined the research at the current, several new methods for processing about zinc-containing dust were introduced. Proposed a new thought of dealing with zinc-containing dust using the sensible heat of molten slag.

Key words zinc-containing dust, recycling, molten steel slag

转底炉炼铁技术改进研究

经文波 薛 逊 刘占华

（北京神雾环境能源科技集团股份有限公司研究院，北京 102200）

摘 要 应用转底炉炼铁技术可以利用钢铁企业内的富余的煤气或低价煤制气和含铁的低品位矿及其镍渣、铜渣、难选矿、复合共伴生矿、尾矿等炼铁，降低了成本，减少环境污染，扩大了炼铁资源应用范围，提高了含铁资源利用率。还可以通过转底炉炼铁工艺改进，减少膨润土黏结剂用量，使用有机黏结剂或用红土镍矿代替，保证生球强度，增加有益元素含量，改善产品品质；改善造球，进一步提高生球料层厚度；提高生球碱度，强化脱硫，改善质

量；降低炉渣黏度，改善渣铁分离效果，提高金属收得率，增加产量；精确控制炉膛各部温度，改善还原过程，改善煤气热能和化学能利用，降低能耗；优化装备，生产还原铁粉高附加值产品。使企业更进一步适应市场需要，提高市场竞争力和经济效益。

关键词 转底炉，炼铁，改进，研究

Rotary Hearth Furnace Ironmaking Technological Improvement Research

Jing Wenbo, Xue Xun, Liu Zhanhua

(Beijing Shenwu Environmental Energy Technology Group Co., Ltd. Research Institute, Beijing 102200, China)

Abstract Application of rotary hearth furnace ironmaking technology can make use of iron and steel enterprise surplus gas or coal gas at low prices and low grade of iron ore, copper and nickel slag, difficult separation and composite associated ore, tailings, to produce iron, reduce costs, reduce environmental pollution, expand ironmaking resources application range, improve the utilization of iron resources can also through the rotary hearth furnace ironmaking process improvement, reduce the dosage of bentonite binder, using organic binder or replace with laterite nickel ore, guarantee the strength of the pellet, increase content of beneficial elements, improve product quality; to improve the balling, further improve the material thickness of the pellet; to improve the pellet alkalinity, strengthen the desulfurization, improve quality; reduce the viscosity of slag, improve the slag iron separation effect, increase the metal yield, increase production; precise control of the various parts of the furnace temperature, improve the reduction process, improve the gas heat energy and chemical energy utilization, reduce energy consumption; optimization of equipment, to produce reduced iron powder, high added value products. To further meet the needs of the market, improve market competitiveness and economic benefits.

Key words rotary hearth furnace, ironmaking, improvement, researcher

氯元素对高炉原燃料冶金性能的影响研究

兰臣臣 张淑会 吕 庆 武兵强

（华北理工大学冶金与能源学院教育部现代冶金技术重点实验室，河北唐山 063009）

摘 要 生产实践证明，高炉内的氯元素主要来源于原燃料，并且主要以 HCl 气体的形式随高炉煤气运动。高炉煤气中的 HCl 不仅危害冶炼设备，还会对高炉原燃料冶金性能产生影响。基于此，研究了 HCl 气体对烧结矿的低温还原粉化性和还原性能以及焦炭的反应性和反应后强度的影响规律。结果表明：HCl 气体可以抑制烧结矿的低温还原粉化，改善高炉块状带透气性。随着煤气中 HCl 气体浓度的增加，烧结矿的 $RDI_{+6.3}$ 指数和 $RDI_{+3.15}$ 指数均增加，$RDI_{-0.5}$ 指数降低。HCl 气体阻碍烧结矿的还原，使烧结矿的还原性下降。随着 HCl 气体浓度逐渐增大，烧结矿的 RI 指数逐渐减小。高炉煤气中 HCl 气体可以劣化焦炭的高温冶金性能。

关键词 氯元素，烧结矿，焦炭，冶金性能

Study on the Effect of Chlorine on the Metallurgical Properties of Raw Material and Fuel of Blast Furnace

Lan Chenchen, Zhang Shuhui, Lv Qing, Wu Bingqiang

(College of Metallurgy & Energy, Ministry of Education Key Laboratory of Modern Metallurgy Technology, North China University of Science and Technology, Tangshan 063009, China)

Abstract The production practice shows that, the chlorine element in blast furnace mainly originates from the raw material and fuel , which forms HCl in the blast furnace and flows with the gas. The HCl in blast furnace gas not only does harm to smelting equipment, but also has an effect on the metallurgical properties of raw material and fuel. The effect of chlorine on the low-temperature reduction degradation property and reduction property of sinter and reactivity and post-reaction strength of coke were studied at laboratory. The results show that: the chlorine element can inhibit the low-temperature reduction degradation property and improve the permeability of lumpy zone of blast furnace. With increasing the concentration of HCl in gas, the $RDI_{+6.3}$ and $RDI_{+3.15}$ of sinter increase, and the $RDI_{-0.5}$ decreases. The chlorine element make the reduction property of sinter decrease. The *RI* of sinter decreases with the increase the concentration of HCl in gas. HCl gas has the promotion effect to the coke quality deterioration.

Key words chlorine, sinter, coke, metallurgical properties

高炉中氯的危害及行为分析

张　伟　王再义　张立国　邓　伟　韩子文　王　亮

（鞍钢股份有限公司技术中心，辽宁鞍山　114009）

摘　要　结合鞍钢及有关企业生产中出现的实际问题，阐述了高炉中氯的影响和危害，并分析了高炉中氯的来源及反应行为。结果表明，高炉中氯的主要来源是由于烧结矿喷洒氯化物，此外通过热力学分析明确了氯化物与碱金属盐类化合物的反应机制，据此探讨了烧结矿喷洒氯化物的利弊关系，并针对高炉中氯的危害控制提出了相关措施。

关键词　高炉，氯化物，碱金属，热力学

Analysis of Harmfulness and Reaction Behavior of Chlorine in Blast Furnace

Zhang Wei, Wang Zaiyi, Zhang Liguo, Deng Wei, Han Ziwen, Wang Liang

(Angang Steel Co. , Ltd. , Anshan 114009, China)

Abstract The influence and harmfulness of chlorine in blast furnace was introduced and the source of chlorine and its reaction behavior was analyzed integrated with the problem occurred in Ansteel and other enterprise. The result showed that, the chlorine in blast furnace mainly came from chloride sprinkled on sinter mine. The reaction mechanism of chloride and alkali metals salt compound was confirmed by thermodynamic analysis. Hereby, the advantages and disadvantages of chloride sprinkled on sinter mine was discussed and the control measure of chlorine in blast furnace was advanced.

Key words blast furnace, chloride, alkali metal, thermodynamic

首钢京唐烧结利用钢铁流程废弃物的研究与实践

裴元东[1,2] 赵志星[2] 安 钢[3] 程峥明[3] 潘 文[2]
罗尧升[3] 石江山[3] 赵景军[3]

（1. 北京科技大学冶金与生态工程学院，北京 100083；2.首钢技术研究院，北京 100043；
3. 首钢京唐钢铁联合有限责任公司，河北唐山 063200）

摘 要 充分利用钢铁流程中的废弃物，对于节能减排、发展低碳生产和循环经济有重要意义。在对京唐钢铁流程各种工业固体废弃物研究的基础上，对其在京唐烧结的使用进行了优化。首先对烧结使用的各种固废性能进行了基础测试，包括化学成分可取性、燃料热值、燃烧高温保持时间、烧结杯测试等。基于试验结果，认为当前条件下适宜京唐烧结配加的各项固废比例分别为：钢渣：<5%，铁皮：2.0%，环境除尘灰：2.0%，旋风灰：1.0%，中断高炉干法灰、烧结电场灰、二次炼钢灰的循环；燃料结构中，焦化除尘灰<15%、焦丁<15%；通过优化固废配加结构，烧结和高炉的有害元素危害均减轻，满足了高炉的用料要求，并创造了效益，为资源的减量化和发展循环、清洁经济做出了贡献。

关键词 烧结，固体废弃物，循环经济，绿色经济

Application and Optimized Research on Waste of Iron and Steel Production Process during Sintering in Shougang Jingtang

Pei Yuandong[1,2], Zhao Zhixing[2], An Gang[3], Cheng Zhengming[3],
Pan Wen[2], Luo Yaosheng[3], Shi Jiangshan[3], Zhao Jingjun[3]

(1.School of Metallurgical and Ecological Engineering, University of Science and Technology Beijing, Beijing 100083, China; 2. Shougang Research Institute of Technology, Beijing 100043, China; 3.Shougang Jingtang Integrated Iron and Steel Co., Ltd., Tangshan Hebei 063200, China)

Abstract It is important significance to make full use of waste of iron and steel production process for energy conservation and emissions reduction, low carbon production and cycle economy. Based on research to all kinds of industrial solid waste from Jingtang iron and steel production process, the waste has been optimized for use in sintering process. At first, all kinds of solid waste on sintering performance has carried out on the basis of tests, including the chemical composition of desirable, calorific value of fuel, burning high temperature holding time, the sintering pot test, etc. The result showed that it was appropriate for current sintering condition with addition of various solid waste ratio respectively: steel slag: < 5%, iron sheet: 2.0%, environmental dust: 2.0%, cyclone dust: 1.0%, dry ash of blast furnace, ash of sintering electric field, the circulation dust of the secondary steelmaking had to be interrupted. For fuel structure, coking dust < 15%, and nut coke < 15%. Based on optimizing solid waste addition, it was greatly mitigated against the dangers of blast furnace and sintering, satisfying the demands of the blast furnace material, and creating the benefit, which has made a contribution with the reduction and the resource development cycle and clean economy.

Key words sintering, solid waste, cycle economy, green economy

2 炼钢与连铸

Steel Making and Continuous Casting

2.1 铁水预处理、转炉电炉/ Hot Metal Pretreatment, Converter, Electric Arc Furnace

Utilization of Steelmaking Slag for Rehabilitation of Marine Environment

MATSUURA Hiroyuki, TSUKIHASHI Fumitaka

(Graduate School of Frontier Sciences, The University of Tokyo, 5-1-5-501 Kashiwanoha, Kashiwa, Chiba 277-8561 Japan)

Abstract Steelmaking slag is one of the main by-products of steelmaking process, and approximately 13 million tons of BOF slag are generated annually in Japan. Ironmaking and steelmaking slags have to be effectively utilized for the development of sustainable steelmaking process. In Japan, most of steelmaking slag is recycled as civil engineering materials and so on. Since the demand of steelmaking slag as construction materials is expected to decrease with the formation of developed society, the recycling use of slags becomes important issues. Two schemes should be considered to solve the issues of steelmaking slag, first one is to reduce the amount of slag generation by the improvement of refining processes and second one is to find the new application way of steelmaking slag in other fields by adding the new functional properties. Recently, steelmaking slags are utilized as the rehabilitation materials for marine environment. In the present paper, the present situation of slag generation and previous and present research projects regarding slag utilization in Japan are introduced and the possibility of utilization of slag for rehabilitation materials of marine environment as a new recycling technology is discussed. Some fundamental experimental results conducted in our research group are also introduced.

Key words ironmaking and steelmaking, slag, recycling, marine environment

Simulate the Transfer Behavior of Phosphorous in the Multi-phase Slag Through the Single Hot Thermocouple Technique

Xie Senlin, Wang Wanlin, Huang Daoyuan, Ma Fanjun

(School of Metallurgy and Environment, Central South University, Changsha 410000, China)

Abstract The mass transfer behavior of phosphorus in the solid $2CaO{\cdot}SiO_2$ phase/liquid phase coexisting multiphase dephosphorization slag at 1623K(1350°C) was simulated through the Single Hot Thermocouple Technique (SHTT). It was found that the transfer rate of phosphorus in the liquid slag was quickly after adding the P_2O_5 powder, with the extension of time, the dissolved phosphorus in the liquid slag was gradually fixed into the directly precipitated solid $2CaO{\cdot}SiO_2$ particle in the form of the ($2CaO{\cdot}SiO_2$-$3CaO{\cdot}P_2O_5$) solid solution, after 900s, the enrichment of phosphorus was accomplished. Besides, it was indicated that the key of accelerating the dephosphorization is to improve transfer rate of phosphorus in the

solid $2CaO·SiO_2$ phase.Those results obtained could provide some fundamental guidance for effective dephosphorization process by using the multiphase slag for hot-metal treatment.
Key words multiphase slag, dephosphorization, ($2CaO·SiO_2$-$3CaO·P_2O_5$) solid solution, transfer behavior

Metal Emulsion Formed by Bottom Gas Flow in Sn Alloy/Oxide System

Liu Jiang, Kim Sunjoong, Gao Xu, Ueda Shigeru, Kitamura Shinya

(Institute of Multidisciplinary Research for Advanced Materials, Tohoku University, Sendai, Japan)

Abstract Metal emulsion produced by passage of gas bubble though interface has potential to increase the surface areabetween molten slag and steel.Chemical reaction efficiencywould be greatly improved by the formation of metal droplets which can disperse in slag phase for some times andthen fall back to metal phase. Some cold systems consisted of immiscible phases have been applied to simulate the emulsion behavior. It was found that the physical properties of system would affect the droplets formation profoundly. Previously, the emulsion in Sn alloy/chloride system was observed by high speed camera by our group. Depending on the gas flow rate, several different formation mechanisms were proposed. In addition formationand sedimentation rate were estimated. In order to get more understanding the Sn alloy/oxide system was employed in this work due to more similar physical properties with actual steel/slag. The oxide with metal droplets in the slag phase during bottom bubbling was sampled. Metal droplets can be separated from the oxide by dissolution into water. After analyzing the number and diameter by microscope, the surface area and mass of droplets were summarized and compared with that in Sn alloy/chloride system.It was found that more large droplets were produced and existed in the upper phase in oxide than that in chloride. Formation rate in oxide was also higher. This can be attributed to the high viscosity of oxide. The effect of gas bubble size will be studied in future.
Key words metal emulsion, bottom gas injection, size distribution, formation rate

Optimization and Practice of 300t Converter Bottom Blowing Technology at Baosteel

Jiang Xiaofang, Mou Jining, Zhang Min, Zhang Geng

(Steelmaking Plant of Baoshan Iron & Steel Co., Ltd., Shanghai 200941, China)

Abstract During the pre- and middle periods of one whole BOF campaign in 300 t converters at Baosteel, bottom tuyeres status is good so that the effect of combined blowing is more obvious. But during the later period of the campaign, with the increase of furnace lining life, the tuyeres would have been blocked or even closed, this would bring a negative impact on the cost and quality of steel. In order to solve these problems, some measures and ways are adopted. Developing a new type of bottom tuyeres improved the metallurgical results of combined blowing during the latter period of the campaign. The converter bottom refractory structure was optimized to prevent the abnormal erosion of refractory nearby the converter bottom corner. According to the different production schedule, converter lining life and bottom blowing conditions, the maintenance mode of bottom tuyeres was determined. Coating technology were implemented to reduce the erosion of bottom tuyeres. Through three years of continuous improvement, the effect of converter combined blowing during the whole campaign, especially in the later period of the campaign, is improved obviously. The ratio of converter combined blowing increases from 79.2% to 100%, the value of [%C]. [%O] at endpoint in BOF decreases from 0.00334 to 0.00228, a decrease of 30.9%, the index of soluble oxygen content at endpoint in BOF decreases 22%, and the refractory consumption for converters maintenance drops from 0.98kg/(t.s) to 0.64kg/(t.s).

Key words converter, bottom blowing technology, bottom tuyeres maintenance

Dynamic Load Characterization of Basic Oxygen Furnaces Using Random Vibration Analysis

Zangeneh Pouya, Al-Dojayli Maher, Maleki Majid, Ghorbani Hamid, Choey Mervyn

(Hatch Ltd., Canada)

Abstract Basic Oxygen Furnaces (BOF) suffer from damages due to excessive dynamic loading and fatigue during their life time as a result of significant vibrations of the furnace during oxygen injection process. Typical BOF vibration responses can be classified as an indeterministic non-stationary vibration, in which the regime of the vibration changes with the continuous operation of oxygen blowing. In order to assess the structural integrity of BOFs, dynamic loads are key design variable and should be determined. In this paper, a methodology is presented to evaluate the dynamic loads generated by the BOF's chemical reactions through random vibration investigation using Finite Element Analysis (FEA). The actual time domain vibration of the system is measured and transformed using Fourier Transformation to obtain the response defined in the frequency domain. Harmonic analysis is conducted to capture the Frequency Response Functions (FRF) of the furnace over a sweep of frequencies covering a wide range of the system natural frequencies, up to 90% of the system mass participation factor. The frequency domain response is used for comparison with FEA dynamic responses to characterize the vibrating load. The calculated loads can be used effectively to better design and/or retrofit BOF systems and their supporting structures to avoid major losses in structural assets.

Key words basic oxygen furnace, random vibration, load, characterization, PSD, signal processing

Research on the Reactivity and Grain Size of Lime Calcined at Extra-high Temperatures via Flash Heating

Wang Xiaoyuan, Xue Zhengliang, Li Jianli

(The State Key Laboratory of Refractories and Metallurgy, Wuhan University of Science and Technology, Wuhan 430081, China)

Abstract In low-carbon energy-efficient basic oxygen furnace (BOF) steelmaking processes, limestone partly or completely replaces the active lime. The effects of limestone calcination temperature (1200 ~ 1500 °C) and time (5 ~ 15 min) on lime reactivity and CaO grain size were investigated. The reactivity was evaluated via titration with hydrochloric acid, and the CaO grain size was analyzed using SEM. The results revealed that the temperature exceeded 1300 °C, the reactivity decrease after an appearance of maximum. The higher temperature is, the earlier the peak of reactivity appears. The CaO grains grow with the increase of temperature and time, which leads to the decrease of reactivity. Notably, the effects of temperature on CaO grain size and reactivity are more remarkable than that of time. To obtain active lime calcined at ultra-high temperature by flash heating, the calcination condition should be at 1300 ~ 1400 °C for 10 ~ 15 min or at the temperature of 1400 ~ 1500 °C for 8 ~ 10 min.

Key words limestone, calcinations, reactivity of lime, flash heating, ultra-high temperature

Recovery of Zinc and Lead from Electric Arc Furnace Dust by Using Selective Chlorination and Evaporation Reaction

Kitao Hiroki, Matsuura Hiroyuki, Tsukihashi Fumitaka

(Department of Advanced Materials Science, Graduate School of Frontier Sciences, The University of Tokyo, 5-1-5 Kashiwanoha, Kashiwa, Chiba 277-8561, Japan)

Abstract Selective chlorination and evaporation reactions could become one of the applicable methods for the treatment of electric arc furnace dust containing considerable amount of zinc and lead. In this study, the selective chlorination and evaporation rate of the Fe_2O_3–$ZnFe_2O_4$–ZnO–PbO mixture or practical electric arc furnace dust with Ar–Cl_2–O_2 gas was measured at 1073 K by gravimetry and the effects of partial pressures of chlorine and oxygen on the Zn and Pb removal rates were investigated. Zinc oxide and lead oxide contained in the oxide mixture were selectively chlorinated and zinc and lead chlorides evaporated to gas phase, while iron oxide remained as oxide in a residue. By chlorinating with Ar–Cl_2 gas, over 99 % of zinc and lead was removed from oxide mixture as well as about 3 % of iron was lost by the chlorination and evaporation reactions. Selective chlorination using Ar–Cl_2–O_2 gas was conducted to decrease the loss of iron and improve the efficiency of the selective chlorination and evaporation reactions. The selectivity of zinc and lead was improved by using O_2 gas mixture. In the case of practical electric arc furnace dust, over 99 % of zinc was removed while about 17 % of iron was also lost by the chlorination and evaporation reactionswith Cl_2–O_2 gas.

Key words electric arc furnace dust, zinc, lead, selective chlorination and evaporation,recycling

Effect of ArcSave® on the EAF Process Improvements and Cost Effectiveness in Both Carbon and Stainless Steel Production

Teng Lidong[1], Hackl Helmut[1], Zhong AlexYuntao[2], Pan Hanyu[2], Sjöden Olof[1]

(1. ABB Metallurgy, Process Automation, ABB AB, Västerås, SE-721 59, Sweden;

2. ABB AB, Beijing 100015, China)

Abstract Electromagnetic stirring (ArcSave®) is a potential and reliable solution for improving the EAF process and reducing the operation cost. Two ArcSave systems have been introduced by ABB in 2014. One ArcSave has been installed on a 90 ton spout tapping furnace (SS-EAF) for stainless steel production in Europe and the other installed on a 90 ton EBT furnace (CS-EAF) for carbon steel production in USA. The effect of ArcSave on the EAF process and cost effectiveness has been studied during the hot test period and the resulting performance has been compared based on the 6-month-data without stirring and 6-month-data with ArcSave. In this paper, the detailed information of the investigation results obtained from these two installations will be summarized and presented. ArcSave speeds up the scrap and FeCr melt-down, accelerates the homogenization of the temperature and chemical composition of the steel bath, forces the metal/slag reactions closer to equilibrium state, increases the decarburization rate, and also improve the operation safety, reliability, and productivity. ArcSave reduces the consumption of electric energy, electrode, oxygen, refractory, de-oxidants, and slag builders and increase the scrap yield which makes the electric arc furnace more cost effectiveness.

Key words electric arc furnace (EAF), electromagnetic stirring (EMS), ArcSave, ferrochromium melting, EBT tapping

双联脱磷转炉强底吹搅拌效果数值分析

汪成义　杨利彬

（钢铁研究总院冶金工艺研究所，北京　100081）

摘　要　进行 300t 脱磷转炉数值模拟，研究了 8 支底吹元件下不同底吹布置方式、底吹流量大小、顶吹参数对熔池混合效果的影响。通过数值模拟研究得出，当底吹元件集中对称布置在炉底 0.42D 圆周上，熔池整体搅拌效果较好。进一步研究各底吹布置方式及复吹参数变化对不同熔池深度搅拌的影响规律，当熔池深度在 $y = -0.8$m 时，8-B 方案“传质更新区”面积达到 93.1%，此时大底吹流量条件将发挥良好的搅拌作用。当底吹流量增加至 2500Nm^3·h^{-1} 值后，熔池有效搅拌区域面积（v＞0.15m/s）未再明显增加，熔池整体的搅拌效果此后提升不大。

关键词　脱磷转炉，数值模拟，底吹布置，速度场，熔池搅拌

Numerical Simulation of Stirring Effect of Strong Bottom Blowing on the Duplex Dephosphorization Converter

Wang Chengyi, Yang Libin

（Iron & Steel Research Institute, Beijing 100083, China）

Abstract　Taking numerical simulation study of a 300 t dephosphorization converter, studied when under eight bottom blowing element, different bottom blowing arrangement, the size of the bottom blowing flow and top blowing parameters on the influence of the molten pool mixing effect. From the research, when the bottom blowing elements concentrated and symmetrically arranged in the furnace bottom of 0.42D on the circumference of a circle, the blending effect of the molten pool will be better. Then research the bottom blowing arrangement and the combined-blowing parameter change on the influence of pool mixing rule of different depth of molten pool by numerical simulation study. When the depth of molten pool $y = -0.8$ m, “the update region of mass transfer” of 8-B plan is 93.1%, and at this time the strong bottom blowing condition will play a good stirring effect. After the bottom blowing flow rate increased to 2500 Nm^3·h^{-1}, the effective stirring area (v>0.15m/s) will not significantly increased and the overall stirring of molten pool effect will be small.

Key words　dephosphorization converter, numerical simulation, bottom blowing arrangement, velocity field, stirring of the pool

100t 转炉大流量底吹脱磷工艺试验研究

杨利彬　曾加庆　王　杰　林腾昌

（钢铁研究总院冶金工艺研究所，北京　100081）

摘　要　针对 100t 转炉高效双渣脱磷工艺，经过热力学及动力学分析及试验得出工艺制度。为了保证冶炼过程脱磷动力学条件，经过系统分析对原有底吹系统进行改造，改造后底吹供气强度达到 0.25Nm^3/t·min。工艺优化后，有效提高了冶炼前期及过程脱磷效率，冶炼终点钢水碳氧积由 0.00335 降低到了 0.00283，避免了钢水过氧化，降

低有效石灰加入量25%，渣量减少20.12%。

关键词　复吹，底吹改造，脱磷，碳氧积

Dephosphorization Technique of 100t Converter Based on Bottom Intensity Blowing

Yang Libin，Zeng Jiaqing, Wang Jie, Lin Tengchang

(Metallurgical Department, Central Iron and Steel Research Institute, Beijing 100081, China)

Abstract　Aiming at the effective dephosphorization technique of 100t converter, based on thermodynamic and dynamical analysis and experiments, relative technique has been developed. Based on evaluation, the bottom system has been optimized. The bottom blowing intensity has been inproved to 0.25Nm^3/(t·min). The results show that the dephosphorization raised effectively，the average measured *w*(C)・*w*(O) at endpoint of whole campaign reduced from 0.00335 to 0.00283, lime comsumption reduced 25%，and the slag weight reduced 20.12%.

Key words　combined blowing, bottom optimization, dephosphorization, *w*(C)・*w*(O) product

鞍钢180吨转炉经济炉龄控制模式实践

李伟东　张晓军　舒　耀　王成青　何海龙

（鞍钢股份有限公司炼钢总厂，辽宁鞍山　114021）

摘　要　通过对180吨转炉经济炉龄的论证，得出转炉最佳经济炉龄约为4500炉，同时得出相应的炉衬维护方案，依据分析结果进行了工业实践。实践结果表明，在保证耐材成本最低化的同时，实际炉龄达到了4401炉，与目标炉龄接近，实际耐材成本指数降低14.5%，全炉役复吹比率从27.3%提高到97.4%，平均碳氧积从0.0030降低到0.0024。

关键词　转炉，经济炉龄，复吹

Practice of Controlling Mode for Economic Campaign Life of 180t Converter

Li Weidong, Zhang Xiaojun, Shu Yao, Wang Chengqing, He Hailong

(General Steelmaking Plant of Angang Steel Co., Ltd., Anshan Liaoning 114021, China)

Abstract　By analyzing the economic campaign life of 180t converter in Angang, the optimal economic campaign life is demonstrated to be about 4500 heats and the corresponding maintenance plan for the converter lining is obtained. The industrial production practice based on the analytical results is carried out and the results show that on condition of the minimum cost for using refractory materials, the actual campaign life is reached to 4 401 heats, the cost index for refractory materials is reduced by 14.5%, the combined blowing rate of the whole campaign life is increased to 97.4% from 27.3% and the average product of carbon content and oxygen content is reduced to 0.0024 from 0.0030.

Key words converter, economic campaign life, combined blowing

拜耳赤泥用于炼钢过程脱磷的实验研究

王瑞敏 张延玲 李显鹏

（北京科技大学钢铁冶金新技术国家重点实验室，北京 100083）

摘 要 本文主要通过赤泥和 CaO 按照一定比例调配作为脱磷剂应用于炼钢过程，借助热力学软件 FactSage 研究加入 Na_2O、Al_2O_3 的 $CaO\text{-}FeO\text{-}SiO_2$ 渣系在不同温度液相区面积的变化，最后根据之前的脱磷实验研究结果，将赤泥应用于吹氧炼钢过程。实验结果表明：根据相图观察得到，在典型炼钢过程的 $CaO\text{-}SiO_2\text{-}FeO$ 渣系，加入 Al_2O_3、Na_2O 会引起渣系在不同温度下液相区面积的增加，因而会降低渣的熔点。拜耳法赤泥和 CaO 混合用作脱磷剂具有良好的脱磷效果和较低的熔点。当赤泥与氧化钙质量比为 1.5 左右，渣铁比为 20%，初始硅为 0 时，脱磷率可达 80%以上。根据之前实验得到的结论，将赤泥用作吹氧炼钢过程脱磷剂也取得了很好的效果，当满足脱磷条件$[\%Si]_{始}=0$时，温度控制在 1350℃左右，脱磷均取得很好的效果(脱磷率可以达到 85%)。

关键词 拜耳法赤泥，液相区，渣系，脱磷

Dephosphorization of Bayer Red Mud Based Flux

Wang Ruimin, Zhang Yanling, Li Xianpeng

(State Key Laboratory of Advanced Metallurgy, University of Science and Technology Beijing, Beijing 100083, China)

Abstract In this paper, red mud and CaO mixed according to a certain proportion were used in steelmaking process as dephosphorizing agent, and the liquid phase region changes of $CaO\text{-}FeO\text{-}SiO_2$ slag with Na_2O and Al_2O_3 added at different temperatures were studied by using the thermodynamic software FactSage, finally according to the previous dephosphorization results, the red mud were applied to oxygen steelmaking process. The experimental results showed that, the observation of phase diagram showed that, the liquid phase regions of the $CaO\text{-}SiO_2\text{-}FeO$ slag system in the typical BOF steelmaking process can be increased with the addition of Al_2O_3 and Na_2O, thus the melting point of slag was decreased. The Bayer red mud and CaO mixed as dephosphorizing agent has good dephosphorization effects and lower melting point. When the mass ratio of red mud and calcium oxide was about 1.5, the ratio of slag and iron was 20% and the initial silicon was 0, the dephosphorization rate was more than 80%. According to previous results, the red mud used for oxygen steelmaking process dephosphorizer also achieved good results. When the initial [%Si] was 0 and temperature controlled was about 1350℃, dephosphorization were obtained good effects (the dephosphorization rate can reach 85%).

Key words Bayer red mud, liquid phase region, slag system, dephosphorization

复吹转炉少渣、高效脱磷冶炼工艺的生产试验研究

王 杰 曾加庆 杨利彬 林腾昌

（钢铁研究总院工艺所，北京 100081）

摘　要　对复吹转炉脱磷过程进行了热力学与动力学分析，并进行了工业试验。工业试验取得了良好的效果：石灰、白云石消耗分别降低 25.00%、30.00%，冶炼前期平均脱磷率 63.97%，冶炼过程平均脱磷率 90.72%。通过对脱磷率影响因素分析，得出转炉冶炼脱磷期控制工艺，R：1.80~2.00，TFe：12.00%~16.00%，T：1350~1400℃左右。

关键词　脱磷率，倒渣率，渣量

Trial Experiments of Slagless and Efficient Dephosphorization Producing Technology of Combined Blowing Converter

Wang Jie, Zeng Jiaqing, Yang Libin, Lin Tengchang

(Metallurgical Department, Central Iron and Steel Research Institute, Beijing 100081, China)

Abstract　Thermodynamics and dynamics analysis were made, and trial experiments were also taken into actions. Results of the trial experiments show that, the lime and dolomite quantity dropped by 25%and 30% respectively, and the average dephosphorization rate of 63.97% during the dephosphorization stage and 90.72% during the total process were got. By means of influencing factors analysis of dephosphorization rate, the control technology of dephosphorization stage was concluded that，the basicity of the slag is 1.8~2.0, the mass fraction of TFe is 12%~16% ,and the temperature of the bath is around 1350~1400℃.

Key words　dephosphorization rate, deslagging ratio, slag quantity

大型转炉少渣冶炼返回渣熔化特性及其应用效果的研究

吴　伟　孟华栋　刘　浏

（钢铁研究总院冶金工艺研究所，北京　100081）

摘　要　为了充分利用 300t 转炉脱碳炉炉渣对其进行轻打水和焖渣处理，之后作为脱磷炉的炉料。本文主要研究了轻打水和焖渣处理后炉渣的熔化特性以及在脱磷炉作为炉料的应用效果。结果表明，从化学成分和岩相分析上看，焖渣的水分高于干渣，两者的全铁含量都较高，熔点相近；干渣中硅酸二钙含量高于焖渣的，而硅酸三钙含量低于焖渣的，两者都含有一定量的铁酸钙，有利于脱磷和化渣。从应用的效果看，通过每炉次加入 3~6t 的干渣或焖渣，每炉可节省石灰 0.7~1t，终点成分控制达到了要求。加入干渣和焖渣的脱磷炉终点炉渣的岩相组成主要是硅酸二钙、RO 相和玻璃相，与不加返回渣得到的脱磷炉终渣成分的相差不大，使用干渣的终渣中含有少量的铁酸钙有利于脱磷。由此表明，采用干渣和焖渣都能达到一定的脱磷效果。

关键词　脱碳炉，返回用渣，熔化特性，脱磷炉，冶金效果

The Melting Characteristics of the Recycling Slag in 300t Less Slag Smelting Converter and Its Application Effects

Wu Wei, Meng Huadong, Liu Liu

(Metallurgical Technology Research Department of Central Iron & Steel Research Institute, Beijing 100081, China)

Abstract The steel slag in the decarburization converter is treated by little water-spreading and heat-stewed methods, which is used as the furnace burden of the dephosphorization converter. The slag characteristics by little water-spreading and heat-stewed methods and its application effects as the furnace burden are researched. The results show the water content in heat-stewed slag is higher than that of little water-spreading slag. There is higher total iron content in heat-stewed and little water-spreading slag and almost same melting point. The dicalcium silicate content in little water-spreading slag is higher than that in heat-stewed slag. The tricalcium silicate in little water-spreading slag is lower than that in heat-stewed slag. There are a lot of calcium ferrite in little water-spreading and heat-stewed slag, which is beneficial to dephosphorization and slag melting. The little water-spreading and heat-stewed slag are added to dephosphorization converter each charge by 3 or 6 tons. The results show that the saving lime is 0.7~1 tons each charge. The petrographic constitute of final slag for the dephosphorization converter by adding the little water-spreading and heat-stewed slag is dicalcium silicate, RO phase and glass phase. There are little calcium ferrite phase in the little water-spreading slag, which is helpful to dephosphorization. So there are better dephosphorization effects for the little water-spreading and heat-stewed slag.

Key words the decarburization converter, returned slag, melting characteristics, the dephosphorization converter, smelting effect

重钢 210t 转炉“留渣+双渣”工艺实践

尹 川 蒲胜亮 刘德宏 刘向东

（重庆钢铁股份有限公司炼钢厂，重庆 400084）

摘 要 文章叙述了重钢 210t 转炉“留渣+双渣”操作工艺在生产过程中通过控制留渣量、双渣倒炉时温度实现了脱磷期低碱度高效脱磷、通过控制脱磷期碱度、双渣倒炉温度降低了脱磷期炉渣带铁量、通过控制双渣后二次开吹氧气流量及氧枪枪位解决了双渣二次吹炼干法除尘易泄爆的问题。在重钢 210t 转炉“留渣+双渣”工艺与单渣法对比后得出了“留渣+双渣”工艺利用转炉冶炼前期熔池温度低等有利条件能够实现少渣量低碱度高效脱磷，从而达到降低降低钢铁料消耗及辅料消耗的目的，但是对转炉煤气回收量有一定的负面影响。

关键词 留渣+双渣，脱磷渣，辅料，钢铁料

Practice on “Slag Residual and Reduced” Operation on 210ton Furnace in Chongqing Steel

Yin Chuan, Pu Shengliang, Liu Dehong, Liu Xiangdong

(Chongqing Iron & Steel Co., Ltd., Steelmaking Plant, Chongqing 400084, China)

Abstract This paper describes the problems of Chongqing Iron & Steel 210t converter “Residual and Reduced” Operation process in the production process such as through control the left amount of slag and double deslagging furnace temperature that realizes dephosphorizing period low alkalinity, through control of dephosphorization alkalinity and double slag pour furnace temperature that reduces the dephosphorization slag with iron content, through control double slag after the second open blowing oxygen flow and oxygen lance position solves the problem that dust is easy to leak and explosion. Comparing with Chongqing Iron & Steel 210t converter “Residual and Reduced” Operation process and single slag method, it comes to the conclusion that “Residual and Reduced” Operation process achieves less slag low alkalinity dephosphorization when it takes advantage of low temperature of converter smelting furnace, so as to achieve the purpose of reducing the consumption of steel material and the consumption of material. But this process take a negative impact on the recovery of LDG.

Key words　slag residual and reduced, dephosphorization slag, auxiliary materials, metal materials

转炉“留渣+双渣”少渣炼钢工艺实践

李伟东　何海龙　刘鹏飞　乔冠男

（鞍钢股份有限公司炼钢总厂，辽宁鞍山　114021）

摘　要　介绍了鞍钢股份有限公司炼钢总厂转炉“留渣+双渣”工艺的关键技术，包括留渣及炉渣固化技术、炉渣流动性控制及高效脱磷技术、快速足量放渣及渣铁分离技术、炉渣返干控制及终渣 FeO 控制技术以及“留渣+双渣”快速生产技术，采用这些技术后，吨钢成本降低 12.19 元。

关键词　转炉，少渣，留渣，双渣，脱磷

Practice of Smelting Process with Less Slag Mode Based on LD Slag Reserving and Duplex Slag Process

Li Weidong，He Hailong，Liu Pengfei，Qiao Guannan

(General Steelmaking Plant of Ansteel Co., Ltd., Anshan Liaoning 114021, China)

Abstract　The key technologies of the smelting process based on the LD slag reserving and duplex slag process in general steelmaking plant of Ansteel have been introduced. These techniques included slag reserving and solidifying, slag fluidity ability control, dephosphorization process with high efficiency, quick and sufficient deslagging process, controlling process of dry slag formation and FeO content in final top slag. Based on these processes, slag reserving and duplex slag process has been formed. After it application, the production cost has been reduced by 12.19 yuan RMB per ton steel.

Key words　LD, less slag, slag reserving, duplex slag, dephosphorization

双渣冶炼过程炉渣成分在线预报模型系统的研究与开发

朱正海[1]　张鹤雄[1]　赵国光[2]　王建军[1]

（1. 安徽工业大学冶金工程学院，安徽马鞍山　243002；
2. 上海梅山钢铁股份有限公司炼钢厂，江苏南京　210039）

摘　要　建立了一种同时基于冶金热力学和冶金动力学的转炉炉渣成分实时计算模型，该模型用于“转炉双渣冶炼过程炉渣成分在线预报模型系统”的开发。模型可实现预测双渣冶炼过程的炉渣成分，以保证转炉冶炼在最佳的时间点完成倒渣操作，减少倒渣时机的判断失误，提高炼钢生产的经济效益。模型系统分为三部分，分别安装在服务器、客户端、离线计算机中，供不同人员使用。

关键词 双渣，炉渣成分，实时预报，模型

Research and Development on Slag Composition Online Prediction Model System of Double Slag Process in BOF

Zhu Zhenghai[1], Zhang Hexiong[1], Zhao Guoguang[2], Wang Jianjun[1]

(1. College of Metallurgy Engineer, Anhui University of Technology, Ma'anshan 243002, China;
2. Steelmaking Plant,Shanghai Meishan Iron & Steel Co., Ltd., Nanjing 210039, China)

Abstract This paper describes a slag composition online prediction model of double slag process in BOF, based on thermodynamics and kinetics of metallurgical. The model is on the purpose of prediction slag composition of the double slag process in BOF. It can make sure that the pour slag operation is done at the best time, reducing error judgment of pour slag time, improving the economic benefit of steelmaking production. The model system is divided into three parts for different user, which are installed on the server, the client and the computer.

Key words double slag，slag composition, online prediction, model

260t 转炉留渣双渣法冶炼低磷 IF 钢半钢控制研究

齐志宇 田 勇 王 鹏 张志文 朱国强 孙艳霞

（鞍钢股份有限公司炼钢总厂，辽宁鞍山 114021）

摘 要 针对低磷 IF 钢生产，存在温度、磷和氧很难同时命中的难题，采用了留渣双渣的冶炼控制工艺，通过合理控制留渣量、一次倒渣温度、一次倒渣时间等技术措施，低磷 IF 钢磷含量合格率达到 100%，终点氧值降低了 112ppm，絮流率降低，提高钢水的质量。

关键词 转炉冶炼，脱磷，双渣

260t Converter Slag Double Slag Process of Smelting Low Phosphorus Steel IF Control of Semi Steel

Qi Zhiyu, Tian Yong, Wang Peng, Zhang Zhiwen, Zhu Guoqiang, Sun Yanxia

Abstract Aiming at the production of low phosphorus steel IF, temperature, oxygen and phosphorus problem is very difficult to hit, the smelting slag double slag process control, through the reasonable control of slag amount, a fall, a temperature of slag slag pour time and other technical measures, the qualified low phosphorus steel phosphorus rate reached IF 100%, the end of oxygen decreased by 112ppm, flocculating rate decreased, improve the quality of molten steel.

Key words converter, dephosphorization, double slag

气化室纯镁喷吹脱硫工艺水模试验研究

李明晖　欧阳德刚　朱善合　王海清　罗　巍

（武汉钢铁（集团）公司研究院，武汉　430080）

摘　要　通过水力学模型试验，分析了气化室纯镁脱硫工艺熔池流动状态，指出了工艺优化的有效途径，研究了工艺参数对气化室纯镁脱硫混匀时间和液面波动高度的影响规律。气化室纯镁脱硫工艺中喷吹气体和镁蒸气从气化室底部以大气泡形式间歇式排出，导致镁气泡反应比表面积小、喷枪摆动剧烈、液面大幅波动甚至喷溅，降低了金属镁的脱硫利用率，限制了金属镁喷吹强度的提高；改善镁蒸气喷入铁水方式是气化室纯镁脱硫工艺优化的有效途径。试验条件下，在混匀时间变化关系曲线上均存在一个最佳喷吹气体流量；在喷吹气体流量较小情况下,混匀时间随喷枪插入深度的增加而减小，而当喷吹气体流量较大时，增大喷枪插入深度混匀时间有所增加；随着喷吹气体流量和喷枪插入深度的增大，熔池液面波动高度不断增加。

关键词　铁水脱硫，纯镁，气化室，水模试验，动力学条件

Hydraulics Simulating Experiment Study on Hot Metal Magnesium Desulphurization by Injection Lance with Gas Chamber

Li Minghui, Ouyang Degang, Zhu Shanhe, Wang Haiqing, Luo Wei

(Research and Development Centre of Wuhan Iron and Steel (Group) Corp, Wuhan 430080, China)

Abstract　Through hydraulic model experiments, the flow state and an effective approach for improving process of hot metal magnesium desulphurization by injection lance with gas chamber was studied. Effects of process parameters on mixing time and surface wave was also been discussed in this paper.

Key words　hot metal desulphurization, pure magnesium, gas chamber, hydraulic model experiment, dynamic conditions

含钒铁水冶炼 CDC03 工艺优化与实践

杨　坤　国富兴　孙福振　梁新维　刘　超　康爱元　李亚厚　段喜海

（河北钢铁集团承德分公司长材事业二部，河北承德　067002）

摘　要　本文分析介绍了承钢半钢冶炼冷轧基料 CDC03 低碳低硅钢工艺实践过程，对冶炼工艺进行了优化和改进，重点探讨了 CDC03 冷轧基料脱氧工艺、转炉出钢挡渣工艺、LF 炉精炼造渣工艺、成分控制、夹杂物控制及钙处理工艺，工艺优化后，大大减少了由于成分不合造成改判的数量，铸坯表面质量明显提高，为后续轧制创造了有利条件。

关键词　挡渣，钙处理，酸溶铝，低碳低硅，Al_2O_3

Optimization and Practice of CDC03 Process for Smelting of Vanadium Containing Vanadium Iron

Yang Kun, Guo Fuxing, Sun Fuzhen, Liang Xinwei, Liu Chao, Kang Aiyuan, Li Yahou, Duan Xihai

(Hebei Iron and Steel Group Chengde Branch Long Material Cause Two, Chengde Hebei 067002, China)

Abstract This paper introduces the bearing steel semi steel smelting base material for cold rolling CDC03 low carbon low silicon steel process practice process, the smelting process was optimized and improved, and focuses on the CDC03 base material for cold rolling deoxidization process, tapping retaining slag process, LF refining slagging process, control components, inclusion control and calcium treatment process, process optimization, greatly reduced due to the compositional differences caused by the judgment of the number of cast slab surface quality is obviously improved, creating favorable conditions for subsequent rolling.

Key words slag, calcium treatment, acid soluble Al, low Carbon and low Silicon, Al_2O_3

半钢脱磷预处理技术的开发及应用

段喜海 孙福振 梁新维 闫维三 杨 坤

（河北钢铁承德分公司长材事业二部，承德 067102）

摘 要 为适应对钢材质量要求的不断提高，促进铁水预处理技术不断发展，承钢开发了新型铁水预处理技术，针对承钢含钒铁水的特殊性，在传统工艺上，开发半钢脱磷预处理技术，降低半钢磷含量，实现转炉少渣、高拉碳冶炼操作，降低了石灰、钢铁料消耗和合金消耗，并有效提高了钢材质量和使用性能，对研发更多低磷品种钢具有重要意义。

关键词 铁水预处理，提钒，半钢脱磷

Chenggang Hot Metal Containing Vanadium New Pretreatment Technology Development and Application

Duan Xihai, Sun Fuzhen, Liang Xinwei, Yan Weisan, Yang Kun

(Hebei Chengde Iron and Steel Branch of the Company Two, Chengde 067102, China)

Abstract In order to adapt to the requirement of steel quality is increasing, and promoted the hot metal pretreatment technology development, ChengGang developing new type of hot metal pretreatment technology, in view of the particularity ChengGang containing vanadium iron, on the traditional process, the development of new steel dephosphorization pretreatment technology, reduce the semisteel phosphorus content, implementation of converter slag and smelting high carbon less operation, reduce the lime, iron and steel material consumption and alloy consumption, and effectively improve the quality of the steel and the use of performance, is of great significance to the research and development of more varieties of low phosphorus steel.

Key words hot metal pretreatment, the desulfurization V, half steel dephosphorization

基于炉口火焰分析的转炉终点预报技术

田　陆　刘卓民

（衡阳镭目科技有限责任公司，长沙　410000）

摘　要　转炉炉口的火焰信息和熔池碳含量及温度之间存在紧密联系。实时采集转炉炉口火焰光强和图像信息，可以对熔池碳含量和温度进行预报。基于光电技术、机器视觉和冶金原理的火焰分析技术在钢厂得到了应用，使用效果表明，当终点碳含量<0.15%，控制范围为±0.02%时，碳命中率大于 90%，温度控制范围为±15℃时，温度命中率大于 90%，这一技术对提高中小转炉控制水平、提升钢材产量和质量具有积极意义。

关键词　转炉，终点控制，火焰，光强，图像

End-point Prediction Technology of BOF Based on Flame Analysis

Tian Lu, Liu Zhuomin

(Ramon Science & Technology Co., Ltd., Changsha 410000, China)

Abstract The flame information is connected with the carbon content and temperature in the converter. We can predict the carbon content and temperature of the converter by gathering the light intensity and flame images online. The flame information technology which is based on photo-electricity technology, machine viewer and metallurgic theory has been used in the steel plant. The experiments indicated that When the carbon content is less than 0.15%, and the control requirement is ±0.02%, the hit rate is more than 90%; when the temperature control requirement is ±15℃, the hit rate is more than 90%. The technology is good for improving the converter control level, steel quality and quantity.

Key words converter, endpoint control, flame, light intensity, image

基于炉气分析的转炉炼钢过程动态控制

田　陆[1,2]　杨斌虎[1,2]　黄郁君[1,2]　王利君[1,2]

（1. 镭目科技有限责任公司，湖南长沙　410007；
2. 北京光科博冶科技有限责任公司，北京　100080）

摘　要　基于冶金热力学、动力学平衡原理，利用当前最先进的转炉炉气激光动态检测技术，镭目公司开发了转炉炼钢全过程动态控制系统。本文论述了转炉炼钢过程动态控制的机理，描述了利用激光气体分析仪进行动态检测的系统组成，分析了影响转炉炼钢过程控制的过程变量，应用实践表明通过炉气分析进行转炉炼钢过程动态控制是可行有效的。

关键词　转炉炼钢，炉气分析，过程控制，动静态模型

Process Dynamic Control of BOF Steelmaking Based on Offgas Analysis

Tian Lu[1,2], Yang Binhu[1,2], Huang Yujun[1,2], Wang Lijun[1,2]

(1. Ramon Science & Technology Co., Ltd., Changsha Hunan 410007, China;
2. Beijin GKBY Science & Technology Co., Ltd., Beijing 100080, China)

Abstract Based on metallurgical theory of thermodynamics and dynamics ,BOF dynamical control system is developed by Ramon co., ltd using the most advanced BOF offgas dynamic d detection. The mechanism of dynamic control of BOF steelmaking in this paper. The system construction of dynamic detection using laser off gas analyzer is described. The process variables of BOF steelmaking process is analysised. The application result shows that BOF control its process using offgas analysis is feasible and effective.

Key words BOF steelmaking, offgas analysis, process control, dynamic & model

转炉出钢过程中增氮的影响因素分析

田志国

（湖南华菱湘潭钢铁集团，湖南湘潭 411101）

摘 要 通过转炉炼钢的生产实践，分析了转炉出钢过程中对钢中氮的影响因素。选择合适的原材料，采用低压力小流量的吹氩工艺和出钢过程中不完全脱氧的方式，可以减少钢液在出钢过程中的增氮量。

关键词 转炉出钢，氮含量，因素分析

The Analysis of Influence Factors in LD Tapping Process of Nitrogen Pickup

Tian Zhiguo

(Xiangtan Iron & Steel Co.,Ltd., Hunan Valin Group, Xiangtan 411101, China)

Abstract Analysis the influence factors in LD tapping process of adding nitrogen by practice 。Choosing appropriate raw material，using lowing press and less flux for blowing Argon，incomplete deoxidization in tapping process，can reduce the quantity of adding Nitrogen in tapping process of steel melt.

Key words LD tapping, nitrogen content, factors analysis

锰矿含碳团块高温熔融自还原试验研究

张 波 胡洵璞 高泽平 王大萍 苏振江

（湖南工业大学冶金工程学院，湖南株洲 412007）

摘　要　本文采用湖南贫锰矿为主要原料，在高温碳管炉内进行锰矿含碳团块高温熔融自还原试验，试验温度控制在1550~1650℃，对影响团块还原度的主要因素进行分析，并对还原过程中金属Mn的挥发情况进行计算。结果表明：锰矿自还原团块内C/O最佳值为1.2，当C/O<1.2时，团块内还原剂量不够，团块还原度较低，而当C/O≥1.2时，团块内还原剂出现过剩，多余的还原剂对团块还原度的提高没有明显作用；还原温度的提高可促进团块还原度增加，但变化幅度不大；助熔剂 CaF_2 可降低团块熔点和黏度，加快团块在前期的还原；金属Mn的挥发呈抛物线状，反应前期挥发量较大，而随着反应的进行，由于液渣量的增加，阻隔了金属Mn的向外扩散，金属Mn的挥发量逐渐减少。提高液渣的形成速度是抑制金属Mn挥发的有效途径。

关键词　锰矿含碳团块，高温，熔融自还原

Study on the Melting Self-reduction Behavior of Manganese Ore Briquette Containing Carbon at High Temperature

Zhang Bo, Hu Xunpu, Gao Zeping, Wang Daping, Su Zhenjiang

(College of Metallurgy, Hunan University of Technology, Zhuzhou 412007, China)

Abstract　Use the lean manganese ore in Hunan province as main raw materials, the self-reduction experiment of manganeseore briquette containing carbon was carried out in the high-temperature carbon tube furnace, and the reduction temperature is 1550~1650℃. The main factors affecting the reduction rate were analyzed, and the volatilization of Mn is calculated in the process of reduction. The results show that the optimal value of C/O in briquette is 1.2, when the C/O<1.2, the amount of reductant is not enough and the reduction degree of briquette is low, and if the C/O≥1.2, the amount of reductant is excess has no obvious effect for improve the reduction degree of briquette;The increase of reduction temperature has effect on improvement of reduction degree, but the effect is not obvious; It can accelerate reduction and decrease reduction time when the appropriate flux CaF_2 is added to the briquettes. The volatilization rate of Mn in self-reduction process form of parabola, the volatilization rate is the largest in the first 3 min, and gradually reduced with the increase the quantity of liquid slag. Therefore, to speed up the formation of liquid slag is an effective way to suppression the volatilization of Mnvolatilization.

Key words　manganese briquettes containing carbon, high temperature, melting self-reduction

高马赫数氧枪喷头在100t转炉的应用

周泉林　刘海春

（唐山不锈钢有限责任公司，河北唐山　063105）

摘　要　针对唐山不锈钢100 t转炉实现自动炼钢后原设计氧枪喷头存在的吹炼不稳定、喷头寿命低等问题，对氧枪喷头及吹氧工艺进行了重新设计与优化。氧枪马赫数由1.9提高到2.15，并通过吹炼过程对L/L0的控制实现了变枪变流量操作，使高马赫数氧枪性能得到充分发挥，调整后喷溅、终渣全铁、喷头寿命等方面均有明显改善，达到预期效果。

关键词　氧枪，马赫数，L/L0

The Application for Lance Tip with High March Number in 100 tons Converter

Zhou, Quanlin, Liu Haichun

(Tangshan Stainless Steel Co., Ltd., Tangshan 063105, China)

Abstract aiming at the problems existed in the old lance tip after 100 tons converter in Tang Shan Stainless steel company has achieved automatic steel making, such as unstable blowing and low lifetime of the lance tip, we redesigned and optimized the lance tip and blowing process. Mach number of new designed lance is increased from 1.9 to 2.15, and has realized the operation for variable lance height and oxygen flow rate by controlling L/L0 during blowing, which have given the lance performance with high Mach number a full play. It is indicated that after such optimization the slopping behavior and total FeO in slag as well as lance tip lifetime have improved much better and has achieved the expectation.

Key words lance, March number, L/L_0

转炉溢渣喷溅控制与研究

吕长海[1,2] 程树森[1] 杨燕春[2]

(1. 北京科技大学冶金与生态学院，北京 100083；
2. 山东莱钢永锋钢铁有限公司，山东齐河 251100)

摘　要 本文论述了溢渣喷溅形成的原因，对喷溅时渣子的各组分含量进行分析，阐述了溢渣与喷溅的形成同加料时机、枪位、氧压、C-O 反应的程度等有直接的关系，对炉渣取样情况进行了详细分析。取样化验发现，喷溅渣中含有大量的粒铁，全铁含量也相对终点渣样偏高，针对此情况制定了相应的溢渣与喷溅控制措施。

关键词 溢渣，喷溅，粒铁，全铁，喷溅渣

Overflow of Converter Slag Splashing Control and Research

Lv Changhai[1,2], Cheng Shusen[1], Yang Yanchun[2]

(1. Beijing University of Science and Technology Institute of Metallurgy and Ecology, Beijing 100083;
2. Shandong Laiwu Yongfeng Steel Co., Ltd., Shandong Qihe 251100)

Abstract Pick to the paper discusses the formation cause of overflow slag splashing, the splash of pulp are each component content analysis, this paper expounds the formation of overflow slag and splash with charging time, gun, oxygen pressure, C - O reaction degree has a direct relationship, and the sampling conditions are analyzed in detail. Sampling tests found that splash slag containing large amounts of iron, total iron content is relatively high end slag samples, for this situation made the corresponding overflow slag and splash of control measures.

Key words slag overflow, splash, grain of iron, all the iron, splash slag

高磷硅锰合金优化还原脱磷研究

朱子宗　周志强　魏晓伟　王　开

（重庆大学材料科学与工程学院，重庆　400044）

摘　要　以 SiCa 合金作为脱磷剂，采用 $CaO\text{-}CaF_2$ 作为覆盖渣，对高磷硅锰合金进行了脱磷实验研究。实验在硅钼棒炉中进行，并采用 ICP 光谱仪和化学分析技术，检测硅锰合金中元素含量。重点考察了处理温度、保温时间、脱磷剂加入量、CaC_2 加入等因素对硅锰合金脱磷行为的影响规律。研究结果表明：为了获得最佳脱磷效果，脱磷温度为 1400~1420℃，脱磷保温时间不宜过长，增大硅钙合金用量可以提高脱磷率。在渣中加入 CaC_2 可以获得更高的脱磷处理温度，提高合金的脱磷效率。

关键词　高磷，硅锰合金，脱磷剂，还原脱磷

Optimal Experiment Research on Reductive Dephosphorization of High Phosphor Silicon-manganese Alloy

Zhu Zizong, Zhou Zhiqiang, Wei Xiaowei, Wang Kai

(College of Materials Science and Engineering, Chongqing University, Chongqing 400044, China)

Abstract　Dephosphorization of silicon-manganese alloy with high phosphorusis is studied under reducing atmosphere with the dephosphorizing agent of SiCa and the slag of $CaO\text{-}CaF_2$. Experiments were conducted in a silicon molybdenum rod furnace and the ICP spectrometer and chemical analysis technique were used to test the content of elements in silicon manganese alloy.　The influence of temperature, holding time, dephosphorizing agent and CaC_2 on the dephosphorizing rate are mainly investigated. The experimental results show that the dephosphorization effect is the best when the temperature is from1400℃ to 1420℃. Prolonged holding time is useless to dephosphorization. The dephosphorization rate increases with the increase of dephosphorization agent. CaC_2 in the slag contributes to help obtain high temperature of dephosphorization and increase the efficiency.

Key words　high phosphor, SiMn alloy, dephosphorizing agent, reducing dephosphorization

煅烧系统压强对石灰活性及微观结构的影响

郝素菊[1]　蒋武锋[1]　张玉柱[1,2]　郝华强[2]

（1. 华北理工大学冶金与能源学院，河北唐山　063009；
2. 东北大学材料与冶金学院，辽宁沈阳　110819）

摘　要　石灰广泛应用于烧结和炼钢等工艺中，其活性直接影响生产效率和产品质量。本研究在高压电炉中煅烧石灰石制备了活性石灰，采用酸碱滴定法测定了石灰的活性度，采用场发射扫描电镜（FE-SEM）对活性石灰的形貌

进行了表征，详细研究了系统绝对压强对石灰活性度、CaO 晶粒大小及孔隙结构的影响。结果表明：高压条件下制备的活性石灰，活性度较高，CaO 晶粒较大，孔隙也增大，并且呈层状结构。对于反应产物体积膨胀的石灰消化反应，反应速率不仅与晶粒大小有关，而且与孔隙结构有关，大孔道的存在和层状结构有利于石灰消化反应的进行。

关键词 压强，活性石灰，活性度，微观孔隙结构，晶粒

Effect of Pressure on Lime's Activity and Micro Structure

Hao Suju[1], Jiang Wufeng[1], Zhang Yuzhu[1,2], Hao Huaqiang[2]

(1. College of Metallurgy & Energy, North China University of Science and Technology, Tangshan Hebei 063009, China; 2. School of Materials and Metallurgy, Northeastern University, Shenyang Liaoning 110819, China)

Abstract Lime was extensively employed in metallurgical production, has significant influence on sintering and steelmaking process. The activity of lime directly affects the production efficiency and product quality. In this paper, lime was prepared by calcimining limestone in a high-pressure electric furnace. The activity of lime was determined by acid-base titration. The morphology of lime was characterized by Field Emission Scanning Electron Microscopy (FE-SEM). Influence of pressure on the activity of lime, size of CaO grain and pore structure were finely studied. The results indicate that the reaction ratio is related to both grain size of CaO and the micro pore structure when investigated the volume expansion of reaction product in lime slaking reaction. As expected, lime produced under high pressure has good performances in the activity, large grain size of CaO, big pore, etc, compared with lime which was made under atmospheric pressure.

Key words pressure, active lime, activity, micro pore structure, grain

武钢 KR 法顶吹扒渣工艺的研究与实践

邓品团 李明晖 丁金发 李胜超

（武汉钢铁股份有限公司炼钢总厂，湖北武汉 430080）

摘 要 本文介绍了武钢二炼钢厂基于 KR 法的顶吹辅助驱渣工艺的水模试验研究结果，叙述了适合该厂实际的 KR 法顶吹辅助驱渣工艺技术的研发过程，详细阐述了该工艺在该厂的工业性应用、优化过程及其良好的实践效果。

关键词 KR 法脱硫，水模试验，顶吹辅助扒渣工艺，工业性应用

转炉高效低成本自动炼钢技术的开发与运用

龙治辉

（重庆钢铁股份有限公司一炼钢厂，重庆 400084）

摘 要 转炉自动炼钢动态控制技术可以提高钢水质量，降低生产消耗，提高生产效率，改善工人劳动条件，是现

代炼钢工艺发展的必然趋势。本文介绍了重钢在传统自动炼钢模型基础上结合自身特点进行的一系列技术开发工作，使自动炼钢技术成功应用于重钢“一罐制”高效化生产组织模式中，模型投用率提高约 70%，转炉双命中率也提高约 15%，并取得了可观的经济效益和社会效益。

关键词 自动炼钢，适应性，一罐制，成本控制

The Exploitation and Appliance of Efficient and Low Cost Automatic Steelmaking Technique on Converters

Long Zhihui

(Chongqing Steel and Iron Co., Ltd., Chongqing 400084, China)

Abstract The dynamic control of converter automatic steelmaking can improve the quality of molten steel，reduce consumption of production，improve the production efficiency and labor conditions.It is the inevitable trend in the development of modern steelmaking process. Based on the traditional automatic steelmaking model，the text introduces the series of technical exploitation work combining with the characteristics of Chonggang.These，which makes the automatic steelmaking model is successfully applied in the high efficient production organization mode of “the system of one can”, model operational rate and converter both-hit rate increase 70% and 15%，which obtains onsiderable economic and social benefits.

Key words automatic steelmaking, adaptability, the system of one can, cost control

转炉滑板法挡渣出钢技术概述

杨 奕 王丽坤 杨玉富

（马鞍山市雨山冶金新材料有限公司，安徽马鞍山 243000）

摘 要 随着钢铁产业的迅猛发展，用户对钢材质量的要求日益提高，而下渣会对转炉炼钢生产及后续工序造成许多影响，转炉滑板法挡渣能有效控制转炉前、后期下渣，提高钢水的洁净度。本文旨在从转炉滑板法挡渣的技术背景、工艺原理、实际使用、今后课题等方面说明转炉滑板法挡渣在转炉炼钢过程中产生的各项优势、可见的经济效益及发展前景。

关键词 转炉，炼钢，下渣，滑板挡渣

Summary of Slag-stopping Tapping Technology by Slide Gate in BOF

Yang Yi, Wang Likun, Yang Yufu

(Ma’anshan Yushan Metallurgy New Materials Limited Company, Ma’anshan Anhui 243000, China)

Abstract With the rapid development of iron and steel industry, user requirements for steel quality are higher and higher, The slag will have many effects on the production and subsequent process of converter steelmaking, slag-stopping tapping technology by slide gate in BOF can effectively control the converter slag, improve the cleanliness of molten steel. This thesis is illustrate slag-stopping tapping technology by slide gate in BOF to the process of converter steelmaking, the advantages, economic benefits and development prospects from the technical background, process principle, practical use, future topics and so on.

Key words converter, steel-making, slag, slag-stopping by slide gate

提钒铁水预脱硫渣扒渣性能研究

郭　凯[1,2]　宋　波[1,2]　杨必文[1,2]

（1. 北京科技大学钢铁冶金新技术国家重点实验室，北京　100083；
2. 北京科技大学冶金与生态工程学院，北京　100083）

摘　要　提钒铁水预脱硫渣为高炉铁水经过提钒后预脱硫处理产生的炉渣。通过对提钒铁水预脱硫渣进行渣性分析发现，提钒铁水预脱硫渣在扒渣温度 1330~1340℃时为非均匀性渣，渣中存在固相质点。利用 FactSage 热力学软件 Equilib 和 Viscosity 模块并结合 Einstein-Roscoe 公式对提钒铁水预脱硫渣 $CaO-SiO_2-MgO-Al_2O_3-FeO$ 在 1300~1460℃进行黏度计算，确定渣中各组元对炉渣黏度影响。通过提钒铁水预脱硫渣和铁水预脱硫渣黏度分析并结合现场生产实践，确定铁水预处理合理扒渣黏度为 2.0~3.0 Pa·s。根据提钒铁水预脱硫渣在 1330~1340℃扒渣时黏度低、铁损大现象，结合炉渣黏度计算结果，采取提高炉渣中 CaO 含量的措施升高炉渣黏度，最终确定提钒铁水预脱硫渣合理成分范围是：*R*=2.7~2.8、MgO=12%~13%、Al_2O_3=7%~8%、FeO=12%~13%。

关键词　提钒铁水预脱硫渣，黏度，扒渣

Exploration the Slag Removed of Vanadium-bearing Hot Metal Desulfuration Slag

Guo Kai[1,2], Song Bo[1,2], Yang Biwen[1,2]

(1. State Key Laboratory of Advanced Metallurgy, University of Science and Technology Beijing, Beijing 100083, China; 2. School of Metallurgical and Ecological Engineering, University of Science and Technology Beijing, Beijing 100083, China)

Abstract The hot metal is treated with vanadium recovery process and then the desulfuration process, the slag we got in desulfuration process is vanadium-bearing hot metal desulfuration slag. The viscosity of it is low. It is different to remove from the surface of the bath. With the laboratory determination on the property of the vanadium-bearing hot metal desulfuration slag, we found that the vanadium-bearing hot metal desulfuration slag is heterogeneous system. There are some slag-particle mixture in the liquid slag. It is not suitable to predict its viscosity by the traditional viscosity calculation model. The effect of CaO, MgO, Al_2O_3 content on the viscosity of a $CaO-SiO_2-MgO-Al_2O_3-FeO$ system at 1300~1460℃were calculated using Einstein-Roscoe combine with Equilib and Viscosity module in thermodynamic software FactSage. Through the studying on the viscosity of vanadium-bearing hot metal desulfuration slag and hot metal desulfurization slag. We make sure the suitable viscosity of slag removed is 2.0~3.0 Pa·s. It is a reasonable viscosity to remove the desulfuration slag. On the basis of the calculating results, we increased the CaO content to raise the viscosity of the slag. The reasonable content of the desulfuration slag of vanadium recovery molten iron is: *R*=2.20~2.3、MgO=13%~14%、Al_2O_3=7%~8%、FeO=12%~13%.

Key words desulfuration slag, viscosity, slag removed

低硫钢种铁水预脱硫过程稳定控制研究

方　敏　徐福泉　王文涛　刘鹏飞　魏　元　宋吉锁

（鞍钢股份有限公司炼钢总厂，辽宁鞍山　114021）

摘　要　针对鞍钢股份炼钢总厂低硫钢种铁水预处理高、喷溅严重、扒损严重等问题，采取了喷枪定位、优化扒渣方法以及应用工业盐等措施，喷溅比例和扒损分别降低了 68%和 51%，同时还解决了高钒钛铁水扒渣困难的问题。

关键词　低硫钢种，脱硫，扒渣

Investigation on the Hot Metal Pretreatment Process Stability Control of the Low Sulfur Steel

Fang Min, Xu Fuquan, Wang Wentao, Liu Pengfei, Wei Yuan, Song Jisuo

(General Steelmaking Plant, Ansteel Co., Ltd., Anshan Liaoning 114021, China)

Abstract　Due to the high ratio of low sulphur steel production, splashing and iron loss during slag skimming problems, by taking the gun positioning, optimization of slag skimming method and application of industrial salt and so on, not only splash proportion and grilled loss is reduced greatly, but also the high vanadium titanium iron slag difficult removal has been solved. Therefore, the production cost and the labor intensity of the workers has been reduced effectively.

Key words　desulphurization, slag skimming, low sulfur steel

鞍钢 260 吨转炉自动化炼钢开发与应用

费　鹏　赵　雷　牛兴明　王鲁毅　徐国义　贯春辉

（鞍钢股份鲅鱼圈钢铁分公司炼钢部，辽宁营口　115007）

摘　要　阐述了鞍钢自主开发了自动化炼钢模型 ACSAS 的背景，介绍了 ACSAS 的静态模型、氧枪模型、静态模型、自学习模型的主要功能及技术改进。通过提高基础数据的准确性，提高设备的可靠性，提高冶炼条件的稳定性，培养高技能的员工队伍等管理措施，保证了自动化炼钢系统的稳定运行。与引进德国蒂森克房伯 OTCBM 模型相比，终点碳的命中率(±0.01%)提高了 2.5%，达到 97.2%；终点温度的命中率([-5℃,+15℃])提高了 10.9%，达到 92.2%；终点碳温双命中率提高了 11.8%，达到 90.3%。熔剂成本下降 4.6 元，磷元素和碳元素超标质量事故下降了 52%，经济效益显著。

关键词　自动化炼钢，静态模型，氧枪模型，动态模型，自学习模型

The Development and Application of Automatic Steelmaking at Ansteel 260 tons Converter

Fei Peng, Zhao Lei, Niu Xingming, Wang Luyi, Xu Guoyi, Jia Chunhui

(Bayuquan Iron & Steel Branch of Angang Steel Co., Ltd., Yingkou Liaoning 115007, China)

Abstract Describes the background of developing automation steelmaking model ACSAS in Ansteel. This paper introduces the main function and technology of static model, oxygen lance model, static model, self-learning model of ACSAS. By improving the accuracy of basic data, the equipment reliability, the stability of smelting conditions, the cultivation of high skilled workforce management measures, to ensure stable operation of the automatic control system. Compared with the introduced Thyssen Krupp OTCBM model from German, the hit rate of end carbon (± 0.01%) increased by 2.5%, reaching 97.2%; the hit rate of the end temperature ([-5℃,+15℃]) increased by 10.9%, reaching 92.2%; the endpoint temperature and carbon double hit rate increased by 11.8%, reaching 90.3%. The flux decline in the cost of 4.6 Yuan, the quality accidents of phosphorus and carbon elements exceed the standard decreased by 52%. ACSAS has significant economic benefits.

Key words automatic steelmaking, static model, oxygen lance model, dynamic model, self-learning model

基于副枪检测的转炉终点磷预报技术

牛兴明　王鲁毅　徐国义　马　宁　贾春辉

（鞍钢股份鲅鱼圈钢铁分公司炼钢部，辽宁营口　115007）

摘　要　分析了影响转炉磷成分的热力学和动力学因素，并对这些因素进行了量化。特别是增加了炉渣温度、炉渣氧势、复吹效果、氧枪搅拌效果的量化。通过选取磷预报的显著因子建立数学公式，实现了终点预报磷与化验分析磷最大偏差为±0.006%，偏差小于±0.003%的比例达到90.7%。该技术在节约能耗、减少温度损失等方面降低成本效果显著。

关键词　副枪检测，磷预报，渣氧势

Prediction Technology for LD End Point Phosphorus Content Based on Sublance

Niu Xingming, Wang Luyi, Xu Guoyi, Ma Ning, Jia Chunhui

(Bayuquan Iron & Steel Branch of Angang Steel Co., Ltd., Yingkou Liaoning 115007, China)

Abstract Analysis of the thermodynamic and kinetic factors influencing phosphorus converter components, and these factors were quantified. Especially in the temperature of slag, slag blowing effect, oxygen potential, oxygen lance of mixing effect are added for quantification. The establishment of mathematical formula by significant factor selection P forecast. And it realizes the endpoint prediction of phosphorus and phosphorus chemical analysis the maximum deviation is + 0.006%, deviation less than 0.003%, the proportion reached 90.7%. The technology plays an important role in energy saving, reduce temperature loss, reduce the cost effect.

Key words　detection by sublance, phosphorus prediction, oxygen content in slay

铁矿石在 260t 转炉生产实践中的应用

李　超　牛兴明　徐国义　尹宏军

（鞍钢股份有限公司鲅鱼圈钢铁分公司炼钢部，辽宁营口　115007）

摘　要　介绍了 260t 转炉采用铁矿石作为原料应用到生产实践的情况，探讨了铁矿石在转炉内的反应机理，比较了转炉应用的各种冷却剂的冷却特性，从理论上分析了铁矿石对化渣的影响。根据本厂生产工艺条件，通过一系列数据分析整理，在克服跑渣喷溅和保证溅渣护炉的前提下，提出了铁矿石的实际操作应用办法，有效论证了铁矿石是转炉炼钢良好的冷却剂和化渣剂，表明铁矿石作为冷却和化渣辅助料在转炉生产实践的可行性和可操作性，随着废钢等冷却辅料减少矿石消耗量的增加，转炉化渣效果较好，有利于生产低磷的高附加值钢种，同时矿石的加入，有利于效缓解废钢紧缺压力，有效降低钢料消耗，节省吹氧时间和氧气消耗等成本支出，能大幅度提高相关经济技术指标。

关键词　铁矿石，冷却剂，化渣剂，转炉

Application of 260 Ton Converter with Iron Ore in Pratical Production

Li Chao, Niu Xingming, Xu Guoyi, Yin Hongjun

(Steelmaking Department in Bayuquan Iron and Steel Branch of Angang Steel Company Limited, Yingkou Liaoning 115007, China)

Abstract　This paper introduces the 260t converter with iron ore as raw material in the practical production, the reaction mechanism of iron ore in converter has been studied also, the cooling characteristics of various coolant for converter has been studied. By the theoretical analysis of the influence of iron ore slag melting, according to the production conditions of the factory, through a series of data analysis, to overcome the slag splashing and ensure the slap-splashing converter-protecting on the premise, it puts forward the practical application way of iron ore, argues that iron ore is well-behaved slagging medium and cooling agent in convert steelmaking. Shows that the iron ore and slag cooling as auxiliary material in the production practice of converter's is feasible and workable, with steel cooling materials reduced, ore consumption increased, the slagging effect is good, it is good at producing high added value and low grade phosphorus steel, at the same time, adding ore can effectively alleviate the pressure of shortage of scrap steel, reduce material consumption, save the time of blowing oxygen and oxygen consumption costs, improve the economic and technical indicators.

Key words　iron ore, cooling agent, slagging agent, converter

首钢转炉一次除尘尘泥生产转炉冷却造渣剂应用研究

武国平

（北京首钢国际工程技术有限公司，北京　100043）

摘 要 转炉炼钢一次除尘采用LT干法方式产生的除尘灰或OG湿法除尘方式产生的污泥，具有粒度细、含铁量高的特点，是可回收的二次资源。介绍了首钢迁钢公司和首钢京唐公司分别采用烧结工艺和冷固球团工艺，通过原料分析、配料计算，使用转炉炼钢尘泥生产出符合转炉使用要求的造渣剂，经炼钢厂使用证明烧结矿和冷固球团均具有很好的化渣效果。处理炼钢尘泥的两种工艺流程可为炼钢厂产生很好的经济效益和环境效益。分析表明，冷固球团工艺是钢铁企业高效回收利用转炉尘泥的发展方向。

关键词 转炉一次除尘，烧结，尘泥，造渣剂，冷固球团

Application Rresearch of Producing Coolant and Slag Former Using Dust and Sludge of Shougang Converter Primary Dedust

Wu Guoping

(Beijing Shougang International Engineering Technology Co.,Ltd., Beijing 100043, China)

Abstract The primary dust which is generated by dry de-dusting of converter gas or sludge which is generated by OG wet de-dusting, with the characteristics of fine particle size, high iron content, is a recyclable secondary resource.This paper describes the sintering process of Shougang Qian'an Iron & Steel Corporation and cold-hardended pellet process of Jingtang company, through the analysis of the raw materials and charge calculation, using dust and sludge of coverter primary dedust produced slag former which meet the requirements of converter. Using slag former in converter show that the sinter and cold-hardened pellets have good effect on slag formation. Two kinds of process for treatment of dust and sludge of coverter primary dedusting can produce good economic and environmental benefits for the steel plant. The analysis shows that the technology of cold-hardended pellet is the development direction of high efficiency of iron and steel enterprises to recycle and utilize the dust and sludge in converter primary dedusting.

Key words converter primary dedust, sinter, dust and sludge, slag former, cold-hardened pellets

滑板挡渣技术在承钢公司提钒和炼钢转炉的应用

黄宏家 高 海 张晓磊 吴 丽

（河北钢铁集团承钢公司热轧卷板事业部，河北承德 067002）

摘 要 滑板挡渣是移植大包滑动水口原理，在传统转炉的出钢口位置安装滑动水口装置，结合红外下渣检测和计算机控制，当出现下渣时，立即关闭滑板以彻底切断钢流达到挡渣的目的。因关闭速度快（0.6s 内），出钢前期和后期下渣量均得到有效控制，钢水洁净度提高，脱氧剂成本也随之降低。滑板挡渣技术应用于提钒转炉为国内首次应用于提钒转炉，下渣量明显减少，钒回收率明显提高，经济效益显著。此外，部分元素在提钒工序氧化进入钒渣，而滑板挡渣技术有效降低了下渣量，使用提钒半钢炼钢后的产品残余元素降低。

关键词 转炉，钒渣，提钒转炉，滑板挡渣

Slag-Stopping Technology by Slide Gate in BOF for Steel-making and Vanadium-Extraction at Chengde Steel

Huang Hongjia, Gao Hai, Zhang Xiaolei, Wu Li

(Hot-rolled Coil Division, Chengde Iron and Steel Company, Hebei Iron and Steel Group, Chengde, Chengde Hebei 067002, China)

Abstract The control principle of slide gate on BOF is similar to that of ladle after being transfered to CC at the beginning and the end of casting. Combined with infared slag detection, PLC and hydraulic system, a swift openning and closing(in 0.6s) is achieved on slide gate, which realizes the effective control of carrier-over slag in earlier and later stage during tapping, avoiding slag outflowing to ladle. Additionally, the purity of steel is improved and the cost of deoxidation agent is reduced. While slide gate slag-stopping technology is utilized on BOF for vanadium extraction, it's the first time to capitalise on this technology. The result was proved remarkable: the yield rate of vanadium slag is extentially higher than any other slag-stopping technology, which facilitates Chengde Steel better economic benefit. Furthermore, due to parts of elements are oxidized as composition of vanadium slag at the process of vanadium-extraction, slide gate slag-stopping technology makes it possible that less vanadium-slag outflow into semi-steel, a byproduct of vanacium-extraction and raw material of steel-making. The content of residual elements of fianl product such as Cr and V are decreased.

Key words converter, vanadium slag, vanadium extraction converter, slide gate

“一罐到底”铁水运输工艺铁水罐配置的优化分析

杨楚荣

（北京首钢国际工程技术有限公司，北京　100043）

摘　要　铁水运输“一罐到底”工艺可通过铁水物质流区段时间因素的解析，分析铁水罐热周转过程和时间，优化铁水罐配置数量。首贵特钢项目高炉-转炉界面和高炉-电炉界面采用“一罐到底”工艺铁水罐的热周转较为复杂，本文采用时序解析得知，配置 11 个铁水罐即可满足铁水运输要求，有效减少铁水罐数量、降低工程投资和生产维护费用。同时，铁水运输的时间因素解析表明，双向铁水运输系统铁水物质流运行协调、有序、连续、高效。

关键词　“一罐到底”，铁水罐配置，优化分析

Quantity Optimization and Analysis of Hot Metal Ladle for “Common Ladle System” Hot Metal Transportation

Yang Churong

(Beijing Shougang International Engineering Technology Co., Ltd., Beijing 100043, China)

Abstract By the method of analysis for time factor of “common ladle system” hot metal transportation process in hot metal flow section, Work cycle and time of hot metal ladle can be analyzed, and also its quantity can be optimized. “Common ladle system” technique has been adapted on BF—BOF interface and BF—EAF interface in Shougui project, and work cycle of hot metal ladle is very complicated. In this paper the analysis for order/time relation shows that eleven hot metal ladles can meet the requirement of hot metal transportation. Thereby hot metal ladles, project investment and maintenance cost can be reduced. Meanwhile, the analysis shows that hot metal flow runs in the status of harmony, order, continuation and high efficiency.

Key words "common ladle system", hot metal ladle quantity, optimization and analysis

高温铁水下长寿命 KR 搅拌头的开发

高 攀[1] 赵东伟[1] 崔园园[1] 李海波[1] 王 飞[2] 田志红[1]

（1. 首钢技术研究院，北京 100043；2. 首钢京唐钢铁联合有限责任公司，河北唐山 063200）

摘 要 通过单因素分析法分析了首钢京唐 KR 脱硫站的实际生产数据，主要分析了温度、脱硫剂组成、搅拌速度、搅拌头尺寸、生产节奏、搅拌头的修补等工艺对搅拌头寿命的影响，最终结论为：KR 进站铁水温度越高，相应搅拌头寿命越低，但由于高温有利于 KR 脱硫反应进行，因此开发了一种耐高温、耐侵蚀的 KR 搅拌头耐火材料，其具备较好的抗热震性能，提高了搅拌头寿命；CaF2 可以加速脱硫渣系对搅拌头耐火材料的侵蚀，因此开发了一种无氟脱硫剂，其应用后搅拌头平均寿命从 220 次左右提高到 280 次左右；搅拌速度由 130r/min 降低至 85r/min 后，搅拌头寿命可以提高 13.8%，搅拌头尺寸增大后，搅拌头平均寿命由 279 次提升到 315 次，因此降低搅拌头转速，增大搅拌头尺寸，最终提高了搅拌头寿命；加快生产节奏，加强搅拌头修补，可以有效地提高搅拌头寿命；通过工艺改进，2015 年 1~5 月，首钢京唐公司 KR 脱硫结束平均硫含量为 0.0008%，同时，KR 搅拌头寿命平均为 300 次，最高 350 次。

关键词 KR，搅拌头，寿命，脱硫工艺

Development of Long Life KR Stirrer under High IronTemperature

Gao Pan[1], Zhao Dongwei[1], Cui Yuanyuan[1], Li Haibo[1], Wang Fei[2], Tian Zhihong[1]

(1. Shougang Research Institute of Technology, Beijing 100043, China;
2. Shougang Jingtang United Iron & Steel Co., Ltd., Tangshan Hebei 063200, China)

Abstract By univariate analysis method, the actual production data of ShougangJingtang KR desulphurization station was analysed, main effects of temperature, desulfurization agent composition, stirring speed, stirrer size, production rhythm, stirrer repairon stirrer life were analysed and ultimately concluded that: the higher the temperature of hot iron, corresponding lower stirrer life, but due to high temperature conducive KR desulfurization reaction, and therefore a high-temperature resistant, erosion resistant refractory for KR stirrer was developed, which have good thermal shock resistance and improves the stirrer life; CaF_2 desulfurization slag system can accelerate refractory erosion of the stirrer, and therefore a fluorine-free desulfurization agent was developed, after its application, average stirrer life expectancy increased from about 220 to 280 times or so; stirring speed reduced from 130r/min to 85r/min, the stirrer life can be improved by 13.8%, after stirrer size increases, the average stirrer life expectancy raised from 279 times to 315 times, thus reducing the stirring speed, increasing the stirrer size and ultimately improve the stirrer life; to accelerate the pace of production, strengthening the stirrer repair, can effectively improve the stirrer life; through process improvement, from January to May 2015, average end sulfur content of ShougangJingtang KR desulfurization is 0.0008%, meanwhile, KR stirrer life is 300 times the average, up to 350 times.

Key words KR, stirrer, Life, desulfurization process

尖晶石类型对刚玉-尖晶石浇注料性能的影响

阮国智　邱文冬　齐晓青　刘光平

（宝钢工程技术集团上海宝钢工业技术服务有限公司，上海　201900）

摘　要　本文研究了不同铝含量的镁铝尖晶石对刚玉-尖晶石浇注料性能的影响。研究结果表明，在高温状态下，不同的镁铝尖晶石由于其铝含量的不同会引起材料体系内部发生不同反应，富铝尖晶石发生 Al_2O_3 的脱溶，而富镁尖晶石则引起二次尖晶石的生成，进而造成不同类型的尖晶石对刚玉-尖晶石浇注料的物理和力学性能产生显著影响，稳定尖晶石可以提高刚玉-尖晶石浇注料的常温和高温强度。

关键词　刚玉-尖晶石，浇注料，镁铝尖晶石，高温抗折强度

Effect of Spinel Types on Properties of Alumina-spinel Castables

Ruan Guozhi, Qiu Wendong, Qi Xiaoqing, Liu Guangping

(Shanghai Baosteel Industry Technological Service Co., Ltd., Baosteel Engineering & Technology Group Co., Ltd.,　Baoshan District, Shanghai 201900, China)

Abstract　Effect of spinel types on the properties of alumina-spinels castable was investigated. The results showed that the addition of various kinds spinel have an effect on the physical and mechanical properties of these refractory castables. The reason is that the Al_2O_3-rich spinel precipitated Al_2O_3 and MgO-rich spinel solid solution Al_2O_3 and formed secondary spinel. As well as, stable spinel in castables during heat treatment process to be more effective for enhancing cold and hot strength.

Key words　alumina-spinels, castable, Spinels, hot modulus of rapture (HMOR)

冷态渣回收利用实践

阎丽珍　蔡士中　苏庆林　王秋坤

修建军　张云鹏　赵世杰　李立刚

（邢台钢铁有限责任公司，河北邢台　054027）

摘　要　冷态渣作为精炼回收渣，具有高碱度、低氧化性等精炼合成渣的优点。通过对冷态渣的成分和回收再利用方案进行研究试验，确定了使用钢种，计算了使用冷态渣工艺的精炼渣脱硫能力和组份，最终实现了在稳定质量的基础上降低造渣料和脱氧剂消耗，降低生产成本的目的。

关键词　LF, 冷态渣, 精炼合成渣, 脱硫

Practice to Reclaiming and Using of LF Cold Slag

Yan Lizhen, Cai Shizhong, Su Qinglin, Wang Qiukun, Xiu Jianjun,
Zhang Yunpeng, Zhao Shijie, Li Ligang

(Steel-making Plant, Xingtai Iron & Steel Corp., Ltd., Xingtai Hebei 054027, China)

Abstract Cold salg is reclaimed from refining slag, as refining synthetic slag, the cold slag also has the merits as high basicity, low oxidizability, etc.This paper studies the composition and project carry on the recycle of cold slag, defined the steel grade that use cold slag, computes the ability of desulphurization and composition of refining slag that use cold slag.As the result of using the cold slag, the quality is stable, the amout of slag and deoxidizing agent is decreased, the production cost is reduced.

Key words LF, cold slag, refining synthetic slag, desulphurization

首钢炼钢系统技术创新进展

李海波[1] 陈　斌[1] 季晨曦[1] 刘　洋[1] 高　攀[1] 刘国梁[1]
崔　阳[1] 朱国森[2]

(1. 首钢技术研究院，北京　100043；
2. 首钢京唐钢铁联合有限责任公司，河北唐山　063200)

摘　要 回顾了首钢总公司迁顺产线、京唐产线、首秦产线近五年来的炼钢技术创新进展。在过去的五年以来，优化了 KR 铁水脱硫的生产工艺参数，理顺了“KR+转炉+RH+连铸”流程生产低硫钢种（[%S]≤0.0030）工艺；围绕转炉“全三脱”的特点进行高效低成本脱磷、控制增氮的工艺研究；开发了 SGRS 工艺，转炉炼钢石灰、轻烧白云石消耗降低 30%以上；针对不同钢种的使用要求，采用相应的措施，使钢中的非金属夹杂物得到良好控制；开发了倒角结晶器技术，通过采用倒角结晶器改善了铸坯冷却及受力状态，铸坯角部横裂纹发生率降低到 0.4%左右；进行了高拉速连铸工艺研究，实现了 SPHC 钢种的拉速从 1.7 m/min 提高到了 2.5 m/min，浇铸 IF 钢的拉速从 1.5 m/min 上升到 2.0 m/min。开发了厚板坯防窄面鼓肚技术，对足辊区的足辊数量、排布方式、喷淋冷却进行优化设计，保证了 400mm 厚连铸坯的质量控制。

关键词 KR，全三脱，SGRS，夹杂物，倒角结晶器，高拉速，特厚板

Advances in Technology Innovation of Shougang Steel-making System

Li Haibo[1], Chen Bin[1], Ji Chenxi[1], Liu Yang[1], Gao Pan[1], Liu Guoliang[1],
Cui Yang[1], Zhu Guosen[2]

(1. Shougang Research Institute of Technology, Beijing 100043, China;
2. Shougang Jingtang United Iron & Steel Co., Ltd., Tangshan Hebei 063200, China)

Abstract The article recalled the steelmaking technology advances of Shougang Corporation Qianan-Shunyi production line,

ShougangJingtang production line, Shouqin production line in the past five years. In the past five years, KR desulphurization parameters were optimized, and “KR+BOF+RH+CC” process producing low-sulfur grades ([%S] ≤0.0030) process was straightened. High efficient low cost dephosphorus around Dec-DeP process and nitrogen pick up control technology were researched. The adoption of SGRS process made BOF lime, dolomite consumption reduced by more than 30%. For different requirements of different types of steel, the non-metallic inclusions in steel was controlled well by the use of appropriate measures. Chamferedmold technology was developed,and industrial experiments showed that the applying of mold with chamfered corners improved the slab cooling and stress state, the occurrence of transverse corner crack reduced to 0.4％or less. Highspeed casting technology was developed, andthe speed of SPHC increased from 1.7m/min to 2.5m/min and the speed of IF steel from 1.5m/min increased to 2.0m/min.Thick slab anti narrow face bulging technology is developed, the number of foot roller zone foot roller, arranged manner, spray cooling were optimized, to ensure the quality control of 400mm thick slab.

Key words KR, DeP-DeC, SGRS, inclusions, chamfered corner mold, high casting speed, extra-heavy plate

炼钢降本工艺的开发与运用

龙治辉 胡 兵

（重庆钢铁股份有限公司一炼钢厂，重庆 400084）

摘 要 炼钢行业已进入微利时代，如何结合本企业自身特点降低生产成本、实现最大利益已成为国内众多钢铁企业争相研究和探索的重要课题，本文介绍了重钢一炼钢厂研究开发的几项较重要的转炉降本工艺。在结合自身特点的基础上，深入分析了含铁资源的回收利用、声纳化渣系统和转炉自动炼钢技术等项目的基本原理和对企业降本的重要贡献。

关键词 降本工艺，自动炼钢，回收利用

The Exploitation and Appliance on Steelmaking Cost Reduction Process

Long Zhihui, Hu Bing

(Chongqing Steel and Iron. Co., Ltd., Chongqing 400084, China)

Abstract Iron and steel industry has already enter in the meager profit times，how to reduce productive cost and obtain maximum profit is important assignment associated to self characters. The text intruduces several kinds of important cost reduction process of converters in NO.1 steelmaking factory of Chonggang. It analyses the basic principle of recovery using ferruginous resource，sonar slag formation system and automatic steelmaking for converters profoundly. Meanwhile，it discusses the important effects of them to business cost reduction.

Key words ferruginous resource, automatic steelmaking, sonar slag formation

尖晶石粒度对刚玉-尖晶石浇注料的性能影响研究

邹 龙[1] 胡四海[2] 王 忠[2] 陈华圣[1] 刘 孟[1] 何明生[1]

（1. 武汉钢铁（集团）公司研究院，湖北武汉 430080;

2. 武汉钢铁（集团）耐火材料有限责任公司，湖北武汉 430082）

摘 要 以电熔白刚玉、烧结尖晶石颗粒、电熔镁砂细粉、电熔尖晶石细粉、电熔尖晶石超微粉、白刚玉细粉、活性 α–Al_2O_3 微粉、纯铝酸钙水泥为主要原料制备了刚玉尖晶石浇注料，研究了尖晶石的引入形式和粒度对试样性能的影响，结果表明：加入 2%的镁砂细粉在浇注料内部原位生成尖晶石形成微裂纹能提高浇注料的抗热震性，加入 4%的烧结尖晶石颗粒、2%电熔尖晶石细粉和 4%电熔尖晶石超微粉时形成粒度梯度，此时浇注料综合性能较佳。

关键词 尖晶石，粒度，浇注料，抗热震性

The Influence of Grain Size for Spinel on the Properties of Corundum-spinel Castables

Zou Long[1], Hu Sihai[2], Wang Zhong[2], Chen Huasheng[1], Liu Meng[1], He Mingsheng[1]

(1. R & D Center of WISCO, Wuhan 430080, China;
2. Refractory Company of WISCO,Wuhan 430082, China)

Abstract Corundum-spinel castables was prepared using tabular alumina 、 sintered spinel particles、fused spinel powder、white fused alumina powder、α-Al_2O_3 powder and pure calcium aluminate cement as main starting materials， and the influence of spinel on the properties of the samples were investigated, The result shows that the introduction of 2% fused magnesium powder samples can create microcrack by formatted spinel in situ, which can improve the thermal shock resistanc of castables. when the grain size of spinel are make up of 4% sintered spinel particle、2% fused spinel powder and 4% fused spinel ultra-micro powder, the combination properties of castables is the better.

Key words spinel, grain size, castable, thermal shock resistanc

基于紧密堆积模型的轻量 Al_2O_3-MgO 质浇注料颗粒级配优化及其影响

邹 阳 顾华志 黄 奥 张美杰

（武汉科技大学省部共建耐火材料与冶金国家重点实验室，湖北武汉 430081）

摘 要 采用数值模拟方法基于 stovall 线性堆积模型计算了不同临界粒径，不同 q 值的混合颗粒紧密堆积时的堆积密度；在此基础上，以微孔刚玉作为骨料，白刚玉粉、α-Al_2O_3 微粉、镁砂细粉、MgO 微粉等作为基质，调整基质粒度组成，分别制备了 q 值为 0.25、0.28、0.31 和 0.34 的轻量 Al_2O_3-MgO 质浇注料，分析了不同 q 值下材料的常温性能、孔径分布和显微结构，并对不同 q 值下的轻量 Al_2O_3-MgO 质浇注料的抗渣性能进行了研究。结果表明适当的 q 值不仅需保证浇注料坯体形成较紧密堆积，还要避免烧结过程中基质微粉过多造成的收缩开裂，同时要考虑生成尖晶石等体积膨胀反应的影响。q 值过大（>0.31），基质堆积不紧密，烧结性能差；q 值过小时（q≤0.25），基质烧结时收缩过大，与骨料边缘脱离，对材料抗渣渗透性能不利。q=0.28 时，浇注料可以同时获得较佳的烧结性能（AP≈14.8%，BD≈3.02g·cm^{-3}，PLC<0.6%）、强度（MOR≈12.4MPa, CCS≈155.5MPa）、孔径结构(D_{50}≈2.07μm)及抗渣侵蚀和渗透性能（I_C≈25.4%,，I_P ≈11.55%）。

关键词 Al_2O_3-MgO 质浇注料，显微结构，颗粒堆积，抗渣渗透，轻量化

转炉煤气尘泥的利用研究

高燕军

（西安西矿环保科技有限公司，陕西西安　710055）

摘　要　传统的转炉烟气除尘系统采用湿法除尘，其产生的粉尘为沉淀的污泥，俗称“红泥”，其中含有的氧化钙已经充分水化，压制成球后不容易破裂，而采用目前的干法电除尘工艺，捕集到的粉尘含有较多石灰粉、氧化铁、少量碳，粒度较细，80%以上的粒度为5～76.4μm，属高细粉状态物质。它的运输方式大部分以机械运输为主，目前国内已有气力输送的实例，但小问题比较多。转炉尘泥的直接利用一直困扰着许多厂家。将转炉尘泥用于转炉炼钢、烧结、制备铁红颜料？究竟怎样的利用方法才是最经济、可靠地利用方法。每个钢厂根据自身情况采用合理的利用方式将大大降低转炉炼钢除尘的成本。

关键词　转炉粉尘，红泥，气力输送

The Research and Utilization of Converter Dust

Gao Yanjun

(Xi’an Xikuang Environmental Protection Co., Ltd., Xi’an 710055, China)

Abstract　Traditional converter flue gas cleaning system using wet process, the dust is precipitated sludge, commonly known as “red mud”, which contains calcium oxide has been fully hydrated and pressed into the ball is not easy to rupture, while the use of the present dry process, trapping the ash containing more lime powder, iron oxide, a small amount of carbon and the grain size is fine, more than 80% of the size is 5 ~ 76.4μm, belongs to the high fine powder material. Its transport mode is mainly based on mechanical transport, and the current domestic pneumatic conveying has been the case, but the problem is more.The direct use of converter dust has been plagued by many manufacturers. The converter is used for converter steelmaking, sintering, the preparation of iron red material? How to use the method is the most economical and reliable method. Each steel mill according to their own situation to adopt a reasonable way to use will greatly reduce the cost of converter steelmaking.

Key words　the converter dust, red mud, pneumatic conveying

鞍钢高炉风口工业水改闭路循环冷却研究

吕　程　张荣军　孙维强

（鞍钢股份有限公司炼铁总厂，辽宁鞍山　114011）

摘　要　对鞍钢高炉风口工业水改闭路循环冷却水工艺进行分析，并将风口小套单独软水闭路循环冷却系统与传统高炉风口小套高压工业净环水开路冷却系统进行技术、经济及安全方面的比较，论证高炉风口小套采用闭路循环软水冷却工艺具有不结垢、冷却强度高、冷却效果好、能耗低、运行安全可靠等诸多优点，符合现代大型高炉发展要

求，能实现高炉高效、优质、长寿生产。

关键词 高炉，风口小套，软水密闭循环，净环水开路循环

Research on the Closed Circulating Cooling Water of Blast Furnace Tuyere Industry Reform

Lv Cheng, Zhang Rongjun, Sun Weiqiang

(Limited by Share Ltd Angang ironmaking plant, Liaoning Anshan 114011, China)

Abstract The process of blast furnace tuyere soft water closed circulating cooling system was analyzed, and the tuyere separate water closed cycle cooling system and traditional blast furnace tuyere high pressure industrial clean circulating water cooling system are compared, the open economy and safety, demonstration of blast furnace tuyere small sleeve adopts closed loop soft water coolingprocess has not. The scale, high cooling strength, good cooling effect, low energy consumption, safe and reliable operation and many other advantages, in line with the requirements of the development of modern large blast furnace, blast furnace can achieve high efficiency and high quality, longproduction.

Key words blast furnace，tuyere，closed loop，open circuit of clean circulation water

2.2 炉外精炼及钢水质量控制/ Second Refining and Liquid Steel Quality Control

Numerical Simulation of Multiphase Phase Flow in Ladle Metallurgy

Li Linmin, Liu Zhongqiu, Li Baokuan

(School of Materials and Metallurgy, Northeastern University, Shenyang 110819, China)

Abstract Argon gas purgingin liquid steel plays a significant role in the ladle metallurgy. One key aspect ofthis processis the bubble transport and slag layer behavior. Bubble movement plays a significant role in the phase structure andcauses the unsteady complex turbulent flow pattern. This is one of the mostcrucial shortcomings of the current two-fluid models. The present work advised a mathematical model using the large eddy simulation (LES) approach coupled with both the Eulerian volume of fluid (VOF) and the Lagrangian discrete phase model (DPM) to simulate the gas stirring process. The argon gas bubble is tracked using the Lagrangian approach. And the liquid-slag-air free surfaces are tracked using the VOF model. The procedure ofbubble coming out of the liquid and getting into the air is modeled using a user-defined function. The results show that the current LES–DPM–VOF coupledmodel can well predict the unsteady bubble movement, slag eye formation,interface fluctuation, and particularly slag entrainment.

Key words numerical simulation, ladle metallurgy, large eddy simulation, discrete phase model

钙处理洁净钢的热力学和工业试验研究

方　文　张立峰　任　英

（北京科技大学冶金与生态工程学院，北京　100083）

摘　要　过去几十年来钙处理在洁净钢生产中取得广泛的应用，关于钙处理的热力学计算也逐步完善。钙处理热力学主要有经典热力学算法以及基于最小自由能的热力学软件计算两种方法。本文对热力学软件 FactSage 的计算方法进行了完整介绍，并用其精确计算钙处理过程。某钢厂现场钙处理试验结果表明，喂入 300m 左右钙铝线可对夹杂物实现良好改性。

关键词　钙处理，热力学计算，FactSage，夹杂物改性

Thermodynamic Consideration and Industrial Trials Study on Calcium Clean Steel

Fang Wen, Zhang Lifeng, Ren Ying

(School of Metallurgical and Ecological Engineering, University of Science and Technology Beijing, Beijing 100083, China)

Abstract　Calcium treatment technologies have been extensively applied in clean steel production, thermodynamic considerations for calcium treatment were also gradually improved. Thermodynamics on calcium treatment mainly consists of two methods, classical thermodynamics and calculated by thermodynamic software based on the minimum of Gibbs free energy. In the current paper, computational thermodynamics by FactSagewas introduced perfectly, and employed for calcium treatment calculation accurately guidingthe calcium treatment operation in industrial production, it was indicated that 300m Ca-Al wire would achieve good modification.

Key words　Calcium treatment, thermodynamics, FactSage, inclusion modification

反应诱发微小异相快速脱磷技术研究

王晓峰　魏　元　王　军　王小善

（鞍山钢铁集团公司，鞍山　114000）

摘　要　本文提出一种全新的洁净钢生产工艺——反应诱发微小异相净化钢水技术，设计了一种具有该种功能的复合球体，并开展了工业现场试验研究。结果表明，反应诱发微小异相净化钢水工艺是一种成本低、高效率、简便易行的钢水净化技术。采用这种工艺可以降低钢水传输过程温降，LF 精炼快速成渣，LF 升温速率提高 2℃/min，处理周期缩短 3～8min。铸坯中磷、全氧最低可以达到 30ppm 和 6ppm，从而实现快速深脱磷以及去除钢液中的细小夹杂。

关键词 洁净钢，脱磷，快速成渣，复合球体

Investigation on the Novel Technology of Dephorphouization Process due to the Dispersed In-situ Phase Induced by the Composite Ball Explosive Reaction

Wang Xiaofeng, Wei Yuan, Wang Jun, Wang Xiaoshan

(Anshan Iron and Steel Group Company, Anshan 114000, China)

Abstract In the present investigation, a novel cleanness steel production technology due to the dispersed in-situ phase induced by the composite ball explosion reaction has been put forward. A composite ball with this function has been designed and the industrial experimental investigations have also been carried out. The results indicate that this novel technique is a low-cost and high efficiency cleanness steel production method, which can realize quick slag forming, fast dephosphorization and fine inclusion removal. Temperature drop during transferring is small. The heat-up rate during LF has been increased for more than 2℃/min and the refining period has been shorten for 3～8min. The phosphorous and total oxygen in the as-cast slab can approach to 30ppm and 6ppm, respectively after its treatment.

Key words clean steel, dephosphorization, quick slag forming, composite ball

钢包和LF生产能力匹配性分析

董金刚[1] 蒋 鹏[2] 王 涛[2] 钟 源[1]

（1. 上海宝山钢铁股份有限公司制造管理部，上海 201900；
2. 上海宝山钢铁股份有限公司炼钢厂，上海 201900）

摘 要 由于第一炼钢单元的LF精炼周期大于钢水浇注时间，两台连铸机不能同时生产LF精炼钢种，限制了LF生产能力和品种钢的生产。为提高LF处理钢生产能力，需要打通2#LF的过跨轨道，实现2#LF、4#RH主要生产3#CC钢种，1#LF、1#RH主要生产1#CC钢种和模铸钢种的LF精炼钢种生产模式，LF精炼钢种生产能力可提高到27炉/天以上，需要底吹钢包10个/天以上，钢包包龄预计从119炉下降到105炉左右，对钢包使用和维修带来挑战。今后钢包技术进步的重点是在提高钢包底吹次数的同时提高钢包包龄，降低钢包使用成本。

关键词 钢包底吹，LF，生产能力，匹配

Matching Analysis of Ladle and LF Production Capacity

Dong Jingang[1], Jiang Peng[2], Wang Tao[2], Zhong Yuan[1]

(1. Manufacture Management Department，Baoshan Iron & Steel Co., Ltd., Shanghai 201900, China;
2. Steelmaking Plant, Baoshan Iron & Steel Co., Ltd., Shanghai 201900, China)

Abstract Due to LF treatment cycle is longer than the steels' casting cycle in the 1st steelmaking plant,two continuous casting machines can not produce LF treatment steels simultaneously ,LF production capacity and grade steels' production are restricted. In order to increase LF production capacity, the 2nd LF cross track must be extended, LF treatment steels'production models will come true that the 2nd LF and 4th RH mainly produce the 3rd CC steels,the 1st LF and 1st RH mainly produce the 1st CC steels and die casting steels. LF production capacity can be increased to more than 27 heats daily, but 10 bottom blowing ladles are taken daily,ladle age will be reduced from 119 heats to about 105 heats, the ladle use and maintenance will be challenged. From now on,the key technology is to improve the ladle bottom heats and ladle age at the same time to degree the ladle cost.

Key words ladle bottom blowing, ladle furnace, production capacity, matching

钢包底吹 RH 内气液流场的离散相模拟

吕 帅 耿佃桥 勾大钊 雷 洪 赫冀成

(东北大学材料电磁过程研究教育部重点实验室，辽宁沈阳 110004)

摘 要 本文基于离散相模型，分别对侧吹氩、底吹氩和侧底复吹氩 RH 真空精炼装置内气液两相流进行数值模拟，获得钢包内的气泡分布和循环流量。模拟结果表明，侧吹条件下，RH 装置的循环流量随侧吹氩量的增大而增大，增大到一定值后出现气体体积饱和现象，使钢液循环流量不增反减；底吹位置设在上升管投影圆的圆心处时，RH 真空精炼装置的循环流量最大，在其他位置时，装置循环流量较小：圆心左侧喷嘴对应的循环流量比圆心上侧对应的循环流量大，而圆心右侧喷嘴对应的循环流量比圆心下侧对应的循环流量小；在相同吹氩量条件下，对比侧吹方式，采用侧底复吹时的 RH 装置循环流量更大，在相同气体增量条件下，底吹氩比侧吹氩对装置循环流量的影响更大。

关键词 RH 真空精炼，数值模拟，侧底复吹，循环流量，湍动能

Discrete Phase Simulation of Gas Liquid Flow in RH with Ladle Bottom Blowing

Lv Shuai, Geng Dianqiao, Gou Dazhao, Lei Hong, He Jicheng

(Key Laboratory of Electromagnetic Processing of Materials, Ministry of Education, Northeastern University, Shenyang 110004, China)

Abstract In present paper, based on the discrete phase model, the side-blowing argon, bottom blowing argon and side-bottom blowing argon in the RH vacuum refining device have been studied, which focus on the bubble distribution and circulation flow rate in ladle. The numerical results show that under the condition of the side-blowing argon, circulation flow rate increases with the side blowing argon flow rate, but when reaches a certain value, the saturation phenomenon will arise and circulation flow rate decreases. When the position of bottom blowing in projection circle's center, the circulation flow rate reach maximum, in other places, the circulation flow rate is small. When the nozzle is on the left side of the center, the circle circulation flow rate is greater than that when the nozzle is on the center of the circle. When the nozzle is on the right side of the circle center, the circle circulation flow rate is smaller than that when the nozzle is under the center of the circle side. On the condition of the same gas flow rate, the circulation flow rate with side bottom blowing is greater than that with side-blowing. On the condition of same increment of gas flow rate, the bottom blowing has greater influence than side for circulation flow rate.

Key words RH vacuum refining, numerical simulation, side-bottom blowing argon, circulation flow rate, turbulent kinetic energy

高压底吹条件下钢液中气泡的物理模拟

王书桓　郭福建　赵定国　高爱民

（华北理工大学冶金与能源学院，河北唐山　063009）

摘　要　通过物理实验，研究了压强、底吹流量对气泡形状、尺寸、数量及“渣眼”尺寸的影响，研究表明：随着底吹流量增加，高压条件下的气泡比常压条件下的气泡更趋向于正圆形，气泡直径平均值变小，气泡数量增多，气泡形状和直径在底吹流量为 4L/h 时变化最为明显；当底吹位置和底吹流量不变时，随着压强的增加，气泡形状逐渐趋向于正圆形，气泡直径逐渐变小，气泡数量逐渐增多，“渣眼”面积逐渐减小。在常压和高压条件下，底吹流量超过 5L/h 之后，气泡形状和直径平均值就没有明显变化了。当底吹流量为 4L/h 时，压强升高到 5 个大气压之后，气泡形状和直径平均值变化幅度就很小了。

关键词　高压，常压，物理模拟，气泡行为

Physical Simulation of Bubble Behavior in Liquid Steel under the Condition of High Pressure and Bottom Blow

Wang Shuhuan, Guo Fujian, Zhao Dingguo, Gao Aimin

(College of Metallurgical and Engineering, North China University of Science and Technology, Tangshan 063009, China)

Abstract　Experiments of pressure and flow rate on the shape,diameter,quantity of bubble and slag eye area indicate that: with the increase of bottom blowing flow rate , under the condition of high pressure air bubbles tend to round than atmospheric pressure under the condition of air bubbles, the average bubble diameter decreases, bubbles quantity increases, when the bottom blowing flow rate of 4 L/h,the change of bubble diameter and shape is very obvious; When at constant flow rate and bottom blowing position, with the increase of the pressure, the bubble shape gradually tend to round, air bubble diameter became smaller, the bubble number increase gradually, "slag eye" area gradually decreases.Under the condition of normal pressure and high pressure, bottom blowing flow after more than 5 L/h, the changing of bubble diameter and shape of the average ate not obvious .When the bottom blowing flow rate is at 4 L/h and pressure is at 0.5MPa, the change of average bubble diameter and shape is very small.

Key words　high pressure, atmospheric pressure, physical simulation, bubble behavior

超低碳 IF 钢生产冶炼工艺研究

王小善　魏　元　王晓峰　齐志宇　王鹏飞　赵志刚

朱国强　高洪涛　李　冰

（鞍钢股份有限公司炼钢总厂，辽宁鞍山　114021）

摘　要　本文以鞍钢260t转炉生产超低碳IF钢为研究对象，重点研究了超低碳IF钢冶炼工艺。研究结果表明：通过原料的精细化管理提高废钢量准确性，优化铁水罐折铁量提高转炉装入铁水比；增加复吹转炉底枪支数和供气流量进行强化冶炼以及部分炉次零位搅拌工艺，降低吹炼终点碳氧积和终点氧含量，可以为RH精炼处理创造较好的初始条件。
关键词　IF钢，顶底复吹转炉，终点氧，终点碳

Investigation on Smelting Process of Ultra-low Carbon Steel Production

Wang Xiaoshan, Wei Yuan, Wang Xiaofeng, Qi Zhiyu, Wang Pengfei, Zhao Zhigang, Zhu Guoqiang, Gao Hongtao, Li Bing

(General Steelmaking Plant of Angang Steel Co., Ltd., Anshan 114021, China)

Abstract　in the present paper, the smelting process of ultra-low carbon steel production has been investigated further by 260t convert in ANSTEEL. The results show that meticulous management of the raw materials can contribute to the improvement of the scrap feeding accuracy. The optimization of hot melt tapping amount can improve its charging in convert. Furthermore, the increasing of the bottom blowing gun amount and argon flow rate can strengthen the smelting process and stirring technique at convert zero point, which can result in the decreasing C-O value and oxygen content at ending point and can contribute to better the initial treatment conditions for RH treatment.
Key words　IF steel，top and bottom blowing converter, ending point oxygen content，ending point carbon content

Ca/S对中碳合金钢中硫化物生成特性的影响研究

赵　啸　何志军　金永龙　庞清海　戴雨翔　么敬文　袁　平

（辽宁科技大学材料与冶金工程学院辽宁省化学冶金重点实验室，鞍山　114051）

摘　要　对Ca/S值在0.042~0.079之间的中碳合金钢进行了冶炼，采用光学显微镜、扫描电镜等实验仪器，并结合图像分析软件Image Pro Plus研究了钙处理后不同Ca/S值对钢中硫化物变性效果的影响。结果表明：当Ca/S值为0.042时，钢样中条状硫化物的长度和宽度最小分别为7.9μm和1.8μm，颗粒状硫化物直径最小为3.1μm，条状硫化物与颗粒状硫化物数量比值约为2.8:1；当Ca/S值为0.079时，条状硫化物的长度和宽度最大分别为12.3μm和2.9μm，颗粒状硫化物的直径最大为4.5μm，条状夹杂物与颗粒状硫化物数量比值约为0.76:1，Ca处理变性效果十分明显。实际冶炼时，应该在满足钢对夹杂物级别要求的前提下，合理控制钢中Ca/S值。
关键词　中碳钢，钙处理，硫化物，变性效果

Research on Sulfide Formation Characteristics Affected by Ca/S Ratio in Medium Carbon Steel

Zhao Xiao, He Zhijun, Jin Yonglong, Pang Qinghai, Dai Yuxiang,Yao Jingwen, Yuan Ping

(School of Materials and Metallurgy, University of Science and Technology Liaoning, Key Laboratory of Chemical Metallurgy Liaoning Province, Anshan 114051, China)

Abstract The medium carbon alloy steels whose Ca/S ratios are between 0.042 and 0.079 were melted in laboratory. Besides, experimental apparatuses such as optical microscope, SEM and Image Pro Plus image analysis software were employed to study the effect of Ca/S value on sulfide modification. It is indicated in the results that the minimum length and width of strip sulfides in sample are respectively 7.9μm and 1.8μm when the Ca/S value is 0.042, while the minimum diameter of granular sulfide is 3.1μm and simultaneously the number ratio of strip and granular sulfide is 2.8:1. When the Ca/S value is 0.079, the maximum length and width of strip sulfides in the sample are respectively 12.3μm and 2.9μm, while the maximum diameter of granular sulfide is 4.5μm. And the number ratio of strip and granular sulfide is 0.76:1. Through the comparison of sulfide size, the conclusion can be made that the modification effect of sulfide by Ca treatment is very significant. Therefore, the ratio of Ca/S value in steel should be reasonably controlled to meet the requirements for inclusion level.

Key words medium carbon steel, ca treatment, sulfide, modification effect

椭圆形和圆形浸渍管 RH 真空精炼过程中夹杂物的行为研究

李 菲[1] 张立峰[1] 王佳力[2] 彭开玉[2] 李树森[2]

（1. 北京科技大学冶金与生态工程学院，北京 100083；
2. 首钢股份公司迁安钢铁公司，迁安 064404）

摘 要 本文以工业实验为基础，研究了椭圆形和圆形浸渍管的两种 RH 真空装置精炼过程中夹杂物形貌、成分、数量及尺寸的变化。实验表明：椭圆形浸渍管由于具有更大的循环流量，可有效提高精炼效率，缩短精炼时间。尤其在精炼前 20min 内，椭圆形浸渍管夹杂物数密度下降 89%，而圆形浸渍管夹杂物数密度下降仅 68%。在夹杂物尺寸方面，椭圆形浸渍管夹杂物前 20min 仍以 2~3μm 为主，而圆形浸渍管夹杂物尺寸有所增大，直到镇静过程面积百分数才有所降低。同时研究发现，椭圆形浸渍管并未改变冶炼终点夹杂物成分及形貌，夹杂物数密度及尺寸也相差不多。出站夹杂物均以点线状纯 Al_2O_3 为主，且椭圆形浸渍管夹杂物在镇静过程中尺寸有所增加。

关键词 椭圆形浸渍管，圆形浸渍管，数密度，面积百分数，夹杂物尺寸

Effect of Oval and Round Snorkels on Inclusions Behavior in the RH Process

Li Fei[1], Zhang Lifeng[1], Wang Jiali[2], Peng Kaiyu[2], Li Shusen[2]

(1. Metallurgical and Ecological Engineering, University of Science and Technology Beijing, Beijing 100083, China; 2. Qian'an Iron & Steel Co., Ltd., Shougang Group, Qian'an 064404, China)

Abstract Morphology, chemical content, quantity and size change of inclusions were studied during different shape snorkels RH refining. Results show that oval snorkel can improve refining effective and short process time for its lager recirculation rate. Especially, number density of oval snorkel inclusion decreased 89% in early 20 minutes, whereas that of round snorkel inclusion decreased just 68%. In the aspect of inclusion size dispersion, 2~3μm inclusions were still the main component in oval snorkel inclusions in early 20 minutes. Size of round snorkel inclusions continued to increase until the standing process. Besides, chemical content and morphology of inclusion were not be changed at the end of refining using oval snorkel. And there were no remarkable effect of snorkel shape on inclusion number density and size dispersion. Inclusions at the end of refining were mainly point and line Al_2O_3. And oval snorkel inclusion size increased during standing process.

Key words oval snorkel, round snorkel, number density, area fraction, inclusion size

不同工艺路线生产低碳铝镇静钢钢水洁净度分析

张 飞[1] 董廷亮[1] 刘占礼[2] 李建设[2] 段云祥[2] 李 杰[1] 李双江[1]

（1. 河北钢铁集团河北钢铁技术研究总院，河北石家庄 052165；
2. 唐山不锈钢有限责任公司，河北唐山 063010）

摘 要 唐钢不锈钢公司生产低碳铝镇静钢采用 BOF-LF-CC 和 DeS-DeC-RH-CC 两种工艺，通过对两种工艺下各工序钢中氧、氮含量的分析和热轧板中夹杂物的分析，得出以下结论：两种工艺均具有较高的氧含量控制能力，需要进一步加强 DeS-DeC-RH-CC 工艺的控制稳定性。冶炼过程中增氮量最大的环节均为转炉出钢过程；但 DeS-DeC-RH-CC 工艺增氮量相对较小，且 RH 还具有一定的脱氮功能，更适合于生产低氮含量的钢种。BOF-LF-CC 工艺生产的热轧板中的夹杂物多为球形 Al_2O_3-CaO-MgO 类，而 DeS-DeC-RH-CC 工艺多为块状纯 Al_2O_3 夹杂物；与 BOF-LF-CC 工艺比较，DeS-DeC-RH-CC 工艺产热轧板中夹杂物数量少，但大尺寸夹杂物的比例较多。

关键词 钢水洁净度；T[O]；氮；夹杂物

Analysis of Different Technological Processes on the Cleanliness of Low Carbon Al-Killed Steel

Zhang Fei[1], Dong Tingliang[1], Liu Zhanli[2], Li Jianshe[2], Duan Yunxiang[2], Li Jie[1], Li Shuangjiang[1]

(1. Hebei Iron and Steel Technology Research Institute，Hebei Iron and Steel Group, Shijiazhuang, Hebei 052165, China；2. Tangshan Stainless Steel Co., Ltd., Tangshan Hebei 063010, China)

Abstract Both the BOF-LF-CC process and the DeS-DeC-RH-CC process are used to produce the LACK steel in Tangshan Stainless Steel. The total oxygen content, nitrogen content and inclusions in LCAK steel produced by BOF-LF-CC and DeS-DeC-RH-CC processes were studied by the way of sampling systematically and comprehensive analysis in steel production. The research results indicated that: both of the two process have a high level of oxygen content control ability, and it's necessary to improve the stability of DeS-DeC-RH-CC process. The converter tapping process is the maximum amount of nitrogen pick-up stage. But DeS-DeC-RH-CC process have a smaller nitrogen pick-up amount, moreover, RH process can remove a part of nitrogen, so it's better to adopt the DeS-DeC-RH-CC processes route to product the low nitrogen content steel. The non-metallic inclusions of the strip produced by BOF-LF-CC processes are mainly spherical Al_2O_3-CaO-MgO inclusions, while the non-metallic inclusions of the strip produced by DeS-DeC-RH-CC processes are bulk or irregularly shaped Al_2O_3 inclusion. Compared with BOF-LF-CC process, DeS-DeC-RH-CC process has less amount of micro-inclusions, but the size of inclusion is larger.

Key words cleanliness of steel, T[O] , nitrogen, inclusion

吹氩方式对钢包混匀效果的影响

郭晓晨[2] 唐海燕[1,2] 张江山[2] 王 杨[2] 刘成松[2] 李京社[2]

（1. 钢铁冶金新技术国家重点实验室，北京 100083;
2. 北京科技大学冶金与生态工程学院，北京 100083）

摘 要 针对某钢厂120t底吹氩钢包建立了相似比为1:3的物理模型，研究了双孔等流量吹氩时的不同吹氩孔布置和吹氩流量对钢包混匀效果的影响，对比探究了双孔一强一弱（流量不等）的吹氩模式下的混匀效果；同时，综合钢渣模拟试验，得出了适合该钢包的吹氩制度。实验结果表明：双孔相同吹氩流量下，较好的吹氩口布置为0.7 R处，夹角180°，吹气流量300~400L/min；双孔一强一弱吹氩模式下，吹氩口布置在0.64 R处，夹角137.3°，一孔吹氩流量为100 L/min，另一孔为500~700 L/min时能获得较好的混匀效果，其混匀时间比相同流量下的混匀时间缩短40%左右。该研究为工厂的工艺改进提供了实验依据。

关键词 钢包，水模拟，混匀时间，一强一弱吹氩方式

Effect of Blowing Argon Style on the Mixing Phenomena of Ladle

Guo Xiaochen[2], Tang Haiyan[1,2], Zhang Jiangshan[2], Wang Yang[2], Liu Chengsong[2], Li Jingshe[2]

(1. State Key Laboratory of Advanced Metallurgy, Beijing 100083,China;
2. School of Metallurgical and Ecological Engineering，University of Science and Technology Beijing，Beijing 100083，China)

Abstract A 120 t physical model of 1:3 similarity ratio was established based on a ladle in a steel mill. The effect of the bottom blowing position and gas flow rate on mixing was investigated. Meantime, the suitable process of bottom blowing argon for this ladle was summarized according to the experiments results of slag simulation.The results show that the optimized blowing gas position is at 0.7 R from the ladle centre, and the angle is 180° when the double holes were at the same flow of 300-400 L/min. While it got a better effect when the double holes were at different flow of 100 L/min and 500-700 L/min with 0.64 R, and the mixing time decreased by 40%. The study provided an experimental basis to the factory for its process improvement.

Key words ladle, water simulation, mixing time, a strong and a weak bottom blowing argon way

RH中夹杂物运动行为的数值模拟研究

王明辉 李京社 杨树峰 刘 威 高向宙 王 杨

（北京科技大学冶金与生态工程学院，北京 100083）

摘 要 钢液中夹杂物的去除对提高钢材质量具有重要意义，夹杂物在钢液中聚集长大、上浮至钢渣界面以及穿越界面

进入渣层受到很多因素的影响。为了检验钢渣界面作用对夹杂物去除的影响，本文以100t RH为研究对象，将PBM模型、夹杂物碰撞模型和夹杂物界面运动数学模型相结合，首次建立了夹杂物碰撞长大、上浮运动、穿越钢渣界面全过程的数值模拟模型。研究发现：(1) 钢渣界面作用对RH中夹杂物去除过程的影响比较明显，冶炼600s时，有无钢渣界面作用的夹杂物去除率差值为 2.56%；(2) 钢渣界面对小粒径夹杂物的阻碍作用大于大粒径；(3) 最佳驱动气体流量为60Nm³/h，该气体流量下钢渣界面夹杂物最少；(4) 对于粒径为16~161.27μm的夹杂物，钢渣界面对夹杂物的阻碍作用随着气体流量增加，不断减小，对于其他粒径的夹杂物，钢渣界面对夹杂物的阻碍作用随着气体流量增加几乎不变。

关键词　夹杂物，运动行为，数学模型，数值模拟

Numerical Simulation on Moving Behavior of Inclusion During Process of RH

Wang Minghui, Li Jingshe, Yang Shufeng, Liu Wei, Gao Xiangzhou, Wang Yang

(School of Metallurgical and Ecological Engineering, University of Science and Technology Beijing, Beijing 100083, China)

Abstract　Inclusion removal in molten steel is very significant to improving steel quality. The processes that the aggregation and growth of inclusion, floating to steel-slag interface and passing through steel-slag interface are influenced by many factors. For testing the influence of steel-slag interface on inclusion removal, in this paper, the mathematical model of inclusion movement at interface and PBM module were combined to build a new numerical model to simulate the processes including inclusion collision and growing up, floating, and passing through steel-slag interface in a 100t RH. The following conclusions are found by the simulation results: (1) Steel-slag interface has an dominant influence on the inclusion removal in RH. The difference of inclusion removal rate between with interface influence and without interface influence is 2.56% when the smelting time is 600 s; (2) Comparing with the inclusion of larger size, the inclusion of small size is easier to be impeded by steel-slag interface; (3) The optimal driving gas flow rate is 60 Nm³/h, which can result in the least inclusion in the molten steel; (4) For the inclusions between 16 μm and 161.27 μm, the impediment of interface decreases as the gas flow rate increases. For the other inclusions, the gas flow rate has little impact on the impediment of interface.

Key words　inclusion, moving behavior, mathematical model, numerical simulation

SWRH82B钢生产中氮含量的控制

康爱元　孙福振　梁新维　杨　坤　吴雨晨　刘兴元　肖元生

（承德钢铁公司长材事业二部，承德　067000）

摘　要　针对高碳钢线材SWRH82B氮含量超标的问题进行了分析，通过对炼钢各工序关键工艺控制点进行调查，提出相应的改进措施，从而使连铸坯质量得到明显改善。

关键词　SWRH82B，钢水增氮，控制

Control of Nitrogen Content in SWRH82B Steel Production

Kang Aiyuan, Sun Fuzhen, Liang Xinwei, Yang Kun, Wu Yuchen,
Liu Xingyuan, XiaoYuansheng

(Chengde Iron and Steel Company, Chengde 067000, China)

Abstract Aiming at the problem of excessive nitrogen content in high carbon steel SWRH82B wire rod were analyzed, through on the key process in each procedure steelmaking control survey, corresponding improvement measures were put forward, so that the quality of continuous casting billet are improved obviously.

Key words SWRH82B, steel increasing nitrogen, control

High-productivity Plants and Innovative Technologies Meet Market Demands and Future Trends

Christian Geerkens, Lothar Fischer, Dr. Jochen Wans, Ronald Wilmes

(SMS Siemag AG, Germany)

Abstract The milestones in continuous casting were set by the SMS group and its predecessor companies (Schloemann, Concast and Mannesmann Demag) and testify to 50 years of leadership in continuous casting. The thin, medium and thick slabs produced on continuous casters from SMS Siemag form the basis for superior steel products.

Today, approx. 95% of steel is produced by means of continuous casting. Being a frontrunner in this technology, SMS Siemag today offers the world's most extensive portfolio of continuous casting machines meeting the respective demands of the market as well as the customers' requirements. The great numbers of continuous casters which we have already successfully commissioned, or will be commissioning are eloquent proof of this successful and trustworthy cooperation with our customers.

Continuous new developments are characteristic for the history of SMS Siemag. Interdisciplinary cooperation within knowledge networks, with a bearing on the tasks in hand, leads to the necessary innovations in plant engineering, automation and process technology and thereby to the solutions demanded by our customers.

In this paper we are presenting several references which show that our customers manufacture high-quality products in their plants. We are also reporting some new developed technologies increasing production, reducing production costs, enhancing quality and saving energy.

Key words slab casting, assistance, compact strip production (CSP®), belt casting technology (BCT®), HD mold, ECO-mode, STEC-Roll®

LF 单联堆冷工艺生产 X70 管线钢研究

初仁生 刘金刚 李战军

（首钢技术研究院，北京 100043）

摘 要 本文炼钢工艺采用单联工艺的方法生产较薄规格的 X70 管线钢（厚度不大于 16.1mm），即转炉→LF 精炼→Ca 处理→连铸→堆冷，在保证产品质量的情况下取消 RH 精炼工艺，降低了成本，增加了生产效率。利用扫描电镜（SEM+EDS）等对某厂 LF 单联堆冷工艺生产的较薄规格的中厚板 X70 管线钢进行全面分析。从钢中气体含量控制和钢中非金属夹杂物控制两个角度分析发现，厚度不大于 16.1mm 的 X70 管线钢可以用 LF 单联工艺进行冶炼。研究表明采用 LF 单联堆冷工艺冶炼的 X70 管线钢成分满足要求，在堆冷条件下，钢中气体成分 H 含量和 N 含量满足要求，非金属夹杂物各类评级均在 1.5 以下，钢板探伤和钢坯内部质量满足要求。同时提出 X70 管线钢非金属夹杂物的控制策略为小尺寸、成分单一的高熔点类钙镁铝酸盐类非金属夹杂物。

关键词 非金属夹杂物，LF 单联，堆冷，控制策略

Study on the Control of Non-metallic Inclusions in Pipeline for Heavy and Medium Plate

Chu Rensheng, Liu Jingang, Li Zhanjun

(Shougang Research Institute of Technology, Beijing 100043, China)

Abstract In the current paper, the control of gas (H_2 and N_2) and inclusions was stuied in the X70 pipeline that the thickness is less than 16mm. The steelmaking process is Pretreatment of hot metal → LD → LF → Calcium treatment → CC → Heap cooling. The steelmaking process for no RH refining process can ensure the product quality situation, so it can reduce the cost and increase the efficiency. The size, composition and type of the non-metallic inclusions were analyzed by aspex explorer automated scanning electron microscope in X70 pipeline samples. The results show that the composition in the smelting process and the quality for the casting billet control satisfied the requirement of the steel. The control of gas content in the plate gets a high level by the technology of heap cooling. The inclusions control for the size, composition and type indicates that calcium aluminates with little magnesia in high melting point can meet the requirement of inclusions control at a excellent level. The ratings for various types of the non-metallic inclusions are 1.5 or less. The control strategy for the inclusions in X70 pipeline is small size, diffuse distribution and little amount of the deformation after rolling. On the contrary, the specific chemical composition of the inclusions is not important, single component in the inclusions is better.

Key words non-metallic inclusions, LF refining process, heap cooling, control strategy

一种智能钢水在线测氢系统

何涛焘　田　陆　文华北　易　兵　万乐仪　陈旭伟

（湖南镭目科技有限公司，湖南长沙　410100）

摘　要 设计了一种由定氢仪表、气动系统、测枪和定氢探头四部分组成的智能钢水在线测氢系统，可以直接测量钢包，中间包及钢锭模中钢液的氢含量。本系统采用平衡分压法实现测量,氮气作为载气在气动系统的推动下进入钢液形成气泡,并连续在气动系统回路中循环。钢液中的氢会持续向氮气气泡中扩散,当载气中的氢含量变化率接近零时测量终止。当仪表测得钢液的平衡氢分压后，用热导传感器检测出混合气体中氢气的含量，然后根据内嵌的计算模型得出钢水的氢含量。现场实验数据证明：该系统氢含量测量范围 0.5～15ppm，测量时间<2min，测量误差≤±0.2ppm，测成率≥95%，具有稳定性高、扩展性强、内嵌模型可调整等优点。本项技术无论从分析精度、重现性还是测成率方面考察均已达到国外同类技术水平。本系统可以作为冶金行业对钢的质量进行实时跟踪、控制和判断的关键依据之一。

关键词 氢分压，在线检测，测氢系统，钢液

The System of Intelligent Online Hydrogen Content Measuring System in Molten Steel

He Taotao, Tian Lu, Wen Huabei, Yi bing, Wan Leyi, Chen Xuwei

(Hunan Ramon Tech. Co., Ltd., Changsha Hunan 410100, China)

Abstract An intelligent online hydrogen content measuring system in molten steel is designed, which is composed of hydrogen measuring instrument, pneumatic system, lance and probe. It can measure the hydrogen content in steel ladle, tundish and ingots directly. A nitrogen carrier gas is injected in the steel melt and recirculated between melt and pneumatic system. It will pick up hydrogen during its passage in the melt. The measurement is stopped when the equilibrium is reached between the hydrogen dissolved in the melt and the hydrogen in the carrier gas. The Ramon system measures the hydrogen content by means of a thermal conductivity detector. It measures the hydrogen concentration in the nitrogen carrier gas. The final hydrogen content is then calculated based on the embedded computing model. The results of experiment show that the measuring range of hydrogen from 0.5 to 15ppm, the measurement time≤2min, the measurement error≤±0.2ppm and the measurement success rate ≥95%. Compared with the traditional microcontroller, this device has higher accuracy, scalability, and adjustable embedded model. This technology has reached the same level of foreign technology. This system can be used as a key standard to the steel quality tracking, control and judgment in metallurgical industry.

Key words hydrogen partial pressure, online detection, hydrogen content measuring system, molten steel

无取向硅钢钙处理改性夹杂物的热力学研究

赵 勇 孙彦辉

（北京科技大学钢铁共性技术协同创新中心，北京 100083）

摘 要 本文对无取向硅钢的钙处理工艺进行了实验室研究，并从热力学角度分析了钙处理对夹杂物的变性程度。实验结果表明，钙处理前夹杂物主要以 Al_2O_3 系夹杂物为主，钙处理后夹杂物向钙铝酸盐系夹杂转变，并有少量 CaS 生成。Al、Ca 脱氧平衡计算表明，理论上生成的钙铝酸盐系夹杂位于 $C_{12}A_7$ 附近的纯液态夹杂区域，但通过 Al-Ca 平衡反应计算得到钢液中生成的钙铝酸盐系夹杂物位于 CA 与 CA_2 中间的固液共存区域。为了尽可能减少 CaS 夹杂物的生成，同时使 Al_2O_3 系夹杂变性成低熔点钙铝酸盐系夹杂，钢液中理想的钙含量应控制在 2×10^{-5}~3×10^{-5} 之间。

关键词 无取向硅钢，钙处理，夹杂物，热力学

Thermodynamics of Inclusion Modification in Calcium Treated Non-oriented Silicon Steel

Zhao Yong, Sun Yanhui

(Collaborative Innovation Center of Steel Technology, University of Science and Technology Beijing, Beijing 100083, China)

Abstract A laboratory study on calcium treatment about non-oriented silicon steel was carried out, aiming at analyzing the degree of modification about inclusions by thermodynamic calculations. The experimental results indicated that Al_2O_3 was the main inclusion before calcium treatment, and this type of inclusions turned into calcium aluminates after calcium treatment with a small amount of CaS precipitating. Furthermore, deoxidation equilibrium calculations of aluminum and calcium revealed that calcium aluminates generated in theory were located in the region of pure liquid inclusion near $C_{12}A_7$, but calcium aluminates generated in molten steel were located in the liquid-solid coexistent region between CA and CA_2 based on the calculation of Al-Ca equilibrium reaction. The ideal calcium content in molten steel should be controlled in the range of 2×10^{-5} to 3×10^{-5} in order to decrease the formation of CaS and transform Al_2O_3 system inclusions into calcium aluminates with low-melting temperature.

Key words　non-oriented silicon steel, calcium treatment, inclusion, thermodynamic

底吹氩钢包钢渣界面行为及夹杂物去除水模型研究

程普红[1]　马国军[1,2,3]　胡黎宁[2,3]　薛正良[1]　赵昊乾[2,3]　黄源升[1]

(1. 武汉科技大学钢铁冶金及资源利用省部共建教育部重点实验室，湖北武汉　430081；
2. 邢台钢铁有限责任公司，河北邢台　054027；
3. 河北省线材工程技术研究中心，河北邢台　054027)

摘　要　本文根据相似原理，采用物理模拟方法研究了在线钢包吹氩时渣-钢界面卷混行为及夹杂物的去除行为。研究了吹气量、渣层厚度和透气砖堵塞对渣-钢卷混的影响及在线底吹氩对夹杂物去除的影响。研究结果表明：在相同工艺条件下，随底吹气量增加钢液面裸露面积增加；随着钢包内钢液量减小吹开裸露面积逐渐减小。在恒定吹气量时，裸露面积比随着渣层厚度的增加而逐渐减小。在吹气量较小时，堵塞50%透气砖，顶部裸露面积和无堵塞相近；但随着底吹气量的增加，顶部裸露面积要稍大于无堵塞情况下钢液面。随着浇铸的进行，油滴脱落临界卷渣气量逐渐减少。采用在线底吹氩可以降低夹杂物进入中间包或铸坯的比例。

关键词　连铸，钢包，夹杂物，卷渣，水模型

Water Model Study on the Behavior of Steel-slag Interface and Inclusions Removal in Ar Bottom Blowing Ladle

Cheng Puhong[1], Ma Guojun[1,2,3], Hu Lining[2,3], Xue Zhengliang[1], Zhao Haoqian[2,3], Huang Yuansheng[1]

(1. Key Laboratory for Ferrous Metallurgy and Resources Utilization of Ministry of Education, Wuhan University of Science and Technology, Wuhan Hubei 430081, China; 2. Xingtai Iron & steel Co., Ltd., Xingtai Hebei 054027, China; 3. Hebei Engineering Research Center for Wire Rod, Xingtai Hebei 054027, China)

Abstract　According to similar principles, slag entrapment behavior of steel-slag interface and inclusions removal in an online bottom argon injection ladle slag was studied by physical simulation method in this paper. The results show that the exposed surface area of steel increases with increasing of the bottom-blown gas flow rate under the same conditions, While the open exposed area of steel decreases with decreasing of the amount of molten steel. At a constant gas flow rate, the exposed area ratio reduced with increasing of slag layer thickness. When it is injected with a lower gas rate, it is no significant difference of the top exposed area between 50% of purging blocking and the non-clogging, while the top exposed area of steel with clogging is slightly larger than that without clogging. The critical gas volumes of oil droplets shedding decrease with proceeding of casting process. Inclusions removal will increase when the process of on-line bottom Ar blowing of ladle is employed.

Keywords　Continuous casting; ladle; inclusions; slag entrapment; water model

RH 循环过程中高熔点钙铝酸盐夹杂物的行为分析

马焕珣[1] 黄福祥[1] 王新华[1] 秦颐鸣[1] 季晨曦[2] 潘宏伟[2]

（1. 北京科技大学冶金与生态工程学院，北京 100083；
2. 首钢技术研究院，北京 100043）

摘 要 以 RH 精炼前进行钙处理的某低碳铝镇静钢为试验钢种，利用 ASPEX 扫描电镜(SEM+EDS)对 RH 精炼处理过程中 1μm 以上的高熔点 $CaO-Al_2O_3$ 夹杂物的形貌、成分、数量和尺寸进行系统研究。结果表明，钙处理工艺将钢中 Al_2O_3 夹杂物有效的变性为含有少量 MgO 的高熔点的 $CaO-Al_2O_3$ 类夹杂物，经过 RH 真空循环，钢中总氧从 RH 进站的 98.7ppm 降低至 32.5ppm，钢中各尺寸范围的夹杂物均能得到明显去除，1~5μm、5~15μm 及 15μm 以上夹杂物的去除率分别达到 70.7%、77.6%和 100%。1~5μm 小尺寸夹杂物和 15μm 以上的大尺寸夹杂物随着 RH 真空循环的进行其数量密度逐渐降低，而 5~15μm 的夹杂物呈现出先降低后增多再降低的变化规律。高熔点 $CaO-Al_2O_3$ 夹杂物的成分和形貌在 RH 真空循环过程中未发生明显变化。

关键词 RH 处理，夹杂物，去除，洁净度

Behavior Analysis of High Melting Point $CaO-Al_2O_3$ Inclusions during RH Treatment Process

Ma Huanxun[1], Huang Fuxiang[1], Wang Xinhua[1], Qin Yiming[1],Ji Chenxi[2], Pan Hongwei[2]

(1. School of Metallurgical and Ecological Engineering, University of Science and Technology Beijing, Beijing 100083, China;
2. Shougang Research Institute of Technology, Shougang Group, Beijing 100043, China)

Abstract This paper focuses on low carbon aluminum killed (LCAK) steel treated by calcium before RH refining process. The morphology, composition, and size of high melting point $CaO-Al_2O_3$ inclusions larger than 1μm were investigated by automated feature analysis (AFA) equipped on ASPEXPSEM Explorer. Results suggest that: The Al_2O_3 inclusions are modified to high melting point $CaO-Al_2O_3$ inclusions containing a small amount ofMgO. Total oxygen value reduces from 98.7 ppm to 32.5 ppm through RH refining process. The inclusions in different sizes are removed efficiently, resulting in the removal rate of inclusions in size of 1~5μm、5~15μm and larger than 15μm reaching to 70.7%、77.6% and 100%. The smaller inclusions in size of 1~5μm and the inclusions larger than 15μm reduce continuously during RH refining process, but the inclusions in size of 5~15μm present decreasing-increasing-decreasing tendency. The composition and morphology of high melting point $CaO-Al_2O_3$ inclusionspresent no obvious change during RH refining process.

Key words RH treatment, inclusions, removal, cleanliness

LF 吹氮合金化工艺试验

刘晓峰 黄志强

（重庆钢铁股份有限公司二炼钢厂，重庆 401258）

摘　要　在理论分析LF吹氮合金化钢水增氮热力学和动力学的基础上，介绍了重钢80t LF吹氮合金化工艺试验，提出了合理控制钢水成分（硫、氧）、氮气流量、吹氮时间等工艺控制措施，指导下一步生产实践。

关键词　LF，氮气，合金化，试验

Blowing Nitrogen Alloying Process Test of LF

Liu Xiaofeng, Huang Zhiqiang

(Second Steel Plant of Chongqing Iron and Steel Company Limited,Chongqing 401258, China)

Abstract　On the basis of theoretical analysis LF blowing nitrogen-alloyed steel by thermodynamics and kinetics of nitrogen introduced on CISC 80 t LF blowing nitrogen alloying process test, proposed the reasonable control of steel components (sulfur, oxygen), nitrogen flow, blowing nitrogen time process control measures, to guide further production practices.

Key words　ladle furnace，nitrogen，alloying，test

铸余渣在炉外精炼中的应用实践

胡昌志　牟炳政　陈露涛　马立华

（重庆钢铁股份有限公司）

摘　要　通过对LF热态钢渣渣系和硫容量的研究，重庆钢铁股份有限公司一炼钢厂采用了相应的技术与措施实现了不同工艺路线铸余渣热态回收至LF精炼炉的循环再利用，不但减少了对环境的污染，并取得了较好的经济效益。

关键词　热态钢渣，硫容量，循环利用，价值

Application of Casting Residue in the Refining

Hu Changzhi, Mu Bingzheng, Cheng Lutao, Ma Lihua

(Chongqing Iron and Limited by Share Ltd.)

Abstract　Through the research of LF hot steel slag slag system and sulfur capacity, Chongqing iron and steel Limited by Share Ltd first steelmaking plant adopts the technology of the corresponding measures to realize the different processes of casting residue thermal recovery to LF refining furnace for recycling, not only reduce the pollution to the environment, and achieved better economic benefits.

Key words　hot steel slag, sulfur capacity, re-utillzation, value

含铌钢增氮原因及控制措施探讨

王鹏飞　朱国强　刘振中　高洪涛　王小善

（鞍钢股份公司炼钢总厂，辽宁鞍山　114021）

摘　要　介绍了鞍钢炼钢总厂四分厂采用“BOF-LF-CC”工艺生产含铌钢氮含量控制情况，分析了转炉和 LF 炉处理过程中钢水增氮的主要影响因素，其中转炉补吹、出钢合金化控制、LF 炉埋弧效果和吹氩搅拌对钢水增氮有较大影响。提出了相应的控制措施，通过优化脱硫目标、提高转炉合金电振速度、加强 LF 炉埋弧效果、优化 LF 炉吹氩工艺等措施，很好地控制了钢水的增氮，使 LF 炉搬出氮含量控制在 0.0045%以下的比率达到 92%以上。

关键词　含铌钢，增氮，转炉炼钢，LF 精炼

Analysis of Mechanism of Nitrogen Increasing in Nb Micro-alloyed Steel and Control Measures

Wang Pengfei, Zhu Guoqiang, Liu Zhenzhong, Gao Hongtao, Wang Xiaoshan

(Steelmaking Plant, Ansteel Co., Ltd., Anshan 114021, China)

Abstract　In this paper, nitrogen increasing phenomenon in niobium micro-alloyed steel produced in Ansteel steelmaking plant adopting BOF-LF-CC process was introduced. The main influence factors causing nitrogen increase in BOF and LF refining process were analyzed. The results indicated that reblowing, alloying during steel tapping, effect of submerged arc and argon stirring had a great influence on nitrogen increasing. Based on above conclusions, corresponding control measures such as optimizing the desulfurization goals, increasing alloy feeding speed of electric vibrators of BOF, improving the effect of submerged arc and adjusting argon blowing argon stirring during LF refining process were used for effectively controlling nitrogen increasing in molten steel. The ratio of heats which nitrogen content was lower than 0.0045% increased above 92%.

Key words　Nb micro-alloying steel, nitrogen increasing, BOF steelmaking, LF refining

应用二级模型冶炼低磷钢的生产实践

雷爱敏　李吉伟　连　庆　滕广义　李　涛

（河北钢铁集团承钢公司，河北承德　067002）

摘　要　本文介绍了承钢长材事业二部在半钢炼钢的条件下，应用二级模型冶炼低磷钢的生产实践。该厂采用的是恒压变枪操作，通过应用二级模型，对枪位控制、加料时机、物料加入量、过程温度控制等重要参数进行优化，以此达到控制终点熔池[C]在 0.10%~16%，［P］在 0.010%~0.020%，完成冶炼低磷钢的目的。

关键词　二级模型，低磷钢

Production Practice of Two-Level Model of Smelting Low Phosphorus Steel

Lei Aimin, Li Jiwei, Lian Qing, Teng Guangyi, Li Tao

(Chengde Iron & Steel Croup Co., Ltd., of Hebei Steel Company, Chengde Hebei 067002, China)

Abstract　This article introduced the production practice of smelting low phosphor steels by using two-level model in the 3rd. Vanadium-extracting and Steel-rolling Plant of Chenggang, under the condition of semi-steel steelmaking. With the process of changing lance with constant pressure, the plant optimized such important parameters of lance position control, feeding time, adding quantity of material, temperature control of process and so on by using two-level model , to control [C] in 0.10%~16% and [P] in 0.010%~0.020% at endpoint, and finally achieved smelting low phosphor steels.

Key words two-level model, low phosphor steel

铝镇静钢深脱硫工艺研究

张新法　钱润锋　李亚厚

（河北钢铁集团承钢公司长材事业二部，承德　067002）

摘　要　本文介绍了铝镇静钢深脱硫工艺的生产实践。通过对吹炼终点和出钢过程操作两个环节进行控制，保证精炼进站钢水合适的温度、成分及稳定的钢、渣含氧量；同时优化钢水脱硫的热力学和动力学条件，即优化钢水渣量、顶渣碱度、顶渣还原性、顶渣流动性、钢水温度、钢水吹氩的搅拌强度等因素，把钢水中的硫含量降低到0.005%以下，使铸坯质量得到改善。

关键词　转炉，LF 炉，深脱硫，造渣

Deep Desulphurization Process of Aluminum Killed Steel

Zhang Xinfa, Qian Runfeng, Li Yahou

(Hebei Iron and Steel Group Company Chenggang Long Products Business Department, Chengde 067002, China)

Abstract The production practice of deep desulfurization foraluminum killed steelprocess is introduced in the study. Through controlling of two operational units (blowing final point? and Tapping process) to ensure the appropriate temperature and contents of molten steel, also stabilize the oxygen content of steel and slag; meanwhile optimize the conditions of thermodynamics and kinetics for desulfurization of molten steel, including optimization of the slag content for molten steel, Top slag alkalinity，Top slag reducibility，flowability, temperature and stirring intensity and so on, lower the sulphur content in molten steel to less than 0.005%, then the quality of casting blank is improved.

Key words converter, ladle furnace, deep desulphurization, slagging

硅铁合金洁净度研究

杜嘉庆　张海杰　张立峰　任　英

（北京科技大学冶金与生态工程学院，北京　100083）

摘　要　硅铁合金在钢铁厂被广泛用于脱氧和合金化。然而，硅铁合金中的杂质元素对钢液纯净度及最终产品质量都会产生一定的影响。本文分析了硅铁合金中主要杂质元素 Al，Ca 和 O 的质量百分含量，分别是 1.07%，0.79% 和 0.16%。用扫描电镜（SEM）观察了合金中相的形貌，并用能谱仪（EDS）分析了杂质元素的含量以及存在相。研究发现，硅铁合金中硅主要以两种形式存在，高硅相和硅铁相。硅铁合金中的杂质主要以金属间相的形式存，氧

化物形式的杂质很少。在硅铁合金中发现了 Si-Ca、Al-Si-Ca、Al-Si-Fe-Ca、Fe-Si-Ti 等杂质相，这些杂质相多以混合相的形式分布在相界面处，少数分布在基体相中。

关键词 硅铁合金，杂质，纯净度

Study of Cleanliness of Ferrosilicon

Du Jiaqing, Zhang Haijie, Zhang Lifeng, Ren Ying

(School of Metallurgical and Ecological Engineering, University of Science and Technology Beijing, Beijing 100083, China)

Abstract Ferrosilicon was widely used in the deoxidation and alloying. However, the impurities in the ferrosilicon have much influence on the cleanliness of the molten steel, as well as the final products. The main impurities elements of the ferrosilicon from one industry were analyzed, contents of Al, Ca and O were 1.07%, 0.79% and 0.16% respectively. Scanning electron microscope (SEM) equipped with Energy Dispersive Spectroscopy system (EDS) were used to evaluate impurity phases in the alloy. It was found that two main phases were existed in the matrix which were high silicon phase and iron-silicon phase. The impurities were present primarily in intermetallic phases in the ferrosilicon. Oxide inclusions could hardly be detected. Phases of Si-Ca, Al-Si-Ca, Al-Si-Fe-Ca and Fe-Si-Ti were found in the phase interface with a mixed phases, and few of them were found in the matrix phases.

Key words ferrosilicon, impurity, cleanness

2.3 连铸工艺/ Continuous Casting Technology

Effect of EMBr on Multiphase Flow and Bubble Entrapment in Steel Continuous Casting

Jin Kai[1], Thomas Brian G.[1], Ruan Xiaoming[2]

(1. Dept. of Mechanical Science and Engineering, University of Illinois at Urbana–Champaign, Urbana, IL, USA; 2. Steelmaking Research Department, Research Institute, Baoshan Iron & Steel Co., Ltd.,Shanghai, China)

Abstract In steel continuous casting, double-ruler electromagnetic braking (EMBr) is often applied to control the flow pattern in the mold. In addition, argon gas is usually injected to prevent clogging, but the bubbles also affect the flow pattern, and may become entrapped to form defects in the final product. To investigate the combined behaviors, plant measurements were conducted and a computational model was applied to simulate turbulent flow of the molten steel and the transport and capture of argon gas bubbles into the solidifying shell in a continuous slab caster, including EMBr. An Eulerian k-ε model of the steel flow was two-way coupled with a Lagrangian model of the large bubbles using a Discrete Random Walk method to simulate their turbulent dispersion. The top surface velocities agreed well with nailboard measurements, and indicated strong cross flow caused by biased flow of Argas due to the slide-gate orientation. Then, the trajectories and capture of over two million bubbles (25 μm to 5 mm diameter range) were simulated using an advanced capture criterion. The number, locations, and sizes of captured bubbles agreed well with measurements, especially for larger bubbles. The relative capture fraction of 0.3% was close to the measured 0.2% for 1mm bubbles, and occurred mainly near the top surface. About 85% of smaller bubbles were captured, mostly deeper down in the caster. EMBr produced similar behavior with slightly lower capture rates.

Key words continuous casting, bubble entrapment, computational model, model validation, EMBr

Modeling of Bubbly Flow in Continuous Casting Mold Incorporating Population Balance Approach

Liu Zhongqiu, Li Linmin, Qi Fengsheng, Li Baokuan

(School of Materials and Metallurgy, Northeastern University, Shenyang 110819, China)

Abstract A generalized Multiple Size Group (MUSIG) model based on the Eulerian-Eulerian approach has been developed to describe the polydispersed bubbly flow inside the continuous-casting mold. The Eulerian-Eulerian approach is used to describe the equations of motion of the two-phase flow. Sato and Sekiguchi model is used to account for the bubble induced turbulence. Luo and Svendsen model and Prince and Blanch model are used to describe the bubbles breakup and coalescence behavior respectively. A laboratory scale mold has been simulated using four different turbulence closure models; with the purpose of critically comparing their predictions of bubble Sauter mean diameter distribution with previous experimental data. Furthermore, the influences of all the interfacial momentum transfer terms including drag force, lift force, virtual mass force, wall lubrication force, and turbulent dispersion force are investigated. The appropriate drag force coefficient, lift force coefficient, virtual mass force coefficient, turbulent dispersion force coefficient are chosen in accordance with measurements of water model experiments. Finally, the MUSIG model is then used to estimate the argon bubble diameter in the molten steel of the mold. The argon bubble Sauter mean diameter generated in molten steel is predicted to be larger than air bubbles in water for the similar conditions.

Key words bubble size distribution, multiple-size-group model, eulerian-eulerian approach, continuous-casting mold

Numerical Modelling of Continuous Casting as a Real Tool for Process Improvement

Pooria Nazem Jalali[1], Pavel Ramirez E. Lopez[1], Thomas Jonsson[1], Christer Nilsson[2]

(1. Swerea MEFOS AB, Aronstorpsvägen 1, SE 97 437 Luleå, SWEDEN,

Materials Science and Engineering Department, KTH University, Brinellvägen 23, Stockholm, SWEDEN;

2. Senior Specialist R & D, Metallurgical Development, SSAB EMEA, 971 88 Luleå, SWEDEN,

Corresponding author: pavel.ramirez.lopez@swerea.se)

Abstract An advanced numerical model, developed by Swerea MEFOS AB, has been used to diagnose and optimize steel casting practices for peritectic steels for a Scandinavian steel producer. The model is based on the commercial code ANSYS-FLUENT 15.0 which solves the Navier-Stokes equations by using an interface tracking technique known as Volume of Fluid (VOF) for the multiple phase system steel/slag. As a result, the model predicts metal flow and slag infiltration as well as their influence on heat flux and solidification under transient conditions. Recent improvements to the model include separation between mould powder and slag film and considering the effect of crystallization on interfacial resistance for a Peritectic mould powder through a User Defined Function (UDF). The casting parameters analysed consist ofcasting speed interlocked with oscillation settings, Submerged Entry Nozzle (SEN) immersion depth and Argon injection flow rate. These practises were optimised by performing parametric studies to evaluate shell growth, lubrication depth, cooling channel heat flux, etc. Application of the model allows the prediction of trends and opportunities for further improvement in the form of guidelines for the process and enhanced operational windows. These have been tested at industrial level showing improvements on surface quality and process stability.

Key words continuous casting, numerical modelling, mould oscillation, immersion depth, argon injection

Mathematical Investigation of Proper Soft Reduction Parameters for CC Bloom of Railway Steel

Liu Ke, Sun Qisong

(Special Steel Technology Dep., Shougang Research Institute of Technology, Beijing 100043, China)

Abstract Soft reduction has been proved to be the most promising way to minimize centre segregation together with the strand centre porosity elimination, through which the negative pressure driven residual melt flow could be balanced by a reasonable reduction amount to compensate the centre solidification shrinkage in the crater mushy zone. The optimum soft reduction amount to bloom casting for given heavy railway steel has been studied in the paper, in which a solidification analysis model and a thermal-mechanical coupled FEM model were developed to determine the reduction amount based on the local shrinkage in the final solidification stage, along with the consideration of the deformation state. The volume shrinkage in the centre interdendritic region has been quantitatively calculated through the evaluation to the changing melt density. The squeezing effect of the soft reduction on the centre mushy zone at various solid fractions was revealed quantitatively as well. The results shown that the theoretical soft reduction amount, from the point of view of the volume shrinkage compensation, should decrease from 12.2 mm to 4.0 mm corresponding to the solid fraction from 0.3 to 0.9. A new definition for the calculation of the internal soft reduction efficiency was proposed and compared with the conventional one. The nomogram for the determination of optimum soft reduction has been presented for the bloom casting at various centre solid fraction upon the implementation of soft reduction. Accordingly, an optimized soft reduction practice was determined for quality high speed railway steel production with clearly improved strand soundness.

Key words continuous casting bloom, soft reduction amount, internal reduction efficiency, coupled FEM model

Nitrogen Control in Continuous Casting for High Strength Microalloyed Steels

Guzela Danilo Di Napoli[1], Vicenzetto Marcus A. Prates[2], Kao Chian Tou[3]

(1. Metallurgical Engineer, Consultant, CBMM, Brazil;

2. Metallurgical Engineer, Market Development Manager, CBMM, Brazil;

3. Metallurgical Engineer, Consultant, CBMM, Brazil)

Abstract Prerequisites for the production of HSLA and API steel grades of the highest standard include cleanliness, internal soundness and defect-free slab surfaces. In this context, this paper presents operating practices necessary to obtain good surface quality in slabs of high strength microalloyed steels. In these steels, nitrogen control throughout the steelmaking process from hot metal to steel during continuous casting is the key to achieving a high quality, defect-free slab.

Key words continuous casting, microalloyed steels, transverse cracks, nitrogen

连铸大方坯凝固末端重压下技术及其应用

朱苗勇　祭　程

（东北大学冶金学院，辽宁沈阳　110819）

摘　要　凝固末端重压下技术是实现均质化、高致密度大断面连铸坯生产的有效手段。本文介绍了凝固末端在线检测、最小压下量参数理论计算、压下区间参数理论计算、凝固末端强电磁搅拌与渐变曲率凸型辊等大方坯重压下关键工艺与装备控制技术。现场应用表明，采用凝固末端重压下技术后，生产铸坯及产品轧材的中心偏析与疏松得到有效控制，铸坯致密度得到显著提升，内部质量改善明显。

关键词　重压下，大方坯连铸，凝固末端，中心疏松，中心偏析

Heavy Reduction Technology and Its Application for Bloom Continuous Casting Process

Zhu Miaoyong, Ji Cheng

(School Metallurgy, Northeastern University, Shenyang Liaoning 110819, China)

Abstract　Heavy reduction of continuous casting strand is an effective method to improve the homogeneity of as-cast with large section size. In the present work, the key technologies of strand heavy reduction including online detection of solidification end, minimum reduction amount and reduction zone calculation method, strong electromagnetic stirring and convex roller with curving surface were introduced. The results of application of heavy reduction to bloom continuous casting show that the center segregation and porosity of bloom and rolled bar were reduced effectively, and the relative density of bloom were improved significantly.

Key words　Heavy reduction, bloom continuous casting, solidification end, center segregation, center porosity

低氟含钛结晶器保护渣结晶及传热性能研究

文光华[1]　高金星[1]　陈力源[1]　杨昌霖[1]　唐　萍[1]　梅　峰[2]　施春月[2]

（1. 重庆大学，重庆　400044；2. 上海宝山钢铁股份有限公司，上海　201900）

摘　要　为了减少连铸保护渣中的氟造成的危害，本研究中利用 TiO_2 替代连铸保护渣中的部分氟化物，开发了一个含氟小于 4%的低氟含钛连铸保护渣（$R_{1.35}$），与工厂现用的中碳包晶钢板坯连铸用高氟连铸保护渣（3D）的结晶与传热性能进行对比研究，并进行了工业试验。研究结果表明：(1) 通过 TTT 曲线分析发现，3D 与 $R_{1.35}$ 的析晶温度区间和结晶孕育时间相近，但 3D 析出的矿相为 $Ca_2(SiO_4)$和 $Ca_4Si_2O_7F_2$，而 $R_{1.35}$ 析出矿相则为 $Ca_2(SiO_4)$ 和 $CaTiO_3$；（2）通过对传热性能的研究发现，$R_{1.35}$ 渣膜中析出的 $Ca_2(SiO_4)$ 和 $CaTiO_3$ 可作为析晶替代物质，能够满足中碳包晶钢板坯连铸对连铸保护渣控热的需求；（3）工业试验过程中，低氟含钛连铸保护渣在结晶器内拔热量正常，保护渣的消耗量合理，无黏结报警现象发生，铸坯表面质量达到现有高氟保护渣 3D 的水平。

关键词 低氟含钛保护渣，传热，中碳钢，板坯裂纹

Investigation on Crystallization and Heat Transfer of Low-Fluoride and Titanium-Bearing Mold Fluxes

Wen Guanghua[1], Gao Jinxing[1], Chen Liyuan[1], Yang Changlin[1], Tang Ping[1], Mei Feng[2], Shi Chunyue[2]

(1. College of Materials Science and Engineering, Chongqing University, Chongqing 400044, China;
2. Baoshan Iron and Steel Company Limited, Shanghai 201900, China)

Abstract In order to reduce the hazard effects of the fluorine in mold fluxes, a low- fluoride (less than 4%) and titanium-bearing mold fluxes ($R_{1.35}$) which replaced a part of fluorine with titanium oxide was prepared in this research, and the industrial tests were done to compare the crystallization and the heat transfer properties of mold fluxes with a high- fluoride-bearing mold fluxes (3D) for the medium carbon steel slab casting. The results show that: (1) From the temperature time transformation curves (TTT curves) of mold fluxes, the crystallization temperature zone and the incubation time of $R_{1.35}$ mold fluxes were similar with that of 3D mold fluxes. The precipitations were $Ca_2(SiO_4)$ and $Ca_4Si_2O_7F_2$ for 3D mold fluxes, but that were $Ca_2(SiO_4)$ and $CaTiO_3$ for $R_{1.35}$ mold fluxes. (2) From the heat transfer properties, the precipitations of $R_{1.35}$ mold flux films ($Ca_2(SiO_4)$ and $CaTiO_3$) should be able to meet the demand of heat transfer for the medium carbon steel slab casting. (3) During the industrial trials, $R_{1.35}$ mold fluxes have a normal heat extraction behavior, appropriate slag consumption and no sticker alarm in the mold. The slab surface quality cast by $R_{1.35}$ is comparable to cast by 3D mold fluxes.

Key words low-fluoride and titanium-bearing mold fluxes, heat transfer, medium carbon steel, slab crack

Vertical Electromagnetic Brake Technique and Fluid Flow in Continuous Casting Mold

Wang Engang, Li Zhuang, Li Fei

(Key Laboratory of Electromagnetic Processing of Materials (Ministry of Education)
Northeastern University, Shenyang 110004, China)

Abstract This paper proposed a new type electromagnetic brake technique that called vertical electromagnetic brake (V-EMBr), which applied to the slab continuous casting process. A three-dimensional mathematical model is developed to describe molten steel flow and the interfacial behavior of molten steel/liquid slag in the mold with vertical electromagnetic brake, considering the coupled effects of electromagnetic field and flow field. The effect of magnetic flux density, casting speed and the submergence depth of the SEN on flow field in mold with V-EMBr are investigated by numerical simulation method. Meanwhile, the effect of current density on the interfacial behavior of molten steel/liquid slag is also investigated with Volume of fluid (VOF) model. The results show that the application of V-EMBr can play an effect control to the molten steel in the mold, and significantly reduce the flow intensity in upper recirculation zone, decrease the impact strength of jet flow on the narrow face and the flow velocity of molten steel in the free surface of mold. At the same time, the height of molten steel/liquid slag interface can be obviously depressed by V-EMBr with increasing of current density, which is beneficial to reduce the possibility of mold flux entrapment.

Key words fluid flow, V-EMBr, continuous casting, mold, level fluctuation

基于"受力检测"的板坯连铸动态辊缝控制系统的开发及应用

胡志刚[1] 李 杰[1] 赵国昌[2] 郑京辉[2] 赵英利[1]

（1. 河北钢铁技术研究总院，河北石家庄 050000；2. 河北钢铁集团舞钢公司，河南舞阳 462500）

摘 要 本文介绍了自主开发的基于"受力检测"的板坯连铸动态辊缝控制系统。该系统包括了扇形段在线标定模块，关键工艺参数监控与报警模块，辊缝实时矫正模块和动态轻压下模块。新开发的扇形段在线标定模块充分考虑了扇形段装配的机械间隙和拉杆在实际生产中因受力变形造成的位移传感器漂移等问题，改进了标定方法，使标定效率提高了约62.5%，连铸机辊缝控制精度提高了约80%。开发的动态轻压下模块基于"压力反馈"原理，并结合热力学计算模型来判断凝固末端位置。开发的辊缝实时矫正模块根据实际受力情况对辊缝值进行实时修正，使显示值与实际值更为接近。辊缝实时矫正模块和动态轻压下模块的结合使用，有利于提高压下效率，改善了板坯的内部质量，其内部中心偏析平均等级由原先的B1.5提高为C2.0。

关键词 辊缝控制，动态轻压下，中心偏析

The Development and Application of Slab Dynamic Roll Gap Control System based on "Force Detection"

Hu Zhigang[1], Li Jie[1], Zhao Guochang[2], Zheng Jinghui[2], Zhao Yingli[1]

(1. Technology Research Institute of HBIS, Shijiazhuang 050000, China;
2. Wugang Company of HBIS, Wugang, 462500, China)

Abstract This article describes the self-developed slab dynamic roll gap control system based on the "force detection". The system includes segment calibration module, critical process parameters monitoring and alarm module, the roll gap real-time correction module and dynamic soft reduction module. The new segment calibration module considers the mechanical gap and the tie rod elastic deformation in the process of production, and improves the calibration method. After application, the calibration efficiency is improved by 62.5%, and the roll gap control accuracy is improved by 80%. Dynamic soft reduction module detects solidification end using the "force feedback" and thermodynamic calculation model. The roll gap real-time correction module can correct the roll gap according to the actual stress situation, making the display value be close to the actual value. The roll gap real-time correction of module and dynamic soft reduction were useful to improving the efficiency of soft reduction, and the center segregation average level of slab were improved from B1.5 to C2.0.

Key words gap Control, dynamic soft reduction, center segregation

倒角结晶器在武钢板坯连铸中的应用

李俊伟 杨新泉 唐树平 张 余 黄 君 徐之浩

（武汉钢铁股份有限公司炼钢总厂，湖北武汉 430080）

摘　要　微合金化钢板坯角横裂和超低碳钢、低碳钢热轧板卷的边部翘皮是直弧形宽板坯连铸机的固有问题，生产含妮、钒、钦、铝的亚包晶低合金钢更是难以避免其出现。为了消除连铸板坯的角部裂纹和翘皮缺陷，提出了通过倒角结晶器减少角部横裂纹，以改变连铸板坯角部传热，提高矫直时连铸坯的角部温度。采用了板坯连铸倒角结晶器技术进行了大量的试验，分析并解决了生产过程中存在的角部渗钢、纵裂和边裂等问题，实现了该技术的稳定连续生产，并取了较大的经济效益。

关键词　连铸板坯，倒角结晶器，角部裂纹，翘皮

Applied of Chamfered Mold on Slab Continuous Casting Production in WISCO

Li Junwei, Yang Xinquan, Tang Shuping,
Zhang Yu, Huang Jun, Xu Zhihao
(WISCO, Wuhan Hubei 430080, China)

Abstract　Microalloyed steel slab corner transverse crack and ultra low carbon steel, mild steel edge of hot rolled plate roll become warped leather is the inherent problems straight curved wide slab continuous casting machine, production conatining niobium, vanadium, chin, aluminum and peritectic low alloy steel is hard to avoid appearing.In order to eliminate the continuous casting slab cracks and side sliver defect, a design of one kind of mould with chamfered corners was programs to modify the transfer the conditions at slab corners and so to elevate slab corner temperature unbending in zone, a lot of industrial trials were made by using the chamfered mould technology, Analyze and solve the existing in the production process of angle steel, the longitudinal crack and edge crack problems, realize the stable and continuous production of the technology, and took larger economic benefits.

Key words　continuous casting slab, chamfered mould, corner crack, side sliver defect

板坯倒角结晶器应用研究

刘国梁[1,3]　马文俊[1]　刘　珂[1]　张利君[1]　罗衍昭[1]
李海波[1]　张丽丽[2]　李中华[2]　倪有金[2]

（1. 首钢技术研究院，北京　100043；
2. 首钢迁钢公司，迁安　064404；
3. 北京科技大学钢铁冶金新技术国家重点实验室，北京　100083）

摘　要　含Nb、V、B等微合金化钢铸坯角横裂发生率较高，传统调水方法很难避开这些钢种第三脆性区，本文进行了倒角结晶器技术研究，数值模拟表明：金属在结晶器上部凝固收缩较快，包晶钢在结晶器上部收缩为2.4%，中碳钢收缩为1.74%；流场模拟发现普通结晶器最大液面波动速度为0.297m/s，波高为5.7mm；倒角结晶器最大液面波动速度为0.273m/s，波高为5mm。工业试验表明：通过采用倒角结晶器，改善了铸坯冷却及受力状态，铸坯角部横裂纹发生率由原来的4.16%降低到0.4%以下。

关键词　倒角结晶器，连铸板坯，角部横裂纹，高温力学性能，数值模拟

The Applying of Mould with Chamfered Corners for Continuous Casting of Slabs

Liu Guoliang[1,3], Ma Wenjun[1], Liu Ke[1], Zhang Lijun[1], Luo Yanzhao[1],
Li Haibo[1], Zhang Lili[2], Li Zhonghua[2], Ni Youjin[2]

(1. Shougang Research Institute of Technology, Beijing 100043, China;
2. Shougang Qian'an Iron & Steel Co., Ltd., Qian'an 064404, China;
3. State Key Laboratory of Advanced Metallurgy,
University of Science and Technology Beijing, Beijing 100083, China)

Abstract The transverse cracking with a higher incidence of steel slabs containing Nb, V, B and other micro-alloyed elements, the traditional cooling methods are difficult to avoid brittle zone III of these steels. In this paper chamfered mold was studied and the numerical simulation results showed that: the solidification shrinkage of the molten steel is faster in the upper part of the mold and reached to 2.4% and 1.74% for peritectic steel and medium carbon steel. The flow field simulation found that level fluctuation speed of the common mold is 0.297m/s, wave height is 5.7mm, and the level fluctuation speed of chamfer mold is 0.273m/s, wave height of 5mm. Industrial experiments show that the applying of mold with chamfered corners improved the slab cooling and stress state, the occurrence of transverse corner crack reduced from 4.16% to 0.4% or less.

Key words chamfered cornermould, continous casting of slab, transvers corner crack, high temperature mechanical properties, numerical simulation

高铝 TRIP 钢结晶器保护渣反应性研究

王　征[1]　崔　衡[1,2]　张开天[1]　陈　斌[3]　庞在刚[3]　罗衍昭[3]　青　靓[4]

（1. 北京科技大学冶金工程研究院，北京　100083；2. 北京科技大学钢铁共性技术协同创新中心，北京　100083；3. 首钢技术研究院，北京　100043；4. 河北省首钢迁安钢铁有限责任公司，迁安　064404）

摘　要　为减弱高铝 TRIP 钢浇铸过程中渣-钢反应，从调节保护渣 Al_2O_3/SiO_2 及 MnO 含量进行热力学和高温渣-钢平衡研究，结果表明：提高渣中 Al_2O_3/SiO_2 比值可明显降低渣-钢反应性，反应前后成分及物性稳定，由 0.16 增至 2.08，ΔG 开始时变化显著，至 1.18 时趋于稳定；保护渣中加入 MnO＞1.5%后，可在一定程度上保护 SiO_2，但反应后保护炸毁保护渣黏度仍然显著增大，故 MnO 的加入对保护渣反应性的改善效果较小。

关键词　高铝 TRIP 钢，结晶器保护渣，反应性

Research on the Reactivity of Mold Fluxes for Al-TRIP Steel

Wang Zheng[1], Cui Heng[1,2], Zhang Kaitian[1], Chen Bin[3], Pang Zaigang[3],
Luo Yanzhao[3], Qing Jing[4]

(1. Engineering Research Institute, University of Science and Technology Beijing, Beijing 100083, China;
2. Collaborative Innovation Center of Steel Technology, University of Science and Technology Beijing,
Beijing 100083, China; 3. Shougang Technology Research Center, Shougang Corporation,
Beijing 100043, China; 4. Hebei Shougang Qian'an Iron & Steel Co., Ltd., Qian'an 064404, China)

Abstract To weaken the reactivity of mold fluxes for Al-TRIP steel,changes of ratio of Al_2O_3/SiO_2 and MnO content were studied. The results showed that the reactivity sharply reduced with the increasing of Al_2O_3/SiO_2 ratio and the composition and physicochemical properties changedminor. When it changed from 0.16 to 2.08, the ΔG had a significant change until it reached 1.18. The addition of MnO (>1.5%) could protect SiO_2 in the slag to a certain extent. However, the viscosity of the slag after steel-slag interfacial reaction would still increase obviously. So the addition of MnO played a small role in weakening the reactivity of mold fluxes.

Key words Al-TRIP steel, mold fluxes, reactivity

板坯结晶器浸入式水口不对称出口影响分析

季晨曦[1] 李林平[2] 崔 阳[1] 田志红[1] 王胜东[3]
张丙龙[3] 朱国森[3]

(1. 首钢技术研究院，北京 100043；2. 北京科技大学，北京 100083；
3. 首钢京唐钢铁联合有限责任公司，河北唐山 063200)

摘 要 连铸过程中水口受到堵塞物沉积、耐材侵蚀等情况的影响，水口内部及出口结构会发生不规则改变，从而导致结晶器流场的改变和偏流的产生，并可能严重恶化板坯质量甚至产生粘结或者漏钢事故。作者对某次板坯连铸漏钢事故分析发现，结晶器浸入式水口两侧出口大小不同，漏钢侧水口尺寸较大。本文分析了该水口所浇注板坯宽度方向的振痕、液位波动，并进行了数值模拟。结果表明：结晶器在线液位检测系统无法识别板坯两个窄面处的液位波动，水口出口较大一侧，表面流速较高，波动较大，冲击窄面压力较小，冲击深度较浅，上回流发展较为充分；水口出口较小一侧，表面流速较低，波动较小，冲击窄面压力较大，冲击深度较深，下回流发展较为充分。

关键词 非对称水口，流场，数值模拟，漏钢

Effect of Asymmetrical Outlet SEN on Flow Field Control in Slab Mold

Ji Chenxi[1], Li Linping[2], Cui Yang[1], Tian Zhihong[1], Wang Shengdong[3],
Zhang Binglong[3], Zhu Guosen[3]

(1. Shougang Research Institute of Technology, Beijing 100043, China;
2. University of Science and Technology Beijing, Beijing 100083, China;
3. Shougang Jingtang United Iron & Steel Co., Ltd., Tangshan Hebei 063200, China)

Abstract Submerged entry nozzle (SEN) is vulnerable to clogs, refractory erosion, which often make it structure irregularly changed. Furthermore, it would change flow filed in the mold, even lead to bias flow. This maybe deteriorate the slab quality or result in sticking or breakout. In this paper, SEN with different size outlets were found in a breakout during slab casting of low carbon Al-killed steel. At the breakout side, there was a larger size SEN outlet. Oscillation marks of the

slab and mould level were analyzed. After that, flow filed was numerically simulated under the breakout condition. And the following results were obtained. Mould level online detecting system cannot distinguish the fluctuation of the slab narrow sides. Outlet with larger size results in higher surface velocity, larger fluctuation, lower impact pressure, and shallower impact depth. Correspondingly, there is a stronger upward flow at this side compared with that of the smaller outlet side.
Key words asymmetrical outlet SEN, flow field, numerical simulation, breakout

连铸水口堵塞、絮流原因分析及预防措施

苏小利 王德义 于海岐 邢维义 吕志勇 殷东明 张宏亮

（鞍钢股份有限公司鲅鱼圈钢铁分公司，辽宁营口 115007）

摘 要 造成该厂连铸水口堵塞、絮流的主要原因为：钢水温度低冻结水口;异物堵塞钢水通道;夹杂物吸附于水口内壁堵塞钢水通道。本文对造成连铸水口堵塞、絮流的原因进行分析，通过制定中包温度目标模型、减少内生一次夹杂和防止二次氧化等措施，水口堵塞、絮流现象明显减少，单浇次水口由平均 2.9 根降低为 1.6 根。单支水口使用时间由原来的 2.5h 提高为 4.5h。铸机恒拉速由原来的 75%提高为现在的 92%以上。
关键词 水口，堵塞，絮流，夹杂物

Analysis and Preventive Measures of Nozzle Clogging or Flocculating

Su Xiaoli, Wang Deyi, Yu Haiqi, Xing Weiyi, Lv Zhiyong,
Yin Dongming, Zhang Hongliang

(Ansteel Bayuquan Iron & Steel Subsidiary, Yingkou Liaoning 115007, China)

Abstract Main causes of this factory casting nozzle clogging or flocculating is: low freezing temperature of molten steel; the molten steel nozzle blockage by other material; inclusion adsorbed on the inner wall of the nozzle channel, and the nozzle blocked finally. This paper analyzes the causes of nozzle clogging or flocculating. These phenomenon is reduced through the development of tundish temperature target model, reducing endogenous inclusion and preventing the molten steel oxidation and other effective measures. Single cast nozzle using quantity by an average of 2.9 is reduced to 1.6. Single nozzle using time is improved from 2.5 hours to 4.5 hours. Constant casting speed rate is increased from the original 75% to the present more than 92%.
Key words nozzle, clogging, flocculating, inclusion

提高汽车钢钢水可浇性研究

高洪涛 魏 元

（鞍钢股份有限公司炼钢总厂，辽宁鞍山 114021）

摘 要 分析了汽车钢钢水在浇注过程中出现絮流现象的原因，通过采取转炉降低出钢氧含量、严格控制出钢下渣量、保证出钢温度、RH 合理控制较低的真空度、减少加铝批次、选择合适的净吹氩时间和提高钢水的静置时间、

连铸提高钢水保护浇注效果等措施，使得在浇注过程中浸入式水口每浇次更换次数由 3.7 次降至 2.3 次，冷轧板夹杂缺陷率由 0.90%降至 0.65%。

关键词 汽车钢，可浇性，夹杂物，絮流

Improve the Pouring of Molten Steel on Steel for Automobile

Gao Hongtao, Wei Yuan

(General Steelmaking Plant, Ansteel Co., Ltd., Anshan Liaoning 114021, China)

Abstract Analysis of the causes of automobile steel of flocculating flow phenomena in the casting process, by lowering the tapping of oxygen converter, strict control to ensure the amount of slag tapping, tapping temperature; RH reasonable control of vacuum degree, reduce the time of low aluminum batch, select the appropriate net argon blowing time and improving steel static the time of continuous casting steel casting; measures to improve the protection effect, so that in the casting process of submerged entry nozzle replacing each cast number by 3.1 to 2.3, cold-rolled sheet inclusion rate from 0.9% to 0.65%.

Key words automobile steel, castability, inclusions, turbulent flow

3#CC 生产能力影响因素分析

董金刚

（上海宝山钢铁股份有限公司制造管理部，上海 201900）

摘 要 3#CC 主要生产厚板钢种，具有生产批量小、浇注速度低、异常终浇多、规格切换多的特点，生产能力不能充分发挥。影响生产能力的主要因素有生产工艺、生产组织、行车负荷和生产异常等。为此，从优化精炼工艺和生产组织、消除行车限制、降低生产异常等方面提出了有关措施，有利于提高 3#CC 的生产能力。

关键词 生产能力，精炼工艺，生产组织，行车负荷，生产异常

Analysis of the Factors Affecting the 3rd CC Production Capacity

Dong Jingang

(Manufacture Management Department, Baoshan Iron & Steel Co., Ltd., Shanghai 201900, China)

Abstract the 3rd CC mainly produces plate steels,it has the characteristics of small batch production, low casting speed, many abnormal end pourings and much specifications change, Its production capacity can not fully play.the main factors affecting the 3rd CC production capacity are production technology, production organization, crane load and abnormal production etc. The measures have been put forward wich are the optimization of refining process and production organization, eliminating crane restrictions, reducing the abnormal production etc. the 3rd CC production capacity will be improved.

Key words production capacity, refining process, production organization, crane load, abnormal production

CSP隧道炉炉温优化控制模型的研究与应用

蔺凤琴[1] 宋 勇[1] 荆丰伟[1] 孙文权[1] 张南风[2]

（1. 北京科技大学高效轧制国家工程研究中心，北京 100083；
2. 湖南华菱钢铁股份有限公司一炼轧，湖南娄底 417009）

摘 要 紧凑式带钢生产线（CSP）隧道炉位于上游连铸机与下游轧机之间，对热坯进行加热和均热之后，向轧机输送高温板坯。隧道炉各区的炉温实测值直接影响板坯的出炉温度和均匀性，从而影响最终产品的成材率和质量。本文介绍的炉温优化模型基于板坯基础炉温策略，依据炉内板坯数量和参数的变化实时拟合综合炉温。此外，根据板坯位置的变化，将板坯理想加热曲线和预测加热曲线进行比对，两者的差值作为输入参数，模型输出炉温修正值，对炉温进行动态修正。该模型在某厂的应用表明，板坯的出炉温度命中率得到了很大提高，同时降低了能耗，节约了成本，提升了企业在行业的竞争力。

关键词 CSP，隧道炉，炉温，优化控制，延迟休炉

Research and Application of Furnace Temperature Optimizing Control for CSP Tunnel Furnace

Lin Fengqin[1], Song Yong[1], Jing Fengwei[1], Sun Wenquan[1], Zhang Nanfeng[2]

(1.National Engineering Research Center for Advanced Rolling Technology, University of Science and Technology Beijing, Beijing 100083, China; 2. No.1 Steelmaking and Plate Rolling Plant of Lianyuan Iron & Steel Co., Ltd., Loudi Hunan 417009, China)

Abstract The tunnel furnace for CSP (Compact Strip Production) is located between the thin slab caster and the rolling mill.It heats and soaks hot slabs from caster ,and transports high temperature slabs to downstream mill.The measured temperature values of furnace zones directly impact disarging temperature and uniformity of slabs, thus impact product rate and quality of finished coils. The furnace temperature optimizing control model refered to this article is based on basic furnace temperature policy ,fits the comprehensive furnace temperature in time according to the changes of quantity and param of slabs in furnace. In addition, a comparison of the slab ideal heat curve and predicting heat curve shows offset of two curves.The model outputs correction of furnace temperature on the basis of the offset,and revises furnace temperature dynamically.The application of this model in a plan substantially impoves hit rate of slab discharging temperature ,and impoves enterprise competitiveness by the reduction of energy and cost.

Key words CSP, tunnel furnace, furnace temperature, optimizing control, delay and stop

板坯凝固末端重压下的模拟研究

董其鹏 张炯明 赵新凯 王 博

（北京科技大学钢铁冶金新技术国家重点实验室，北京 100083）

摘　要　连铸坯低压缩比轧制是厚板生产的发展趋势，然而探伤不合一直是限制此技术应用发展的主要原因。因此，本研究在轻压下技术的基础上提出凝固末端重压下技术，欲充分利用铸坯凝固末端高温、大的温度梯度等有利条件，施加大的压下量，以达到有效控制铸坯内部质量的目的。本文先利用 TherCast 软件对铸坯凝固过程进行了模拟研究。模拟结果显示了铸坯凝固过程中表面和内部温度变化曲线，以及铸坯内部固相率和钢水静压力的变化。确定铸坯中心固相率后，为了分析压下位置，选取了三个压下位置对铸坯实施 10mm 的压下量。研究发现，压下位置在 f_s=0.86 以后时，对铸坯内部偏析无改善效果，但此时压下有利于改善内部疏松，并且产生内部裂纹可能性较小。最后，研究了两种不同的重压下模式对铸坯坯壳变形和等效应变的影响，发现相同压下量，单辊压下比扇形段压下坯壳变形量稍大，而等效应变稍小。

关键词　板坯，凝固，内部质量，重压下，模拟研究

Simulation Study on the Heavy Reduction at the End of Slab

Dong Qipeng, Zhang Jiongming, Zhao Xinkai, Wang Bo

(State Key Laboratory of Advanced Metallurgy, University of Science and Technology Beijing, Beijing 100083, China)

Abstract　The development tendency for the production of heavy plate is rolling with small reduction ratio, however, the technology is mainly limited by ultrasonic test defects. Therefore, in order to control internal quality of slab effectively, a novel technology named heavy reduction at the end of solidification is presented here, basing on the soft reduction process, taking full advantage of high temperature and large temperature gradient to implement large reduction amounts. Firstly, the solidification process of slab was simulated using TherCast. Results showed the evolutions of surface temperature, central temperature, central solid fraction and ferrostatic pressure. Then, to analyze the influence of reduction position, three reduction positions were chosen to implement heavy reduction with reduction amount of 10mm. The study found that implementing heavy reduction at the position with large solid fraction (0.86 here) cannot improve the segregation, but is good for the improving of porosity. Moreover, the possibility for the internal crack is very small. Finally, two different models of heavy reduction were simulated to study the effect on the deformation of shell and equivalent strain. Results showed that if implemented the same reduction amounts, the deformations of shell for the model with single roller are bigger, but with smaller equivalent strain.

Key words　slab, solidification, internal quality, heavy reduction, simulation study

连铸板坯二冷区辊式电磁搅拌的数值模拟研究

易　军　麻永林　李振团　陈重毅

（内蒙古科技大学材料与冶金学院，内蒙古包头　014010）

摘　要　利用 ANSYS 软件对连铸板坯二冷区辊式电磁搅拌进行数值模拟，分别计算了频率为 5Hz、7Hz、10Hz；安匝数为 400A/84 匝、360A/91 匝、320A/105 匝及铸坯厚度为 150mm、180mm、200mm 的条件下，铸坯内部磁场变化情况。结果表明：在电流为 400A 时，频率越大，磁感应强度越小，但产生的电磁力越大；在频率为 10Hz 时，安匝数为 400A/84 匝和 320A/105 匝在相同时刻的磁感应强度值和电磁力基本相同，但大于 360A/91 匝的情况；在电流为 360A，频率为 10Hz 时，铸坯厚度越薄，沿铸坯宽度方向中心截面的最大磁感应强度和最大电磁力越大。

关键词　辊式电磁搅拌，连铸板坯，数值模拟

Numerical Simulation Research on Roll Type Electromagnetic Stirring in Secondary Cooling Zone of Continuous Casting Slab

Yi Jun, Ma Yonglin, Li Zhentuan, Chen Zhongyi

(School of Material and Metallurgy, Inner Mongolia University of Science and Technology, Baotou 014010, China)

Abstract The roll type electromagnetic stirring in secondary cooling zone was simulated by ANSYS soft. Magnetic field variation of inner of continuous casting slab was calculated at which frequency is 5Hz, 7Hz, 10Hz, the number of ampere-turns is turns of 400A/84, 360A/91 and 320A/105, and casting slab thickness is 150mm and 180mm, 200mm. The results indicate that when the current is 400A, the greater the frequency, the smaller the magnetic induction intensity, but electromagnetic force increases, when the frequency is 10 Hz, ampere-turns is turns of 400A/84, 360A/91, the magnetic induction intensity and electromagnetic force are basically equal at the same time. But in the case of turns of 360A/91, when current is 360A and the frequency is 10 Hz, the thickness of the slab is thinner, and the maximum magnetic induction intensity and the maximum electromagnetic force is greater along the direction of center section of the slab width.

Key words roll type electromagnetic stirring, continuous casting slab, numerical simulation

宝钢股份 5 号连铸机改造新技术的应用

吕宪雨　蒋晓放　曹德鞍　姚建青　常文杰

（宝山钢铁股份有限公司炼钢厂，上海　200941）

摘　要　主要介绍宝钢炼钢厂改造后的 5 号板坯连铸机采用的新技术，这些技术包括塞棒控流、中间包连续测温、结晶器电磁搅拌、结晶器电动缸振动、整体式水口、连续弯曲连续矫直的辊列技术等。通过这些新技术的应用，5 号连铸机的产能得到了较好的发挥，板坯质量得到了很好的改善。同时，通过这些新技术的应用，5 号连铸机生产的品种也得到了很好的拓展，为公司精品规模化生产奠定了良好的基础，尤其是对表面质量和内部质量要求极为严格的高品质钢种的生产起到了至关重要的作用。

关键词　塞棒控流，中间包连续测温，结晶器电动缸振动，电磁搅拌

New Technologies Adopted in No.5 Continuous Caster Revamping at Baosteel Co., Ltd.

Lv Xianyu, Jiang Xiaofang, Cao Dean, Yao Jianqing, Chang Wenjie

(Steelmaking Plant，Baoshan Iron & Steel Co., Ltd., Shanghai 200941, China)

Abstract The new technologies adopted in No.5 continuous caster Revamping at Baosteel Co., Ltd are mainly introduced, which include the flow controlling with stopper rod, continuous temperature measuring in tundish, mold electromagnetic stirring, mold electric cylinder oscillation, integral type submerged entry nozzle, continuous bending and unbending carve technology, etc. The productivity of No.5 continuous caster was significantly increased and the slab quality was greatly improved after using these new technologies. Meanwhile, the steel grades that could be produced in No.5 CC was also extended, which provides a good opportunity to establish the base for excellent steels mass production and especially plays an important roll in producing those high quality steel grades demanded the strict controlling on surface and inner quality.

Key words flow controlling with stopper rod, continuous temperature measuring in tundish, mold electric cylinder oscillation, electromagnetic stirring

薄板坯连铸中间包温度优化工艺实践

高福彬[1] 李建文[1] 关会元[2]

（1. 河北钢铁集团 邯钢公司技术中心，河北邯郸 056015;
2. 河北钢铁集团 邯钢公司三炼钢厂，河北邯郸 056015）

摘 要 主要对邯钢连铸连轧厂薄板坯连铸机中间包温度波动大的问题进行了探讨。通过采取优化精炼冶炼工艺、稳定精炼终点温度，加强钢包周转、提高 A 类钢包使用率，优化中间包烘烤制度，缩短钢水到站等待时间等措施使薄板坯连铸中间包温度波动明显减少。

关键词 连铸，中间包温度，温度波动，温度控制

Practice on Process for the Thin Slab Continuous Casting Tundish Temperature

Gao Fubin[1], Li Jianwen[1], Guan Huiyuan[2]

(1. Technique Center, Handan Iron and Steel Company, Hebei Iron and Steel Group, Handan Hebei 056015, China;
2. Three Steel Making Plant, Handan Iron and Steel Company, Hebei Iron and Steel Group, Handan Hebei 056015, China)

Abstract Mainly discussed the thin slab continuous casting tundish temperature fluctuations at Continuous casting and rolling plant of Hangang.By adopting optimized refined smelting process,stabilize the temperature of refined smelting,strengthen ladle turnaround, improve class A ladle usage, optimization of tundish baking system, shortening the time of molten steel to wait for measures such as the thin slab continuous casting tundish temperature fluctuations significantly reduced.

Key words continuous casting, tundish temperature, temperature fluctuation, temperature control

水平振动电极电渣重熔过程多物理场和凝固过程研究

王 芳 任冬冬 王 强 李宝宽

（东北大学材料与冶金学院，辽宁沈阳 110819）

摘　要　振动电极的意义在于更有效的实现电极和熔融渣液之间的相对运动、增加传热能力，这是降低能耗、提高凝固质量的有效方法。实验与理论计算表明，通过采用振动电极，可以有效提高相对运动速度，并且振动电极方法容易实现工程化，具有较强的实用性。本文主要研究竖直振动与水平振动的电极对电渣重熔过程的影响，通过计算不同频率与振幅下的水平振动情况下的物理场的分布，来确定频率、振幅等工艺参数对电渣重熔过程的影响。结果表明：振动电极可以有效提高电极与熔融渣液之间的相对运动，从而增加了渣池的传热能力，增大熔速。增加振动频率、增大振动幅度，都能够有效增大电极与熔融渣液之间的相对运动，从而使液滴变小，渣池中平均温度梯度降低，熔池趋于浅平，渣池温度变的相对均匀，有利于凝固。

关键词　电渣重熔，振动电极，温度场，流场，电磁场

The Research on Distribution of Multi-physics Field and Solidification under the Horizontal Vibrating Electrode in ESR Process

Wang Fang，Ren Dongdong, Wang Qiang，Li Baokuan

(School of Materials & Metallurgy, Northeastern University, Shenyang 110819, China)

Abstract　The vibration electrode method is easy to implement and has strong practicability. It is reported that vibration electrode promote the relative movement between electrode and melting slag, and it could promote heat transfer capacity. Therefore the vibration electrode is the method which decrease energy consumption and increase solidification quality. This project focus on the influence of vertical and horizontal vibrating electrode to ESR process, To research the distribution of physics field in vertical and horizontal vibrating electrode under different frequency and amplitude. The results how that vibrating electrode can effectively increase the relative motion between the electrode and the molten slag fluid, and promote heat transfer capacity, increasing the melting rate. The increasing of frequency and amplitude promote the relative movement between electrode and molten slag. Therefore it make the droplets smaller, the average temperature of molten slag lower and molten pool more shadow. The molten slag temperature would become more uniform and it is to beneficial to solidification.

Key words　the ESR system, vibration electrode, temperature filed, flow filed, electromagnetic field

单向凝固技术在大型扁钢锭制造中应用现状

耿明山[1,2]　刘　艳[1,2]　曹建宁[1]　李耀军[1,2]

（1. 中冶京诚工程技术有限公司，北京　100176;

2. 北京市铸轧工程技术研究中心，北京　100176）

摘　要　本文介绍了国内外单向凝固技术基本原理和技术特点，分析了日本和国内单向凝固钢锭的开发应用现状，分别阐述了单向凝固钢锭的锭型、工艺、凝固组织和轧制特厚板性能，总结了单向凝固技术难点和相应措施。单向凝固钢锭无 V 形偏析、疏松结构和非金属夹杂物集聚。单向凝固钢锭适用于生产制造大单重高品质特厚板，用该种工艺生产的钢板表现出良好的力学性能、均匀性、清洁度和加工性能。

关键词　单向凝固，扁钢锭，特厚板，锭型，水冷模具

Application Status of Unidirectional Solidification Technology in Large Flat Ingot Manufacture

Geng Mingshan[1,2], Liu Yan[1,2], Cao Jianning[1], Li Yaojun[1,2]

(1. MCC Capital Engineering & Research Incorporation Limited, Beijing 100176, China;
2. Beijing Cast-Rolling Engineering Research Center, Beijing 100176, China)

Abstract The basic principle and technical characteristics of the unidirectional solidification technology at home and abroad were introduced in this paper, the development application situations of unidirectional solidification ingot in Japan and domestic were analyzed, the unidirectional solidification ingot mould, produce process, solidification structure and rolling plate performance were elaborated respectively, unidirectional solidification technology difficulty and corresponding measures are summarized.The unidirectional solidification ingots were found to be free from V segregates,the loose structure and the accumulation of non-metallic inclusions.It was realized that the unidirectional solidification ingot was suitable for production of high grade heavy gauge plates of the large unit weight,the qualities of the plates produced by the process were found to show superior mechanical properties, homogeneity, cleanliness and workabilities.

Key words unidirectional solidification, flat ingot, ultra-heavy plate, ingot shape, water-cooled mould

基于电磁超声的连铸坯液芯凝固末端位置在线测量

田志恒 田 陆 田 立 周永辉 刘润秋 肖小文

（衡阳镭目科技有限责任公司，湖南衡阳 421001）

摘 要 介绍基于电磁超声的连铸坯液芯凝固末端位置在线测量的原理、组成，结合某钢厂板坯连铸机的生产实际情况，设置了 5 对电磁超声传感器来在线测量连铸坯液芯凝固末端位置，选取其中的 3～4 对在线测量了连铸坯在不同拉速情况下的液芯凝固末端位置并与动态轻压下相结合。测量结果表明，系统能根据拉速等工艺条件的改变，准确、实时地测量到连铸坯液芯凝固末端位置；测量数据能较好地与动态轻压下结合，显著改善连铸坯内部质量。

关键词 液芯，凝固末端位置，电磁超声，在线测量

Online Measurement of Solidification End Point of Casting Slabs Based on the Electromagnetic Acoustic Transducer

Tian Zhiheng, Tian Lu, Tian Li, Zhou Yonghui, Liu Runqiu, Xiao Xiaowen

(Hengyang RAMON Science & Technology Co., Ltd., Hengyang 421001, China)

Abstract It introduces the principle and composition of online measurement of solidification end point of casting slabs based on the electromagnetic acoustic transducer, combines with the actual production situation of steel slab continuous casting machine, sets 5 pairs of electromagnetic ultrasonic sensors to measure the solidification end point of casting slabs online. Selecting 3 to 4 pairs of electromagnetic ultrasonic sensors to measure the solidification end point of casting slabs

online at different casting speed, and combine with dynamic soft reduction. The measurement results show that the system can measure the solidification end point of casting slabs accurately and real-time according to the change of the casting speed and so on process conditions. The measurement data can combine with dynamic soft reduction well, improve the internal quality of continuous casting slab significantly.
Key words　liquid core, solidification end point, electromagnetic acoustic transducer, online measurement

CLSM-HT 实验钢样凝固组织形貌的特征研究

王　伟　侯自兵　陈　晗　邓　刚　文光华　唐　萍

（重庆大学材料科学与工程学院冶金工程系，重庆　400044）

摘　要　共聚焦激光扫描显微镜高温系统（Confocal laser scanning microscope high temperature system, CLSM-HT）广泛用于研究金属凝固过程中相变、第二相析出行为及钢液中夹杂物的聚集等现象，但对共聚焦激光扫描显微镜高温系统下实验钢样凝固组织研究很少。为了扩展共聚焦激光扫描显微镜高温系统在热态模拟方面的应用，就很有必要研究试验钢样凝固组织的形貌特征。因此，本文以 70 号钢为试验钢种，研究该 CLSM-HT 不同凝固阶段冷却速度下试样凝固组织的形貌特征。研究发现试样整个观察面凝固组织分布不均匀，内部与外部凝固组织存在差异，同时存在柱状晶与等轴晶；对比同一试样的外部区域和内部区域，发现外部二次枝晶间距的小于内部二次枝晶间距（外部为内部之比范围为 0.837～0.946），且两区域的差异程度随凝固阶段冷却速度的增大而增大。
关键词　共聚焦激光显微镜高温系统，凝固组织，枝晶间距，冷却速度

Characteristics for Solidified Structure Morphology of Hot-model Experimental Steel Samples Using CLSM-HT

Wang Wei, Hou Zibing, Chen Han, Deng Gang, Weng Guanghua, Tang Ping

（College of Materials Science and Engineering，Chongqing University, Chongqing 400044, China）

Abstract　Confocal laser scanning microscope high temperature system, which is abbreviated as CLSM-HT, is widely used for study of the phase transition during solidification process, the behavior of the second phase precipitation and inclusion gathered in molten steel and the like. However, the solidified structure morphology of experimental steel samples using CLSM-HT has not been studied specially now. Then, it is very necessary to study the solidified structure morphology of experimental steel samples to extend the application of CLSM-HT for hot-model experiment. Therefore, high carbon steel (0.70% C) as test material, the study about solidified morphology of different cross section under different solidifying cooling rate (5 ~ 25℃ / min) is carried out. It is found that the solidified morpholog distribution of the specimen is not uniform in the whole viewing surface, presence of columnar crystals and the equiaxial crystals at the same time, and there is a difference between the central area of the specimen and the outer region on its solidification structure. Then, the same same observation surface is divided into two parts: the outer region and the inner region. Outer secondary dendrite arm spacing is less than the inner secondary dendrite arm spacing (the ratio of outer region to inner region is 0.837 ~ 0.946). And the different extent between two regions is increased with the increase of cooling rate during solidification stages
Key words　CLSM-HT, solidification structure, dendrite arm spacing, cooling rate

Study on New Mold Flux for Heat-Resistant Steel Continuous Casting

Qi Jie, Liu Chengjun, Liu Yingying, Jiang Maofa

(Key Laboratory for Ecological Metallurgy of Multimetallic Ores (Ministry of Education), Northeastern University, Shenyang Liaoning, China)

Abstract In order to restrain the slag-metal interface reaction in the continuous casting process of heat-resistant steel, the new mold fluxes with CaO-Al_2O_3-Li_2O-CeO_2 system were proposed. The fundamental properties of the new mold fluxes such as the melting temperature, the viscosity and the crystallization were investigated systematically. It was found that the appropriate content of CeO_2 was determined as 5~10wt%, the melting temperature show good stability with increasing CeO_2, indicating that the mold fluxes could show steady smelting properties after continuously absorbing inclusions containing cerium in the continuous casting process. CeO_2 show significantly good properties of decreasing the viscosity and restraining the crystallization process. After adding CeO_2, the crystalline phases were shifted from CaO and $LiAlO_2$ to $CaCeAlO_4$ and $LiAlO_2$, the precipitation of CaO in the crystallization process was restrained. With a content of less than 14wt%, Li_2O can reduce the melting temperature and viscosities before the breaking point obviously, but the fluxing action of Li_2O became ineffective when its content exceeded 14wt%. The results hold great theoretical significance for developing mold fluxes for heat-resistant steel.

Key words heat-resistant steel, mold flux, continuous casting, viscosity, crystalline phase

Influence of Monoclinic Zirconia Powder on the Properties of Zirconia Metering Nozzle Materials

Ouyang Junhua, Zhao Yuxi, Cui Xiujun, Qiu Wendong, Liu Guangping

(Shanghai Baosteel Industry Technological Service Co., Ltd., Baosteel Engineering & Technology Group Co., Ltd., Shanghai 201900, China)

Abstract By means of pressureless sintering process, the zirconia ceramic materials were prepared using calcia partially stabilized zirconia and monoclinic zirconia as raw materials, Y_2O_3 and CeO_2 powder as composite stabilizer, PVA as binder. The effects of monoclinic zirconia grain size distribution, molding pressure and granulation on the sintering, mechanical strength, thermal shock resistance and phase composition of the specimens sintered at 1700℃ and 1720℃ for 3h in an air atmosphere were investigated. The results indicate that zirconia materials with good thermal shock resistance can be produced by this method. At the same time, monoclinic zirconia powder is finer, and the more contributions can be made to stability and densification of zirconia because of the transformation of monoclinic zirconia to cubic zirconia.

Key words zirconia, monoclinic zirconia, thermal shock resistance, metering nozzle

板坯连铸中间包结构优化的研究

秦 聪 余作朋 吴艳青 刘占礼 徐 烨

（唐山不锈钢有限公司，河北唐山 063105）

摘　要　结合不锈钢公司板坯中间包生产现状，利用水模实验对中间包内钢水流动特性进行研究。结果表明：原型中间包结构存在明显的短路流，钢水停留时间偏短，死区体积分率偏高，不利于夹杂物上浮去除。采用移动挡墙、改变挡坝高度的控流方式时，由于钢水流经路径增长和全混流体积分率增加，响应时间增长，死区体积分率显著减小至14%以下，有利于夹杂物上浮去除。

关键词　连铸，中间包，结构优化，水模实验

Study on Optimization of Tundish Configuration in Slab Continuous Casting

Qin Cong, Yu Zuopeng, Wu Yanqing, Liu Zhanli, Xu Ye

(Tangshan Stainless Steel Co.,Ltd., Converter Workshop, Production Technical Department, Tangshan 063105, China)

Abstract　Fluid flow characteristics of the slab tundish was studied by water modeling experiment based on the practical production of Stainless Steel Co., Ltd. The results showed that there is obvious short circuited flow that resulted in a short residence time and higher dead volume ratio，which is harmful to removal of inclusions. By using of remove weirs and change height of dams，the flow path is prolonged and the volume fraction of complete mixing flow can be increased. As a result，the minimum residence time is obviously increased，and the dead zone volume fraction is decreased up to lower than 14%，which is favorable for the removal of inclusions and increase of yield ratio of molten steel.

Key words　continuous casting, tundish, configuration optimization, water modeling experiment

大型扁钢锭锭型的设计原则

项　利[1]　耿明山[2]

（1. 钢铁研究总院连铸技术国家工程研究中心，北京　100081；
2. 中冶京诚工程技术有限公司，北京　100176）

摘　要　国内关于大型扁钢锭锭型设计原则的论述鲜有报道，本文在查阅大量外文文献的基础上论述了关于大型扁钢锭设计原则，重点分析了钢锭锭型参数对钢锭质量的影响。介绍了国内大单重扁钢锭生产现状，提出国内特厚板用大型扁钢锭的锭型设计。

关键词　扁钢锭，锭型，设计原则，特厚板

Design Principle of Large Flat Ingot Shape

Xiang Li[1]，Geng Mingshan[2]

(1.National Engineering Research Center of Continuous Casting Technology, China Iron and Steel Research Institute Group, Beijing 100081, China;
2. MCC Capital Engineering & Research Incorporation Limited, Beijing 100176, China)

Abstract Research works on the design principle of large flat ingot shape have been seldom reported in homeland. Based on large amount of foreign literature in aboard, the design principle of large flat ingot on ingot quality were summarized, the influences of ingot shape factor on the quality of ingot were analysed.The manufacture present situation of large flat ingot at home wereindtroduced, the design parameter of large flat ingot for extra-heavy plate were clarified.

Key words flat ingot, ingot shape, design principle, ultra-heavy plate

CSP 流程结晶器内钢水到冷却水的水平传热热阻模型

牛 群 程树森 乐庸亮

（北京科技大学冶金与生态工程学院，北京 100083）

摘 要 钢水在结晶器内的凝固传热近似分为通过结晶器的水平传热和拉坯方向传热。钢水在结晶器内停留的时间较短，钢液要在较短的时间内放出大量的热，结晶器内散失的热量大部分是通过水平传热散失的。结晶器的传热的快慢与铸坯的质量密切相关。本文在前人的基础上建立了 CSP 流程中小漏斗型结晶器内钢水到冷却水的水平传热热阻模型，计算了热阻模型中各部分热阻占总热阻的比例。计算发现：当气隙层厚度为 10μm 时，保护渣热阻、凝固坯壳热阻、气隙热阻分别占钢水到冷却水总热阻的 59.48%、15.28%、10.42%，三者是结晶器水平传热的限制性环节；当气隙层厚度大于 58μm 时，气隙层热阻将大于保护渣热阻；随着冷却水流速的增加，结晶器铜板到冷却水的综合热阻降低，当水速大于 7 m/s 时，热阻变化不大。

关键词 结晶器，传热，热阻

The Model of Horizontal Heat Transfer Resistance of Mold From Molten Steel to Cooling Water in CSP Process

Niu Qun, Cheng Shusen, Yue Yongliang

(School of Metallurgical and Ecological Engineering, University of Science and Technology Beijing, Beijing 100083, China)

Abstract Solidification conduct heat of molten steel in mold approximately classifies into two categories:horizontal heat transfer through the mold and the heat transfer along the casting direction. The molten steel's residence time is short and meanwhile gives off a lot of heat. Most of heat in the mold is dissipated through horizontal heat transfer. The speed of heat transfer in the mold is closely related to the quality of the slab. Resistance model of small funnel mold from liquid steel to the cooling water in CSP process is built up on the basis of predecessors' research and calculates the thermal resistance model of the proportion of the total thermal resistance of each part.When the air gap thickness is 10micrometre, the thermal resistance of mold flux resistance, solidified shell and the air gap accounts for 59.48%, 15.28% and 10.42% of total thermal resistance, respectively and those are therestriction factors of horizontal heat transfer.When the airgap thickness is more than 58micrometre,air gap's thermal resistance will be bigger than the thermal resistance of mold flux.With the increase of cooling water flow rate, the comprehensive thermal resistance of the mold copper plate to the cooling water decreased, and when water flow rate is bigger than 7 metre per second, the thermal resistance shows little change.

Key words mold, heat transfer, thermal resistance

吹氩下不同水口类型的结晶器流场和温度场的数值模拟

丁　晖　王延峰　金汉青

（武汉钢铁股份有限公司炼钢总厂四炼钢连铸车间，湖北武汉　430083）

摘　要　以武钢四炼钢连铸车间铸坯断面为 230mm×1300mm 的结晶器为研究对象，采用 1.5m/s 的拉速，利用 Fluent 软件建立钢液-氩气-保护渣三相流模型研究改变水口倾角、水口扩张角和吹氩量对结晶器流场及温度场的影响。结果表明：适当的吹氩量为夹杂物和气泡碰撞、长大、上浮及其去除提供了良好的物理环境，有利于提供夹杂物的去除率。水口采用适当的扩张角，液面波动较小，注流冲击点较小,有利于夹杂物上浮去除，有利于钢水坯壳的均匀生长，也有助于均匀结晶器的温度，促进钢水过热耗散，改善化渣条件。

关键词　扩张角，吹氩，数值模拟，高拉速，水口

Simulation on the Flow Field and Temperature Field of the Mould with Different Nozzle Type under the Blowing Argon

Ding Hui，Wang Yanfeng，Jin Hanqing

(Continuous Casting Workshop of Fourth Steelmaking plant, Steelmaking Plant of WISCO, Wuhan 430083, China)

Abstract The influence of flow field and temperature field in slab continuous casting mould by changing nozzle angle、divergence angle and the amount of argon was described using the Fluent to based on steel-argon-slag there phase model, which taking mould for sectional width230mm×1300mm of WISCL steel plant and the speed is 1.5m/s as research object. The results show that the appropriate amount of argon for inclusion and bubble collision, grow up, floating and removal provides good physical environment, which conducive to the removal rate of inclusions. using appropriate divergence angle,level fluctuation is smaller, the impact point injection flow is small, which is benefic to the inclusion removal and a uniform growth for steel billet shell, benefic the crystallizer molten steel overheating temperature, promo--ting steel dissipation, improving quick melting of mould powder.

Key words　divergence angle, argon blowing, numerical simulation, high speed casting, nozzle

重钢一炼钢厂 1 号连铸机不漏钢实践

刘　渝　邓吉祥

（重庆钢铁股份有限公司一炼钢厂，重庆　400084）

摘　要　对重钢一炼钢厂 1 号板坯连铸机各类漏钢事故的分析，发现黏结漏钢是主要因素。进一步分析发现结晶器

液面波动大、SEN潜入深度不合适、生产条件保障不到位及保护渣性能波动等容易造成黏结漏钢的发生，通过采取系统性相应措施后，未发生漏钢事故，实现了1号连铸机至今连续生产400余万吨不漏钢。

关键词 不漏钢，黏结，原因，措施

Without Breakout in Practive of No.1 Continuous Caster in Chongqing Iron and Steel Company

Liu Yu, Deng Jixiang

(Chongqing Iron & Steel Co., Ltd., Chongqing 400084, China)

Abstract After analyzed on the cause of the accident of breakout of No.1 slab continuous caster in the chongqing steel , found the bonding breakout is the major factor . Further analysis found that the water level of mould fluctuations ,the dive depth of SEN was inappropriate, production conditions were not in place and the quality fluctuation of mould fluxes , prone to bonding breakout ,through the systemic measures, without breakout,continuous processing more than four million ton of No.1 slab continuous caster have been achieved.

Key words without breakout, bonding, casue, measures

方圆坯铸机生产实践

潘统领　白　剑　温荣宇　王英林

（鞍钢股份有限公司炼钢总厂，辽宁鞍山　114021）

摘　要 为了生产多规格高质量铸坯，鞍钢新建一条方圆坯铸机生产线，针对生产初期遇到的问题，通过采用增强起步二冷冷却强度、调整结晶器铜管锥度、调整二冷水嘴等技术措施，铸坯缩孔率评级绝大部分在1.0级以下，铸坯质量满足下道工序要求。

关键词 方圆坯, 改进, 实践

Square/Round Billet Casting Machine Production Practice

Pan Tongling, Bai Jian, Weng Rongyu, Wang Yinglin

(Angang Steel Co., Ltd., Steel Mills, Anshan 114021, China)

Abstract In order to produce more high quality casting, Anshan iron and steel to build a square billet casting machine production line, In view of the production problems early, through the adoption of enhanced started secondary cooling strength, adjust the crystallizer copper tube taper, adjust the cold water technical measures, such as casting shrinkage rate for the most part rating below 1.0, casting blank quality meet the requirement of next working procedure.

Key words square/round billet，improvement，practice

影响钢包自开率的原因分析与解决方法

李亚厚　杨　坤　梁新维　孙福振　国富兴

（承德钢铁公司长材事业二部技术科，承德　067000）

摘　要　连铸环节若钢水不自动开浇，须实施烧眼操作，此操作会造成钢中氧含量升高，夹杂物增多，影响铸坯质量。严重时甚至会造成连铸机停浇事故，造成生产节奏紊乱。承德钢铁公司长材事业二部通过统计分析引流剂状况、水口清理、钢包状况、钢水在钢包停留时间对 120t 钢包自动开浇的影响，通过制定并落实各种措施，将钢包的自开率由原来的93.6%提高至97.3%。从而为稳定生产节奏，提高钢水纯净度提供了有力的保障。

关键词　引流剂，自开率，纯净度

Influence of Automatic Free-opening Causal Analysis and Resolution Methods

Li Yahou, Yang Kun, Liang Xinwei, Sun Fuzhen, Guo Fuxing

(Chengde Iron and Steel Company the Second Long Products Business Technical Department, Chengde 067000, China)

Abstract　If the continuous casting not free-opening,must implement the buring nozzle,this will caused by increased oxygen conten in the steel,inclusion increased,effect of casting billet quality,serious when even cause continuous casting machine casting stop accidents,resulting in the production rhythm disorder.Chengde iron and steel company long products business department through the statistical analysisChengde Iron and steel company long material cause second through statistical analysis of flow agent status, cleaning nozzle, the condition of the ladle　and molten steel in the ladle residence time of 120t ladle automatic cast influence, through the development and implementation of various measures, the ladle free opening rate by original 93.6% up to 97.3%. In order to stabilize the production rhythm, provides a guarantee to improve the purity of molten steel.

Key words　drainage agent, free-opening rate，purity

大型特厚板坯料水冷模铸技术

刘　艳　曹建宁　耿明山　李耀军　王永涛　白　冰

（中冶京诚工程技术有限公司，北京市铸轧工程技术研究中心，北京　100176）

摘　要　连铸板坯由于受最大厚度尺寸的限制，无法满足厚度超过 150mm 的特厚板轧制需求，为解决大单重特厚板坯料供应问题，出现了真空板坯焊合技术，但采用该项新技术生产的品种钢特厚板得不到相关行业认证，因此大型扁钢锭仍然是特厚板坯料的最佳来源。随着扁钢锭吨位增大，钢锭上部缩孔较深，影响成材率，同时疏松等级高，偏析严重，因此大型扁钢锭的质量问题成为发挥特厚板轧机能力的瓶颈。本文介绍了中冶京诚工程技术有限公司自主研发的特厚板坯料水冷制造技术，采用该技术可以生产 50t 级以上大型特厚板坯料，产品质量高于普通铸锭，生产成本远低于电渣重熔钢锭，采用水冷坯料轧制的特厚板，性能优异，表明该项技术及成套装置具有广阔的市场推广前景。

关键词 特厚板，水冷，钢锭

Technology for Production Supper-thickness Plate Slabs with Intensive Water Cooling

Liu Yan, Cao Jianning, Geng Mingshan, Li Yaojun, Wang Yongtao, Bai Bing

(MCC Capital Engineering & Research Incorporation Limited, Beijing Casting-Rolling Engineering Research Center, Beijing 100176, China)

Abstract Continuous casting slabs cannot fully satisfy the requirement for rolling the more than 150 mm supper plates due to the limits of maximum thickness. To solve the problem of supplying large single plate slabs, a new technique that two continuous casting slabs are welded under vacuum has emerged, but the products using this technique cannot be certified by related industry, so the large rectangular ingots are still the mainly source to manufacture supper-thickness plates. With the increasing of ingot weight, the porosity is so deep on the upper of ingot that the metal yield is reduced, and the organization density is relatively poor and segregation is serious. So the large-rectangular ingot quality becomes the bottleneck to play heavy plate mill capacity. This paper introduced a new technology researched by MCC Capital Engineering & Research Incorporation Limited (CERI) which be used to cast large ingot with intensive water cooling. More than 50t supper-thickness plate slabs can be produced and the product quality is better than that of traditional ingot, the cost is much lower than that of Electroslag Remelting ingot. The mechanical property of plate with water cooling slab is excellent. It shows that the technology and the equipment will have broad market prospects.

Key words super thickness plate, water cooling, ingot

2.4 品种及铸坯质量控制/ Steel Grade Development and Quality Control

Hot Ductility Behavior and Slab Crack Performance of Microalloyed Steels During Continuous Casting

JANSTO Steven G.

(Technical & Market Development Department, CBMM North America, Inc., Pittsburgh, USA)

Abstract This research shows that high carbon equivalent microalloyed steels which exhibit inherently lower hot ductility, through the hot tensile test as measured by percent reduction in area (%RA), still exhibit sufficient ductility to satisfactorily meet the unbending stress and strain gradients existing in the straightening section of most industrial casters. Operational results indicate only 10%~15%RA as measured by the hot tensile test, exhibits superior ductility and no cracks are generated through the unbending section of the caster. The published %RA data over the past several years has grossly overstated the minimum ductility required for crack-free casting of Nb-bearing steels by two to threefold, as well as other microalloyed steel grades. The research study clearly connects the relationship between the steelmaking and caster operation and carbon content as the primary drivers of the hot ductility behavior and resultant slab quality. The overall composition is not the primary driver. However, the carbon level of the steel and solidification considerations are the primary drivers affecting surface quality such as slab cracking. The location of the equiaxed-columnar grain transition zone below the surfacesignificantly affects the sub-surface residual strain gradient. This paper also introduces strain energy as a better measure of the hot ductility behavior than %RA which better explains the incongruence between %RA and propensity for slab cracking in microalloyed steels. This research is based solely upon industrial produced heats with an emphasis on the effect of process parameters on hot ductility behavior and successful production of crack-free slabs.

Key words continuous casting, equiaxed-columnar grain transition zone, hot ductility

Research on Behavior of Secondary Phase Particles Precipitation and Dissolution in Micro-alloyed Steel in Process of Continuous Casting

Zhang Qiaoying, Zhang Jian, Liu Guoping, Wang Litao

(Maanshan Iron & Steel Co., Ltd., Maanshan Anhui 243000, China)

Abstract The behavior of secondary phase particles precipitation in micro-alloyed steel in process of continuous casting has been studied by numerical simulation. The secondary phase precipitation possibility was analysed in molten steel, solidification process, and austenite. The secondary phase precipitation rule and interactant rule were found in molten steel, solidification process, austenite+ferrite and ferrite. The paper is the condition of dynamics of the secondary phase precipitation.

Key words micro-alloyed steel, secondary phase particles, precipitation

Inclusion Control in 55SiCr Spring Steels

Zhao Baojun[1], Huang Zongze[2], Yao Zan[2], Wang Geoff[1], Jiang Zhouhua[3]

(1. School of Chemical Engineering, The University of Queensland, Brisbane, Australia;
2. Baoshan Iron and Steel Co., Ltd., Shanghai, China;
3. School of Materials and Metallurgy Northeastern University, Shenyang, China)

Abstract 55SiCr is one of the important spring steels widely used in automobile manufacturing. The presence of micro inclusions in the steel can significantly influence a number of spring properties such as fatigue strength. It is important to understand the mechanisms of the inclusion formation so that size and proportion of the inclusions can be controlled. Inclusions in two 55SiCr spring steels deoxidized by different processes have been analysed by electron probe X-Ray microanalysis. The sizes and compositions of the inclusions in two steels are significantly different. The chemistry of the inclusions observed in 55SiCr steels can be represented by the SiO_2-Al_2O_3-CaO-MgO system. The liquidus temperatures and viscosities relevant to the refining slags and inclusions have been discussed. It would be possible to use these fundamental properties to characterize and manage the inclusions in the steels produced by different deoxidation processes.

Key words 55SiCr, spring steel, inclusion, phase equilibrium, viscosity

Inclusions in Al-Ti Deoxidized Ca/Mg Treated Microalloyed Steels

Li Guangqiang[1], Zheng Wan[1], Wu Zhenhua[1], Hiroyuki Matsuura[2], Fumitaka Tsukihashi[2]

(1. The State Key Lab. of Refractories and Metallurgy, Wuhan University of Science and Technology, Wuhan 430081, China; 2. Depart. of Advanced Materials Science, Graduate School of Frontier Sciences, the University of Tokyo, 277-8561, Japan)

Abstract Inclusions are very critical to the manufacturing process and the performance of steels. Microalloyed high strength steels such as pipeline steel usually deoxidized with Al and alloyed with Ti. Due to the complicated interaction between Al, Ti, O, S, Mn, and Ca or Mg in liquid steel, it is difficult to control the composition, size and quantity of inclusions. To clarify the formation mechanism of beneficial inclusions and make use of their benefit effects on microstructure refinement of steels, the inclusions' properties and the relationship between inclusions and the microstructures of Al-Ti complex deoxidized Ca or Mg treated microalloyed steels were investigated. Large amount of finely dispersed complex inclusions composed oxides core and MnS shell in steels with relative low Al can be obtained comparing with the relative smaller amount of Al_2O_3-CaO or Al_2O_3-MgO inclusions in high Al steels. The Al_2O_3-TiO_x-CaO or Al_2O_3-TiO_x-MgO oxides cores are more conducive to the precipitation of thin MnS layer on their local surfaces. The inclusions with complex structure of oxides core and the MnS surface shell are important to the nucleation of interlocking AFs on them in steels. The AFs quantity is much more and the grains are much finer in low Al steel than that in high Al steel.

Key words steelmaking, inclusions, oxides, deoxidation, microstructure

基于力学行为分析的连铸坯初始缺陷形成机理研究

林仁敢[1,2] 孟祥宁[1] 朱苗勇[1]

（1. 东北大学材料与冶金学院，沈阳 110819；2. 宝钢特钢有限公司，上海 200940）

摘 要 连铸结晶器内弯月面初凝坯壳强度和塑性较小，受结晶器振动影响显著，且形成的初始缺陷会在后续加工过程中继续扩展成为严重的质量缺陷。本文基于材料力学理论，针对弯月面初凝坯壳建立简支梁力学模型，分析结晶器振动过程中初凝坯壳力学状态。研究表明，结晶器振动正滑脱段内，当结晶器上振速度大于 1.6m/min 时，坯壳初始表面裂纹将形成于弯月面处距钢水自由液面 1.9~5.6 mm 范围内。凝固前沿一次枝晶因坯壳弯曲而搭桥形成瞬时封闭结构，导致枝晶间吸附的气泡、富集溶质、夹杂等无法通过与外部液相交流而去除，从而诱发坯壳缩孔、皮下气泡以及夹杂等凝固缺陷。负滑脱时期内，当结晶器下振速度超过 6m/min 时，坯壳初始内裂纹将形成于弯月面处距钢水自由液面 3.1~4.3 mm 范围。

关键词 振动结晶器，弯月面，初始缺陷，一次枝晶

Formation Mechanism of Initial Defects in Continuous Casting Slab Based on Analysis of Mechanical Behavior

Lin Rengan[1, 2], Meng Xiangning[1], Zhu Miaoyong[1]

(1. School of Materials and Metallurgy, Northeastern University, Shenyang 110819, China;
2. Baosteel Special Metals Co., Ltd., Shanghai 200940, China)

Abstract As the low strength and ductility, initial solidifying shell in continuously oscillating mold is impacted significantly. The induced initial defects would be expanded and propagated in the subsequent handling process and evolved into serious defects that are able to restrict the development of steel products. In this study, a mechanical model of meniscus shell based on the principle of materials mechanics was developed for describing the change of stress state during mold oscillation. The results show that the initial surface crack would be formed in the distance from 1.9 mm to 5.6 mm below free surface of steel when the mold oscillation velocity exceeds 1.6 m/min during positive strip time in the process of continuous casting low-carbon steel. Because inter-dendritic bridging is induced in solidification front by the bending deformation of initial solidifying shell, an approximate enclosed space is formed to interdict the contact with the outside liquid steel temporarily. The inter-dendritic solute concentration, shrinkage, surface blowhole and inclusion would be produced in the solidification structure in positive strip time. Besides, the subsurface crack would be formed in the distance from 3.1 mm to 4.3 mm below free surface of steel when the mold oscillation velocity oversteps 6 m/min during negative strip time.

Key words oscillating mold, meniscus shell, initial defects, primary dendrite

冷轧薄板钢铸坯的钩状坯壳特征研究

邓小旋[1] 季晨曦[1] 田志红[1] 崔 阳[1] 朱国森[2]

(1. 首钢技术研究院，北京 100043；
2. 首钢京唐钢铁联合有限责任公司，河北唐山 063200)

摘 要 铸坯表层的大尺寸非金属夹杂物是引起冷轧薄板线性缺陷的主要原因。冷轧薄板钢种由于液相线高，连铸结晶器弯月面的“钩状”坯壳相对发达，较易捕获上浮的夹杂物、气泡或保护渣，因此研究铸坯钩状坯壳特征有利于找出其影响因素，减少轧板表面缺陷。本文利用一种特殊腐蚀液侵蚀得到冷轧薄板钢铸坯全周长方向的钩状坯壳特征，发现铸坯宽度方向钩状坯壳在边部和中心处较深，窄面钩状坯壳在铸坯边部较深。通过逐层腐蚀方法得到铸坯角部钩状坯壳的三维形貌，研究发现拉速越大，铸坯宽面、窄面和角部的钩状坯壳均变浅，其捕获夹杂物能力也变弱。此外，铸坯表层 0.5mm 处大于 150 μm 的夹杂物/气泡较易聚集在钩状坯壳处。因此，高拉速连铸有利于降低被钩状坯壳捕获的夹杂物数量。

关键词 钩状坯壳，冷轧薄板，连铸坯，非金属夹杂物

Hook Character in Continuously-cast Steel Slabs for Cold-rolled Grades

Deng Xiaoxuan[1], Ji Chenxi[1], Tian Zhihong[1], Cui Yang[1], Zhu Guosen[2]

(1. Shougang Research Institute of Technology, Beijing 100043, China;
2. Shougang Jingtang United Iron & Steel Co., Ltd., Tangshan Hebei 063200, China)

Abstract Large-sized non-metallic inclusions on the surface layer of continuously-cast slabs are responsible for slivers defects for cold-rolled steel grades. The cold-rolled steel grades, such as low carbon (LCAK) and ultra-low carbon aluminum killed (ULCAK) steel slabs, with relatively high liquidus temperature are prone to hooks which entrapped macro-inclusions, argon bubble and mold flux. So hooks should be fully characterized to better the understanding of the entrapping behavior of inclusions. In this paper, Two LCAK slabs and two ULCAK slabs are sampled to investigate the frozen hooks around the slab perimeter. It is found that the larger hook depth was observed near the slab corner and slab center in the wide faces, while deeper hook was observed near the slab corner in the narrow faces. The three dimensional morphology of hooks near the slab corner were obtained by step etching method. The result showed that larger casting speed result in shallower hook depth for all location of slabs. In addition, inclusions larger than 150 μm on the surface layer of slabs are almost entrapped by hooks. Accordingly, high speed casting may lower the hook depth and thus reduce the number of macro-inclusions entrapped by hook structure.

Key words hook, cold rolled steel, continuously-cast slab, non-metallic inclusions

连铸板坯裂纹缺陷的控制实践

汪洪峰 王 勇

（宝钢股份梅山炼钢厂，江苏南京 210039）

摘 要 文章从钢的凝固行为、高温力学性能、作用在连铸坯上的各种应力、应变、化学成分以及微合金元素析出行为等角度出发，总结分析了各主要因素对连铸坯产生裂纹的影响，对铸坯裂纹的形成条件、机理进行了综述和总结，从理论与实际的结合上提出了相应的预防及控制措。

关键词 连铸板坯，裂纹，形成原因，控制措施

Production Practice of the Cracks of Continuous Casting Slab

Wang Hongfeng, Wang Yong

(Meishan Steelmaking Plant, Baoshan Iron & Steel Co., Ltd., Nanjing 210039, China)

Abstract The causes and the factors affecting the formation of the cracks in continuously cast strand are summarized through analyzing the solidification behaviors and high temperature mechanism of steel, stress and strain acting on continuously east strand, chemical component, presence behavior of micro alloying elements. The corresponding preventive and control measures are proposed from the point of the theory and practice.

Key words continuous casting slab, cracks, causes, control practice

硅钢非稳态浇铸铸坯洁净度分析

康 伟[1] 栗 红[1] 赵晨光[2] 孙 群[2] 吕志升[1] 曹亚丹[1]

（1. 鞍钢集团钢铁研究院，辽宁鞍山 114001；2. 鞍钢股份炼钢总厂，辽宁鞍山 114021）

摘 要 为研究非稳态浇铸对硅钢铸坯洁净度的影响，对 BOF→RH→连铸生产的硅钢非稳态及稳态铸坯进行取样分析，主要分析头坯、换水口坯、尾坯及稳态坯的 TO 含量、氮含量及夹杂物类型及分布。结果表明：头坯 TO 含量、氮含量比其他铸坯高，铸坯内存在较大尺寸的 Al_2O_3 夹杂；换水口坯 TO 含量、氮含量相对较高，铸坯内部出现大尺寸 $Al_2O_3 \cdot MgO$ 复合夹杂；尾坯 TO 含量、氮含量接近稳态坯，但铸坯内部出现较大尺寸 Al_2O_3 夹杂；依据

TO 含量、氮含量及夹杂物分析，铸坯质量由高到低依次为：稳态坯>尾坯>换水口坯>头坯。

关键词 硅钢，非稳态浇铸，铸坯，洁净度

Research on Cleanliness of Silicon Steel Slab Produced in Unsteady Casting Process

Kang Wei[1], Li Hong[1], Zhao Chenguang[2]，Sun Qun[2]，Lv Zhisheng[1]，Cao Yadan[1]

(1. Ansteel Group Iron and Steel Reasearch Institute, Anshan 114001, China；
2. General Steelmaking Plant of Ansteel Anshan 114021, China)

Abstract In order to research the impact of unsteady casting process on Silicon steel slab, the cleanness of Silicon steel slab of unsteady casting conditions produced by the BOF-RH-CC process was investigated，analyzed TO content、nitrogen content and inclusions of first slab, SEN-changing slab, steady slab, last slab. The results show that TO content, nitrogen content in the first slab are higher than other slab，there are large size Al_2O_3 inclusions in the slab；TO content、nitrogen content in the SEN-changing slab are relatively high，there are large size Al_2O_3.MgO composite inclusions in the slab；TO content、nitrogen content in the last slab close to the steady slab，here are large size Al_2O_3 inclusions in the slab；according to the analysis of TO content、nitrogen content and inclusions，the slab quality from high to low is：steady slab> last slab > SEN-changing slab > first slab.

Key words silicon steel, unsteady casting, slab, cleanliness

Ti-Al 脱氧夹杂物尺寸对晶内针状铁素体形核的影响

黄 琦[1] 姜 敏[1] 王新华[1] 胡志勇[2] 杨成威[3]

（1. 北京科技大学冶金与生态工程学院，北京 100083；2. 南京钢铁公司，江苏南京 210035；
3. 武汉钢铁研究院，湖北武汉 430000）

摘 要 本文研究了 C-Mn 钢中 Ti-Al 复合脱氧夹杂物的尺寸对晶内针状铁素体(Intragranular Acicular Ferrites: IAF）形核的影响。发现同种成分的夹杂物，其尺寸对 IAF 的形核有较大影响。利用 ASPEX 自动扫描电镜大面积地分析了夹杂物的成分以及尺寸，并利用“三元云相图”描述夹杂物在三元相图中的主要成分分布。采用场发射扫描电镜（FE-SEM）分析了有效形核核心和非有效核心的成分及尺寸分布特征。结果表明，小尺寸夹杂物（小于 1μm）容易在晶界处集聚，不利于 IAF 的形核，反而促进晶界铁素体的生成。当夹杂物尺寸超过某一临界值时（4μm），随着夹杂物尺寸的不断增大，夹杂物诱导 IAF 形核的能力会逐渐降低，最后为 0（大于 6.5μm）。

关键词 Ti-Al 复合脱氧，三元云相图，夹杂物尺寸，晶内针状铁素体

The Effects of Ti-Al Deoxidation Inclusion Size on Intragranular Acicular Ferrite nucleation

Huang Qi[1], Jiang Min[1], Wang Xinhua[1], Hu Zhiyong[2], Yang Chengwei[3]

(1. School of Metallurgical and Ecological Engineering, University of Science and Technology Beijing, Beijing 100083, China;
2. Nanjing Iron & Steel United Co., Ltd., Nanjing 210035, China;
3. Research Institute of Wuhan Iron and Steel (Group) Co., Ltd., Wuhan 430000, China)

Abstract The influences of Ti-Al complex deoxidation inclusions on nucleation of Intragranular Acicular Ferrite (IAF) in C–Mn steel have been investigated. It was found that the size inflicted suffering on the nucleation of IAF for these particles with same chemical composition. ASPEX automatic SEM has been used to numerously analyze the size and inclusion composition and the he "ternary cloud phase diagram" was used to accurately and visually describe the distribution of predominant chemical compositions of inclusions. Field-Emission SEM with ultra-high resolution was used to analyze the size distributions of effective nuclei and inert particles. The results show that the small inclusion (less than 1μm) is more likely to aggregate at grain boundaries, consequently promoted the formation of Grain Boundary Ferrite (GBF) and restrained the nucleation of IAF. When the inclusion size exceeds a threshold (about 4 μm), the possibility of IAF nucleation starts to decrease gradually. Consequently, the inclusions cannot nucleate IAF any more when the size is larger than 6.5 μm in present study.

Key words Ti-Al deoxidation inclusion, ternary cloud phase diagram, size of inclusion, IAF

邯钢硬线钢中心碳偏析的控制

董廷亮[1] 张守伟[2] 李 杰[1] 翟晓毅[2] 李博斌[1]

（1. 河北钢铁集团河北钢铁技术研究总院，河北石家庄 052165；
2. 河北钢铁集团邯钢公司一炼钢厂，河北邯郸 056015）

摘 要 邯钢一炼钢小方坯连铸机采取控制中间包钢水过热度在10~35℃，控制二冷比水量为0.8 L/kg，控制拉速为1.8m/min等优化措施，使55号硬线钢中心碳含量由0.648%降低到0.573%，中心碳偏析指数由1.16降低到1.03，中心碳偏析显著降低，中心疏松和中心缩孔也得到明显改善。

关键词 硬线钢，小方坯连铸，中心碳偏析

Control of Central Carbon Segregation of Hard Wire Steel in Hansteel

Dong Tingliang[1], Zhang Shouwei[2], Li Jie[1], Zhai Xiaoyi[2], Li Bobin[1]

(1. Hebei Iron and Steel Technology Research Institute, Hebei Iron and Steel Group, Shijiazhuang Hebei, 052165, China;
2. No.1 Steelworks, Handan Iron and Steel Company, Hebei Iron and Steel Group, Handan Hebei, 056015, China)

Abstract The method of controlling the molten steel heat to 10~35℃ in tundish, controlling second cooling water ratio to 0.8 L/kg and controlling casting speed to 1.8m/min were carried out on billet caster in Han-steel No.1 Steelworks, the result was central carbon content of 55# hard wire steel decreased to 0.573% from 0.648%, the center carbon segregation index decreased to 1.03 to 1.16, the center carbon segregation decreased obviously, simultaneously center porosity and central pipe were improved obviously.

Key words hard wire steel, billet caster, center carbon segregation

连铸拉速和加热炉制度对 GCr15 轴承钢碳偏析的影响

李双江[1]　肖国华[2]　师艳秋[2]　赵劲松[2]　李　杰[1]

（1. 河北钢铁技术研究总院，河北石家庄　050000；
2. 河北钢铁集团邯钢公司，河北邯郸　056015）

摘　要　本文在现场生产试验的基础上，采用系统的分析检测方法，研究了 200mm 方断面 GCr15 轴承钢连铸坯生产 ϕ55mm 棒材产品过程，连铸拉速和加热炉制度对轴承钢碳化物偏析的影响。结果表明，连铸拉速越大，铸坯的宏观碳偏析越严重，适当低的连铸拉速有利于控制铸坯的宏观碳偏析。在二冷水量不变及中间包钢水过热度为 28℃时，连铸拉速控制在 0.9 m/min 比较适宜，此时铸坯碳的宏观偏析指数在 1.08 左右。加热炉的加热温度是影响轴承钢产品碳化物偏析的重要因素。200mm 方连铸坯比较适宜的加热温度和加热时间为 1200℃和 4.5 h，此时棒材产品碳化物带状为 2.5 级，液析为 0.5 级别，满足钢产品的质量要求。

关键词　GCr15 轴承钢，碳偏析，连铸拉速，加热炉制度

Effect of Continuous Casting Speed and Heating Institution to Carbon Segregation Control of GCr15 Bearing Steel

Li Shuangjiang[1], Xiao Guohua[2], Shi Yanqiu[2], Zhao Jinsong[2], Li Jie[1]

(1. Hebei Centre Iron & Steel Technology Research Institute, Shijiazhuang Hebei 050000, China;
2. Handan Iron & Steel Co., Ltd., Hebei Iron & Steel Group, Handan Hebei 056015, China)

Abstract　Based on production experiment and systematic detection and analysis, the effect of continuous casting speed and heating institution to carbon segregation control of GCr15 bearing steel have been studied in steel making process, and the continuous casting billet of 200 mm square was rolled to ϕ55mm bar in this process. The results show that macro carbon segregation of casting billet is more serious with casting speed higher, and it is benefit for controlling moderately lower casting speed to carbon segregation. At the two cold water being inalterability and tundish superheat controlled at 28℃ respectively, the casting speed at 0.9m/min is appropriate, while the macro carbon segregation index in the casting billet is about 1.08. The heating temperature of reheating furnace is an important factor to affect carbide segregation of bearing steel products. It is suitable for heating temperature and heating time are control at 1200 and 4.5 h, while the carbide strip of bar product is 2.5 grade, and liquation is 0.5 level. It satisfy products requirements.

Key words　GCr15 bearing steel, carbon segregation, continuous casting speed, heating institution

浸入式水口内夹杂物碰撞吸附行为

赵定国　李　新　王书桓　张文祥

（华北理工大学冶金与能源学院，河北唐山　063009）

摘　要　连铸浇注过程中浸入式水口常出现水口结瘤，不仅降低了连铸生产效率和水口寿命，还影响了铸坯质量。研究了钢液中夹杂物在浸入式水口壁面结瘤过程，建立了夹杂物与水口内壁间的物理碰撞吸附模型。模型分析结果表明，当夹杂物半径小于 2.5μm 时，Al_2O_3 夹杂物临界碰撞停留速度大于 $1cm·s^{-1}$，临界剪力大于 $1N·m^{-2}$；TiO_2 夹杂物临界碰撞停留速度大于 $4cm·s^{-1}$，临界剪力大于 $10N·m^{-2}$；夹杂物越小越容易结瘤，含钛铝镇静钢更容易结瘤。模型的分析结果与实际生产结果相吻合，具有较好的适用性。

关键词　浸入式水口，夹杂物，碰撞吸附，临界碰撞停留速度，临界剪力

The Behavior of Collision and Adsorption for Inclusion in Submerged Entry Nozzle

Zhao Dingguo, Li Xin, Wang Shuhuan, Zhang Wenxiang

(School of Metallurgical and Engineering, North China University of Science and Technology, Tangshan 063009, China)

Abstract　There are nodes of inclusions often on the Submerged Entry Nozzle (SEN) wall during continuous casting process. It not only reduces the production efficiency of continuous casting and the life of SEN, but also affects the quality of casting billet. Inclusions in molten steel were studied on the SEN wall nodular process, and the physical collision and adsorption model about the inclusions and the SEN wall was established. The results shown that when the radius of Al_2O_3 inclusions was less than 2.5μm, the critical collision velocity was more than $1cm·s^{-1}$, the critical shear force was more than $1N·m^{-2}$. When the radius of TiO_2 inclusions was less than 2.5μm, the critical collision velocity was more than $4cm·s^{-1}$, the critical shear force was more than $10N·m^{-2}$. The smaller the inclusion is, the easier nodulation is, and the nodulation of steel containing titanium element is easier. The model results fitted actual production results, so the model has better applicability.

Key words　submerged entry nozzle, inclusion, collision and adsorption, critical collision velocity, critical shear force

板坯角部横裂纹与热轧卷边部缺陷遗传性探讨

马　硕　陈玉鑫　王章岭　宋佳友

（首钢京唐钢铁联合有限公司炼钢部，河北唐山　063200）

摘　要　含碳量 0.08%～0.21%的微合金化钢连铸板坯角部横裂纹和由此引起的热轧钢卷边部缺陷是长期以来困扰国内外钢铁企业的难题。倒角结晶器在首钢京唐公司全面应用生产后，板坯角部横裂纹所引起的钢卷裂纹得到明显改善，控制在 1%以下。本文利用生产数据对倒角结晶器条件下板坯缺陷与热轧钢卷缺陷的关系进行了探讨。分析认为，直角结晶器管线钢热轧钢卷裂纹和翘皮发生率均与板坯角横裂有正向对应关系；倒角结晶器条件下翘皮、裂纹缺陷与板坯角横裂的正向对应关系得到消除。文章尝试提出角部横裂纹缺陷遗传性模型，实践中通过模型可以评估板坯角部横裂纹缺陷，检出准确性、连铸或热轧工况是否发生了变化。

关键词　微合金化钢，结晶器，翘皮，裂纹

Discussion for what Role Corner Cross Cracks Play in Side Defects on Hot Coil

Ma Shuo, Chen Yuxin, Wang Zhangling, Song Jiayou

(Steel-making Plant, Shougang Jingtang United Iron & Steel Co., Ltd., Tangshan 063200, China)

Abstract It is a well-known problem some side defects on coil of micro-alloyed steel inherit from the slab corner cross cracks. The chamfering mold could help improving corner cracks on slab and side overlay defects on coil. This reappeared in SGJT with rate of occurrence under 1%. This paper has discussed what role chamfering mold play in hot coil defects by decreasing slab cracks in pipeline. By analyzing the plant data, it was found that the occurrence of seam defect has a plus relationship with corner crack by usual mold which would disappear by chamfering mold. It was revealed by an inheritance model between corner cross cracks and side defects how to distinguish wrong process of continuous casting, slab QC or hot rolling.

Key words micro-alloyed steel, mold, crack, hot coil, seam

铸坯表面纵裂影响因素研究概况

李德军　许孟春　于赋志

（鞍钢集团钢铁研究院，辽宁鞍山　114009）

摘　要　本文对连铸坯表面纵裂的形成机理及特征进行了介绍，在此基础上对钢中 C、S、Als 及微合金元素 Nb 的成分控制与拉坯拉速、钢水过热度、水口插入深度、液面波动等工艺操作上容易引起连铸坯纵裂的影响因素的研究概况分别进行了阐述，并对在成分控制及工艺操作上防止铸坯纵裂缺陷的产生提出了相应的控制措施。

关键词　连铸坯，表面纵裂，影响因素

Research Situation on the Influencing Factors of the Longitudinal Surface Crack of Casting Slab

Li Dejun, Xu Mengchun, Yu Fuzhi

(Angang Group Iron and Steel Research Institute, Anshan Liaoning 114009, China)

Abstract The formation mechanism and characteristics of the longitudinal surface crack of continuous casting slab are described, and based on this, the research situations on the influencing factors easily causing longitudinal crack on the continuous casting slab are discussed, such as the composition control of C、S、Als and microalloy Nb in steel, and the process operation of slab drawing speed, superheat of molten steel, nozzle submerged depth and level fluctuation. Finally, the corresponding control measures preventing the longitudinal crack on the casting slab surface are proposed according to the composition control and the process operation.

Key words continuous casting slab, longitudinal surface crack, influencing factor

重压下对高碳钢碳偏析的影响研究

于赋志[1] 廖相巍[1] 许孟春[1] 李德军[1] 张 宁[1]
张 崇[2] 张树江[3] 魏 元[4] 罗建华[4] 栾华兵[4]

（1.鞍钢集团钢铁研究院，辽宁鞍山 114001; 2．鞍山钢铁集团公司战略规划部，辽宁鞍山 114001;
3. 鞍钢集团设计研究院，辽宁鞍山 114001; 4．鞍钢股份有限公司炼钢总厂，辽宁鞍山 114001）

摘 要 为改善高碳钢大方坯中心偏析，进行了凝固末端重压下工艺试验。通过对试验大方坯、中间坯与线材的碳偏析分析，研究了不同连铸工艺参数下大方坯的碳偏析行为。结果表明，在单点压下的情况下，0.8m/min 时，压下量 20mm 的效果最佳，铸坯最大 C 偏析指数为 1.10，而重压下铸坯中心线的最大 C 偏析指数为 1.05，盘条的最大 C 偏析指数由 1.22 降低到 1.15 以下。

关键词 大方坯, 重压下, 高碳钢, 偏析

Study on the Influence of Heavy Reduction on the Carbon Segregation of High Carbon Steel

Yu Fuzhi[1]，Liao Xiangwei[1]，Xu Mengchun[1]，Li Dejun[1]，Zhang Ning[1]，
Zhang Chong[2]，Zhang Shujiang[3]，Wei Yuan[4]，Luo Jianhua[4]，Luan Huabing[4]

(1. Ansteel Group Iron & Steel Research Institute，Anshan 114001，China；
2. Department of Strategy & Project of Anshan Iron & Steel Group, Anshan 114001, China；
3. Ansteel Group Design & Research Institute, Anshan 114001, China；
4. General Steelmaking Plant of Ansteel Co., Ltd., Anshan 114001, China)

Abstract The application of heavy reduction at the end of solidification on high carbon steel bloom was investigated to improve the center segregation in bloom. According to the analysis of carbon segregation of bloom, the workpiece and tire cord steel, the carbon segregation behavior was studied with different bloom continuous casting process parameters. The results showed that the effect of reduction was the best under the condition of the bloom velocity at 0.8m/min and the thickness reduction of 20mm in the case of single-point reduction at present. Then the maximum C segregation index was 1.10 in bloom,1.05 in workpiece and from 1.22 down to 1.15 or less in wire rod.

Key words bloom, heavy reduction, high carbon steel, segregation

连铸板坯角部横裂纹成因探析

阳祥富

（宝山钢铁股份有限公司炼钢厂，上海 201900）

摘　要　针对连铸坯角部发生横裂纹问题，通过采用金相显微镜、扫描电镜、能谱仪及化学分析等手段，并结合现场生产工艺，查明了裂纹产生的原因。在此基础上，通过优化连铸保护渣、控制液面波动、适当降低结晶器冷却强度及调整结晶器振动参数等手段，有效地降低了连铸坯角部横裂纹的产生。

关键词　连铸坯，角部横裂纹

Formation Reasons Study on the Transverse Corner Cracks in Continuous Casting Slab

Yang Xiangfu

(Steelmaking Plant, Baoshan Iron & Steel Co., Ltd., Shanghai 201900, China)

Abstract　The transverse corner cracks in continuous casting slab have been analyzed by using metallographic microscope, SEM, EDS and chemical analysis. The formation reasons of cracks have been found out combined with the production processes. Upon this, the improvement solutions have been proposed, including optimizing mold fluxes, controlling liquid surface fluctuation, decreasing the cooling intensity and adjusting the vibration parameters of casting mould. After applying these solutions, the transverse corner cracks of continuous casting slab have been reduced effectively.

Key words　continuous casting slab, transverse corner crack

轴承钢不同连铸工艺偏析程度对比研究

王　伟[1,2]　赵铮铮[1,2]　逯志方[1]　孟耀青[1]　赵昊乾[1]

（1. 邢台钢铁有限责任公司，河北邢台　054027；

2. 河北省线材工程技术研究中心，河北邢台　054027）

摘　要　采用淬火+回火方法，研究了不同连铸工艺轴承钢钢坯的偏析情况。结果表明，不同工艺下连铸坯边部偏析均较轻，混晶区和中心区域差别较大。低拉速和轻压下的采用对轴承钢连铸坯偏析影响大，低过热度相对没有表现出明显的影响。从混晶区到铸坯中心，偏析严重区域出现了一次共晶碳化物，并伴随着 TiN、MnS 等夹杂物，对产品疲劳寿命造成了影响。

关键词　偏析，一次共晶碳化物，连铸工艺

Investigation Different Technology of Continuously Cast of Bearing Steel Segregation

Wang Wei[1,2], Zhao Zhengzheng[1,2], Lu Zhifang[1], Meng Yaoqing[1], Zhao Haoqian[1]

(1. Xingtai Iron & Steel Corp., Ltd., Xingtai 054027, China;

2. Wire Engineering Technical Research Center of Hebei, Xingtai 054027, China)

Abstract　By use the method of quench and temper, we investigation the segregation of different technology of continuously cast of bearing steel. The result show that, different technology continuously cast bloom has soft segregation by the edges, but major difference on mixed zone and center. The low drawing speed and soft-reduction technology has more effect on bearing steel bloom, the low superheat degree have no obvious effect. From mixed zone to center of bloom, eliquation has appear in the zone of serious segregation, and some impurity like TiN, MnS are turn up with eliquation, which have bad influence to product fatigue life.

Key words segregation, eliquation, continuously cast technology

杆体对水口内夹杂物吸附的数值模拟

王书桓　张文祥　赵定国

（华北理工大学冶金与能源学院，河北唐山　063009）

摘　要　利用 Fluent 软件建立三维有限体积模型，采用多相流模型对水口内钢液的流动特性与夹杂物吸附规律进行数值模拟，重点研究向水口内插入吸附杆，吸附杆表面粗糙度，吸附杆半径，吸附杆结构以及夹杂物的尺寸对夹杂物吸附的影响。模拟结果表明：随着吸附杆表面粗糙度的增大，吸附率到达一个最大值 20.15%；随着杆体半径越大，吸附率增高；随着夹杂物粒径增大，吸附率降低；杆体结构是影响水口夹杂物吸附率的重要因素。

关键词　水口，夹杂物，吸附率，数值模拟

Numerical Simulation of Absorbing Inclusions with Rod in the Nozzle

Wang Shuhuan, Zhang Wenxiang, Zhao Dingguo

(North China University of Science and Technology School of Metallurgical and Engineering, Tangshan Hebei 063009, China)

Abstract By using the FLUENT software,simulation of absorbing inclusions with rod in the nozzle.Research on the influence of adsorption rod surface roughness,adsorption rod size, adsorption rod structure and inclusion size on the adsorption rate.The results showed that, with the increase of adsorption rod surface roughness, under the experimental conditions study, the maximum absorption rate of adsorption rod is 20.15%;As the radius of adsorption rod size, the greater the adsorption rate increased; Increases with the diameter of inclusions, adsorption rate lower; The adsorption rod structure influence on the adsorption rate of the most significant.

Key words nozzle, inclusion, absorption rod, numerical simulation

宽厚连铸板坯角部裂纹控制技术

何宇明[1]　胡　兵[1]　周　宏[1]　王　谦[2]　张　慧[3]

（1. 重庆钢铁股份有限公司，重庆　401258；2. 重庆大学，重庆　400044；

3. 北京钢铁研究总院，北京　100081）

摘　要　直弧型连铸机自身特点，生产宽厚板坯时容易出现角部横裂纹，也容易出现角部纵裂纹。通过偏喷嘴，二

冷动态配水，大倒角结晶器等技术，板坯温度避开钢的第三脆性区；大倒角结晶器，直弧型连铸机对弧，板坯宽窄面鼓肚预防等技术降低了坯壳所受到的应力，两方面的综合效果使板坯角部裂纹形成的几率明显降低，基本解决板坯角部横裂纹和纵裂纹问题。

关键词 宽厚板坯，直弧型连铸机，角部裂纹，大倒角结晶器，对弧技术

Generous Slab Corner Crack Control Technology

He Yuming[1], Hu Bing[1], Zhou Hong[1], Wang Qian[2], Zhang Hui[3]

(1. Chongqing Iron and Steel Co., Ltd., Chongqing 401258, China;
2. Chongqing University, Chongqing 400044, China
3. Beijing Iron and Steel Research Institute, Beijing 100081, China)

Abstract Straight characteristics arc continuous caster, the production of transverse corner cracks in generous slab of falling, and prone to Angle and longitudinal crack. By partial nozzle, secondary cooling dynamic water distribution, large chamfer mould technology, such as the slab temperature to avoid the third brittle zone of steel; big chamfer mould, straight arc continuous caster of arc, the slab width drum belly prevention technology reduces the stress of the shell is, two aspects of the comprehensive effect to reduce the risk of the slab corner crack formation, basically solved the slab corner transverse cracks and longitudinal crack problem.

Key words wide and thick slab, straight mold type caster, corner cracking, big chamfer mould, technology of arc bulging

高氧搪瓷钢的生产实践

黄成红[1] 罗传清[1] 林利平[1] 秦世民[1] 宋乙锋[2]

（1. 武钢股份炼钢总厂；2. 武钢研究院）

摘 要 本文叙述了武钢高氧搪瓷钢的成分设计思路、工艺选择、成分控制技术和用户使用情况。武钢的生产实践表明，高氧搪瓷钢成分设计合理，冶炼工艺成熟，产品质量能够满足用户需求。

关键词 高氧，搪瓷钢，实践

Behavior of Non-metallic Inclusions in a Continuous Casting Tundish with Channel Type Induction Heating

Wang Qiang[1], Li Baokuan[1], TSUKIHASHI Fumitaka[2]

(1. School of Materials and Metallurgy, Northeastern University, Shenyang 110819, China;
2. Department of Advanced Materials Science, Graduate School of Frontier Sciences, The University of Tokyo, Chiba, Japan)

Abstract A transient 3D mathematical model has been developed for the motion of inclusions in a continuous casting tundish with channel type induction heating. The Euler-Lagrange approach was employed, and the inclusion trajectories were obtained by numerical solution of the motion equation including gravity, buoyancy, drag, lift, added mass, Brownian, electromagnetic pressure and thermophoretic forces. Besides, the collision and coalescence of inclusions and adhesion to the lining solid surfaces were also taken into account. The Brownian, Stokes and turbulent collision were clarified, separately. The effect of induction heating power on the inclusion behavior was demonstrated. The results indicate that the electromagnetic pressure force significantly promotes the removal of inclusions, especially for the bigger inclusion. Although the thermophoretic force goes against the removal of inclusions, its influence is negligibly small. The removal ratio of inclusions in the tundish with induction heating increases from 67.45% to 96.43% while the power varies from 800kW to 1200kW. The collision and coalescence should be included when model the inclusion motion, because it can promote the removal of inclusion. The turbulent and Brownian collision becomes more active with the increasing power, while the Stokes collision is just opposite.

Key words tundish, induction heating, inclusion motion, collision and coalescence, removal ratio

X80 管线钢中 Ca 处理氧硫化物夹杂的机理研究

徐 光[1] 姜周华[1] 李 阳[1] 胡汉涛[2] 马志刚[2]

（1. 东北大学材料与冶金学院，辽宁沈阳 110819； 2. 宝山钢铁股份有限公司，上海 200941）

摘 要 模拟现场 X80 管线钢精炼过程喂钙线生产工艺，在实验室条件下精炼 X80 管线钢，控制 Al_2O_3 夹杂物钙处理变性程度，分析了复合夹杂物中氧化物和硫化物的产生及变化情况。夹杂物检测结果发现，钙处理后，当复合夹杂物中 CaO 和 Al_2O_3 摩尔比为 1:1 时，夹杂物中未发现硫；当复合夹杂物中 CaO 和 Al_2O_3 摩尔比为 2:1 时，夹杂物中有硫化物出现，但含量比较小；当复合夹杂物中 CaO 和 Al_2O_3 摩尔比为 3:1 时，夹杂物中出现大量的硫化物。实验结果表明，CaS 的产生与 Al_2O_3 夹杂物变性程度关系密切。夹杂物的面扫描分析显示，精炼过程中 CaS 有两种产生方式，一种为 Ca 和 S 直接反应生成，然后聚合长大，并被氧化物包裹；另一种为 Al、S 等与钙铝酸盐夹杂物发生反应，在夹杂物边缘形成 CaS 活性区域。前者 CaS 多在夹杂物核心位置，而后者多分布在复合夹杂物的边缘。

关键词 X80 管线钢，钙处理，CaS

Mechanism of Oxysulfide in X80 Pipeline Steel Treated by Ca

Xu Guang[1], Jiang Zhouhua[1], Li Yang[1], Hu Hantao[2], Ma Zhigang[2]

(1. School of Materials and Metallurgy, Northeastern University, Shenyang 110819, China;
2. Baoshan Iron and Steel Co., Shanghai 200941, China)

Abstract Simulated the Ca-wire feeding production process of X80 pipeline steel on the field situation, X80 pipeline steel was refined under the laboratory controlling the degree of calcium treatment of Al_2O_3 inclusion, and the generating states of sulfide and oxide were also analyzed. The testing result of inclusion shows that, sulfur can't be found in the complex inclusions when the mole ratio of CaO and Al_2O_3 is 1:1 in the inclusion after feeding calcium. A little sulfur can be found when the ratio of CaO and Al_2O_3 is 2:1 and a great deal of sulfur appears when the ratio is 3:1. According to the result of experiment, it has a close relationship between the generation of CaS and the degree of calcium treatment of Al_2O_3 inclusion. Result of surface scanning indicates that, it has two ways for the generation of CaS in the refining process. One kind is generated with the direct reaction between Ca and S, then aggregation and growing up. And the other is the reaction between Al, S and calcium aluminates, which generated an active area of CaS. The first kind of CaS distributes in the core of inclusions in most cases and the latter distributes at the edge of inclusion for the most time.

Key words　X80 pipeline steel, calcium treatment, CaS

超低碳钢微小夹杂缺陷形成分析

班必俊

（宝山钢铁股份有限公司，上海　201900）

摘　要　冶炼产生的夹杂物在冲压拉伸过程中容易暴露导致缺陷，如裂纹、凸点、印痕等缺陷，对加工制品质量影响较大，甚至一些微小的夹杂也是致命的。为了探究此缺陷产生的机理，采用 Laser-OES、 EPMA 等手段分析夹杂成分及分布，利用凝固定律计算夹杂在钢水中被捕捉的位置，分析夹杂产生的原因。通过研究分析本文的微小夹杂系三氧化二铝类夹杂，推断在连铸浇铸过程中由于侵入式水口结瘤脱落黏附在结晶壳上产生了皮下夹杂，经过轧制成冷轧钢板，在制罐变形过程中皮下微小夹杂凸显在钢罐表面上，造成制品印痕缺陷。

关键词　微小夹杂缺陷，印痕，凝固壳

Analysis of Micro Inclusions Defects Formed in Ultra Low Carbon

Ban Bijun

(Baoshan Iron & Steel Corporation Ltd., Shanghai 201900, China)

Abstract　Inclusions of steelmaking exposed easily leads to defects in the stamping and stretching process, such as micro inclusions defects, salient point, marks, which impact on the quality of the processed product, and even some small inclusions is fatal. In order to study the mechanism of the defects, inclusion composition and distribution is analyzed by EPMA, Laser-OES and other method; the captured position of the inclusions in the molten steel of mold is calculated by solidification law; analysis of the causes of inclusions generated. Through this paper's research and analysis of micro inclusions that are aluminum oxide inclusions, it can be deduced that subcutaneous inclusions form in solidified shell in casting process due to nozzle clogging adhesion. After rolled into the cold rolling steel plate, subcutaneous micro inclusions are exposed on the surface of steel can in the canning process, which cause the product mark defect.

Key words　micro inclusions, mark, solidified shell

中厚板边直裂缺陷的产生及控制

马静超　成旭东　许红玉

（河北钢铁集团邯钢公司三炼钢厂，河北邯郸　056000）

摘　要　大倒角结晶器技术是一种极其有效的解决板坯角部横裂的技术，但在中厚板轧制过程中钢板表面容易产生通长或断续的边直裂现象。通过金相组织、扫描电镜等方式检测认为，边直裂的产生是由于大倒角铸坯角部在轧制过程中角部与表面延伸率不同，角部向内折叠造成的，通过先纵轧后展宽、将结晶器倒角改为圆弧状及铸坯角部轻压下措施，可大幅度抑制角部边直裂缺陷的产生。

关键词 大倒角结晶器，边直裂，折叠，中厚板

Analysis and Control of Vertical Edge Crack in Middle and Thick Plate

Ma Jingchao, Cheng Xudong, Xu Hongyu

(3rd Steelmaking Plant Handan Iron & Steel Company, Hebei Iron & Steel Group, Handan Hebei 056000, China)

Abstract Chamfered corner mould was a new and effective technique to solve the corner cracks in billet, but it was easy to cause continuous or intermittent defect of vertical edge crack. vertical edge crack was analyzed by means of metalgraphy and SEM，the reason of the crack was that corner overlapped to inner because of different elongation between corner and inner during rolling process. By the application of longitudinal rolling first then broadwise direction; fillet of mould changed to arc shape and soft reduction to billet technologies, vertical edge crack had been controlled effectively.

Key words chamfered corner mould, vertical edge crack, overlap, middle and thick plate

承钢小板坯连铸机铸坯表面纵裂的控制

钱润锋 国富兴 孙福振 潘爱龙 李亚厚 张新法

（河北钢铁集团承钢公司长材事业二部，河北承德 067002）

摘 要 连铸板坯表面纵裂起因于结晶器弯月面初生凝固坯壳厚度的不均匀性. 针对承德钢铁公司小板坯连铸机的实际生产情况，分析了影响板坯表面纵裂的各种因素，发现连铸板坯纵裂与钢水成分、结晶器冷却强度、浸入式水口、拉坯速度、结晶器液面波动、结晶器保护渣、钢水过热度、结晶器质量等诸因素密切相关。通过采取相应的措施，可使连铸板坯纵裂指数有一定改善。

关键词 连铸板坯，表面纵裂，结晶器，钢水成分，过热度

Small Surface Longitudinal Crack of the Slab Caster Chenggang Control

Qian Runfeng, Guo Fuxing, Sun Fuzhen, Pan Ailong, Li Yahou, Zhang Xinfa

(Hebei Iron and Steel Group Company Chenggang Long Products Business Department, Chengde Hebei 067002, China)

Abstract Continuous casting slab surface longitudinal crack due to mold meniscus primary solidified shell thickness nonuniformity. Small slab continuous caster in chengde iron and steel company's actual production situation, analyzed the influence factors of surface longitudinal crack of slab, longitudinal crack on continuous casting slab and steel components, mould cooling intensity, invasive nozzle, casting speed, mould liquid level fluctuation, degree of superheat of liquid steel slag, and closely related to the factors such as the mold quality. By taking corresponding measures, can make the continuous casting slab longitudinal crack index have to improve.

Key words continuous casting slab, surface longitudinal crack, mould, molten steel composition, degree of superheat

风电轴承钢 GCr15SiMn 中钛含量控制实践

江成斌　罗　辉

（宝钢特钢有限公司，上海　200940）

摘　要　风电技术装备是风电产业的重要组成部分，也是风电产业发展的基础和保障。其中风电轴承的难点技术是针对主轴轴承的要求无故障运转长，并具有极高的可靠度。宝钢特钢有限公司于 2012 年研制出用于风力发电主轴轴承用的高碳铬轴承钢 GCr15SiMn 锻棒，具有高的纯净度、较低的有害元素含量（如钛和氧）。其中钛含量要求控制≤30ppm。

本文根据宝钢特钢炼钢厂 40t 炼钢产线在风电轴承钢冶炼过程对钛含量控制上的生产实践，生产中发现除原料控制外还有其他因素如钢液氧位、脱氧元素等影响冶炼过程中钛含量控制。通过采用控制电炉出钢钛含量，精炼使用低钛合金、低碱度精炼渣、合理使用脱氧剂等一系列控制钛措施，使冶炼过程钛含量得到有效抑制。生产的风电轴承钢 GCr15SiMn 钢的钛含量控制 20ppm 以下，兼顾氧含量控制在良好水平。

关键词　高碳铬轴承钢，GCr15SiMn，风力发电机，钛

Practice and Control for the Titanium Content of Bearing Steel GCr15SiMn in Wind Power Generation

Jiang Chengbin, Luo Hui

(Baosteel Special Steel Co., Ltd. Steelmaking plant, Shanghai 200940, China)

Abstract　Wind power equipment is an important part of the industry of wind power, and it is the basis and guarantee for the development of the industry of wind power. The difficult technology in wind power bearing is that the requirements of spindle bearing fault-free operation time and the reliability of spindle bearing requirements has very good. The high damage rate of gear box bearing requirements has high load bearing capacity design, etc. Baosteel Special Steel Co., Ltd. has developed for wind power main shaft bearing with high carbon chromium bearing steel GCr15SiMn forging bar in 2012. It has high purity, low content of harmful elements (such as titanium and oxygen). It controls that titanium content is less than or equal to 30ppm.

In this paper, according to the 40 tons of steelmaking line of Baosteel Special Steel Co., Ltd. has the production practice of titanium content control in the wind power bearing steel smelting process. By using the titanium content before the tapping of EAF, using low titanium alloy and low basicity slag in refining, using of rational deoxidizing agent, etc. We effectively control the titanium content in smelting process. The titanium content of the wind power bearing steel GCr15SiMn steel is controlled below 20ppm, and the oxygen content is controlled at good level.

Key words　high carbon chromium bearing steel, GCr15SiMn, wind power generation, titanium

含钛 H13 热作模具钢中大尺寸析出物特征及热稳定性研究

谢　有[1]　成国光[1]　孟晓玲[1]　屈志东[1]　陈　列[2]　张燕东[2]　赵海东[2]

（1. 北京科技大学钢铁冶金新技术国家重点实验室，北京 100083；
2. 西宁特殊钢公司，青海西宁 810005）

摘 要 本文对含钛H13钢铸态钢锭中大尺寸析出物的存在形式及热稳定性进行了研究。研究发现，未添加Ti的H13钢中大尺寸析出物包括富V析出物，呈长条状或树枝状，以及V-Mo-Cr析出物，呈典型的共晶形貌。V-Mo-Cr类析出物1000℃下保温3h即可溶解，而富V大尺寸析出物热稳定性相对较高，1000℃及1100℃下保温长时间虽然有溶解趋势，但即使在1100℃下保温6h仍无法达到消除的效果。H13+0.015%Ti钢铸锭中大尺寸一次析出物除了富V及V-Mo-Cr析出物之外，还存在一定量的数微米级Ti-V类析出物，成分以Ti为主，V含量变化较大，在1000℃及1100℃下没有发现溶解趋势，热稳定性高。计算表明，在凝固末期，富V析出物、富Cr相M_7C_3及富Mo相M_6C依次析出，而Ti的添加则明显提高了一次析出物的析出温度。H13钢中添加Ti时应该特别注意此类大尺寸热稳定性高的析出物的控制。

关键词 Ti微合金化，大尺寸析出物，热稳定性，H13钢

Large Precipitates in Titanium-containing H13 Tool Steel

Xie You[1], Cheng Guoguang[1], Meng Xiaoling[1], Qu Zhidong[1], Chen Lie[2], Zhang Yandong[2], Zhao Haidong[2]

(1. State Key Laboratory of Advanced Metallurgy, University of Science and Technology Beijing, Beijing 100083, China; 2. Xining Special Steel Co., Ltd., Xining Qinghai 810005, China)

Abstract The characteristics of large precipitates in titanium-containing H13 tool steel were studied in this paper. The results showed that a large number of Ti-rich precipitates with different amounts of vanadium existed in H13+0.015% Ti steel besides the primary carbide, the V-rich ones with a long strip or dendritic shape and the V-Mo-Cr ones with a typical eutectic morphology which existed in H13 tool steel without Ti addition. The thermal stability of the three kinds of large precipitates was estimated through water quenching treatment at 1000℃ and 1100℃ after different holding time, that is, 0.5h, 3h and 6h. V-Mo-Cr precipitates were unstable and disappeared just after 3h at 1000℃. V-rich precipitates had a dissolved tendency after 3h even though they still existed even after 6h at 1100℃. However, the large precipitates with high content of Ti were stable and no significant change occurred at test temperature. The calculation by Thermo-Calc showed that V-rich phase, Cr-rich phase M_7C_3 and Mo-rich phase M_6C precipitated successively at the end of solidification in H13. The addition of Ti increased the precipitation temperature of primary precipitates. The large precipitates with high content of Ti should be paid more attention in Ti-containing H13 steel.

Key words Ti-microalloyed steel, large precipitates, thermal stability, H13 tool steel

优质园艺工具用钢55MnB热轧钢带的研制开发

佟 岗 刘万善 赵彦灵 张本亮 曹黎猛 骆思超 唐 辉

（宁波钢铁有限公司，宁波 315807）

摘 要 介绍了宁钢钢铁有限公司根据用户和市场需求，自主进行优质园艺工具（出口园林工具园林剪）用钢55MnB质量设计和工艺设计，采用铁水预处理→转炉炼钢→LF精炼炉→RH真空处理（钙处理）→板坯连铸→热装热送→热连轧制→缓冷出厂的工艺路线，试制成功了符合用户质量需求的55MnB热轧钢带，经用户机械加工、热处理及热处理后的性能检验， 硬度完全能够达到用户制作出口园林剪的要求。

关键词　园艺工具，园林工具，园林剪，55MnB，热轧钢带

Research and Development of the High-quality Hot Strip Steel 55MnB for Garden Tools

Tong Gang, Liu Wanshan, Zhao Yanling, Zhang Benliang, Cao Limeng, Luo Sichao, Tang Hui

(Ningbo Iron & Steel Co., Ltd., Ningbo 315807, China)

Abstract　In this paper introduced Ningbo steel Co., Ltd. completed quality design and process design of the high-quality hot rolled strip 55MnB for garden tools independently ,based on user demand and market demand . The technological path is preliminary desulphuzation of hot metal→converter steelmaking→Ladle Furnace→vacuum treatment by RH(Ca treatment)→slab continous casting→hot delivery and hot charging→steel rolling→slow cooling. Trial the high-quality hot strip steel 55MnB successfully which is line with the user demand for quality. The machinability,heat treatment, performance testing after heat treatment and hardness can fully meet user requirements for the production of export garden shears

Key words　gardening tools, garden tools, garden shears, 55MnB, hot rolled strip

邯钢 CSP 工艺下低碳钢中非金属夹杂物来源及性质的研究

范　佳[1]　巩彦坤[1]　聂嫦平[2]　李建文[1]

（1. 河北钢铁集团邯钢分公司，河北邯郸　056015；2. 北京科技大学冶金学院，北京　100000）

摘　要　本文以邯钢三炼钢-连铸连轧产线所生产的 SPHCZ 铸坯为研究对象，以大样电解实验、扫描电镜检测为主要手段，对其内部的非金属夹杂物含量、尺寸、种类及分布规律进行研究，并对产生各类夹杂的根源进行了分析，从而为制定面向 CSP 工艺下低碳钢产品的经济、合理的夹杂物去除工艺提供了重要的参考。

关键词　CSP，大样电解，低碳钢，非金属夹杂物，来源

Research of Source and Nature of Non-metallic Inclusions of the Low Carbon Steel in CSP Technology of Han-steel

Fan Jia[1], Gong Yankun[1], Nie Changping[2], Li Jianwen[1]

(1. Hebei Iron and Steel Company Limited Handan Branch, Handan 056015, China;
2. Beijing University of Science and Technology Institute of Metallurgy, Beijing 100000, China)

Abstract　in this paper, as the research object to the SPHCZ casting blank produced by the production line of the third steel-CSP of Han-steel, by main way of the electrolysis test and the scanning electron microscope testing,the content size species and distribution law of it are researched, and the source about generating various of inclusions is analyzed. Thus, the important references are provided to drawing up the economical and reasonable technology about removing the inclusions faced to the low carbon steel production in CSP technology.

Key words CSP, electrolysis, low carbon steel, non-metallic inclusions, source

2.5 钢铁近终形连铸/ Near Net-Shape Continuous Casting

Defect Formation Processes and Mechanisms during Twin Roll Strip Casting of Steel

Ferry Michael1, Xu Wanqiang (Martin)

(School of Materials Science and Engineering, The University of New South Wales, Sydney NSW 2052, Australia)

Abstract Twin roll strip casting is a net-shape low carbon emission frontier technology for fast, efficient and direct casting of molten steel (and other alloys) into final products in the form of 1~2 mm thick strip. Due to its high cooling rate (up to 2000℃/s) and complex process during solidification, surface defects such as cracks and dents may form on the surface of the strip during casting. Therefore, their formation mechanisms need to be understood so as to eliminate these defects and generate high quality commercial steel strip products by the optimization of the production process.

In this work, defect formation in low carbon steel strip produced in a twin roll strip casting pilot plant were examined by cross sectioning through the strip thickness, with the microstructure and crystallography of each cross section studied using optical microscopy, SEM and EBSD techniques. The microstructure and crystallography of the defects on the strip surface change significantly during sectioning from surface to the strip centre. In this way, the microstructures and their crystallography of the defects were successfully revealed in 3D. It was found that the atmosphere, surface texture and temperature profile of the mould (twin rollers) in the casting compartment play an important role in defect formation. Accordingly, their mechanisms of formation and relationship with key processing variables have been revealed and discussed herein. These results lay a solid foundation for the elimination of these defects in the future production of strip-cast steel products.

薄带连铸的亚快速凝固组织演变与控制

杨院生[1] 胡壮麒[1] 于 艳[2] 方 园[2]

（1. 中国科学院金属研究所，辽宁沈阳 110016；2. 宝钢研究院，上海 201900）

摘 要 研究了薄带连铸 304 不锈钢在冷却速度为 100~3000 K/s 范围内的亚快速凝固组织演变，通过理论推导和实验验证得到亚快速凝固初凝相析出的判据，当凝固速率高于 4.2 cm/s 时，初生相为亚稳奥氏体相，小于 4.2 cm/s 时初生相为铁素体。原位观察到铁素体相的析出和长大过程以及铁素体转变为奥氏体的固态相变过程，对于不同的凝固速度，304 不锈钢中存在块状铁素体、枝晶状铁素体、板条状铁素体、胞状铁素体、条带状铁素体等不同的铁素体形态，并提出了不同形态铁素体的形成机制。增加亚快速凝固冷却速度，可显著减少亚快速凝固薄带中的 Cr 和 Ni 的微观偏析。通过亚快速凝固控制实现了 304 不锈钢的薄带连铸，薄带直接冷轧后表面质量显著提高，内部组织均匀，综合力学性能达到了用常规工艺流程生产的不锈钢薄板的要求。

关键词　薄带连铸，亚快速凝固，不锈钢，组织演变

Sub-rapid Solidification Microstructure Evolution and Control for Strip Casting

Yang Yuansheng, Hu Zhuanglin, Yu Yan, Fang Yuan

Rolling Technologies for Direct Strip Casting at Ningbo Steel

Shimpei Okayasu

(Primetals Techndogies Japan, Ltd.)

Abstract　The direct strip casting process has attracted attention as a near net shape technologies. BAOSTEEL GROUP CORPORATION and Primetals Technologies Japan, Ltd. have jointly developed Twin Roll Caster for carbon steel based on the direct strip casting technologies. The strip casting demo plant, with 800mm diameter and 1340mm long casting rolls, was installed in Ningbo Steel and its operation was started up in 2014. A stable castability of 1340mm wide strips and long term sequence casting of more than 180ton was succeeded. This paper presents the downstream equipment of the direct strip casting such as an In-line reduction mill, Strip cooling system and Carrousel reel type coiler. Our discussion focuses on stable rolling technologies including Dynamic Pair Cross and strip temperature control technologies.

Thin Slab Direct Rolling Modeling of Nb Microalloyed Steels

Pereda Beatriz[1], Uranga Pello[1], López Beatriz[1], Rodriguez-Ibabe Jose M.[1], Liu Zhongzhu[2], Stalheim Douglas[3], Barbosa Ronaldo[4], Arantes-Rebellato Marcelo[5]

(1. CEIT and TECNUN, University of Navarra, San Sebastian, Spain; 2. CITIC Metal Co., Ltd, Chaoyang, Beijing, China; 3. DGS Metallurgical Solutions Inc., Vancouver, WA, USA;
4. Metallurgical and Materials Engineering Dept., Universidade Federal de Minas Gerais, Belo Horizonte, Brazil;
5. CBMM, Sao Paulo, Brazil)

Abstract　The production of Nb microalloyed coils by thin slab casting and direct rolling (TSDR) technologies shows some metallurgical characteristics that significantly differentiate this process from conventional casting and rolling of thick slabs. When improved combinations of strength and thickness are sought, modeling tools designed ad hoc to predict the austenite microstructural evolution provide an insightful input to process and composition optimization. This paper shows the latest simulation tools developed by CEIT and CBMM for different TSDR configurations available in the market. These models predict grain size distributions taking into account softening-precipitation interactions, refinement due to static and dynamic recrystallization and strain accumulation during last rolling passes. Based on the metallurgical analysis, parameters such as Recrystallization Limit Temperature (RLT) and Recrystallization Stop Temperature (RST) are also defined. A wider

knowledge of all the mechanisms involved in these grades will open new possibilities to produce new final grades by TSDR technologies with optimized processing routes, microstructures and, as a consequence, final mechanical properties.
Key words thin slab direct rolling, Nb microalloyed steels, microstructural evolution, rolling mill layouts

Microstructures and Mechanical Properties of theMn-N Alloyed Lean Duplex Stainless Steel Processed by Twin-roll Strip Casting

Zhao Yan, Liu Zhenyu, Wang Guodong

(The State Key Laboratory of Rolling and Automation, Northeastern University, Shenyang Liaoning, China)

Abstract Lean duplex stainless steel, in which nickel is replaced by manganese and nitrogen, is one of the most important trends for developing high strength stainless steels.However, manufacturing thin gage strips of lean duplex stainless steelhas been one of the most challenging technologies in the production of stainless steels. In the present work, lean duplex stainless steel with the composition of Fe-18.7Cr-6.3Mn-0.38N was fabricated by twin-roll strip casting,in which serious pore defects can be effectively avoided as compared to conventional ingot casting. The solidification structure observed by OMindicates that no solidified dendrites have beenformedthrough surface to center of the cast strip, and the coarse equi-axedgrainswere formed in the center.The cast strip was hot rolled and cold rolled to the final thickness of 0.5mm, with no obvious edge cracks having been formed during both hot and cold rolling.After the annealing treatment at 1050°C for 5min following the cold rolling, distribution of Mn and Cr in the cast strip can be further homogenized.The cold rolled and annealed cast strip exhibited an excellent combinationof strength and ductility, with the ultimate tensile strength and elongation having been measured to be 1000MPa and 65%, respectively.TEM micrographs clearly showed that strain induced martensitic transformation had occurred during plastic deformation.
Key words duplex stainless steel, Mn-N alloying, twin-roll strip casting, solidification structure, mechanical properties

用 MATLAB 拟合并预测带钢厚度的温度补偿系数

张 军

（宁波钢铁有限公司，浙江宁波 315800）

摘 要 根据现有的基础温度补偿数据，用 MATLAB 拟合构造一个合适的函数，用这个函数预测更多的温度补偿数据，解决了多功能仪在薄带连铸试验机组超高温下的温度补偿数据缺失导致厚度测量不准问题。
关键词 MATLAB，拟合，多功能仪，温度补偿

Fitting and Calculating Strip Temperature Compensation Coefficient with MATLAB Software

Zhang Jun

(Ningbo Iron & Steel Co., Ltd., Zhejiang Ningbo 315800, China)

Abstract According to the existing base temperature compensation data, construct a proper function with the MATLAB software, use this function to predict more temperature compensation data, to solve the multi-function meter thickness measurement problems because of ultra-high temperature compensation data missing in Ningbo strip casting line.

Key words　MATLAB, fitting, the multi-function meter, temperature compensation

双辊薄带连铸界面热流的数学模型研究

张　琦[1]　于　艳[2]　方　园[2]

（1. 青岛理工大学机械学院，山东青岛　266520；
2. 宝钢技术中心前沿技术研究所，上海　201900）

摘　要　熔池与结晶辊间的界面热流影响着铸带的组织性能和表面质量，因此，本文采用数值模拟的方法，获得了连铸工艺参数和界面热流对结晶辊出口处薄带表面温度的影响，并获得界面热流与连铸工艺参数间的定量关系式，可以得到如下结论：(1) 界面热流、液位高度与熔池出口处薄带表面温度呈线性反比变化；拉速、钢水过热度、辊缝与熔池出口处薄带表面温度呈线性正比变化。(2) 拉速是影响熔池出口处薄带表面温度的最主要的因素，其次为界面热流和辊缝，最后为液位高度和钢水过热度。此外，本文建立了界面热流与连铸工艺参数间的关系模型可以为连铸工艺参数的制定提供参考。

关键词　双辊薄带，界面热流，数值模拟

Interface Heat Flux Mathematical Model of Twin Roll Strip Continuous Casting

Zhang Qi[1], Yu Yan[2], Fang Yuan[2]

(1. Department of Mechanical Engineering, Qingdao Technological University, Shandong Qingdao 266520;
2. Advanced Technology Institute, Technology Center of Baosteel, Shanghai 201900)

Abstract　Interface heat flux between the pool and the rolls affects the structural properties and surface quality of the twin roll cast strip, so, in this paper, the numerical simulation method is used to build the mathematical model of flow and heat transfer of pool, and the effect of the interface heat flux and continuous casting process parameters on the strip surface temperature at the roll outlet is acquired, then the MINITAB is used to build the relationship between the strip surface temperature and the interface heat flux and continuous casting process parameters. Because the strip surface temperature is easy to measure, then the quantitative relationship between interface heat flux and continuous casting process parameters can get by reverse derivation which provide a reference for custom casting process parameters.

Key words　twin roll strip continuous casting, interface heat flux, numerical simulation

双辊薄带铸轧电磁侧封分体式磁极结构设计

杜凤山　孙慕华　杨　波　许志强　吕　征

（燕山大学国家冷轧板带装备及工艺工程技术研究中心，秦皇岛　066004）

摘 要 薄带钢双辊铸轧整体式磁极电磁侧封装置存在电磁感应强度过大、磁极底部发热严重、同一熔池深度磁场分布不均的缺点。针对整体式磁极的缺点，本文设计并优化了一种新型的分体式磁极电磁侧封机构，采用分体式磁极与线圈绕组克服了熔池底部磁感应强度集中，磁极过饱和等问题。通过三维模型电磁场有限元模拟，分析了分体式磁极侧封结构熔池内磁场、感应电流、电磁压力分布情况，验证了分体式磁极侧封结构的可行性。通过进一步对分体式磁极结构进行优化，得出磁极与熔池之间的距离和磁极的大小是影响侧封的关键因素，优化后的分体式磁极结构侧封效果优于整体式磁极。

关键词 双辊薄带铸轧，电磁侧封，分体式磁极

The Design of Split Electromagnetic Poles for Side Sealing Structural for Twin Roll Strip Casting

Du Fengshan, Sun Muhua, Yang Bo, Xu Zhiqiang, Lv Zheng

(National engineering research center for equipment and technology of C.D.R., Yanshan University, Qinhuangdao 066004, China)

Abstract For the Electromagnetic side seal of twin roll strip casting process, drawbacks were existed in integral pole: the larger electromagnetic induction intensity, the seriously fever at bottom of the pole, the uneven magnetic field distribution in the same depth of molten pool. For the shortcomings of Integral pole, This paper designs a new split electromagnetic poles for side sealing structural configuration, to overcome the issues of integral pole. Through three-dimensional finite element simulation, to analyze the magnetic field, induced current, electromagnetic pressure distribution in the molten metal pool and to verify the feasibility of the split pole side seal structure. By further optimizing the structure of the split pole, the key factor is the size of the poles and the distance between the molten metal pool and poles. And the split pole structure is better than the Integral pole.

Key words twin roll strip casting, electromagnetic side seal, split electromagnetic poles

薄带连铸布流系统的优化设计

张捷宇

（上海大学，上海 200072）

摘 要 薄带连铸技术是当今世界冶金及材料研究领域的一项前沿技术。该技术改变了传统冶金工业薄带材的生产过程，将连续铸造、轧制等过程串为一体，具有生产工序简化、低成本、低能耗的优点；同时可以利用薄带连铸技术的亚快速凝固效应，细化铸带组织，改善材料性能。被认为是最具发展前景的近终形连铸技术之一。目前，薄带连铸过程中薄带坯的最终质量很难得到保证。如何控制薄带坯的凝固组织和改善铸坯质量，是薄带连铸亟需解决的难题。

薄带连铸过程狭窄的熔池、较快的拉速和较高的冷却强度使薄带连铸熔池内的流动、凝固行为十分复杂。本文采用物理模拟和数学模拟的研究方法，在对薄带连铸熔池内钢液传输特性分析和凝固行为研究的基础上，对比不同的布流系统，优化设计出的布流系统可以保证熔池液面波动合理、出流均匀、温度分布均匀；采用 VOF 方法研究了钢液开浇时的流动和液面波动情况，模拟结果与水模型测量结果基本一致。通过模拟工艺参数的变化，来控制薄带连铸流、热和凝固过程的宏观传输行为，最终实现改善铸坯内部组织，提高薄带质量。

关键词 薄带连铸，传输特性，液面波动，凝固组织

Simulation Study on Melt Delivery System of Twin-roll Strip Casting Process

Zhang Jieyu

(Shanghai University, Shanghai 200072, China)

Abstract The strip casting process is an advanced technique which can directly produce the coiled strip from the melt by combining the casting and the hot rolling into a single step. Compared to strips made from the conventional slab or thin slab casting process, the features of the strip casting process have advantages of the reducing the production cost as well as the investment cost due to the elimination of the entire hot rolling process. Furthermore, it can produce better properties due to sub-rapid cooling conditions. Among the various strip casting processes applied to steelmaking, the twin-roll strip casting is regarded as the most prospective technology of near-net-shape casting. However, the practical application of the strip casting process on an industrial scale is still hindered by the crucial control of the casting process and strip quality.

The strip casting is a very complicated process, in which the flow, heat transfer and solidification can be complete in the narrow curve pool and in less than on second. In this work, the fluid transfer phenomena and solidification characteristics in the pool of twin-roll strip casting are studied by using physical and mathematical simulations. The optimized delivery system can produce desirable flow patterns and even temperature distribution in the casting pool. A Volume of Fluids (VOF) approach was employed to simulate the transient fluid flow and level fluctuation during the initial pouring stage. The predicted surface profile agreed well with the measured values in water model. These simulation reslts can help us to control the twin-roll strip casting process and improve the quality of products in practice.

Key words strip casting, transfer phenomena, level fluctuation, solidification microstructure

Microstructure and Mechanical Properties of Sub–rapidly Solidified Fe–Mn Based Alloy Strip

Yang Yang[1], Song Changjiang[1], Duan Lian[1], Zhang Yunhu [2], Zhai Qijie [1]

(1. State Key Laboratory of Advanced Special Steels, Shanghai University, Shanghai, China;
2. Helmholtz-Zentrum Dresden-Rossendorf, 01314 Dresden, Germany)

Abstract The purpose of this paper was to investigate the microstructure and the mechanical properties of Fe-Mn binary alloys and Fe-Mn-C ternary alloys prepared by using sub-rapid solidification technique. The results demonstrated that the ε-martensite and α'-martensite would be directly formed in the sub-rapidly solidified Fe-Mn and Fe-Mn-C samples. And the lath width of ε-martensite and α'-martensite was lower than that of sample produced by conventional method. The mechanical properties of Fe–Mn and Fe-Mn-C samples prepared by sub–rapidly solidification were even higher than those by hot rolled and heat treated. This work discusses the relationship between the microstructure and the mechanical properties of the strips.

Key words sub-rapid solidification, TWIP steel, mechanical property, microstructure

大块非晶合金板材连续铸造技术研究

周秉文　杨洪硕　蒋博宇　房　园　殷实鉴　张兴国

（大连理工大学材料科学与工程学院，辽宁大连　116024）

摘　要　非晶合金具有多种优异的性能，如高强度、高硬度、良好的耐腐蚀性、耐磨能力、优异的磁性能和生物相容性等，但是块体非晶合金受临界尺寸及制备工艺的制约，一直未能在工业上广泛应用。本文主要研究连续铸造块体非晶合金板材所需的凝固条件，以及初步探索 $Cu_{40}Zr_{50}Al_{10}$ 非晶合金板材连续铸造技术。本文通过 Procast 数值模拟技术获得了石墨水冷铜模复合铸型内熔体温度场变化，以及连铸的拉坯速率对非晶合金形成的影响。采用真空双室水平连铸设备制备出表面质量良好，尺寸为 300mm（长）×50mm（宽）×6mm（厚）的 $Cu_{40}Zr_{50}Al_{10}$ 合金板材，合金凝固组织包含大量非晶态结构，且该合金强度超过 1290MPa。

关键词　块体非晶合金，数值模拟，真空水平连铸

Study on Continuous Casting Technology for Bulk Metallic Glassy Plate

Zhou Bingwen, Yang Hongshuo, Jiang Boyu, Fang Yuan, Yin Shijian, Zhang Xingguo

(School of Materials Science and Engineering, Dalian University of Technology, Dalian 116024, China)

Abstract　Glassy alloy has many good properties, such as high strength, high hardness, high corrosion resistance, good wear resistance, good magnetic properties and biocompatibility. However, it has not been widely used in industry, because the preparation methods and critical dimension limited its development. A comprehensive 3D temperature field model has been carried out to simulate a horizontal continuous fabrication slab blank caster for the $Cu_{40}Zr_{50}Al_{10}$ bulk metallic glasses. The model used in this study takes into account the coupled temperature field and solidification aspects of the process and is verified by comparing with the continuous casting experiment. By Procast numerical simulation verify that the cooling intensity of vacuum multi-mold horizontal continuous casting equipment exceed the critical cooling rate of some glassy alloys. A 50mm*6mm*300mm $Cu_{40}Zr_{50}Al_{10}$ board with good high surface quality has been prepared by continuous casting, and the fracture strength of this board is over 1290 MPa.

Keywords　bulk metallic glasses，vacuum multi-mold horizontal continuous casting，numerical simulation

大块非晶合金板材连续铸造技术研究

Study on Continuous Casting Technology for Bulk Metallic Glassy Plate

3　轧制与热处理

Steel Rolling and Heat Treatment

Development of the Unified Rolling Force Model Using von Karman Equation on Elastic Foundation Method

Guo Remnmin

(V. P. Research and Development, Tenova I2S, Yalesville, Connecticut, USA)

Abstract For the past decades, rolling models have been applied based on the rolling mill types. There is no unified rolling force/power model in the industrial and academic fields. The original von Karman equation was derived from the force equilibrium condition without any special assumptions. Hence, the unified solution should exist if no special assumptions are introduced in the solving routine. This article proposes a unified rolling force/power model by solving von Karman equation on the elastic foundation whose unit spring constant is obtained from Hertzian contact equation. This new developed solution can be applied to all rolling mill types.

Key words slip friction, elastic foundation, stress-strain curve, rolling force, rolling model, rolling mill

微量Nb对热镀锌双相钢组织性能的影响

梁 轩 刘再旺 黄学启

（首钢技术研究院，北京 100043）

摘 要 基于淬火回火的双相钢热镀锌退火工艺，本文试验研究了微量Nb对热镀锌双相钢显微组织和力学性能的影响。结果表明，随着热镀锌或及合金化时间延长，材料的强度下降、塑性升高，相比于热镀锌退火，热镀锌合金化退火的温度更高、时间更长，因此双相钢的强度下降更明显；微合金元素Nb延迟回火时马氏体中固溶碳的析出及延迟位错马氏体的回复再结晶过程，可以提高双相钢的回火稳定性，从而改善热镀锌双相钢的力学性能；Nb微合金化的GA双相钢可比碳锰钢提高抗拉强度50MPa以上，从而得到具有优良性能的高强度镀锌双相钢板。

关键词 微合金，Nb，双相钢，显微组织

Effect of Nb on the Microstructure and Mechanical Properties of Galvanized Dual Phase Steel

Liang Xuan, Liu Zaiwang, Huang Xueqi

(Shougang Research & Development Center, Beijing 100043, China)

Abstract Based the quenching & tempering process，the effect of Nb on the microstructure and mechanical properties of galvanized dual phase steel was investigated. The results showed that the niobium can retard the precipitation of solute carbon and recrystallization of dislocation martensite, and therefore improves the mechanical properties of galvanized dual phase steel.

Key words microalloying, Nb, dual phase steel, microstructure

The Application of Object-oriented Technology to Coil Temperature Control in Hot Strip Mill

Shen Yuchun

(Process Information Section Electrical & Control Department, China Steel Corporation, Taiwan, China)

Abstract Coil temperature control always plays a very important role to coil quality in Hot Strip Mill. Due to the limitation of computer capability in the past, it is hard to get the detail coil processing history of each sample point on run out table. Now, the modern computer technology has already advanced enough to apply Object Oriented Design on Coil temperature control. This not only makes past impossible possible but also raises various applications. The applied Object Oriented Design contains from object function analysis to the way objects chain created and reflected to each sample object on coil, to make the coil temperature prediction hit the target.

Key words hot strip mill, coil temperature control, object orient design

新一代中厚板正火控冷装置的开发与应用

何春雨[1]　张明山[2]　余　伟[1]　宋国栋[2]

（1. 北京科技大学高效轧制国家工程研究中心，北京　100083;
2. 山东钢铁股份有限公司济南分公司，山东济南　250132）

摘　要　正火控制冷却技术是中厚板正火热处理线的重要工艺手段，对于控制正火后钢板的组织性能有着重要的影响。介绍了新一代中厚板正火控冷装置的工艺及设备特点。新一代正火控冷装置采用 0.4~0.5MPa 的冷却水与淬火机系统共用淬火机低压段冷却水，减少水处理系统的投资，设备布置紧凑，重量比传统设备降低 30%，占地面积降低 20%，设备最大高度降低 50%。可自由组合不同类型的冷却喷头实现不同强度冷却速度。在工业生产中达到了以下应用效果：作为新产品开发的平台，可实现 4~120mm 厚度范围钢板的正火控制冷却，正火控冷钢板的性能合格率到 96.5%以上；可在较少合金含量的情况下实现钢板高强度、高韧性和优良的焊接性能的正火钢板生产；作为性能控制手段，可用于挽救轧后部分性能不合格钢板。正火控制冷却技术在正火类厚板及特厚板生产中具有良好的应用价值。

关键词　中厚板，正火，控制冷却，低成本，减量化

Research and Application of New Generation Plate Normalizing Control Cooling Technology

He Chunyu[1], Zhang Mingshan[2], Yu Wei[1], Song Guodong[2]

(1. National Engineering Research Center for Advanced Rolling Technology,
University of Science and Technology Beijing, Beijing 100083, China;
2. Shandong Iron and Steel Co., Ltd., Jinan 250132, China)

Abstract Normalizing controlled cooling technology is an important means of technology on plate normalizing heat treatment line. It has an important impact on microstructure and properties of steel after normalizing. Introduced characteristics of a new generation plate normalizing cooling device. A new generation of normalizing cooling device using 0.4~0.5MPa cooling water. Quenching machine can be shared with the low-pressure section of the cooling water quench system. Reduce investment, equipment layout compact water treatment system, the weight is reduced by 30% compared to conventional equipment. Covers an area of reduced 20%. The maximum height of the device by 50%. Different types can be combined to achieve different intensity of cooling nozzle cooling rate. Applications effect are achieved in industrial production. As a platform for new product development, 4~137mm thickness plate can be controlled cooling after normalizing. The plate qualified rate can be reached more than 96.5%. The plate can obtain high strength, high toughness and excellent weldability with low alloy content after normalizing control cooling process. The technology also can improve the performance of steel which substandard after hot rolling normalizing controlled cooling technology has a good application value in heavy plate normalizing production.

Key words plate, normalizing, control cooling, low cost, alloy reduction

冷轧不锈钢连续退火过程带钢温度、晶粒和力学性能预测模型

豆瑞锋[1] 温 治[1] 张 雄[2] 周 钢[3] 李 志[4] 冯霄红[4]

（1. 北京科技大学机械工程学院，北京 100083；2. 西安石油大学，西安 710065；
3. 中国恩菲工程技术有限公司，北京 100038；
4. 重庆赛迪冶炼装备系统集成工程技术研究中心有限公司，重庆 400013）

摘 要 建立了冷轧不锈钢带钢退火过程的温度预测数学模型，在全面系统地等温退火和连续退火实验研究的基础上，利用参数拟合方法，得到了430铁素体不锈钢的再结晶动力学模型、晶粒长大模型、硬度模型和强度模型等。现场实验结果表明：带钢温度预测数学模型的计算精度达到3%(命中率90%以上)。晶粒尺度、力学性能模型的计算结果表明：带钢温度达到最高退火温度之前，由于再结晶的作用，其硬度、晶粒尺寸和屈服强度迅速降低，而延伸率则快速升高。带钢晶粒的长大会提高延伸率，但同时降低了屈服强度。本文所建立的冷轧不锈钢带钢退火过程的温度、晶粒和力学性能预测数学模型能够为冷轧不锈钢退火过程工艺优化提供理论参考。

关键词 冷轧不锈钢，连续退火，卧式炉，力学性能，预测模型

Prediction Model for Temperature, Grain Size and Mechanical Properties of Cold Rolled Stainless Steel in Continuous Annealing Process

Dou Ruifeng[1], Wen Zhi[1], Zhang Xiong[2], Zhou Gang[3], Li Zhi[4], Feng Xiaohong[4]

(1. School of Mechanical Engineering, University of Science and Technology Beijing, Beijing 100083,China;
2. Xi'an Shiyou University, Xi'an 710065, China;
3. China ENFI Engineering Corporation, Beijing 100038, China;
4. CISDI R&D Co., Ltd., Chongqing 400013, China)

Abstract The temperature prediction model for cold rolled stainless steel annealing was built. Based on the systematically study on isothermal annealing experiments and continuous annealing experiments, the recrystallization kinetics, grain growth, hardness and strength models were established. The verification based on onsite data indicates that: the temperature prediction model has an error of 3% with an accuracy of 90%. The grain size and mechanical properties prediction models show that: before the strip temperature reaching the highest point of annealing temperature, because of the recrystallization, the hardness, grain size and yield strength decrease dramatically, but the elongation increases sharply. The grain growth after recrystallization will simultaneously increase the elongation and decrease yield strength slightly. The strip temperature, grain size and mechanical properties prediction models built in this research can provide theoretical reference value for optimization of cold rolled stainless steel continuous annealing process.

Key words cold rolled stainless steel, continuous annealing, horizontal furnace, mechanical properties, prediction model

板带热连轧负荷分配新方法研究与应用

李维刚

（武汉科技大学信息科学与工程学院，武汉 430081）

摘　要 提出一种“压下模式+轧制力模式”相结合的板带热连轧机组在线负荷分配新方法。根据压下模式负荷分配计算确定精轧机组厚度分配的初始值，根据轧制力模式负荷分配的CLAD算法进行轧制力按比例分配迭代计算，最后采用压下模式分配计算确定的压下率分布范围对轧制力模式分配结果进行限幅处理。新方法已成功应用于国内某热连轧生产线改造项目，在线应用表明该方法具有轧制稳定性好、操作工干预少等优点，在减小精轧各机架压下率波动的同时尽量满足轧制力按目标比例分配的要求，有效提高带钢精轧过程的轧制稳定性。

关键词 带钢热连轧，负荷分配，轧制力比例，轧制稳定性

Research and Application of a New Method of Load Distribution for Hot Strip Mill

Li Weigang

(College of Information Science and Engineering, Wuhan University of Science and Technology, Wuhan 430081, China)

Abstract A new method of online load distribution for finishing trains of hot strip mill was proposed, which combined reduction mode and rolling force mode of load distribution coefficient method. The new method determined the initial value of thickness distribution based on the reduction mode of load distribution, and then used the CLAD algorithm for the rolling force mode of load distribution to make rolling force per pass reduce by a certain percentage, lastly used the reduction range determined by reduction mode to limit the results of the rolling force mode of load distribution. The new method has been successfully applied in Ningbo Steel 1780 hot rolling production line, and the online application validates the new method with the advantages such as fast computing speed and excellent convergence performance. It reduces the fluctuation range of reduction per pass in finishing mill and most likely makes rolling force per pass decrease in proportion, effectively improves rolling stability.

Key words hot strip mill, load distribution, roll force ratio, rolling stability

智能制造技术在钢产品生产中的应用探讨

胡恒法

（梅钢公司制造管理部，南京 210039）

摘 要 智能化制造技术是在大数据和物联网支撑下新出现的一个具有良好前景的新技术，有望解决用户对产品低成本、高质量、个性化的要求，又满足企业大规模生产的需求，以提高企业的竞争能力。本文探讨了智能化制造技术在钢产品生产中的应用。

关键词 智能制造，大数据，新技术

Discussion on Application of Intelligent Manufacturing Technology in Steel Industry

Hu Hengfa

(Manufacturing Management Department of Meishan Iron & Steel Co., Ltd., Nanjing 210039, China)

Abstract Intelligent manufacturing technology has a good prospects of new technology, appeared in the supporting of big data and Internet networking, expected to solve the requirements of users of products with low cost, high quality, personalized, and to meet the needs of large-scale production enterprises, in order to improve the competition ability of the enterprise. This paper discusses the application of intelligent manufacturing technology in the production of steel products.

Key words intelligent manufacturing, big data, new technology

基于边界积分方程法的轧机辊间压扁模型及其应用

肖 宏[1] 任忠凯[1] 员征文[1,2]

（1. 燕山大学国家冷轧板带装备及工艺工程技术研究中心，河北秦皇岛 066004；
2. 徐工集团江苏徐州工程机械研究院，江苏徐州 221004）

摘 要 辊间压扁理论是板形控制理论的重要组成部分，针对半无限体模型在计算辊间压扁时的误差问题，给出一种计算辊间压扁的新模型。该模型将轧辊视为有限长度的半无限体，基于边界积分方程法建立了一个精确的辊间压扁模型。基于该压扁模型建立新型六辊轧机辊系变形模型，所得结果与有限元法、半无限体模型和费普尔公式进行对比，发现新模型所得结果更加准确。计算了中间辊窜辊量的影响，发现新模型所得辊间压力和压扁优于其他两模型，特别是在轧辊边部区域比其他两模型更符合实际。

关键词 辊间压扁，辊系变形，有限长半无限体，边界积分方程，六辊轧机

Roll Flattening Model and Its Application by Boundary Integral Equation Method

Xiao Hong[1], Ren Zhong kai[1], Yuan Zhengwen[1,2]

（1. National Engineering Research Center of Cold Strip Rolling Equipment and Technology, Yanshan University, Qinhuangdao 066004, China;
2. Jiangsu Xuzhou Construction Machinery Research Institute, XCMG, Xuzhou 221004, China）

Abstract Roll flattening theory is a primary part of the plate shape theory, to improve the accuracy of roll flattening calculation based on semi-infinite body model, a new roll flattening model is proposed. In this model, the roll barrel is considered as a finite length semi-infinite body, an accurate roll flattening model is established by the boundary integral equation method. Based on the new roll flattening model, a new 6-Hi mill deformation model is established and verified by FEM. The new model is compared with Foppl formula and semi-infinite body model in different roll shifting value. The results show that the pressure and flattening between rolls calculated by the new model are more precise than other two models, especially near the two roll barrel edges.

Key words roll flattening, roll deformation, finite length semi-infinite body, boundary integral equation method, 6-hi mill

5m 厚板精轧机轧制力偏差问题仿真研究

孙建亮[1] 谷尚武[1] 刘宏民[1] 张志忠[2]

（1. 燕山大学国家冷轧板带装备及工艺工程技术研究中心，河北秦皇岛 066004；
2. 宝山钢铁股份有限公司厚板厂，上海 201900）

摘 要 本文针对某厂5mCVC-Plus厚板轧机两侧轧制力偏差过大的问题，首先对轧机的结构和控制特点进行分析，初步确定产生轧制力偏差的原因；然后基于ANSYS/LS-DYNA建立厚板轧制过程有限元模型，仿真模拟多种轧制工况，模拟了不同非对称因素对轧制力偏差的影响，确定了影响轧制力偏差的主要因素是轧辊交叉、轧辊倾斜和非对称弯辊；最后提出了轧制力偏差控制措施，即尽可能减小非对称量，控制轧机的非对称因素在合适范围内；定期检修时严格控制辊系与牌坊的间隙和预埋量，控制效果良好。

关键词 轧制力偏差，厚板精轧机，非对称弯辊，辊系空间交叉

Simulation of Rolling Force Deviation of Heavy Plate Mill

Sun Jianliang[1], Gu Shangwu[1], Liu Hongmin[1], Zhang Zhizhong[2]

(1. National Engineering Research Center for Equipment and Technology of Cold Rolling Strip, Yanshan University, Qinhuangdao 066004, China; 2. Baoshan Iron & Steel Company Limited Plate Mill, Shanghai 201900,China)

Abstract According to the large rolling force deviation problem of 5m CVC-Plus plate mill, First, the structure and control characteristics of the rolling mill were analyzed, the causes of rolling force variation were preliminarily determined; Second, the

finite element model for plate rolling process was made based on ANSYS/LS-DYNA, the influence of different asymmetric factors on the rolling force variation were simulated, the main factors affecting rolling force deviation are roll cross, roll tilt and asymmetric bending; Finally, the rolling force deviation control measures are put forward, that is, as far as possible, reducing the asymmetric, and controlling asymmetric factors of rolling mill in a suitable range; in the preventive maintenance, the clearance between the mill house and bearing chock and embedded parts should be controlled strictly, and the control effect is good.
Key words rolling force deviation, heavy plate rolling mill, asymmetric bending, roll cross

厚板 TMCP 钢板终冷温度均匀性改善策略

刘 斌 肖桂林 吴扣根

（宝山钢铁股份有限公司厚板部，上海 200941）

摘 要 分析了厚板 TMCP 钢板在加速冷却过程中，沿长度方向终冷温度分布不均的原因，并根据现场生产实践，针对不同问题给出了相应的处理措施，取得了良好的效果。
关键词 厚板，控制冷却，温度均匀性

Improvement of Stop Cooling Temperature Uniformity for TMCP Plate

Liu Bin, Xiao Guilin, Wu Kougen

(Plate Department，Baoshan Iron & Steel Co., Ltd., Shanghai 200941, China)

Abstract The cause of nonuniform temperature distribution along length direction of plate in the process of accelerated cooling control for TMCP plate was analysed, and the corresponding corrective measurement to the different problems were given and effective results were also achieved.
Key words heavy plate, controlled cooling, temperature uniformity

Dynamic Transformation of Austenite to Ferrite During the Torsion Simulation of Strip Rolling

John J. Jonas, Clodualdo Aranas Jr.

(Materials Engineering, McGill University 3610 University St. Montreal, Canada H3A 0C5)

Abstract Torsion simulations of strip rolling were carried out on a 0.06%C-0.3%Mn-0.01%Si steel over the temperature range 930℃ to 1000℃. Pass strains of 0.4 were applied at a strain rate of $1s^{-1}$. Interpass times of 0.5s, 1s, 1.5s and 3s were employed to determine the mean flow stresses (MFS's) applicable to strip rolling. By means of double differentiation, critical strains of 0.04 and 0.11 were established for the initiation of dynamic transformation and dynamic recrystallization, respectively. It is shown that the dynamic transformation of austenite to ferrite takes place above the Ae_3 temperature, resulting in significantly lower than expected MFS values. The nucleation of ferrite reduces the rolling load and modifies the microstructure. Shorter interpass times produce larger decreases in MFS than longer ones because of reduced recovery during short interpass intervals.
Key words Strip mill simulation, interpass time, dynamic transformation

硅含量对热轧低碳贝氏体钢组织和性能的影响

袁 清[1,2] 徐 光[1] 王 利[2] 何 贝[1] 周明星[1]

（1. 武汉科技大学耐火与冶金国家重点实验室，高性能钢铁材料及其应用湖北省协同创新中心，湖北武汉 430081；
2. 宝钢集团有限公司汽车用钢开发与应用技术国家重点实验室，上海 201900）

摘　要　对三种不同硅含量的低碳贝氏体钢进行轧制实验，对轧制后试样进行组织检验和拉伸实验，对比分析其显微组织和各项力学性能。结果表明,Si 含量从 1.0 wt.%增加到 1.5 wt.%时，微观组织基本相同，均为粒状贝氏体+M/A 岛，屈服强度、抗拉强度略有增加，延伸率基本相同，强度的增加主要来源于 Si 的固溶强化作用。Si 含量从 1.5wt.%增加到 2.0wt.%时，显微组织产生明显变化，含硅 2.0 wt.%钢的显微组织为粒状贝氏体+板条马氏体+等轴铁素体，屈服强度及抗拉强度显著增加，但相比于含 Si 量 1.5wt.%的钢，延伸率、强塑积均下降，强度的提高主要是固溶强化、相变强化等的综合作用。综合比较三种低碳贝氏体钢，若只考虑钢的强度因素，则 Si 的添加量应达到 2.0 wt.%；若考虑钢种塑性及强塑积，则 Si 的添加量应为 1.5 wt.%。

关键词　硅，低碳贝氏体钢，粒状贝氏体，M/A 岛，显微组织，力学性能

Effect of Si Content on the Microstructure and Mechanical Properties of Hot Rolled Low Carbon Bainite Steel

Yuan Qing[1,2], Xu Guang[1], Wang Li[2], He Bei[1], Zhou Mingxing[1]

(1. The State Key Laboratory of Refractories and Metallurgy; Hubei Collaborative Innovation Center for Advanced Steels, Wuhan University of Science and Technology, Wuhan 430081, China;
2. State Key Laboratory of Development and Application Technology of Automotive Steels (Baosteel Group), Shanghai, 201900, China)

Abstract　Rolling experiments of three low carbon bainite steels with different silicon contents were performed on laboratory roll mill. The microstructure observation and tensile tests were carried out after rolling. The microstructure and mechanical properties were compared and analyzed. The results show that when Si content increased from 1.0 wt.% to 1.5 wt.%, the microstructure was basically the same, consisting of granular bainite + M/A island. The yield strength and tensile strength increased slightly, while the elongation was basically the same. When Si content increased from 1.5wt.% to 2.0wt.%, the microstructure changed significantly, consisting of granular bainite, lath martensite and axial ferrite. Yield strength and tensile strength increased significantly, but elongation and the product of strength and elongation were decreased. The improved strength mainly results from the combined effects of solid solution hardening and transformation hardening. Comprehensive comparison of three low carbon bainite steels, it shows that if only the strength is considered, the Si content should be about 2.0 wt.%. If elongation and comprehensive propertied are considered, the amount of Si should be about 1.5 wt.%.

Key words　Si, low carbon bainite steel, granular bainite, M/A island, microstructure, mechanical properties

超快冷条件下时效温度对高强冷轧板组织性能影响的研究

胡靖帆[1] 宋仁伯[1] 王林炜杰[1] 蔡恒君[1,2]

（1. 北京科技大学材料科学与工程学院，北京 100083；
2. 鞍钢股份有限公司冷轧厂，辽宁鞍山 114021）

摘 要 在200℃/s的退火超快冷条件下，研究了高强冷轧薄板在退火温度为820℃时，不同时效温度对低合金高强钢组织和性能的影响。实验结果表明：随着时效温度的升高，实验用钢的抗拉强度呈现下降趋势，屈服强度呈现上升趋势，当时效温度从240℃提高到320℃时，抗拉强度从893MPa下降到了773MPa，下降了13%，屈服强度从496MPa增加到了577MPa，提升了16%。延伸率呈现上升趋势。随着时效温度的升高，马氏体体积分数逐渐下降，当时效温度从240℃提高到320℃时，马氏体体积分数从49.2%下降到了40.9%，随着时效温度的升高，铁素体的晶粒尺寸呈现一定的增大趋势，从3.8μm增加到了4.6μm。不同时效温度下的拉伸断口形貌基本为等轴状的韧窝，当时效温度从240℃升高到320℃时，韧窝尺寸从1.4μm增大到了2.9μm，随着时效温度的升高，韧窝变的大而且深，说明材料的塑性随着时效温度的升高有所提升。

关键词：超快冷，时效温度，高强钢，组织，性能

Effect of Aging Temperature on Microstructure and Properties of Low Alloy High Strength Steel under Ultra-fast Cooling Condition

Hu Jingfan[1], Song Renbo[1], Wanglin Weijie[1], Cai Hengjun[1,2]

(1. School of Materials Science and Technology, University of Science and Technology Beijing, Beijing 100083, China; 2.Cold Rolling Department of Ansteel, Anshan 114021, China)

Abstract The effect of aging temperature on microstructure and mechanical properties of high strength cold rolling steel was studied when the annealing temperature was 820℃ under the ultra-fast cooling condition of 200℃/s. The results show that with the increasing of the aging temperature, the tensile strength of the steel is decreased, and the yield strength is increased. When the aging temperature increased from 240 to 320℃,the tensile strength decreased from 893MPa to 773MPa by 13%, and the yield strength increased from 496MPa to 577MPa by 16%. The elongation increased with the increasing of aging temperature. With the increasing of aging temperature, the volume fraction of martensite decreased gradually. when the aging temperature increased from 240℃ to 320℃, the volume fraction of martensite decreased from 49.2% to 40.9%, and the grain size of ferrite increased from 3.8μm to 4.6μm. The morphology of the tensile fracture surface at different aging temperature is the equiaxed dimples. When the aging temperature increased from 240℃ to 320℃,the size of dimple increased from 1.4μm to 2.9μm. With the increase of aging temperature, the dimples become larger and deeper, which indicates that the plasticity of the steel increased.

Key words ultra-fast cooling, aging temperature, high strength steel, microstructure, properties

回火时间对核电压力容器用钢的组织及性能的影响

李传维　韩利战　刘庆冬　顾剑锋　晏广华　张伟民

（上海交通大学材料科学与工程学院材料改性与数值模拟研究所，上海　200240）

摘　要　以核电压力容器用 SA508Gr.3 钢为研究对象，研究了显微组织和第二相随回火时间的变化规律及其对性能的影响。结果表明，在 650℃回火时，粒状贝氏体中的 M/A 岛首先分解为铁素体基体和堆积的长条状 Fe_3C 碳化物；随后长条状的碳化物长大球化，M/A 岛分解形成的铁素体发生合并结晶为大的铁素体板条；当回火时间进一步延长，Fe_3C 碳化物数量减少并粗化，新生成的 M_2C 碳化物弥散分布在铁素体基体中。随着回火时间延长，材料的强度逐渐下降，冲击性能先上升至极值后逐渐恶化。结果表明，回火过程中基体组织的软化效应、合并再结晶，M-A 岛分解程度，碳化物的类型、形貌、尺寸和分布是影响材料力学性能的重要因素，其中晶界大颗粒碳化物能够显著降低晶界的结合强度，改变裂纹非稳定扩展的临界断裂应力，是 SA508 Gr.3 钢冲击韧性达到峰值后下降的主要的原因。

关键词　核电压力容器用钢，回火时间，组织，性能

The Effect of Tempering Time on Microstructure and Mechanical Properties of a Reactor Pressure Vessel Steel

Li Chuanwei, Han Lizhan, Liu Qingdong, Gu Jianfeng,Yan Guanghua, Zhang Weimin

(Institute of Materials Modification and Modeling, School of Materials Science and Engineering, Shanghai Jiao Tong University, Shanghai 200240, China)

Abstract　This paper indicates microstructure and their influence on the properties of the nuclear reactor power pressure vessel steel during tempering at 650°C. The results show that: the M/A (Martenite/Austenite) constituents with higher carbon content in the granular bainite are first broken up into the ferrite lathes and accumulated Fe_3Ccarbides. Later, the ferrite lathes which decomposed from martensite island undergoes merged crystallization and broaden. Fe_3C grew bigger and spheroidized. At the same time, the carbon atoms diffused to the matrix, so the dispersed carbides, M_2C, were found in the matrix. When the tempering time was 5h, the material strength is decreased to reach a balance state, while the material impact toughness reaches the plateau. With the extension of tempering time, the bainiteferrites are merged; the carbides in the grain boundary are further coarsened, although the material strength declines slightly, the material impact performance deteriorates seriously. The analysis suggests that the softening effective, the merge and the crystallization of the basal body organization, the decomposition degree of the M-A island, the type, shape, dimension and distribution of the carbide are important factors influencing the material mechanical performance of the material.

Key words　reactor pressure vessel (RPV) steel, tempering time, microstructure, mechanical properties

Study of Clad Rolling Technology on Super-heavy Plates

Chen Zhenye[1], Li Jiangxin[1], Lin Zhangguo[1], Wei Ming[2],

Liu Hongqiang[1], Zhang Yunfei[1], Liu Dan[2]

(1. Central Iron and Steel Technology Research Institute of Hebei, Shijiazhuang Hebei 052000, China;

2. Wuyang Iron and Steel Company HBIS, Pingdingshan Henan 462500, China)

Abstract After surface cleaning and weld combining to Q345D continuous casting billets, we can get compound billets. By vacuum-pumping the compound billet cave, packing and heating to 1220~1260℃, the compound billets are rolling during 1160~986℃on one of the medium plate production lines of Hebei Iron and Steel Group(HBIS). The products are Q345D super-heavy plates. The experiment include transverse tensile, the tensile of Z direction, scan of tension fracture, the microstructure observation in the bonding interface, clod bending and ultrasonic inspection etc. Examination results show that all mechanical properties of the Q345D compound plate meet the requirements of GB/T1591-2008 and GB/T6396—2008. The average tensile strength on Z direction of Q345D compound plate is 537MPa, and the area reduction is from 45% to 52%. The cold bending property is good and the ratio of the bonding interface is higher than 99%. The tensile specimen fracture is mixing fracture and a small amount of toughness fracture. From the trial products and examination results, we can get two conclusions. Firstly, the vacuum composite rolling craft can achieve metallurgical combination of same composite interface material, and this craft can replace ingot casting rolling process for super-heavy plates. Secondly, It's reasonable to explore dissimilar metal clad plate rolling craft on the basis of the super-heavy plate composite rolling craft research.

Key words super-heavy plate, compound billet, rolling, mechanical property, microstructure

回火工艺对 S450EW 力学性能及组织影响

王俊霖 刘志勇 陈吉清 梁 文 陶文哲 黄大伟

（武汉钢铁(集团)公司研究院，湖北武汉 430080）

摘 要 对铁路车辆用耐候钢 S450EW 薄板进行了回火模拟试验，并进行力学性能测试，同时结合扫描电镜（SEM）、金相显微镜（OM）对显微形貌进行分析。研究回火工艺对 S450EW 钢板力学性能的影响规律，以及对 S450EW 组织的影响。研究发现，经工艺参数优化，随回火温度的提高，试验钢的强度下降，屈强比下降；钢的屈强比与组织中软硬相的强度差有关，强度差越大时，屈强比越低，在本组试验中，回火前的组织较为单一，基本以贝氏体组织为主，该组织的屈强比较高，在较高的回火温度可以有效的增加组织中软相—准多边形铁素体的含量，从而降低其屈强比。在显著降低其强度同时，使钢板获得理想的屈强比性能，达到获得良好的冲压性能的目的。

关键词 S450EW，回火，屈强比

Effect of Annealing Temperature on Structure and Mechanical Properties of S450EW

Wang Junlin，Liu Zhiyong, Chen Jiqing, Liang Wen, Tao Wenzhe, Huang Dawei

(Research and Development Centre of Wuhan Iron and Steel (Group) Corp., Wuhan Hubei 430080, China)

Abstract It is done the simulating test of annealing for S450EW sheet, mechanical properties were test, combined with scanning electron microscopy (SEM), optical microscope (OM) the microstructure were analyzed. The influence of annealing temperature on the mechanical properties and microstructure of S450EW steel are studied. The study found that with the process parameters optimization, with the increase of the annealing temperature, the strength of the test steel declined, yield strength ratio decreased; steel yield strength ratio is related to percentage of soft phase in the microstructure , the strength difference is bigger, bend strong ratio is lower. in this experimental group, before tempering , microstructure contain bainite only , and the yield strength is relatively high, at high temperature tempering can effectively increase the content of soft phase quasi polygonal ferrite thereby, reducing the yield strength ratio. Finally, the ideal ratio of yield and yield of the steel plate is obtained.

Key words　S450EW, annealing, yield ratio

A Hybrid Damage Mechanics Model to Predict the Impact Toughness of Structural Steel

Münstermann Sebastian[1], Golisch Georg[2], Wu Bo[2]

(1. Forschungszentrum Juelich GmbH, Germany; 2. Department to Ferrous Metallurgy, RWTH Aachen University, Germany)

Abstract　The toughness transition behavior of bcc steel results from the strong competition between two possible fracture mechanisms. At low temperatures, cleavage fracture is observed, while ductile fracture happens at elevated temperatures. In the transition phase, ductile fracture is followed by cleavage fracture. Charpy impact toughness properties of bcc steel are predicted with a newly-developed hybrid damage mechanics model. An elastic visco-plastic plasticity model with stress-state dependent yielding and isotropic hardening forms the basis of the new model. Since temperature, strain rate and the material's strain hardening properties are the major influences on the microscopic failure mechanisms, the plasticity core of the model considers these influencing parameters. Thus, failure under adiabatic loading conditions can be predicted. Additionally, the effect of damage on the material's plasticity is described with a scalar damage variable D coupled into the yield potential, and both the cleavage and the ductile fracture mechanisms are considered in the corresponding damage evolution law. Stress triaxiality and normalized Lode angle are the major influencing parameters that the hybrid damage evolution law takes into account. Material parameter identifications and successful model application in terms of Charily impact toughness tests are demonstrated. As the model has got a phenomenological character, its applicability for the safety assessment of engineering structures can be expected.

Key words　charpy impact toughness, damage mechanics, stress triaxiality and normalized lode angle, failure at high strain rates, adiabatic conditions, MBW model

马氏体薄带钢张力水淬残余应力的数值模拟

林　潇　张清东

(北京科技大学机械工程学院，北京　100083)

摘　要　利用有限元法和热弹塑性理论，对马氏体薄带钢淬火过程中的温度场、组织场和应力场进行数值模拟，重点分析了带钢张力及横向初始温度分布对带钢淬火残余应力的影响。模拟发现，带钢横向初始温度分布对纵向残余应力在宽度方向上的分布具有重要影响，温差越大，应力分布的不均匀程度越大。基于以上分析，选取具有一定横向初始温差的带钢作为研究对象，研究张力大小及分布对其淬火纵向残余应力的影响规律。结果表明：施加与温度分布规律相同的分布张力可补偿带钢横向温差引起的纵向残余应力在宽度方向上的分布不均，而均布张力对残余应力不均匀程度的影响较小。结合生产实际，建议在导向辊上使用正凸度辊形以控制带钢的板形缺陷。本文研究结果可为生产实际提供指导。

关键词　马氏体薄带钢，水淬，残余应力，数值模拟，张力，横向温差

Numerical Simulation on Residual Stress of Martensitic Thin Steel Strip in Tension During Water Quenching Process

Lin Xiao, Zhang Qingdong

(School of Mechanical and Engineering University of Science and Technology
Beijing, Beijing 100083, China)

Abstract In this paper, the temperature field, microstructure field, and stress field of martensitic thin strip during quenching are numerical simulated using FEM and thermoplastic theory. The effects of tension and transverse initial temperature distribution on residual stress during quenching process are analyzed specially. Based on simulating results, the effect of transverse initial temperature difference on transverse distribution of longitudinal residual stress can not be ignored. The larger the temperature difference, the greater the degree of non-uniform stress distribution. Based on the above analysis, the influence of value and distribution of tension on longitudinal residual stress in martensitic steel strip with a initial temperature difference is studied. The results show that the uneven distribution of longitudinal residual stress caused by non-uniform of transverse initial temperature can be diminished by distributed tension with the same distributing characteristic as temperature. In combination with the production practice, guide roll with positive convexity is advised to control flatness defects of strip. This can provide guidance for practical production.

Key words martensitic thin steel strip, water quenching, residual stress, numerical simulation, tension, transverse temperature difference

“四辊轧机”与连铸相结合的技术构想

陈金富　满春红

（东北特钢集团北满特殊钢有限责任公司，齐齐哈尔　161041）

摘　要　“四辊轧机”是本文提到的新轧机构想，它的构成方式、孔型系统、表面变形和应力状态都与两辊轧机存在很大区别。针对“四辊轧机”的自身特点，结合连铸坯内部疏松、缩孔及偏析缺陷产生的原理，提出采用“四辊轧机”与连铸相结合的新技术，解决连铸坯内部质量问题。文中对“四辊轧机”的特点做了详细介绍，对轻压下技术应用存在的问题和“四辊轧机”与连铸相结合的技术进行了对比，创造性提出了“四辊轧机”与连铸拉坯速度相匹配，在连铸坯液芯位置上实施大压下量变形，从而改善连铸坯内部质量的新方法，提出了“四辊轧机”确定大压下量变形位置和压下量大小的理论根据。同时，对“四辊轧机”连轧机组与连铸坯结合，形成后续产品延伸，打造能源节约型工艺流程生产线，提出了自己的想法。

关键词　两辊轧机，四辊轧机，连铸坯

Technical Conception of the Combination of “Four Rollers Mill” and Continuous Casting

Chen Jinfu, Man Chunhong

(Dongbei Special Steel Group Beiman Special Steel Co., Ltd., Qiqihaer 161041, China)

Abstract "Four Rollers Mill" is a new mill concepion which is mentioned in this paper. It is a very big difference between forming type, pass system, surface deformation, stress state and the two rollers mill. According to the characteristics of the "Four Rollers Mill", combine principle of continuous casting billet internal porosity, shrinkage hole and segregation defects, advance the new technology which adopts the combination of "Four Rolls Mill" and continuous casting, solve the billet internal quality problems. The characteristics of "Four Rollers Mill" is introduced in detail, contrast soft reduction technology application issues with the technology of combining "Four Rolls Mill" and continuous casting, creatively put forward to the matching of "Four Rolls Mill" and continuous casting casting speed, implement the high pressure deformation at liquid core position of continuous casting slab, so as to improve the the new method of billet internal quality, advance that "Four Rollers Mill" determines high pressure volume deformation position and the theoretical basis of pressure volume. At the same time, combine "Four Rolls Mill" continuous rolling mill and continuous casting billet, form follow-up product extension, create energy saving process flow production line, put forward their own ideas.

Key words two rollers mill, four rollers mill, continuous casting bille

The Thermal-Mechanical Coupled FEM Analysis on Continual Tube Rolling MPM Deformation Process

Zou Shuliang[1], Li Pengxue[1,2], Tang Dewen[1,2], Liu Bing[2]

(1. Hunan Key Laboratory of Nuclear Facilities Emergency Safety Technology and Equipment, Hengyang 421001, China; 2. School of Mechanical Engineering, University of South China, Hengyang 421001, China)

Abstract Based on the characteristics of MPM deformation process, a thermal-mechanical coupled model of this process was established by the three-dimensional elastic-plastic finite element method. Some important parameters such as the rolled product size in each stand, the temperature variation curves of the key nodes on the tube, the rolling force and torque of each roller, and the equivalent stress and strain of the tube were analyzed using this model. The analysis results are as follows: The external diameter and wall thickness are alternate reduced between the pass bottom position and the roll gap position and tend gradually to finished size; The temperature of nodes, contacting with mandrel or rolls, decrease for heat transfer between tube and mandrel, tube and rolls, rube and surroundings, while the temperature of internal nodes increase for plastic work and decrease following for conduction within tube; The flow law of metal is that the metal in the bottom of pass is radial compressed, and flow towards the roll gap position or along the axial direction. Comparison between simulation solutions and experiment results shows a good agreement, which means that this model is capable of simulation MPM deformation process as well as forecasting product quality.

Key words MPM, finite element model, continual tube rolling, simulation

冷轧 4830mm 横向条纹分析及对策

陶　涛

（宝山钢铁股份有限公司冷轧厂，上海　201900）

摘　要　某五机架四辊冷连轧机是从德国 SMS 公司引进的具有 20 世纪 80 年代初水平的无头轧制机组，可进行全连续和常规轧制。通过对入口设备和出口滚筒剪的改造，基本所有规格、钢种都采用全连续轧制。支承辊轴承采用的动静压油膜轴承 54－72"，静压主要用于轧机停机启动或低速运行过程中抵消其自重减少摩擦，便于启动过程正常实现。由于对静压认识的不足在轧机出现了4830mm轧机启动横向条纹，导致大量缺陷产生并流出，产生了严重的后果。通过对静压对轧机启动的影响进行探讨和分析，制定了对策措施，有效防止静压异常导致的横向条纹发生。
关键词 冷连轧机，动静压，4830mm，横向条纹

Analysis and Solution to 4830mm Transverse Stripes at Cold Rolling Mill

Tao Tao

(Cold Rolling Plant of Baosteel Co., Ltd., Shanghai 201900, China)

Abstract 5 stand four roll cold tandem mill is endless rolling machine imported from German SMS in twentieth Century at the beginning of the 80's, it can carry out the fully continuous and normal rolling. Through the revamping of equipment in the entrance and durm flying shear in the exit, basically all specifications、steel grades can be rolled using fully continuous rolling.Dynamic and static pressure oil film bearing 54 - 72 " is used for back-up roll bearing, static pressure is mainly used for offsetting the weight and reducing the friction in the process of restarting after stop or low speed , easy to restart. Due to lack of understanding of the static pressure, 4830mm transverse appers at colding mill, resulting in a large number of defects and outflow and resulting in serious consequences, according to the discussing and analysis on the effect of static pressure to restarting after stop ,then solution is formulated,transverse stripes caused by the abnormal static pressure is prevented effectively.
Key words 2030mm cold rolling, dynamic and static, 4830mm, transverse stripes

Development and Application of Steel for Marine Engineering in Baosteel

Gao Shan, Zhang Caiyi, Lu Xiaohui

(Baosteel Research Insititue, Shanghai 201900, China)

Abstract In order to meet the needs of marine engineering, the marine engineering steel has been a strategic product in the area of heavy plate since 2005 in Baosteel. Baosteel's marine engineering steel includes all grades steel in the specification of Classification Society, ASTM and API, EN etc. The marine structural steel are approvaled by ABS, DNV, CCS, GL, LR, BV Classification Society, and has been used by Singapore SembCorp marine,South Korea's Hyundai Heavy industries, Daewoo shipbuilding and marine, Yantai CIMC Raffles Offshore Engineering Co., Ltd., Shanghai Zhenhua Heavy Industries (Group) Co., Ltd., China Merchants heavy industry (Shenzhen) Co., Ltd., Dalian shipbuilding Co., Ltd., and other builders, to construct self elevating offshore platform, semi submersible offshore platform, 30 million tons of marine FPSO . In the future, Baosteel will focus on the development of ultra high strength steel, well-welding steel , corrosion resistance steel and so on.
Key words Baosteel, marine engineering steel, development, application

Study on Microstructure and Properties of High Strength Shipbuilding Steel with High Crack-Arrest Toughness

Lu Xiaohui， Gao Shan， Zhang Caiyi

(Baoshan Iron ＆ Steel Co., Ltd., Shanghai 201900, China)

Abstract The effect of chemical composition and process on high strenth shipbuilding steel has been studied in this paper.The lab research reveals that the addition of Al、Cr、Ni elements to the Nb、Ti microalloyed C-Mn steel will promote the formation of acicular ferrite microstructure.And strength will be improved significantly，at the same time the low temperature toughness is also improved and maintain high values,the Charpy impact value at -60℃ is over 300J.According to lab research，the EH47 grade steel the yield strength of which is over 460MPa has been trial-produced in the mill.And its mechanical properties，especially low temperature crack-arrest toughness have been analysed. The results of temperature gradient double-tension crack-arrest test show that the brittle crack arrest toughness（K_{ca}） of the 50mm thick industrial trial-produced plate is over $9000N/mm^{3/2}$,which is much bigger than the standard of classification society，$K_{ca}>6000N/mm^{3/2}$ at -10℃.Variable energy instrumented DWTT has been explored to measure the crack-arrest toughness，the result is $K_{ca}=7240N/mm^{3/2}$,but the relationship between them needs much research.

Key words crack-arrest, shipbuilding steel, double-tension test, DWTT

Analysis of Intragranular Acicular Ferrite in Low Carbon Steels with Zr and Mg Co-additions

Li Xiaobing, Yu Zhe, Min Yi, Liu Chengjun, Jiang Maofa

(Key Laboratory for Ecological Metallurgy of Multimetallic Mineral (Ministry of Education), Northeastern University, Shenyang 110819, China)

Abstract The formation of intragranular acicular ferrite (IAF) was investigated by adding Zr and Mg simultaneously into Al–killed low carbon steel. Four steels containing various Mg and Zr were refined with vacuum induction furnace, and the characteristics of the inclusion and microstructure were studied scanning electron microscopy equipped with an energy–dispersive X–ray spectrometer. The formation of IAF is demonstrated to be enhanced by the additions of Mg and Zr in the steels, and the highest volume fraction of IAF is obtained in 0.0024% Mg-0.030% Zr steel. The nuclei inclusions are identified as Al_2O_3–MgO–MnS and Al_2O_3–MgO–ZrO_2–(MnS) multiphase particles with size in the range of 1.5–3.0μm. Due to the number density of inclusions for 1.5–3.0μm is relatively low, thus the volume fraction of IAF is low as 18% even in 0.0024% Mg-0.030% Zr steel. Nevertheless, it is observed that the nucleation potential of inclusions is found to increase in the sequence of Al_2O_3–MgO and Al_2O_3–MgO–ZrO_2.

Key words oxide metallurgy, inclusions; Zr–Mg deoxidized steel, intragranular acicular ferrite, microstructure

Characterizing the Interface Stresses between an Elastic Rough Ball and a Smooth Rigid Plate under Lubrication

Wu Chuhan[1], Zhang Liangchi[1], Li Shanqing[2], Jiang Zhenglian[2], Qu Peilei[2]

(1. School of Mechanical and Manufacturing Engineering, The University of New South Wales, NSW 2052, Australia; 2. Baoshan Iron & Steel Co., Ltd., Shanghai 200941, China)

Abstract This paper presents a new method for characterizing the interface stresses between an elastic rough ball and a smooth rigid plate under lubrication. Both micro asperity deformation and macro elastic deformation of the ball were considered. A statistical analysis was carried out to integrate the random asperity deformation with the asperity-lubricant interaction. The surface roughness effect on the hydrodynamic pressure of the pressurized lubricant entrapped at the contact interface was incorporated into the analysis. An iterative approach, with the aid of fast Fourier transform (FFT), was used to predict the interface stresses and the deformed surface profile of the ball. It was found that the method can successfully predict the effects of surface roughness and multi-scale deformation on the contact stresses. It is expected that this method can be used to reveal the property variation of lubricant using ball-on-disk tests.

Key words interface stress, statistical analysis, multi-scale deformation, FFT

热轧带钢边部翘皮缺陷的成因调查

安守滨　李　波

（宁波钢铁有限公司制造管理部，浙江宁波　315807）

摘　要　扫描电镜和能谱分析发现,宁钢热轧带钢边部翘皮缺陷的主要成分是 FeO。据此推断，炼钢工序板坯的角裂纹、气孔缺陷，热轧工序的立辊轧制均可导致热轧带钢边部翘皮。本文重点讨论了立辊轧制对热轧带钢边部翘皮缺陷的影响。调查发现，孔型立辊 E1 可以抑制狗骨形凸起，提高侧压效率和边部质量。但使用异常可导致大批量边部翘皮缺陷发生，同时也造成了立辊磨损异常，降低立辊使用效率和寿命。虽然小侧压量对预防边部翘皮缺陷有利，但应综合考虑生产率和成材率。本文提出避免孔型立辊 E1 异常工作及磨损的措施，包括确保出炉板坯的平坦度、保证轧制线和辊道精度、确保立辊轧机平稳咬钢；优化轧制工艺和模型，针对不同钢种制作立辊道次压下量与边部缺陷等级曲线，优化道次侧压量负荷分配，提高精细化操作水平等。

关键词　边部翘皮，立辊轧制，产品缺陷，热轧带钢，“狗骨状”变形

Mechanism Investigating on the Hot Rolled Strip Edge Shell Defects

An Shoubin, Li Bo

(Ningbo Iron & Steel Co., Ltd., Ningbo Zhejiang 315807, China)

Abstract Analyzed by Scanning Electron Microscopy and Energy Dispersive Spectroscopy，we found the Edge shell defects main component is FeO of Ningbo steel hot rolled strip. So we deduced that the Edge shell defects caused by Corner Cracks、edge hole of slab and Vertical rolling process. In this paper, we especially discussed the influence of Vertical rolling to the Edge shell defects. The investigation result is that Vertical roll with groove can control the Dog bone shape, increase Vertical rolling efficiency and edge quality. But, while the Vertical roll with groove works wrong, it will cause a plenty of Edge shell defects and abnormal edge wear. It also reduce the Vertical rolling efficiency and shorten the life of the Vertical roll. Although small Edge reduction can prevent Dog bone shape , but should be integrated into the productivity and yield. In this paper, we propose some measures to avoid the Vertical roll E1 abnormal wears and works. Including :ensuring the heated slab flatness, ensuring the rolling line and roller table precision, ensuring the Vertical rolling stabilization. Optimizing the rolling process and model. Finding the relation of pass edge reduction with edge defect level according to the different steel type . Optimizing the pass edge reduction distribution ,And improving the level of delicacy operation etc.

Key words edge shell, vertical rolling, product defect, hot rolled strip, “dog bone” deformation

U 形卷取工艺对薄规格低碳铝镇静钢热镀锌板力学性能均匀性的影响

胡燕慧　张　浩　刘再旺　李　维　李　洁　宋鹏心　滕华湘

（首钢技术研究院薄板所，北京　100043）

摘　要　为了提高 0.6mmDX51D+Z 通卷力学性能均匀性，对热轧卷进行了提高头尾卷取温度即 U 形卷取试验。利用带钢力学性能在线检测设备 EMG-IMPOC，对 0.6mmDX51D+Z 通卷力学性能进行检测，根据头尾与中部的力学性能差异来设定 U 形卷取工艺。对实施 U 形卷取工艺前后的 DX51D+Z 头、中、尾力学性能进行了对比。采用金相显微镜对碳化物形貌进行观察，采用试样电解和 ICP 方法对 AlN 析出物数量进行检测，对 DX51D+Z 头尾力学性能改善的原因进行了分析。通过实施 U 形卷取工艺，改善了低碳铝镇静钢热镀锌板通卷组织、析出物和与力学性能均匀性。

关键词　低碳铝镇静钢，热镀锌板，带钢力学性能在线测量系统，U 形卷取

The Effect of U-shape Coiling Temperature Method on Uniformity of Mechanical Property for Hot-Dip Galvanized Low Carbon Aluminum Killed Steel Sheet

Hu Yanhui, Zhang Hao, Liu Zaiwang, Li Wei, Li Jie, Song Pengxin, Teng Huaxiang

(Strip Technology Department of Shougang Research Institute of Technology, Beijing 100043, China)

Abstract　To optimize uniformity of mechanical properties for 0.6mm DX51D+Z, the U-shape coiling temperature tests were carried out on hot rolling line. The mechanical properties of 0.6mm DX51D+Z were tested from head to tail by EMG-IMPOC online measuring system and the U-shape coiling temperature was set based on the results. The mechanical properties of head tail and central parts with and without U-shape coiling temperature were tested. The micrograph of carbides were observed by metalloscope, and quantity of AlN was tested by electrolyzing and ICP analysis. The reasons for optimizing uniformity of mechanical property were analyzed. The U-shape coiling temperature method is effective to optimize uniformity of microstructure, precipitation and mechanical properties of hot-dip galvanized low carbon aluminum killed steel sheet.

Key words　low carbon aluminum killed steel, continuous hot-dip galvanized steel sheet, online measuring system for mechanical properties of steel sheet, U-shape coiling temperature

P 对双层焊管高温钎焊后组织与性能的影响

李　雯

（上海梅山钢铁股份有限公司技术中心，南京　210039）

摘　要　本文通过对比不同P元素含量对双层焊管产品不同退火工艺后强度的影响，证明P对退火后晶粒细化有一定作用，通过对上述不同P含量不同退火工艺的冷轧板模拟钎焊工艺后的组织性能影响发现，在高温模拟钎焊工艺下，材料奥氏体化已十分充分，冷却后组织与钎焊前退火组织的相关性不大，P虽然可以对退火板起到一定细晶强化作用，但高温钎焊后强度得不到显著提升，反而会使韧性明显下降。

关键词　P，双层焊管，钎焊，组织与性能

Effects of P on Microstructure and Properties of Brazing Welded Double Coif Tube

Li Wen

(Technology Center of Meishan Iron & Steel Co., Nanjing 210039, China)

Key words　Phosphorus, welded double coif tube(WDCT), brazing, microstructure and properties

1970 冷连轧机振动与抑制研究

陈　兵　吕　涛

（北京科技大学机械工程学院，北京　100083）

摘　要　针对现场冷连轧机的振动原因以及带钢表面振纹进行了现场跟踪测试，通过对比垂直系统及主传动系统振动优势频率，可以得出轧机扭振是引起轧机垂振的原因。通过计算机仿真分析建立轧机垂直系统及主传动系统模型，研究了轧机的固有特性，与现场测试所得到的垂直系统及主传动系统振动优势频率对比，运用ABAQUS谐响应以及瞬态响应分析技术，分析了轧机在扭矩激励下的谐响应以及瞬态响应，验证了轧机振动测试结果的准确性。最后从轧制乳化液入手对抑制冷连轧机振动进行了试验研究。通过试验研究了现场乳化液的摩擦性能，得到了各种浓度乳化液的油膜强度和摩擦系数值。通过对比油膜强度和接触区压应力分析乳化液摩擦因素及油膜强度对系统振动的影响，得到抑制轧机振动的方法。

关键词　冷连轧机，振动，固有频率，抑制，乳化液，浓度

Research on Vibration Inhibition Experiment of 1970 Cold Rolling Mill

Chen Bing, Lv Tao

(School of Mechanical Engineering, University of Science and Technology Beijing, Beijing 100083, China)

Abstract　For the vibration reason of cold rolling mill and the chatter marks on steel strip, continuous tracking measurement is carried out . Compared the main vibration frequency of the vertical system and the main drive system. It found torsional vibration is the cause of vertical vibration. Through computer simulation analysis to establish the model of vertical system and the main drive system of the mill, Studied the intrinsic properties of the mill and compared with field test results. Using ABAQUS analyzed the harmonic response and the transient response of mill in the excitation of torque, As a result, it validated the accuracy of the vibration testing. Last end from emulsion friction and take a research on

vibration inhibition experiment of cold rolling mill, obtained the film strength and the friction coefficient of various concentrations emulsion via experiment. Comparing oil film strength and contact area pressure, analyzing the effects of friction factors and film intensity on vibration system and the effective counter measures are put forward.
Key words cold rolling mill, vibration, natural frequencies, inhibition, emulsion, coentration

Optimising Plate Surface Quality and Flatness

S. Samanta, J. Hinton, I. Robinson

(Primetals Technology Ltd., Europa Link, Sheffield, S9 1XU, UK,
A Joint Venture of Siemens, Mitsubishi Heavy Industries and Partners)

Abstract The end users of the steel plate are demanding better quality in terms of surface cleanliness and flatness. In the current competitive market, the knowledge and understanding of the complete process route is vital for plate producers to deal with surface and flatness issues. During all operations of plate production, the hot steel surface is exposed to air with which it reacts to form an oxide surface layer (scale). This paper describes how the influence of the scale layer during production extends the entire production process, from the Reheat Furnace through Primary and Mill Descaling practices to Intensive Cooling and Hot Levelling and finally to the Cooling Bed. When rolling thin and wide plate, flatness becomes a major issue where the importance of close coupled process control in the Plate Mill Line is illustrated. The paper highlights how the interaction of surface scale in post-rolling Intensive Cooling can lead to non-homogeneous surface heat transfer which may result in subsequent flatness issues. Finally the use of Levelling technology is briefly described which leads to the best flatness and minimum residual stress in the final product.
Key words steel plate rolling, MULPIC®, scale metallurgy, levelling, plate cooling, flatness control

Microstructure and Hot Deformation Behaviour of High-Cr Cast Iron / Low Carbon Steel Composite

Gao Xingjian[1], Jiang Zhengyi[1,2], Wei Dongbin[1], Jiao Sihai[3]

(1. School of Mechanical, Materials and Mechatronic Engineering, University of Wollongong, Wollongong, Australia; 2. School of Materials and Metallurgy, University of Science and Technology Liaoning, Anshan, China; 3. Research Institute, Baoshan Iron and Steel Co., Ltd., Shanghai, China)

Abstract A hot diffusion-compression bonding process was developed to fabricate a novel laminated composite consisting of high-Cr cast iron as the core and low carbon steel as the cladding on a Gleeble 3500 thermo mechanical simulator at a temperature of 1150℃ and a strain rate of 0.01 s^{-1}. Interfacial bond quality was examined by microstructural characterisation, and hot deformation behaviour of the composite was studied by finite element modelling. Experimental results show that the metallurgical bond between the cast iron and carbon steel was achieved under the proposed bonding conditions. The interface has a good bonding quality because no defects such as micro crack and lack of bonding were detected. The intergrowth between the cast iron and carbon steel was observed, and a carbide-free zone was detected near the interface on the cast iron side. These phenomena indicate that the cast iron and carbon steel were bonded together by diffusion of elements. After well cladding by the low carbon steel, the brittle high-Cr cast iron can be severely deformed at high temperature with crack-free. This significant improvement should be attributed to a decrease of crack sensitivity due to stress being relieved by the soft cladding and an enhanced flow ability of the cast iron by a simultaneous deformation with ductile steel.
Key words hot deformation behaviour, high-Cr cast iron, brittleness, laminated metal composite

特厚复合钢板开发

李文斌 原思宇 李广龙 王 勇

（鞍钢集团钢铁研究院，辽宁鞍山 114009）

摘 要 采用低合金连铸板坯，经过表面清理、真空焊接及室式炉加热、宽厚板轧机轧制的工艺生产 200mm 特厚复合钢板。用探伤、拉伸、剪切及冷弯等试验检验其结合度和力学性能，利用光学显微镜和扫描电镜等分析特厚钢板的组织及拉伸断口。结果表明：采用该工艺生产的特厚复合钢板结合性良好，能够满足 GB/T 7734—2004 Ⅰ级探伤要求；结合部位组织和基体组织均为铁素体+珠光体组织；复合钢板的厚度方向断面收缩率达到 35%以上。

关键词 特厚，复合钢板，真空焊接，组织性能，结合

Research on the Bonding Properties of Clad Heavy Steel Plate

Li Wenbin, Yuan Siyu, Li Guanglong, Wang Yong

(Iron & Steel Research Institute of Angang Group., Anshan 114009, China)

Abstract Using low alloy casting billet, heavy steel plate rolling mill produced 200mm thick clad steel plates after surface cleaning, vacuum welding and chamber furnace heating. Bonding degree and mechanical properties were tested with testing, tensile, shear and cold bending tests.Microstructure and the tensile fracture of heavy steel plate were analyzed by means of optical microscope and scanning electron microscope. Results show that the thick steel plate produced by the technology has good bonding properties and satisfy the GB/T 7734-2004 Ⅰ flaw detection demand .The bonding site structure and matrix structure are both ferrite plus pearlite.Clad.The Clad steel plate's percentage reduction of area with through-thickness direction achieves more than 35%.

Key words heavy, clad steel plate , vacuum welding, structure property, bonding

2180mm 冷连轧机板形目标曲线的应用与优化

贾生晖[1] 李洪波[2] 包仁人[2] 覃忍冬[2] 褚玉刚[1] 刘海军[1]

（1. 武汉钢铁股份有限公司，湖北武汉 430080；

2. 北京科技大学机械工程学院，北京 100083）

摘 要 板形目标曲线是冷连轧机板形控制系统的重要组成部分，某 2180mm 冷连轧机采用了从国外引进的板形控制系统，对适用于该轧机的目标曲线应用与优化方法长期缺乏了解，为此本文通过分析轧机一级、二级板形控制模型，解释了当前目标曲线的调用过程，并结合温度测量数据说明当前所采用的标准目标曲线存在温度补偿不足及调用方式不合理的问题，因此决定采用按宽厚比方式对目标曲线进行分类与调用，并据此对当前的目标曲线进行了优化。

关键词 冷连轧机，目标曲线，宽厚比，温度测量

Application and Optimization of Flatness Setpoint Curve in 2180mm Cold Strip Mill

Jia Shenghui[1], Li Hongbo[2], Bao Renren[2], Qin Rendong[2], Chu Yugang[1], Liu Haijun[1]

(1.Wuhan Iron & Steel Co., Ltd., Wuhan Hubei 430080, China;

2. School of Mechanical Engineering, University of Science and Technology Beijing, Beijing 100083, China)

Abstract Setpoint curve is an important component of flatness control system. The flatness control system of 2180mm is introduced from abroad. Most engineers of this line know little about the setting method and optimization of setpoint curve. The Level 1 and Level 2 flatness control system is analyzed in this paper, and the calling procedure of setpoint curve is given out. With the temperature measuring results, the unreasonable problems in calling method and temperature compensation are discussed. Width-thickness ratio is used for classification of the setpoint curve, which is an improvement of the flatness control system.

Key words tandem cold rolling mill, setpoint curve, width-thickness ratio, temperature measurement

610MPa 级水电钢减量化生产工艺研究

李新玲 应传涛 乔 馨 王若钢 丛津功

（鞍钢股份有限公司鲅鱼圈钢铁分公司厚板部，辽宁营口 115007）

摘 要 为降低 610 MPa 级水电钢的生产成本，鞍钢鲅鱼圈厚板部 5500 生产线采用降低合金成分，以“TMCP+回火”的工艺方式代替常规调质工艺生产,本文通过对 TMCP 及回火工艺研究发现：提高入水温度及返红温度可提高产品力学性能，改善钢板板形；采用合理的回火工艺回火，产品屈服及抗拉强度会进一步提高。有效的提高了产品的一次性能合格率及板形合格率。

关键词 610MPa 级水电钢，低合金成分，入水温度，返红温度，回火工艺

Abstract In order to reduce the production cost of 610 MPa grade hydropower steel, the “TMCP+ tempering” technology instead of conventional quenching and tempering technology which was applied in producing reducing adding content of alloying elements was produced in Ansteel 5500.From this paper, to improve start accelerated cooling temperature and final cooling temperature can improve the product mechanical properties, and improve the steel plate shape. Adopting reasonable tempering process, product yield and tensile strength will be further improved, and Performance qualification rate and steel plate shape will be obviously improved also.

Key words 610 MPa grade hydropower steel, low alloy composition, start accelerated cooling temperature, final cooling temperature, tempering process

超快冷条件下含 Ti 低碳微合金钢析出行为研究

李小琳[1] 高 凯[1] 李军辉[1] 王昭东[1] 邓想涛[1] 李顺超[2] 王国栋[1]

（1. 东北大学轧制技术及连轧自动化国家重点实验室，辽宁沈阳 110819；
2. 宝山钢铁股份有限公司，上海 201900）

摘 要 采用热模拟试验机研究了含 Ti 低碳微合金钢的连续冷却转变行为，利用 OM 和 TEM 对经超快冷工艺处理后实验钢的显微组织和析出行为进行观察和分析。结果表明，变形可以使 CCT 曲线整体向左上方移动，而且铁素体和珠光体的相变区间增大，马氏体相变推迟；当冷速小于 5℃，主要发生铁素体转变，且温度区间为 600~720℃，当冷速在 5~30℃/s 范围内，主要以贝氏体转变为主，温度区间为 500~630℃；实验钢变形后以 80℃/s 和 20℃/s 的冷却速度冷至 600℃后保温 20min 后空冷，显微组织均由铁素体和细小的粒状贝氏体组成，其中铁素体相中析出物平均尺寸分别为 5.0 和 8.2 nm，利用 Orowan 机制对析出强化量进行计算得出，析出强化量分别为 89.3 和 52.4 MPa。
关键词 微合金钢，连续冷却转变行为，透射电子显微镜，纳米析出物

The Precipitation Behavior of the Ti-bearing Low-carbon Steel with the Method of Ultra Fast Cooling

Li Xiaolin[1], Gao Kai[1], Li Junhui[1], Wang Zhaodong[1],
Deng Xiangtao[1], Li Shunchao[2], Wang Guodong[1]

(1. State Key Laboratory of Rolling and Automation, Northeastern University, Shenyang 110819, China;
2. Baoshan Iron and Steel Co., Shanghai 201900, China)

Abstract The continuous cooling transformation behavior of the Ti-bearing low carbon steel was investigated by means of thermal simulation experiments. The microstructure and precipitation behavior of tested steel treated with different cooling rate after deformation have been investigated by using optical microscopy and field-emission-gun transmission electron microscopy. The results show that deformation makes the overall CCT curve of the tested steel move to the bottom left, which greatly expands the ferrite and pearlite transformation area, while inhibit the martensite transformation. The presence of ferrite is restricted to the cooling rates below 5℃/s, the transformation zone is restricted to the temperature range of 600~720℃. The bainite was observed to spread over a wide range of cooling rates from 5~30℃/s in the temperature range of 500~630℃. The microstructure of the steel treated with the cooling rate of 80 and 20℃/s after deformation mainly consists of ferrite and tiny-sized bainite. The average size of the precipitates in ferrite has been determined to be 5.0 and 8.2 nm, respectively. The contribution of the precipitation hardening to the yield strength have been identified to be 89.3 和 52.4 MPa, respectively, based on the Orowan mechanism.
Key words micro-alloyed steel, continuous cooling transformation behaviors, TEM, nano-precipitation

冷镦钢 SCM435 奥氏体连续冷却转变曲线研究

杨 静[1] 罗建华[2] 李桂艳[1] 赵宝纯[1] 黄 磊[1] 王晓峰[1]

（1. 鞍钢集团钢铁研究院，辽宁鞍山 114009；
2. 鞍钢股份有限公司炼钢总厂，辽宁鞍山 114042）

摘 要 利用 Gleeble-3800 热模拟试验机采用热膨胀法结合硬度测试研究了冷镦钢 SCM435 的连续冷却转变过程，通过光学显微镜、电子探针分析了不同冷速下 SCM435 的组织转变行为。结果表明：SCM435 在低于 1℃·s^{-1} 的冷速速率下的组织主要为铁素体+珠光体+贝氏体，在 1～10℃·s^{-1} 的冷速范围内组织主要为贝氏体和少量珠光体+铁

素体或马氏体，在冷速大于 10℃·s^{-1} 时组织为马氏体。

关键词 连续冷却转变，冷镦钢，显微组织

Continuous Cooling Transformation Curve of Undercooling Austenite in SCM435 Cold Heading Steel

Yang Jing[1], Luo Jianhua[2], Li Guiyan[1], Zhao Baochun[1], Huang Lei[1], Wang Xiaofeng[1]

(1. Iron & Steel Research Institute of Angang Group, Anshan 114009, China;
2. Medium Heavy Plate Plant of Ansteel Ltd., Anshan 114042,China)

Abstract The continuous cooling transformation curve of SCM435 cold heading steel was measured by means of Vickers-hardness measurement and thermal dilation method on the Gleeble-3800 thermo-simulation machine. The microstructure was observed and analyzed by optical microscopy (OM), EPMA. The results show that the microstructure of the steel consists of ferrite+pearlite+bainite under the cooling rate 1℃·s^{-1},the main microstructure of steel cosists of bainite and a small amount of ferrite+pearlite or martensite at the cooling rate from 1℃·s^{-1} to 10℃·s^{-1},and the microstructure of steel cosists of martenstie when the coiling rate more than 10℃·s^{-1}.

Key words dynamic CCT diagram, cold heading steel, microstructure

The Occurrence of Separation in High Toughness Linepipe Steel During Drop Weight Tear Test

Liang Xiaojun, Ding Jianhua, Jiao Sihai

(Research Institute, Baoshan Iron & Steel Co., Ltd., Shanghai 201900, China)

Abstract As an effective and significant method for evaluating crack arrestability, drop weight tear test (DWTT) results become even more important for those high toughness steels with large thickness since it is more difficult to assure enough ductile fracture in a severe environment or with a larger thickness. Separation is a commonly seen phenomenon during drop weight tear testing in high toughness linepipe steels. A small numbers of separations are believed not to affect the running ductile fracture resistance, while severe separation may be harmful for fracture or burst of pipes. However, the occurrence of separation may be related to multiple factors. It is helpful to find methods to eliminate separations with a clear understanding about separation for development of high grade high toughness linepipe steels. In this paper, a quantitative relationship was built up betweent microstructures and separations and according to the quantitative analysis data and stress analysis, the essential reasons causing the occurrence of separation in high toughness plate linepipe steels are discussed and summarized.

Key words linepipe steel, DWTT, separation, bainite, stress distribution

Effect of Ti Content on Microstructure and Properties of Hot-rolled Low Carbon Steels for Enamelling

Wang Shuangcheng, Wei Jiao, Sun Quanshe

(Research Institute, Baoshan Iron & Steel Co., Ltd., Shanghai 201900,China)

Abstract Two kinds of hot-rolled low carbon steel plates that contain different contents of titanium element were designed in laboratory. The microstructure and mechanical properties before and after the enamelling were contrastively investigated, and the precipitates in the samples were analysed by TEM and EDS. The results show that there is higher volume fraction of TiC precipitates in the sample containing higher content of titanium element. These precipitated particals, on the one hand,can be the hydrogen traps and improve the fishscaling resistance, on the other hand, can increase the yield strength of the steel plates. For the samples containing higher content of Ti, the yield strength after enamelling can reach more than 400MPa, and no fishscales were found at the surfaces of the samples after two-side enamelling.

Key words hot -rolled low carbon steel , enamelling, precipitate , Titanium content

A Study of the Microstructure and Properties of Dual-Phase Steels Processed Using A Continuous Galvanizing Line Simulation: The Benefits of Optimized Hot and Cold Rolling and the Addition of Vanadium

Yu Gong[1], Anthony J. DeArdo[1,2]

(1. Basic Metals Processing Research Institute Mechanical Engineering and Materials Science University of Pittsburgh; 2. Centre for Advanced Steels Research, Materials Engineering Laboratory Department of Mechanical Engineering, P.O Box 4200 FI-90014 University of Oulu, Oulu, Finland)

Abstract This paper reports recent research into methods of improving the strength and formability of high strength Dual-Phase steels after processing using continuous galvanizing line (CGL) simulations. In this study changes in both composition and processing were explored. The base composition was a low carbon C-Mn-Si steel which would exhibit good spot weldability. To this steel were added two levels of Cr and Mo for strengthening the ferrite and increasing the hardenability of the intercritically formed austenite. Also, these steels were produced with and without the addition of vanadium in an effort to further increase the strength. Since earlier studies revealed a relationship between the nature of the starting cold rolled microstructure and the response to CGL processing, the variables of hot band coiling temperature and level of cold reduction prior to annealing were also studied. Finally, in an effort to increase strength and ductility of both the final sheet (general formability) and the sheared edges of cold punched holes (local formability), a new thermal path was developed that replaced the conventional GI ferrite-martensite microstructure with a new ferrite-martensite-tempered martensite and retained austenite microstructure. The new microstructure exhibited a somewhat lower strength but much high general and local formabilities. In this paper, both the physical and mechanical metallurgy of these steels and processes will be discussed. This research has shown that simple compositions and processes can result in DP steels with so-called Generation III properties.

Key words galvanizing, high strength dual phase steels, vanadium addition, hole expansion, high ductility

利用历史数据回归轧制力模型参数

叶红卫

（宝山钢铁股份公司设备部，上海 200941）

摘 要 本文详尽地介绍了宝钢热轧二热轧的三电技术改造中热轧轧制力模型的基本情况以及模型参数确定的具体方法，通过利用历史数据回归轧制力模型参数的方法保证了二热轧三电改造的一次性通板成功，模型精度也在较短的时间内迅速到项目指标，大大缩短了模型调试的爬坡期。

关键词　L2，轧制力模型，回归分析，模型参数

The Regression for Parameters of Rolling Force Model with Historical Data

Ye Hongwei

(Baoshan Iron and Steel Co., Equipment Dep., Shanghai 200941, China)

Abstract　This paper detailed introduces the method of regression parameters of rolling force model for revamping of the hot rolling mill of Baosteel, and the accuracy of the model also can be quickly to project indicator in a relatively short period of time.

Key words　L2, rolling force model, optimization, model parameters

MULPIC®: In-line Intensive Cooling for Plate Mills

G. Garfitt[a], I. Whitley[b]

Primetals Technology Ltd.[1], Europa Link, Sheffield, S9 1XU, UK

([1]A Joint Venture of Siemens, Mitsubishi Heavy Industries and Partners

[a]*gary.garfitt@primetals.com*, [b]*ian.whitley@primetals.com*)

Abstract　The In-line intensive cooling has become the most powerful process in modern plate mills in order to produce higher strength and value added grades. The key to success is higher cooling rate, better cooling accuracy and uniform cooling across the length and width. Many existing Plate Mills have insufficient cooling capability or no cooling equipment at all. These Plate Mills need to add or improve their cooling capacity in order to respond to the end customer demand. However, every plant has a unique layout and hence it is difficult to incorporate a standard machine to all these customers. This paper illustrates several design techniques that has been adopted to incorporate a standard cooling machine in to existing Plate and Plate Steckel Mills.

Since 2002, Primetals Technologies has commissioned many MULPIC® (MULti Purpose Interrupted Cooling) machines in plate mills around the world. Until recently, most of these installations were standard MULPIC® machines of 24m in length. However, some of the existing plate producers are looking for performance improvements by adding high flow density headers in front of existing laminar cooling sections. This hybrid configuration provides several challenges for process control as well as engineering. In this paper we discuss these challenges and the benefits associated with such a hybrid cooling machine and the steps taken to ensure a smooth and rapid design, installation, start-up and fast achievement of the results. Many new customised design features, such as, flexible hoses, new flow control valve and hybrid cooling model have been introduced to improve performance and ease integrating MULPIC® in to existing systems. This enhances the product range but minimises the costs and disruption to the existing production line during installation and start-up.

Key words　plate cooling, multi-purpose interrupted cooling, intensive cooling, MULPIC, ACC, DQ, heat treatment

Improvement of Mechanical Properties of Cold Rolled High Strength DP-Steels Using Laboratory Scale Process Simulations

Roik Jan[1], Yan Bo[2], Zhang Suoquan[2], Zhu Xiaodong[2], Jiao Sihai[2], Bleck Wolfgang[1]

(1. Steel Institute, RWTH Aachen University, Aachen,Germany;

2. R&D Center, Baoshan Iron and Steel Corporation Ltd., Shanghai, China)

Abstract Laboratory process simulations are carried out to design the processing parameters of two high strength cold rolled DP steels. The hot rolling process is simulated by means of a Thermo-Mechanical-Treatment Simulator. Subsequently, specimens are cold rolled and intercritically annealed using a micro rolling mill and an annealing simulator.Mechanical properties are determined with mini-tensile test specimens. Different hot rolling conditions aiming for bainitic or ferritic microstructures are adjusted by applying low and high coiling temperatures, respectively. The ferritic hot rolled microstructure contains a fraction of bainite and martensite which is aligned in bands parallel to the rolling direction. This banded microstructure is inherited to the final product,while for the bainitic hot rolling microstructure no banding is observed after processing. Both hot rolling procedures lead to a fine or ultrafine grained DP microstructure after cold rolling and intercritical annealing. The higher microalloyed grade has a grain size of 1.2 μm and contains 60 vol% martensite, whereas the other grade has a grain size of 2 μm and 70 vol% martensite. Both DP steels exhibit UTS of 1090 MPa and total elongations $A_{12}$of 13.5 %. Nevertheless, a slightly lower strength-ductility balance is observed for the ultrafine-grained DP steel.

Key words dual phase steels, laboratory process simulation, hot rolling, intercritical annealing

Competitive Formation of Reverted Austenite with Cu-rich Precipitate and Alloyed Carbide in a HSLA Steel during Intercritical Heat Treatment

Liu Qingdong, Gu Jianfeng, Li Chuanwei, Han Lizhan

(Institute of Materials Modification and Modelling, School of Materials Science and Engineering, Shanghai Jiaotong University, Shanghai 200240, China)

Abstract The austenite reversion in a Cu-containing 3.5wt.% Ni high-strength low-alloy (HSLA) steel with intercritical heat treatment was investigated by electron back-scatter diffraction (EBSD), transmission electron microscopy (TEM) and atom probe tomography (APT). It was found that the reverted austenite was formed predominantly located at prior austenite grain boundaries (PAGBs), martensite packet or block boundaries and martensite lath boundaries (LBs) during intercritical tempering at 675℃ for 1 h. The reverted austenite was rich in austenite stabilizers Ni, Mn, Cu and C. Cu-rich precipitates were also detected by TEM distributed at PAGBs and martensite LBs as well as in ferritic matrix. C-rich precipitates, which were simultaneously enriched with austenite stabilizers Ni and carbide formers Cr and Mo, were found by APT. Since austenite reversion and formation of alloyed carbides are both triggered by initial C segregation, there is a competitive diffusion of austenite stabilizers and carbide formers during the further proceeding of austenite reversion and carbide precipitation, leading to the formation of paraequilibrium C-rich precipitates. On the other hand, Cu precipitation is greatly associated with the clustering and segregation of austenite stabilizers Ni and certainly Cu, which inevitably induces a competitive diffusion of the related atoms in participating in austenite reversion and Cu precipitation. As a result, no Cu-rich precipitates are detected in and around the reverted austenite.

Key words reverted austenite, Cu-rich precipitate, C-rich precipitate, intercritical tempering, atom probe tomography, high-strength low-alloy steel

The Microstructure of Dislocated Martensitis Steel

Morris, J.W.Jr.[1], Kinney, C.[1], Qi Liang[1], Pytlewski, K.R.[1], Khachaturyan, A.G.[1], Kim, N.J.[2]

(1. Department of Materials Science and Engineering, University of California, Berkeley, CA 94720, USA; 2. Graduate Institute of Ferrous Technology, POSTECH, Pohang, Korea)

Abstract One of the most fascinating microstructures found in martensitic steels is that of Fe-12Mn, which spontaneously produces an ultrafine-grained microstructure on quenching, with an effective grain size in the micron range and an exceptional resistance to cleavage fracture. The source, and, in fact, the details of this microstructure have been unclear for decades. However, recent research using a combination of TEM, EBSD and theoretical analysis offers new insight. As will be discussed, the microstructure is a laminate of thin plates of dislocated martensite laths that with predominantly twinned boundaries, and an invariant plane of the type $\{011\}_{\alpha'} \| \{111\}_{\gamma}$. Each plate contains all 6 of the martensite variants that are compatible with the martensite packet that includes the plate. The structure of the plate follows from the fact that it forms from the parent austenite through the intermediary of a planar, hexagonal ε-martensite phase. The overall microstructure is the simplest compatible structure that fully relaxes the shear strain of the martensite transformation.

Key words dislocated martensitic steel, martensitic transformation, microstructure

Clean and Efficient Burner Systems for Heat Treating Furnaces

Dr. Joachim G. Wünning[1], Dr. Li Jun[2]

(1. WS Wärmeprozesstechnik GmbH, Renningen, Germany;
2. Baosteel Group Corporation Research Institute R&D Center)

Abstract The reduction of waste gas losses is often the most effective and economic way to increase the efficiency of industrial furnaces. Regenerative burners achieve highest efficiency but one has accept a certain expenditure for cyclic switching and exhaust gas suction. This might not be justified for smaller burner sizes and furnaces. A new gap flow recuperative burner reaches almost the same efficiencies with a recuperative system. To increase the working temperature range, the gap flow heat exchanger can be combined with a ceramic heat exchanger. One important aspect which has to be considered is thermal NO-formation, especially when preheating the combustion air. The use of flameless oxidation technology enables lowest NOx emissions despite high air preheat temperatures.

Key words combustion, heat recovery, preheated air, low NOx, flameless oxidation

The Development of Armor Plate with Super High Strength 2000MPa in Baosteel

Zhao Xiaoting[1], Yao Liandeng[1], Wang Xiaodong[2], Gu Zhifei[2]

（1. Baoshan Iron & Steel Co., Ltd., Shanghai, China；
2. Inner Mongolia First Machinery Group Co., Ltd., Baotou, China）

Abstract The research development of armor plate with super high strength 2000MPa in lab was introduced in detail, involving the results of mechanical properties, cutting, cold bending and ballistic test. The results display that the strength of the test steel is up to 2000MPa with excellent ballistic behavior. In addition, the test steel has good cutting and cold-forming performance with suitable cutting machine and cold forming process.

Key words armor plate with super high strength, cutting, bending, ballistic behavior

回火温度对 DQ 工艺高强钢板组织和性能的影响

杨云清[1] 杜 江[1] 张朋彦[2]

（1. 华菱湘潭钢铁有限公司，湖南湘潭 411101；2. 东北大学，辽宁沈阳 110819）

摘 要 本文主要研究钢板奥氏体在未再结晶区 60%的变形量、终轧 860℃直接淬火后，不同的回火温度对钢板力学性能及组织的影响。600~680℃回火试验研究表明，试验钢板具有较好的回火稳定性、性能良好；回火索氏体的形貌没有发生太大变化；试验钢的最佳回火温度应为 640~660℃。研究结果为制定 DQ-T 钢板的生产工艺提供了依据。

关键词 DQ，回火温度，组织，力学性能

Effect of Tempering Temperature on Microstructure and Properties of DQ High Strength Steel

Yang Yunqing[1], Du Jiang[1], Zhang Pengyan[2]

(1.Valin Xiangtan Iron & Steel Co., Ltd., Xiangtan 411101, China;
2. Northeastern University, Shenyang 110819, China)

Abstract This paper mainly studies the deformation of austenite in the steel, non recrystallizationzone 60% finishing 860℃ after direct quenching effect of different tempering temperature on Microstructure and mechanical properties. 600~680℃ of tempering experiments show that, test plate with tempering stability is good, good performance; the morphology of tempered sorbite did not change much; the best tempering temperature of tes steel should be 640~660℃. The results provide a basis for the production process of DQ-T steel plate.

Key words DQ, tempering temperature, microstructure, mechanical properties

首钢迁钢 2160 热轧超快冷技术的开发与应用

江 潇[1] 王学强[1] 牛 涛[2] 吴新朗[1] 董立杰[1] 王淑志[1]

（1. 首钢股份公司迁安钢铁公司，河北迁安 064404；2. 首钢技术研究院，北京 100043）

摘 要 超快速冷却是热轧控轧控冷技术（TMCP）新的发展方向，为轧制工艺优化和产品合金减量化提供了新的手段。迁钢 2160 热连轧将轧后冷却的前 4 组层流集管替换为 3 组超快冷集管，新增吸水井满足 8000m^3/h 超快冷用水需求，应用高密喷嘴和缝隙喷嘴实现高速均匀冷却，采用新旧控制系统并行、高低压模式自动切换方式实现温度控制精度的稳定，设备上线提前做好准备工作并科学安排工程实施节点，在调试过程采用先低压后高压的步骤，保障生产平稳过渡后逐步将超快冷工艺用于生产实践。本文最后以 25.4mmX70 管线钢这一典型品种规格为例介绍了迁钢 2160 热轧超快冷技术在产品提质降本中的实践应用。

关键词 热轧，超快冷，冷却控制系统，管线钢，控轧控冷

The Development and Application of Ultra-fast Cooling Technology in Shougang Qiangang 2160 HSM

Jiang Xiao[1], Wang Xueqiang[1], Niu Tao[2], Wu Xinlang[1], Dong Lijie[1], Wang Shuzhi[1]

(1. Shougang Qian'an Iron & Steel Company, Qian'an Hebei 064404, China;
2. Shougang Research Institute of Technology, Beijing 100043, China)

Abstract Ultra-fast cooling (UFC) was a newtrendof hot rolling TMCP, which provided a new method to optimize rolling process and reducemicro alloy content. Qiangang 2160 HSM reconstructedits first 4 laminar cooling groups into 3 ultra-fast cooling groups, built a new well to ensure 8000m^3/h water supply capacity, adopted high-density and slit type nozzles to raise cooling rate, developed a parallel cooling temperature control system and high-low water pressure switch mode to maintain the control accuracy, prepared and planed carefully for equipment set-up, implemented low-pressure and high-pressure during commission in turn toguarantee production steady, then put the technology into practice. At the end, an example of X70 25.4mm was given to introduce the application of Qiangang 2160 HSM UFC technology.

Key words hot rolling, ultra-fast cooling, cooling control system, pipe steel, TMCP

高效控制轧制的理论、开发与应用前景

余 伟 李高盛 刘 涛 蔡庆伍

（北京科技大学高效轧制国家过程研究中心，北京 100083）

摘 要 常规控制轧制受三阶段轧制温度的限制，生产效率受影响，也无法抑制再结晶奥氏体晶粒长大。高效控制轧制是将轧制过程的材料温度、应变与材料力学行为实时耦合的新工艺方法，细化奥氏体组织的同时，缩短控轧周期。依据Q345钢的高温性能特点，采用有限元法对高效控制轧制过程进行了模拟，并用实验室轧制模拟以及工业中厚板轧机试验的方法对模拟结果进行验证。研究结果表明：高效控制轧制有利于增加坯料心部应变量，最大增加了61.35%。模拟和轧制结果显示，轧制新工艺还可以减小特厚板心部晶粒尺寸，晶粒度级别提高了1.5级,进一步提高钢板中心区域冲击功。该工艺不仅可以用于中厚板生产，还可以用于生产大中规格棒材。

关键词 轧制，有限元，组织，力学性能，生产效率

Theory, Research and Application Perspective of High Efficient Controlled Rolling

Yu Wei, Li Gaosheng, Liu Tao, Cai Qingwu

(National Engineering Research Center of Advanced Rolling of USTB, Beijing, 100083, China)

Abstract The efficiency of normal controlled rolling process is limited by three phase rolling temperature transition, meanwhile, the recrystallized austenite grain growth cannot be inhibited. The high efficient controlled rolling is a new process which real-timely couples the temperature, strain and mechanical behavior of the deformed metal, further fines grain size and shortens the rolling cycle time. The high efficient controlled rolling of Q345B steel was numerically modeled by FEM based on the test material high temperature performance. The results were verified by laboratory rolling experiment and industrial plate rolling production. The results indicated that the high efficient controlled rolling can increase the strain in the core of rolled piece, mostly by 61.35%. The calculation and experiment proclaim that the grain size in the ultra-heavy plate core were refined by this new rolling process, grain grade improve 1.5 grade level. The toughness of heavy plate is also raised. The new process is used to produce not only medium and heavy plate, but also medium and large diameter bar.

Key words rolling, FEM, microstructure, mechanical properties, production efficiency

RAMON 电磁超声中厚板探伤

——一种非接触式自动钢板探伤技术的实现

周永辉 张均辉 王君君

（衡阳镭目科技有限责任公司，湖南衡阳 421000）

摘 要 超声波无损探伤技术是工业中应用非常广泛的一种检测技术,它具有对被测工件没有损伤、操作简单、性能稳定、可靠性高等优点,可以对工件从生产到在役的各个环节进行全程检测。随着超声检测理论日益成熟，以及现代大规模集成电路和计算机技术的快速发展，超声波检测技术以其快速、准确、高效、无污染、低成本等特点，广泛应用于机械制造、石油化工、航空航天等工业部门，成为保证工程质量、确保设备安全的一种重要手段。而电磁超声与传统的压电超声技术相比，它具有非接触、不需耦合剂、检测速度快、精度高、易激发各种超声波形、表面的油污等非导电材料不影响检测等优点，能更好地适应高温、高油污等恶劣场合。RAOMN 电磁超声中厚板探伤系统可以满足轧钢厂中厚钢板以及铝厂铝合金板的在线检测。本文介绍 RAMON 电磁超声中厚板探伤系统的构成及其测试数据。

关键词 电磁超声，EMAT，探伤

RAMON Electromagnetic Ultrasonic Flaw Detection System for Medium-thick Plate

——Realization of Non-contact Automatic Flaw Detection Technology for Steel Plate

Zhou Yonghui, Zhang Junhui, Wang Junjun

(RAMON Science & Technology Co., Ltd., Hengyang 421000, China)

Abstract Ultrasonic nondestructive detection technology is widely used in industry, it has the advantages of non-destructive to the detected object, easy operation, stable performance, high reliability, etc. and can fully detect every link from production to in service. With the gradual maturity of ultrasonic detection theory, and the rapid development of modern large-scale integrated circuit and computer technology, ultrasonic detection technology is widely applied to machine manufacturing, petrochemical engineering, aerospace engineering etc. with its rapid, accurate, high efficient, non-pollution, low cost etc. features, which has to be an important approach to guarantee the quality of engineering and the safety of equipment. Compared with traditional piezoelectric ultrasonic technology, the electromagnetic ultrasonic has the advantages of non-contact, no need of coupling medium, fast detection, high precision, easy to stimulate various ultrasonic waveforms, no influence on detection for the surface oil contamination and other non-conducting materials, which can better adapt to high temperature, heavy oil contamination and other harsh occasions. RAMON electromagnetic ultrasonic flaw detection system for medium-thick plate satisfies the online detection of medium-thick steel plate in rolling mill and aluminum alloy plate in aluminum plant. This article introduces the components and testing data of RAMON electromagnetic ultrasonic flaw detection system for medium-thick plate.

Key words electromagnetic ultrasonic, EMAT, flaw detection

兼顾热轧硅钢边裂与边降的工艺优化

赵 阳[1] 赵 林[2] 郭 薇[1] 王凤琴[1] 李 飞[1] 王秋娜[2] 李 彬[2]

（1. 首钢技术研究院，北京 100041；2. 河北省首钢迁安钢铁有限责任公司，河北迁安 064404）

摘 要 热轧硅钢带钢存在边裂与边降两大控制难点。边裂缺陷通常是由边部温降所致，控制边裂需采用边部加热器提高中间坯边部温度；边降缺陷与带钢边部金属流动性成正相关，中间坯边部温度越高，热轧硅钢产品的边降越差，造成经济效益损失。针对上述问题，本文深入研究了边部加热器的工作原理，重点分析了边部加热器工艺参数与温降和边降的对应关系，通过实验对比优化，在控制硅钢边裂的同时，提高了热轧硅钢边降指标，解决了长期困扰硅钢的边裂与边降的矛盾问题。

关键词 热轧硅钢，边裂，边降，工艺优化

Process Optimization Helpful to Edge Cracking and Edge Drop of Hot-rolled Silicon Steel

Zhao Yang[1], Zhao Lin[2], Guo Wei[1], Wang Fengqin[1], Li Fei[1], Wang Qiuna[2], Li Bin[2]

(1. Shougang Research & Development Institute, Beijing 100041, China;
2. Shougang Qian'an Iron & Steel Co., Ltd., Qian'an Hebei 064404, China)

Abstract Edge cracking and edge drop are two control points of hot-rolled silicon steel. Edge cracking is caused by temperature drop of strip edge. In order to control edge cracking of hot-rolled silicon steel, we need to increase edge temperature of slab, by using edge heater. Edge drop is caused by lateral metal flow. It is inevitable for the edge drop to increase along with the increase of slab edge temperature, which causes lots ofeconomic loss. To solve this problem, this paper studies the principle of edge heater. Through the study of process parameters of edge heater, this paper gives a method for solving the contradiction of edge cracking and edge drop.

Key words hot-rolled silicon steel, edge cracking, edge drop, process optimization

钙元素对 Mg-2Al 合金显微组织、宏观织构和力学性能的影响

刘 鹏 江海涛 康 强

（北京科技大学国家高效轧制中心，北京 100083）

摘　要　本文设计了不同质量分数的含钙镁合金 Mg-2Al-xCa（x=0.2, 0.5, 1.0），通过光学显微镜观察了合金显微组织，利用扫描电镜观察了第二相析出物的分布形貌，利用能谱仪分析了第二相粒子的化学成分，通过 X 衍射技术研究了合金的宏观织构，此外，还研究了钙元素对合金室温力学性能各向异性的影响。实验结果表明：随着钙元素的增多，合金的晶粒尺寸不断细化，其中 Mg-2Al-0.2Ca，Mg-2Al-0.5Ca 和 Mg-2Al-1.0Ca 合金的平均晶粒尺寸分别为 18μm、16μm、15μm。这是因为钙元素加入到 Mg-2Al 合金中形成了大量 $Mg_{17}(Al, Ca)_{12}$ 和 $(Mg, Al)_2Ca$ 第二相粒子，最终导致了晶粒细化。不同钙元素含量对合金织构影响规律不同，Mg-2Al-0.2Ca 合金呈现典型基面织构特征，Mg-2Al-0.5Ca 呈现双峰织构特征，Mg-2Al-1.0Ca 呈现带状织构特征。双峰织构有利于减小合金各向异性，其中 Mg-2Al-0.5Ca 合金各向异性最小，平均强度可达 235MPa，平均延伸率为 20%。

关键词　Mg-Al-Ca，显微组织，织构，各向异性

The Influence of Calcium Elements on Microstructure, Macro Texture and Mechanical Properties of Mg-2Al Magnesium Alloys

Liu Peng, Jiang Haitao, Kang Qiang

(National Engineering Research Center of Advanced Rolling, University of Science and Technology Beijing, Beijing 100083, China)

Abstract　The Mg-2Al-xCa (x=0.2, 0.5, 1.0) magnesium alloys with different calcium elements content were designed in this paper. The microstructure was observed by optical microscopy (OM). The second phase distribution was observed by the scanning electron microscopy (SEM), and the chemical composition of second phase was analyzed by the energy disperse spectroscopy (EDS). The macro texture of alloys was tested by the X-ray diffraction (XRD) technology. Besides, the influence of calcium element on anisotropy of mechanical properties was investigated. The experiments results shown that the grain size was greatly refined with the increase of the addition of calcium elements, and the average grain size of Mg-2Al-0.2Ca, Mg-2Al-0.5Ca and Mg-2Al-1.0Ca was 18μm, 16μm and 5μm respectively. The reason of this phenomenon was due to the formation of $Mg_{17}(Al, Ca)_{12}$ and $(Mg, Al)_2Ca$ second phases when the Mg-2Al alloy was added into the calcium element. Moreover, the basal texture characteristics varied from the addition of different calcium content. Mg-2Al-1.0Ca alloy exhibited the typical basal texture characteristics. Mg-2Al-0.5Ca alloy exhibited the bimodal texture and the Mg-2Al-1.0Ca alloy exhibited the banding texture. Additionally, the bimodal texture was beneficial to decrease the anisotropy of mechanical properties. The Mg-2Al-0.5Ca alloy had the smallest anisotropy with the average strength of 235MPa and the average elongation of 20%.

Key words　Mg-Al-Ca, microstructure, texture, anisotropy

基于加工图的高硅钢温轧制工艺的研究

程知松　魏　来　蔡庆伍　李　萧

（北京科技大学冶金工程研究院，北京　100083）

摘　要　通过热模拟实验研究高硅钢在变形温度 400~650℃，应变速率 0.001~10s^{-1} 下的中温塑性，作出其流变应力本构方程，得到其平均变形激活能 Q = 446.05kJ/mol，同时绘制出应变为 0.2 和 0.3 时的加工图。从图上可以看出，在 620~650℃、应变速率 5~10s^{-1} 时，处于加工图上的峰值区，功率耗散率达 0.23，具备最佳的加工条件。进一步观察压缩试样的显微组织以及通过实验室温轧验证，确定在此工艺参数组合下能够得到板形良好、无裂纹的温轧板。

关键词　高硅钢，温轧，加工图

Warm Rolling Process of High Silicon Steel Based on Processing Map

Cheng Zhisong, Wei Lai, Cai Qingwu, Li Xiao

(Metallurgical Engineering Institute, University of Science and Technology Beijing, Beijing 100083, China)

Abstract The hot deformation behaviors of High silicon steel have been studied by hot compression test at 400~650℃ and at the strain rates of 0.001~10s^{-1}. The peak flow stress during hot compression was modeled by a hyperbolic sine constitutive equation and an average deformation activation energy of 446.05kJ/mol was calculated. The processing map of strain 0.2 & 0.3 were analyzed according to the dynamic materials model. The results showed that the optimal deformation processing parameters were the temperature range of 620~650℃ and the strain rate range of 5~10s^{-1}, with a peak efficiency of power dissipation of 0.23. Furthermore, the deformed microstructures were observed, and the process was validated by warm-rolling. High silicon steel with the thickness of 0.52mm, bright surface, few edge cracks can be successfully fabricated under the combination of process parameters.

Key words high silicon steel, warm rolling, processing map

冷轧连续退火炉炉辊结瘤原因分析及控制

曾冰林 李 源 邱 波 陈晓伟

（武钢股份冷轧总厂，武汉 430083）

摘 要 本文分析了武钢某冷轧连续退火炉炉辊结瘤的形貌特点、处理过程及化学成分，发现炉辊结瘤的诱因是炉内保护气体氧化性增强，使带钢中的 Mn、P 等元素及清洗段残留的 Na 元素被氧化形成氧化物。明确了炉内气氛异常以及清洗段能力差是导致炉辊结瘤频繁发生的根本原因，根据退火炉及清洗段的设备特点采取相应措施后，炉辊结瘤现象得到有效控制。

关键词 冷轧，连续退火炉，结瘤，炉内气氛

Roller Nodules Cause Analysis and Control of Cold-Rolled Continuous Annealing Furnace

Zeng Binglin, Li Yuan, Qiu Bo, Chen Xiaowei

(WISCO Ltd. CRM, Wuhan 430083, China)

Abstract This article analyzes the furnace roller nodular morphology characteristic, treating process and chemical composition of a cold rolling continuous annealing furnace of wisco, founds the cause of furnace roller nodules is the increase of the shielding gas in the furnace oxidizing , Mn,P and other elements from strip and Na element from cleaning section residues was oxidized to form oxides. Shows the root cause of the furnace roller nodules is the furnace atmosphere abnormalities and cleaning section poor ability, according to the equipment characteristics of annealing furnace and cleaning section the corresponding measures are taken, furnace roll nodules has been controlled effectively.

Key words cold-rolled, continuous annealing furnace,nodules,furnace atmosphere

唐钢酸洗破鳞机的工艺参数优化

王学慧

（河北钢铁技术研究总院，河北石家庄 052000）

摘 要 唐钢冷轧厂的破鳞机长时间不能正常的投入使用，严重影响了酸洗的产品质量、增加了酸耗，通过对破鳞机进行大量的研究，在破鳞机张力、循环水、设备、管理等方面进行改进和调整，摸索出一套适合本厂破鳞机的工艺参数，使该设备能正常的投入使用，提高产品质量，达到了预期的目的。

关键词 酸洗，破鳞机，工艺参数，优化

Optimization on Processing Parameters of Tang Steel's Picking Scale Breaker Equipment

Wang Xuehui

(Hebei General Research Institute of Iron and Steel Technology, Shijiazhuang Hebei 052000, China)

Abstract Ti was not work on scale breaker equipment of Tang Steel cold rolling plate Co., Ltd. long time. Product quality of picking was serious influence and Increase HCL consume. Ti was researched of a lot to scale breaker equipment ,Ti was improved and adjusted at tension and circulating water and equipments and manage of scale breaker equipment, and find out one set processing parameters suitable of scale breaker equipment, this equipments can working and raise product quality of picking, expect purpose was carry out.

Key words picking, scale breaker equipment, processing parameters, optimization

模拟焊后热处理对 14Cr1MoR 钢组织和力学性能的影响

赵燕青[1] 吴艳阳[2] 刘宏强[1] 张 鹏[1] 熊自柳[1] 安治国[1] 黄世平[1]

（1. 河北钢铁技术研究总院，河北石家庄 050000；
2. 河北钢铁集团舞阳钢铁有限责任公司，河南平顶山 462500）

摘 要 研究了模拟焊后热处理对 14Cr1MoR 钢组织和力学性能的影响。结果表明：采用正火（加速冷却）＋回火生产的试验钢，显微组织以贝氏体为主，贝氏体板条的宽度范围为 100~200nm，板条边界及板条内均分布有析出物颗粒，析出物颗粒的尺寸范围为 50~100nm，而经过模拟焊后热处理的试验钢得到贝氏体和多边形铁素体的混合组织，贝氏体板条宽度增加，板条宽度范围为 500~600nm；组织中的析出物明显长大，析出物颗粒的尺寸范围为 200~500nm，析出物主要是 Cr 和 Mo 的碳氮化物析出，并且以 Cr 的碳氮化物析出为主。模拟焊后热处理对钢的力学性能有明显的影响。交货态试验钢的屈服强度、抗拉强度分别为 496MPa 和 610MPa，延伸率为 21.33%；–20℃冲击功为 216J，试验钢经模拟焊后热处理之后，屈服强度下降到 406MPa，抗拉强度降低至 586MPa，延伸率

升至 29%，–20℃冲击功下降至 46J。
关键词 14Cr1MoR 钢，模拟焊后热处理，显微组织，力学性能

Effects of Simulated Post Welding Heat Treatment on Microstructure and Mechanical Properties of 14Cr1MoR Steel

Zhao Yanqing[1], Wu Yanyang[2], Liu Hongqiang[1], Zhang Peng[1], Xiong Ziliu[1], An Zhiguo[1], Huang Shiping[1]

(1. Floor Index of Hebei Iron & Steel Technology Research Institute Pilot Base Inspection Center Shijiazhuang 050000, China; 2. Wuyang Iron and Steel Co., Ltd., Pingdingshan 462500, China)

Abstract The effects of Simulated Post Welding Heat Treatment on Microstructure and Mechanical Properties of 14Cr1MoR Steel were investigated. The results show that the 14Cr1MoR steel which is processed by the normalization with accelerated cooling + tempering has the microstructures of dominant bainite, the width of bainite lath is about 100~200nm, precipitates particles are distributed on bainite lath boundary and bainite lath, the width of precipitates particles is about 100~200nm.The mixed microstructure of bainite and polygon ferrite was obtained after simulated post welding heat treatment, the width of the bainite lath increased, the width of bainite lath is about 100~200nm, the size of the precipitates obviously increased, the size of the precipitates is about 200~500nm, the precipitation particles was Cr carbonitride and Mo carbonitride, and the Cr carbonitride was mainly. The effects of simulated post welding heat treatment on mechanical properties of the tested steel is obviously, the tested steel after normalization with accelerated cooling + tempering heat treatment, the yield strength is 406 MPa, tensile strength is 586 MPa, elongation is 29%, Charpy impact energy is 46J at –20℃, the tested steel after simulated post welding heat treatment, the yield strength decreased to 406 MPa, tensile strength decreased to 586 MPa, elongation increased to 29%, Charpy impact energy decreased to 46J at –20℃.
Key words 14Cr1MoR steel, simulated post weld heat treatment, microstructure, mechanical properties

热轧微合金钢高温变形抗力研究及数学模型优化

刘 勇 刘富贵 郭 韬

（攀钢集团攀枝花钢钒有限公司热轧板厂，四川攀枝花 617062）

摘 要 利用 MMS-200 热力模拟试验机对 510MPa 级含钒低合金钢在 850~1100℃温度区间变形抗力进行研究，分析试验中变形温度、变形速率和变形程度对变形抗力的影响规律，并根据实验研究结果优化和完善热连轧变形抗力数学模型，该品种轧制力预报误差由 8%~10%降低到 5%以下，有效提高了过程控制轧制力数学模型的预报精度。
关键词 含钒低合金钢，热模拟，变形抗力，轧制力，数学模型

Micro Hot Rolled Alloy Steel High Temperature Deformation Resistance Research and Mathematical Model of Optimization

Liu Yong，Liu Fugui, Guo Tao

(Pangang Group Panzhihua Steel & Vanadium Co., Ltd. Hot-rolled Plate Factory,
Panzhihua Sichuan 617062, China)

Abstract By the MMS-200 testing machine for thermal simulation, researched the deformation resistance in 850~1100℃ temperature for vanadium low-alloy steel whose tensile strength is 510MPa and analyzed deformation temperature, deformation speed and deformation extent to influence the deformation resistance in the test. According to the experimental result, the mathematical model of deformation resistance in hot continuous rolling has been optimized and improved. The prediction error for rolling force is reduced from 8%~10% to below 5%. The prediction accuracy of the mathematical model for process control has been improved effectively.

Key words low-alloy steel with vanadium, thermal simulation, deformation resistance, rolling force, mathematical model

冷轧低碳铝镇静钢表面暗斑及指甲痕缺陷分析

高小丽[1] 刘 朋[1] 于 洋[1] 王文广[1] 孙超凡[1] 周 旬[2]

（1. 首钢技术研究院，北京 100043；2. 首钢京唐钢铁公司，河北 063200）

摘 要 针对某钢厂低碳铝镇静钢冷硬卷表面的点状暗斑及指甲痕缺陷，采用 SEM、显微硬度仪等检测手段对此类缺陷进行了详细表征，发现暗斑处有异常粗大的晶粒，指甲痕缺陷正好发生于粗晶/细晶界面处，粗晶区与细晶区显微硬度有明显差别。讨论了暗斑与指甲痕的伴生机理，发现这两种缺陷的根源为带钢表层异常粗大的晶粒。研究了粗晶的形成机理并梳理了钢厂实际工艺，认为高温卷取过程落入了两相区，卷取前夹送辊压力下带钢表面存在小变形，这为后续缓冷过程中表层晶粒异常长大创造了条件。提出适当降低卷取温度、降低冷却速度提高相变点的措施使卷取过程尽量远离两相区，同时降低夹送辊压力防止临近变形蓄积能量。新措施实施后，该类缺陷发生率大幅下降，成品性能也没有受到明显影响，效果良好。

关键词 低碳铝镇静钢，表面，粗晶，指甲痕

Cold Rolled Low Carbon Aluminum Killed Steel Surface Dark Spots and Nail Mark Defects

Gao Xiaoli[1], Liu Peng[1], Yu Yang[1], Wang Wenguang[1], Sun Chaofan[1], Zhou Xun[2]

（1. Shougang Research Institute of Technology, Beijing 100043, China；
2. Shougang Jingtang United Iron & Steel Co., Ltd., Hebei 063200, China）

Abstract The point-like dark spots and nail mark defects in LCAK steel chilled strip surface were characterized in detail using SEM and microhardness detection.We found abnormal coarse grains in the dark spotsarea, nail mark defects just occur in interface area of coarse grain and fine grains,and the microhardness of coarse grains and fine grains areawas significantly different. We discussed the mechanism of dark spots associated with nail marks, and found the main cause of both was abnormal coarse grains in the strip surface layer.The formation mechanism of coarse grains were analysed,and we found that high coiling temperature made the strip surface easy to fall into two-phase region before coiling, and little deformation carried by coiling pinch rolls created the conditions for abnormal grain growth in subsequent slow cooling process.The way controlling this kind of defects is lowering coiling temperature, reducing the cooling speed for increasing the phase transition point to make the coiling process as far as possible away from the two-phase region, and reducing the pressure ofpinch roll to prevent the energy accumulation of the critical deformation. After using the new measures, incidence rate of such defects declined substantially, and the product performance has not been significantly affected.

Key words LACK steel, surface, coarse grain, nail mark

高强冷轧板超快冷条件下组织转变与力学性能的研究

王林炜杰[1] 宋仁伯[2] 赵征志[1] 于三川[2] 高 喆[2] 胡靖帆[2]

（1. 北京科技大学冶金工程研究院，北京 100083；
2. 北京科技大学材料科学与工程学院，北京 100083）

摘 要 利用金相显微镜、TEM 等方法分析了不同保温温度和冷却速度对 C-Si-Mn-Nb 系高强冷轧板显微组织、力学性能和第二相粒子析出行为的影响。结果表明：50℃/s 冷速下，保温温度低于 800℃时出现带状马氏体，强度高，延伸率较低，超过 840℃时铁素体体积分数显著下降，并出现 M-A 和残奥组织，在 820℃时，马氏体组织弥散分布在铁素体基体上，观察到尺寸在 7~9nm 之间的 Nb（C，N）析出物，其综合力学性能最佳，抗拉强度达到 826MPa，延伸率为 20.2%。400℃/s 冷速下的显微组织在保温温度为 780~860℃时都由马氏体和铁素体两相构成，随保温温度的升高，抗拉强度提高，延伸率下降明显，780℃保温时其综合力学性能最佳，抗拉强度为 985MPa，延伸率为 14.2%，观察到尺寸在 5~7nm 之间的 Nb（C，N）析出物。

关键词 高强钢，超快冷，力学性能，显微组织，第二相粒子

Microstructure Evolution and Mechanical Properties of High Strength Cold Rolled Sheet under Ultra-fast Cooling Rate

Wang Linweijie[1], Song Renbo[2], Zhao Zhengzhi[1], Yu Sanchuan[2], Gao Zhe[2], Hu Jingfan[2]

(1. Research Institute of Metallurgy Engineering, University of Science and Technology Beijing, Beijing 100083, China; 2. School of Materials Science and Engineering, University of Science and Technology Beijing, Beijing 100083, China)

Abstract Effects of annealing temperature and cooling rate on microstructure, property and precipitation behavior of second-phase particle for the C-Si-Mn-Nb ultra high strength steel were studied by metalloscope and TEM. The results showed that zonal martensite, high strength and low elongation was observed when the cooling rate was 50 ℃/s and the holding temperature was lower than 800℃. At the same cooling rate, M-A and retained austenite was observed when the holding temperature was higher than 840℃. And the best mechanical property appeared when the temperature was 820℃, the tensile strength and elongation exceeded 826 MPa and 20.2%. At that temperature, martensite dispersed in ferrite and the second phase particles were Nb(C,N) precipitates which sizes were 7~9nm. When the cooling rate was 400℃/s and the holding temperature came from 780℃ to 860℃, the microstructure consisted of martensite and ferrite. The tensile strength enhances and elongation reduced obviously with holding temperature increasing. The best mechanical property appeared when the temperature was 780℃, the tensile strength and elongation exceeded 985 MPa and 14.2%, and Nb（C,N） precipitates which sizes were 5~7nm were observed.

Key words high strength steel, ultra-fast colding, mechanical property, microstructure, second-phase particle

冷轧罩式炉锈蚀缺陷研究

庚 丽 许 新 杨军荣 韩 炯

（鞍钢股份有限公司冷轧厂，辽宁鞍山 114021）

摘 要 由于罩式炉退火钢卷后在 5~10 月份经常会产生锈蚀等缺陷，影响用户的外观使用。我们在鞍钢冷轧 1780 机组排查时找不到锈蚀产生的根本原因需很长时间，对锈蚀缺陷量控制不利，因此我们根据分析几个最长产生锈蚀的源头，通过现场实验，研究不同工序下锈蚀的形貌，对照这个锈蚀形貌可以发现锈蚀产生的根源，结果表明，不同的锈蚀源头产生的锈蚀形貌和颜色不同，不同工序经过罩式炉退火后产生的锈蚀也不同。

关键词 罩式炉，水，油，锈蚀

Study on the Rust Defect of Cold Rolled Batch Annealing

Geng Li, Xu Xin, Yang Junrong, Han Jiong

(The Cold Rolling Plant of Ansteel Co., Ltd., Anshan Liaoning 114021, China)

Abstract The bell type furnace annealing coil after 5-10 months often have defects such as corrosion, affect the user's appearance. The fundamental reason why we in the investigation of Ansteel 1780 units can not find the corrosion production needs for a long time, the corrosion defect control adverse, so we according to the source analysis of the longest corrosion, through field experiments, study of the morphology of different processes of corrosion, the corrosion morphology can be found in the root control, corrosion result show that the corrosion morphology and corrosion source color different, different processes after annealing furnace after corrosion is also different.

Key words batch annealing, water, oil, rusting

鞍钢 1450mm 冷连轧机薄料板形研究

张福义[1] 刘英明[1] 王有涛[1] 李富强[1] 英钲艳[1] 辛 鑫[2]

（1. 鞍钢股份有限公司冷轧厂，辽宁鞍山 114021；
2. 鞍钢股份有限公司产品发展部，辽宁鞍山 114021）

摘 要 冷轧带钢轧制过程中，薄料带钢的板形较难控制，鞍钢 1450mm 冷连轧机以生产薄料为主，为改善带钢板形，根据实际生产总结出改善板形的措施：提高轧辊精度，缩短轧辊轧制周期，优化窜辊补偿量、架间平均张力，优化乳化液温度，使生产的带钢边浪及二肋浪等浪形缺陷明显减少，板形得以大幅度改善。

关键词 薄板，板形控制，冷连轧机

The Study of Sheet Shape for 1450mm Tandem Cold Mill of Ansteel

Zhang Fuyi[1], Liu Yingming[1], Wang Youtao[1], Li Fuqiang[1],Ying Zhengyan[1], Xin Xin[2]

(1. Cold Rolled Strip Steel Mill of Angang Steel Co., Ltd., Anshan Liaoning 114021, China;
2. Strip Development Department of Angang Steel Co., Ltd., Anshan Liaoning 114021, China)

Abstract The flatness of sheet is hard to control in the process of cold rolling. Cold tandem mill of 1450mm is given priority to with producting sheet. In order to improve the flatness of cold rolled strips, the measures are adopted based on practical production experience. The roll grinding is improved, rolling period is shorten, intermediate roll shifting compensation, mean tension and the temperature of emulsion is improved. Edge wave and the double-rib wave are reduced obviously, sheet shape is improved.

Key words sheet, flatness control, tandem cool mill

邯钢热轧-冷轧带钢“亮带-鼓包”攻关与工艺优化

李冠楠[1]　贾改风[1]　席江涛[2]

（1. 河北钢铁集团邯钢公司技术中心，河北邯郸　056015；
2. 河北钢铁集团邯钢公司连铸连轧厂，河北邯郸　056015）

摘　要　邯钢CSP薄规格软钢不同程度出现了纵向亮带，影响了CSP-冷轧的正常生产秩序。通过寻找形成亮带的直接原因，优化卷取阶段拉力分配制度，优化原始辊型凸度、轧辊均匀磨损控制、优化板型控制等措施的实施，改善了邯钢CSP薄规格低碳钢横断面厚度分布情况，降低局部高低的出现率，避免带钢低温阶段发生冷加工变形，CSP亮带-冷轧鼓包现象大幅度降低，提高了产品等级。

关键词　亮带，薄规格，工艺优化

Research and Process Optimization of Handan Iron and Steel Hot Rolling-cold Rolling Steel “Bright Band-drum Kits”

Li Guannan[1], Jia Gaifeng[2], Xi Jiangtao[2]

(1. Technique Centure, Handan Iron and Steel Co., Ltd. Corporation, Hebei Steel and Iron Group, Handan Hebei 056015, China; 2. CSP, Handan Iron and Steel Co., Ltd. Corporation, Hebei Steel and Iron Group, Handan Hebei 056015, China)

Abstract Handan Iron and Steel CSP thin gauge mild steel varying degrees vertical bright band, affecting the normal order of production of the CSP-Cold Rolling Plant. Analyze the reasons, optimization the coiling stage Rally distribution system, optimization of the original roll crown, roll uniformity wear control, optimize the implementation. Improve CSP thin gauge low carbon steel cross-sectional thickness distribution, reduce the local high and low rate of occurrence, avoid strip cryogenic stage of cold deformation. CSP band-Cold-rolled drum kits substantial reduced, product grade improved.

Key words bright band, the thin gauge, process optimization

345MPa 高强度级别特厚钢板板形研究

王亮亮 潘庆轩 潘凯华 罗 军 乔 馨 李靖年 于金洲 曹春玉

（鞍钢股份公司鲅鱼圈钢铁分公司，辽宁营口 115007）

摘 要 针对 345MPa 强度级别特厚钢板板形不平度超差问题，通过分析加热、轧制、矫直等工序的影响因素，得到了此类钢板的最佳生产工艺，成功将钢板不平度由原来的 5mm/m 降低至 2mm/m 以下，达到协议标准，满足客户需求。

关键词 高强度，特厚钢板，不平度，加热，轧制，矫直

Abstract Contraposed the problem of 345MPa high strength heavy plate’s profile shape exceed standard, according to analyze the factor of heating、rolling、leveling, have found the optimal method, and finally reduced the unflatness to 2mm/m from 5mm/m, meeting the customer’s request.

Key words high strength，heavy plate，unflatness，heating，rolling，leveling

锚具用夹片热处理硬度不均分析

翟进坡[1,2,3] 于 浩[1] 陈继林[1,2] 王利军[1,2] 李 军[1,2]

（1. 北京科技大学材料科学与工程学院，北京 100083；2. 邢台钢铁有限责任公司，河北邢台 054027；3. 河北省线材工程技术研究中心，河北邢台 054027）

摘 要 锚具用夹片采用 20CrMnTi 盘条制造，正常检验过程中发现：夹片热处理后出现硬度不均现象。本文对夹片毛坯的化学成分、显微组织、晶粒度及夹片成品的硬度、显微组织进行了分析，分析结果表明：热处理工艺不当是造成夹片硬度不均的原因。

关键词 锚夹片，20CrMnTi，硬度不均，热处理工艺

Analyzing of Nonuniform Hardness for Anchorage Clamp Plate after Heat Treatment

Zhai Jinpo[1,2,3], Yu Hao[1], Chen Jilin[1,2], Wang Lijun[1,2], Li Jun[1,2]

(1. Material Science and Engineering Institute, Beijing University of Science and Technology, Beijing 100083, China；2. Xingtai Iron and Steel Co., Ltd., Xingtai 054027, China；
3. Hebei Wire Engineering Technology Research Center, Xingtai 054027, China)

Abstract 20CrMnTi wire rod can be used to manufacture the anchorage clamp plate. In the normal production, the test results show that the hardness of anchorage clamp plate is nonuniform. In this paper, the chemical composition, microstructure, grain size of clamp plate billet and the microstructure, hardness of the finished clamp plate product were studied and analyzed. The results show that improper heat treatment technics causes nonuniform hardness for anchorage clamp plate.

Key words anchor cramp, 20CrMnTi, nonuniform hardness, heat treatment process

高速极薄冷轧带钢酸轧机组的工艺润滑系统研究与分析

毛召芝

（中冶京诚工程技术有限公司，北京　100176）

摘　要　随着钢铁行业步入产品结构调整的深水期，目前国内冷轧企业特别是民营企业新建的冷轧带钢机组主要突出高速和极薄两个技术特点。本文通过对国内某典型全国产化高速极薄酸轧联合机组工艺润滑系统的设计理念出发，详细分析了高速极薄机组工艺润滑系统自身的技术特点，希望为类似机组和工程技术公司从事该系统设备维护和设计研发的人员提供些许思路。

关键词　酸轧机组，工艺润滑，高速，极薄

The Research and Analysis of Emulsion System of High-speed and Thin Strip Pickling-mill Stand Process

Mao Zhaozhi

(Steel Rolling Division, Capital Engineering & Research Incorporation Limited, Beijing 100176, China)

Abstract　With the iron and steel industry into the product structure adjustment of the deep water，At present，domestic enterprises especially cold rolling mill of new private enterprises mainly highlight the two technical features of high speed and thin. Based on a typical national production design process of emulsion system of high speed and thin strip pickling-mill stand，The technical features of high speed and thin technology emulsion for detailed analysis of the system itself，Hope to provide certain ideas for this system equipment maintenance and design researching of engineer technical company and similar mill stand group.

Key words　picking-mill stand，process lubrication，high-speed，very thin

KOCKS 旋转轧机　新型的无缝管延伸轧管方法

Jörg Surmund, Erich Bartel, Patrick E. Connell

(KOCKS 公司，希尔登 40721，德国)

摘　要　当前无缝管生产领域内的延伸机组年生产能力超过了 30 万吨。为了满足中小型企业对设备生产能力的较

低要求，KOCKS 公司开发了旋转轧机 KRM，该设备的年生产能力为 15 万吨。KRM 旋转轧机的设计成熟，生产管材质量高，管材成材率高，同时相比于纵轧延伸机组的设备投资更低。

关键词 KOCKS 四辊旋转轧机，斜轧延伸轧管

KOCKS Rotation Mill An Innovative Elongation Method for Seamless Tubes

Jörg Surmund, Erich Bartel, Patrick E. Connell

(KOCKS GmbH, Hilden 40721，Germany)

Abstract The market for elongation systems of seamless tubes is largely driven by high production methods capable of producing in excess of 300,000 t/a. In an effort to satisfy the requests of numerous customers for a process capable of producing in the medium range of up to 150,000 tpy, Friedrich Kocks GmbH & Co KG has developed the Kocks Rotation Mill (KRM). This machine is a proven, yet innovative elongation method that allows the production of high quality tubes with excellent yield, all with a capital cost that is much lower than longitudinal elongation equipment.

Key words Kocks rotation mill, cross roll elongation

超快冷工艺 800MPa 级高强钢的组织与性能

陶军晖 马玉喜 杜 明 宋 畅 习天辉 徐进桥 郭 斌

（武汉钢铁（集团）公司研究院，武汉 430080）

摘 要 针对武钢热轧总厂二分厂超快冷改造前后生产的 5～12 mm 厚度规格 800MPa 级高强钢的力学性能进行了全面检测，分析了超快冷工艺生产的试验钢金相组织及应用情况。结果表明：对于厚度规格小于 10mm 的 800MPa 级高强钢，超快冷工艺对其屈服强度和抗拉强度提高幅度不大，而对冲击韧性有较大改善。对于厚度规格大于 10mm 的 800MPa 级高强钢，超快冷工艺对其屈服强度和抗拉强度有较大提高，而对冲击韧性提高幅度不大。超快冷工艺生产的试验钢具有优良的力学性能、成型性能和焊接性能，并成功应用于工程机械吊车吊臂和泵车伸缩臂等零件的生产。

关键词 超快冷，800MPa，高强钢，组织，性能

Microstructure and Performance of 800MPa High Strength Steels Produced by Ultra-Fast Cold

Tao Junhui, Ma Yuxi, Du Ming, Song Chang, Xi Tianhui, Xu Jinqiao, Guo Bin

(Research and Development Center of WISCO, Wuhan 430080, China)

Abstract The mechanical performance of 800MPa high strength steels (thickness specifications:5~12mm) which were produced before and after the UFC transformation at the second factory of hot-rolled mill of WISCO were comprehensively tested, and the microstructure and applications of the test steels done by the UFC were also analyzed. The results show that for the 800MPa high strength steels of thickness specifications of less than 10mm the UFC has little effect on their yield strength and tensile strength but greatly improves their toughness, while for the steels of thickness specifications of more than 10mm the UFC greatly improves their yield strength and tensile strength but has little effect on their toughness. The test steels produced by the UFC has excellent mechanical performance, formability and weldability and were also successfully used in production of the booms of cranes or pumping trucks and other parts.

Key words ultra-fast cold, 800MPa, high strength steel, microstructure, performance

中厚板控制冷却过程温度场的计算模型

彭宁琦　汤　伟　汪贺模　钱亚军

（湖南华菱湘潭钢铁有限公司，湖南湘潭　411101）

摘　要　根据中厚板轧线 L2 级记录的大量现场数据以及传热学原理，采用有限差分、神经网络、分类自适应等方法，建立了适应中厚板冷却过程的温度场计算模型。离线模拟结果表明：预矫温度和返红温度计算值与测量值误差在±10℃内的比例分别为 96.1%和 92.0%，标准差分别为 4.9℃和 5.6℃。并通过建立的模型对厚度和集管组数变化时空冷、水冷和返红 3 个过程厚度方向上的温度场和平均冷速进行了分析，分析结果为理解和制定合理的中厚板控冷工艺提供了依据。

关键词　温度场，中厚板，有限差分，神经网络，自适应

Calculation Model of Temperature Field for Medium and Heavy Plate during Controlled Cooling Process

Peng Ningqi, Tang Wei, Wang Hemo, Qian Yajun

(Hunan Valin Xiangtan Iron and Steel Co., Ltd., Xiangtan Hunan 411101, China)

Abstract According to large amounts of rolling field-data recorded by L2 system in wide heavy plate mill plant and basis of heat transfer theory, the practical calculation model of temperature field for medium and heavy plate during controlled cooling process was developed by methods of finite difference, artificial neural network and sorted self-learning. Off-line simulation results show that ratio of absolute error less than 10℃ between measured temperatures and computed ones are respectively 96.1 % and 92.0 % for pre-leveller temperature and stop cooling temperature, and standard deviation are respectively 4.9℃ and 5.6℃. Then using the further model, the temperature distribution along thickness and the average cooling rates on air-cooled, water-cooled and re-reddened phases in different plate thicknesses and pipe arrangements were analyzed, such can be used to understand and establish proper technological parameters of controlled cooling process.

Key words temperature field, medium and heavy plate, finite difference method, neural network, self-learning

F550 船板钢中心偏析遗传性研究

季益龙[1]　刘建华[1]　陈　方[2]　刘　建[1]

（1. 北京科技大学冶金工程研究院，北京　100083；
2. 马钢工程技术集团有限公司，安徽马鞍山　243000）

摘　要　本研究采用了原位统计分布分析技术，分别对加热处理前后的 F550 船板钢铸坯以及轧材中的 C、P、S、Mn 四种元素进行了偏析度分析。实验结果显示，铸坯中心存在着宽度为 5mm 左右的偏析带，与 Mn 元素相比，P、C、S 的偏析较为严重；在经过加热处理后，中心偏析得到明显改善，但不能完全消除，这表明中心偏析具有遗传性；轧制后 C、S、P、Mn 的偏析度均得到不同程度改善，其中 S、Mn 的偏析趋于消除，但 C、P 的偏析带依然明显，且偏析带宽度由 5mm 扩展为 10mm。

关键词　F550 船板钢，中心偏析，遗传性，原位统计分布分析

Research on Centerline Segregation Heredity of F550 Shipbuilding Steel

Ji Yilong[1], Liu Jianhua[1], Chen Fang[2], Liu Jian[1]

(1. Engineering Research Institute, University of Science and Technology Beijing, Beijing 100083, China;
2. Masteel Engineering and Technology Group Limited Company, Maanshan Anhui 243000, China)

Abstract　In this paper,the technology of original position statistic distribution analysis was applied to analyze the element segregation of C, P, S and Mn in the F550 shipbuilding steel slab before and after heat treatment and rolled-products. The results show that there is a 5mm-thick segregation band in the center of the slab. The segregation of P, C and S is more serious than C. Although the segregation was obviously ameliorated after heat treatment, it could not be eliminated completely by heating. This indicates that centerline segregation is hereditary. After rolling, the segregation was improved in different degree. The band of S, Mn was nearly invisible, while that of C,P was still obvious. And thewidth of the center segregation band extended to 10mm after rolling.

Key words　F550 shipbuilding steel, centerline segregation, hereditary, original position statistic distribution analysis

热处理方式对中碳抗酸容器钢性能的影响

赵新宇　邹　扬　刘　洋　秦丽晔　赵　楠　隋鹤龙　樊艳秋

（首钢技术研究院，北京　100043）

摘　要　随着含硫油气的开发，对应用于湿 H_2S 环境下的容器类钢种的抗氢致裂纹要求越来越高。中碳抗酸容器钢是指碳含量在 0.1%~0.3%之间的具备抗湿硫化氢酸性环境的压力容器用钢。通过研究不同热处理方式：TMCP、正火和正火加回火对中碳抗酸容器钢各项性能的影响，全文得到以下结论：拉伸性能方面，TMCP 最高，正火与正火加回火试样基本相当，低于 TMCP 试样 10~50MPa 左右；低温冲击性能方面，TMCP 韧脆转变温度最低为-60℃左右，正火与正火加回火试样韧脆转变温度有所升高，在-40~-50℃之间；抗酸性能方面，正火抗酸性能最优，TCMP 其次，正火加回火试样抗酸性能最差。

关键词　中碳抗酸容器钢，热处理，TCMP，正火，回火，氢致裂纹

Effect on Properties of Medium Carbon Vessels Steel with Resistant Sour of Style of Heat Treatment

Zhao Xinyu, Zou Yang, Liu Yang, Qin Liye, Zhao Nan, Sui Helong, Fan Yanqiu

（Shougang Technology and Research Institute, Beijing 100043, China）

Abstract With the exploitation of gas mixed with sulfur, the requirement of hydrogen induced cracking (HIC) resistance for medium carbon vessels serving at environmental with hydrogen sulfide solution become more rigorous. Medium carbon vessel steels for resistance to sour are the steels pressure vessel steels with the capacity of resistance to environment of acid waterish hydrogen sulfide, which the element of carbon is designed at 0.1%~0.3%. Various styles of heat treatment are taken such as TMCP (Thermo Mechanical Control Process), normalizing and tempering after normalizing to research on properties of medium carbon vessel steel. The conclusions are following: Aspects on properties of stretching, the strength of samples under TMCP is highest. However the strength of samples under normalizing and tempering are equivalent, which is lower than that under TMCP almost 10-50MPa. Meanwhile aspects of impact, the ductile-brittle transition temperature of samples under TMCP is excellence of -60℃. And that of samples under normalizing and tempering increases to -40℃ or -50℃. Aspects on properties of resistance to sour environment, the properties of that under normalizing are best, that under TMCP take second place and that under tempering are worst.

Key words medium carbon vessel steel, heat treatment, TCMP, normalizing, tempering, HIC

Assessment of Cold Rolling Lubricants by 3-D Stribeck Curves and Roll Bite Mimicking Tests

Smeulders Bas

(Research Scientist, Quaker Chemical, The Netherlands)

Abstract In order to account for the plastic surface deformation occurring in cold rolling, a novel, 3-D Stribeck curve is proposed. This curve is characterised by three 'anchor points' which represent (1) the friction in the boundary regime, (2) the friction in the elastohydrodynamic regime, and (3) the friction experienced in a thin lubricant film. Two operational numbers define the lubrication regime, representing (1) film formation in the inlet zone, governed among others by the process speed and (2) in-bite film formation facilitated by plastic surface deformation, governed among others by the sliding speed in the contact. Lubrication in the rolling process can be influenced by any of the 3 anchor points and 2 film formation mechanisms, all of which are influenced by the lubricant composition. The ideas behind the 3-D Stribeck curve can be used in a more targeted lubricant development. The 3-D Stribeck curve can be measured in the laboratory, preferably for emulsions. Additional test methods were developed to investigate boundary lubrication of emulsions under severe conditions, mimicking frictional power intensity and flash temperatures in the roll bite. Good correlations are found with performance in a pilot mill and a production mill.

Key words plastic surface deformation, boundary lubrication, EHL, thin-film lubrication, wear

提升 40kg 级薄规格船板韧性的轧制工艺研究

刘泽田[1,2] 麻永林[1] 陆 斌[2]

（1. 内蒙古科技大学材料与冶金学院，内蒙古包头 014010；
2. 内蒙古包钢钢联股份有限公司薄板坏连铸连轧厂，内蒙古包头 014010）

摘 要 本文设计了采用厚规格连铸坯生产薄规格船板的工艺技术，并研究了两种不同轧制工艺下生产的 40Kg 级薄规格船板的力学性能和显微组织的差异。结论表明，自开坯工艺会显著影响到钢板的低温冲击韧性和微观组织。

不合适的自开坯工艺使钢板混晶严重，并产生较多尺寸较大的异常晶粒，对钢板的低温冲击韧性不利。而采用大压下量、少道次数完成自开坯，会减轻钢板的混晶程度，消除尺寸较大的异常晶粒，显著提升钢板的低温冲击韧性。
关键词 薄规格，高强船板，自开坯工艺，低温韧性

Research of Rolling Process for Improving Toughness of 40kg Class Thin Hull Structural Steels

Liu Zetian[1,2], Ma Yonglin[1], Lu Bin[2]

（1. School of Materials and Metallurgical，Inner Mongolia University of Science and Technology，Baotou 014010, China; 2. Compact Strip Production Plant, Baotou Iron & Steel (Group) Co., Ltd., Baotou 014010, China）

Abstract Rolling process was designed that how to make thin plates used thick casting slabs in this article. The difference of mechanical property and microstructure between EH40 made under two different rolling processes was also researched. The results showed that the self-cogging process had obvious influences on the low temperature toughness and microstructure of plates. An unreasonable self-cogging process can cause terrible mixed crystal and some bigger grains, which have bad effect to the low temperature toughness. If less passes were used in self-cogging process by applying bigger draft every pass, good microstructure was gained. The grain was uniform and small, and plates have excellent low temperature toughness.
Key words thin pates, high strength hull structural steels, self-cogging process, low temperature toughness

轧后冷却方式对调质态 Q690 屈强比的影响

张苏渊 邹 扬 刘春明 王海宝 顾林豪

（首钢技术研究院宽厚板所，北京 100043）

摘 要 采用调整调质态 Q690 钢板轧制后冷却方式的试验，考察轧后水冷、空冷和炉冷对试验钢屈强比的影响。试验结果表明：采用不同的轧后冷却方式，经两相区调质处理后，力学性能满足 690 级别高强钢的要求，但是屈强比（YS/TS）不同。采用轧后水冷的试样，冷却速度达到 35℃/s，调质后软相组织铁素体含量最低，约为 10%，YS/TS 最高，为 0.937。采用轧后空冷的试样，轧态组织中铁素体含量增加，调质后软相组织含量也增加，为 11.8%，YS/TS 降低为 0.913。采用轧后炉冷方式的试样，轧态组织中铁素体最多，调质后软相组织含量最高，为 15%，YS/TS 值最低，为 0.897。
关键词 冷却方式，调质态，屈强比，Q690

Effects of Cooling Mode after Rolling on the Quenched and Tempered Q690's Yield Ratio

Zhang Suyuan, Zou Yang, Liu Chunming, Wang Haibao, Gu Linhao

（Shougang Research and Technology Institute, Beijing 100043, China）

Abstract　Adjusting the cooling mode after rolling of the quenched and tempered Q690 , effects of water cooling, air cooling and furnace cooling on the yield ratios has been studied. The test results show that: using different cooling mode after rolling, the yield ratio (YS/TS) are different, although the mechanical properties can meet the 690 level after quenched and tempered in diphase area.　Samples used water cooling after rolling, the cooling rate reaches to 35℃/s. After quenching and tempering in diphase, the ferrite content was the lowest, about 10%, with the highest YS/TS, 0.937. The sample application of air cooling after rolling has more ferrite content，11.8%, YS/TS decreased to 0.913. The highest ferrite's content was 15% with the lowest YS/TS, 0.897.

Key words　cooling mode, quenched and tempered, yield ratio, Q690

热处理钢板下表面凹坑控制与管理技术

李　伟

（宝钢股份公司厚板部，上海　200941）

摘　要　本文对造成热处理钢板下表面凹坑的原因进行分析，并针对炉辊结瘤这一主要原因进行了详细的说明，提出了相应的技术改进措施和管理改进措施，同时对实施前后的效果进行对比。

关键词　热处理炉，下表凹坑，结瘤

The Controlling and Management Technology of the Pits in Low Surface of Heat Treatment Plate

Li Wei

(Baosteel Heavy Plate Mill, Shanghai 200941, China)

Abstract　In this paper, we analyses the causes of the pits in the low surface of the heat treatment plate, and describe the main reason of nodulation of furnace-roller in detail, put forward the corresponding technical improvement measures and management measures, and make a comparison between the measures being carried out and before.

Key words　heat treatment furnace, pit in low surface, nodulation of furnace-roller

E690 海洋工程用钢的相变规律与再结晶行为研究

刘智军　宋仁伯　秦　帅

（北京科技大学材料科学与工程学院，北京　100083）

摘　要　本文在实验室条件下试制了 E690 海洋工程用钢，对其高温变形后的相变规律和再结晶规律进行了研究，为后期热轧时控轧控冷提供参考。结果表明：该成分体系下，在 1000℃施加 60%的工程应变后，试验钢在 0.1~30℃/s 冷速下都有贝氏体生成，在冷速为 5~30℃/s 冷速范围内有板条马氏体生成，且随着冷速增加，板条马氏体数量增加；试验钢的流变应力受变形温度和应变速率的影响显著，随着变形温度的升高和应变速率的降低，流变应力逐

渐减小。在热变形过程中，温度越高，应变速率越低，越容易发生动态再结晶。通过线性回归的分析方法获得试验钢热变形热激活能 Q 为 244.5779 kJ/mol，最后得到了热变形本构方程。

关键词 E690 海洋工程用钢，相变，再结晶

Study on Phase Transformation Law and Recrystallization Behavior of Offshore Platform Steel E690

Liu Zhijun, Song Renbo, Qin Shuai

(School of Materials Science and Engineering, University of Science and Technology Beijing, Beijing 100083, China)

Abstract This paper obtained the E690 offshore platform steel under laboratory conditions, and studied the phase transformation law and recrystallization behavior of the forged steel. These will provide a conference for thermal mechanical control process (TMCP). Results showed that the test steel with 60% deformation at 1000℃, bainite occurred in a wide range of 0.1~30℃/s, while lath martensite(LM) occurred in the rage of 5~30℃/s and increased with the temperature rose. The flow stresses of steel were strongly influenced by deformation temperature and strain rate during high temperature compression. It decreased with the increase of deformation temperature and the decrease of strain rate. In the process of thermal deformation, the higher temperature and lower deformation rate is, the more likely they were to develop dynamic recrystallization. The hot deformation activation energy of steel was calculated by regression analysis method，which was 244.5779 kJ/mol. At last, we got the constitutive equation of tested steel.

Key words E690 off-shore platform steel, phase transformation, recrystallization

热处理炉烧嘴常见故障分析及处理

马洪旭

（烟台宝钢钢管有限责任公司，山东烟台 265500）

摘 要 本文介绍了热处理炉及烧嘴燃烧系统，对热处理炉烧嘴燃烧系统常见故障原因进行了分析，并归纳总结出具体处理方案和给出烧嘴故障快速诊断与处理流程，大幅度的降低了热处理炉烧嘴故障率。

关键词 热处理炉，燃烧系统，烧嘴，故障诊断

Heat Treatment Furnace Burner Fault Analysis and Treatment

Ma Hongxu

(Yantai Baosteel Steel Tube Co., Ltd., Yantai Shandong 265500, China)

Abstract This paper introduces the heat treatment furnace and burner combustion system, to heat treatment furnace burner combustion system common faults are analyzed, and summarizes the specific solutions and give burner fault quick diagnosis and treatment process, greatly reduces the heat treatment furnace burner failure rate.

Key words heat treatment furnace, burner, combustion system, fault diagnosis

合金减量化 Q345B 钢板热轧后弛豫影响分析

温志红　黄远坚　李成良　周中喜

（宝钢集团广东韶关钢铁有限公司，广东韶关　512123）

摘　要　在 Q345B 钢板减量化生产过程中，对于轧后需弛豫才能进入超快冷系统的情况，进行了不同冷却速度的热模拟试验，制作 CCT 曲线，并结合生产实例进行分析，分析结果如下：（1）进行 Mn 减量化后，由理论计算可知每降低 0.1%Mn，A_{r3}将降低 8℃。（2）热模拟试验结果表明，在钢板空冷弛豫过程中，5℃/s 与 0.25℃/s 冷却速度相比，A_{r3}温度将降低 65℃。（3）实例分析表明，薄规格钢板由于粗大先共析铁素体数量的增加造成强度的不足。（4）通过系列措施使钢板的开冷温度应大于相变开始温度 A_{r3}，解决钢板强度不足问题，同时钢板韧性达到–40℃要求。

关键词　弛豫，超快冷，减量化

Analysis of Relaxation Effect after Finishing Rolling on the Alloy Redution of Q345B Steel Plate

Wen Zhihong, Huang Yuanjian, Li Chengliang, Zhou Zhongxi

(Baosteel Group Guangdong Shaoguan Iron and Steel Co., Ltd., Shaoguan Guangdong 512123, China)

Abstract　A variety of cooling rate of thermo-mechanical simulate experiments were conducted and the experimental results were made into the continuous cooling transformation(CCT) diagrams, aiming at that steel enter the ultra-fast cooling system having experienced the process of relaxation after rolling, in the application of reducing rolling technology. The data were analyzed, combined with the production practice, and the conclusions were as follows: (1) Theoretical calculation shows that A_{r3} temperature will be reduced about 8℃,every 0.1% (weight) reduction in Mn; (2) Thermo-mechanical experiments indicates that A_{r3} temperature of the cooling rate of 5℃/s will be reduced about 65℃ compared with the cooling rate of 0.25℃/s in the process of relaxation of steel plate; (3) Production practice Analysis indicated that the insufficient strength of thin gauge plate due to the increase of massive proeutectoid ferrite; (4) A series of measures was taken which makes the initial cooling temperature of plates higher than A_{r3} temperature to tackle the problems of insufficient strength of plates and to satisfy the demand of -40℃ impact toughness.

Key words　relaxation, ultra fast cooling, alloy reduction

基于光度立体的带钢表面缺陷三维检测方法

王　磊　徐　科

（北京科技大学高效轧制国家工程研究中心，北京　100083）

摘　要　现有表面缺陷在线检测系统一般采用两维灰度图像检测方式，在氧化铁皮等伪缺陷的干扰下，缺陷的误报严重。本文提出一种基于光度立体视觉的钢板表面三维缺陷在线检测方案，通过绿色、蓝色两台激光线光源从不同角度对带钢表面照明，用一台彩色线阵3CCD摄像机高速扫描带钢表面，从采集到的彩色图像中分离出绿光和蓝光通道，根据三维重建算法得到带钢表面的三维信息，根据表面三维信息检测缺陷，从而降低由于伪缺陷造成的误报率。通过对热轧带钢表面辊印和冷轧带钢表面压痕缺陷进行表面三维检测试验，结果表明该方法可以准确检测到带钢表面三维缺陷。

关键词 带钢，表面缺陷，三维检测光度立体

3D Defect Detection of Steel Strips Based on Photometric Stereo

Wang Lei, Xu Ke

(National Engineering Research Center for Advanced Rolling Technology,
University of Science and Technology Beijing, Beijing 100083, China)

Abstract 2D detection with gray level images ispopularly employed in current surface inspection systems of steel strips, andleads to high false detection of defects because of interference of pseudo-defects, such as scales. A3Ddefect detection technique of steel plates based on photometric stereo is developed. Steel strips are illuminated with green and blue linear lasers from different directions.One color line-scan camera with 3CCD isemployed to scan themoving surface of steels. Color imagescaptured by the CCD camera are separated into green channel and blue channel, which are used for 3D reconstruction of steel strips. Defects are detected according to 3D information of steel surfaces,and false detection rate because of pseudo-defects is reduced. The technique wasexamined withroll imprints of hot-rolled steels and indentations of cold-rolled steels, andthe results showed that 3D defects of steel stripswere detected accurately.

Key words steel strips, surface defects, 3D detection photometric stereo

低碳钢板轧制起梗的机理分析

罗石念 佟 岗

（宁波钢铁有限公司，浙江宁波 315807）

摘 要 低碳钢SPHC热卷在供单机架冷轧厂生产0.30mm以下薄带钢时，表面出现批量性的起梗缺陷。本文系统地分析热轧、冷轧工艺过程对起梗缺陷的诱发影响规律，找出了其关键的内在影响机制。冷轧工序薄板起梗缺陷与带钢局部高点（硬度不均）导致材质内部横向张力分布不均、致使延伸不均这一内在板形问题有关，本文对这种延伸不均的产生机理进行了解析，并针对现场实际装备条件提出了解决措施。为系统解决冷轧薄板起梗缺陷难题，还研究了其与热轧带钢凸度和楔形值的相关关系，得出控制较大的热卷凸度值，并有效降低楔形值，可抑制起梗。生产应用实践表明，通过优化热连轧机组窜辊、弯辊工艺与轧制计划编排，防止出现过大的局部高点，设定带钢断面楔形与凸度比值小于2，可显著抑制冷轧薄带钢起梗、局部延伸不均缺陷，大幅提高产品实物质量水平。

关键词 薄带钢，局部应力，起梗，应力分布

Investigation of Ridge Buckle of Low Carbon Sheet Steel

Luo Shinian, Tong Gang

(Ningbo Iron & Steel Co., Ltd., Ningbo Zhejiang 315807, China)

Abstract The hot-rolled sheet SPHC of Ningbo Iron & Steel Corporation were used to produce the thin cold sheet less than 0.35mm in the single-stand cold rolling mill, which lead to the frequent occurrence of ridge buckle defect. Based on a large number of statistic data on the position and characteristics of the defect, the systematic analysis of the factors affecting the ridge buckle were discussed, especially, the key source element were found. The results show that this kind of ridge buckle defect were firmly related to the partial high spot（heterogeneous hardness） of the cold rolled strip and the hot rolled strip. The analysis of the reason of the partial high spot have been made, and the coping measures on the basis of local equipment in Ningbo Iron & Steel Corporation have been raised accordingly. In order to solve the ridge buckle problem, the relationship between the ridge buckle and the crown and wedge of the hot band were investigated. Controlling the high crown value and low wedge value of the hot band is very beneficial to the restrain of ridge buckle of cold sheet. The temperature gradient and the difference in phase transformation along the width of the hot-rolled strip bring about the formation of the compressive residual stress at the hot-rolled strip edge during the water cooling process. By adopting a series of available measures from hot rolling scheduling to product processing, the transverse section shape of hot-band were optimized. The crown value were controlled in a proper range and the wedge value were less than 50% crown value, as a result, the swelling and edge ridge buckle of the cold sheet faded away and the rate of finished cold rolled sheet were enhanced a lot.

Key words thin sheet, local stress, ridge buckle, stress distribution

冷轧带钢拉矫机破鳞参数优化与实验研究

陈 兵 何名成

（北京科技大学机械工程学院，北京 100083）

摘 要 拉伸弯曲矫直机作为带钢轧制过程中的重要设备，对高效除鳞和改善板形起着重要作用。本文主要就拉矫机两组弯曲辊组在总插入深度值相同的情况下前后插入深度值合理设置的问题进行了研究。首先对拉矫机矫直部分进行了简化并建立了有限元模型，调换前后弯曲辊组插入深度值来设置不同的仿真工况，通过仿真结果可以得出在总插入深度值一样的情况下，后弯曲辊组的插入深度值大于前弯曲辊组插入深度时可以得到更好的破鳞效果，有利于拉矫机的破鳞。同时前弯曲辊组插入深度值不能过小。然后进行了试验研究，对采用不同工艺参数进行拉矫的带钢进行取样，通过酸洗试验的方式，比较不同工艺参数情况下试样酸洗时间的长短，同样得出了相同的结论。

关键词 拉矫机，参数优化，插入深度，破鳞

Research on Parameter Optimization & Experiment of Descaling in Tension Leveler for Cold Rolled Strips

Chen Bing, He Mingcheng

（School of Mechanical Engineering, University of Science and Technology Beijing，Beijing 100083, China）

Abstract Asa kind of key equipment in the process of strip rolling, the tension leveler machine plays important role in efficient descaling and improving the shape of the strip.Reasonable setting of the insertion depth of the two bending rolls with the same total insertion depth has been studied in this paper. First the straightening section of tension leveler is simplified and finite element model is established. Different simulation conditions are set up by swapping the insertion depth of two groups of bending roll. Simulation results show that, in the case of the same total insertion depth, efficient descaling can be get when the insertion depth of the behind bending roll is bigger than the former. And the insertion depth of the former cannot be too small. Than experiment study is carried out. The samples are selected from strips with different process parameters. Through the pickling test, the same conclusion is drawn by comparing the pickling time of samples with different process parameters.

Key words tension leveler, parameter optimization, insertion depth, descaling

特厚 TMCP 态 EH40 船板生产开发

刘 源[1] 鲁 强[1] 杨 军[1] 王 华[2] 陈 华[2]

（1. 鞍钢股份有限公司鲅鱼圈钢铁分公司，辽宁营口 115007；
2. 鞍钢股份有限公司产品发展部，辽宁鞍山 114009）

摘 要 通过微合金化成分及 TMCP 工艺的合理制定，利用 300mm 厚连铸坯料，在 5500mm 宽厚板轧机上成功开发了 100mm 厚规格的高强船板 EH40–Z35。产品的综合性能、组织分析结果表明：微观组织结构合理，具有良好的低温韧性、抗层状撕裂性、低冷裂纹敏感性，可以完全满足九国船级社规范的要求。

关键词 EH40–Z35 高强船板，100mm 厚板，TMCP 工艺

Development of TMCP High-strength Ship Steel in 5500 Heavy Plate Mill

Liu Yuan[1], Lu Qiang[1], Yang Jun[1], Wang Hua[2], Chen Hua[2]

(1.Bayuquan Iron & Steel Subsidiary Company of Angang Steel Co., Ltd., Yingkou Liaoning 115007, China;
2. Product Development Department of Angang Steel Co., Ltd., Anshan Liaoning 114009, China)

Abstract The 100mm thickness high-strength ship steel EH40-Z35 heavy plate was developed successfully by using appropriate micro-alloying and TMCP process parameters in 5500 heavy plate mill. The results show that the product has suitable microstructure and low temperature toughness, good resistance to lamellar tearing, low cold cracking sensitivity, and meets the certified specifications specified by nine country classification societies.

Key words EH40-Z35 high-strength ship steel, 100mm Heavy steel plate, TMCP process

中厚板辊式淬火机系统开发及应用

刘 涛[1] 余 伟[1] 许少普[2] 张立杰[1] 李彦彬[1] 唐郑磊[2]

（1. 北京科技大学高效轧制国家工程研究中心，北京 100083；
2. 南阳汉冶特钢有限公司，河南南阳 474500）

摘 要 辊式淬火机是中厚板现代化热处理线的关键设备，是当前高强度中厚板淬火生产的首选设备形式。为此，开发了一套中厚板辊式淬火机系统，淬火机由本体装置及辅助系统组成，完成淬火机喷水装置、钢结构框架、提升机构、供水系统、传动系统及液压系统等设计开发，并开发了淬火机控制系统。开发的辊式淬火机能实现连续淬火、连续加摆动淬火及正火控冷等工艺。所开发的淬火机已成功地应用于新产品的开发及生产，通过在国内某钢铁公司中厚板生产中的应用，结果表明：淬火冷却强度大，能满足耐磨钢、高强钢、压力容器钢及油罐钢等钢种淬火生产；

产品厚度规格适用范围宽，能实现6~100mm厚度范围内钢板的淬火工艺；淬火后钢板具备较好的板形及冷却均匀性；正火控冷工艺模型控制精度98.46%在±18℃内。

关键词 中厚板，辊式淬火机，开发，应用

Development and Application of Plate Roller Quenching Machine

Liu Tao[1], Yu Wei[1], Xu Shaopu[2], Zhang Lijie[1], Li Yanbin[1], Tang Zhenglei[2]

(1. National Engineering Research Center for Advanced Rolling Technology, University of Science and Technology Beijing, Beijing 100083, China; 2. Nanyang Hanye Special Steel Co., Ltd., Nanyang Henan 474500, China)

Abstract Roller quenching machine is the key equipment of the medium and heavy plate heat treatment line, and it is the first choice for the high strength steel production. Therefore, we develop a plate roller quenching machine system which is composed of the device body and auxiliary systems, and complete system design and development such as the water jetting device and the steel frame structure and the lifting mechanism and the water supply system and the transmission system and the hydraulic system and the control system. Roller quenching machine can realize continuous quenching and swing quenching and normalizing control cooling process. The developed quenching machine has been successfully applicated for the new product development and production, according to the application in a domestic iron and steel company, the results show that: high quenching cooling intensity can be acquired，and it can meet the quenching production requirements of wear resistant steel and high strength steel and pressure vessel steel and oil tank steel; wide thickness range of the application for the product specifications can be acquired such as 6~100mm thickness; the good flatness and cooling uniformity can be acquired after quenching; the model control precision of normalizing control cooling can be acquired such as 98.46% within ±18 ℃.

Key words plate, roller quenching machine, development, application

多普勒激光测速仪与脉冲发生器联合测速在平整延伸率控制中的应用

李赢 李超 李明伟 李红雨

（鞍钢股份有限公司冷轧厂，辽宁鞍山 114021）

摘 要 简单介绍了冷轧平整的目的和意义，说明了延伸率的定义和测量方法。描述了使用激光测速仪或者脉冲发生器单一形式测速的优缺点，找到了在加减速等扰动情况下带钢延伸率出现较大波动的原因，并通过不同速度区间选用不同的测速方式，优化延伸率控制系统，解决了在加减速或出现其他扰动时的延伸率波动问题。

关键词 平整机，延伸率，激光测速仪，脉冲发生器

Research on the Combining Speed Detective System Working on Elongation Control System of Cold Rolling Skin Pass Mill

Li Ying, Li Chao, Li Mingwei, Li Hongyu

(The Cold Rolling Plant of Ansteel Co., Ltd., Anshan Liaoning 114021, China)

Abstract The objective and significance of cold skin pass rolling were simply introduced; the definition and measuring method were explained. The elongation control system of cold rolling skin pass mill was described; the speed feed forward control and feedback control in the control system were detailed explicated. Aiming at the fluctuation problem of elongation during acceleration, deceleration and the appearance of other disturbances, a combining detective system of encoder and laser was brought forward. The result indicates that the dynamic response of elongation control was obviously accelerated.
Key words skin pass mill, elongation, laser, encoder

承钢 800MPa 复合强化 Q550E 的开发

邢俊芳 张叶欣 陆凤慧

（河北钢铁股份有限公司承钢分公司，河北承德 067001）

摘 要 针对国标中低合金耐低温冲击 Q550E 高强钢板的性能要求，发挥承钢自身钒钛优势，采用先进的 V、Nb、Ti 复合微合金强化设计思路，并对现有 TMCP 工艺优化，同时制定质量控制点要求。试制成品性能满足标准要求的 Q550E，其中屈服强度 690～735MPa，抗拉强度达 790～830MPa，-40℃夏比 V 型冲击功≥75.61J。热轧卷板 Q550E 的成功试制，为高强度级别的低合金结构钢板 Q690E 提供了试轧可能。
关键词 复合强化，钒钛强化，成分设计，层流冷却
Abstract According to the national standard performance requirements for low alloy and low-temperature impact resistance of high strength steel Q550E, use Chengang Vanadium and Titanium advantages, use advanced V, Nb, Ti composite micro alloy strengthening design, optimize the existing TMCP process, and formulate the requirements for quality control. The trial product performance to meet the requirements of the Q550E standard is among 690 to 735MPa for yield strength, 790 ~ 830 MPa for tensile strength and above 75.61J for - 40 degree Charpy V-notched impact energy. Low alloy structural steel plate Q690E may try rolling for the successful trial production of Q550E.
Key words composite reinforcement, vanadium and Titanium enhanced, composition design, laminar cooling

承钢 1780 节约支撑辊换辊时间

李子文 高少华 韩立伟 柴培斌 徐 良 张瑞波

（河北钢铁集团承钢公司热轧卷板事业部轧钢作业区，河北承德 067000）

摘 要 承钢公司热轧卷板事业部 1780 精轧机组共 7 架四辊轧机，根据轧辊能力、轧制计划、制定各架次轧辊更换吨位，其中，F1~F6 更换吨位为 20t，F7 更换吨位为 10t。每月大概需要 1.5 次更换支撑辊，而支撑辊更换往往需要很长时间，影响检修进度，降低产能释放。所以本文研究如何节约支撑辊更换时间，可以减少支撑辊更换占用的大量检修时间，节约换辊时间，使分厂的检修计划制定更加容易，缩短了检修时间，可以提高生产作业率。
关键词 支撑辊更换，产能释放，缩短时间，提高作业率

Chenggang Company 1780 Save Backup Roll Roll Change Time

Li Ziwen, Gao Shaohua, Han Liwei, Chai Peibin, Xu Liang, Zhang Ruibo

(Hot-rolled Coil, Hebei Iron and Steel Group Company Chenggang Plate Group Assignments Section of Steel Rolling, Chengde 067000, China)

Abstract ChengGang company 1780 finishing mill group of hot rolled coil plate division, a total of seven four-roll rolling mill, according to rolling plan, to develop various vehicles roll ability, change the tonnage, among them, the F1~F6 replacement tonnage of 20 tons, F7 replacement capacity of 10 tons. Replace the backup roll about a month to 1.5 times, and supporting roller replacement often takes a long time, affect the maintenance schedule, reduce the capacity to release. So this paper studies how to save the backup roll change time, can reduce the support roller replacement takes up a lot of maintenance time, time of the roller, make it easier to factory maintenance plans, shorten the maintenance time, can improve the production operation efficiency.

Key words backup roll change, capacity to release, shorten the time, improve the operation efficiency

邯钢低碳软钢边部横向折印缺陷攻关与工艺优化

刘红艳　刘永强

（河北钢铁集团邯钢公司，河北邯郸　056015）

摘　要　针对低碳软钢平整横向折印缺陷，对 SPHC 碳含量按照偏下限控制、对 SPCC 加入适当含量的硼元素，适当提高终轧温度、降低卷取温度，提高层流冷却速率来提高带钢自身提高屈服点，带钢横折缺陷明显改善；采用-20μm 方程曲线辊形工作辊上机，选用了合适的轧制力、延伸率、弯辊力及调平值，扩大了弯辊力的调整范围，提高了弯辊对横折和表面质量的控制效果，对折皱的改善能够满足质量要求。最终横折缺陷率由攻关前的月平均 2.31%，降低到工艺优化后的月平均 0.12%，改善效果十分明显。

关键词　低碳软钢，横向折印，屈服点，弯辊力

The Research and Process Optimization of Low-carbon Mild Steel Edge Transverse Defects Pincher Handan Iron and Steel

Liu Hongyan, Liu Yongqiang

(Handan Iron and Steel Co., Ltd. Corporation，Hebei Steel and Iron Group, Handan Hebei 056015, China)

Abstract For low carbon mild steel flat transverse defects pincher, in accordance with the carbon content of SPHC partial lower control of SPCC add the appropriate content of boron, increasing the finish rolling temperature, reducing the coiling temperature, increase the laminar cooling rate and improve the yield point of the strip itself,strip cross breaks defect significantly improved. using -20μm Equation curve roller-shaped work roller machine, the choice of suitable rolling force, elongation, bending force and leveling value, expanding the adjustment range of the bending force, improve the bending and surface quality of the cross-fold control effect, to improve the wrinkles can meet the quality requirements. The final horizontal folding defect rate from the previous month of the average of 2.31%, and reduce the average monthly average of 0.12% after the optimization, improve the effect is obvious.

Key words low carbon steel, transverse folding, yield point, bending force

改善冷轧板表面质量的研究

杨 哲 杨 柳 王清义 武建琦

（河北钢铁集团邯郸钢铁集团有限责任公司冷轧厂，邯郸 056015）

摘 要 本文主要从冷轧带钢表面质量的影响因素和原理入手，利用系统分析方法，对冷轧带钢表面黄带、锈蚀的影响因素进行探讨，对比了终冷台生产工艺、钢卷库区存放、平整液、空气吹扫系统的优缺点，总结了带钢表面质量攻关采取的措施和取得的效果。具有良好的经济效益，值得借鉴。

关键词 黄带，锈蚀，平整液，空气吹扫，影响因素

Research of Improving on the Cold-rolling Strip Surface Quality of Handan Steel

Yang Zhe, Yang Liu, Wang Qingyi, Wu Jianqi

(Hebei Iron & Steel Group Handan Iron & Steel Group Co., Ltd., Handan 056015, China)

Abstract The paper mainly commence with analyzing the factor and principle of influence on the Cold-rolling Strip' surface quality, as the tool that used the system analysis methods, simply state the factor of yellow staining and rustiness of the Cold-rolling Strip, and introduce the advantages and disadvantages of the final coil cooling , the coil storage, the temper solution and the blow-off device. Eventually, summed up the step and effect of improvement. It has good economic profit and is good for reference.

Key words yellow staining，rustiness，temper solution，blow-off device，factor of influence

钢轨矫直引起断面尺寸不合格原因分析

滕 飞 顾双全 刘晓燕

（河北钢铁集团邯钢公司大型轧钢厂，河北邯郸 056005）

摘 要 针对钢轨矫直过程中钢轨轨高减小、钢轨轨底宽度增大、钢轨轨头宽度的减小、对称度情况变化及腹腔高度减小的现象，对钢轨矫直引起断面尺寸不合格的原因进行分析。钢轨在轧制时必须考虑钢轨矫直后断面尺寸的变化。钢轨矫直过程的矫直机参数设置不合理容易造成钢轨对称度不合格和腹腔高度不合格。为钢轨的轧制过程、矫直过程提供指导。

关键词 钢轨，矫直，不合格

Analysis of Rail Lineament Size Uneligibility Caused by Straightening

Teng Fei, Gu Shuangquan, Liu Xiaoyan

（HBIS Handan Corporation of Heavy Section Mill, Handan Hebei 056005, China）

Abstract According to rail lineament size change after rail straightening which include rail height decrease、rail width increase、head width decrease ect,Lineament size uneligibility reason were analyzed in the process of rail straightening.Rail rolling must take into rail lineament size change after rail straightening consideration.The uneligibility of symmetry and flank height caused by defective straightening parameter.It provided the scientific basis for optimizing rolling and straightening process.

Key words rail, straightening, uneligibility

Effect of Tempering Temperature on the Microstructure and Mechanical Properties of V Modified 12Cr2Mo1R Steel Plate

Liu Zili[1,2], Liu Chunming[1], Li Xianju[2], Zhang Hanqian[2]

(1. Northeastern University, Shengyang Liaoning, China;
2. Baoshan Iron & Steel Co., Ltd., Shanghai, China)

Abstract A V modified 12Cr2Mo1R steel plate for pressure vessel was developed to improve the strength of the plate. In this paper ,the mechanical properties and microstructures of the newly developed steel plate treated with the same normalizing process but different tempering temperatures were studied by using the mechanical testing machine, the optic microscope and the SEM(Scanning Electron Microscope). Generally, with increasing the tempering temperature from 700 ℃ to 740℃,the yield strength and tensile strength decreased sharply, while the elongation and reduction properties increased.The toughness of the plate increased with the increasing tempering temperature.The amount of the carbides extracted from specimens with different tempering states was also studied by chemical method,the result indicated that the changing trend of the amount of the carbides had a direct correlation with the mechanical properties.

Key words 12Cr2Mo1R, modified, pressure vessel, carbide

薄板坯连铸连轧过程控制系统在线改造研究

何安瑞[1] 宋 勇[1] 邵 健[1] 荆丰伟[1] 赵海山[2] 周 杰[2]

(1. 北京科技大学冶金工程研究院，北京 100083;
2. 马鞍山钢铁股份有限公司一钢轧，安徽马鞍山 243011)

摘 要 介绍了薄板坯连铸连轧生产线原控制系统存在的问题。针对这种特殊工艺流程的改造需求，设计了基于网关服务器的新老过程控制系统并行运行方案，实现了过程控制系统的在线改造新系统采用了多项模型新技术，完善了控制功能，克服了原系统存在的不足新系统的试生产实绩表明其用户界面友好，生产过程稳定，产品质量满足用户要求。
关键词 连铸连轧，过程控制，在线改造，数学模型

Online Upgrading of Process Control System for Thin Slab Continuous Casting and Rolling

He Anrui[1], Song Yong[1], Shao Jian[1], Jing Fengwei[1], Zhao Haishan[2], Zhou Jie[2]

(1. Engineering Research Institute, University of Science and Technology Beijing, Beijing 100083, China;
2. No.1 Steel Making and Rolling General, Maanshan Iron & Steel Co., Ltd., Maanshan 243011, China)

Abstract The problem of the old control system was introduced on the continuous casting and rolling product line. 1n view of the upgrading demand for this special technological process, the parallel running solution of new system and old system was developed based on the gateway server, and online upgrading of the process control system was realized. Several new model technologies were developed in the new system. Some control functions were improved and the deficiency of the old system was reduced. The actual production data of the new system showed its user interface was friendly, the production process was stable and the product quality can meet the user requirements.
Key words continuous casting and rolling, process control system, online upgrading, mathematical model

连铸板坯轧制过程中的潜在质量风险管理

谢翠红 王国连 李广双 王玉龙

（首秦金属材料有限公司，河北秦皇岛 066326）

摘 要 本文以板坯质量管理实践为基础，充分总结质量实践管理过程的技术要点，并在实践中充分应用,有效解决了钢板边裂的责任划分，钢板边裂发生率由2014年的2.14%降低到2015年上半年的1.19%；通过异常爪裂发生原因的分析，得出加热工艺是影响爪裂的重要因素，严格控制入炉时间间隔，使得爪裂发生率得以有效控制，爪裂废品率由2014年的0.01%降低到2015年上半年的0.004%。通过对轧制钢板缺陷类型的总结分析，将潜在的风险因素转化为技术控制要点输入到常规的质量控制循环中，形成了以钢板缺陷为起点的闭环质量管理体系，同时实现了螺旋式的质量提高，轧制钢板对应原料性板坯缺陷发生率由2014年0.54%降低到2015年上半年的0.25%。
关键词 板坯质量，潜在风险，原料性缺陷

Potential Quality Risk Management of Slab in the Rolling Process

Xie Cuihong, Wang Guolian, Li Guangshuang, Wang Yulong

(Shouqin Metal Materials Co., Ltd., Qinhuangdao 066326, China)

Abstract This paper bases on the practice of slab quality management and summarizes the technical points of quality management process, the division of responsibility for the split of plate edge crack has been solved effectively, the occurrence rate of plate edge crack was reduced from 2.14% in 2014 to 1.19% in the first half of 2015; Through the analysis of the causes of abnormal claw crack, the heating process was an important factor to influence the crack, the time interval of the furnace was controlled strictly, the occurrence rate of the jaw crack was effectively controlled, and the scrap rate was reduced from 0.01% in 2014 to 0.004% in the first half of 2015. The potential risk factors were transformed into the conventional quality control loop by summarizing and analyzing the defect types of rolling mill,it has been applied in the practice of the closed loop quality management system based on the defect of steel plate, and realizes the quality of the spiral type, the occurrence rate of the raw material of the slab was reduced from 0.54% in 2014 to 0.25% in the first half of 2015.

Key words slab quality, potential risks, raw material defects

KOCKS 旋转轧机 新型的无缝管延伸轧管方法

Jörg Surmund，Erich Bartel, Patrick E. Connell

（德国 KOCKS 公司）

摘 要 当前无缝管生产领域内的延伸机组年生产能力超过了 30 万吨。为了满足中小型企业对设备生产能力的较低要求，KOCKS 公司开发了旋转轧机 KRM，该设备的年生产能力为 15 万吨。KRM 旋转轧机的设计成熟，生产管材质量高，管材成材率高，同时相比于纵轧延伸机组的设备投资更低。

关键词 KOCKS 四辊旋转轧机，斜轧延伸轧管

轧辊辊形及辊温测量最新技术进展
——曲线直显式智能辊形测量仪的研究与实现

刘根社 刘 迪 宗 坤 刘 强 徐 梦

（商丘魁斗计量测控有限公司）

摘 要 本文简述了板带钢轧辊目前常用的辊形测量仪器，分析了它们各自的优缺点，介绍了一种针对各种测量仪器的不足而研制的最新专利产品——曲线直显式高精度智能辊形仪。该仪器具有体小质轻、参数可调、自动采集和存储数据、自行运算处理、直接显示辊形曲线和辊温曲线等智能性；具有仪器分辨率 0.001mm、精度误差小于 0.003mm 的测量准确性；具有用“U”盘传输数据到上位计算机进行进一步分析和二次开发的实用性；具有成本价格低廉的经济性等特点；仪器经多家钢厂实际使用，效果很好。一致认为该仪器是一种极有推广应用价值的便携式的高精度智能辊形测量仪器。

关键词 专利，辊型测量，智能辊形测量仪

4　表面与涂镀

Metal Coating Technology

炼铁与原料

炼钢与连铸

轧制与热处理

★ 表面与涂镀

钢材深加工

钢铁材料

能源与环保

冶金设备与工程

冶金自动化与智能管控

冶金分析

信息情报

Innovation in Surface Coating Technology

Kwang Leong Choy

(Institute for Materials Discovery, University College of London, London, WC1E 7JE United Kingdom)

Abstract This contribution presents an overview of the use of chemical vapour deposition (CVD) and physical vapour deposition (PVD) for the manufacture of high quality structural and functional coatings onto steel. The scientific and technological significant of these methods will be discussed and reviewed. In addition, the emerging, innovative and cost-effective non vacuum aerosol assisted chemical vapour deposition (AACVD) will be highlighted. AACVD is a variant of the CVD process. It has the capability to deposit high purity nanostructured thin and thick films onto steel well controlled structure and composition at molecular scale level and at low processing temperatures. The fundamental aspects of AACVD including process principle, deposition mechanism, reaction chemistry, kinetics, and mass transport phenomena will be presented. The relationships of the process/structure/property of the coatings manufactured by the AACVD based methods will be described. The microstructure and properties of the nanostructured materials produced by AACVD method will be described and compared with conventional CVD and PVD coating techniques. The use of the promising non vacuum deposition AACVD based technology for the creation of potential high value added steel products will be presented. The technical viability and the potential of the AACVD process to be scaled up or large area and large scale production will be presented.

Key words chemical vapour deposition(CVD), AACVD, steel substrates, coating

Technology Trends in the Surface Coatings Industry and How They Address the Issue of Sustainability

Lowe Chris

(Becker Industrial Coatings Ltd., Liverpool, UK)

Abstract Sustainability is a growing movement in Western Society with the definition from the Brundtland Â Report being:Sustainable development is development that meets the needs of the present without compromising the ability of future generations to meet their own needs.

The European Coil Coating Industry is developing strategies that will make it more sustainable and the paint suppliers are key players in that development. Projects from improved energy efficiency of Coil Coating lines to improved energy efficiency of buildings through targeted paint development are being progressed. The source of raw materials for paints is an issue which will become more and more of am issue as fossil fuels more expensive and ultimately in short supply, hence there is some focus on Bio-sourced resins and solvents. Finally industry and transport contribute significantly to pollution, can the large surface area covered by precoated metal act as a counter balance.

Key words coil coating，sustainability

Pickling & Galvanizing of Thin Hot Strip A Comparison of Two Processes

W. Karner*, E.Blake, A.Stingl, J. Hykel

(Andritz Metals, 1120 Vienna, Austria)

Abstract By the introduction of processes for the direct production of thin hot strip material like CSP and ESP the need arises to find economical solutions for the pickling and subsequent galvanizing of these products. It is attractive to couple the pickling and the galvanizing process to be performed in one line the so called PGL, a combination of pickling with hot dip galvanizing. For some products it may also be interesting to combine pickling with electrolytic galvanizing (PEGL). In this paper the two processes are compared and their respective merits for different products are evaluated.properties remained fairly high.

Key words hot strip, pickling, PGL, PEGL

Self Fluxing Electrolytes for High Speed Tin Electrodeposition

Levey Carol

(Global Marketing Manager Steel Mills, Dow Chemical, Marlborough, Massachusetts, USA)

Abstract Approximately 65% of tinplate lines are vertically oriented and utilize an electrolyte base consisting of phenol-sulfonic acid and ethoxylated naptholene sulfonic acid (PSA/ENSA). A significant advantage of PSA/ENSA is that the diluted electrolyte can be used to facilitate fluxing of the tinplate. Replacement electrolytes, such as those based on the non-toxic and more environmentally sustainable methanesulfonic acid (MSA), are considered non-self-fluxing and require a separate fluxing agent which adds to the financial and logistical burden during replacement as both conversion capital and tank allotment is typically limited. This paper describes a novel fluxing agent that can be used on MSA and sulfuric acid based electrolytes to permit operation in a self-fluxing configuration, simplifying both conversion and reducing cost of operation.

Key words tinplate, methanesulfonic acid, self-fluxing, sulfuric acid

Correlations Between Chemical State, Dispersion State, and Anticorrosive Properties of Graphene Reinforced Waterborne Polyurethane Composite Coatings

Li Jing, Li Yaya, Yang Zhenzhen, Yang Junhe

(School of Materials Science and Engineering, University of Shanghai for Science and Technology, Shanghai 200093, China)

Abstract Grapheneoxide (GO) / functionalized graphene (FG) / graphene oxide after mild reduction(RGO) reinforced waterborne polyurethane（PU） composite coatings were fabricated on galvanizedsteel surfaces.The anticorrosive properties of the graphene reinforced PU composite coating were characterized. The dispersion state of graphene in PU matrix varied with the tailored chemical state of graphene. The correlations between the chemical state, dispersion state of graphene, and anticorrosive properties of the composite coatings were discussed. The superior anticorrosive properties of the graphene reinforced PU composite coatings were achieved by the addition of 0.2 wt% of RGO. The Electrochemical Impedance Spectroscopy (EIS) results showed that the under-painting corrosion didn't occur after 235 h immersion in 3.5 wt% NaCl electrolyte and the impedance modulus at 0.1Hz remained at 10^9 Ω, barely changed for 235 h.The GO reinforced PU composite coatingsshowed similar anticorrosive properties. The FG reinforced PUcomposite coatings exhibited a conductive percolation threshold of 0.1 wt%, while their anticorrosive properties remained fairly high.

Key words anticorrosive properties, graphene, composite coatings

The Current State of the Art in Laboratory Corrosion Test Methods

Fowler Sean

(Q-Lab Corporation, USA)

Abstract The salt spray laboratory corrosion test was introduced as an ASTM standard more than 100 years ago, and even today it is the most commonly performed method. The people who developed the standard recognized that the test did not accurately replicate real world corrosion in most cases, but they considered the test critical because it enabled repeatable and reproducible results. This made it a critical quality control test in the automotive and other industries. The method was also adaptable for a variety of metallic substrates by acidifying the salt solution.

Wet and dry cyclic tests were developed for better replication of real world corrosion. This innovation allowed more accurate differentiation of relative performance levels among materials and protective coatings, but these tests did not supplant the traditional salt spray test in popularity. The likely reason is that the wet and dry cyclic tests were not repeatable or reproducible, so they were not reliable for use in a quality control environment.

During the last 25 years, the automotive industry has developed wet and dry cyclic tests with modernized requirements that allow them to accurately replicate real world corrosion mechanisms and appearance. Better test chamber technology and test performance requirements have also significantly improved the reproducibility of these tests, so they can be used for material development and in quality control applications. This presentation will review these developments and provide some data to illustrate the importance of these new test performance requirements.

Key words corrosion test, laboratory, cyclic test

An Advanced Electrolyte and Flux System Used to Convert Phenol Sulfonic Acid Electrotin Lines to Methane Sulfonic Acid

Gary J. Lombardo

(Global Technical Director, Plating Quaker Chemical Corporation)

Abstract With ever increasing demand to reduce cost and environmental impact, more tinplate producers are looking to switch away from Phenol Sulfonic Acid (PSA) based electrolytes. The tried and tested choice when switching, is the Methane Sulfonic Acid (MSA) based electrolyte. One hurdle to converting an existing PSA electrotin line (ETL) to MSA is the haze problem caused by MSA build up in the flux tank. This haze problem necessitates adding extra rinse tanks or using excessive rinse water to prevent the blue haze caused by MSA in the flux tank. This is because most PSA ETL's have only two tanks after plating, one dunk tank, and one flux tank. Some tinplate producers have installed expensive post rinse flux applicators to get around this problem when using MSA. This paper describes a novel MSA electrolyte and flux system. The flux can be used in the flux tank of all existing PSA lines. The flux is totally compatible with the MSA electrolyte and can be counter flowed back into the plating tank. This unique flexibility negates the need to make expensive modifications to the PSA ETL when switching to MSA.

Key words methane sulfonic acid, phenol sulfonic acid, ferrostan

硅钢表面涂覆及绝缘性研究

任予昌　田志辉

（武钢股份公司冷轧总厂，湖北武汉　430080）

摘　要　由于低碳经济以及节能降耗的需要，冷轧无取向电工钢正向低铁损方向发展。铁损是电机在≥50Hz 交变磁场下磁化时所消耗的电能，通过分析铁损的产生原因，本文引出了表面绝缘性的重要性。本文通过对比不同产线生产的硅钢片绝缘性的差异，研究了表面涂覆与带钢绝缘性之间的关系；通过电子探针分析带钢表面涂覆情况，了解了表面绝缘层受破坏或涂覆不良时的原因，并提出了提高表面绝缘性的对策措施，得到了很好的效果，完成了硅钢表面绝缘性的研究。

关键词　硅钢，表面涂覆，绝缘性

Research of the Silicon Steel Surface Coating and Insulation

Ren Yuchang, Tian Zhihui

(The Cold-Rolling Mill of WISCO, Wuhan Hubei 430080, China)

Abstract　Due to the low carbon economy and the needing of saving energy and reducing consumption, non Oriented Silicon Steel is low iron loss of direction. Iron loss is at 50 hz alternating magnetic field or machine magnetizing electricity consumed, by analyzing the causes of the iron loss, this paper raises the importance of the surface of the insulation. In this paper, by comparing the differences between the production of silicon steel sheet production line insulation, studied the relationship between the surface coated with insulation; Strip surface coated by electron probe analysis, found the reasons of surface insulation damage or coated with bad, and puts forward the measures to improve the surface insulation measure, obtained the very good effect, completed the research of silicon surface insulation

Key words　silicon steel, surface coating, insulation

一种热镀锌板表面点状缺陷研究

蒋光锐[1]　刘李斌[1,2]　李　洁[1]　李明远[1]

（1. 首钢技术研究院，北京　100043；

2. 北京科技大学新金属材料国家重点实验室，北京　100083）

摘　要　采用扫描电镜、电子探针和激光共聚焦显微镜分析了热镀锌板镀层表面一种肉眼几乎不可见的点状缺陷。这种点状缺陷的直径不超过 500μm，缺陷位置的镀层没有超过表面粗糙度水平的厚度起伏，缺陷位置的镀层中含有细小的铁铝化合物以及富铝的膜状氧化物。分析表明，该缺陷是嵌入镀锌板表面的薄膜状氧化铝以及镶嵌其中的细锌渣颗粒。

关键词 热镀锌板，表面，点状缺陷，激光共聚焦显微镜，电子探针

Study of A Dot Defecton Surface of Hot-Dip Galvanized Steel Sheet

Jiang Guangrui[1], Liu Libin[1,2], Li Jie[1], Li Mingyuan[1]

(1. Shougang Research Institute of Technology, Beijing 100043, China; 2. State Key Laboratory for Advanced Metals and Materials, University of Science and Technology Beijing, Beijing 100083, China)

Abstract A kind of dot defect naked eye can hardly see on surface of hot-dip galvanized steel sheet was invested by scan electron microscope (SEM), electron probe microanalysis (EPMA) and laser confocal scanning microscopy (LCSM). The diameter of this kind of dot defect is less than 500 micrometer. No fluctuation of coating thickness more than the roughness exists around the defect. Furthermore, there exists some fine Fe-Al compound and film-like oxidesriched in al. It indicates that kind of dot defect comes from a mixture of fine dross and film-like alumina which lay on the surface of hot-dip galvanized sheet.

Key words hot-dip galvanized sheet, surface, dot defect, laser confocal scanning microscopy, electron probe microanalysis

Study on Stamping Properties of Galvalume Coating

Ding Zhilong

(Technology Center of Baosteel Meishan Iron & Steel Co., Ltd., Nanjing 210039, China)

Abstract As hot-dip material of steel substrate, Galvalume alloy has good corrosion resistance. But the Galvalume coating will have crack when it is deformed, so it is very necessary to study the reasons. In this paper, the reason of the coating cracks have been found by using stamping test and 180 degree bending test for different coating thickness（weight）hot-dipping galvalume steel sheet . The coating of hot-dipping galvalume and transition layers have been tested by SEM and EDS to study crack reasons. The results as follows: coating thickness（weight）is the main reason of crack, coating lateral deformation bending and tensile stress tendency will become smaller with the decrease of coating thickness.With the result of SEM and EDS, Fe, Al and Si are the main elements in the crack position. In the deformation process, Galvalume alloy transition layers crack more easily relatively to the coating .So the transition layers thickness is the main reason of stamping crack. With the decrease of Galvalume alloy transition layers thickness, the coating crack trend become smaller. Combining Galvalume coating production process, coating thickness and alloy layer thickness have optimized to improve Galvalume coating products forming performance by demonstrating crack reason using actual stamping results.

Key words galvalume, coating thickness, transition layer, forming performance

冷轧板表面粗糙度对磷化膜微观形貌的影响

李子涛 周世龙 张 武 张 军 刘永刚

（马鞍山钢铁股份有限公司汽车板推进处，安徽马鞍山 243000）

摘 要 使用粗糙度仪、扫描电镜等试验设备研究了不同粗糙度的冷轧 DC06 钢板的磷化膜微观形貌，结果表明：

随着 Ra 值增加，钢板表面凹坑变大、变深，当 RPc 值增加，钢板表面凹坑变得更加密集；Ra、RPc 值增加均有利于磷化成膜，即使 Ra 值较低，保持较高的 RPc 也能获得良好的磷化膜。

关键词 冷轧板，表面粗糙度，磷化膜，微观形貌

Influence of Surface Roughness of Cold-rolled Steel Sheet on Surface Morphology of Phosphate Coatings

Li Zitao, Zhou Shilong, Zhang Wu, Zhang Jun, Liu Yonggang

(Maanshan Iron & Steel Co., Ltd., Auto Sheet Strategic Business Unit Cihu New District, Maanshan Anhui 243000, China)

Abstract Roughness meter and SEM were used to study influence of surface roughness of cold-rolled steel sheet on surface morphology of phosphate coatings. The results showed that the more and deeper pits were observed with increasing Ra and RPc.Both the increased Ra and RPc were beneficial to develop phosphate coatings with fine and uniform crystalline grain.Even if the Ra value is low, high RPc can be also useful to obtain good phosphate coatings.

Key words cold-rolled steel sheet, surface roughness, phosphate coatings, surface morphology

永磁密封技术在热镀锌中的应用研究

李 静[1] 李小占[1] 赵忠东[2] 陈 健[3]

（1. 北京科技大学冶金工程研究院，北京 100083；2. 河北钢铁邯郸分公司能源中心，河北邯郸 056015；3. 营口京华钢铁有限公司，辽宁营口 115005）

摘 要 本文提出了一种新型的永磁封流技术和封流原理，并构建实验平台。通过三维有限元方法对永磁封流过程进行了数值模拟计算，得出永磁封流过程中永磁转子转速、镀槽与永磁转子间距、锌液高度等永磁封流参数对锌液所受电磁力的影响规律。结果表明，相对于传统的电磁封流技术，永磁封流技术控制简单，节约能源，具有良好的应用前景。

关键词 热镀锌，永磁封流，有限元，数值模拟

Application and Research of Finite Element Numerical Simulation on Permanent Magnet Sealing in Hot-dip Galvanizing

Li Jing[1], Li Xiaozhan[1], Zhao Zhongdong[2], Chen Jian[3]

(1. University of Science & Technology Beijing, Beijing 100083, China; 2. Handan Iron and Steel Group Co., Ltd., Handan 056015, China; 3. Minmetals Yingkou Medium Plate Co., Ltd., Yingkou 115005, China)

Abstract This paper presents a new type of permanent magnetic sealing technology and sealing principle, and builds an experimental platform. In order to achieve the correction of the permanent magnet rotor speed, the space between the plating bath and the permanent magnet rotor, and the zinc liquid height, the numerical simulation is studied by the three-dimensional finite element method. The results show that permanent magnetic sealing technology control is simple, saving energy, compared with the traditional electromagnetic sealing technology, and has a good prospect in application.

Key words hot-dip galvanizing, permanent magnet sealing, finite element, numerical simulation

Q420qENH 桥梁钢模拟海洋环境腐蚀行为研究

李 琳 艾芳芳 陈义庆 侯华兴 高 鹏 钟 彬 肖 宇

（鞍钢集团钢铁研究院，辽宁鞍山 114009）

摘 要 本文采用 Q420qENH 低碳贝氏体耐候桥梁钢和对比试样传统耐候钢 09CuPCrNi，通过周期浸润和浸泡腐蚀试验来模拟海洋环境下桥梁钢的腐蚀行为，利用试样微观腐蚀形貌观测和电化学分析等手段对其进行了腐蚀行为研究。结果表明：低碳贝氏体组织的 Q420qENH 在模拟海洋环境耐腐蚀性能优于传统耐候钢 09CuPCrNi。周期浸润腐蚀试验后 Q420qENH 形成的锈层均匀致密、保护性较强。模拟海洋环境中，Q420qENH 和 09CuPCrNi 试样电化学阻抗值先增大后减小，最后增大至相对稳定值。

关键词 模拟海洋，周期浸润试验，耐候桥梁钢

Study on Corrosion Behavior of Q420qENH Bridge Steel in Simulated Ocean Environment

Li Lin, Ai Fangfang, Chen Yiqing, Hou Huaxing, Gao Peng, Zhong Bin, Xiao Yu

(Iron & Steel Research Institute of Angang Group, Anshan Liaoning 114009, China)

Abstract In this paper, low carbon bainite weathering bridge steel Q420qENH were used, comparing with the traditional weathering steel 09CuPCrNi. The corrosion behavior of bridge steels in ocean environment was simulated through cyclic and longtime immersed corrosion test. And the morphology of samples after corrosion test was observed. Meanwhile, the electrochemical measures were researched. The result indicated that the corrosion resistance of Q420qENH was better than 09CuPCrNi in simulated ocean environment. The rust layer was uniformity and compact，which has a strong protection after cyclic immersed corrosion test. EIS of Q420qENH and 09CuPCrNi increased first and decreased then，finally raised comparatively stabilization in simulated ocean environment.

Key words simulated ocean environment, cyclic immersion corrosion test, weathering bridge steel

冷轧酸洗钝化技术的生产应用

张 毅[1] 马 昊[2] 张 虎[2] 刘 昕[3]

（1. 路宝利（北京）科技有限公司，北京 100070；
2. 山钢股份济南分公司薄板厂冷轧车间，山东济南 250100；
3. 中国钢研科技集团有限公司，北京 100081）

摘 要 通过对冷轧厂酸洗机组工艺的研究分析，结合酸洗下游工序和酸洗商品板客户对表面质量和防锈性能的要求，在济钢薄板厂酸洗机组进行了酸洗钝化剂的连续使用，实践证明，在酸洗机组漂洗段使用酸洗钝化剂后，酸洗卷表面质量显著提高，有效抑制了水锈、黄斑和停车斑等缺陷，成功实现了无油钝化防锈酸洗商品卷的批量生产，节省了酸洗机组的防锈油成本，同时也节约了商品卷客户的脱脂成本和废水处理量，填补了国内冷轧行业的技术空白，具有较好的市场前景和推广价值。

关键词 酸洗钝化剂，酸洗，缺陷，无油防锈

Application of Picking Passivation Technology in Cold Rolling Production

Zhang Yi[1], Ma Hao[2], Zhang Hu[2], Liu Xin[3]

(1. Lubritalia S.p.a(Beijing) Ltd., Beijing 100070, China;
2. Cold Rolled Workshop in Sheet Plant of Jinan Iron & Steel Co., Ltd., Jinan 250100, China;
3. China Iron & Steel Research Institute Group, Beijing 100081, China)

Abstract Through the study of pickling process in the cold rolling mill, combined with the lower procedure of the pickling process and the customer requirements for surface quality and rust resistance for pickling plate products, the pickling passivation agent was continuously used and now is still used in the pickling line in sheet plant of Jian Iron & Steel Co., LTD.. Practice has proved that the surface quality of the pickling products was improved significantly after using the passivation agent in the acid pickling line. The defects such as watermark, yellow spot and roll stop mark had been successfully controlled. So the cost of the anti-rust oil of the pickling line was saved, the cost of degreasing and waste water processing was also saved for the pickling coil customers. Meanwhile, this technology filled in the blank in domestic cold-rolled industry technology, and has good market prospect and promotion value.

Key words pickling passivation agent, pickling, defect, oil-free rustproof

Ti 添加对热浸镀 55wt.%Al-Zn-Si 镀层中析出相的影响

罗 群[1] 李 谦[1,2] 张捷宇[1] 李 麟[1] 周国治[1]

（1. 上海大学省部共建高品质特殊钢冶金与制备国家重点实验室，上海 200072；
2. 上海大学材料基因工程研究院，上海 200444）

摘 要 为了探究 Ti 添加导致热浸镀 55wt.%Al-Zn-Si 镀层锌花减小的机理，本文通过实验对比观察 55wt.%Al-Zn-Si 和 55wt.%Al-Zn-Si-Ti 镀层合金层的形貌和各物相的成分。为了减少杂质元素如 V、Ce 的影响和便于析出相的检测，本文利用高纯的原料锭配制了 55wt.%Al-Zn-1.6Si-1Fe-0.7Ti 合金，研究 Fe 和 Ti 过饱和后的析出相种类。SEM 和 EDS 的结果表明，55wt.% Al-Zn-Si-Ti 镀层中通常可以检测到颗粒尺寸大于 10μm 的金属间化合物，而 55wt.%Al-Zn-Si 镀层的金属间化合物颗粒尺寸小而均匀。配制的合金中，Fe 和 Ti 过量将导致析出两种金属间化合物，分别为 $TiAl_3$ 和 Fe_4Al1_3。利用 Al-Zn-Si-Fe-Ti 相图数据库计算 610℃的 55wt.% Al-Zn-1.6Si 等温截面，结果显示过量 Ti 导致镀液中析出 $TiAl_3$ 相，Fe 含量超过饱和溶解度导致镀液中析出 Fe_4Al1_3 相，计算得到的 $TiAl_3$ 和 Fe_4Al_{13} 的成分与实验结果吻合。根据计算结果分析 Ti 添加导致锌花减小的机理为：Ti 添加使镀

液中析出 $TiAl_3$ 相($Ti_{25}Al_{60}Si_{15}$)，消耗镀液中的部分 Si，使镀液中有效 Si 含量降低，从而使 Fe 的饱和溶解度降低，析出更多的 Fe_4Al_{13}。$TiAl_3$ 和 Fe_4Al_{13} 含量的增加使镀液中的形核密度增加，细化晶粒。

关键词 55wt.%Al-Zn-Si 镀层，金属间化合物，合金层，热力学，相图计算

The Effect of Ti Addition on the Precipitated Phases in Hot-dip 55wt.%Al-Zn-Si Coating

Luo Qun[1], Li Qian[1,2], Zhang Jieyu[1], Li Lin[1], Zhou Guozhi[1]

(1. State Key Laboratory of Advanced Special Steels, Shanghai University, Shanghai 200072, China;
2. Institute of Genomic Material, Shanghai University, Shanghai 200444, China)

Abstract In order to study the mechanism of reduction of spangle size for 55wt.% Al-Zn-Si-Ti coating, the alloy layers of the hot-dip 55wt.%Al-Zn-Si coatings and 55wt.%Al-Zn-Si-Ti coatings were comparatively investigated by scanning electron microscopy (SEM) and energy dispersive X-ray spectrometer (EDS). Aiming to avoid influence of the impurity elements (V and Ce) and easily identify the intermetallic compounds in the alloy, the high purity 55wt.% Al-Zn-1.6Si-1Fe-0.7Ti alloy was prepared. The results showed that the size of Fe-Al-Si intermetallic compounds in 55wt.% Al-Zn-Si-Ti coatingwas always larger than 10 μm. Many big intermetallic particles were observed at the center of the spangle, but the intermetallic particles in the 55wt.% Al-Zn-Si coatings were small and distributed uniformly. The calculated isothermal sectionof 55wt.% Al-Zn-1.6Si at 610 °C showed that the Ti addition made $TiAl_3$ precipitated and the high Fe content led to the precipitation of Fe_4Al_{13}. The calculated compositions of $TiAl_3$ and Fe_4Al_{13} were consistent with that determined from experiment. The mechanism of grain reduction caused by Ti addition can by summarized as follows: the addition of Ti makes the $TiAl_3$($Ti_{25}Al_{60}Si_{15}$) formed, then the Si content would decrease in the bath, leading to the amount of precipitated Fe_4Al_{13} increase. Both the Fe_4Al_{13} and $TiAl_3$ could act as the heterogeneous nucleation sites of the primary Al dendrites and refine the grains.

Key words 55wt.%Al-Zn-Si coatings, intermetallic compound, alloy layer, thermodynamics, phase diagram calculation

La 对热浸镀 55%Al-Zn-Si 合金熔池中底渣的影响

许 晋[1] 李 钦[1] 吴 岳[1] 李 谦[1,2]

（1. 上海大学省部共建高品质特殊钢冶金与制备国家重点实验室，上海 200072;
2. 上海大学材料基因工程研究院，上海 200444）

摘 要 Galvalume 镀液中添加 La 元素可以提高镀层的耐腐蚀性、镀层组织均匀性、降低合金层厚度，但是，当这种元素的添加量不合适时，会使镀液中的渣相增多，粘附于镀层表面的渣相会损害镀层的质量，使镀层性能下降，严重时会影响生产的运行。本文通过配渣的方式来模拟实际熔池中成渣情况，首先研究单独添加 La 进入 55%Al-Zn-Si 熔池中的成渣现象，再观察 Fe、La 两种元素之间的相互作用对镀液中渣相的影响，分析 Al-Zn-Si-Fe-La 体系中的渣相析出行为。采用 X-ray 衍射分析(XRD)、扫描电子显微镜电镜(SEM)、聚焦离子束(FIB)、透射电子显微镜(TEM)等手段对比研究了含 La 镀液中析出渣相的成分、形貌组织、物相结构以及它们之间的相互关系，探索了 La 在镀液中底渣生成的作用规律。研究表明 La 的添加不仅使熔池中析出不同形貌的含 La 渣相($Al_4Zn_4Si_{1.5}La$ 和 Al_2Zn_2La)，也促进 τ_5、τ_6 相形成，且 Fe、La 的协同作用也促使 $Al_4Zn_4Si_{1.5}La$ 在更低的 La 含量下析出。

关键词 热浸镀 Galvalume 合金，La 元素，底渣

Effect of La on the Dross Formation in the Hot-dip 55%Al-Zn-Si Bath

Xu Jin, Li Qin, Wu Yue, Li Qian

Abstract With adding La into the bath of Galvalume, it can improve the corrosion resistance and microstructure uniformity of the coating, reduce the thickness of the alloy layer. However, if the amount of La was not in the suitable range, the amount of dross would increase in the bath. The quality and properties of the coating would be degraded if the dross were picked up on the coating surface. The dross problems could bring harm to the operation of the coating line. Therefore, the paper adopted the method of adding La directly into the liquid alloy to simulate the dross formation in the bath. First, the author added La into the Al-Zn-Si bath to observe the dross formation in the bath. Then, added La and Fe to observe the effect of Fe and La on the dross formation in the complicated system. X-ray diffraction (XRD), scanning electron microscope (SEM), focused ion beam(FIB) and transmission electron microscope (TEM) were used to investigate the chemical composition, morphology, microstructure and the correlation of the dross formation in the bath. The impacting mechanism of Fe and La on the dross formation in the bath was studied: with La adding into the bath, it will not only precipitate $Al_4Zn_4Si_{1.5}La$ and Al_2Zn_2La which have different features, but also promote formation of τ_5, τ_6 phase. Moreover, $Al_4Zn_4Si_{1.5}La$ is affected by Fe to grow at lower La content.

Key words hot-dip galvalume alloy, La, bottom dross

Effect of Post-annealing on the Microstructure and Mechanical Properties of Cold-sprayed Zn-Al Coatings

Liang Yongli[1,2,3], Zhang Junbao[2], Yang Xiaoping[2]

(1. University of Science and Technology of China, Hefei 230026, China;
2. Advanced Technology Division, Research Institute, Baoshan Iron & Steel Co., Ltd., Shanghai 201900, China; 3. Shenyang National Laboratory for Materials Science, Institute of Metal Research, Chinese Academy of Sciences, Shenyang 110016, China)

Abstract A compact Zn-Al coating was successfully deposited by cold spraying technique. The effect of annealing temperature on the microstructure, microhardness and wear properties of the coatings was explored. The typical morphology of as-sprayed coating was characterized as intensively deformed particles with elongated shape. After annealing treatment, more homogeneous microstructure was achieved at 250°C while an eutectoid lamellar structure started to form when annealing temperature was over 300°C. An obvious anneal-hardening effect was observed especially at the annealing temperature range of 200 to 350°C. The wear resistance of annealed coatings showed an opposite trend with the variation of their microhardness. The lowest friction coefficient was obtained after annealed at 250~300°C. Generally, The best wear properties with both relatively small weight loss rate and friction coefficient were achieved after annealing at 250°C for 120 min.

Key words cold spray, Zn-Al coating, microstructures, hardness, wear properties

蓝膜及蓝膜应用

张浙军

（威海蓝膜光热科技有限公司，山东威海 264200）

摘　要　蓝膜也称超级蓝膜，是一种太阳光谱选择性吸收涂层，是沉积在金属卷材表面，用来收集太阳能的吸收涂层，属新一代太阳能利用技术。其对太阳辐射能具有极高的吸收率，且其自身发射率很低，可有效提高太阳能光热转换效率。

欧美约90%以上的太阳能热水系统采用蓝膜作为系统的核心部件，蓝膜可将阳光中的热能充分吸收，并最终传递给水。此类系统广泛应用于家庭热水、公用热水、医药、化工、养殖等诸多领域。

蓝膜太阳能采暖系统是将阳光中能量充分吸收后转化为热能，再将热能传送到所需区域进行采暖使用。采暖系统可分为主动采暖系统和被动采暖系统。主动太阳能采暖系统，可由蓝膜产品与空气源热泵等主动得热组件共同组成，采用将蓝膜金属板所得太阳能提供给热泵进行提升，以达到适宜的室内供热温度。蓝膜被动采暖系统如同彩钢建材一样安装于建筑墙面，可广泛应用于仅在白天需要采暖的厂房、学校、商场等建筑。

关键词　蓝膜，太阳能的吸收涂层，采暖系统，金属

日照钢铁连续酸洗镀锌生产线工艺特点

高仁辉　郑志刚

（日照钢铁控股集团有限公司，山东日照　276806）

摘　要　本文重点介绍了日照钢铁连续酸洗镀锌生产线的工艺特点。两条生产线分别由奥地利安德里茨公司和意大利达涅利公司设计，采用了国际上最先进的生产工艺和关键设备。日照钢铁连续酸洗镀锌生产线的工艺和装备达到了国际先进水平，具备了一定的“以热代冷”能力。本文还系统阐述了连续酸洗镀锌生产线相对于传统镀锌的技术优势和成本优势，并指出了产线运行、生产和质量管理的关键在于“稳定”。

关键词　热轧薄板，酸洗，热镀锌，工艺特点

Technology Characteristics of Continuous Pickling and Galvanizing Line of Rizhao Steel Co., Ltd.

Gao Renhui, Zheng Zhigang

(Rizhao Steel Holding Group Co., Ltd., Rizhao 276806, China)

Abstract　The technology characteristics of continuous pickling and galvanizing line of RiZhao steel holding group company was introduced emphatically in the paper. The No.1 product line was designed by Andritz in Austria and another line was designed by Danieli in Italy, which adopts the most advanced technology and key equipments. The technology and equipment of the line comes to the international advanced level, which can produce hot rolling galvanized sheet replaces the cold rolling galvanized sheet. This paper also described the technique and cost advantage of continuous pickling and galvanizing line relative to the common galvanizing line, and point out that the key of line operation, production and quality management is “Stability”.

Key words　hot rolling sheet，pickling, hot dip galvanizing, technology characteristics

大气暴晒和实验室加速腐蚀环境下耐候钢的耐候性能研究

王志奋　刘建荣　周元贵　刘　敏　王俊霖

（武汉钢铁(集团)公司研究院，湖北武汉　430080）

摘　要　研究了大气暴晒和实验室加速腐蚀环境下耐候钢和结构钢的耐腐蚀性能。结果表明，由于合金元素作用造成致密锈层，在两种环境下耐候钢的耐腐蚀性能均优于结构钢。通过研究两种环境下耐候钢和结构钢的锈层情况，结果表明实验室环境能较好的模拟大气环境，但是无法全面真实的反映实际大气环境腐蚀情况。

关键词　耐候钢，X 射线衍射，SEM，锈层

The Corrosion Resistance of Weathering Steels in Atmospheric Exposure and Laboratory-accelerated Corrosion Tests

Wang Zhifen, Liu Jianrong, Zhou Yuangui, Liu Min, Wang Junlin

(Research and Development Center, Wuhan Iron and Steel (Group) Corporation, Wuhan 430080, China)

Abstract The corrosion resistances of weathering steel and structure steel in atmospheric exposure and laboratory-accelerated corrosion tests have been investigated. The results showed that the corrosion rate of weathering steel was lower compared with structure steel due to the alloying additions. The chemical composition and morphology of the rust layers of weathering steel and structure steel have been studied in the two experiments. The results showed that the laboratory test preferably simulated the site circumstance. But the difference between the two tests was in existence.

Key words　weathering steel, X-ray diffraction, SEM, rust

表面粗糙度对热镀锌自润滑钢板表面润滑及耐蚀性能的影响

杨家云　戴毅刚

（宝山钢铁股份有限公司研究院，上海　201900）

摘　要　在表面粗糙度分别 0.82μm、1.23μm、1.68μm 和 2.07μm 的 4 种热镀锌钢板上涂覆自润滑涂层制得热镀锌自润滑钢板，采用表面性能分析仪和中性盐雾试验研究了热镀锌钢板表面粗糙度对热镀锌自润滑钢板表面润滑及耐蚀性能的影响。结果显示无涂层时，钢板表面动摩擦因数较大，均在 0.35 以上，且随其表面粗糙度的增加迅速增加。当涂覆自润滑涂层后，自润滑钢板表面动摩擦因数均显著降低至 0.15 以下并在摩擦测试初期几乎保持不变，表明自润滑钢板表面粗糙度在摩擦测试初期对其表面动摩擦因数无影响。但在摩擦测试后期，自润滑钢板表面动摩

擦因数逐渐增大，且增大幅度随着表面粗糙度的增加而增加。同时结果显示表面粗糙度对自润滑钢板耐蚀性能无影响。

关键词 表面粗糙度，自润滑涂层，热镀锌钢板，动摩擦因数，耐蚀性能

Effect of Surface Roughness on Surface Lubrication and Anticorrosion Performance of Galvanized Steel Sheet with Self-lubricated Coating

Yang Jiayun, Dai Yigang

(Research Institute, Baoshan Iron & Steel Co., Ltd., Shanghai 201900, China)

Abstract 4 kinds of galvanized steel sheets with different surface roughness were used to prepared self-lubricated steel sheets with self-lubricated coating, and the effects of surface roughness on surface lubrication and anticorrosion performance of the self-lubricated steel sheets were investigated by friction coefficient tester and neutral salt spray test. The results showed that dynamic friction coefficient of the galvanized steel sheets without self-lubricated coating were high and increased with increase of surface roughness of the sheets. After coated with elf-lubricated coating, dynamic friction coefficient of the sheets reduced remarkably and almost kept unchanged at first stage of the friction test. This proved that the surface roughness of the self-lubricated steel sheets had a little relationship with the dynamic friction coefficient at first stage of the test. However, the dynamic friction coefficient of the sheets increased with increase of surface roughness at the end of the test. Moreover, the results also showed that the anticorrosion performance of the self-lubricated steel sheets had a little relationship with the surface roughness of the sheets.

Key words surface roughness, self-lubricated coating, galvanized steel sheet, dynamic friction coefficient, anticorrosion property

唐钢镀锌锌花尺寸及均匀性控制的实践

孙　力　刘大亮　韩　冰

（河北钢铁集团唐钢公司，河北唐山　063016）

摘　要 在建材市场，不同的国家和地区及不同使用部位对于锌花要求不同，如南美洲地区要求大锌花，而欧洲一般喜欢无锌花，外露雨水管或工业通风要求大锌花，彩涂基板则要求光整无锌花。本文主要研究了在无小锌花装置条件下，利用锌液成分调整对锌花尺寸进行控制，并成功开发出大锌花、小锌花产品。通过对加热制度以及镀后工艺、气刀参数调整，带钢表面锌花均匀性有了较明显提高。何在一条镀锌线上满足不同用户对锌花不同要求，利用锌液中 Sb 含量的变化对锌花尺寸进行调节，同时利用加热制度、镀后冷却风机以及速度的调整对锌花均匀性进行调节。经过多次试验试产，在同条生产线达到了不同锌花尺寸的控制目的，给接单和镀锌生产组织提供极大方便。通过试验对比分析得到以下结论：随着 Sb 含量的降低，锌花尺寸在逐步减小，但是锌花尺寸的降低过程中有平台不是线性关系， Sb 含量降低至一定水平后，存在锌花状态不稳定；含量降低下线有限度。锌花大小除于 Sb 含量有关外，同工艺速度、冷却速率等因素有关，是个工艺耦合过程。发挥移动风机和气刀作用，提高板面结晶线，避免锌层均匀性，可以有效保证带钢中部与边部的锌花尺寸差异，保证锌花均匀性。

关键词 小锌花，锌液成分，移动风机，形核

The Study of Controlling Particles Size and Uniformity with Spangle for Tangsteel CGL

Sun Li, Liu Daliang, Han Bing

(Thangshan Steel, HBIS, Tangshan 063016, China)

Abstract by controlling composition of galvanizing liquid to adjust spangle size without spangle equipment, this way can produce big and mini spangles. On the other hand, by adjusting heating cycle and process, uniformity of spangle approve evidently.

Key words mini-spangle, composition of galvanizing, moving fan, nucleation

合金化工艺对 DP590 合金化镀层组织和性能的影响

王贺贺[1]　齐春雨[1,2]　江社明[1]　李远鹏[1]　俞钢强[1]　张启富[1]

（1. 中国钢研科技集团有限公司先进金属材料涂镀国家工程实验室，北京　100081；
2. 北京首钢冷轧薄板有限公司，北京　103042）

摘　要　为研究热镀锌合金化工艺参数对 DP590 双相钢合金化行为的影响，采用 SEM、粗糙度仪分析了不同工艺条件下合金化镀层的表面形貌、镀层结构和表面粗糙度。结果表明，随着合金化温度升高和时间延长，合金化镀层表面 ζ 相减少，δ 相增多，出现小孔；480 ℃时 DP590 钢板镀层表面粗糙度随处理时间的延长而增加，500 ℃和 520 ℃处理时镀层表面粗糙度时随时间延长而减小；当合金化时间为 25 s 时，镀层表面粗糙度随温度升高而降低，而 20 s 时，随合金化温度升高，表面粗糙度呈先升高后降低趋势；温度升高，时间增加，合金化程度加深，当合金化温度为 520 ℃，合金化时间为 25 s 时，DP590 钢板热镀锌合金化镀层中出现 Γ 相，耐粉化性能变差。

关键词　双相钢，热镀锌合金化，表面形貌，镀层结构，表面粗糙度

Effect of Alloying Process on Structure and Properties of DP590 Galvannealed Coatings

Wang Hehe[1], Qi Chunyu[1,2], Jiang Sheming[1], Li Yuanpeng[1], Yu Gangqiang[1], Zhang Qifu[1]

(1. National Engineering Laboratory of Advanced Coating Technology for Metals, Central Iron and Steel Research Institute, Beijing 100081, China;
2. Beijing Shougang Cold Rolled Sheet Co., Ltd., Beijing 103042, China)

Abstract To analyze the effect of alloying parameters on the alloying behavior of DP590 dual phase steel plate, SEM, surface roughness measuring instrument were applied to investialloyingte the surface morphology, cross-sectional structure and surface roughness of the alloying coatings under different alloying processes. The results show that with alloying temperature rising and alloying time extending, the content of ζ phase decreases and the content of δ phase increases, and small holes appear on the surface of the coating. When alloying temperature is 480℃, surface roughness of the coating on DP590 steel plate increases with the extending of time. While ALLOYING temperature is 500℃ and 520℃, the surface roughness decreases with time extension. When alloying time is 25 s, the surface roughness decreases with the rising of temperature. While alloying time is 20 s, with the rising of alloying temperature, the surface roughness decreases first and then increases. With alloying temperature rising and alloying time extending, alloying degree deepens, and Γ phase appears in the coating when alloying temperature is 520℃ and alloying time is 25 s, which would do harm to the powdering resistance galvannealed of coatings of DP590.

Key words dual phase steel, galvannealing alloying, surface morphology, cross-sectional structure of coating, surface roughness

热镀锌钢管表面黑斑分析

刘 昕 江社明 张启富

（中国钢研科技集团有限公司先进金属材料涂镀国家工程实验室，北京 100081）

摘 要 通过扫描电镜和能谱分析钝化镀锌钢管表面形貌及表面成分分布情况，结合实验室耐湿热性能检测，寻找钝化镀锌钢管表面出现黑斑的影响因素，从而为提高产品质量提供方法及建议。结果表明，钝化镀锌钢管表面黑斑的形成主要是由于钝化膜耐湿热性能较差引起，同时铅在镀层内的不均匀分布加速了黑斑的出现。

关键词 镀锌，黑斑，耐湿热性，铅

Analysis of Black Spot on Hot Dip Galvanized Steel Pipe

Liu Xin, Jiang Sheming, Zhang Qifu

(National Engineering Lab of Advanced Coating Technology for Metals China Iron & Steel Research Institute Group, Beijing 100081, China)

Abstract Surface morphology and the distribution of elements were observed by scanning electron microscopy(SEM) and energy disperse spectroscopy(EDS), wet heat resistant property of the passivation film was tested in the laboratory to find the influence factors of the black spot phenomenon of the hot dip galvanized steel pipe. So as to provide methods and suggestions to improve the product quality of the hot dip galvanized steel pipe. Results show that the formation of the black spot on the galvanized steel pipe is mainly due to the poor wet heat resistant property of the passivation film, and uneven sistribution of lead in the zinc coating accelerated the appearance of black spots.

Key words galvanize, black spot, wet heat resistance, lead

连铸结晶器铜板电镀 Cu-Ni 技术的应用研究

白新波

（宝钢工程集团 宝钢工业技术服务有限公司，上海 201900）

摘　要　连铸拉坯时，结晶器铜板表面镀层不断磨损，磨损到一定程度后，需下线对铜板进行电镀修复。为保证镀层与结晶器铜板有良好的结合力，电镀之前，要把铜板上镀层铣除干净，去除残余镀层过程中不可避免会铣掉部分基材，经过若干次处理后铜板逐渐变薄直至报废。本文探讨了一种恢复结晶器铜板厚度的电镀方法，延长其寿命和报废期限。通过循环伏安曲线，对焦磷酸盐体系中Cu-Ni合金的共沉积过程进行了探讨，结果表明，在焦磷酸体系中可以实现Cu-Ni共沉积。采用恒电流电沉积的方法制备出了Cu-Ni合金镀层，结果表明镀层的形貌、结合力和导电性良好，Cu-Ni合金镀层与基体铬锆铜相比，硬度和耐腐蚀性都有显著提高，这为该镀层在铜板上的应用、延长铜板使用寿命提供了可能。

关键词　连铸结晶器，Cu-Ni镀层，共沉积，焦磷酸盐，性能

Application Research of Cu - Ni Electrodeposition Technology on Continuous Casting Mould

Bai Xinbo

(Baosteel Industry Technological Services Co., Ltd., Baosteel Engineering Group Co., Ltd., Shanghai 201900, China)

Abstract　In the course of work, the surface coatings of continuous casting mold would wear and get thinner. It is need to repair the mold by plating when the coating wears to a certain extent. Prior to electroplating, the residual coating should be cleaned with a mill, portion of the substrate will be milled out inevitably in the process. After several times, the mold gets thinner and thinner gradually until retirement. An electroplating method to recover the thickness of continuous casting mould was discussed. The process of co-deposition of Cu-Ni alloy was discussed by cyclic voltammetry. The results show that copper and nickle can be co-deposited in pyrophosphate. And Cu-Ni alloy coating were prepared by constant current deposition. Morphology, electrical conductivity and adhesion and of the coating are good. Compared with copper plate (Cu-Cr-Zr), the hardness and corrosion resistance of Cu-Ni alloy deposit Obtained by method of electro-deposition are improved significantly. It provides a possibility of application on mould and prolongs the service life of the mould.

Key words　continuous casting mould, Cu-Ni coating, co-deposition, pyrophosphate, property

冷轧钢板表面粗糙度控制实践

贾生晖[1]　邢　飞[1]　刘海军[1]　卢劲松[1]　王伟湘[2]　李名钢[1]　赵　刚[1]
朱生卫[1]　胡厚森[1]　金　力[1]

(1. 武汉钢铁股份有限公司冷轧薄板总厂，湖北武汉　430083；
2. 武汉三联特种技术有限公司，湖北武汉　430000)

摘　要　通过冷轧生产时，表面粗糙度数据对比分析，得出不同变形率、润滑条件、轧辊毛化方式及辊面粗糙度均匀条件下，对应的带钢表面粗糙度情况，摸索出最佳的工艺路线和工艺参数组合，在最终成品板面粗糙度及均匀性满足客户的要求方面做出了积极的尝试。

关键词　表面粗糙度，轧辊毛化，工艺参数

Roughness Control of Cold Rolling Steel in Practice

Jia Shenghui[1], Xing Fei[1], Liu Haijun[1], Lu Jinsong[1], Wang Weixiang[2], Li Minggang[1], Zhao Gang[1], Zhu Shengwei[1], Hu Housen[1], Jin Li[1]

(1. Wuhan Iron & Steel Co., Ltd., Wuhan Hubei 430083,China;
2. Wuhan Sanlian Special Technology Co., Ltd., Wuhan Hubei 430000, China)

Abstract After cold rolling, roughness data of production was compared and analysised, we knew the relationship between the corresponding strip surface roughness and deformation rate, lubricating condition, different roll roughen way under the condition of uniform and roller surface roughness etc.. We've groped for the best combination of process route and process parameters. We make positive attempts in the final surface roughness of strip to meet customer requirements better.

Key words surface roughness, roll roughen way, process parameters

易拉罐拱底白斑分析

阮如意 顾 婕 张 清 郭文渊

（上海宝钢包装股份有限公司，上海 201908）

摘 要 金属易拉罐灌装饮料中国特有的凉茶饮料，在经过121℃、30min的高温蒸煮杀菌工艺后，往往在未涂覆涂料的罐底会出现各种不同的缺陷，其中罐底白斑是目前影响面最大的质量问题。由于罐底白斑的出现影响产品的外部美观，产品质量下降，影响消费者购买欲望，影响企业的形象和经济效益。本文采用电子扫描电镜从表面分析并结合EDS能谱分析产品白斑的原因，从分析结果来看白斑出现的主要原因是铝合金发生腐蚀的产生，并受铝材表面的平整度，化学钝化层致密性，杀菌水质这三个主要因素的影响。本文提出改善措施，帮助工厂和客户提高产品质量。

关键词 易拉罐，白斑，改善措施

Study of the White Spot Defect on Can Dome

Ruan Ruyi, Gu Jie, Zhang Qing, Guo Wenyuan

(Shanghai Baosteel Packaging Co., Ltd., Shanghai 201908, China)

Abstract Metal beverage cans filling unique Chinese herbal tea, after retort sterilization process at 121℃, in 30min, there will be many defects in the dome area which was not coated. By far, the white spot defect was the most serious quality problem. Since the white spot affect the can appearance, bring down the product quality, influence customers desire to buy. Corporate image and economic benefits. In this paper, the white spot defect area was investigated by scanning electron microscope (SEM), energy dispersive spectrometry(EDS) and Digital Microscope System. The forming reason of defects were analyzed systemically. The results show that aluminum surface flatness, chemical passivation layer density and sterilization water quality are three main factors. This paper proposes improvement measures, to help plant and customers improve products quality.

Key words cans, white spot, improvement

含盐量对DH36钢在海水全浸区腐蚀行为的影响

程 鹏 黄先球

（武汉钢铁（集团）公司研究院，湖北武汉 430080）

摘 要 针对DH36钢在海水全浸区的腐蚀问题，在实验室用人造海水模拟了全浸区的腐蚀环境，采用全浸腐蚀试验方法研究了DH36钢在模拟海水全浸区的腐蚀行为，通过SEM观察腐蚀形貌，利用回归分析建立了溶液含盐量与DH36钢腐蚀之间的关系。结果表明，DH36钢在海水全浸区的腐蚀十分严重，并且随着海水含盐量的升高，DH36钢的腐蚀速度呈现先增大后减小的趋势。

关键词 DH36钢，海水腐蚀，含盐量，全浸区

Influence of Salinity on Corrosion Behavior of DH36 Steel in Seawater Full Immersion Zone

Cheng Peng, Huang Xianqiu

(Research and Development Center of WISCO, Wuhan 430080, China)

Abstract Aimed at the corrosion problem of DH36 steel in the seawater full immersion zone, corrosion environment of full immersion zone has been simulated with artificial seawater in laboratory. The corrosion behaviors of DH36 steel in the full immersion zone were investigated by using full immersion tests. The morphology of corrosion products was observed by SEM. Using regression analysis to establish the relationship between salinity and the corrosion of DH36 steel. The results showed that the corrosion of DH36 steel immersed in full immersion zone corroded seriously, the corrosion rate of DH36 steel firstly increased and then decreased with the increase of water salinity.

Key words DH36 steel, seawater corrosion, salinity, full immersion zone

高温炉辊新型热障涂层结构与性能研究

王 倩 徐建明

（宝钢工程集团宝钢工业技术服务有限公司，上海 201900）

摘 要 采用等离子喷涂技术制备用于高温炉辊的新型复合热障涂层，以Ni基合金涂层打底、ZrO_2-Y_2O_3涂层作为工作层、以ZrO_2-Y_2O_3与纯Al_2O_3混合粉涂层盖面，通过金相显微镜、显微硬度计、拉伸试验机、磨损试验机等对涂层的微观组织及性能进行了观测分析。结果表明，新型复合热障涂层与常规热障涂层相比，涂层微观组织更加致密，具有更高的显微硬度，更好的耐磨损、抗冲击性能，与基体结合强度、抗热震性能良好。

关键词 热障涂层，结构与性能，高温炉辊

Research on Structure and Properties of Thermal Barrier Coatings for High-temperature Hearth Roll

Wang Qian, Xu Jianming

(Baosteel Engineering Group, Baosteel Industry Technological Service Co., Ltd., Shanghai 201900, China)

Abstract The late-model composite thermal barrier coatings for high-temperature hearth roll were prepared by plasma spraying technology, what were nickel-base alloy coating as bottom layer, ZrO_2-Y_2O_3 coating as working layer, and the coating of ZrO_2-Y_2O_3 and pure Al_2O_3 mixed-powder as top layer. Microstructure and properties of the coatings were examined by metallographic microscope, microhardness tester, tensile testing machine, abrasion tester, and so on. The results show that the late-model composite thermal barrier coatings are more compaction and lower porosity compared to that of normal thermal barrier coatings; the microhardness is higher; the wear resistance and shock resistance are enhanced; the bond strength to base and thermal shock resistance of late-model composite thermal barrier coatings are very well.

Key words thermal barrier coatings, structure and properties, high-temperature hearth roll

钢板表面水性钝化防锈技术

唐艳秋[1]　安成强[2]

（1. 沈阳防锈包装材料有限责任公司，辽宁沈阳　110032；
2. 东北大学金属防护技术工程研究中心，辽宁沈阳　110004）

摘　要 通过对当前钢板涂覆防锈油进行防锈的影响分析，结合市场用户对钢板表面处理的要求，提出了钢板无油防锈的技术需求。对无油防锈方法进行了阐述，提出实现钢板无油防锈的最佳方法是表面水性钝化处理，并以冷轧钢板脱脂过程表面水性钝化处理和热轧钢板酸洗后表面水性处理为例，提出了实施的工序和过程，通过加速腐蚀试验分析了表面水性钝化处理后的板面耐蚀性变化，得出钝化处理能使钢板表面耐蚀性明显提升，在高温、高湿和凝露等苛刻环境中存放时间更长。表面耐蚀性提升后，可抵御钢板在包装前存放过程、开包后使用过程的锈蚀影响。结合气相防锈纸包装，可提供产品在储运期间的防锈保护，钢板防锈期比单纯去掉防锈油延长 39%，提升了无油防锈包装效果，可用于要求钢板表面不涂防锈油的产品防锈包装。

关键词 防锈，防锈油，钝化，钢板

Rust Prevention with Surface Passivation of Steel Sheet

Tang Yanqiu[1], An Chengqiang[2]

（1. Shenyang Rustproof Packaging Material., Co., Ltd., Shenyang Liaoning 110032, China;
2. Metal Protection Technology Engineering Research Center of Northeastern University，
Shenyang Liaoning 110004, China ）

Abstract This article put forward the technical requirements for steel sheet oil-free rustproof by analyzing the influence of the rust preventive oil method for steel sheet and user's requirements of sheet surface treatment. Oil-free rustproof methods were detailed, and put forward the water-based surface passivation treatment is the best way. It puts forward the applying process in cold rolled steel degreasing and hot plate after pickling. By the analysis of changes on steel sheet surface corrosion resistance with the accelerated corrosion test, it is concluded that water-based passivation treatment can make the steel plate surface corrosion resistance improved significantly, and the storage time longer even in harsh environments, such as high temperature, high humidity and condensation. The enhanced corrosion resistance can defense the corrosion in steel storage process before packaging and after using. Combined with volatile corrosion inhibiting paper, it can prevent the corrosion better during storage and transportation, and long the protection period 39% than just get rid of rust preventive oil. Water-based passivation can increase the antirust effect of steel sheets and can be applying in non-oiled steel products.

Key words antirust, rust preventive oil, passivation, steel sheet

基于电机振动的辊涂设备传动结构设计分析

陈方元 朱 志 褚学征

（中冶南方工程技术有限公司，湖北武汉 430223）

摘 要 本文对影响带钢辊涂涂层质量的因素进行分析综述，涂料的物理性能、辊涂工艺、设备精度及振动是影响涂层质量的重要因素。其中电机振动是辊涂设备振动的重要源头，基于两种方案的辊涂设备传动结构，通过有限元方法，对电机振动在辊涂机刚度及涂层精度等方面的影响进行了详细分析。对辊涂生产过程电气化控制及设计结构合理、高精度辊涂设备有一定的指导意义。

关键词 辊涂工艺，辊涂设备，振动，有限元分析

Roller Coating Technology and Equipment Affect the Quality of the Coating on the Strip Discussion

Chen Fangyuan, Zhu Zhi, Chu Xuezheng

(WISDRI Engineering and Research Incorporation Ltd., Wuhan 430223, China)

Abstract This paper reviews the factors affecting the quality of the roller coating, which includes physical properties of the coating, roller coating technology, equipment precision and vibration, which mainly comes from motor. Based on two different structure of transmission system of roller coating equipment, the paper analyzes the effects, in detail, of vibration of the motor, which affects the rigidity of roller coating equipment and the precision of coating, with FEA method. It is helpful for the electrical control of roller coating and is instructive to design a roller coating equipment with reasonable structure and high precision.

Key words roller coating technology, roll coater, vibration, finite element analysis

武钢智能化电泳模拟线在汽车板涂装实验上的应用

庞 涛 龙 安 白会平 马 颖 张思思

（武钢研究院，湖北武汉 430080）

摘　要　本文介绍了由武钢自主设计并建成的国内钢厂首个全流程智能化电泳涂装试验线，说明了该试验线的工作特点及优势、主要设备配置和性能参数，列出了试样要求及试样漆膜性能参数，测试表明，本试验线与同类工业化电泳生产线产品性能相当。对该试验线在武钢汽车板涂装试验上的应用前景予以展望。

关键词　电泳涂装，模拟，智能化，汽车板

Application of the Intelligent Electrophoresis Simulation Line in the Automobile Panel Coating Experiment

Pang Tao, Long An, Bai Huiping, Ma Ying, Zhang Sisi

(Research and Development Centre of Wuhan Iron and Steel Coap, Wuhan Hubei 430080, China)

Abstract The first full process intelligent electrophoresis coating test line built and designed by WISCO independently among domestic steel mills was introduced. In this paper, the working principle, characteristics and advantages, equipment and equipment performance parameters of the test line were described . Test shows that performance of this test line equals to industrial electrophoresis production line. At last, the future application in automobile plate coating test of this test line was prospected.

Key words electrophoresis, simulation, intelligent, automobile panel

提高超深冲钢板表面粗糙度的工艺分析

冯冠文　杜小峰　蔡　捷　黄道兵　周文强

（武钢研究院，湖北武汉　430080）

摘　要　钢板表面粗糙度是指具有微小间距的峰和谷组成的微观几何轮廓，钢板在汽车厂使用期间，表面粗糙度会对润滑油的含津能力和涂漆后的光亮度产生影响，是汽车冲压成型和涂装质量的重要参数之一。本文选取的是汽车厂主要关注的粗糙度参量 Ra 和 RPc，针对武钢为了达到汽车厂对钢板表面粗糙度的要求，以武钢超深冲钢板为原料，对平整工艺进行调整试验的分析总结。通过改变平整轧辊粗糙度，结合实测数据分析了板面粗糙度与轧辊粗糙度间的关联规律，发现在一定范围内（Ra≤4.0μm）增加轧辊粗糙度利于提高钢板粗糙度；通过调整的平整入出口张力来达到较大的轧制力，也可以获得理想的表面粗糙度。文章提出现场生产中增加钢板粗糙度的解决方案，满足客户要求，获得了良好的现场与市场应用效果。

关键词　超深冲钢板，粗糙度，平整工艺

Process Analysis on Increasing Surface Roughness of Extra Deep Drawing Steel

Feng Guanwen, Du Xiaofeng, Cai Jei, Huang Daobing, Zhou Wenqiang

(The R&D Center of WISCO, Wuhan 430080, China)

Abstract Surface roughness of steel is defined as microscopic geometry which is composed with small spacing peaks and troughs. On the application of steel for automotive factory, the surface roughness has great influence on the bearing capacity of lubricating oil and the luminance after coating. Therefore, the surface roughness is one of the critical parameters for steel drawing and application. In this paper, we studied the influence of Ra and RPc on extra deep drawing steel of WISCO. The roughness of equalizing rolling mill was changed to study the relation between the roughness of strip and roll. We find that the strip roughness increased when roll roughness raised within a certain range(Ra≤4.0μm); and the expected surface roughness would be gained by adjusting the tension of skin miller entry and exit for more draught pressure. The solution on increasing surface roughness mentioned in this paper met requirements of customers and had good performances on the market.

Key words extra deep drawing steel, surface roughness, skin passed mill technology

脱脂工艺对热镀锌钢板表面特性的影响

张剑萍 杨家云

（宝山钢铁股份有限公司研究院，上海 201900）

摘 要 分别采用水、有机溶剂和碱性脱脂剂水溶液对热镀锌涂油钢板进行脱脂清洗处理，并通过三维显微镜、辉光光谱仪和电化学工作站研究了不同脱脂工艺对热镀锌钢板表面形貌、元素成分及表面反应活性的影响。结果显示不同脱脂工艺对热镀锌板表面形貌无影响，且三种脱脂工艺均能够有效清洁热镀锌板表面油污。水和有机溶剂脱脂工艺能够保留热镀锌板表面氧化层，但碱脱脂工艺会破坏其表面氧化层，因此会影响热镀锌钢板表面的反应活性。电化学数据表明采用有机溶剂脱脂工艺可以保持热镀锌板表面的氧化膜状态不发生变化，因此其表面反应活性相对较低；采用水和碱脱脂工艺会改变热镀锌板表面氧化膜结构，会提高热镀锌表面反应活性。

关键词 脱脂工艺，热镀锌钢板，表面元素，电化学，表面反应活性

Effect of Degreasing Process on Surface Property of Galvanized Steel Sheet

Zhang Jianping, Yang Jiayun

(Research Institute, Baoshan Iron & Steel Co., Ltd., Shanghai 201900, China)

Abstract The galvanized steel sheets with antirust oil were degreased by neutral water, organic solvent and alkaline aqueous solution, respectively, and the effects of degreasing process on surface morphology, surface elements and surface reaction activity of the galvanized steel sheets were investigated by 3D microscope and electrochemical workstation. The results showed that three kinds of degreasing methods could effectively clean the antirust oil on the surface of galvanized steel sheets and had a little relationship with surface morphology of galvanized steel sheets. The electrochemical results proved that the solvent-degreasing process could keep the oxide film on the surface of galvanized steel sheet, which led to low surface reaction activity of the sheet. The neutral water-degreasing and alkaline-degreasing process would ruin the oxide film of galvanized steel sheet, and this would improve surface reaction activity of the sheet.

Key words degreasing process, galvanized steel sheet, surface element composition, electrochemistry, surface reaction activity

热镀锌沉没辊沟槽印缺陷研究

金鑫焱[1] 钱洪卫[2]

（1. 宝山钢铁股份有限公司研究院，上海 201900；
2. 宝山钢铁股份有限公司冷轧厂，上海 201900）

摘 要 利用配备能谱的扫描电镜分析了热镀锌沉没辊沟槽印缺陷的镀层表面、溶锌后基板表面以及截面的形貌和成分。发现了该缺陷具有锌层局部偏厚、锌渣残留、基板表面与带钢运行方向呈一定角度的轻微擦伤、Zn-Fe 合金相爆发组织等微观特征。从而证实了沉没辊沟槽印缺陷的形成过程：当带钢与沉没辊接触时，沉没辊沟槽内积聚的锌渣在带钢表面造成轻微擦伤，严重的位置局部形成 Zn-Fe 合金相爆发组织，残留的锌渣及生成的爆发组织导致锌层局部偏厚，从而在出锌锅后表现出与沉没辊沟槽位置对应的沟槽印缺陷。

关键词 热镀锌，沉没辊，沟槽印缺陷

Study on the Sink Roll Groove Mark Defect on Hot Dip Galvanized Steel Sheet

Jin Xinyan[1], Qian Hongwei[2]

（1. Baosteel Research Institute, Shanghai 201900, China；
2. Baosteel Cold Rolling Plant, Shanghai 201900, China）

Abstract A groove mark defect on hot dip galvanized steel sheet was studied by SEM equipped with EDS. The morphologies and the chemical compositions of the coating surface, substrate surface after zinc coating removal and cross section were characterized. It was found that the groove mark defect had a little thicker coating. The substrate showed slight scratches whose directions had a certain angle to the rolling direction. Both zinc dross and outburst structures of Zn-Fe intermetallic phase were found in the scratches. The formation mechanism of the sink roll groove mark defect was approved. It is the accumulated zinc dross in the groove causes the slight scratches on the substrate, which results in the formation of outburst structure. It is the zinc dross and the outburst structure in the scratches influence the coating thickness and appears to be a sink roll groove mark defect finally.

Key words hot dip galvanizing, sink roll, groove mark defect

5 钢材深加工

Further Processing for Rolled Products

炼铁与原料

炼钢与连铸

轧制与热处理

表面与涂镀

★ 钢材深加工

钢铁材料

能源与环保

冶金设备与工程

冶金自动化与智能管控

冶金分析

信息情报

先进短流程-深加工新技术与高强塑性汽车构件的开发

干 勇[1,2] 李光瀛[2] 马鸣图[3] 毛新平[4] 罗 荣[2]

（1. 中国工程院，北京 100088； 2. 中国钢研科技集团有限公司钢铁研究总院，北京 100081；
3. 中国汽车工程研究院股份有限公司，重庆 401122；4. 武汉钢铁集团公司，湖北武汉 430083）

摘 要 在近年来国内外薄板坯连铸连轧 TSCR 和先进热成型处理 AHFT 技术发展的基础上，提出了先进的短流程与深加工技术相结合的工艺途径，为高强塑性汽车构件的生产制造，开发一种高效率、低能耗、低排放、低成本的新工艺。首先采用先进短流程工艺，包括以 CSP、FTSR、ISP 半无头轧制为主要特征的第二代 TSCR 技术和以 ESP 无头轧制为主要特征的第三代 TSCR 技术，生产高强度薄规格汽车板作为热冲压成型的原料；而后采用先进热成型处理 AHFT 技术，对短流程薄规格热轧酸洗板进行热冲压与热处理相结合的深加工，制作高强塑性汽车构件。本文简要介绍了短流程薄板坯连铸连轧 TSCR 的技术发展，先进热成型处理 AHFT 的强塑化工艺与主要技术特征。讨论了采用短流程 TSCR 新工艺生产先进高强度钢 AHSS 薄规格汽车板，为先进热成型处理 AHFT 提供热冲压成型板料的主要技术关键。介绍了在超短流程的双辊薄带连铸工艺线上开发高强塑性 TWIP 钢的试验进展。短流程 TSCR 新工艺与先进热成型处理 AHFT 深加工新技术相结合，开发与生产超高强塑性汽车构件，需要钢铁与汽车行业的密切合作。这项短流程-深加工新技术不仅可以满足新一代汽车对轻量化节能减排和抗冲撞安全的要求，而且可以显著降低汽车板构件生产制造过程中的能耗与温室气体排放。

关键词 先进高强度钢 AHSS，短流程，深加工，薄板坯连铸连轧 TSCR，超薄带，先进热成型处理 AHFT，高强塑性 HSD，汽车构件

Development of Advanced Compact Steel Process and Deep Working Technology for High-Strength-Ductility Auto-Parts

Gan Yong[1,2], Li Guangying[2], Ma Mingtu[3], Mao Xinping[4]， Luo Rong[2]

(1. Chinese Academy of Engineering, CAE, Beijing 100088, China;
2. Central Iron & Steel Research Institute, CISRI, Beijing 100081, China;
3. China Automotive Engineering Research Institute, CAERI, Chongqing 401122, China;
4. Wuhan Iron and Steel (Group) Corporation, WISCO, Wuhan 430083, China)

Abstract An advanced manufacturing route for high-strength-ductility auto-parts is proposed on the basis of thin slab casting & rolling TSCR new technologies and advanced hot forming treatment AHFT in deep working. The new process is a combination of the innovated compact strip production with sheet-steel deep working for high-efficiency, low-energy-consumption, low-emission and low cost. First, the high-strength thin-gauge AHSS strips are produced through TSCR lines, incl. semi-endless-lines of CSP, FTSR, ISP and endless-strip-production ESP line, to deliver hot-stamping blanks. Then, high-strength-ductility auto-parts are manufactured through advanced-hot-forming-treatment AHFT by integrated hot-stamping with heat-treatment. In the present paper, the recent development of TSCR technology is briefly summarized， the main features of AHFT is shortly described. The key technologies of TSCR for thin-gauge AHSS strips production to deliver sheet-blanks for AHFT are discussed. The new R&D progress in TWIP steel through ultra-compacted twin-roll strip-casting line is introduced. The combination of TSCR with AHFT technology for high-strength-ductility auto-parts needs intimate cooperation between steel-producer and auto-maker. The new compact process of TSCR-AHFT can not only promote weight-lightening, energy-saving, GHG-emission-decreasing and crash-safety during auto-running, but can also decrease energy-consumption and GHG-emission during auto-steel & auto-parts producing.

Key words　advanced high strength steel AHSS, compact steel process, deep working，thin-slab casting & direct rolling TSCR，ultra-thin-gauge-strip，advanced hot forming treatment AHFT，high-strength-ductility HSD，auto-parts

高强钢辊压技术在轻量化汽车车身上的应用

Pietro Passone

（Emarc S.p.A., Italy）

摘　要　随着环境和能源对汽车工业严苛要求即将到来，实现单车油耗大幅度降低已经是迫在眉睫的任务。在采用新能源以及动力传动系统之后，实现车身轻量化是实现以上目标的重要途径。

钢铁还是未来车身的主要应用材料。利用高强度实现料厚减薄是减重的主要原理。除了材料，引用新的设计手段、制造工艺与前者结合，才能实现减重、性能提升与保持成本的统一。因此辊压成为一种重要的选择。

笔者结合多年的设计、工艺以及产线技术经验，在欧美多款重要车型上实现了很大的突破并实现量产。此文将介绍一种系统性采用高强钢辊压方法实现车身减重的技术，并以案例的方式进行了详细说明。

关键词　高强钢，辊压，轻量化，汽车

Steel Developments for the Packaging Industry

Jones David

(PACS Consulting, UK, Formerly of Corus Packaging Plus & British Steel Tinplate)

Abstract　The steel and metallurgical developments to support lightweighting, performance improvements and improved line efficiencies for the can-making and packaging industries in general are discussed. The paper concentrates on cans produced by the drawn wall and ironing (DWI) process for both beer and beverage, and food applications. It also covers the metallurgical developments to support the growth of steel easy-open ends for the food can market. The steel chemistries and processing routes for these applications are compared demonstrating the versatility of steel in meeting the different performance requirements. In both cases, the improvements and resulting gauge reductions have been achieved by close co-operation between the steel suppliers, the equipment and tooling suppliers and the can-makers. Specialist techniques like Finite Element Analysis (FEA) have been used to speed up the development process. The benefits of this approach are demonstrated by improved performance despite the reductions in steel thickness. Quality control procedures to achieve high can-making line efficiencies are reviewed. Other developments are discussed which include the growth of pre-coated polymer coated steels and their application for the manufacture of aerosol containers and components. Future trends are proposed which would see a move away from just weight and cost reduction towards greater differentiation in terms of different can sizes, shaped containers, improved decoration and increased use of polymer coated steels.

Key words　steel DWI beverage cans, polymer coated steel, easy-open ends, ductile and high strength steels, finite element analysis

Better Steel, Better Life Need Forming Innovation

Dipl. Ing. Thomas Krueckels

(Dreistern GmbH & Co KG, Germany)

Abstract Major forming innovations are driven by global and powerful trends. Such trends are a challenge to the automotive and other industries; they also represent an opportunity for material and machine suppliers. The rollforming industry has realized these opportunities and developed a number of innovations which can efficiently contribute to cope with these challenges by cost-effective processes for light-weight components and highly flexible production technologies for customized mass production.

Key words forming innovation, rollforming industry, cost-effective, highly flexible

更好的钢铁更好的制品开启美好梦想

毛海波 窦光聚

（中钢集团郑州金属制品研究院有限公司，河南郑州 450001）

摘 要 介绍我国金属制品行业的发展现状：从小到大创世界规模之最，品种和质量数量满足国家建设所需，并成为最大的线材制品出口国，形成了比较完备的质量标准体系，生产原料和装备基本满足制品需求。提出行业由大到强的新梦想：转型升级占领国际高端市场，增加品种、拓宽应用创造美好生活，开发智能装备提高产品质量和改善工作条件，和钢铁企业联合开发满足产业升级的新品种线材。结合自身企业实际提出企业发展的思路和工作重点：作高端弹簧钢丝生产的领跑者；研发和推广金属制品节能环保智能装备；提高信息检测服务能力，业务范围推向国际。

关键词 线材，金属制品，制品大国，制品强国，无铅热处理，弹簧钢丝，智能装备，信息检测服务

Better Iron and Steel and Better Products to Start Happy Dream

Mao Haibo, Dou Guangju

(Sinosteel Zhengzhou Research Institute of Steel Wire Products Co., Ltd., Zhengzhou 450001, China)

Abstract To introduce development present situation of China metal products industry: being maximum scale in the world from little to big, the variety, quality and quantity satisfy the needs of construction of China, have become maximum steel wire products export country, more perfect quality standards system come into being, production raw materials and equipment meet products requirement basically. New dream of the industry from big to strong is put forward: by transformation and upgrading to occupy international high-end market, by adding variety and widening application to create happy life, by developing intelligent equipment to improve products quality and work conditions, joint with iron and steel enterprises to develop new steel wire rod variety for the industry upgrading. Combining with self enterprise practice to put forward development thought and key jobs: to act as pacemaker of high-end spring steel wire production; to develop and popularize energy saving and environment protecting and intelligent equipment for metal products; to increase information and detection service capability, scope of service extend to international.

Key words wire rod, metal products, big products country, strong products country, heat treatment free lead, spring steel wire, intelligent equipment, information and detection service

A New Residual Stress Measurement Method for Roll Formed Products

Sun Yong[1], Li Yaguang[2], Liu Zhaobing[3], Daniel W.J.T.[1], Li Dayong[2], Ding Shichao[1]

(1.School of Mechanical and Mining Engineering, The University of Queensland, St Lucia, QLD 4072, Australia; 2. Institute of PFKBE, School of Mechanical Engineering, Shanghai Jiaotong University, Shanghai, China; 3.Ningbo SaiRolf Metal Forming Co., Ltd., Ningbo, China)

Abstract The residual stress in a roll formed product is generally very high and un-predictable. This is due to the occurrence of redundant plastic deformation in the roll forming process and it cancause various product defects. Although the residual stress of a roll formed product consists of longitudinal and transverse residual stress components, the longitudinal residual stressplays a key role in causingproduct defects of a roll formed product and therefore, the longitudinal residual stress is of mostconcern to the roll forming scholars and engineers. However, how to inspect the residual stress of a product quickly and economically as a routine operation is still a challenge. This paper introduces a new residual stress measurement method to study the longitudinal residual stressvariation through layers of the thickness of a roll formed product. The new method is also extendable to a product with similar processing such as a rolled sheet.The detailed measuring procedure is given and discussed. The residual stress variation through eachlayer can be derived based on the variation of curvature in different layers and steps. This new method has been explored and validated by experimental study on a roll formed square tube. The new method is expected to be a routine testing method to monitor the quality of a product that has been formed and that could have a great impact on the roll forming industry.

Key words roll forming, residual stress, measurement method

超低碳钢 Qst32-3 静态再结晶行为的研究

刘 超[1] 孟军学[2] 崔 娟[1] 陈继林[1]

（1. 邢台钢铁有限责任公司，河北邢台 054027；
2. 邢台钢铁线材精制有限责任公司，河北邢台 054027）

摘 要 以超低碳钢 Qst32-3 为研究对象，分别探讨不同拉拔变形量(35%~55%)及退火工艺参数(加热温度：680℃和 720℃，保温时间：10~300min)对其静态再结晶的影响，并分析了 Qst32-3 钢静态再结晶的组织演化规律。结果表明，Qst32-3 的静态再结晶为晶界弓出形核机制。在相同的退火参数条件下，随拉拔变形量增大，再结晶晶粒尺寸减小。当变形量一定时，随再结晶温度升高和时间延长，再结晶更为充分，且组织中碳化物分布的均匀性增加。

关键词 Qst32-3，静态再结晶，拉拔，退火

Research on Static Recrystallization of Ultra-low Carbon Steel Qst32-3

Liu Chao[1], Meng Junxue[2], Cui Juan[1], Chen Jilin[1]

(1. Xingtai Iron & Steel Co., Ltd., Xingtai 054027, China;
2. Xingtai Steel Precisiong Engineering Company, Xingtai 054027, China)

Abstract Effect of different drawing deformation(35%~55%) and annealing parameter(temperature: 680℃ and 720℃; time: 10~300min) on static recrystallization was investigated based on the ultra-low carbon steel Qst32-3. Furthermore, the microstructure evolution law of ultra-low carbon steel was also analyzed. The results show that grain boundary bulging is the main nucleation mechanism of Qst32-3. The recrystallization grain size decreases with the increase of drawing deformation under the same annealing parameters. The full development of recrystallization was obtained with the increase of temperature and time under the same drawing deformation. In addition, the carbide uniformity was also improved.

Key words Qst32-3, static recrystallization, drawing process, annealing process

轨交用高强难燃镁合金的开发

徐世伟[1] 唐伟能[1] 陆 伟[2] 庄建军[2] 潘 华[1] 蒋浩民[1] 张鲁彬[2]

（1. 宝钢集团中央研究院，上海 201900；
2. 宝钢金属有限公司，上海 200940）

摘 要 本文系统地介绍了宝钢集团近年来在轨交用高强难燃镁合金研发方面的进展。随着我国轨交建设的蓬勃发展以及国家对于节能减排的重视，轨交车辆的轻量化迫在眉睫，轻量化材料的使用是其中的一个重要手段，而镁合金作为最轻质的结构金属材料具有广泛的应用前景。宝钢集团围绕着轨交轻量化材料的需求，开发出了屈服强度大于400MPa、燃烧温度大于700℃的新型低成本挤压镁合金。该系列镁合金不含任何稀土和其他贵重元素，耐腐蚀性能优于商用AZ91D镁合金,但是密度却低于AZ91。目前该系列新型镁合金已经成功实现了产业化试制，制备出了直径大于300mm的半连续铸锭，实现了优于商用AZ31的挤压速度，并成功制备出了大型轨交挤压型材。宝钢集团还针对轨交领域的应用需求，开发出了面向该系列新型镁合金挤压型材的表面处理工艺等用户技术。

关键词 镁合金，轨交，高强，难燃，用户技术

Development of High Strength and Flame-retardant Magnesium Alloys for Rail Transit Vehicles

Xu Shiwei[1], Tang Weineng[1], Lu Wei[2], Zhuang Jianjun[2],
Pan Hua[1], Jiang Haomin[1], Zhang Lubin[2]

(1. Research Institute (R & D Center), Baosteel Group Corporation, Shanghai 201900, China;
2. Baosteel Metal Co., Ltd., Shanghai 200940, China)

Abstract In this study, we systematically introduced the research and development progress of high strength and flame-retardant magnesium alloys for rail transit vehicles in Baosteel Group Corporation. In recent years, there is a vigorous development of China's rail transit construction, meanwhile, the light weight of the rail transit vehicles becomes a important topic due to the national energy saving strategy. And the usage of light weight materials, such as magnesium alloy, which is the lightest metal structure material, is one of the important means. Accordingly, Baosteel Group Corporation developed some new-type magnesium alloys with common elements, which could show a high yield strength over 400MPa after hot extrusion and a high ignition temperature over 700℃. These new-type magnesium alloys show better corrosion resistance but lower density than the commercial AZ91D magnesium alloy. Recently, we have successfully produced the direct-chill (DC) cast ingots of these new-type alloys with the diameter over 300mm, and these DC ingots were also successfully subjected to the direct extrusion under a faster ram speed than the commercial AZ31 magnesium alloy. Some complicated extrusion profiles were obtained. Furthermore, in order to expand the application range, the surface treatment of these new-type magnesium alloys were also studied in Baosteel Group Corporation.

Key words magnesium alloys, rail transit vehicles, high strength, flame-retardant, techniques

拉丝乳化液润滑性能评估实验机的研制

胡东辉[1]　苏　岚[2]

（1.宝钢金属有限公司，上海　201900；2. 北京科技大学，北京　100083）

摘　要 研制了一种新型的湿拉乳化液润滑性能评估专用的摩擦实验机。以法莱克斯摩擦实验原理为基础，通过对高碳钢丝的湿式拉拔工艺进行有限元模拟计算，新型摩擦实验机可以模拟湿拉工艺条件。通过在摩擦实验机上测试，可以对比评估不同状态或不同类型的乳化液的拉拔润滑性能，也可用于对乳化液进行优化后拉拔润滑性能否改善的评估。

关键词 湿式拉拔，乳化液，拉拔润滑性能，摩擦实验机

Development of Test Machine for Wire Drawing Lubrication Performance Evaluation

Hu Donghui[1], Su Lan[2]

(1. Baosteel Metal Co., Ltd., Shanghai, 201900, China;
2. Beijing Science and Technology University, Beijing 100083, China)

Abstract A new friction test machine for wet wire drawing lubricant performance evaluation is developed.Through finite element simulation about wet wire drawing process,and based on Falex test principle, the key parameters of test machine shall simulate the wet wire drawing process. The new friction test machine can be used for the lubricant performance evaluation of different type of lube. It can also be used for the evaluation of lube improvement.

Key words wet wire drawing, lube, lubrication performance, friction test machine

帘线钢 72A 氧化铁皮状态对机械破鳞的影响及机理研究

张 鹏[1] 刘宏强[1] 刘艳丽[1] 刘红艳[2] 赵燕青[1]

（1. 河北钢铁技术研究总院，河北石家庄 050000；
2. 河北钢铁集团邯钢公司技术中心，河北邯郸 056015）

摘 要 实验得到帘线钢 72A 在不同加热温度下保温三分钟时的氧化铁皮生成情况，并构建了数学模型，分析了相同保温时间时，不同加热温度对帘线钢 72A 单位面积氧化铁皮生成情况的影响，以及不同加热温度对生成氧化铁皮结构的影响，进一步分析了对其机械破鳞性的影响，文中也对机械破鳞机理进行了探讨。试验结果表明，氧化铁皮结构对 72A 成品的机械破鳞性能影响不大，而氧化铁皮厚度对其机械破鳞性能影响较大，并且存在一个厚度的最佳值；机械破鳞的机理是在外力的作用下，基体与氧化铁皮界面处的附加应力超过氧化铁皮的许用应力，导致在界面处产生微裂纹，而裂纹沿界面的扩展导致了氧化铁皮的脱落，因此，可以通过调整帘线钢 72A 的吐丝温度、斯太尔摩线的冷却制度以及辊道的速度等来获得易于机械除鳞的氧化铁皮结构与厚度，方便下游客户的使用。

关键词 72A 帘线钢，氧化铁皮，机械破鳞，加热温度

Influence and Mechanism of Steel Cord 72A Oxide Scale Situation on the Mechanical Descalability

Zhang Peng[1], Liu Hongqiang[1], Liu Yanli[1], Liu Hongyan[2], Zhao Yanqing[1]

(1. Hebei Iron & Steel Technology Research Institute, Shijiazhuang Hebei 050000, China;
2. Technique Centure, Handan Iron and Steel Co., Ltd. Corporation, Hebei Steel and Iron Group, Handan Hebei 056015, China)

Abstract From the experiment that the steel cord 72A heat preservation with three minutes at different heating temperature, we know the iron oxide generation case, construct a mathematical model, and analyze the generation of iron oxide per unit area of steel cold 72A under different heating temperature with the same time , furthermore we analyze the influence to oxide scale structure under different heating temperature. In this paper we also probes into mechanical descalability mechanism. The result shows that it is not the structure but the thickness of the iron oxide that effects the mechanical descaling much more greater, and there exists a best thickness; the mechanism of mechanical descaling is that, when the additional stress of interface between the matrix and the iron oxide is larger than the allowable stress of iron oxide, there will generate some microcracks between the interface, and the growing of crack along the interface leads to the shedding of iron oxide. therefore, we can adjust the cord steel through the laying temperature of 72A 、cooling system as well as the speed of the Stelmor line's roller table for ease of mechanical descaling oxide iron structure and thickness, convenient for the use of the customers.

key words 72A steel cord, iron oxide, mechanical descaling, heating temperature

汽车用超高强钢辊压件的局部热成形工艺研究

彭雪锋[1]　韩静涛[1]　晏培杰[2]

（1. 北京科技大学，北京　100083；2. 上海宝钢型钢有限公司，上海　201900）

摘　要　随着汽车轻量化的快速发展，超高强钢在汽车上的应用成为现代汽车行业的发展趋势。但超高强钢在室温下屈强比大、塑性差等特点，给传统辊压成形工艺及设备带来了前所未有的挑战。以汽车用超高强钢方管为例，本文提出一种弯角局部感应加热辊压成形技术来制备超高强钢方管，阐述了其成形机理，并通过弯角力学性能、金相组织和显微硬度分析了成形温度、成形速度、弯角半径以及线圈形状等主要参数对热辊压成形技术的影响。研究结果表明，随着加热温度的升高，弯角处力学性能得到了明显的改善；随着成形速度的减慢，位错得到了完全的回复；弯角处不同程度的塑性形变造成中性层硬度的变化；通过合理的选择线圈形状可明显优化工艺设计。该技术推广了超高强钢在汽车上的应用。

关键词　超高强钢方管，热辊压成形，局部感应加热

Study on Local Hot Forming Process of Ultra-High Strength Roll-Formed Parts for Automobile

Peng Xuefeng[1], Han Jingtao[1], Yan Peijie[2]

(1. University of Science and Technology Beijing, Beijing 100083, China;
2. Shanghai Baosteel Section Steel Co., Ltd., Shanghai 201900, China)

Abstract　With the rapid development of lightweight of automobile, it has become the development trend of modern automotive industry to apply ultra-high strength steel in automobile body parts. However, ultra-high strength steel at room temperature has high yield ratio and low ductility, which brings a huge challenge to the traditional roll forming process and equipment. Taking the ultra-high strength square tube for Automobile as an example, in this paper, a local hot roll forming technology at corner for ultra-high strength square tube is proposed. The basic forming mechanism of the system is clarified. And the influence of forming temperature, forming speed, fillet radius and style of the coil on hot roll forming technology is analyzed by the mechanical properties, microstructure and microhardness of corner. Experimental results show that as the temperature rises, mechanical properties of the corners have been significantly improved. With the slower speed of roll forming, the dislocation has been full recovery. The changes of the layer hardness are caused by the different degree of plastic deformation at bending corners. Moreover, the reasonable choice of coil can significantly optimize process design. The application of ultra-high strength steel in automobile will be promoted by hot roll forming technology.

Key words　ultra-high strength square tube, hot roll forming, local induction heating

Study on High Volume Flexible Manufacturing by Using the Sheet Metal Roll Forming Process

Albert Sedlmaier, Thomas Dietl

(Data M Sheet Metal Solutions GmbH, Valley 83626, Germany, datam@datam.de)

Abstract Increasing model variety, more vehicle customization and shorter innovation cycles have a strong impact on productivity and quality. The number of platforms and body types are permanently increasing inside a plant, leading to high tool and set-up costs in the body in white section of the factory. Moreover, as legislation sets mandatory emission reduction targets for new cars and trucks, new lightweight design strategies for the "Body-inWhite" (BIW) need to be found like ultra-high tensile materials with important springback and load optimized design of parts leading to complex geometries. Often parts of the BIW belong to a family differing only in a single geometrical parameter like overall length or width, or even only in material thickness. Hence,the companies tend to replace manufacturing processes like press forming due to their lack of flexibility by other more flexible production methods.

We present a new manufacturing system using the sheet metal roll forming process to overcome many of these challenges. Traditional roll forming lines only produce parts with constant cross sections. A new 3D roll forming line for complex structural automotive parts opens new ways for production. It also gives more flexibility and higher product quality.In this study, an overview is given for possible families of parts being produced with this new production technique. The basic machine concept and its current application is shown.

Key words flexible manufacturing, mass production, 3D-roll forming, sheet metal forming, truck beam

基于响应面法的高强钢方矩管滚弯成型参数优化研究

晏培杰[1] 赵亦希[2] 吴振刚[1] 马越峰[1]

（1. 上海宝钢型钢有限公司，上海 201107； 2. 上海交通大学，上海 200240）

摘 要 针对滚弯成型工艺中高强钢方矩管的内壁凹陷缺陷，开展了工艺参数优化研究。首先考虑了方矩管管坯材料非均匀性，基于 ABAQUS/Explicit 平台建立了高精度的高强钢方矩管滚弯成型仿真模型。应用响应面方法对辊轮直径、压边宽度和圆角半径这三个参数进行了回归拟合，分析了各因素间的交互作用影响。基于响应面回归方程进行了参数优化实验，得到了相应成型条件下的最优参数，并验证了其有效性。

关键词 滚弯工艺，分区材料，凹陷深度，响应面优化

Optimization of Technical Parameters of Roll Bending of High-strength Rectangular Tube Based on Response Surface Method

Yan Peijie[1], Zhao Yixi[2], Wu Zhengang[1], Ma Yuefeng[1]

(1. Shanghai Baosteel Section Co., Ltd., Shanghai 201107, China;
2. Shanghai Jiaotong University, Shanghai 200240, China)

Abstract In the roll bending of high-strength rectangular tube, there will be some depressions on the inner bending regions. In order to solve these problems, based on the heterogeneity of material of rectangular tube, the article build a three-roll bending model by ABAQUS/Explicit. Through regressing fitting of diameter of roll,width of edges and fillet radius, the article analyses the interactive relationship between these three parameters. Parameters are optimized and verified through experiment based on the response surface regression equation.

Key words roll bending, different regions, depth of concavity, response surface optimization

浅谈冷弯成型中的边部浪形控制

盛珍剑

（武汉钢铁江北集团冷弯型钢有限公司）

摘　要　随着冷弯产品结构的不断优化，市场上越来越对冷弯辊式成型产品的兴趣在不断升温，特别是在效率，质量等方面有极大的兴趣；但是，一些冷弯辊式成型的产品难以取代或者达到折弯和冲压产品的外观质量，主要集中在一些大型的薄壁，开口型钢等产品，辊式成型中对于边部浪形控制技术方面存在很多问题，特别是对原料要求比较严格，一般原料公司家难以生产出对应的原料。本文主要就几个典型的产品浪边质量控制方法，从原料质量要求，浪边控制的理论依据，下山法技术，调车技术等方面，以闭口肋，F180×4.0 等为例，讲叙如何在生产过程中减少边部浪形的出现，以便于提高大型薄壁产品的整体质量，也为将来控制边部浪形提供一些技术基础。

关键词　边部浪形（浪边），薄壁产品，下山法，原料外形

Analysis of the Wave Shape Control of the Edge Part in Cold Roll Forming

Sheng Zhenjian

(WISCO JIANGBEI COLD-FORMED CO., LTD.)

Abstract　With the continuous optimization of structure of the cold-formed productions ,the market take more and more interest in cold-formed productions continues , especially in efficiency and quality. However, some cold-formed productions is difficult to replace the appearance quality which is used to produce by bending or pressing , mainly concentrated in some large thin-walled, open type steel productions, there are many problems in roll forming for edge wave shape control process, especially more strictly to the requirements of raw materials, it is difficult to produce the corresponding raw materials for mang raw materials companies. This paper will introduce several typical products quality control methods of the edge waves , example as the quality of raw material requirements, edge wave control of the theoretical basis, down mountain method, shunting technology, as closed rib, the F180 * 4.0, introduce how in the production process reduced the presence of edge wave shape, in order to improve large thin-walled products, the overall quality, but also for the future control of edge wave shape to provide some technical basis.

Key words　edge wave shape, thin-walled, down mountain method, raw material form

7×(3+9+15×0.245)+0.245HT 钢帘线的开发

崔世云　刘臣　顾军

（江苏宝钢精密钢丝有限公司，江苏海门　226100）

摘　要　介绍了 7×（3+9+15×0.245）+0.245HT 钢帘线新产品的整个研发过程，在满足钢帘线成品性能要求的前提下，包含了半成品黄丝直径的选择、钢帘线产品的扭转控制、对钢帘线股绳预变形量的控制、钢帘线成品的扭转

目标值确定等研发过程设计。实验表明，随着半成品黄丝直径的增大，压缩比增大，成品帘线反而破断拉力降低；通过控制前道容易控制的 7×（3+9+15×0.245）扭转，可以很好地控制成品的扭转，并且具有一定的多项式关系；股绳 3+9+15×0.245 的预变形量对成品帘线成型质量有很大的影响，随着预变形量的增大，芯股缩进量变小，具有一定的线性关系；通过模拟客户生产工艺确定了扭转目标值。按上述控制开发出的样品各项性能均满足标准，用于 40.00R57 全钢巨型工程机械子午线轮胎胎体和带束层，轮胎成品性能满足国家标准要求，产品经矿山使用验证未出现因钢丝帘线变化引起的不良反馈。

关键词 工程胎，钢帘线，胎体，带束层

The Development of 7×(3+9+15×0.245)+0.245HT Steel Cord

Cui Shiyun, Liu Chen, Gu Jun

(Jiangsu Baosteel Fine Wire Co., Ltd., Haimen Jiangsu 226100, China)

Abstract The development of 7×(3+9+15×0.245) +0.245HT steel cord was introduced, which includes the diameter of half product, the control of the steel cord torsion, the control of the pre deformation, and the torsion of the product. Results show that when the diameter of half product increased, the compression ratio increased, and the breaking load of steel cord reduced; The 7×(3 +9+15×0.245)+0.245HT steel cord torsion can be controlled by 7×(3+9+15×0.245) torsion , and these is a polynomial relationship between them; The deformation of 3+9+15×0.245 cord has a great influence on the forming quality of the finished product. With the increasing of pre deformation, inner core strand indentation is small, which has a certain linear relationship; by simulating the production process of customer to determine the torsion target. The finished product properties of specimens meet the standards, and can be used for 40.00R57 all steel giant OTR radial tire carcass and belt. The performance of finished tire to meet the requirements of the national standard, products have no adverse feedback on the Mines.

Key words OTR, steel cord, carcass, belt

相变诱导塑性钢的非弹性回复行为研究

孙蓟泉　牛　闯　滕胜阳　阚鑫峰

（北京科技大学冶金工程研究院，北京　100083）

摘　要 相变诱导塑性钢由于存在马氏体相变，其回弹行为包括弹性回复和非弹性回复两部分，可以通过位错理论加以解释。对两种强度级别的 TRIP 钢进行单向循环加载实验，通过数据拟合建立弹性模量模型，其随着应变的增大而呈指数形式递减。将该模型引入模拟软件中进行拉弯回弹实验仿真，并通过与实际拉弯回弹实验的对比，可以证明变弹性模量模型比恒定弹性模量模型具有更高的预测精度，其仿真结果与实验值更为吻合。

关键词 相变诱导塑性，非弹性回复，弹性模量，拉弯回弹

Research on Inelastic Recovery Behavior of Transformation-induced Plasticity Steels

Sun Jiquan, Niu Chuang, Teng Shengyang, Kan Xinfeng

(Metallurgical Engineering Research Institute, University of Science and Technology Beijing, Beijing 100083, China)

Abstract Because of the transformation to Martensite, the springback behavior of transformation-induced plasticity (TRIP) steels contain elastic recovery and inelastic recovery, which can be explained by dislocation theory. This paper carried out uniaxial cyclic tension tests on TRIP steels of two strength levels, and established an elastic modulus model by fitting experimental data, in which the elastic modulus decreased in exponential form as strain increased. The elastic modulus model was introduced into Dynaform5.9 for stretch bending simulation, the comparison results with actual experiments proved that the model with variable elastic modulus had better prediction accuracy than that with constant elastic modulus, and the simulation result was more consistent with experimental value.

Key words TRIP effect, inelastic recovery, elastic modulus, stretch bending

矫直机辊缝偏差现象对策与措施的研究

卜彦强

（首钢股份公司迁安钢铁公司，河北迁安　064400）

摘　要　本文根据迁钢公司2号热轧卷开平生产线矫直机矫直辊传动侧和操作侧辊缝差值偏大的现象，采用鱼骨图分析法从人、机、料、法、环五个方面分析了辊缝差值偏大的原因，并结合现场实际采取相关措施，最终使得辊缝偏差保持在工艺要求范围内。为了防止辊缝差值偏大现象的再次发生，对矫直机的日常管理和维护提出一些建议和要求，通过这些措施的实施，大幅度减少了设备故障时间，明显提高了产品的成材率，经济效益显著。

关键词　矫直机，辊缝，偏差，鱼骨图

Research on the Straightener Roll Gap Deviation Phenomenon Countermeasures

Bu Yanqiang

(Shougang Joint-stock Company Qian'an Iron and Steel Company, Qian'an Hebei 064400, China)

Abstract In this passage, according to the big difference phenomenon in QIANGANG 2# HRC Straightened and production line straightening machine's straightening roller drive side and operation side roll gap, using the method of fishbone diagram from workers, machines, material, methods, ring five aspects analyzes the reasons of roll gap's big difference and connecting with the relativity measures actually to make the roll gap deviation to keep within the scope of the technical requirements. In order to prevent the roll gap big difference phenomenon from happening again, we put forward some suggestion and requirements about daily management and maintenance for the straightening machines. Through the implementation of these measures, it can reduces the equipment failure time greatly, significantly improve the yield and the economic benefit.

Key words straightening machine, roll gap, deviation, fishbone diagram

减薄 202 钢罐口味问题浅析

顾 婕 郭文渊 张 清

（上海宝钢包装股份有限公司，上海 201908）

摘 要 本研究针对 202 钢罐口味问题开展进行，明确了钢罐发生口味问题的原因在于罐身内壁未被内涂层/密封胶有效覆盖，暴露的金属元素被内容物产品腐蚀，析出到内容物中并积累到一定的数量值，宏观上表现为饮料的非正常口感。发生金属析出的位置集中在罐身的身钩半径区域。研究从钢罐自身质量控制、灌装工艺和口味问题风险评估三个方面提出解决方案，通过保证罐口直壁区合适的内涂膜厚度，保证良好的卷封结构，以及控制卷封厚度为 1. 12±0.05mm，卷封顶隙≤0.10mm 等关键参数，可以有效阻止身钩半径区域因卷封形变而暴露的金属元素析出到内容物产品中。此外，采用合适的测试液，严格控制测试液位的高度，可以及时反映罐内涂层，特别是罐口涂层的质量，是防止口味问题的有效预防措施。钢罐包装碳酸饮料和啤酒，可以在保质期内达到金属离子零析出的水平。

关键词 钢罐，金属析出，二重卷封

Analysis on Metal Pick-up of 202 Steel Can

Gu Jie, Guo Wenyuan, Zhang Qing

(Shanghai Bao Packaging Co., Ltd., Shanghai, 201908, China)

Abstract The research is focus on 202 steel can metal pick-up problem, clarified that the metal pick-up is caused by product reacting with the metal substrate at a site(s) of exposed metal. In particular the forming of the Body Hook during seaming can cause a degree of lacquer cracking in this region, which results in potential exposure of the metal substrate. The exposed metal can be corroded by the content products, consumer could taste unusual flavor when the precipitation quantity accumulated to a certain value.The The plans to solving metal pick-up problem are: (1)ensure the suitable coating thickness of can straight wall; (2) ensure the good seaming structure, and control seam thickness is 1.12 ± 0.05mm, the seamgap is no more than 0.10 mm; (3)using of appropriate ME test solution, and strictly control the height of ME test liquid. Steel cans packaging carbonated drinks and beer could achiever metal zero precipitation in warranty period, the flavor of the content of will not be affected, that reflects the steel cans is adaptability packaging form for beer and carbonated drinks

Key words steel can, metal pick-up, double seaming

基于等腰梯形变截面扭转构件的加工制作关键控制技术

王超颖 李文杰 虞明达

（宝钢钢构有限公司，上海 201900）

摘　要　本文对等腰梯形变截面扭转构件加工制作的全过程进行了专门的论述，介绍了其扭转成型的原理、深化设计的要求及具体方法，从下料、划线、单块零件板成型、构件装配、焊接、检测等方面分析了构件加工过程中的重难点，并提出了具体的技术处理方法，对类似空间复杂曲面结构构件的加工制作有一定的参考价值。

关键词　等腰梯形变截面扭转构件，扭转成型，深化设计，加工制作

The Key Control Technology of Fabricating and Making Based on the Isosceles Trapezoid Cross-section of Torsional Component

Wang Chaoying, Li Wenjie, Yu Mingda

(Baosteel Structure Co., Ltd., Shanghai 201900, China)

Abstract　This paper specially discusses the whole process about the fabricating and making of the isosceles trapezoid cross-section of torsional component, introduces the principle of reverse molding, the requirements and concrete methods of detailed design, analyzes the difficult point from the aspect of blanking、ruling、forming of the single block parts plate、assembly、welding and inspection of the component in the component machining process, and proposes specific technical solution. This paper has the certain reference value for the processed of similar spatial complicated curved structure.

Key words　the isosceles trapezoid cross-section of torsional component, teverse molding, detailed design, fabricating and making

新一代硅片切割用结构线的开发

顾春飞　闵学刚　顾　军　李庆峰　刘冠荣　杨志伟　张建刚

（江苏宝钢精密钢丝有限公司，江苏海门　226100）

摘　要　随着光伏产业的快速发展，高效率、低成本已成为太阳能硅片切割领域的趋势。作为硅片切割的重要辅料，钢丝的切割效率及砂浆的使用成本至为关键。目前，传统的直钢丝已无法满足行业的需求，结构线由于其特殊的结构，可有效增加钢线的浆料携带率，提高钢线的切割效率，降低砂浆的使用成本。本文通过对新一代硅片切割用结构线的特点及切割原理的研究，对结构线的变形技术及变形量的定量测试方法的开发，成功研发了结构线产品。对0.115 规格结构线基本性能进行了检测，其直径为 115±3μm，强度为 3600±100MPa，两组波高分别为 175±10μm 及 155±10μm，达到了稳定的工艺要求。跟踪其切割效果发现，同等条件下，0.115 结构线较直钢丝切割速度提升了 22%，砂浆使用成本降低了 25%；0.115 结构线的 A++率提高了 8 个百分点，B 线痕率降低了 7 个百分点，C 线痕降低了 2.5 个百分点，切割效果较好，有效验证了结构线已可以替代直钢丝应用于硅片切割。

关键词　结构线，变形技术，浆料携带率，切割效率

The Development of New Generation of Silicon Wafer Cutting Structured Wire

Gu Chunfei, Min Xuegang, Gu Jun, Li Qingfeng, Liu Guanrong, Yang Zhiwei, Zhang Jiangang

(Jiangsu Baosteel Fine Wire Co., Ltd., Haimen Jiangsu 226100, China)

Abstract With the rapid development of photovoltaic industry, high efficiency and low cost has become a trend in the field of solar wafers cutting.As important materials of silicon wafer cutting, the cutting efficiency and cost of slurry are crucial. At present, the traditional straight wire has been unable to meet the needs of industry. Structured wire, because of its special structure, which can effectively increase the slurry- carrying rate and cutting efficiency, decrease the cost of slurry. This article did a research on the characteristic and cutting principle of the new generation structured wire for silicon wafer cutting.Through the exploitation of the structured wire deformation technology and deformation quantitative testing method, the structured wire products have be successfully developed. The properties of 0.115 structured wires were tested, and the diameter for 115±3μm，intensity for 3600±100MPa, two groups of wave height are 175±10μm and 155±10μm, reached the stable process requirement.Compared with straight wire, the cutting speed is increased by 22%, cost of slurry is reduced by 25% under the same condition. The A++ rateis increased by 8 percent, B line mark rate is reduced by 7 percent, C line mark is reduced by 2.5 percent, the effect ofstructured wireis better than straight wire. The results effective verify structured wire which has been applied in the silicon wafer cutting could replace the straight wire.

Key words structured wire, deformation technology, slurry- carrying rate, cutting efficiency

316L 不锈钢丝冲压过程中表面缺陷研究及工艺优化

谭 瑶[1] 宋仁伯[1] 王宾宁[1] 陈 雷[1] 郭 客[2,3] 王忠红[3]

（1. 北京科技大学材料科学与工程学院，北京 100083；2. 辽宁科技大学，辽宁鞍山 117022；
3. 鞍钢集团矿业设计研究院，辽宁鞍山 114004）

摘 要 通过对 316L 不锈钢丝试样的微观组织和表面缺陷进行观察、对表面应力状态进行分析、对试样力学性能进行测试，探究冲压后产品的表面缺陷成因。结果表明：不合格试样的表面缺陷主要有牙齿印、凹坑和擦伤；进口原料试样的抗拉强度为 671.4MPa，屈服强度为 272.9MPa，延伸率为 56.9%，而国产原料试样的抗拉强度为 672.9MPa，屈服强度达到 287.7MPa，延伸率仅为 47.9%；国产钢丝冲压后组织更粗大且分布不均匀，其晶粒尺寸为 20.9~29.7μm；对试样表面不同位置的应力状态分析表明两拉一压的应力状态使其塑性较差，在变形过程中的协调性差。随着走线速度的提高，退火时间缩短，钢丝的断后伸长率呈现下降趋势，屈服强度和抗拉强度升高，但过高的走线速度会导致退火不完全达不到软化效果，故在退火温度为 1050℃时，最佳走线速度为 6m/min。

关键词 316L 不锈钢丝，表面缺陷，走线速度

Surface Defects in the Process of Stamping and Process Optimization for 316L Stainless Steel Wire

Tan Yao[1], Song Renbo[1], Wang Binning[1], Chen Lei[1], Guo Ke[2,3], Wang Zhonghong[3]

(1. School of Materials Science and Engineering, University of Science and Technology Beijing, Beijing 100083, China; 2. University of Science and Technology Liaoning, Anshan 117022, China; 3. Angang Group Mining Engineering Corporation, Anshan 114004, China)

Abstract The main purpose of the study was to analyze the surface defects in the process of stamping and process optimization for

316L stainless steel wire by observing the microstructure and surface defects, analyzing surface stress and mechanical performance. The results show that pits, teeth-mark-like and bruise were the main surface defects of the unqualified specimens. The specimen of the imported raw material has tensile strength of 671.3MPa, yield strength of 272.8MPa and the elongation of 56.8%. In comparison, the specimen of domestic raw material has tensile strength of 672.9MPa, yield strength of 287.6MPa and the elongation of 47.8%. The grain of the domestic material was coarser and more inhomogeneous, ranging from 20.9μm to 29.7μm. The surface of the specimen had bad plasticity and uniformity on the surface with the tensile stress state in two dimensions and one compressive stress state in one dimension. The elongation became lower, the yield strength and the tensile strength values were higher with the improvement of the working speed and the shortening of the time. However, the raw material will be softened inadequately with excessive wire-line speed. Therefore , the optimum working speed was 6m/min while the annealing temperature was 1050℃.

Key words　316L stainless steel wire, surface defects, wire-line speed

微观组织与织构特征对无取向硅最终成品板磁性能的影响

孙　强　米振莉　李志超　党　宁

（北京科技大学冶金工程研究院，北京　100083）

摘　要　通过对 3 种不同成分的无取向硅钢成品退火板进行微观组织观察以及分别使用 XRD 和 EBSD 进行宏观织构和微观织构观察，研究了微观组织与织构特征对无取向硅钢成品板磁性能的影响。结果表明：无取向硅钢再结晶组织对其磁性能有影响，晶粒尺寸越大，无取向硅钢的磁性能越好，1.35Si-0.25Mn-0.28Al 的再结晶平均晶粒尺寸达 51.6 μm，铁损值达 3.577 W/kg。Si 和 Al 元素有利于平均晶粒尺寸的增大，Mn 含量的提高有利于减少夹杂物对晶粒长大的限制。无取向硅钢的织构分布影响了成品板的磁感应强度。无取向硅钢再结晶织构主要由强的γ织构（特别是{111}<112>织构）和弱的立方织构以及高斯织构等组成。有利织构中，立方{100}<001>织构和旋转{100}<011>立方织构含量较高，提高了磁感应强度值，1.35Si-0.25Mn-0.28Al 钢的磁感应强度达 1.739 T。

关键词　无取向硅钢，织构，组织，磁性能

Effect of Characteristics of Microstructure and Textures on Magnetic Properties in final Products of NGO Steel

Sun Qiang, Mi Zhenli, Li Zhichao, Dang Ning

(Engineering Research Institute, University of Science and Technology Beijing, Beijing 100083, China)

Abstract　Microstructure of three non-oriented electrical steels with different compositions was observed by metallographic analysis. The micro-texture and macro-texture of the steels were analyzed by EBSD and XRD. The effect of characteristics of microstructure and textures on magnetic properties of non-oriented electrical steel was studied. The results indicate that recrystallized microstructure in non-oriented electrical steel has a big effect on iron loss which is decreased by the expanding average grain size. The average grain size is 51.6 μm and the iron loss is 3.577 W/kg in 1.35Si-0.25Mn-0.28Al steel. Effect of varied elements on the microstructure was analyzed. The Si and Al contents can boost the growth of grains and increased Mn content could reduce the effect of inclusions on limiting the growth of grains. Magnetic flux density of non-oriented electrical steels is effected by texture distribution. Recrystallized textures are comprised of strong γ-fiber textures, weak cube texture, weak Goss texture and others. Among favorable components, the proportions of cube texture and rotated texture were large while almost no copper and brass textures existed in annealed samples. The more favorable textures exist, the larger the flux density is in annealed sheets. The flux density of 1.35Si-0.25Mn-0.28Al steel attained 1.739 T.

Key words non-oriented silicon steel, microstructure, texture, magnetic properties

汽车控制臂挤压铸造数值模拟及工艺优化

邢志威[1] 杨海波[1] 陈学文[2]

（1. 北京科技大学机械工程学院，北京 100083；2. 广州金邦有色合金有限公司）

摘 要 以汽车控制臂为研究对象，运用 MAGMASOFT 对其挤压铸造工艺进行数值模拟，准确预测到铸件在凝固过程中出现缩孔、缩松缺陷的位置。结果表明：铸件在上支臂顶端厚壁处和上支臂中部存在缩孔、缩松缺陷。对上支臂顶端进行局部挤压，当挤压力为 130MPa 时，该区域热节消除；在上支臂中部加设冷却水路，使控制臂整体满足顺序凝固原则，缩孔、缩松缺陷得到消除。

关键词 挤压铸造，数值模拟，缺陷预测，局部挤压

Squeeze Casting Technics Numerical Simulation and Process Optimization of Automobile Control Arm

Xing Zhiwei[1], Yang Haibo[1], Chen Xuewen[2]

(1. School of Mechanical Engineering,University of Science and Technology Beijing,Beijing 100083,China;
2. Guangzhou Kinbon Non-Ferrous Alloy Metals Co., Ltd.)

Abstract Taking the automobile control arm as the research object, using MAGMASOFT to simulate the squeeze casting process, the position of the shrinkage cavity and shrinkage defects in the casting process is accurately predicted. The results show that there exist shrinkage defects in the end and middle of the upper arm. The top of the upper arm is partially squeezed, extrusion pressure is 130MPa, the hot spot is eliminated; In the middle of the upper arm, the cooling water is added, so the control arm as a whole to meet the order of solidification principle, and shrinkage defects are eliminated.

Key words squeeze casting, numerical simulation, defect prediction, local extrusion

基于虚功原理布置屈曲约束支撑方法的研究

陈 云

（上海力岱结构工程技术有限公司，上海 200086）

摘 要 屈曲约束支撑的不同布置方式对结构的耗能减震有很大的影响。本文提出一种基于虚功原理布置屈曲约束支撑的方法，并采用有限元软件 Etabs 对一栋 5 跨 8 层的框架结构进行了验证对比分析，主要分析不同的布置方式

对结构侧移、顶层的位移时程以及对结构的附加阻尼比的影响。通过对比发现：虚功原理布置屈曲约束支撑的方法可以有效地减小结构侧向位移以及结构的动力响应，并且可以为结构提供更多的附加有效阻尼比，因此可以充分地发挥支撑的耗能能力。

关键词　屈曲约束支撑，耗能减震，虚功原理，附加阻尼比，有限元

Research on Arrangement Way of Buckling-restrained Brace Based on Virtual Work Principle

Chen Yun

(Lead Dynamic Engineering Co., Ltd., Shanghai 200086, China)

Abstract　Different arrangement ways of buckling-restrained braces（BRB）in frame structures have great influence on anti-seismic performance with energy dissipation. This paper proposes an arrangement way based on virtual work principle. Based on a 5-span and 8-layer frame structure model, A verification and contrast analysis was carried out by finite element software Etabs. Influence of the structural lateral drifting, displacement--time history at the top floor and additional damping ratio for the frame were compared in different brace arrangement ways. The result shows that the proposed ways based virtual work principle could effectively decrease structural lateral drifting and dynamic response while increasing additional damping ratio. Therefore, the above arrangement ways could make full use of dissipation ability of braces.

Key words　buckling-restrained brace（BRB）, energy dissipation, virtual work principle, additional damping ratio, finite element method

一种超高强度半奥氏体沉淀硬化不锈钢的合金设计及其对冷加工性能的影响

郭燕飞　李居强　陈　涵　赵星明

（中钢集团郑州金属制品研究院有限公司，河南郑州　450001）

摘　要　0Crl2Mn5Ni4Mo3Al 为节镍型半奥氏体沉淀硬化不锈钢，是介于马氏体沉淀硬化不锈钢与奥氏体沉淀硬化不锈钢之间的一种过渡型沉淀硬化不锈钢。它既具有马氏体沉淀硬化不锈钢的超高强度，又具有奥氏体沉淀硬化不锈钢的高韧性及良好的成型加工性。本文通过其所含合金元素在亚稳态奥氏体组织中的作用，对其合金化原理进行了讨论和分析。通过不锈钢相图分析了不同成分钢中奥氏体的稳定性，表明成分越偏上限，钢中奥氏体稳定性越高，并指出奥氏体的稳定性与层错能有关，成分越偏上限钢的层错能越高，钢的塑性越好越适合冷变形。成分下限尽管相变诱发塑性，但相变速率较快，致使随着形变的累积变形主要在马氏体组织内进行，因此塑性反而下降。

关键词　半奥氏体沉淀硬化，CrNi 当量，M_s，层错能

The Alloy Design of A Super-high Strength Semi-austenitic Precipitation Hardening Stainless Steel and the Effect to the Cold Working Properties

Guo Yanfei, Li Juqiang, Chen Han, Zhao Xingming

(Sinosteel Zhengzhou Research Institute of Steel Wire & Steel Wire Products Co., Ltd., Zhengzhou 450001, China)

Abstract 0Crl2Mn5Ni4Mo3Al is low nickel semi-austenitic precipitation hardening stainless steel, which belongs to the transition type between martensitic precipitation-hardening steel and austenitic precipitation hardening stainless steel. It has Super-high strength such as martensitic precipitation-hardening steel, and high ductility and Excellent processability such as austenitic precipitation hardening stainless steel. Through the effects of alloying elements to the metastable austenite structure, the alloying principle of 0Crl2Mn5Ni4Mo3Al is analyzed and discussed. Analyzing the stability of the austenite structure in steels with different chemical compositions, it show that the more closer to the upper limit of the chemical composition, the better the stability of the austenite structure, and the better the plasticity and cold deformation ability with stacking fault energy increasing.At the lower limit of chemical compositions, due to the transformation rate increasing, with the accumulation of deformation, which is mainly within the martensitic structure, the plastic deformation ability decreased despite the transformation induced plasticity.

Key words semi-austenitic precipitation hardening, CrNi equalweight, M_s, stacking fault energy

UV-LED 光源和 UV 汞灯光源比较
——UV-LED 光源用于印铁的研究

夏科峰

（上海宝钢包装股份有限公司，上海 201908）

摘 要 本文比较了 UV 汞灯光源和 UV-LED 光源不同的光谱特性，并用这两种光源对 A 公司生产的 UV 印铁光油进行固化测试和性能测试；UV-LED 光谱分布是狭窄的单一波峰，并且随着距离变大，光强快速下降，5cm 高度时的光强是 1cm 高度的 33%；UV-LED 光源对涂层表面固化不完全；LED 固化后的涂层加工性能良好。

关键词 UV-LED 光源，光强分布，加工性能

Comparison of UV Mercury Lamp Curing Light Unit and UV-LED Light Curing Unit
—The Research for Metal Printing of UV-LED Light Curing Unit

Xia Kefeng

(Shanghai Baosteel Packaging Co., Ltd., Shanghai 201908, China)

Abstract UV mercury lamp light curing unit and UV-LED light curing unit with different spectral distribution and the two light units of A company's production of UV Metal decorating varnish for curing test and performance test; UV-LED spectral distribution is narrow single peak, and with distance, the rapid decline in the intensity, the height of 5 cm intensity is 1cm height of 33%; UV-LED light source to surface coating curing incomplete; LED curing coating has good processing properties.

Key words UV-LED, spectral distribution, processing properties

SWRCH45K 螺栓帽头开裂原因分析

李世琳[1,2] 李 龙[1,2] 李宝秀[1,2] 李永超[1,2] 郭明仪[1,2]

（1. 邢台钢铁有限责任公司，河北邢台 054027；

2. 河北省线材工程技术研究中心，河北邢台 054027）

摘 要 客户使用我公司冷镦钢 SWRCH45K 牌号热轧盘条制作的外六角螺栓帽头开裂率达 80%。对开裂螺栓帽头进行化学成分分析，满足 JIS 3507 标准要求；对其进行宏观形貌和金相组织检验，非金属夹杂物级别满足标准要求，裂纹由表面向内曲折延伸，显微组织为铁素体+球化体+粗片状珠光体，存在不均匀的塑性变形区。结果表明：SWRCH45K 螺栓帽头开裂的原因是冷镦前球化组织不良，存在粗片状珠光体；当珠光体呈层片状分布时，冷镦过程中，塑性变形主要集中在铁素体内，而珠光体中的渗碳体层片方向与冷镦时施加的剪切应力方向呈近 45º时，变形渗碳体具有流变形特征，渗碳体层片方向与剪切应力方向呈 90º时，具有扭折或剪切变形特征，但仍与基体变形不一致，对基体造成分割，加剧应力集中，超出塑性变形极限引起开裂。

关键词 SWRCH45K 螺栓，帽头开裂，球化退火，组织

Analysis on Cracking Reason of SWRCH45K Bolt Head

Li Shilin[1,2], Li Long[1,2], Li Baoxiu [1,2], Li Yongchao[1,2], Guo Mingyi[1,2]

(1. Xingtai Iron and Steel Corporation Limited, Xingtai Hebei 054027, China;

2. Hebei Engineer Technology Research Center for the Wire Rod, Xingtai Hebei 054027, China)

Abstract The hexagon bolt head made by SWRCH45K hot-rolled wire rod cracked and the ratio of cracking was 80%. The cracked bolt head is chemical composition analyzed, it meets the requirement of JIS 3507 standard. Macro-examination and microstructure were carried out on the cracked bolt head, non-metallic inclusion meets the requirement of standard, the crack stretchs from surface to inside flexural, where microstructure is F+particle cementite+ thick pearlite and non-uniform plastic deformation occurred. The result show that the cracking reason of the SWRCH45K bolt head is the thick pearlite existed in the microstructure before cold heading, the plastic deformation mainly occurred in ferrite during the cold heading process, and when the directions of eutectoid lamellae and the one shearing stress applied at an angle of 45℃, the cementite lamellae have the flowed characteristic ,when the directions of eutectoid lamellae and the one shearing stress applied at an angle of 90℃, the cementite lamellae have contored or shearing characteristic, and its deformation is not uniform with the α-Fe matrix, which separated each other and caused stress concentration that overstepped the plastic deformation in the cold heading process, brought the bolt head to crack.

Key words SWRCH45K bolt, bolt head rupture, spheroidizing annealing, microstructure

电磁监测卫星伸杆机构弹性卷筒成形工艺研究

李占华　韩静涛

（北京科技大学，北京　100083）

摘　要　电磁监测卫星主要用于实时监测空间电磁环境状态变化，对预测地震发生有重要意义。其监测需利用伸杆机构将探测载荷伸展至远离星体的位置，弹性卷筒用作为伸杆机构的核心部件，具有展开自驱动、展开长度长（可达15m以上）、驱动力稳定、指向性高、刚度高、重量轻、储能高等优点，成为近年来的发展趋势，但其成难难度很高。本文首次提出采用拉压复合连续弯曲成形的概念用于成形此构件，并对其成形原理进行了阐述，对其成形工艺进行了分析。成形方法兼有压弯力矩小与拉弯稳定性高的特点，并通过正反向弯曲减小残余应力，进一步增加构件稳定性。可使弹性卷筒构件在无热处理情况下直接应用，而仍表现出很高的稳定性。成形结果连续可调、易控，可满足各部位等曲率成形的要求；是一种高效、节能的成形方法。本文利用此方法进行弹性卷筒的成形试验研究，对所得样件性能参数进行了测试，构件在长距离展开下，仍可表现出良好的直线度与弯曲刚度，较高且稳定的驱动力，可较好满足卫星的应用需求。

关键词　电磁监测卫星, 伸杆机构, 弹性卷筒, 工艺研究

Study on the Stacer Forming Process of the Deployable Boom Device Applied in the Electromagnetic Monitoring Satellite

Li Zhanhua, Han Jingtao

(University of Science and Technology Beijing, Beijing 100083, China)

Abstract　The electromagnetic monitoring satellite is used to monitor the state of the space electromagnetic environment in real time, which is significant for predicting the occurrence of earthquakes. The deployable boom device should be used to deploy the detector to the position away from the satellite when monitoring. As the core component of the device, the advantages of the stacer are self-driven, long deployed length (up to 15 meters), stable driving force, high directivity, high rigidity, light weight and high energy storage. It has become the development trend in recent years. However the formation of the stacer is very difficult. In this paper, a novel continuous bending forming by stretch bending and press bending is applied to produce the stacer. The forming principle is described, and the forming process is analyzed.　The method is the combination of small press bending moment and high stretch bending stability. As the residual stress is reduced by positive and negative bending, the stability of component is increased further. It is shows high stability for the stacer used without any heat treatment. The forming can be adjusted continuously and controlled easily, which can meet the requirements of the equal curvature forming. So it is an efficient and energy saving forming method. The stacer is finished by this method, and the performance parameters of the obtained samples are tested. In a long distance expansion, the component can perform excellent straightness and bending rigidity in a long distance expansion, and it has higher and stable driving force. The stacer can meet the application needs of the satellite preferably.

Key words　electromagnetic monitoring satellite, deployable boom device, stacer, process research

双金属复合板压弯及辊式矫直特性研究

郝东佳　吴迪平　秦　勤

（北京科技大学机械工程学院，北京　100083）

摘　要　双金属复合板因其独特的性能与价格优势而具有极其广泛的应用前景，其相关的加工制造技术的研究正越来越受到人们的关注。本文针对不锈钢-普碳钢复合板，首先建立了不同受力条件下钢板弯曲变形的平面应变弹塑性仿真分析模型，对比分析了碳钢板与复合板、压弯与纯弯过程变形区应力应变场的异同，系统分析了压弯条件下中性层的偏移现象、影响因素与变化规律，并对复合板压弯弹复后的残余曲率大小、影响因素、变化规律及正反压弯时残余曲率的不同进行了研究。其次，本文还建立了九辊矫直过程多体接触、弹塑性平面应变动态仿真分析模型，研究了九辊矫直过程对原始缺陷的矫直能力，并发现了复合板矫直存在一定的方向性的事实。

关键词　双金属复合板，中性层，残余曲率，矫直

Study on Bending and Roller Leveling Property of Bimetal Composite Plate

Hao Dongjia, Wu Diping, Qin Qin

(School of Mechanical Engineering, University of Science and Technology Beijing, Beijing 100083, China)

Abstract　Bimetal composite plate has an extremely wide range of applications because of its unique performance and price advantages. The researches of related manufacturing technologies are also more and more concerned. A FEM simulation model of stainless steel-carbon steel composite plate has been suggested to discuss bending deformation under different loading conditions. The strain and stress field of the carbon steel have been compared with that of the composite plate to investigate the differences of them. The phenomenon and change rule of neutral layer offset have been analyzed. And the change rule of residual curvature has also been investigated. Moreover, a nine-roller leveling plane strain dynamic simulation model has also been established to research the leveling ability of composite plate' initial defects. The results show that roller leveling has a good leveling effect on different initial defects with a suitable reduction process, as well as the direction of composite plate in the process of leveling exists.

Key words　bimetal composite plate, neutral layer, residual curvature, leveling

高强激光拼焊板 B 柱成型数值模拟及工艺优化

刘立现　李晓刚　金　茹　李春光　陈　庆　赵　宁

（首钢技术研究院，北京　100043）

摘　要　汽车 B 柱加强板是激光拼焊板在汽车车身上应用的典型零件之一，对提高车身碰撞安全性能有着重要的作用。本文利用有限元分析软件 AutoForm 对差厚（1.8mm/1.2mm）高强激光拼焊板 B 柱进行了冲压成型数值模拟，

研究了 DP 钢和 TRIP 钢对 B 柱成型过程中出现的开裂和焊缝移动等成型缺陷的影响，分析了产生缺陷的原因。研究表明，与 DP 钢相比，由于同强度级别的 TRIP 钢的 r 值和 n 值更高，因此其生产的 B 柱拼焊板具有更好的成型质量；并且通过优化 B 柱加强板圆角设计和合理布置拼焊板薄侧和厚侧材料的拉延筋，可以改善板料在 B 柱加强板冲压成型过程中的流动，使各区域的应变均匀化，使开裂缺陷和焊缝移动得以很好的改善。通过模拟仿真分析可以在新车型开发设计和试模阶段为零部件结构和工艺优化提供指导，减少了模具开发周期，降低了生产成本。

关键词 先进高强钢，激光拼焊板，数值模拟，焊缝移动

Numerical Simulation and Process Optimization on High Strength Tailored Welded Blank for B-pillar

Liu Lixian, Li Xiaogang, Jin Ru, Li Chunguang, Chen Qing, Zhao Ning

(Shougang Research Institute of Technology, Beijing 100043, China)

Abstract B-pillar reinforcement with tailored welded blanks is a typical part in the automotive applications, which plays an important role on improving crash safety of vehicles. In this paper, the forming process of high strength tailored blanks for B-pillar with different thickness (1.8mm/1.2mm) was simulated by finite element software AutoForm. The influence of DP and TRIP steels on the forming defects of cracks and weld line movement was investigated and the cause of defects was analyzed. The research showed that, compared with the DP steel, due to the higher r and n value, the TRIP with the same level of tensile value was more suitable for the production of B-pillar reinforcement. By optimizing the fillet of B-pillar and setting the drawbeads reasonably, the cracks and weld line movement could be controlled and improved, because it could improve the flow of sheet in B-pillar stamping process and uniform the strain of each region. Simulation analysis can provide guidance for the structure and process optimization of components in the design and the tryout phases, which can reduce mold development cycle and production costs.

Key words advanced high strength steel, tailored welded blank, numerical simulation, weld line movement

钢结构电渣焊的常见缺陷检测和探讨

朱卫民 刘志瑞

（宝钢钢构有限公司，上海 201999）

摘 要 本文针对钢结构加工内隔板采用电渣焊时出现的一种常见缺陷，即焊缝熔宽不足及熔合边界偏离，论述了为判定该类型缺陷在箱型截面构件应用中采取的直探头探伤方法及内部缺陷的评定标准和计算方式。本文是结合实际生产经验，参照标准，为进一步实现该类焊缝探伤作业的标准化而进行的探索。

关键词 钢结构，电渣焊，直探头探伤

Common Defect Detection of Steel Electroslag Welding

Zhu Weimin, Liu Zhirui

(Baosteel Construction Co., Ltd., Shanghai 201999, China)

Abstract　This paper in view of the steel structure processing within the partition using electroslag welding appeared a kind of common defects, i.e. lack of weld width and the fusion boundary deviation, to determine the type of defect is discussed in the box section member application straight probe detection methods and evaluation criteria and calculation of internal defects. This article is based on the actual production experience, with reference to the standard, in order to further implement the standardization of the weld inspection work of exploration.

Key words　the steel structure, electroslag welding, straight probe detection

HFW 管线管焊缝冲击功“两高一低”现象研究

李书黎[1]　张　毅[1]　李　烨[1]　冷洪刚[1]　魏　君[1]
陈　浮[2]　刘占增[2]

（1. 武钢集团江北钢铁有限公司钢管厂，湖北武汉　430415；
2. 武钢研究院，湖北武汉　430080）

摘　要　高频感应焊（HFW）管线管焊缝冲击韧性存在不稳定现象，在进行同一组的三个试样夏比冲击试验中，焊缝冲击功值出现“两高一低”情况。通过扫描电镜分析了冲击试样残样的断面，冲击功低值的试样断面更平整且断面有残留氧化物夹杂。结合 HFW 焊管制管中焊接及热处理工艺特点，分析了焊缝冲击功不稳定的原因以及焊接过程中氧化物夹杂形成的原因，提出了控制氧化物夹杂的措施。

关键词　HFW 焊管，冲击功不稳定，氧化物夹杂

Study on the Strong Instability Impact Strength of HFW Welded Pipe

Li Shuli[1], Zhang Yi[1], Li Ye[1], Leng Honggang[1], Wei Jun[1], Chen Fu[2], Liu Zhanzeng[2]

(1. Steel Pipe Plant of Wisco Wuhan Jiangbei Iron and Steel Co., Ltd., Wuhan Hubei 430415, China;
2. Design & Research Institute of WuhanIron & Steel Groups, Wuhan Hubei 430080, China)

Abstract　The strong instability of impact strength of HFW welded pipe is appeared occasionally. Charpy impact text of weld samples turn out to be two high single value of impact energy and one low value. The fracture surface of the charpy impact samples are analyzed by SEM. The fracture surface owns low impact strength shows more flat sections. At the same time, the inclusions which proved to be residual oxide inclusions are found on the fracture surface. The influence factors of the strong instability of impact strength are confirmed on the analyze of the welding and heat treatment of the HFW welded pipe procedure, the control measures of oxide inclusions are advised as well.

Key words　HFW welded pipe, strong instability impact strength, inclusions

基于力学试验机的新型拉伸夹具设计与探讨

王超颖　李永忠

（宝钢钢构有限公司，上海　201900）

摘　要　力学拉伸试验是评判材料物理性能的一个重要指标，针对之前测量试样必须大于等于400mm这一不足之处，设计研发了一套满足拉伸设备的可重复使用的新型拉伸试验夹具，通过对多组材料厚度方向评定数据的测量分析，得出结论：新型夹具在保证测量精度的前提下有效的克服了拉伸试验机测量范围难以满足一般标准试样的问题，且所测数据的偏差较小，数据离散型较小。

关键词　拉伸试验，新型拉伸夹具，设计

Design and Research of New Tensile Testing Jig Based on the Mechanical Testing Machine

Wang Chaoying, Li Yongzhong

(Baosteel Structure Co.,Ltd., Shanghai　2001, China)

Abstract　Mechanical tensile test is an important indicator that judging the material physical properties.In order to overcome the deficiency that the length of testing samples shall be more than 400mm, we research a suite of new tensile testing jigs can be repeatedly used for the stretching machine. By means of analyzing multiple sets of data about material thickness direction, we made a conclusion. While ensuring the measurement accuracy , the new tensile jig effectively overcome the problem that the mechanical testing machine can not meet the scope of standard sample, and the testing data of the deviation and discretion is smaller than before .

Key words　tensile test, new tensile testing jig, designing

碳钢的阳极极化过程研究

陈海林[1]　贺立红[1]　邵远敬[1]　叶理德[1]　王英英[2]　夏先平[2]　蔡　炜[1]

（1. 中冶南方工程技术有限公司技术研究院；2. 华中科技大学材料学院）

摘　要　用线性电位扫描法研究了碳钢在不同浓度电解质条件下的阳极极化过程，再进一步用恒电位法、恒电流法分别研究了碳钢在不同浓度电解质条件下电解行为。发现碳钢阳极极化过程依次经历活化区、钝化区、过钝化区及极限区，不存在活化-钝化过渡区；在过钝化区电极表面发生析氧反应，使得阳极电流呈指数增加；当阳极电流达到最大值后开始降低并最终稳定在某个范围内，电极表面进入极限区，除继续发生析氧反应外，还会发生基体溶解反应，并在电极表面覆盖大量氧化产物；进一步分析发现，电解质浓度越低，电位越高，基体溶解反应越容易进行，而极限区的存在是由于OH^-离子存在浓度极限。

关键词　碳钢，过钝化，恒电位

The research of Anodic Polarization Process of Carbon Steel in Alkali Solution

Chen Hailin[1], He Lihong[1], Shao Yuanjing[1], Ye Lide[1], Wang Yingying[2], Xia Xianping[2], Cai Wei[1]

(1. Technology Research Division of WISDRI Engineering & Research Incorporation Limiited Wuhan, China;

2. Huazhong University of Science and Technology, Wuhan, China)

Abstract Anodic polarization process of carbon steel in solution was studied by linear potential scan.Electrolysis behavior was also analyzed under differential electrolyte concentration and differential electric potential or current by electrolysis at constant potential or current. They were found that anodic polarization process of carbon steel in solution included activated region, passive region, transpassive region, and terminal region except of transition zone between activated and passive. In transpassive region, oxygen evolution reaction which caused that anodic current increased sharply occurred on the electrode surface. Besides oxygen evolution reaction, dissolution reaction of matrix would start whose oxidation products covered on the surface in terminal region. Anodic current decreased, then come to be steady after reaching the peak by the effect of overburden. Furher research showed that dissolution reaction of matrix occurred easily when the solution concentration was low and anodic electric potential was high, and OH^- concentration limit cause the existing of terminal region.

Key words carbon steel, transpassive, constant potential

薄壁厚隔板箱型柱四面丝极电渣焊工艺技术

高国兵　王　衍　刘春波　贺明玄

（宝钢钢构有限公司，上海　201999）

摘　要　箱型柱结构在建筑结构中应用日益广泛，节点区域内部隔板为满足四面全熔透要求，往需采用两面电渣焊；因工程设计要求，对隔板采用四面电渣焊，其装配、焊接及工艺设计要求均有不同；本文结合四面丝极电渣焊特点，具体分析电渣焊诸影响因素之间相互关系及其对熔池宽度的影响程度，并根据工程实际进行试验验证，提出了箱型柱制作四面丝极电渣焊工艺规范及工艺控制措施，确保箱型柱隔板焊接质量得到有效控制。

关键词　箱型柱结构，隔板，丝极电渣焊，工艺规范，工艺控制

Filament Electroslag Welding in the Four Sides of the Box Column which has Thick Web and Thin Flange

Gao Guobing, Wang Yan, Liu Chunbo, He Mingxuan

(Baosteel Constructure Co., Ltd., Shanghai 201999, China)

Abstract Box structure is increasing widely used in building structure. The box's clapboard in node area are required to meet the requirements of full penetration. We also use two sides electroslag welding to finish the work. According engineer's requirement, four sides electroslag welding is very different to two sides electroslag welding. The author consider about the characteristics of four sides filament electroslag welding, analyzes the various factors affecting the electroslag welding and how does it affect the width of molten pool. And according to the experiment of engineering practice, we made four sides filament electroslag welding process specifications and process control measures of box column to ensure the welding quality of box column separator are under control.

Key words box column structure, clapboard, filament electroslag welding, electroslag welding process specifications, process control

退火温度对1.3%Si无取向硅钢成品板磁性能的影响

孙 强 米振莉 李志超 党 宁

（北京科技大学冶金工程研究院，北京 100083）

摘 要 本研究测量了不同退火温度条件下的无取向硅钢成品板的铁损值和磁感应强度，使用EBSD、XRD和OM技术研究了1.3%Si无取向硅钢在不同退火温度条件下的宏观织构、微观取向和微观组织。分析了退火温度对再结晶组织和织构的影响；讨论了退火温度与无取向硅钢成品板磁性能的关系。实验结果表明：无取向硅钢的退火温度对再结晶组织和成品板磁性能有影响，随着退火温度的上升，再结晶晶粒平均尺寸增大且铁损值下降；不利织构含量上升且有利织构含量下降，导致磁感应强度值下降。γ纤维织构是再结晶织构中的优势组分，高斯{110}<100>织构强度也较高。退火温度对再结晶织构也有影响，随着退火温度上升，γ织构的含量不断上升，退火温度的上升降低了立方{100}<100>织构和旋转立方{100}<110>织构但增加了高斯{110}<100>织构的强度。

关键词 退火温度，无取向硅钢，铁损值，磁感应强度

Influence of Annealing Temperature on Magnetic Properties of Final Products of 1.3%Si Non-oriented Electrical Steel

Sun Qiang, Mi Zhenli, Li Zhichao, Dang Ning

(Engineering Research Institute, University of Science and Technology Beijing, Beijing 100083, China)

Abstract Iron loss and Flux density in final products of non-oriented electrical steel was measured. Macroscopic texture, microscopic orientation and microstructure in 1.3%Si non-oriented electrical steel was analyzed by EBSD, XRD, OM techniques. The relationship between annealing temperature and magnetic properties was discussed. The effect of annealing temperature on recrystallized texture and microstructure was analyzed. The experimental results indicate that, the annealing temperature has a large effect on the recrystallization microstructure and iron loss in products. When the annealing temperature was increased, the average grain size of recrystallization microstructure was increased and the iron loss of annealed sheets was decreased. The content of favorable textures was decreased and the content of unfavorable textures was increased, which caused lower flux density in final products. The γ-fiber texture was dominant component in the recrystallized texture in 1.3%Si non-oriented electrical steel. The annealing temperature also has a large influence on texture distribution. With higher annealing temperature, the intensity of γ-fiber texture was increased; the cube {100}<100> texture and rotated cube {100}<110> texture were weakened while the Goss texture was obviously strengthened.

Key words annealing temperature, non-oriented electrical steel, iron loss, flux density

退火织构对SAF2507双相不锈钢r值的影响

荀晓晨[1] 罗照银[2] 陈雨来[1] 李静媛[2]

（1. 北京科技大学冶金工程研究院，北京 100083；

2. 北京科技大学材料科学与工程学院，北京 100083）

摘 要 通过拉伸试验和XRD等实验手段对冷轧退火后的双相不锈钢SAF2507的塑性应变比r值以及其退火织构对r值的影响进行了研究。结果表明，实验钢冷轧退火板表现出了较低的平均塑性应变比r_m值，其值约为0.77和明显的r_{45}～r_{60}峰值，这主要与其在冷轧退火后形成的织构有关。ODF图显示，织构主要存在于铁素体相中，其主要的织构组分为{001}<110>织构、{112}<110>织构和{111}<112>织构，并没有显著的γ纤维再结晶织构产生。奥氏体相中的织构强度较弱，主要有Cu织构{112}<111>，Brass织构{110}<112>和GOSS织构{110}<100>几种类型。铁素体相的{001}<110>织构是造成r_m值较低的主要原因。同时，铁素体相的{001}<110>织构、{112}<110>织构和{111}<112>织构的共同作用使得板材的r值与θ的关系曲线呈倒"V"型，即在θ为45°～60°的范围内存在峰值。此外，奥氏体相的{110}<001>织构也对钢板的r值产生了一定程度的影响。

关键词 SAF2507双相不锈钢，织构，r值

Effect of Annealing Texture on the r Value of SAF2507 Duplex Stainless Steel

Xun Xiaochen[1], Luo Zhaoyin[2], Chen Yulai[1], Li Jingyuan[2]

(1. Metallurgical Engineering Research Institute, University of Science and Technology Beijing, Beijing 100083, China; 2. School of Materials Science and Engineering, University of Science and Technology Beijing, Beijing 100083, China)

Abstract The plasticstrainratio r value of SAF2507 duplexstainlesssteel and the effect of annealing texture on the r valuewas investigated by the tensile test and X-ray diffraction.It was observed thatthe test steel annealing of cold rolled sheet showed lower $r_m$value which is about 0.77 and the obvious r_{45}～r_{60} peak point,which is related with the annealing texture. The ODF showed that the texture mainly existsin the ferrite phase whose maincomponentsis{001}<110>texture, {112}<110>textureand {111}<112> texture, but not the significantγ fiber recrystallization texture.The intensity of austenitic texture is lower than ferritic,of which main components is copper texture {112}<111>,Brass texture {110}<112> and GOSS texture {110}<100>.The ferritic {001}<110> texture is the main reason of resulting in the low r_m.At the same time, theinteractionofferritic {001}<110>texture,{112}<110>texture and {111}<112> texture results in the peak value of when the θ is in the rangefrom45to60deg..Besides, the austenitic{110}<001> texture have a certain degree of impact on the value of the test steel sheet.

Key words SAF2507 duplex stainless steel, texture, r value

314奥氏体不锈钢丝冷拔过程中断裂行为研究

王宾宁[1] 宋仁伯[1] 陈 雷[1] 谭 瑶[1] 郭 客[2,3] 王忠红[3]

（1. 北京科技大学材料科学与工程学院，北京 100083；2. 辽宁科技大学，辽宁鞍山 117022；3. 鞍钢集团矿业设计研究院，辽宁鞍山 114004）

摘 要 通过对企业生产的314不锈钢丝试样的微观组织、断口形貌及夹杂物进行观察，对试样力学性能进行测试，分析拉拔变形前和拉拔变形后钢丝的强度，塑性以及导致钢丝断裂的原因。结果表明：ϕ3.0mm的钢丝试样抗拉强度为1096.7MPa，断后伸长率为10.6%，随着冷拉拔的进行，ϕ2.15mm的钢丝试样抗拉强度上升到1148.2MPa，断

后伸长率急速下降至 6.2%。冷拉拔变形使得钢丝强度增大，塑性变差，由于道次变形量过大导致拉拔过程容易发生钢丝断裂。对变形前后钢丝的组织以及断口分析表明：试样组织为基体为奥氏体，富铬相 M23C6 或 M7C3 则弥散分布于表面，其形状主要呈长条状，富铬相的大量析出显著降低钢丝的塑性，易引起晶间腐蚀的发生；314 不锈钢丝的断裂属于以脆断为主的混合断裂。断口无明显塑性变形，纤维区存在明显沿晶断裂，还有少量球形和花瓣状的韧窝；此外，试样中存在较多灰色块状夹杂物，EDS 分析表明，夹杂物为硅酸盐及氧化铝，尺寸为 6.5μm 左右。
关键词 314 不锈钢丝，冷拔，断裂，夹杂物

The Fracture Behavior of 314 Austenitic Stainless Steel Wire During the Cold Drawing Process

Wang Binning[1], Song Renbo[1], Chen Lei[1], Tan Yao[1], Guo Ke[2,3], Wang Zhonghong[3]

(1. School of Materials Science and Engineering, University of Science and Technology Beijing, Beijing 100083, China; 2. University of Science and Technology Liaoning, Anshan 117022, China; 3. Angang Group Mining Engineering Corporation, Anshan 114004, China)

Abstract With the observation of the microstructure, fracture morphology and inclusions,as well as the test of the mechanical properties of 314 stainless steel wire, the strength and plasticity of before and after drawing deformation along with the cause of the breakage were analyzed.The results showed that, in terms of ϕ 3.0mm steel wire sample,the tensile strength were 1096.7MPa. As the cold drawing proceed, the tensile strength of ϕ2.15mm steel wire sample went up to 1148.2MPa, and the elongation after fracture dropped from 10.6% to 6.2% dramatically.Cold drawing deformation enhanced the strength of the steel wire, while worsen the plasticity. The breakage was liable to occur during the drawing process as a result of the too much rolling reduction per pass.The analysis result of organization and fracture of steel wire before and after drawing deformation indicatedthat the matrix was austenite with the dispersedly distributedlong strips of Cr-rich phases, $M_{23}C_6$ or M_7C_3, on the surface. The mass precipitation of Cr-rich phases decreased the plasticity of steel wire significantly.The breakage of 314 steel wire was a kind of brittleness based mixed fracture.There were no distinct plastic deformation in the fracture, whilst obvious intergranular fracture with trace of spherical and petaloiddimple were existed in the pars fibrosa.Moreover, the sample contained many grey block inclusions, which identified by EDS as silicate and alumina, in size of approximately 6.5μm.
Key words 314 stainless steel wire, cold drawing, fracture, inclusion

410 不锈钢丝抗拉强度异常分析及退火工艺研究

陈　雷[1]　宋仁伯[1]　杨富强[1]　王斌宁[1]　谭　瑶[1]　郭　客[2,3]　王忠红[2]

（1. 北京科技大学材料科学与工程学院，北京　100083；
2. 鞍钢集团矿业设计研究院，辽宁鞍山　114021；
3. 辽宁科技大学，辽宁鞍山　114051）

摘　要　对在生产过程中抗拉强度出现异常的 410 不锈钢丝进行分析，同时对 410 不锈钢丝在不同退火工艺下的组织与性能进行检测和观察，分析不同退火工艺参数对其组织与性能的影响规律。结果表明：出现抗拉强度异常的钢丝中马氏体与铁素体含量存在差异，差值约为 23.27%，平均晶粒尺寸差值约为 5μm；退火后，当退火温度相同时，

随退火时间降低，抗拉强度增大，断后延伸率减小，晶粒尺寸减小；当退火时间相同时，随退火温度升高，抗拉强度减小，断后延伸率增大，晶粒尺寸增大。

关键词　410 不锈钢丝，抗拉强度，退火工艺

The Analyze of the Tensile Strength and the Study on Annealing of 410 Stainless Steel Wire

Chen Lei[1], Song Renbo[1], Yang Fuqiang[1], Wang Binning[1], Tan Yao[1], Guo Ke[2,3], Wang Zhonghong[2]

(1. School of Materials Science and Engineering, University of Science and Technology Beijing, Beijing 100083, China; 2. Ansteel Group Mining Engineering Corporation, Anshan 114021, China; 3. University of Science and Technology Liaoning, Anshan 114051, China)

Abstract　The anomaly of 410 stainless steel wire in tensile strength during manufacturing process, then the microstructure and properties of 410 stainless steel wire with different annealing process were observed and measured. The influences of different annealing process on the microstructure and properties of 410 stainless steel were analyzed. The results show that the content of martensite and ferrite is different between the abnormal steel wires in tensile strength, the difference is about 23.27%, and the difference of grain size is about 5μm. After annealing, when the annealing temperature is the same, the tensile strength increased, the elongation improved and the grain size lower as the annealing time shorter. When the annealing time is the same, the tensile strength decreased, the elongation reduced and the grain size bigger as the annealing temperature higher.

Key words　410 stainless steel wire, tensile strength, annealing temperature

基于有限元模拟的钢丝拉拔过程变形均匀性分析

毛小玲　苏　岚　米振莉

（北京科技大学冶金工程研究院，北京　100083）

摘　要　利用有限元软件 ABAQUS 建立了钢丝拉拔工艺过程的分析模型，研究了工作锥角和道次压缩率对拉拔过程变形均匀性的影响，提出并分析了如何通过改变工作锥角和压缩率改善心部变形不足的问题。研究表明：钢丝拉拔过程中，当压缩率一定时，随着工作锥角的增大，变形区域应变分布会逐渐由均匀变为不均匀，且心部与边部应变分布差异越来越大，出现心部变形不足；当工作锥角一定时，随着压缩率的增大，变形区应变分布会由不均匀向均匀转变，心部变形不足逐渐消失。

关键词　钢丝拉拔，有限元，变形均匀性，工作锥角，压缩率

Deformation Uniformity Analysis of Wire Drawing Process Based on Finite Element Simulation

Mao Xiaoling, Su Lan, Mi Zhenli

(Research Institute of Metallurgical Engineering, USTB, Beijing 100083, China)

Abstract The finite element software ABAQUS was introduced to establish the deformation model of steel wire during the drawing process. The influence of die cone angle and compression rate on deformation uniformity of drawing process was investigated. This research also supplied an approach to improve the insufficient deformation in core area by setting appropriate die cone angle and compression ratio.When the compression ratio was constant, with the increasing of die cone angle, deformation behavior will gradually change from homogeneous to heterogeneous. At the same time, the difference between core and edge area extended, and the insufficient deformation in core area occurred;Whendie cone angle was constant, with the increasing of compression ratio, the conclusion was adverse.
Key words wire drawing, FEM, deformation uniformity, die cone angle, compression ratio

冷轧深冲薄板成型极限图（FLD）的实验研究

陈海斌 米振莉 吴海鹏

（北京科技大学冶金工程技术研究院，北京 100083）

摘 要 利用电蚀网格法，在 Zwick/Roell BPB600 板材成型性试验机上进行冷轧深冲薄板的成型实验，采用 VIALUX 的 AutoGrid 软件对网格进行识别和分析，测试深冲板的成型极限图(FLD)。实验表明，L 钢厂的冷轧板的成型性能与 H 钢厂的冷轧板有较显著的差异。成型极限图 FLD 的测试对 L 钢厂改进冷轧板的生产工艺提供了理论依据。
关键词 冷轧板，成型极限图，工艺改进

Experiment Researchon the Forming Limit Diagrams(FLD) of Cold Rolled Steel Sheet

Chen Haibin, Mi Zhenli, Wu Haipeng

(National Engineering Research Center for Advanced Rolling Technology,University of Science and Technology Beijing,Beijing 100083,China)

Abatract As shown in this paper, the forming limit diagrams(FLD) of cold rolled steel sheet are tested by means of the electroetching grid method based on the forming experiment of cold rolled steel sheet carried on with Zwick/Roell BPB600 sheet forming testing machine and the grid identification and analysis made with the AutoGrid of VIALUX. Experiments show that the forming property of the cold rolled steel sheet produced by L plant is obviously different from that produced by H plant. It can be concluded that the testing of the forming limit curves(FLC) offers the theoretical foundation for the drawing of the deep drawing and forming process of cold rolled steel sheet.
Key words cold rolled steel sheet, forming limit diagrams, process improvement

总压缩率对镀锌钢丝扭转性能的影响

母俊莉

（宝钢集团南通线材制品有限公司，江苏南通 226000）

摘　要　钢丝在拉拔过程中，随着金属塑性变形的加剧，导致钢丝力学性能发生显著变化，在保证钢丝抗拉强度满足工艺设计要求的前提下，减少总压缩率对钢丝扭转性能有大幅度提高。

关键词　总压缩率，镀锌钢丝，扭转

The Influence of Total Compression Rate on the Torsion Property of Galvanized Steel Wire

Mu Junli

(Baosteel Nantong Wire Product Co., Ltd., Nantong Jiangsu 226000, China)

Abstract　In the process of drawing the wire, since the plastic deformation of the metal increasing, it will cause that the mechanics performance of the steel wirehas changed significantly. On the premise of ensuring the tensile strength of the steel wire tomeet the requirements of process design, reducing the total compression rate will increase the torsion performance of the steel wire significantly.

Key words　stotal compression rate, galvanized steel wire, torsion

13.9 级紧固件用冷镦钢开发

徐正东　张　弛

（南京宝日钢丝制品有限公司，江苏南京　210038）

摘　要　高强度螺栓连接是继铆接、焊接之后发展起来的一种新型连接形式，它具有承接能力高、受力性好、耐疲劳、不松动、较安全及施工简便、可拆换等优点，是世界各国研究开发的热点。

关键词　高强度，紧固件，疲劳试验

13.9 Fasteners for Cold Heading Steel Development

Xu Zhengdong, Zhang Chi

(Nanjing Baori Wire Product Co., Ltd., Nanjing 210038, China)

Abstract　the high strength bolt connection is developed following the riveting and welding after a new connection form, it has to undertake high ability, good stress of, fatigue resistance, not loose, is safe and convenient construction, removable and advantages, is the focus of research and development in the world.

Key words　high strength，fastener，fatigue test

6 钢铁材料

Steel Products

炼铁与原料

炼钢与连铸

轧制与热处理

表面与涂镀

钢材深加工

★ 钢铁材料

能源与环保

冶金设备与工程

冶金自动化与智能管控

冶金分析

信息情报

6.1 汽车用钢/Automotive Steel

Knowledge Based Steel Design for Improved Performance

Bleck Wolfgang

(IEHK Steel Institute, RWTH Aachen University, Aachen, Germany)

Abstract Modern steel concepts aim at improved properties by tailoring the microstructure. Three trends are of major interest in nowadays materials development activities: the decrease of the structural length of microstructural constituents towards nanosized quantities, the increase of alloying level and by this the interaction of different elements in enriched zones, extensive use of numerical materials design over various length scales and through long process chains.

Key words AHSS (advanced high strength steels), HMnS (high Mn steels), bainitic steels, virtual microstructure, damage tolerance

Nb-Microalloying in Next-Generation Flat-Rolled Steels

SPEER John G, ARAUJO Ana L, MATLOCK, David K, DE MOOR Emmanuel

(Advanced Steel Processing and Products Research Center, Colorado School of Mines, 1500 Illinois St., Golden, Colorado, 80401, USA)

Abstract Extensive efforts are underway worldwide to develop new steels with substantial fractions of retained austenite, for lightweight automobile manufacturing and other applications requiring improved combinations of strength and formability. It is likely that microalloying can provide product enhancements in these emerging products, such as Q&P, TBF, medium-Mn TRIP, etc. And this paper examines the expected behavior of niobium using inferences based on published AHSS literature and principles of Nbmicroalloying.Some benefits of Nb in terms of microstructure refinement and precipitation strengthening have been reported, and increased retained austenite levels are shown here in Q&P hot-rolling simulations. The potential influences of Nb are complex due to the sensitivity of Nb dissolution and precipitation to chemical composition and processing; differences in the expected role of Nb are pointed out with respect to different product forms produced via hot-rolling or annealing after cold-rolling, and microstructures with or without substantial quantities of primary ferrite. Some issues that warrant further examination are identified, as a deep understanding of Nbmicroalloying and other fundamental behaviors will be needed to optimize the performance of these next-generation steels.

Key words niobium, high strength steel, Q&P, sheet, plate

冷轧高强汽车板的高应变速率行为及纳米析出特征

康永林[1,2] 李声慈[1] 朱国明[1]

（1. 北京科技大学材料科学与工程学院，北京 100083；

2. 材料先进制备技术教育部重点实验室，北京　100083）

摘　要　汽车用钢需求高速增长，同时也向着高强度化方向发展。钢在高应变速率条件下的力学性能及变形行为与静态条件下的相比具有显著区别；通过高速拉伸实验，分别研究了 DP780 双相钢和 DC53D+ZF 钢两种材料的高应变速率行为，分析了应变速率对钢的力学性能、组织形变的影响，并建立了相应的本构方程。根据获得的本构方程，对 DC53D+ZF 钢板的矩形梁碰撞进行了模拟计算。结果表明，碰撞过程中应变速率的影响较大，采用所推导 J-C 模型的仿真结果与试验获得的材料真应力-应变曲线的计算结果最为接近。以 980MPa 级 Nb 微合金化（C-Si-Mn-Cr-Nb）冷轧超高强双相钢为例，分析了 Nb 析出历程及相应粒子形貌，可以通过在不同冶金工艺阶段的析出特征进行相应的工艺控制，以得到目标组织及纳米析出粒子的分布。

关键词　汽车用钢，高应变速率，碰撞，纳米析出

High Strain Rate Behavior and Nano-precipitation of Cold Rolled High Strength Automobile Steel Sheets

Kang Yonglin[1,2], Li Shengci[1], Zhu Guoming[1]

(1.School of Materials Science and Engineering, University of Science and Technology Beijing, Beijing 100083, China;

2. Key Laboratory for Advanced Materials Processing (MOE), Beijing 100083, China)

Abstract　The demand for automotive steel is rapidly increasing, and it develops toward a higher strength. Compared with the static conditions, mechanical properties and deformation behavior of steel at high strain rate are significantly different. High-speed tensile test were used to study the high strain rate behavior of DP780 dual phase steel and DC53D+ZF steel, effects of strain rate on mechanical properties and microstructural deformation were analyzed, and corresponding constitutive equation was established, which was used to simulate the collision of rectangular beam for DC53D+ZF steel. The results show that strain rate has great influences on the collision, the simulated results by JC model was the closest to the results by experimental true stress-strain curve. 980MPa grade Nb micro-alloyed (C-Si-Mn-Cr-Nb) cold-rolled ultra high strength dual-phase steel was used to analyze the precipitation and corresponding particle morphology. Process can be controlled base on the characterization of precipitation at different stages of the metallurgical process in order to obtain the required mucrostructure and distribution of nano-precipitated particles.

Key words　automotive steel, high strain rate, collision, nano-precipitation

轴承钢精炼中大型夹杂物来源的示踪研究

刘　浏[1]　范建文[1]　王　乐[1]　王　品[2]

（1. 中国钢研科技集团有限公司，北京　100081；2. 湖北新冶钢有限公司，湖北黄石　435001）

摘　要　采用在 LF 精炼初渣中添加示踪剂的方法，研究确定了轴承钢大型夹杂物的来源为：初期脱氧夹杂，一直悬浮在钢中，未与炉渣接触，不含 BaO 成分，在钢材中残留量占全部氧化物夹杂的 10%~15%；内生夹杂，由渣钢反应生成，随精炼进行，含 BaO 夹杂的比例升高。无渣冶炼如 RH 可抑制该类夹杂生成，在钢材中残留量约占 25%~40%；卷渣夹杂，由钢渣搅动生成，绝大多数含有 BaO 成分，随精炼进行尺寸逐渐减小，残留在钢材中的比例约占 50%~60%。因此，降低钢中大型夹杂物的技术措施是：严格控制脱氧前钢水 a_0，降低 T.O；尽可

能避免或减弱渣钢反应强度，或降低精炼渣的还原势；优化钢水搅拌强度，减少卷渣并促进微细夹杂物聚合上浮。

关键词 轴承钢，夹杂物，转炉，炉外精炼

Study on Generation Mechanisms of Large Inclusions during Bearing Steels Refining Process by Tracer Method

Liu Liu[1], Fan Jianwen[1], Wang Le[1], Wang Pin[2]

(1. Central Iron & Steel Research Institute, Beijing 100081, China;
2. Hubei Xinyegang Co., Ltd., Huangshi Hubei 435001, China)

Abstract The generation mechanisms of large size oxidation inclusions during whole refining process have been studied by the method of adding $BaCO_3$ tracer agent into the initial ladle refining slag. The results show that the large size oxidation inclusions in the bearing steels mainly originate from primary de-oxidation reaction, intrinsic reaction and refining slag mixing. Each type of the inclusions accounts for 10%~15%, 25%~40% and 50%~60% of all the residual oxidation inclusions respectively. The primary de-oxidation inclusions always linger in molten steel without contact with slag, therefore they do not contain BaO, MgO, CaO and so on. The intrinsic reaction inclusions originate from interface reactions between slag and molten steel, so that the BaO content of the inclusions increases with the refining time. For the refining process without slag, such as RH process, the formation of these inclusions can be hold back. The slag-mixing inclusions originate from the slag entrapped into the molten steel by the disturbance between slag and molten steel, so most of them contain BaO and the size of inclusions decreases with the refining time. The methods for decreasing the number of large size inclusions are as follow: confining strictly the oxygen activity (a_0) before deoxidizing process to decrease T.O content in molten steel, weakening even avoiding the interface reactions between slag and molten steel, decreasing the reduction intensity of LF slag as possible as we can, and optimizing the molten steel stirring intensity to decrease the entrapped slag and promote the tiny inclusions coagulating and floating upwards.

Key words bearing steel, non-metallic inclusion, converter steel making, secondary refining

The Influence of Pre-deformationon Tensile Properties of Hydrogen-chargedaustenitic Stainless Steels

Lee Young-Kook, Park Il-Jeong, Ji Hyunju

(Department of Materials Science and Engineering, Yonsei University, Seoul 120-749, Korea)

Abstract The effect of pre-deformation before hydrogen charging on tensile properties of hydrogen-charged austenitic stainless steels, such as STS 310S and STS 316L, was investigated through the slow strain rate tensile tests (SSRTs) and thermal desorption analysis (TDA). Theelongations of both steels decreased due to atransition of ductile-to-brittle fracture by hydrogen charging. For STS 310S, the degree of the reduction in elongation by hydrogen charging decreased with increasing pre-strain. The TDA revealed that the migration of hydrogen atoms to twin boundaries was hindered by pre-strain. Therefore, the reason why pre-strain suppressed the fracture transition was that pre-strain hindered both mechanical twinning during SSRT sand hydrogen atomsdelivery to twin boundaries. For STS 316L, the pre-strain degraded the resistance to hydrogen embrittlement. The formation of ε-martensitewas accelerated by pre-strain in hydrogen-charged specimens during SSRTs. The TDA elucidated that hydrogen atoms migrated from grain boundaries and dislocations to not only mechanical twins but γ/ε interfaces in pre-strained specimens during the SSRTs. Accordingly, the elongation loss by pre-strain resulted from both the increase in fraction of ε-martensite with pre-strain and the segregation of hydrogen atoms to γ/ε interfaces.

Key words hydrogen embrittlement, pre-deformation, austenitic stainless steel, mechanical twin, ε-martensite

大规格弹簧钢无槽连续粗轧有限元仿真的研究

张业华[1]　王保元[2]　戴江波[2]　徐大庆[1]

（1．武钢集团襄樊钢铁长材有限公司，湖北襄阳　441100;
2．中冶南方武汉钢铁设计研究院有限公司，湖北武汉　430083）

摘　要　为了解决襄樊钢铁长材有限公司新建年产 30 万吨弹簧扁钢生产线，投产初期轧制不连续及大规格扁钢质量不能满足客户要求等问题，本文利用全球第一家非线性大型有限元 MARC 软件，基于旋转轧辊刚性面接触摩擦引领钢坯运动、钢坯热轧塑性变形的热力耦合有限元法,建立了钢坯在加热至1100℃的粗轧轧制仿真有限元模型，为弹簧扁钢无槽连续轧制提供了理论计算的手段，并结合相关生产工况实际，针对弹簧扁钢主要轧制工艺参数进行仿真，揭示了襄钢新建弹簧扁钢生产线易造成堆钢和产品质量不稳定的原因，依据研究分析结果完善了轧制工艺制度，经过实际验证，大规格弹簧扁钢无槽轧制连续性和产品质量得到了较大的改善，同时降低了工人劳动强度，提高了经济效益。

关键词　钢坯，轧制，连续，无槽，有限元

Research on Finite Element Simulation of the Big Size Spring Steel in Continuous Grooveless Rolling

Zhang Yehua[1], Wang Baoyuan[2], Dai Jiangbo[2], Xu Daqing[1]

(1. Wuhan Iron and Steel Group Xiangyang Steel Co., Ltd., Xiangyang 441100,China;
2. WISDRI Wuhan Iron and Steel Design Institute,Wuhan 430083,China)

Abstract Xiangfan iron and Steel Co., Ltd built the production line of annual 300000 tons spring steel in 2013. In order to solve the discontinuity and quality instability problem of the billet rolling , The Author use large finit element MARC software for the simulation calculation.The basis of the rolling process parameters of large size spring steel, the finite element model was establishe. the calculation results reveal the reasons about the billet heap and the low precision products, the rolling program file about continuous rolling pass and technology of the rolling mill was revised and perfected, After actual verification,the continuity and stability of the large size spring flat steel is improved greatly, and the unit operation efficiency increased in continuous slotless rolling.

Key words steel billet，rolling，continuous，grooveless，finite element

胀断连杆用 V-N 微合金锻钢的开发

张贤忠　周桂峰　陈庆丰　熊玉彰

（武钢研究院，湖北武汉　430081）

摘 要 胀断工艺被广泛应用于汽车发动机连杆加工，针对连杆胀断工艺，开发了一种新的高强度 V-N 微合金锻钢，通过成分和工艺设计，能够得到利于胀断加工的网状铁素体和珠光体组织。和传统的胀断连杆用钢 C70S6 相比，开发的 V-N 微合金锻钢有更高的强度和优越的疲劳性能，而且，连杆胀断表面显示出脆性断裂特征，满足连杆胀断工艺的要求。

关键词 微合金钢，组织，力学性能，疲劳性能，胀断

Development on V-N Microalloyed Steel Used in Fracture Splitting Connecting Rods

Zhang Xianzhong, Zhou Guifeng, Chen Qingfeng, Xiong Yuzhang

(Reseach and Develoment Center of WISCO, Wuhan 430081, China)

Abstract Fracture splitting technology is widely used in automotive engine connecting rod processing. According to the characteristics of connecting rods fracture splitting processing, developed a new kind of high strength of V-N microalloyed forging steel. By chemical composition and process design, a network of ferrite and pearlite could obtain, it was benefit to fracture splitting. Compare with C70S6 which normally used in fracture splitting rods, the developed microalloyed steel was superior in mechanical properties and fatigue strength. Furthermore, the fracture splitting surface showed distinct brittle fracture character and favorable fracture splitting properties.

Key words microalloyed steel, mechanical properties, fatigue strength, microalloyed steel, fracture splitting

Effects of Structure Refinement on Properties of Multiphase AHSS

Nina Fonstein

(CBMM consultant, USA, per request from Brazil)

Abstract The presented paper reviews, based on literature and own experimental data, some basics for the following phenomena:

(1) Yield strength increase in DP steels at ferrite grain refinement, although classical Hall-Petch equation, well describing those changes, does not make an original physical sense, when grain boundaries are no longer are barriers for transferring plastic deformation due to abundance of mobile dislocations;

(2) Strain hardening increase at structure refinement of DP steels whereas in single-phase materials it decreases at diminishing grain size;

(3) Enhancement of TRIP effect (austenite stability) at refinement of austenite whereas grain refinement usually decreases hardenability of austenite at cooling after full austeitization;

(4) Critical influence of grain refinement on impact of precipitation hardening on ductile-brittle transition temperature (DBTT);

(5) Impact of grain size on bainitetransformation kinetics etc.

Discussing different ways of structure refinement, unique advantages of niobium microalloyingof all types of AHSS (DP, TRIP, Carbide Free Bainitic steels as well as medium Mn, martensitic steels and steels processed using Q&P cycle) will be convincingly illustrated by numerous examples.

Key words structure refinement, grain size, multiphase AHSS, microalloying, niobium

On the Role of Aluminium in Segregation and Banding in Multiphase Steel

Ennis Bernard[1,2], Mostert Richard[1], Jimenez-Melero Enrique[2], Lee Peter[2]

(1. Tata Steel, PO Box 10000, 1970 CA IJmuiden, The Netherlands;
2. The School of Materials, The University of Manchester, Oxford Road, Manchester, M13 9PL, UK)

Abstract In this work we have concentrated on the effect of aluminium on the formation of segregation during casting and the resultant effect on the microstructure and properties of multiphase steel. Different processing conditions were applied on the basis of model calculations in order to promote and suppress banding in segregated and non-segregated steel. It is shown that Al segregates preferentially to the solid during solidification, in contrast to the other elements which segregate preferentially to the liquid and increasing Al content leads to a increased partitioning of the other chemical elements. It has been shown experimentally, and in good agreement with the predicted model calculations, that the effect of segregation second phase banding can be suppressed. The tensile behaviour of all the variants is identical up to the yield point, but the non-banded samples show a much higher tensile strength with similar levels of ductility to the banded samples.

Key words steel, casting, segregation, phase transformation, thermomechanical processing

冷却速度对汽车用非调质冷镦钢组织和性能的影响

丁 毅 安金敏

（宝山钢铁股份有限公司研究院长材研究所，上海 201900）

摘 要 本文研究了冷却速度对汽车转向系统用非调钢组织和性能的影响。试验结果表明，试验钢的显微组织主要由铁素体珠光体构成，珠光体片间距随着冷却速度的加快而逐渐变窄。当冷却速度为 0.5℃/s 时试验钢种具有大量均匀的析出物，使试验钢种在 0.5℃/s 冷却的力学性能高于在 2℃/s 时冷却的力学性能。

关键词 非调钢，冷却速度，珠光体片间距，析出物

Effect of the Cooling Speed on the Microstructures and the Properties in Micro-alloyed Steel for Car

Ding Yi, An Jinmin

(Research Institute, Baoshan Iron & Steel Co., Ltd., Shanghai 201900, China)

Abstract Effect of the cooling speed on the microstructures and the properties in micro-alloyed cold heading steel for automatic steering system has been researched in this paper. The microstructures of the researched steel are of ferrite and pearlite. The distance in the pearlite is decreasing with the cooling speed is accelerated. A large number of homogeneous precipitates are observed when the cooling speed is 0.5 ℃/s, as a result the properties with the cooling speed of 0.5 ℃/s are better than that with the cooling speed of 2 ℃/s.

Key words micro-alloyed steel, cooling speed, distance in pearlite, precipitate

Development of "Hot Forging Non-quenched and Tempered Steel"

Yu Shihchuan, Liang Shengleo, Kuo Chunyi

(Metallurgical Department, China Steel Corporation, Kaohsiung Taiwan, China)

Abstract This article describes development and improvement of forging steels which are free of quenching and tempering. By adding V, Ti and other alloy and controlling rolling technology, the strength and toughness have been promoted. Furthermore, by means of controlled cooling after forging, the mechanical properties could be equivalent to conventional quenched and tempered carbon steel. It is not only energy saving and reduction of carbon emissions but also reducing process cost by 10% to 15%.

Key words hot forging, non-quenched and tempered, environment friendly steels

汽车用BN型易切削钢的开发

张 帆 曾 彤 任安超 丁礼权

（武汉钢铁（集团）公司研究院，湖北武汉 430080）

摘 要 热力学计算表明，BN在奥氏体温度区间开始析出；[Al]含量增加，硼氮平衡浓度积降低，过高的[Al]含量，会导致AlN优先于BN析出。试验结果表明，与基础钢相比，BN的添加对力学性能无害，并且能显著提高钢材的切削性能。

关键词 易切削钢，BN，切削性能

Development of BN-type Automotive Cutting Steel

Zhang Fan, Zeng Tong, Ren Anchao, Ding Liquan

(Research and Department of WISCO, Wuhan Hubei 430080, China)

Abstract Based on thermodynamic analysis, the paper shows that BN separate out at Austenitizing temperature, and along with [Al] increasing, the equilibrium concentration of [B][N] decrease, excess content of [Al] may cause AlN priority precipitation than BN. Experiment results showed that, compared with foundation steel, BN does no harm to mechanical properties, and can significantly improve the machinability of steel.

Key words free cutting steel, boron nitride, machinability

一种齿轮钢连铸坯枝晶显示方法

柳洋波 佟 倩 孙齐松 丁 宁 温 娟

（首钢技术研究院，北京 100043）

摘　要　本文提出了一种全新的齿轮钢连铸坯枝晶的显示方法。工艺步骤如下：（1）将齿轮钢连铸坯试样在奥氏体化温度保温 30~60min，然后炉冷却到珠光体转变温度，并保温 2h，最终炉冷到室温；（2）用砂轮打磨，去除试样表面氧化铁皮和脱碳层，然后研磨抛光；（3）将试样浸入 4%的硝酸酒精溶液 5s，然后用清水冲洗表面，并吹干；（4）利用光学显微镜观察枝晶形貌。采用本显示方法，能将齿轮钢连铸坯的枝晶清晰地显示出来，并可以测量一次枝晶间距和二次枝晶间距。本方法和现有方法相比取得以下效果：（1）齿轮钢连铸坯枝晶显示的清晰、准确。（2）采用 4%的硝酸酒精溶液侵蚀，省去了复杂的腐蚀试剂的配制，便于检测人员操作的规范和快捷。

关键词　齿轮钢，连铸坯，枝晶显示，等温退火

A New Method for Dendrite Revelation of Continuous Casting Billet of Gear Steel

Liu Yangbo, Tong Qian, Sun Qisong, Ding Ning, Wen Juan

(Shougang Research Institute of Technology, Beijing 100043, China)

Abstract　A new method for dendrite revelation of continuous casting billet of gear steel was presented in this paper. Process steps were as follows: (1) The continuous casting billet specimens were austenited for 30-60min, then cooled in the furnace to pearlite transformation temperature, and hold for 2h, eventually cooled to room temperature in the furnace; (2) The specimens were ground to remove oxide scales and decarburization layers, then polished; (3) The specimens were immersed in 4 vol% nital for 5s, then flushed and dried; (4) Dendrite morphology was observed by using an optical microscope. The morphology of dendrite was clearly observed by adopting the new method, and the primary dendrite spaces and the secondary dendrite spaces could be measured. Comparing with traditional methods, two advantages were shown as follows: (1) The morphology of dendrite is clear and accurate. (2) It is convenient to etch with 4 vol% nital rather than some complex etches.

Key words　gear steel, continuous casting billet, dendrite display, isothermal annealing

高强度风电螺栓用钢的低温冲击韧性研究

陈继林[1,2]　刘振民[1,2,3]　崔　娟[1,2]　张治广[1,2]　蔡士中[1,2]

（1. 邢台钢铁有限责任公司，河北邢台　054027；2. 河北省线材工程技术研究中心，河北邢台　054027；3. 北京科技大学材料学院，北京　100083）

摘　要　结合实际轧制过程，运用金相显微镜、SEM、TEM，及低温冲击功试验，研究了高强度风电螺栓用钢 B7 的低温冲击韧性研究，结果表明：采取低温轧制，材料组织均匀性较好，多边形铁素体均匀分布；有助于位错密度的改善;在低温轧制工艺下产品内部大角度晶界角度要偏小；产品晶粒度较高，达 8 级。实验证明：通过低温轧制，内部组织的改善有效降低了材料韧脆转变温度，并提高组织晶粒度，最终实现了 B7 产品低温冲击韧性有所提高。

关键词　高强度，低温冲击韧性，低温轧制，晶粒度

Research on Low Temperature Impact Toughness of High Strength Wind Power Bolt Steel

Chen Jilin[1,2], Liu Zhenmin[1,2,3], Cui Juan[1,2], Zhang Zhiguang[1,2], Cai Shizhong[1,2]

(1. Xingtai Iron and Steel Corp., Ltd., Xingtai Hebei 054027, China;2. Hebei Engineering Research Center for Wire Rod, Xingtai Hebei 054027, China;3. University of Science & Technology Beijing, Beijing 100083, China)

Abstract Combined with the actual rolling process, the low temperature impact toughness of high strength steel B7 was studied by means of optical microscope, TEM, SEM and low temperature impact test. The results show that: low temperature rolling, microstructure uniformity of polygonal ferrite, uniform distribution; help to improve the dislocation density; in the low temperature rolling process under the angle of large angle grain boundary to small grain products; a higher degree, up to 8, Experiments show that: the low temperature rolling, improvement in the internal organization to effectively reduce the ductile brittle transition temperature, and increase the grain size, finally realized the impact toughness of B7 products has increased.

Key words high strength, low temperature impact toughness, low temperature rolling, grain size

Evolution of Microstructures in a Nanotwinned Steel Investigated by Nanoindentation and Electrical Resistivity Measurement

Zhou Peng[1], Liu Rendong[2], Wang Xu[2], Huang Mingxin[1]

(1. Department of Mechanical Engineering, The University of Hong Kong, Pokfulam Road, Hong Kong, China; 2. Iron and Steel Research Institute, Ansteel Group, Anshan, China)

Abstract A twinning-induced plasticity (TWIP) steel was subjected to cold rolling followed by recovery heat treatment, forming a nanotwinned steel. During the recovery annealing, the nanotwins are thermally stable so that they remain in the sample after recovery annealing. On the contrary, the dislocation density was reduced greatly after recovery annealing. The nanotwinned steel possesses a high yield strength, high ultimate tensile strength and good ductility. The nanohardness tests and electrical resistivity measurements illustrate that the dislocation density increases dramatically in the nanotwinned steel during the tensile test while the twin volume fraction remains constant. The transmission electron microscopy (TEM) observation confirms the results from nanoindentation and electrical resistivity tests. The maximum volume fraction of deformation twins has been reached during the heavy cold rolling so that no more new deformation twins can be generated in the nanotwinned steel during tensile test. Therefore, the work hardening behaviour of the nanotwinned steel is mainly provided by the accumulation of dislocations.

Key words nanotwinned steel, TWIP steel, nanotwins, nanoindentation, electrical resistivity

The Experimental Investigation on Chain-die Forming U-Profiled AHSS Products

Sun Yong [1], Li Yaguang [2], Zhang Kun [3], Han Fei [3], Liu Zhaobing [4], Daniel W J T(Bill) [1], Li Dayong [2], Shi Lei [3], Ding Shichao[1]

(1. School of Mechanical and Mining Engineering, University of Queensland, St Lucia, QLD 4072, Australia;
2. Institute of PFKBE, School of Mechanical Engineering, Shanghai JiaoTong University, Shanghai, China;
3. Baosteel Research Centre, Baosteel, Shanghai, China;
4. Ningbo SaiRolf Metal Forming Co., Ltd., Ningbo, China)

Abstract Advanced high strength steel (AHSS) iswidely used in the automotive industry because of its advantages for weight reduction and for the resulting improvement in crashworthiness and safety. However it is still a challenge to fabricate AHSS products at room temperature due to its high strength and limited elongation. This greatly limits the new material's application and the spread of its use. Chain-die forming has been proposed and developed as an extension to roll forming in fabricating AHSS products. It has shown its superiority by reducing redundant plastic strains during forming and nearly zero residual stresses can be achieved through increasing the length of deformation zone, via increasing the apparent roll radius. In this paper, an experimental investigation is aiming to demonstrate the working principle and investigate the characteristics of Chain-die forming AHSS products with various material properties and geometrical dimensions which can further expand the application of Chain-die forming and accelerate its commercialization, particularly in automotive industry. As the longitudinal strain, residual strain and springback are regarded as the crucial factors in the quality evaluation of AHSS products, the relationships and comparisons of these factorsof Chain-die formed AHSS products with various material properties and geometrical dimensions are presented. Finally, some concluding remarks for further study are made from the experimental investigation.

Key words Chain-die forming, AHSS, longitudinal strain, redundant plastic deformation, springback

Microstructures and Tensile Deformation Behavior of a Cryogenic-rolled Mn-Al TRIP Steel

Hu Zhiping[1], Xu Yunbo[1], Tan Xiaodong[1], Yang Xiaolong[1], Peng Fei[1], Yu Yongmei[2], Liu Hui[3], Wang Le[3]

(1.The State Key Laboratory of Rolling and Automation, Northeastern University, Shenyang, China;
2. School of Mechanical Engineering, Shenyang University of chemical technology,Shenyang Economical & Technological Development Zone, Shenyang, Liaoning, China; 3. Research Institute of Iron and Steel of Shandong Iron and Steel Group, Jinan, Shandong 250101, China)

Abstract In this work, acryogenic rolling process was appliedto a 3Mn-1.5Al TRIP steel. Microstructure characterization was carried out by means of optical microscope, scanning electron microscope (SEM) equipped with electron probe microanalysis (EPMA) and transmission electron microscope (TEM). Mechanical properties were obtained by uniaxial tension tests. The microstructure characterization results show that the steel presents a complex microstructure composed of three phases (ferrite, martensite and retained austenite). The volume fractions of austenite before and after a deformation were determined by X-ray diffraction (XRD). The XRD results indicate that the amount of retained austenite reaches up to 18vol% and the TRIP effect occurs quite apparently in the specimen with optimizingintercritical annealing temperature of 730°C. The mechanical properties results displays that the tested steel presentsadequateultimate tensile strength of 916MPa and excellent elongation of 30% in the optimal intercriticalannealing temperature. The retained austenite strain hardening behavior (continuous TRIP effect) is observed in the steel. The outstanding combination of strength and ductility with the product of strength and elongation (PSE) reaching up to 27GPa% around indicates that the steel has a bright application prospect.

Key words cryogenic rolling, Mn-Al TRIP steel, microstructures and properties, retained austenite

奥氏体化和等温转变温度对高碳钢盘条组织性能的影响

吴礼文[1] 唐延川[1] 赵自飞[1] 康永林[1]
曹长法[2] 王广顺[2] 徐 凯[2]

（1. 北京科技大学材料科学与工程学院，北京 100083；
2. 青岛钢铁控股集团有限责任公司，山东青岛 266043）

摘 要 采用离线奥氏体化和两段式盐浴等温热处理方式，奥氏体化温度分别为900℃、920℃、940℃和960℃，保温10min后，迅速转入冷却盐浴炉中，冷却浴温度设定为480℃，冷却10s后转入恒温盐浴炉，恒温浴温度分别设定为480℃、500℃、520℃以及550℃。共16组热处理条件，探讨此成分高碳钢盘条最佳的热处理工艺。其中920℃奥氏体化、500℃盐浴等温热处理盘条，抗拉强度1374MPa，断面收缩率40.6%，索氏体化率94.5%，珠光体片层间距138nm，具有较高的强度和均匀的组织，综合性能最好。随着等温转变温度越高，索氏体化率下降，珠光体片层间距增加，抗拉强度降低，断面收缩率也降低。奥氏体化温度对抗拉强度影响不大，主要影响珠光体球团尺寸大小和断面收缩率。利用热力学计算，分析VC在珠光体盘条钢中的析出强化行为。

关键词 高碳钢盘条，盐浴热处理，组织性能关系，VC析出行为

Effect of Austenitizing and Isothermal Transformation Temperature on the Microstructure and Properties of High Carbon Steel Wire Rod

Wu Liwen[1], Tang Yanchuan[1], Zhao Zifei[1], Kang Yonglin[1], Cao Changfa[2], Wang Guangshun[2], Xu Kai[2]

(1.School of Materials Science and Engineering University of Science and Technology Beijing, Beijing 100083, China;
2. Qingdao Iron and Steel Holding Group Co., Ltd., Qingdao 266043, China)

Abstract Offline austenitizing and two stages of salt bath isothermal heat treatment are used in the experiment, with austenitizing temperatures of 900℃, 920℃, 940℃ and 960℃. after 10min incubation, wire rod is rapidly transferred to a cooling bath at 480℃, keeping 10 minutes, and then transferred to thermostatic bath with temperature of 480℃, 500℃, 520℃ and 550℃. Wire rod under 920℃ austenitizing and 500℃ salt bath isothermal heat treatment is of good strength and uniform microstructure, with tensile strength 1374MPa, reduction 40.6%, Patenting rate 94.5%, and pearlite lamellar spacing 138nm. As the isothermal transformation temperature is higher, patenting rate, tensile strength and reduction decrease, and pearlite lamellar spacing increases. Austenitizing temperature has little effect on the tensile strength, mainly affecting the size of pearlite pellet and reduction. Finally, thermodynamic calculations is used to analyze VC precipitation hardening behavior of the pearlite steels.

Key words high carbon steel wire rod, salt-bath heat treatment, structure-property relationships, VC precipitation

Evaluation of Hydrogen Embrittlement Susceptibility of Four Martensitic Advanced High Strength Steels Using the Linearly Increasing Stress Test

Venezuela Jeffrey[1], Liu Qinglong[1], Zhang Mingxing[1], Zhou Qingjun[2], Atrens Andrej[1]

(1. The University of Queensland, Division of Materials, School of Mining and Mechanical Engineering, St. Lucia, 4072 Australia;

2. Baoshan Iron & Steel Co., Ltd., Research Institute, Shanghai 201900, China)

Abstract Martensitic advanced high strength steel (MS-AHSS) belongs to a new class of steels that are currently used in the manufacture of lightweight yet crashworthy cars. One issue with using high-strength steels is hydrogen embrittlement (HE), which causes the degradation of the mechanical propertiess of the steel. This study investigated the influence of hydrogen on the mechanical and fracture properties of four commercially-produced martensitic advanced high strength steels (MS-AHSSs), using linearly increasing stress tests (LISTs).The four steels had tensile strengths above 950 MPa. Electrochemical hydrogen charging was carried out in 0.1 M NaOH at cathodic potentials of −1200, −1500 and −1800 $mV_{Ag/AgCl}$. The LIST sused two appliedstress rates. The four MS-AHSSs had microstructures consisting of martensite and ferrite.Hydrogen had little effect on the threshold and tensile stress but decreased ductility. HE susceptibility increased with increasing steel strength, more negativecharging potential, and decreasing applied stress rate. Two grades of MS-AHSS showed little HE susceptibility, whilst the two strongest ones exhibited some susceptibility.HE susceptibility was manifest by the reduction of ductility, and the change from ductile cup-and-cone fracture to macroscopically-brittle shear fracture. The LISTs were proven capable of assessing HE susceptibility in MS-AHSS.

Key words automotive advance high strength steel, hydrogen embrittlement, SEM

Study on Hydrogen Permeation in DP and Q&P Grades of AHSS under Cathodic Charging and Simulated Service Conditions

Liu Qinglong[1], Venezuela Jeffrey[1], Zhang Mingxing[1], Zhou Qingjun[2], Atrens Andrej[1]

(1. The University of Queensland, Division of Materials, School of Mining and Mechanical Engineering, St. Lucia, 4072 Australia；

2. Baoshan Iron & Steel Co., Ltd., Research Institute, Shanghai 201900, China)

Abstract The hydrogen permeation behaviour of DP and Q&P grades of advanced high strength steels was investigated using the permeability experiments (1) under cathodic charging at different potentials in 0.1 M NaOH solution, (2) in 3% NaCl solution at the free potential simulating the corrosion behaviour of non-galvanized steel in service and (3) at zinc potential for the corrosion condition of the galvanized steel during actual service.Under cathodic charging at different potentials in 0.1 M NaOH solution, hydrogen permeation decreased with less negative charging potential, reflected by lower values of permeation rate and diffusion coefficient. At the most severe charging condition of −1.700 $V_{Hg/HgO}$ in the 0.1 M NaOH solution, the permeation rate for 1200 DP-GI steel was 1.53 μA/cm and for 980 QP steel was 0.492 μA/cm. The values for 1200 DP-GI and 980 QP steels at the least severe charging condition of −1.100 $V_{Hg/HgO}$ were 0.178 μA/cm and 0.111 μA/cm, respectively. The hydrogen permeation rate under simulated service conditions in 3% NaCl solution was significantly lower than the value under the least negative cathodic charging potential of −1.100 $V_{Hg/HgO}$ in NaOH solution.The Zn coating on the 1200 DP-GI steel could protect the substrate steel from corrosion, however, once the Zn coating was corroded under actual service, more hydrogen would be introduced into and permeate through the steel.

Key words steel, hydrogen permeation, corrosion

Advanced Technology to Anneal Current and Future AHSS

Eric Magadoux[1], Stephane Mehrain[2]

(1. Research & Development Engineer, Fives Stein, France;
2. Vice President, Sales & Marketing, Fives Stein, France)

Abstract Water Quench technology has been used on certain continuous annealing lines for more than 30 years now to produce HSS grades and the market tendency for higher strength steel grades is accelerating today.

Various AHSS grades have been developed for which yield strength and other properties are also considered, to ensure formability. The production of AHSS grades in a continuous annealing line requires a strict cooling control to limit strength deviation within the coil to improve homogeneity and formability. To reach the required AHSS characteristics, Wet Flash Cooling® has been developed, offering a more flexible control of the strip cooling cycle, including initial and final strip temperatures and modulation of cooling rate.

This technology has been successfully operating in a large capacity industrial continuous annealing line at a major steel plant for 6 years following its start up in 2009. This paper describes the new challenges of quenching technologies to respond to the evolving AHSS production needs.

Key words water cooling, cooling flexibility, martensite, Q&P

Determining the Respective Hardening Effects of Twins, Dislocations, Grain Boundaries and Solid Solution in a Twinning-induced Plasticity Steel

Liang Zhiyuan[1,2], Li Yizhuang[1,2], Huang Mingxin[1,2]

(1. Shenzhen Institute of Research and Innovation, The University of Hong Kong, Shenzhen, China;
2. Department of Mechanical Engineering, The University of Hong Kong, Hong Kong, China)

Abstract Manganese-rich austenitic twinning-induced plasticity (TWIP) steels with high strength and exceptional ductility have received much attention in the past two decades. Tremendous efforts have been made to explore their unusual hardening behavior which involves complex dislocation-twin interaction, dislocation-dislocation interaction, dislocation-grain boundary interaction and dislocation-solid atom interaction. Nevertheless, the individual hardening effects of twins, dislocations, grain boundaries and solid solution to the high strength of TWIP steels are still unclear. In order to answer this question, the flow stress of a TWIP steel wasexperimentally decomposed into the respective contributions of twins, dislocations, grain boundaries and solid solution in the present study. For the determination of forest hardening, synchrotron X-ray diffraction experiments with line profile analysis were carried out to measure the dislocation density. It is found that the yield stress of the present TWIP steel is controlled by solid solution and grain boundary hardening,which contribute to 238.3 and 238.5 MPa of the flow stress, respectively. After yielding, the work-hardening rate is dominated by dislocation evolutionwhich accounts forup to 922 MPa at a true strain of 0.4, equal to ~60% of the flow stress.In comparison, twins contribute to 118 MPa at the same true strain, equal to ~8% of the flow stress. In other words, twins only have minor effect on the flow stress, in contrast to the current understandings in the literature.

Key words TWIP steel, twin, dislocation, grain boundary, solid solution, synchrotron X-ray diffraction

一种 Fe-Mn-Al-C 低密度高强钢的组织性能研究

赵　超　宋仁伯　张磊峰　杨富强

（北京科技大学材料科学与工程学院，北京　100083）

摘　要　本文对一种成分为 Fe-10.4Mn-9.7Al-0.7C(wt.%)的 Fe-Mn-Al-C 系低密度高强钢的热轧钢板及其在 800°C~1000°C 温度范围热处理后的力学性能和微观组织进行了研究。其密度约为 6.8g/cm^3，比传统钢低约 13%。研究结果表明，本文所研究的钢热轧后组织由铁素体、奥氏体以及 κ 碳化物三种物相组成，并且组织呈现明显的带状，热轧板的抗拉强度为 1106MPa，规定塑性屈服强度为 1008MPa，断后伸长率为 11.0%。热轧板在 800°C 保温 1h 再水淬后组织中 κ 碳化物增多，塑性降低，而 900°C 及 1000°C 保温 1h 再水淬后 κ 碳化物几乎完全溶解，塑性有所改善。热轧板在 900°C 下保温 1h 再水淬后抗拉强度为 855MPa，断后伸长率达到 36.6%，强塑积达到 31GPa·%，达到了最佳的强度和塑性组合。当保温温度升高到 1000°C 后，组织明显粗化，塑性降低。

关键词　低密度，高强钢, Fe-Mn-Al-C 钢，热处理

Microstructure and Mechanical Properties of a Low-density High-strength Fe-Mn-Al-C Steel

Zhao Chao, Song Renbo, Zhang Leifeng, Yang Fuqiang

(University of Science and Technology Beijing, Beijing 100083, China)

Abstract　The mechanical properties and microstructure evolution of a hot-rolled Fe-Mn-Al-C steel with chemical composition Fe-10.4Mn-9.7Al-0.7C(wt.%) before and after heat treated at 800°C~1000°C were investigated. The density of the present steels is 6.8 g/cm^3, which is about 13% lower than traditional steels. Results show that, the hot-rolled steel has a banded structure, and consists of ferrite, austenite and κ carbide. The ultimate tensile strength of the hot-rolled steel is 1106 MPa, the yield strength is 1008 MPa, and the elongation to failure is 11.0%. After held at 800°C for 1 h followed by water quenching, κ carbide increases, and ductility decreases, while after held at 900°C or 1000°C, all the κ carbide is almost dissolved, and ductility is improved. The best combination of strength and ductility is obtained when the steel is held at 900°C for 1 h followed by water quenching, where the ultimate strength is 855 MPa, the elongation to failure is 36.6%, and the product of strength and ductility reaches 31 GPa·%. When the holding temperature is raised to 1000°C, the microstructure coarsened apparently, and the ductility decreases.

Key words　low density, high-strength steel, Fe-Mn-Al-C steel, heat treatment

Controlling of Abnormal Austenite Grain Growth in Banded Ferrite/Pearlite Steel by Cold Deformation

Xiangguang Zhang[1], Kiyotaka Matsuura[2], Munekazu Ohno[2], Goro Miyamoto[3], Tadashi Furuhara[3]

(1. Formerly Graduate Student, Hokkaido University. Now at Institute for Materials Research,

Tohoku University, Sendai, Japan; 2. Division of Materials Science and Engineering,
Faculty of Engineering, Hokkaido University, Sapporo, Japan;
3. Institute for Materials Research, Tohoku University, Sendai, Japan)

Abstract Inhibition of austenite grain growth above the eutectoid temperature in steels is of great importance in the steel production processes. The occurrence of abnormal grain growth of austenite during heating process in steels is a serious problem because the mechanical properties of steels are degraded due to duplex distribution or extremely coarse austenite grain structures caused by the abnormal grain growth. In the authors' recent work, it was found that the grain-coarsening temperature, T_c, (the temperature above which abnormal grain growth occurs) is much lower in a ferrite/pearlite (F/P) banded 0.2 pct carbon steel than in a non-banded steel although annealing at high temperatures is needed to eliminate the F/P banded structure. Instead, in this study we focus on the effects of cold deformation on the abnormal grain growth of austenite. Here it is shown that the abnormal grain growth that occurred in the F/P banded steel at low temperatures compared with non-banded steel can be completely inhibited by applying cold deformation before austenitizing. These should be attributed to the uniform distribution of AlN precipitates caused by the cold deformation. This provides a guideline for the industry to control the abnormal austenite grain growth effectively.

Key words austenite, banded structure, abnormal grain growth, cold deformation, precipitate

Bauschinger Effect and its Influence on Residual Stress of a Twinning Induced Plasticity Steel

Luo Z. C.[1,2], Huang M. X.[1,2]

(1. Shenzhen Institute of Research and Innovation, University of Hong Kong, Shenzhen, China;
2. Department of Mechanical Engineering, University of Hong Kong, Hong Kong, China)

Abstract The Bauschinger effect (BE) of a twinning induced plasticity (TWIP) steel sheet was measured by shear plus reverse shear experiment. Digital image correlation (DIC) technique was used to get the corresponding stress-strain curve and then the exact yield stress was determined. Kinematic hardening of TWIP steel was about one-third of total hardening. Further, the backstress contributes about 13 percent to the total flow stress and prone to saturate with increasing strain. The effect of BE on residual stress was evaluated by the commercial finite element analysis software (ABAQUS) by using the case of cup drawing. The results showed that residual stress after forming was smaller when taking BE into consideration.

Key words bauschinger effect, TWIP steel, cup drawing, digital image correlation

不同应变速率下高强 IF 钢的力学性能特征研究

赵征志[1,2] 梁江涛[1] 闫 远[2] 郗洪雷[2] 唐 荻[1,2] 苏 岚[1,2]

（1. 北京科技大学钢铁共性技术协同创新中心，北京 100083；
2. 北京科技大学冶金工程研究院，北京 100083）

摘 要 针对汽车用高强无间隙原子钢（IF 钢），采用德国 Zwick HTM 16020 高速拉伸试验机对实验钢在在应变速率为 $10s^{-1}$、$100s^{-1}$、$400s^{-1}$、$800s^{-1}$ 下进行高速拉伸实验，得到应力应变曲线，对其变形行为进行了分析。结果表明：与准静态拉伸相比，高速拉伸下 170P1 钢的强度、塑性等力学性能均发生明显的变化，在高速拉伸下屈服强度和抗拉强度均大于准静态时。随应变速率的增大实验钢的断后延伸率先增大后减小，在应变速率 $400s^{-1}$ 时达到最大值62%，而均匀延伸率一直处于比较稳定的范围，在应变速率为100s^{-1}时达到最大值33%。170P1实验钢在高速拉伸下断口呈现出明显的塑性断裂特征，宏观上表现出明显的剪切唇区、放射区、纤维区三区特征。微观上表现出等轴韧窝和拉长韧窝的特征，随应变速率的增大，韧窝的形状和大小都有变化。可见，170P1钢各宏观力学性能在动态载荷下的应变速率敏感性，是高应变速率变形过程中微观机制发生改变的结果。

关键词 高强IF钢，动态载荷，应变速率，变形行为

Mechanical Behaviour Research of High Strength IF Steel under Different Strain Rate Tensile Loading

Zhao Zhengzhi[1,2], Liang Jiangtao[1], Yan Yuan[2], Xi Honglei[2], Tang Di[1,2], Su Lan[1,2]

（1. Collaborative Innovation Center of Steel Technology, University of Science and Technology Beijing, Beijing 100083, China；2. Engineering Research Institute, University of Science and Technology Beijing, Beijing 100083, China）

Abstract High-strength IF steel was studied. The strain-stress curves were measured using Zwick HTM 16020 high-speed material tensile mechine at strain rates of 10s^{-1}、100 s^{-1}、400 s^{-1} and 800 s^{-1}.The results indicate that compared with the quasi static stretching, plasticity and other mechanical properties of 170P1 steel under high-speed tensile strength are obvious changes, and at high speed tensiler the yield strength and tensile strength are greater than the quasi static. Along with the increase of the strain rate of experiment steel, the break extension first increases then decreases. The peak reaches 62% in the strain rate of 400 s^{-1}, and the range of uniform elongation has been in a relatively stable, maximum of which reaches 33% in the strain rate of 100 s^{-1}. 170P1 experimental steel under high tensile fracture shows the characteristics of clear plastic fracture. Shear lip area, radiation area and fiber area were studied from the macro and the characteristics of shaft toughening nest and spin toughening nest were studied from the micro. The shape and size of the nests changed with the increase of strain rate. It is thus clear that the change of high strain rate deformation mechanisms is the reason for strain rate sensitivity under macro mechanical properties under dynamic load of 170P1 steel.

Key words high strength IF steel, dynamic load, strain rate, deformation behavior

50CrVA淬硬性偏低原因分析

罗贻正　祝小冬

（方大特钢科技股份有限公司，江西南昌　330012）

摘　要 通过使用同批次原材料在相同工艺温度条件下，使用不同淬火介质做淬火实验对比，结合GB/T 1222—2007《弹簧钢》标准中规定的实验室条件下的淬火结果，分析了导致50CrVA弹簧钢淬硬性偏低的原因，从而确定导致此批材料淬火后硬度偏低是由于淬火介质冷却能力不足所致。

关键词 50CrVA，淬火介质，硬度偏低

Analysis of the Low Hardness of 50CrVA

Luo Yizheng, Zhu Xiaodong

(Fangda Special Steel Technology Co., Ltd., Nanchang Jiangxi 330012, China)

Abstract By use the same batch of raw materials in the same process temperature conditions, adopt of different quenchant to compare the result of quenching test, refer to the laboratory conditions quenching result of according to the GB/T 1222—2007 《Spring Steel》 standard, analysis of the reason which caused to 50CrVA spring steel has low hardenability, and determine the reason of this batch of material from the lower hardness after quenching is due to inadequate cooling power of quenchants.

Key words 50CrVA, quenchant, low hardness

线材表面氧化铁皮柔性化控制

王宁涛[1,3] 陈继林[1,3] 翟进坡[1,2,3] 刘 超[1,3] 刘振民[1,2,3]

（1. 邢台钢铁有限责任公司技术中心，河北邢台 054027； 2. 北京科技大学材料科学与工程学院，北京 100083；3. 河北省线材工程技术研究中心，河北邢台 054027）

摘 要 为改善线材表面氧化铁皮质量，结合实际轧制生产，利用金相显微 OM、SEM 分析线材表面氧化铁皮形态及物相，研究了冷却速率对氧化铁皮厚度及组分的影响，结果表明：冷速为 0.7℃/s，氧化铁皮厚且疏松，易脱落，冷速 1.6℃/s 时，氧化铁皮薄且致密，表面质量良好。总结分析了化学成分、吐丝温度及冷却速率对氧化铁皮的影响，并提出了线材表面氧化铁皮柔性化控制的应用。

关键词 线材，氧化铁皮，冷却速率，微观组织，柔性化

The Flexible Control of Scale on the Wire Surface

Wang Ningtao[1,3], Chen Jilin[1,3], Zhai Jinpo[1,2,3], Liu Chao[1,3], Liu Zhenmin[1,2,3]

(1. Xingtai Iron and Steel Corp., Ltd. Technical Center, Xingtai Hebei 054027, China;
2. Department of Materials University of Science and Technology Beijing, Beijing 100083, China;
3.Hebei Engineering Research Center for Wire Rod, Xingtai Hebei 054027, China)

Abstract The influence of cooling rate on the scale thickness was studied by OM, SEM to analysis the morphology and phase of scale on wire surface for improving the quality of the wire surface. Results showed that:the scale was Thick and loose which easy to fall off when the cooling speed is 0.7 ℃/s,the scale was thin and dense which has good surface quality when the cooling speed is 1.6 ℃/s.the influence of scale by the chemical composition, spinning temperature and cooling rate was Analysed,and put forward the flexible control of scale on the wire surface.

Key words wire, scale, cooling rate, microstructure, flexible

Relationship between Hardness and Ferrite Morphology in VC Interphase Precipitation Strengthened Low Carbon Steels

Zhang Yongjie[1], Miyamoto Goro[2], Shinbo Kunio[2], Furuhara Tadashi[2]

(1. Department of Metallurgy, Tohoku University, Sendai, Japan;

2. Institute for Materials Research, Tohoku University, Sendai, Japan)

Abstract　Interphase precipitation is the phenomenon that when steels containing strong carbide-forming elements are transformed from austenite into ferrite, alloy carbides are precipitated in sheets as a result of periodic nucleation at migrating interphase boundary. Nano-sized carbides formed through interphase precipitation are used to strengthen low carbon steels due to its large amount of precipitation strengthening. The effects of transformation temperatures and alloy elements on the hardness of ferrite strengthened by VC interphase precipitation in low carbon steels were investigated in the present study. It is found that when the transformation temperature becomes lower, the transformed microstructure changes from equi-axed grain boundary ferrite into acicular Widmanstatten or bainitic ferrite. Such transition also corresponds to the decrease in hardness at lower transformation temperatures due to the absence of VC interphase precipitation in acicular ferrite as confirmed by three-dimensional atom probe analyses. On the other hand, by lowering the Mn content, the formation of acicular ferrite is promoted while Si addition does not have large influence on the ferrite morphology. In contrast, higher N content suppress the formation of acicular ferrite, especially at lower transformation temperature at which intra-granular ferrite is extensively formed instead. The hardness in different alloys can be well understood qualitatively by considering the morphology of ferrite, indicating that the control of ferrite morphology is extremely important to obtain better mechanical properties with interphase precipitation for industrial use.

Key words　interphase precipitation, hardness, ferrite morphology, alloying element

Lightweighting: Impact on Grade Development and Market Strategy

Nauzin Jean-Paul

(FIVES KEODS)

Abstract　Lightweighting implies to take into account needs of the carmaker such as crash behavior and stiffness. These two main functions lead to conclude that developing grades for structural part (with higher Tensile strength) is not the only way. The replacement of steel used for the big inner and outer panels is mandatory if steelmakers don not want to lose this market because of the efficiency of aluminium.

Key words　lightweighting, flat carbon steel, car design, market strategy

带状组织对 20CrMoH 齿轮钢淬火膨胀影响的数值模拟

杨　超　安金敏

（宝山钢铁股份有限公司研究院，上海　201900）

摘　要　运用热膨胀相变仪对 20CrMoH 齿轮钢试样的淬火膨胀性能进行测定，研究了带状组织对钢各向异性的影响，通过 Deform 软件并考虑材料的各向异性参数来模拟带状组织对齿轮钢淬火性能的影响。结果表明：含有带状组织的 20CrMoH 钢淬火时膨胀量呈现明显各向异性，且垂直于轧制方向的膨胀量明显大于平行轧制方向，针对 ϕ4 mm×10 mm 试样横向和纵向膨胀量相差 11μm,而有限元模拟表明齿轮钢淬火后横纵向试样膨胀量相差 13μm，实测值和模拟值比较吻合。

关键词　带状组织，齿轮钢，各向异性，数值模拟

Influence of Banded Microstructure on Quenching Dilatometry in 20CrMoH Gear Steel and Numerical Simulation

Yang Chao, An Jinmin

(Research Institute, Baoshan Iron & Steel Co., Ltd., Shanghai 201900, China)

Abstract Dimensional anisotropy associated with banded structure of 20CrMoH gear steel was investigated via thermal dilatometer and the influence of banded structure on quenching anisotropy of gear steel was validated by means of numerical simulation considering anisotropy parameters of steel with deform software. Results showed that 20CrMoH gear steel with banded structure display different dilatational characteristics along the rolling direction (RD) and the transverse direction (TD)during quenching.Besides,transverse expansion is significantly greater than that of longitudinal. The dilatometric difference of longitudinal and transverse specimens for cylinder of 4mm diameter by 10mm length is 11μm,whereas the value via the finite element simulation is 13μm, the test results were basically consistent with the simulated results.

Key words banded structure, gear steel, anisotropy, numerical simulation

12.9 级高强度紧固件用钢的调质工艺研究

赵浩洋[1] 翟瑞银[1,2] 金 峰[1]

(1. 宝山钢铁股份有限公司，上海 201900;
2. 汽车用钢开发与应用技术国家重点实验室，上海 201900)

摘 要 汽车行业的发展对紧固件提出了更高的要求，高强度紧固件已成为重要的发展方向。本文研究了高强度紧固件用钢 SCM440 调质处理工艺。通过常温拉伸试验，检测不同调质工艺处理后材料的力学性能，并利用金相显微镜、扫描电子显微镜观察材料的组织，分析组织对性能的影响规律。结果表明，SCM440 钢生产 12.9 级紧固件，合适的调质工艺为 850℃保温 80min 淬火，480℃保温 3h 回火。调质处理后 SCM440 的屈服强度和抗拉强度分别为 1163MPa、1242MPa，断面收缩率为 55%，综合力学性能较好，达到了 12.9 级紧固件的要求。研究表明，SCM440 调质处理时淬火温度过高，容易导致晶粒粗化，屈强比降低；回火温度越高，碳化物越容易聚集长大，强度越低。

关键词 SCM440，调质工艺，显微组织，力学性能

Quenching and Tempering of Steel for 12.9-Grade High Strength Fasteners

Zhao Haoyang[1], Zhai Ruiyin[1,2], Jin Feng[1]

(1. Baoshan Iron & Steel Co. ,Ltd., Shanghai 201900, China;
2.State Key Laboratory of Development and Application Technology of Automotive Steels, Shanghai 201900, China)

Abstract With the improvement of auto industry, fasteners should satisfy much higher requirements. The development of high strength fasteners is one of the most important directions currently. In this paper, quenching and tempering of SCM440 was investigated. The effect of heating processing (quenching temperature and tempering temperature) on the mechanical was researched by tensile test. Microstructures were observed by optical microscopy and scanning electron microscopy. The effects of microstructure on mechanical properties was analyzed. The results show that the heat treatment process was optimized as quenching after holding at 850℃ for 80 minutes and then tempering at 480℃ for 3 hours. The yield strength, tensile strength and reduction of area can be as high as 1163MPa, 1242MPa and 55%, respectively. And the better comprehensive mechanical properties could be obtained, meeting the standard of 12.9-Grade fasteners. The increasing quenching temperature can lead to coarsening of grain and decrease of yield ratio.With the increase of tempering temperature, carbides have the tendency to aggregation and coarsening, resulting in the decrease of strength.

Key words　SCM440, quenching and tempering, microstructure, mechanical properties

分段冷却工艺对55SiCr弹簧钢脱碳和组织的影响

金　峰　姚　赞

（宝山钢铁股份有限公司研究院，上海　201999）

摘　要　本文利用 Gleeble1500 热模拟试验机模拟了汽车悬架用 55SiCr 弹簧钢在轧后分段冷却条件下的脱碳和最终冷却组织。利用金相显微镜、显微硬度计、扫描电子显微镜等对试验结果进行分析，研究不同冷却条件对弹簧钢脱碳和显微组织的影响规律。结果表明，轧制结束后的 55SiCr 弹簧钢在高温段停留的时间决定了其表面脱碳的深度，随着高温段降温速度的提高，试样表面脱碳的程度减小，当冷却速度达到 1℃/s 时，降温过程不再产生全脱碳；当降温速度达到 5℃/s 时，为保证得到力学性能优良的索氏体组织，快速降温终点应控制在 650℃以上；当盘条的降温速度超过 5℃/s 后，组织控制相对困难，即使冷却后期采用保温措施也容易在最终组织中出现马氏体，导致盘条无法进行后续加工。所以，在 55SiCr 弹簧钢盘条的轧后冷却中，应先快速冷却至 650℃后采取保温措施使相变完成，可以得到无全脱碳、基体为索氏体的良好显微组织。

关键词　55SiCr 弹簧钢，全脱碳，显微组织，热模拟，分段冷却

The Effect of Step-Cooling Process on Decarburization and Structures of 55SiCr Spring Steel

Jin Feng, Yao Zan

(Research Institute, Baoshan Iron & Steel Co., Ltd., Shanghai 201999, China)

Abstract　The ferrite decarburization behavior and microstructure of 55SiCr spring steel at room temperature for automotive suspensions was investigated using Gleeble1500 thermal-mechanical simulator in this paper. The effect of cooling condition after hot rolling on the decarburization and microstructure was studied by optical microscope, scanning electron microscopy and microhardness tester. The results showed that the decarburization layer depth depended on the holding time at high temperature. The cooling rate at high temperature increased, the degree of decarburization decreased. When the cooling rate was larger than 1℃/s, the complete decarburization would not occur. When the cooling rate at high temperature was equal to 5℃/s and terminal temperature of fast cooling was higher than 650℃, a sorbite microstructure with good mechanical property would be obtained. When the cooling rate was larger than 5℃/s, the martensite microstructure would be form which was harm for subsequent processing performance. Therefore, the cooling process of

55SiCr spring steel showed be: cooling fast to 650℃ and then keep the temperature or slowing down the cooling rate as low as 0.5℃/s to obtain the sorbite microstructure.

Key words 55SiCr spring steel, decarburization, microstructure, thermal-mechanical simulator, step-cooling process

中厚板精密冲裁机理及工艺优化研究进展

赵 震 朱圣法 庄新村

（上海交通大学塑性成形技术与装备研究院模具CAD国家工程研究中心，上海 200030）

摘 要 精密冲裁是在普通冲裁基础上发展起来的一种先进的精密塑性成形工艺。精冲过程中，材料在三向压应力作用下变形，极大改善了材料的塑性变形能力，最终可得到尺寸精度高且质量稳定的优质零部件。聚焦于该工艺复杂的成形机理，本文围绕精冲材料的成形性能分析、成形过程韧性断裂预测、成形工艺优化等关键技术，介绍了上海交通大学在中厚板精密冲裁领域的部分研究进展。

关键词 精密冲裁，成形性能分析，韧性断裂，工艺优化

The Research Advances of Mechanism and Process Optimization for Medium-thick Sheet Metal Fine-blanking

Zhao Zhen, Zhu Shengfa, Zhuang Xincun

(National Engineering Research Center of Die & Mold, Institute of Forming Technology and Equipment, Shanghai Jiaotong University, Shanghai 200030, China)

Abstract Fine-blanking is an advanced metal forming process derived from conventional blanking process. During theprocess, the material deforms under triaxial compressive stress state, which greatly improves the material plastic deformation capacity, and ultimately products manufactured by fine-blanking process have higher dimensional accuracy and stable quality. Focusing on the forming mechanism of fine-blanking, a comprehensive review for research advances of medium-thick sheet metal fine-blanking processin Shanghai Jiao Tong University is presented, including the formabilityanalysis of thematerials, ductile fractureprediction as well as the process optimization.

Key words fine-blanking, formabilityanalysis, ductile fracture, process optimization

Twinning Mechanism in TWIP Steel: From Single Crystalline Micro-pillar to Bulk Sample

Huang M.X., Liang Z.Y.

(Department of Mechanical Engineering, The University of Hong Kong, Pokfulam Road, Hong Kong, China)

Abstract Twinning-induced plasticity (TWIP) steels have excellent combination of strength (1000 MPa) and ductility (60%) and are potential lightweight materials for automotive applications. The excellent mechanical properties of

TWIP steels are due to the formation of intensive deformation twins during deformation. Understanding the twinning mechanisms in TWIP steels is essential for the successful application of TWIP steels in automotive industry. The first part of this work is to employ sub-micon and micron-sized single crystalline pillars to investigate the nucleation and growth mechanism of deformation twins. It is found that the nucleation and growth of deformation twins are due to emission and glide of successive partial dislocations. The twin thickness can range from nanometres to micrometres. A physical model is proposed to simulate the nucleation and growth of deformation twins and the model predictions agree well with experimental observations. The second part of this work investigates the effect of high strain rates on the deformation mechanism of bulk samples. By synchrotron X-ray diffraction experiments, the present work demonstrates that a higher strain rate leads to a lower dislocation density and a lower twinning probability. This unexpected suppression of dislocation and twin evolution have been attributed to the temperature increase due to dissipative heating at high strain rate deformation. A physically-based model has been proposed to simulate the evolution of dislocation and twin densities and to model the stress-strain relation. The modelling results agree well experimental findings.

抗拉强度 980MPa 汽车用超高强钢延迟断裂行为研究

周庆军　黄　发　王　利

（宝山钢铁股份有限公司，上海　201900）

摘　要　超高强钢具有较高的氢脆敏感性，延迟断裂是其最常见的失效形式之一。随着超高强钢在汽车上的应用不断增多，延迟断裂成为钢厂及车厂共同关注的材料性能之一。本文以 980 MS、980 DP 和 980 QP 三种抗拉强度为 980 MPa 的冷轧超高强汽车用钢为对象，采用 U 弯、慢应变速率拉伸和恒载荷试验方法研究了三种不同组织的超高强钢延迟断裂行为，并探讨了钢中氢含量与其开裂行为之间的关系。

关键词　超高强钢，汽车用钢，延迟断裂，氢

Experimental Study on Delayed Fracture of TS 980MPa Grade Steels for Automotive Applications

Zhou Qingjun, Huang Fa, Wang Li

(Baoshan Iron & Steel Co., Ltd., Shanghai 201900, China)

Abstract　Baosteel is the main auto sheet producer in China. In recent years, great progress has been made on the research and development of ultra high strength steel (UHSS). As is known, ultra high strength steels are susceptible to hydrogen embrittlement, delayed fracture is the most common form of hydrogen induced failure. In this paper, three kinds of TS 980MPa grade ultra high strength steel for automotive applications, i.e.980MS, 980DP and 980QP steel were selected, the delayed fracture resistance was evaluated using U-bending test, slow strain rate test and constant load tensile test. And the relationship between hydrogen content and cracking were studied and discussed.

Key words　ultra high strength steel, auto sheet, delayed fracture, hydrogen

An Investigation on the Tensile Property of 10 pct Mn TRIP/TWIP Dual Phase Steels

He Binbin[1], Luo H.W.[2], Huang Mingxin[1]

(1. Department of Mechanical Engineering, The University of Hong Kong, Hong Kong, China;
2. State Key Laboratory of Advanced Metallurgy & School of Metallurgical and Ecological Engineering, University of Science and Technology Beijing, Beijing 100083, China)

Abstract The mechanical behavior of a 10 pct Mn dual phase steel which belongs to the 3rd-generation AHSS for automotive application is investigated. The 10 pct Mn steels were intercritically annealed from the different amount of partial martensite microstructures. The tensile property for the 10 pct Mn steel with larger amount of initial martensite is significantly better than the counterpart with lower one. The X-ray diffraction (XRD) results show that the austenite volume fraction decreased significantly both in small and large strains for the 10 pct Mn steel with larger amount of initial martensite, confirming the operation of transformation induced plasticity (TRIP) effect in 10 pct Mn steel during tensile test. The significant decrease of austenite volume fraction at small and large strain may be ascribed to the sharp different mechanical stability of the large and small austenite in 10 pct Mn steel. Interestingly, the austenite volume fraction remains still at the medium strain. The transmission electron microscopy (TEM) observation suggests an operation of twinning-induced plasticity (TWIP) effect. Therefore, it is proposed that the TWIP effect may dominate at the medium strain to link the retained austenite grains with different mechanical stability, resulting in an excellent combination of strength and ductility of the 10 pct Mn steel with larger amount of initial martensite.

Key words 3rd-generation AHSS, medium Mn steel, stability of austenite

铌微合金化 Q&P 钢组织与力学性能研究

吝章国[1] 熊自柳[1] 刘宏强[1] 张 琳[2] 李守华[3] 孙 力[2]

(1. 河北钢铁技术研究总院，石家庄 052165；
2. 唐钢公司技术中心，河北唐山 063016；
3. 邯钢公司技术中心，河北邯郸 056000)

摘 要 主要采用连续退火热模拟机模拟 Q&P 工艺过程，对 Nb 微合金化 Q&P 钢组织与性能进行了分析与研究。结果表明：加入铌元素后，Q&P 钢热处理后原始奥氏体晶界不明显，马氏体簇板条间距明显变小，钢中奥氏体尺寸细小弥散，奥氏体体积含量和奥氏体含碳量分别达到了 7.6%和 1.42%，强塑积 27280 MPa·%；促进铁素体析出而抑制珠光体析出，淬火时易于形成少量细小的铁素体，向奥氏体排碳，增加奥氏体的稳定性，利于在马氏体板条间获得细小的薄膜状残余奥氏体；在配分过程中铌元素抑制渗碳体析出，稳定残余奥氏体，随着配分时间增加，含铌钢 Q&P 钢抗拉强度和伸长率先增加后降低，配分时间 500s 时，抗拉强度达到最大值，配分时间 700s 时伸长率达到最大值。

关键词 Q&P 钢，铌微合金化，奥氏体，组织，性能

Research on Microstructure and Mechanical Property of Niobium Micro Alloying Q&P Steel

Lin Zhangguo [1], Xiong Ziliu [1], Liu Hongqiang[1], Zhang Lin [2], Li Shouhua[3], Sun Li [2]

（1. Hebei Iron and Steel Technology Research Institute, Shijiazhuang 052165, China ;
2. Technology Center of Tangsteel Iron and Steel Co., Ltd., Tangshan 063016, China;
3. Technology Center of Handan Iron and Steel Co., Ltd., Handan 056000, China）

Abstract The microstructures and mechanical property of Niobium micro alloying Q&P steel were anysised and resached through simulating the process of Q&P by Continue annealing simulator machine.The result showed that original austenite grain boundary of Q&P steel were not obvious, gap of martensite clusters became fine, grain size of austenite was small and disperse, volum percentage and carbon content of austenite were 7.6% and 1.42% respectivly, and tension strength by elongation reached 27280 MPa·%, after adding Niobium. The niobium elements boosted ferrite transformation and depressed cementite precipition, were good for getting thin film austenite in quenching, and delayed the time of cementite precipitaion and defer the reply of dislocation in martensite. With time of cementite precipitation increasing, Tension strength and elongation of Q&P steel with Nb increase and then decrease, and tension strength achieved max at 500s, and elongation achieved max at 700s.

Key words Q&P steel, Niobium micro alloying, austenite, microstructure, mechanical property

材料晶粒尺寸及变形率对钢板加工变形后表面粗糙度的影响

班必俊

（宝山钢铁股份有限公司，上海　200940）

摘　要　用于冲压、拉伸用的冷轧钢板，在加工后钢板容易产生表面粗糙化现象，影响加工件的外观质量，为了研究导致钢板加工变形后表面粗糙化的机理，本文注重研究了钢板材料晶粒度及钢板拉伸变形率对材料表面粗糙化的影响，通过采用冲压、拉伸模拟试验研究基本摸清了晶粒尺寸及拉伸变形率和钢板表面粗糙化的关系。通过冲压模拟试验得出了罐体表面粗糙度主要受钢板晶粒度影响较大，晶粒越大，冲压后罐体表面就越粗糙；通过拉伸模拟试验得出了拉伸变形率对变形后的钢板粗糙程度有较大影响，拉伸变形率越大拉伸后的表面越粗糙；一定量电镀能够降低罐体粗糙程度，但不能消除表面凹凸，特别是粗糙罐体表面更难消除凹凸不平。

关键词　晶粒尺寸，粗糙化，变形率

The Effect of Grain Size and Deformation Rate on Surface Roughness after Steel Sheet Forming

Ban Bijun

(Baoshan Iron & Steel Corporation Ltd., Shanghai 200940, China)

Abstract Cold rolled steel for stamping and drawing is easy to cause surface roughening after processing the steel plate, which will affect the appearance quality of processing parts. In order to study the mechanism of steel sheet surface roughening after forming, this paper focuses on the research of the effects of material grain size and steel tensile deformation rate on the steel sheet surface roughening. The experimental study found out the relationship between material grain size and surface roughening by stamping and stretching simulation. As a result, surface roughness is increased as grain size increases by stamping simulation; Tensile deformation rate increases surface roughness obviously by tensile test; Can body surface roughness can be reduced by a certain amount plating but not be eliminated surface uneven, especially the rough surface of the can body which is difficult to be eliminated uneven.

Key words grain size, surface roughening, deformation rate

热镀锌钢板几种使用开裂模式浅析

马雪丹 郑建平

（宝钢股份制造管理部，上海 201900）

摘 要 本文研究了几种热镀锌钢板在使用中出现的不同开裂模式，包括翻边开裂、折弯开裂、疲劳开裂和颈缩开裂。因基板材质的差异，开裂的机理各有不同。翻边开裂一般出现在高强钢结构件的加工过程中，提高材料的扩孔性能是解决问题的关键，从材质机理上讲就是要获得较细小的晶粒和均匀的组织，减少带状组织，并且尽量降低以硫化物为代表的夹杂物成分，从而提高材料扩孔率。当材料需要180度折弯时，采用高延伸率的冲压用钢可能会出现开裂情况，而晶粒较细小、碳化物分散的CQ钢才是既经济又好用的选择。热镀锌钢板疲劳开裂不仅仅是材料性能本身的问题，还与整机零件设计、材料的服役条件、使用环境息息相关。锌铁合金钢板表面镀层的铁含量往往会成为钢板开裂的影响因素。

关键词 热镀锌钢板，开裂，翻边，折弯，疲劳，颈缩

Galvanized Steel Sheet Used in Several Modes of Cracking

Ma Xuedan, Zheng Jianping

(Department of Manufacturing and Management, Baosteel Iron & Steel Co., Ltd., Shanghai 201900, China)

Abstract In this paper, several different cracking modes of hot-dip galvanized steel sheet are talked about, including flange cracking, bending cracking , fatigue cracking and necking cracking. Due to differences in the substrate material, the mechanism of cracking is different. Flanging cracks usually appear in high-strength steel structure of the process, improve reaming performance materials are the key to solving the problem, from the material terms of the mechanism is to obtain finer grain and uniform structure, reducing the band structure, and to minimize the sulfide inclusions components represented thereby improving hole expansion rate material. When the material have a 180-degree bend finer grain size, carbide dispersion of CQ steel is economical and easy to use options. The surface quality of galvannealed sheet would be a factor when necking cracking .

Key words galvanized steel sheet,cracking, flange, bend, fatigue, necking

宁钢 510L 热轧汽车板表面质量控制

佟　岗　罗石念　唐小勇

（宁波钢铁有限公司，浙江宁波　315807）

摘　要　宁波钢铁公司在开发 510MPa 级热轧汽车大梁钢 510L 的过程中，根据用户使用质量要求，对 510L 带钢表面氧化铁皮在粗轧除鳞及精轧、层流冷却等生产工序中的演变行为进行了研究，找出了带钢表面形成红色氧化铁皮的内在原因。在钢板的成分控制方面，针对硅元素含量进行了减量设计，同时适量提锰以弥补钢板强度损失。此外，为了进一步消除 510L 带钢的红色氧化铁皮缺陷，进行了调整轧制工艺的工业性试验，得出板坯加热制度、粗轧除鳞、开轧及终轧温度、卷取工艺对 510L 带钢红铁皮的产生亦具有重要影响。通过优化热轧工艺参数，粗轧采用高除鳞温度、精轧高温终轧、降低卷取温度，使钢板中 FeO 的共析反应得到控制，从而有效抑制 510L 红色氧化铁皮生长，带钢表面红铁皮发生率大幅降低。

关键词　汽车板，氧化铁皮，硅含量，热轧工艺

Control of Surface Quality of 510L Hot Rolled Sheet in Ning-Steel

Tong Gang, Luo Shinian, Tang Xiaoyong

(Ningbo Iron & Steel Co., Ltd., Ningbo Zhejiang 315807, China)

Abstract　During the process of 510MPa hot-rolled auto sheet in Ningbo Iron & Steel Corporation, the structure, evolution rules of oxide scales have been conducted. The evolution of red scale on hot band during descaling and hot rolling were analyzed and the formation mechanism of red scale was described. The silicon content was reduced, in the mean time, the Mn content was increased to maintain the strength. The trial hot-rolling indicated that the slab reheating temperature, descaling process, finish rolling temperature and coiling temperature played an important role in the formation of red scales and the optimized hot rolling parameters were proposed. After adopting the technology of descaling at higher temperature and coiling at lower temperature, The eutectoid reaction kinetics from FeO can be hindered, as a result, the red scale on 510L sheet was effectively removed and the surface quality was greatly enhanced.

Key words　automobile sheet, oxide scale, silicon content, hot rolling parameters

连退带钢速度对 600MPa 级冷轧双相钢组织及力学性能的影响

詹　华[1]　刘永刚[1]　肖洋洋[1]　潘洪波[2]　程　凯[1]

（1. 马鞍山钢铁股份有限公司汽车板推进处，安徽马鞍山　243000；

2. 安徽工业大学工程研究院，安徽马鞍山　243002）

摘　要　结合马钢 2130mm 连续退火机组特点，利用 ULVAC-CCT-AY-II 型多功能连续退火模拟器模拟了 600MPa 级冷轧退火双相钢的生产过程。采用金相、彩色金相组织观察方法以及力学性能测定方法，研究了在不同退火速率条件下实验用双相钢的组织形貌和力学性能特点。结果表明，随着连退带钢速度的增加，马氏体含量逐渐增加，马氏体分布逐渐变得弥散细小，实验钢抗拉强度略有增加，屈服强度和屈强比均为先增加后降低的趋势；随着连退带钢速度由 80 增加到 137m/min，实验钢伸长率逐渐降低，当带钢速度超过 137m/min 后，伸长率迅速增加；n 值随着带钢速度的增加呈逐渐增加的趋势，当带钢速度超过 100m/min 后增加趋势更加明显。当带钢速度超过 100m/min 后实验钢屈强比降低、n 值增加，实验钢具有较好的综合力学性能。

关键词　双相钢，连续退火，带钢速度，组织性能，屈强比

Effect of Continuous Annealing Strip Speed on Microstructure and Properties of 600MPa Grade Cold-rolled Dual-phase Steel

Zhan Hua[1], Liu Yonggang[1], Xiao Yangyang[1], Pan Hongbo[2], Cheng Kai[1]

(1. Auto Sheet Strategic Business Unit of Ma'anshan Iron and Steel Company，Ma'anshan 243000, China;
2. Engineering Research Institute, Anhui University of Technology, Ma'anshan 243002, China)

Abstract　Based on the characteristics of 2130mm Continuous Annealing Line (CAL) in Masteel,the ULVAC-CCT-AY-II annealing simulator was used to simulate the production process of 600MPa grade dual phase (DP) steel. The characteristics of microstructure and mechanical properties in dual phase steel under different annealing strip speed was studied by using optical, color optical microscopy and mechanical test.The research results show that with the increase of thestrip speed,the martensite content increases andmartensite became distribution and small; the tensile strength increase slightly,yield strength and yield ratio were increased at first and then decrease. The elongation decreases with the strip speed increased from 80 to 137m/min and then increased while the strip speed exceed 137m/min.With the increasing ofstrip speed, n value increases especiallywhen the strip speed over 100m/min. The yield ratio decreases and n valueincreases when the strip speed over 100m/min, and test steel has good comprehensive mechanical properties.

Key words　dual-phase steel, continuous annealing, strip speed, microstructure and properties, yield ratio

超高强马氏体钢 TMCP-DP 工艺中碳配分有效性研究

巨　彪　武会宾　唐　荻　车英健　赵　娇

（北京科技大学高效轧制国家工程研究中心，北京　100083）

摘　要　利用 Gleeble-3500 热模拟试验机对提出的 TMCP-DP（热机械轧制-直接碳配分）工艺中 DP 过程进行了热模拟研究，以与一步法 Q&P 工艺对比碳配分的有效性。实验结果表明，随着 DP 过程冷速的逐渐降低，所得实验钢的残余奥氏体含量先增加后减少，与一步法 Q&P 同一配分温度下配分时间逐渐增加时所得结果规律一致，且在较高残余奥氏体含量时对应残余奥氏体相的碳含量也较高，证明 DP 过程中碳配分的发生。

关键词　超高强钢，碳配分，残余奥氏体

Effectiveness Research of Carbon Partitioning on TMCP-DP Processed Ultrahigh Strength Martensitic Steel

Ju Biao, Wu Huibin, Tang Di, Che Yingjian, Zhao Jiao

(National Engineering Research Center for Advanced Rolling, University of Science and Technology Beijing, Beijing 100083, China)

Abstract Gleeble-3500 was utilized to conduct the TMCP-DP (Thermo-mechanical rolling process-Direct partitioning) and one-step Q&P (Quenching &Partitioning) process.The results show that with the decrease of cooling rate in DP, the volume fraction of obtained retained austenite increased to higher value at first, and then decreased, and the change of which with the increase of partitioning time is similar in one-step Q&P. Higher carbon content of austenite was detected at higher retained austenite volume fraction after both processes, which proved the occurrence of carbon partitioning.

Key words ultrahigh strength steel, carbon partitioning, retained austenite

780 MPa 级热镀锌双相钢疲劳性能的研究

刘华赛[1]　邝　霜[1]　谢春乾[1]　滕华湘[1,2]

（1. 首钢技术研究院薄板研究所，北京　100043；
2. 北京科技大学材料科学与工程学院，北京　100083）

摘　要 采用拉-拉疲劳试验研究了连续热镀锌 780MPa 级冷轧双相钢的高周疲劳性能，绘制了其疲劳寿命-应力幅曲线，并采用扫描电镜对其疲劳断口和样品表面进行了观察，分析了其在高周疲劳情况下的疲劳断裂行为，同时采用透射电镜对其断裂后的断口附近微观结构进行了观察。结果表明，连续热镀锌冷轧双相钢 DP780 在应力比 R 为 0.1，加载频率为 15 Hz 的条件下的疲劳极限为 250 MPa；通过对疲劳断口和样品表面的观察发现疲劳裂纹萌生于铁素体和 MA 岛的界面处，并且微裂纹沿着界面扩展最终形成宏观裂纹；疲劳断口附近的微观组织观察显示在未开裂的铁素体与 MA 岛界面处塞积了大量的位错，由此可以判断铁素体中位错在界面处塞积并撞击铁素体/MA 岛界面是导致热镀锌 DP780 疲劳失效的主要原因。

关键词 双相钢，热镀锌 DP780，疲劳断裂，相界面

Fatigue Properties of 780 MPa Grade Hot-dip Galvanized Dual Phase Steel

Liu Huasai[1], Kuang Shuang[1], Xie Chunqian[1], Teng Huaxiang[1,2]

(1. Strip Technology Departments, Shougang Research Institute of Technology, Beijing 100043, China;
2. School of Materials Science and Engineering, University of Science and Technology Beijing, Beijing 100083, China)

Abstract 780 MPa grade continuous hot-dip galvanized dual phase steelwasstudied, and S-N curve of DP780 was drawn

using tension-tension fatigue tests.The fracture morphologies of DP780 were observed, and the fracture behavior was also analyzed using scanning electron microscopy. The microstructure of DP780 near the fatigue fracture surface were observed using transmission electron microscopy. The fatigue limit ofhot-dip galvanizedDP780is 250MPa with the loading frequency of 15 Hz at a loading ratio of 0.1. SEM surface analyses show that the crack initiation zonesare on the interface of ferrite and martensite-austenite islands of the steel, and the micro-crack propagates along the interface of ferrite and martensite-austenite islands, macroscopic crack finally formed at these interfaces. TEM observations near the fracture surface show that there are a lot of dislocations piled up at the interface of ferrite and martensite-austenite islands. It can be concluded that during cyclic loading processthe dislocations in ferrite impacted the interface of ferrite and martensite-austenite islands continuously, and finally micro cracks occurred at these interfaces.
Key words dual phase steel, hot-dip galvanized DP780, fatigue fracture, phase boundary

控轧控冷工艺在低锰（0.75%~0.85%）HRB400E 盘螺生产中的应用

马正洪　张玺成　钱　萍

（山西建邦集团轧钢厂，山西临汾　043400）

摘　要　利用高线控轧控冷技术轧制无微合金低锰（0.75%~0.85%）细晶高强度 HRB400E 钢筋，轧制温度区间为部分奥氏体未结晶区，阐述了在控轧控冷过程中温度与水冷控制对晶粒度尺寸的影响，轧后风冷对成品内部金相组织构成的影响，通过抽样检测，表明控制加热温度在 995℃左右，进精轧温度 800℃左右，此工艺下生产的无微合金低锰钢筋的晶粒尺寸已相当接近超细晶微合金化高强度钢筋的极值范围（5~10μm）。经过理论分析结合现场实际生产情况得到了无微合金低锰细晶高强度钢筋的最佳金相组织构成百分比，对于今后生产其他类型的特钢产品起到了很好的借鉴作用。此项轧制细晶高强度钢筋的控制控冷技术是建立在无微合金的基础上的，属于国内一流水平，同时每年可为公司创造近亿元的利润。
关键词　控轧控冷，无微合金化，细晶高强度，珠光体含量

The Application of Controlled Rolling and Cooling Process in Low Manganese (0.75%~0.85%) HRB400E Rebar Coil Production

Ma Zhenghong, Zhang Xicheng, Qian Ping

(Shanxi Jianbang Group , Linfen Shanxi 043400, China)

Abstract Using TMCP of high-wire rolling to roll HRB400E rebar coil with no micro-alloy and lower Mn content(0.75%~0.85%) , rolling temperature range is not part of the austenite crystalline region, Analysised the effection about heating temperature and water cooling on the grain size, cooling effect on the microstructure of finished structure after rolling, Through sampling and testing, when controlled the heating temperature about 995℃ and the temperature into NTM about 800℃, no micro-alloy grain size of the low-manganese steel has been quite close to the ultrafine-grained micro-alloyed high strength steel extreme range (5 ~ 10μm). And through theoretical analysis combined with the practical production obtained the best microstructure without micro alloy low manganese fine grained high strength steel structure the percentage, to a very good reference for the future production of other types of steel products. This technology is based on the non - alloy steel, which is a domestic first-class level, and can create tens of millions of dollars every year.

Key words TMCP, no micro alloying, fine grain and high strength, pearlite content

600MPa 级 Si、Mn 系热镀锌双相钢热处理工艺研究

崔磊 詹华 陈乐 马二清

(马钢股份汽车板推进处，安徽马鞍山 243000)

摘 要 本文介绍了一种 600MPa 级 Si-Mn 系热镀锌双相钢工业试制状况。采用金相、SEM 组织观察方法以及力学性能测定方法，研究了不同退火温度下生产线所得双相钢的组织形貌和性能特点，结果表明：热镀锌双相钢中易出现贝氏体组织，随着退火温度的提高，贝氏体含量、屈服强度和屈强比升高。工业试制得出，镀层表面质量及粘附性良好，厚带钢可在较低的带速下于 790℃退火得到综合力学性能良好的产品。

关键词 双相钢，热镀锌，组织形貌，力学性能，镀层

Study on the Heat Treatment of Si-Mn 600MPa Grade hot-dip Galvanized Dual-phase Steel

Cui Lei, Zhan Hua, Chen Le, Ma Erqing

(Auto Sheet Strategic Business Unit, Ma'anshan Iron & Steel Co., Ltd., Ma'anshan Anhui 243000, China)

Abstract An industrial experiment of development Si-Mn600MPa grade hot-dip galvanized dual-phase steel was introduced. Themicrostructure and mechanicalproperties of different critical annealing temperature was investigated by OM, SEM and mechanical property testing method. The results show that the absence of bainite is usual; both the bainite content, the ultimate tensile strength and yield ratio increase with the increasing of critical annealing temperature. The industrial experiment indicate that the surface quality and adhesion of coating is good; at a lower speed the thick strip can get good synthesis mechanical property in 790℃ critical annealing temperature.

Key words dual phase steel, hot-galvanize, microstructure, mechanical property, coating

热轧态组织对合金结构钢 42CrMo 球化率的影响

程吉浩[1] 刘文斌[1] 闵利刚[2] 郭 斌[1]

(1. 武钢研究院，湖北武汉 430081；2. 湖北大帆汽车零部件有限公司，湖北麻城 438300)

摘 要 本文通过调整热轧卷取温度制备了三卷不同热轧态组织的 42CrMo 钢，采用相同的工艺对三卷试验钢进行球化退火，研究了不同热轧态组织对合金结构钢 42CrMo 球化率的影响。结果表明：（1）卷取温度从 720℃下降到 580℃，42CrMo 钢热轧态组织依次为珠光体、伪珠光体、贝氏体组织，同时钢中先析出铁素体的量逐步减少。（2）原始组织越接近平衡态，球状碳化物越难以形成。

关键词 热轧态，42CrMo，球化率

The Effect of Hot Rolled Microstructure on Nodularity of Fine Blanking Steel 42CrMo

Cheng Jihao[1], Liu Wenbin[1], Min Ligang[2], Guo Bin[1]

(1.Research and Development Center of WISCO，Wuhan 430081, China;
2. Hubei Dafan Metal Products Co., Ltd., Macheng 438300, China)

Abstract Three volume 42CrMo test steel with different hot rolled structure was manufactured by adjusted coiling temperature. Then take the same annealing process for the three volume test steel. The results show that: (1)With the coiling temperature decreased from 720℃ to 580℃，the microstructure transformation is pearlite, pseudopearlite, bainite, the volume of proeutectoid ferrite was decreased. (2)The more the original microstructure was close to the equilibrium state，the more difficult the globular carbide was formed.

Key words hot rolled, 42CrMo, nodularity

北美汽车材料的应用现状与发展趋势探讨

郑 瑞[1] 郄 芳[2]

（1. 首钢技术研究院，北京 100043; 2. 首钢规划发展部，北京 100043）

摘 要 近年来，北美地区汽车产业增长势头强劲，严苛的油耗法规推动了高强钢和铝合金等轻量化材料的显著增长，也使轻量化非钢材料对钢铁材料的冲击比全球其他地区都要强烈。钢铁企业正在深挖钢材的潜力，联手汽车制造商开发质量更轻、强度更高的钢材及其加工应用技术，致力于实现车身及零部件各方面的轻量化，努力维持钢在汽车材料市场的主导地位。本文介绍了高强钢和轻量化非钢材料在北美的应用情况，概述了钢铁企业积极应对替代材料冲击的措施。面对其他轻量化材料更大规模的来袭，钢材需要加快自我突破式的研究和技术进步，从而使钢铁材料的综合性能不断跃上新的台阶，为汽车轻量化起到越来越主导的作用。

关键词 北美，轻量化材料，高强钢，铝合金，镁合金，碳纤维

Discussion on the Application Status and development Trend of Automotive Materials in North America

Zheng Rui[1]， Qie Fang[2]

(1.Shougang Research Institute of Technology, Beijing 100043, China; 2. Shougang Planning Development, Beijing 100043, China)

Abstract In recent years, automotive industry in North American grew strongly, stringent fuel regulations promoted significant growth of high-strength steel and aluminum and other lightweight materials,also made more impact of lightweight non-steel on steel materials than the rest of the world. Iron and steel enterprises were digging the potential of steel, joined hands with automotive manufacturers to develop lighter, higher-strength steel and its processing technology, committed to the implementation of the lightweight body and parts, strove to maintain the dominant position of steel in the automotive materials. The application of high strength steel and lightweight non-steel materials in North America were introduced, the measures of iron and steel enterprises to actively respond to the impact of the substitute materials were summarized. In the face of other lightweight materials more large-scale strikes, steel need to speed up the self breakthrough research and technology progress, so that the comprehensive performance of steel materials continued to a new level, played a more and more dominant role.

Key words North America, lightweight materials, high strength steel,aluminum alloy, magnesium alloy, carbon fiber

450MPa级铌微合金化车轮钢的组织与性能

刘 斌[1] 王 孟[1] 王立新[1] 赵江涛[1] 杨海林[2]

（1. 武钢研究院，湖北武汉 430080；2. 武钢热轧总厂，湖北武汉 430081）

摘 要 通过在低碳锰钢的基础上添加微合金化元素铌，结合控轧控冷工艺，成功开发出组织为铁素体和珠光体的细晶粒汽车车轮用热轧钢，该钢钢质纯净、晶粒细小、具有优良的延伸凸缘性能和良好的综合力学性能，成功应用于车轮轮辋和轮辐等零件。

关键词 车轮用钢，热轧，铌微合金化，延伸凸缘性能

Microstructure and Properties of Nb Microalloyed Wheel Steel with 450 MPa Tensile Strength

Liu Bin[1], Wang Meng[1], Wang Lixin[1], Zhao Jiangtao[1], Yang Hailin[2]

(1. Research and Development Center of WISCO, Wuhan 430080, China;
2. Hot Rolling Plant of WISCO, Wuhan 430081, China)

Abstract Using TMCP technology, fine-grain hot rolled wheel steel with a microstructure of pearlite and ferrite was successfully developed by adding Nb on the base of carbon manganese steel. Owing to its pure materials, fine grain size, excellent abrasion resistance and comprehensive mechanical properties, the steel was used to make parts such as the wheel rim and spokes.

Key words wheel steel, hot rolled, Nb-microalloyed, stretch-flange ability

回火温度对冷镦钢SCM435组织及性能的影响

陈继林[1,2] 刘振民[1,2,3] 崔 娟[1,2] 张治广[1,2] 蔡士中[1,2]

（1.邢台钢铁有限责任公司，河北邢台 054027；2.河北省线材工程技术研究中心，
河北邢台 054027；3. 北京科技大学材料学院，北京 100083）

摘 要 研究了回火温度对微观组织、力学性能以及低温冲击韧性的影响。结果表明：随着回火温度的升高，马氏

体晶界及晶面逐渐有碳化物析出，组织中碳化物由片状连续不均匀分布变为颗粒状弥散分布；抗拉强度与屈服强度都随着回火温度的升高而下降，断面收缩率及断后伸长率随着回火温度的升高而增加；在350~450℃温度间，冲击功随回火温度升高稳定增加，回火温度在550℃以上时，冲击功呈急速升高状态，SCM435钢经油淬后在550~650℃区间回火能够同时满足强度和冲击功的要求。

关键词 回火温度，SCM435，组织，力学性能，冲击韧性

Effect of Tempering Temperature on Microstructure and Mechanical Properties of Cold Heading Steel SCM435

Chen Jilin[1,2], Liu Zhenmin[1,2,3], Cui Juan[1,2], Zhang Zhiguang[1,2], Cai Shizhong[1,2]

(1. Xingtai Iron and Steel Corp., Ltd., Xingtai Hebei 054027, China;2. Hebei Engineering Research Center for Wire Rod, Xingtai Hebei 054027, China; 3. University of Science & Technology Beijing, Beijing 100083, China)

Abstract The effects of tempering temperature on microstructure, mechanical properties and cryogenic impact toughness were investigated. The results show that: the carbide precipitation on the grain boundary and crystal plane with tempering temperature increased, and carbide from flake continuous uneven distribution to particle dispersion distribution; Tensile strength and yield strength are decreasing with reduction of area and the elongation are increasing with the tempering temperature increasing; Between 350-450℃, impact energy increase stability with the tempering temperature increase, when the tempering temperature above 550℃, impact energy went into a state of rapid rise, SCM435 steel after oil quenching, tempering within 550-650℃ can meet the requirements of strength and impact energy at the same time.

Key words tempering temperature, SCM435, microstructure, mechanical properties, impact toughness

20Cr齿轮钢异金属夹杂的分析研究

丁礼权[1] 梁正宝[2] 张 帆[1] 罗国华[1]

（1. 武钢研究院，湖北武汉 430080；2. 武钢股份科技管理部，湖北武汉 430083）

摘 要 针对某公司生产的20Cr钢齿轮零件表面产生若干鼓包的现象，采用宏观观察、金相显微镜、扫描电镜和Oxford能谱仪等对试样进行观察和分析。结果表明，20Cr钢齿轮零件表面产生若干鼓包是一种异金属夹杂物，产生该缺陷的主要原因是由于冶炼操作不当，原料未完全熔化或浇注系统中掉入异金属所致。通过加强对炼钢所用废钢、合金等材料的监管、严格执行合金加入时机和顺序、控制好钢液温度、加强对炼钢设备的巡检和维护等措施可减少钢中异金属夹杂缺陷的产生。

关键词 20Cr齿轮钢，鼓包，异金属，金相组织

Analysis and Research the Foreign Metal of 20Cr Gear Steel

Ding Liquan[1], Liang Zhengbao[2], Zhang Fan[1], Luo Guohua[1]

(1. Research and Development Center of WISCO, Wuhan 430080, China;
2. Science and Technology Management of WISCO, Wuhan 430083, China)

Abstract A number of bulges occurred at the surface of 20Cr gear steel , bulges samples were analyzed by using macroscopic observation, metallographic microscope, scanning electron microscope and Oxford spectrometer. The results show that, the bulge occurred at the surface of 20Cr gear steel is a foreign metal. The main reason of the defect is due to improper smelting operation, raw material not completely melted or pouring into the system caused by foreign metal. The defect of foreign metal in steel is maybe reducing, through strengthening control of used steel, alloy and other materials, implementing strictly the timing and sequencing of the alloy to join, controlling the liquid steel temperature, strengthening measures and maintenance of steelmaking equipment .

Key words 20Cr gear steel, bulge, foreign metal, microstructure

高强度 TRIP 钢窄搭接电阻焊焊接工艺与性能研究

成昌晶 计遥遥 张 武 詹 华 刘永刚 单 梅

（马鞍山钢铁股份有限公司汽车板推进处，安徽马鞍山 243000）

摘 要 高强度 TRIP 钢有高的屈服强度、抗拉强度和延展性以及冲压成形能力等优点，广泛应用于汽车制造中。对不同焊接条件下 HC420/780TR 焊接接头进行性能检测和微观分析，结果表明：1.8mm 厚 HC420/780TR 与 1.8mm 厚 HC420/780TR 搭接焊缝未熔合长度约 2mm，接头杯突性能较差；降低接头碳当量，减弱冷裂纹倾向，让 1.8mm 厚 HC420/780TR 与 1.8mm 厚 SPCC 过渡焊，焊缝未熔合部分减少至 0.5mm，杯突性能有所改善；对 1.8mm 厚 HC420/780TR 与 1.8mm 厚 SPCC 过渡焊后进行接头热处理，降低焊接应力，焊缝未熔合长度进一步减少至 0.3mm，杯突性能完全合格。

关键词 高强度 TRIP 钢，窄搭接电阻焊，焊接接头，热处理

Research on Narrow Lap Welding of High Strength TRIP Steel

Cheng Changjing, Ji Yaoyao, Zhang Wu, Zhan Hua, Liu Yonggang, Shan Mei

(Auto sheet strategic business unit of Ma’anshan Iron and Steel Company, Ma’anshan Anhui 243000, China)

Abstract High strength TRIP steel is widely used in automobile manufacturing with the advantages of high yield strength, tensile strength,ductility and stamping forming ability. Mechanical properties and micro analysis of welded joints of HC420/780TR under different welding conditions are studied.The result show that in the narrow lap welding between HC420/780TR and HC420/780TR,the incomplete fusion length is 2.0mm and the cup performance is bad; In the narrow lap welding between HC420/780TR and SPCC with the depress of CE of the joint, the incomplete fusion length is 0.5mm and the cup performance is improved; In the narrow lap welding between HC420/780TR and SPCC with post weld heat treatment, welding stress is reduced, the incomplete fusion length is 0.3mm and the cup performance is fully qualified.

Key words high strength TRIP steel, narrow lap welding, welding joint, heat treatment

U75V 钢轨轨底开裂原因分析

熊 飞 张友登 王志奋

（武汉钢铁(集团)公司研究院，湖北武汉 430080）

摘　要　U75V 钢轨在轧制过程中轨底出现裂纹，本文通过宏观观察，金相检验等方法对裂纹产生的原因进行分析，检验结果显示在轨底裂纹部位未观察到明显冶金缺陷，缺陷部位表层组织为回火索氏体+少量铁素体，与正常部位表层组织珠光体+铁素体不同。从上述分析结果可知轨底表面有淬硬层存在，淬火层组织已自回火，缺陷的产生与该部位被淬火有关。

关键词　裂口，冶金缺陷，淬硬层，内应力

The Cracking Reason Analysis of the U75V Rail Base

Xiong Fei, Zhang Youdeng, Wang Zhifen

(Research and Development Center, Wuhan Iron and Steel (Group) Corporation, Wuhan 430080,China)

Abstract　There are some cracks which are produced at the process of rolling at the base of U75V rail. Macro observation and Metallographic examination are used to analysis how the crack produced in this paper. We find out that there is no obvious metallurgical defects at the site of crack which is produced at the rail base.The microstruction of the surface at the defect section is tempered sorbite and little ferrite,while the microstruction of the surface at the normal section is pearlite and ferrite.Based on the result above,we know that there is hardened layer at the surface of the rail base,and the quench hardened case has self temperinged,this is the reason why the crack produced.

Key words　crack, metallurgical defect, hardened layer, innerstress

82A 高碳钢线材生产实践

曹树卫[1,2]

（1. 安阳钢铁集团有限责任公司生产管理处，河南安阳　455004；
2. 北京科技大学材料科学与工程学院，北京　100083）

摘　要　82A 高碳钢线材是生产钢丝绳、胎圈钢丝、胶管钢丝的重要原料。安阳钢铁集团有限责任公司通过对化学成分设计、冶炼与连铸工艺、加热工艺、轧制工艺及冷却工艺等合理调整与优化，在 100t 转炉与高速线材轧机上成功生产了ϕ5.5mm 规格的 82A 高碳钢盘条。冶炼过程中，炉前出钢温度大于 1600℃，精炼时间不小于 50min；连铸采用全保护浇注，结晶器电磁搅拌；轧钢过程中，加热采用侧进侧出步进梁式加热炉，控轧采用在精轧机组水冷段后增加 4 机架减定径机组，控冷采用带有“佳灵”装置的大风量高风压延迟冷却型斯太尔摩线。结果表明，82A 试验钢盘条力学性能良好，同一炉号、同一公称直径及同一轧制制度的盘条抗拉强度的波动范围不大于 70MPa，盘条金相组织索氏体含量达到 90%以上，完全满足用户的使用要求。

关键词　高碳钢线材，化学成分，力学性能，金相组织

Production Practice of 82A High Carbon Wire Rod

Cao Shuwei[1,2]

(1. Production Management Department of Anyang Iron & Steel Group Co., Ltd., Anyang Henan 455004, China ;
2. School of Materials Science and Engineering, University of Science and Technology Beijing, Beijing 100083, China)

Abstract　82A high carbon wire rod is an important raw material of producing steel wire rope and steel wire for hose and bead wire. Based on design of chemical composition, reasonable adjustment and optimization of smelting and continuous casting process, reheating process, rolling process and cooling process etc., Anyang Iron And Steel Group Co., Ltd. produced successfully ϕ5.5mm high carbon steel wire rod grade C82DA for conversion to wire by 100t converter-high speed wire rod mill. In the course of smelting, the tapping temperature is more than 1600℃, and the refining time is no more than 50min; It uses full protective casting, and electromagnetic stirring of mould for continuous casting; In the course of steel rolling, it uses side charge and side discharge walking-beam furnace, it uses adding 4 stands reducing-sizing block after finishing mill group water-cooling section for control rolling, and uses blast volume blast pressure delay cooling Stelmor line with "Jialing" installation for control cooling. The result illustrates that for 82A test steel, the wire rod is in good mechanical performance with the range of tensile strength deviation of less than 70MPa for same heat, same nominal diameter and same rolling system. The sorbitic metallographic structure of rod reached more than 90%. It can fully satisfy the requirements of customers.

Key words　high carbon wire rod, chemical composition, mechanical property, microstructure

590MPa级车轮用热轧卷板的研制

闵洪刚　张立龙

（本钢热轧高强钢研发所，辽宁本溪　117000）

摘　要　介绍了本钢2300热连轧机组590MPa级车轮用RS590热轧高强度卷板的产品设计、冶炼、轧制工艺和性能情况。结果表明在低碳、硅锰系成分基础上，添加少量的Nb和Cr元素，结合控轧控冷，成功开发了590MPa级的车轮用RS590热轧卷板。所开发的RS590车轮用热轧卷板具有高的疲劳性能、高的成形性能和良好的焊接性能，不仅可用于制作车轮的轮辋，还可用于制作高扩孔率要求的乘用车车轮轮辐。

关键词　590MPa级，车轮用，热轧卷板

Development of 590MPa Grade Hot Rolled Strip Used for Automotive Wheels

Min Honggang, Zhang Lilong

(Hot Rolled High Strengh Steel Development Department of Bengang, Benxi Liaoning 117000, China)

Abstract　This paper introduces product design、smelting technics、rolling technics and performance of 590MPa RS590 hot rolled strip used for automotive wheels produced on Bengang 2300 continuously hot-rolled production line. The results indicated that useing the lower carbon content, Si,and Mn adding a small amount of Nb and Cr elements and appropriate controlling rolling and controlled cooling technique, can succeed develops 590MPa grade RS590 hot rolled strip used for automotive wheels. RS590 hot rolled strip used for automotive wheels has excellent fatigue properties , high performance and the good welding performance, which not only can be used for making wheel rim but also be used for making high expanding ratio wheel disc of passenger cars.

Key words　590MPa, used for wheels, hot rolled strip

承钢二高线 PF 线保温通廊应用实践

陈文勇　黄志华　曾　鹏　闫紫红　于青松

（河北钢铁股份有限公司承德分公司，河北承德　067002）

摘　要　介绍了承钢公司二高线车间 PF 线保温通廊的概况、结构选择和应用要点，同时介绍了关键工艺参数的设计方法，实践证明，该设计方法科学合理，应用效果良好，提高了线材性能，缩短了时效期，经济效益明显增高。

关键词　高速线材，保温通廊，性能，时效期

Application and Practice of Insulating Propylaea PF Line in the Second High-speed Wire Plant in Chengde Iron and Steel Company

Chen Wenyong, Huang Zhihua, Zeng Peng, Yan Zihong, Yu Qingsong

(Chengde Iron and Steel Company, Hebei Iron and Steel Group Co., Ltd., Chengde Hebei 067002, China)

Abstract　The structure selection and application of insulating propylaea are introduced which are used in the high-speed wire PF line in chengde Iron and Steel Company.At the same time,the design methods about the key parameters are described.The production shows that that the design methods are reasonable and better effects are obtained.The properties of high-speed wire is improved and the ageing period of high-speed wire is shortened.The economical benefit rises obviously.

Key words　high-speed wire, insulating propylaea, property, ageing period

退火温度对 600 MPa 级冷轧双相钢微观组织和力学性能的影响

邝春福　郑之旺　常　军

（攀钢集团研究院有限公司钒钛钢研究所，四川攀枝花　617000）

摘　要　将 C-Mn 钢加热至 770℃、800℃和 830℃均热 120s 后，快速冷却至 300℃进行过时效处理以模拟连续退火工艺。通过金相显微镜、扫描电镜和拉伸测试等技术，研究了退火温度对 600MPa 级冷轧双相钢微观组织和力学性能的影响。结果表明：退火温度为 770℃和 800℃时，铁素体晶粒内部存在大量未溶解的碳化物颗粒；退火温度增加至 830℃时，铁素体中的碳化物颗粒充分溶解，同时马氏体晶粒尺寸和体积分数提高。另外，在 770~830℃范围内，随着退火温度提高，双相钢屈服强度和屈强比逐渐降低，抗拉强度明显增加，伸长率呈先增加后降低的趋势，*n* 值变化不大。

关键词　冷轧双相钢，退火温度，力学性能，固溶 C 原子

Effect of Annealing Temperature on the Microstructure and Mechanical Properties of a 600 MPa Cold Rolled Dual Phase Steel

Kuang Chunfu, Zheng Zhiwang, Chang Jun

(Pangang Group Research Institute Co., Ltd., Vanadium and Titanium Containing Steels Division, Panzhihua Sichuan 617000, China)

Abstract In the paper, a C-Mn steel was soaked at 770℃, 800℃ and 830℃ for 120s, and then rapidly cooled to 300℃ to simulate the continuous annealing process. The influence of the annealing temperature on the microstructure and mechanical properties of the 600MPa cold rolled dual phase steel was investigated by using the microscope, SEM, tensile test and so on. The results showed that, a lot of undissolved carbide particles were retained in the ferrite when the dual phase steel was annealed at 770℃ and 800℃. With an increase in the annealing temperature to 830 ℃, the carbide particles located in the ferrite were dissolved fully, and the grain size and volume fraction of martensite were enhanced. By increasing the annealing temperature from 770℃ to 830℃, the yield strength and the yield ratio decreased gradually, the elongation tended to increase first and then to decrease, the tensile strength increased obviously, and the *n*-valued remained almost unchanged.

Key words cold rolled dual phase steel, annealing temperature, mechanical properties, solute C atom

承钢 1780 生产线高强薄规格汽车用钢开发

陆凤慧　包　阔　邢俊芳　胡德勇

（承钢热卷事业部，河北承德　067002）

摘　要　承钢 1780 半连轧生产线在开发高强薄规格汽车用钢，通过 Nb、V、Ti 等微合金化的成分设计，合理的控轧控冷的轧钢工艺，优化精轧机组的负荷分配，通过高精度的轧钢设备管理，实现细晶强化、固溶强化、析出强化的有机结合，达到轧制负荷合理，轧制稳定，板型良好，性能优异，实现以热代冷，满足汽车行业高强度、复杂变形和低成本的用户要求。

关键词　成分设计，高强薄规格，轧机稳定性，板形，以热代冷

不同炉型退火对 0Cr13C 氧化铁皮酸洗行为影响

王利军　翟进坡　冯忠贤　王宁涛　陈继林　李宝秀

（邢台钢铁有限责任公司河北省线材工程技术研究中心，河北邢台　054027）

摘　要　0Cr13C 退火材其表面氧化铁皮非常难于酸洗，本文对不同退火炉型生产的 $0Cr_{13}C$ 退火材酸洗质量进行了对比，分析了不同炉型氧化铁皮厚度、成分及物相结构。经对比发现：1 号炉、2 号炉和 3 号炉退火材表面氧化铁

皮厚度不同及$(Fe_2O_3)_3\cdot(Cr_2O_3)_2$尖晶石结构相造成酸洗难易不一的主要原因，另外，$Fe_2O_3$含量不同也是影响酸洗效果的因素之一。

关键词 0Cr13C，退火，氧化铁皮，酸洗

Effect of Different Furnaces on the Acid-pickling Behavior of the Oxide Scales on 0Cr13C Steel

Wang Lijun, Zhai Jinpo, Feng Zhongxian,
Wang Ningtao, Chen Jilin, Li Baoxiu

(Xingtai Iron & Steel Co. , Ltd. , Hebei Engineering Research Center for Wire Rod, Xingtai 054027, China)

Abstract It is difficult to pickle the oxide scale of 0Cr13C annealing stainless steel. The acid-pickling quality of different 0Cr13C stainless steels was compared from the thickness、composition and phase structure. The compared result shows that oxide scale thickness and $(Fe_2O_3)_3\cdot(Cr_2O_3)_2$ Spinel structure is the main factor to effect the quality of acid-pickling, Ecept that, different Fe_2O_3 content is another factor effecting the acide-pickling quanlity.

Key words 0Cr13C, annealing, oxide scale, acid-pickling

XGML33Cr 钢制螺栓塑性偏低原因分析

王利军 翟进坡 王宁涛 冯忠贤 陈继林 张印甲

（邢台钢铁有限责任公司河北省线材工程技术研究中心，河北邢台 054027）

摘　要 通过试验和分析，研究了 XGML33Cr 调质处理后断面收缩率和断后伸长率偏低的原因。研究结果表明：上贝氏体组织是造成断面收缩率和断后伸长率偏低的原因。通过末端淬火试验和不同规格 XGML33Cr 淬火试验，确定了现有库存钢坯适宜轧制盘条规格。另外，为大规格免退火冷镦钢成分优化提供了方向。

关键词 XGML33Cr，塑形指标，免退火钢，成分优化

Analysis on Low Plastic Property of Bolt Produced by XGML33Cr

Wang Lijun, Zhai Jinpo, Wang Ningtao,
Feng Zhongxian, Chen Jilin, Zhang Yinjia

(Xingtai Iron & Steel Co. , Ltd. , Hebei Engineering Research Center for Wire Rod, Xingtai 054027, China)

Abstract We have analyzed the reason of the low reduction of area and elongation of XGML33Cr after hardening and tempering. The research result shows that upper bainite is the main reason of the low reduction of area and elongation. We have confirmed the maximum size of wire rod through end quenching test and quenching test of different size wire rod. Otherwise ,the paper provides a good advice of component optimization.

Key words XGML33Cr, plastic property, on line annealed, component optimization

热轧带肋钢筋盘条 HRB400E 产品开发

彭　雄　肖　亚　王绍斌
向浪涛　汪　涛　戴　林

（重庆钢铁股份有限公司，重庆长寿　401258）

摘　要　本文介绍了重钢热轧带肋钢筋盘条 HRB400E 产品开发过程，包括生产技术难点、成分设计思路、生产工艺流程、炼钢工艺控制要点、轧制工艺方案、产品质量及金相组织状态等。试制表明，HRB400E 成分设计与工艺控制合理，产品质量稳定，抗震性能良好，HRB400E 产品开发取得了圆满成功。

关键词　热轧带肋钢筋盘条，HRB400E，抗震钢筋，产品开发

Development of Hot Rolled Ribbed Bar HRB400E

Peng Xiong, Xiao Ya, Wang Shaobin, Xiang Langtao, Wang Tao, Dai Lin

(Chongqing Iron & Steel Co., Ltd., Longevity Chongqing 401258, China)

Abstract　The development process of hot rolled ribbed bar HRB400 has been introduced including technical difficulties, composition design, the production technological process, key points of steelmaking process control, rolling process, product quality and the microstructure state, etc., The composition design and process control is reasonable of HRB400E according to the consequence of trial production and the product quality is stable, seismic behavior is also well and the product development of HRB400E has achieved a great success.

Key words　hot rolled ribbed bar, HRB400E, seismic ribbed bar, production development

优质碳素钢盘条 80 钢生产工艺研究

彭　雄　王绍斌　肖　亚
向浪涛　汪　涛　戴　林

（重庆钢铁股份有限公司，重庆长寿　401258）

摘　要　优质碳素钢盘条 80 钢复产试制期间，研究了 80 钢盘条生产工艺控制要点。为满足 80 钢优良的拉拔性能，要求 80 钢钢质纯净、有害杂质元素少、铸坯质量良好；钢坯加热时应适当降低加热温度，控制炉内气氛为弱还原气氛，减少脱碳；开轧及精轧入口温度不能太高，轧后应迅速快冷至索氏体相变区域，提高索氏体化率。通过研究试制，重钢 80 钢盘条使用性能良好，索氏体化率达 90%以上。

关键词　优质碳素钢盘条，80 钢，生产工艺，索氏体化率，拉拔性能

Study on Production Technology of 80 Hot Rolled Quality Carbon Steel Wire Rod

Peng Xiong,Wang Shaobin, Xiao Ya,
Xiang Langtao, Wang Tao, Dai Lin
(Chongqing Iron & Steel Co., Ltd., Longevity Chongqing 401258, China)

Abstract The key point control has been researched of 80 steel wire rod through the process of reproduction trial produce. To get the excellent drawing property, it asks for pure steel, less harmful impurity elements and good slab quality,low heating temperature of billet, weak reducing atmosphere to reducing decarbonization. The temperature of initial rolling and finishing rolling can't be too high and it should be quickly cooled to sorbite phase change temperature to improve the sorbite rate. In the end，the using performance of 80 steel wire rod is excellent and the sorbite rate get above 90% according to the effect of trial production.

Key words quality carbon steel wire rod, 80 steel, manufacturing technique, sorbite rate, drawing property

热轧光圆钢筋 HPB300 生产实践

王允　彭雄　向浪涛　汪涛　卿俊峰　戴林
（重庆钢铁股份有限公司，重庆长寿　401258）

摘　要　本文介绍了重钢热轧光圆钢筋 HPB300 产品试制过程，包括成分设计思路、生产工艺流程、炼钢工艺控制、铸坯质量、轧制工艺方案、红坯料型控制、产品质量及金相组织状态。试制结果表明，HPB300 铸坯及成品质量良好，产品试制取得了圆满成功，实现了稳定生产。

关键词　热轧光圆钢筋，HPB300，生产实践

Production Practice of Hot Rolled Plain Bar HPB300

Wang Yun, Peng Xiong, Xiang Langtao,
Wang Tao, Qing Junfeng, Dai Lin
(Chongqing Iron & Steel Co., Ltd., Longevity Chongqing 401258, China)

Abstract The production process of hot rolled plain bar HPB300 has been introduced including composition design, production process, steelmaking process control and the quality of casting billet, rolling process, red blank control, product quality and microstructure. After the research, it can be produced steady and the quality of slab and finished product is also good. Product development has achieved a great success.

Key words hot rolled plain bar, HPB300, production practice

邢钢 QB30 冷镦钢的开发与生产

李永超[1,2] 陈继林[1,2] 戴永刚[1,2] 郭明仪[1,2] 张治广[1,2]

（1. 邢台钢铁有限责任公司，河北邢台 054027; 2. 河北省线材工程技术研究中心，河北邢台 054027）

摘 要 QB30 冷镦钢常用于生产 10.9 级高强度紧固件。本文介绍了邢钢 QB30 冷镦钢的生产工艺和组织性能。通过成分设计、炼钢、轧钢生产工艺优化，盘条的化学成分、冷镦性能、夹杂物、脱碳层等均达到控制目标。用户使用结果表明，QB30 盘条能够满足生产高强度缸盖螺栓的指标要求。

关键词 QB30，冷镦钢，生产过程，工艺控制

Development and Production of QB30 Cold Heading Steel Wire Rod

Li Yongchao[1,2], Chen Jilin[1,2], Dai Yonggang[1,2], Guo Mingyi[1,2], Zhang Zhiguang[1,2]

(1. Xingtai Iron and Steel Co., Ltd., Xingtai 054027, China;
2. Hebei Engineering Research Center for Wire Rod, Xingtai 054027, China)

Abstract QB30 cold heading steel is usually used to produce 10.9 high strength fasteners. The production process and microstructure property of QB30 are introduced in this paper. Through optimizing steelmaking and rolling process ,the properties such as chemical composition, cold heading performance , inclusion and decarburization all achieve thecontrol goal.According to customers, it is showed that the QB30 can meet the demand for production ofhigh strength cylinder head bolt.

Key words QB30, cold heading steel, production process, process control

汽车用弹簧扁钢表面缺陷分析

祝小冬 罗贻正 刘 斌

（方大特钢科技股份有限公司，江西南昌 330012）

摘 要 本文阐述了弹簧扁钢表面经常出现的三种缺陷形貌特征及产生的原因。线状缺陷大部分系轧制划伤形成，少部分由表面卷渣形成。翘皮类缺陷由方坯表面微裂纹和轧制严重划伤形成。裂纹类缺陷由方坯表面存在大量气孔形成。

关键词 线状缺陷，翘皮，裂纹

Analysis of the Surface Defects of Spring Steel Flat Bars

Zhu Xiaodong, Luo Yizheng, Liu Bin

(Fangda Special Steel Technology Co., Ltd., Nanchang 330012, China)

Abstract This paper describes the reason and shape for three kinds of surface defects of the spring steel flat bars. Most linear defects originated from hot-rolling scratch, a few surface defects originated from rolling slag. Surface unwarping defects by the billet surface micro-crack formation and rolling serious scratch. Lots of pores existence of the billet surface caused the crack in the spring steel flat bars.

Key words line defect, surface unwarping, crack

汽车螺旋悬架弹簧用钢 55SiCrA 组织和性能研究

张　博　郭大勇　高　航　马立国　王秉喜

（鞍钢集团钢铁研究院，辽宁鞍山　114001）

摘　要　采用热膨胀法结合显微金相与硬度法，在 LINSEIS L78 RITA 相变仪上测定了 55SiCrA 弹簧钢的临界点温度（A_{c1}、A_{c3}、A_{r1}、A_{r3} 和 M_s）和连续冷却转变曲线（CCT），研究了冷却速度对组织和硬度的影响规律。在此基础上，进行了控轧控冷工业试验。结果表明：55SiCrA 的 A_{c1}、A_{c3}、A_{r1}、A_{r3} 和 M_s 温度分别为 746℃、775℃、694℃、717℃和 255℃；当冷速≤2℃/s 时，转变产物为少量铁素体、珠光体，珠光体硬度随冷速增大而增大；当冷速≥5℃/s 时，转变产物为马氏体、珠光体；当冷速≥20℃/s 时，转变产物为马氏体，硬度随冷速增大而增大；现场控轧控冷的试验钢抗拉强度达到 1163 MPa，伸长率为 13%，面缩率为 49%，综合力学性能良好，满足了用户的使用要求。

关键词　55SiCrA 弹簧钢，汽车悬架螺旋弹簧，CCT 曲线，显微组织，力学性能

Study on Microstruture and Property of 55SiCrA Steel Used for Automotive Suspension Coil Spring

Zhang Bo, Guo Dayong, Gao Hang, Ma Liguo, Wang Bingxi

(Iron and Steel Institute of Angang Group, Anshan Liaoning 114001, China)

Abstract By using the thermal expansion method combined with metallography-hardness method, the continuous cooling transformation curves (CCT) and the critical points temperature (A_{c1}, A_{c3}, A_{r1}, A_{r3} and M_s) of 55SiCrA spring steel was measured by LINSEIS L78 RITA transformation instrument. The effects of cooling rate on microstructure and hardness of the steel were analyzed. Controlled rolling and controlled cooling experimentations of the steel were carrired out on the basis of above works. The results show that the critical points temperature A_{c1}, A_{c3}, A_{r1}, A_{r3} and Ms is 746℃、775℃、694℃、717℃ and 255℃, respectively. The austenite transformation products are a few of ferrite and pearlite when the cooling rate is less than or equal to 2℃/s. The hardness of pearlite increase gradually with increasing of the cooling rate. The austenite transformation products are martensite and pearlite when the cooling rate is more than or equal to 5℃/s. The austenite transformation products are martensite when the cooling rate is more than or equal to 20℃/s. The hardness of martensite increase gradually with increasing of the cooling rate. The tested steel can reach good comprehensive mechanical properties by the controlled rolling and controlled cooling in the industrial trial-manufacture process. The tensile strength is 1163 MPa, the elongation rate is 13%, the reduction of area is 49%. The tested steels were used very well by downstream manufacturers, meeting their requirements.

Key words 55SiCrA spring steel, automotive suspension coil spring, CCT curves, microstructures, mechanical properties

热冲压钢 22MnB5 等温转变曲线测定

魏国清

（鞍钢股份有限公司冷轧厂，辽宁鞍山 114000）

摘 要 奥氏体冷却转变曲线是用以研究钢的性能及制定钢的热处理工艺必不可少的依据，因此本文也采用了金相-硬度法测定热冲压马氏体钢 22MnB5 的过冷奥氏体等温转变曲线(TTT 曲线)和过冷奥氏体连续冷却曲线（CCT 图），并对曲线进行了分析和讨论，为生产中根据不同的性能要求，为生产实践和新工艺的制定提供了参考依据。

关键词 22MnB5 钢，金相-硬度法, TTT 曲线, CCT 曲线, 热成型

The TTT Diagram of the Hot-stamping Steel 22MnB5

Wei Guoqing

(Cold Rolling Department of Ansteel, Anshan 114000, China)

Abstract The changing curve of austenitic is essential for studying the performance of steel and developing heat treatment process of steel，Therefore，The TTT diagram and CCT diagram of the hot-stamping steel 22MnB5 were measured by metallograph-hardness method，and then according to the different requirements for properties in production，The curve is analyzed and discussed，All those provid the references for productive practice and establishing new technics.

Key words 22MnB5 steel, metallographic-hardness, TTT curve, CCT curve, hot-stamping

400MPa 级别高强 IF 钢镀锌板研制与开发

孙海燕 石建强 李高良 程 迪 王连轩

（河北钢铁股份有限公司邯郸分公司，河北邯郸 056015）

摘 要 介绍了邯钢 400MPa 级别高强 IF 钢镀锌汽车板的研发机理和生产工艺特点。通过化学成分、热连轧及冷轧退火工艺设计，生产制备了满足 EN10346 和 GB2518 对应牌号要求的产品，力学性能检测显示产品在 r 值达到 1.7 以上的前提下，抗拉强度实际超过了 400MPa，且抗拉强度过程能力指标 P_{pk} 达到了 1.26。金相组织检测为铁素体，晶粒比较均匀，晶粒尺寸约为 15μm。

关键词 高强 IF，镀锌，力学性能，P_{pk}

Development of High Strength IF Galvanizing Steel for Grade of 400MPa

Sun Haiyan, Shi Jianqiang, Li Gaoliang, Cheng Di, Wang Lianxuan

(Handan Iron and Steel Group Co., Ltd. of HBIS, Handan Hebei 056015, China)

Abstract 400MPa high strength IF steels for automobile use were introduced mainly on the development mechanism and production process for Hansteel Company. The final product matches EN10346 and GB2518 corresponding needs. The value *r* reaches above 1.7 and meanwhile the tensile strength more than 400MPa. The process controlling index P_{pk} is 1.26 for tensile strength. The micrographs show that there is only ferrite which presents uniform grains and the diameter about 15μm.

Key words high strength IF, galvanizing, mechanical properties, P_{pk}

800MPa 级别冷轧热镀锌双相钢力学性能控制研究

刘志利 孙海燕 石建强 程 迪 杜艳玲

（河北钢铁股份有限公司邯郸分公司，河北邯郸 056015）

摘 要 通过工厂生产和模拟镀锌工序热历程制备了高强度冷轧热镀锌双相钢，利用扫描电镜（SEM）观察了双相钢组织及其微观结构，探讨了不同退火温度对双相钢力学性能和组织的影响规律。研究结果表明，试制的冷轧双相钢具有高的强度和良好的延伸率，符合欧标 EN10346 中对 800MPa 级别热镀锌双相钢力学性能要求。随着退火温度的升高，双相钢中晶粒的尺寸有所细化，马氏体比例有上升的趋势，但温度较高时该趋势弱化。镀锌双相钢显微组织由暗黑色的铁素体基体和弥散分布的亮白色马氏体组成。

关键词 双相钢，热镀锌，显微组织，力学性能

Research on Controlling of Mechanical Properties of 800MPa Cold-rolled Dual Phase Steel for Hot-dip Galvanization

Liu Zhili, Sun Haiyan, Shi Jianqiang, Cheng Di, Du Yanling

(Handan Iron and Steel Group Co., Ltd. of HBIS, Handan Hebei 056015, China)

Abstract High strength cold dual phase steel for hot-dip galvanization was produced by steel plant and experiment by simulating galvanizing thermal history. The microstructure was observed and analyzed by scanning electron microscopy (SEM) and transmission electron microscopy (TEM). The effect of different annealing temperatures on the microstructure and mechanical properties of dual-phase steel was also discussed. The experimental results show that the dual-phase steel possesses excellent strength and elongation that match EN10346 800MPa standards. The grain thinning trend and martensite content growing will be obvious along with annealing temperature going up. But the tendency will be weak when temperature high. The microstructure shows that furvous ferrite comes with dispersive white martensite.

Key words dual-phase, hot-dip galvanizing, microstructure, mechanical properties

6.2 电工钢/Electrical Steel

Advantages and Potentials of E^2Strip-A Subversive Strip Casting Technology for Production of Electrical Steels

Wang Guodong

(State Key Laboratory of Rolling and Automation, Northeastern University, Shenyang 110819, China)

Abstract In this paper, the advantages and potentials of E^2Strip, i.e. ECO-Electric Steel Strip Casting, a subversive strip casting technology for production of electrical Steels created and developed by State Key Laboratory of Rolling and Automation (RAL)· Northeastern University (NEU)· China are reported. The recent progress in the controlling of microstructure, texture, precipitation and magnetic properties of non-oriented, grain-oriented and high-silicon electrical steels is briefly outlined. Furthermore, the new specially designed compact and ecological production routes respectively for non-oriented, grain-oriented and high-silicon electrical steels are also mentioned.

Key words strip casting, non-oriented electrical steel, grain-oriented electrical steel, high silicon electrical steel, texture

Desired Characteristics of Non-Oriented Electrical Steels for Use in Next Generation, Energy Efficient, Electrical Power Equipment

Moses Anthony John

(Wolfson Centre for Magnetics, School of Engineering, Cardiff University, Wales, UK)

Abstract Non-oriented electrical steel is by far the most widely used soft magnetic material. It has evolved over the past 100 years and is likely to remain prominent for decades to come despite the challenge for greater market share from emerging materials. However complex ways in which the steels are magnetised in modern electrical equipment are rapidly becoming more common in many applications so it is useful to review the parameters which control their magnetic properties and consider the growing need for optimising the magnetic properties under more challenging magnetisation conditions such as high flux density, rotational magnetisation or non-sinusoidal flux density.

The presentation reviews the reasons for the dominating influence of composition (silicon content), grain size, internal stress, impurities, sheet thickness and texture. Although the generality of the well know method of analysing losses in terms of hysteresis, classical eddy current and excess losses is often questioned today, it is still a useful approach for attempting to understand the factors influencing the losses in non-oriented steels. However, the effectiveness of the approach under these more challenging magnetising conditions is discussed.

It is likely that users will demand far more versatility from non-oriented materials in the future in terms of effective operation under wide ranges of magnetisation conditions which make the relevance of established Standard ways of grading materials at a fixed frequency and flux density questionable when selecting best material for a given purpose. The inference of this situation on material development is briefly discussed.

Finally the possible impact of future developments in controlled textures or resistivity gradients through the sheet thickness on the performance of energy efficient non-oriented materials operating on under more complex magnetising conditions is considered.

Key words magnetic materials, electrical steel non-oriented silicon steel

中国电工钢产能与技术进步及需求研究

陈 卓

（中国金属学会电工钢分会，湖北武汉 430080）

摘 要 近十年来，我国电工钢产业高速发展，产能不断扩充，技术水平快速提升，但电工钢产能结构性过剩、产品质量稳定性、市场环境等方面值得关注和改善，本文针对中国电工钢的产能、技术进步、下游行业需求等进行分析，同时，就中国电工钢产业存在的突出问题提出建议，并对未来电工钢需求进行预测。

关键词 电工钢，产能，技术进步，下游行业需求及预测

Research on the Capacity, Technological Progress and Demand of Electrical Steel in China

Chen Zhuo

(Electrical Steel of Chinese Society for Metal, Wuhan 430080, China)

Abstract Over the past ten years in China, electrical steel industry developed fast, production capacity continued to expand and technology level grew fast as well. However there are so many outstanding issues worthing to attention such as electrical steel production capacity structural surplus,product quality stability, market environment and other aspects.This paper analyzes the production capacity, technological progress, downstream industry demand and so on,put forward suggestions as outstanding problems in Chinese electrical steel industry,and predict the future demand for electric steel industry.

Key words electrical steel, production capacity, technological progress, downstream industry demand and forecast

新材料发展趋势及铁基非晶合金现状

周少雄

（中国钢研科技集团安泰科技股份有限公司，北京 100081）

涂敷工艺对硅钢无铬环保涂层微观形貌的影响

王双红[1,2] 龙永峰[2] 贾 亮[2]

（1. 沈阳大学 机械工程学院材料科学系，辽宁沈阳 110044;

2. 沈阳防锈包装材料有限责任公司，辽宁沈阳　110033）

摘　要　冷轧无取向硅钢涂层板用于制造各种电气设备。目前，无取向硅钢涂层板主要通过在其表面涂覆绝缘涂料固化成绝缘涂层的方法制造。绝缘涂层的微观形貌及成分直接影响硅钢涂层板的质量和性能，微观形貌与涂敷工艺密切相关。本文研究了几种已应用的无铬环保绝缘涂层，利用扫描电子显微镜及能谱仪分析了不同涂敷工艺所制备的绝缘涂层的微观形貌及成分组成。结果表明：光辊制备的涂层形貌较刻槽辊更易受基板的轧制条纹、表面缺陷影响。刻槽辊制备的涂层以龟裂片状纹理为主，光辊制备的涂层以明暗相间条状纹理为主，龟裂纹较少。已应用的几种无铬环保涂层主要含 O、P、Fe、Al、Si、N、C 等元素，并通过 $NaO·nSiO_2$、CaO、ZnO 等增强涂层的致密度、耐热性及绝缘性。

关键词　涂敷工艺，绝缘涂料，环保，无铬，无取向硅钢

Effect of Coating Process on Morphology of Chromate-free Insulating Film on Silicon Steel

Wang Shuanghong[1, 2], Long Yongfeng[2], Jia Liang[2]

(1. Mechanical Engineering, Shenyang University, Shenyang 110044, China;
2. Shenyang Rustproof Packaging Material Co., Ltd., Shenyang 110033, China)

Abstract　At present, the manufacturing method of the non-oriented silicon steel with insulating coating was mainly through rolling and solidifying the coating on the surface of the silicon steel. Morphology of the insulating coating had an effect on quality and performance of non-oriented silicon steel, and morphology was closely related with coating process. This paper studied some business environment-friendly insulating coating and analyzed morphology and component of the insulating coating prepared by different process using SEM/EDS. The results showed that morphology of the insulating coating by smooth roller process was more susceptible to the surface defects and rolling stripes than by grooved roller process, the insulating coating by grooved roller process was the turtle lobes texture based coating, the insulating coating by smooth roller process was the bright and dark striped texture based coating. Some business environment-friendly insulating coating mainly consisted of O, P, Fe, Al, Si, N, C with $NaO·nSiO_2$ or CaO or ZnO.

Key words　coating process, insulating film, environment-friendly, chromate-free, non-oriented silicon steel

模拟工况条件下电工钢片磁性能测量方法及其应用

王晓燕[1]　马　光[2]　杨富尧[2]　孔晓峰[3]

（1.国家硅钢工程技术研究中心，湖北武汉　430080; 2. 国网智能电网研究院，北京　102211;
3.国网金华供电公司，浙江金华　321017）

摘　要　电工钢是用来生产电机和变压器铁心的主要原材料。现有的国际标准 IEC 60404-2 、IEC 60404-3 和国家标准 GB/T 3655—2008、GB/T 13789—2008 规定电工钢片用爱泼斯坦方圈和单片磁导计在频率 50Hz 保证正弦磁通密度峰值为 1.5T 或 1.7T 条件下评判其等级。由于一些原因，电机和变压器铁心的空载损耗与利用爱泼斯坦方圈或单片磁导计测量的电工钢损耗的数值往往不一致，本文总结了其中的原因。这些导致差异的因素在铁心设计过程中被量化为工艺系数，即铁心单位质量的空载损耗与电工钢片置于爱泼斯坦方圈或单片磁导计内在工作磁通密度下测

量的损耗的比值。本文讨论了模拟不同工况下，测量电工钢损耗的几种方法，并阐述了相关测量方法在评估电机和变压器铁心损耗方面的应用。

关键词 铁心，电工钢，工艺系数，损耗

Measurement Method and Application for Magnetic Properties of Electrical Steel under Modeling Working Conditions

Wang Xiaoyan[1], Ma Guang[2], Yang Fuyao[2], Kong Xiaofeng[3]

(1. National Engineering Research Center for Silicon Steel, Wuhan 430080, China;
2. The Smart Grid Research Institute of State Grid Corp. of China, Beijing, 102211, China;
3. Power Supply Company of State Grid Corp. of China, Jinhua, 321017, China)

Abstract Electrical steel is the main raw material to make the core of motor and transformer. The international standards now available IEC 60404-2 , IEC 60404-3 and state standard GB/T 3655—2008 , GB/T 13789—008 stipulate that the electrical steel sheet should be excited in the Epstein frame or single sheet tester under frequency 50Hz. The grade is judged by the magnetic flux density at 1.5T or 1.7T and ensuring the waveform sinusoidal. The no load loss of motor and transformer is not accord with it measured by the Epstein frame or single sheet tester. This paper summarizes the reasons. These differences are quantified as building factor, the ratio of the per unit weight no load loss of the core to that of the electrical steel sheet in the Epstein frame or single sheet tester measured at working magnetic flux density. This paper discusses several measurement methods of electrical steel loss under different modeling working conditions and elaborates the application of assessing the core loss of motor and transformer with corresponding measurement methods.

Key words core, electrical steel, building factor, loss

化学原位沉积结合高温烧结制备新型金属软磁复合材料

王 健 樊希安 吴朝阳 李光强

（武汉科技大学钢铁冶金及资源利用省部共建教育部重点实验室，武汉科技大学省部共建耐火材料与冶金国家重点实验室，湖北武汉 430081）

摘 要 为了降低金属软磁复合材料的损耗，探究制备金属软磁复合材料的新工艺，本文利用化学原位沉积结合高温烧结制备了两种新型金属软磁复合材料。通过调节化学原位沉积工艺中正硅酸乙酯的加入量研究调控SiO_2绝缘层的厚度，探讨了不同正硅酸乙酯加入量对磁粉芯绝缘层厚度、电性能及磁性能的影响规律。结果表明：（1）通过化学原位沉积包覆得到了Fe/SiO_2、$FeSiAl/SiO_2$核壳结构复合粉末，Fe/SiO_2核壳结构复合粉末经放电等离子得到了Fe/SiO_2复合磁粉芯，$FeSiAl/SiO_2$核壳结构复合粉末经压制成型、高温烧结得到了Fe_3Si/Al_2O_3复合磁粉芯。（2）调节正硅酸乙酯的加入量可以有效调控绝缘层的厚度。（3）随着正硅酸乙酯加入量的增加，Fe/SiO_2及Fe_3Si/Al_2O_3复合磁粉芯的有效磁导率的频率稳定性越来越好，电阻率逐渐增大，损耗降低。

关键词 金属软磁复合材料，化学原位沉积，放电等离子烧结，电性能，磁性能

Novel Soft Magnetic Composites Materials Prepared by In-situ Chemical Deposition Process Combined with High Temperature Sintering

Wang Jian, Fan Xi'an, Wu Zhaoyang, Li Guangqiang

(The Key Laboratory for Ferrous Metallurgy and Resources Utilization of Ministry of Education, Wuhan University of Science and Technology, The State Key Laboratory of Refractories and Metallurgy, Wuhan University of Science and Technology, Wuhan 430081, China)

Abstract In order to reduce the core loss of soft magnetic composites materials and investigate new technique to prepare soft magnetic composites materials, two novel soft magnetic composites materials were prepared by in-situ chemical deposition process combined with high temperature sintering in this paper. The added TEOS contents during the in-situ chemical deposition process were varied to controlling the thickness of insulating layer. The influences of added TEOS contents on the thickness of insulating layer, electrical and magnetic properties for soft composite cores have been investigated systemically. The results were as follows: (1) Fe/SiO_2 and FeSiAl/SiO_2 core-shell structured particles were synthesized by in-situ chemical deposition process. The Fe/SiO_2 soft composites cores were prepared by spark plasma sintering and the Fe_3Si/Al_2O_3 soft composites cores were prepared by high pressure compaction combined with high temperature sintering. (2) The thickness of insulating layer could be controlled effectively by varying the added TEOS contents. (3) The electrical resistivity of Fe/SiO_2 and Fe_3Si/Al_2O_3 soft composite cores increased with increasing the TEOS contents, while the effective permeability and core loss changed in the opposite direction. The effective permeability of soft composite core with larger TEOS content exhibited better frequency stability at high frequencies than other soft composite cores.

Key words soft magnetic composites materials, in-situ chemical deposition process, spark plasma sintering, electrical properties, magnetic properties

二次冷轧法制备 Fe-6.5wt.%Si 薄带的再结晶织构及磁性能

姚勇创　沙玉辉　柳金龙　张 芳　左 良

（东北大学材料各向异性与织构教育部重点实验室，沈阳　110819）

摘　要　本文采用二次冷轧压下率为 60%~80%的二次冷轧法制备出厚度为 0.10~0.20mm 的 Fe-6.5wt.%Si 薄带，沿轧向的磁性能为：B_8=1.474~1.495T，$P_{10/50}$=0.654~0.938W/kg。采用 X 射线衍射技术分析了冷轧及再结晶织构，冷轧形成位于{111}⟨112⟩的强γ织构，α织构相对较弱。经过高温退火后，形成由强η (⟨001⟩//RD)及较弱γ (⟨111⟩//ND) 组成的再结晶织构，不同于通常高温退火后获得的再结晶强γ和弱η织构。冷轧压下率增大导致再结晶γ织构略微增强，但η织构仍然占主导。分析认为，较大的冷轧前晶粒尺寸、冷轧道次间的回复作用以及冷轧强{111}⟨112⟩织构，使得高硅钢经过 60%~80%冷轧后，剪切带的数量以及储能有利于η晶粒形核并在初次再结晶完成时取得尺寸优势，进而在随后的晶粒长大过程中发展成为主导再结晶织构组分。

关键词　Fe-6.5wt.%Si 薄带，冷轧压下率，织构，磁性

Recrystallization Texture and Magnetic Properties of Two-Stage Cold Rolled Fe-6.5wt.%Si Thin Sheel

Yao Yongchuang, Sha Yuhui, Liu Jinlong, Zhang Fang, Zuo Liang

(Key Laboratory for Anisotropy and Texture of Materials (Ministry of Education), Northeastern University, Shenyang 110819, China)

Abstract The 0.10~0.20mm thick Fe-6.5wt.%Si thin sheets were fabricated through two-stage cold rolling process with the second stage cold rolling reductions of 60%, 70% and 80%. The magnetic properties along rolling direction are as follows: B_8=1.474~1.495T, $P_{10/50}$=0.654~0.938W/kg. Deformation and recrystallization textures were analyzed by means of X-ray diffraction technology. The cold rolling texture is characterized by strong γ fiber (⟨111⟩//ND) with peak at {111}⟨112⟩ and weak α fiber (⟨110⟩//RD). After high temperature annealing, the recrystallization texture consists of strong η fiber (⟨001⟩//RD) and relatively weak γ fiber, that is different from the ordinary results in which γ fiber is the dominant recrystallization texture component. With the increasing cold rolling reduction, the recrystallization γ fiber increases slightly, but recrystallization η fiber is still the dominant texture component. In the present study, the combination of large grain size before cold rolling, recovery between cold rolling passes and strong cold rolling {111}⟨112⟩ texture meliorate the cold rolling microstructure, namely, the number and stored-energy of shear bands between 60% and 80% cold rolling are favorable for η grains to nucleate and gain the size advantage after primary recrystallization, which is beneficial to make η fiber become the dominant texture component after grain growth.

Key words Fe-6.5wt.%Si thin sheet, cold rolling reduction, texture, magnetic properties

Recrystallization Texture Transition in Fe-2.1wt.%Si Steel by Different Warm Rolling Reduction

Shan Ning, Sha Yuhui, Xu Zhanyi, Zhang Fang, Liu Jinlong, Zuo Liang

(Key Laboratory for Anisotropy and Texture of Materials (Ministry of Education), Northeastern University, Shenyang 110819, China)

Abstract The competition among Goss ({110}<001>), Cube ({001}<100>) and {111}<112> recrystallization texture components depending on warm rolling reduction was investigated in Fe-2.1wt.%Si hot band, which is characterized by strong Cube at center layer. The deformation and recrystallization textures were both analyzed by X-ray diffraction technique. Warm rolling textures are composed of strong α fiber and relatively weak γ fiber with a peak at {111}<112> for all reductions. Deformed Cube texture component can remain under 70% reduction. Cube deformation bands may be the main contributor to the deformed Cube texture component. In-grain shear bands are well developed above 80% reduction. Recrystallization texture consists mainly of Cube and Goss components for all reduction, while only weak γ fiber with a peak at {111}<112> appears at 90% reduction. Depending on the rolling temperature of 500 °C, Cube, Goss and Cube, Goss components dominate the recrystallization texture in sequence with the warm rolling reduction increasing from 60% to 90%. This recrystallization texture transition with warm rolling reduction can be understood in terms of the number and nature of nucleation sites for various texture components. A variety of final recrystallization textures are thus proposed for non-oriented silicon steel by designing texture and microstructure of hot band and warm rolling reduction.

Key words texture, silicon steel, warm rolling, reduction

Texture and Structure Control in Ultra-low Carbon Steels by Non-conventional Sheet Manufacturing Processes

Leo A.I. Kestens

(Ghent University, Department of Materials Science and Engineering, Technologiepark 903, B9052 Ghent, Belgium)

Abstract The sheet manufacturing process, which involves various solid-state transformations such as phase transformations, plastic deformation and thermally activated recovery processes, determines the texture of the finished product. The conventional process of flat rolling and annealing only offers limited degrees of freedom to modify the texture. After annealing a {111} recrystallization fibre in BCC alloys is commonly obtained. For flux carrying purposes, however, other texture components are required than the ones achievable by conventional processing.

In the present paper it is shown that by severe plastic deformation of ultra-low carbon steel a texture can be obtained with increased intensity on the {001} fibre, which is of interest for magnetic applications. A similar texture is observed at the surface of Mn added low carbon steels by a mechanism of orientation selection, which involves the crystal anisotropy of the metal/vapour surface energy. It is also demonstrated that the conventional {111} deep drawing texture of IF steel can be obtained by a process of ultra-fast annealing (with a heating rate of more than 1000°C/s). Ultra-fast annealing also allows to significantly reduce the grain size of the finished product as compared to the conventionally processed IF steel.

高牌号冷轧无取向硅钢冷轧前后热处理工艺研究

苗　晓　张文康　王新宇

（太原钢铁（集团）有限公司技术中心，太原　030006）

摘　要　本文研究了采用两次冷轧工艺生产高牌号无取向冷轧硅钢时，热轧板分别在常化和不常化条件下，中间退火工艺对无取向硅钢晶粒大小、晶体织构和磁性能的影响。随中间退火温度的升高，二次冷轧前晶粒和成品晶粒增大，成品中不利织构组分减弱，磁性能得到改善。热轧板经过常化时的磁性能明显好于未经常化时的磁性能，但中间退火温度较高时常化对磁性能的有利作用减弱。

关键词　无取向硅钢，退火，晶粒大小，晶体织构，磁性能

Effect of Heat Treatment Before and After Cold-rolling on High Grade Non-oriented Silicon Steel

Miao Xiao, Zhang Wenkang, Wang Xinyu

(Technical Center, Taiyuan Iron and Steel (Group) Co., Ltd., Taiyuan 030006, China)

Abstract Effects of intermediate annealing on grain size, crystallographic texture and magnetic properties of high grade cold-rolled non-oriented silicon steel produced by two-pass cold-rolling and the hot-rolled plate with or without normalization were studied. With increasing intermediate annealing temperature the grain size before the second pass

cold-rolling and finished product increased, the unfavourable texture components in finished product decreased, therefore the magnetic properties improved. The normalization improved magnetic properties, but the effect of normalizing on magnetic properties weakened when the intermediate annealing temperature was higher.
Key words non-oriented silicon steel, annealing, grain size, crystallographic texture, magnetic properties

高效电机用无取向电工钢的研制

杜光梁　冯大军　石文敏　杨　光

（国家硅钢工程技术研究中心，湖北武汉　430080）

摘　要　在现普通冷轧高牌号无取向硅钢的基础上，降低 Si 和 Al 的总含量，添加少量辅助元素，组合常化处理和成品退火工艺，通过中试和大生产试制试验，研制出了高效电机用无取向电工钢。

关键词　实验钢，对比钢，高效，研制

The Development of Non-oriented Electrical Steel for High-efficiency Motor

Du Guangliang, Feng Dajun, Shi Wenmin, Yang Guang

(National Engineering Research Center For Silicon Steel, Wuhan 430080, China)

Abstract Based on existing common cold-rolling high-grade non-oriented silicon steel，a non-oriented electrical steel for high-efficiency motor was developed in laboratory and mass production by reducing Si and Al element，adding a small amount of supplementary elements，and optimizing normalizing process and finished annealing technology.

Key words experimental steel, correlation steel，high-efficiency，development

铁基非晶带材的制备技术现状及应用展望

李晓雨　王　静　庞　靖　李庆华　江志滨　王　玲

（青岛云路新能源科技有限公司，山东青岛　266232）

摘　要　钢非晶合金作为一种新材料，具有优异的软磁、耐蚀、耐磨、高强度及韧性等性能越来越受到人们的广泛重视。随着科学技术的进步，非晶产业化发展迅速，已被成功应用于传感器、开关电源、通讯领域、电力电子领域和抗电磁干扰等领域。但与传统材料相比，非晶合金还存在一些自身缺陷，本文对铁基非晶带材的制备技术现状，及在配电、电机、磁粉芯和非晶涂层等方面的进一步拓展应用作了重点介绍。

关键词　非晶宽带，制备技术，应用拓展

Research Status of Preparation Methods and Application Prospect for Fe Based Amorphous Ribbons

Li Xiaoyu, Wang Jing, Pang Jing, Li Qinghua, Jiang Zhibin, Wang Ling

(Qingdao Yunlu Energy Technology Co., Ltd., Qingdao Shandong 266232, China)

Abstract As a kind of new materials, amorphous alloy has been received much attention, with excellent soft magnetic, corrosion resistance, wear resistance, high strength and toughness. With the development of technology, amorphous alloy as a rapidly developing industry, has been successfully applied in sensor, switching power supplies, communications, power electronics and anti-electromagnetic interference. However, compared with traditional materials, amorphous alloy still has some deficiencies. Research status of preparation methods for Fe based amorphous ribbons and applications in distribution transformers, motors, magnetic powder core and amorphous coating are discussed in this paper.

Key words amorphous ribbons, preparation methods, applied extension

Strain Hardening Behavior During Multistep Nanoindentation of FeSiB Amorphous Alloy

Lashgari H.R.[1], Cadogen[2] J.M., Chu D.[1], Li S.[1]

(1. The School of Materials Science and Engineering, University of New South Wales, Sydney, Australia;
2. School of Physical, Environmental and Mathematical Sciences,UNSW Canberra at the Australian Defence Force Academy,Canberra, BC 2610, Australia)

Abstract The intention of the present study is to investigate the mechanical properties of the $Fe_{80.75}Si_8B_{11.25}$ amorphous alloys produced by melt-spinning process by the means of nanoindentation method. Two types of instrumented nanoindentation were conducted on the specimen: (1) singlestep nanoindentation (loading-unloading) and, (2) multistep nanoindentation (incremental loading-unloading). The strain hardening behavior was observed during multistep nanoindentation of $Fe_{80.75}Si_8B_{11.25}$ amorphous alloys which is quite uncommon in metallic glasses. It was found that multistep nanoindentation (incremental loading-unloading) increased the hardness value up to 20% when compared to singlestep nanoindentation (loading-unloading). This matter shows a unique strain hardening behavior in Fe-based amorphous alloys that can be attributed to the arrestment and entanglement of the propagating shear bands during incremental loading-unloading process. In addition, atomic force microscopy (AFM) of the indented surfaces revealed that surface steps around the indent in amorphous alloy being as a result of the shear band activation had extended to a lesser extent to the surface after multistep nanoindentation when compared to singlestep one. In other words, the amount of material pile-up was higher in multistep nanoindentation as compared to single step one. This matter exhibits further strain localization around the indent after incremental loading-unloading process, leading to the interaction and arrestment of the shear bands during reactivation at the same location.

Key words Fe-based amorphous alloys, strain hardening, nanoindentation, shear bands

Non-oriented Electrical Steels for Eco-friendly Vehicles

Kim Jae-hoon, Kim Jae-seong, Kim Ji-hyun, Park Jong-tae

(POSCO Technical Research Laboratories, POSCO, Pohang, Korea)

Abstract This paper presents the characteristic of newly developed electrical steel and motor performance analysis for HEV/EV. This material is developed and optimized for high frequency operation to reduce the core losses in traction motors to increase fuel efficiency. Four types of electrical steels are introduced, which are optimized for high flux density (PNHF), high frequency low core loss (PNF), high punchability (PNSF) and high strength (PNT) to meet different specifications from different types of traction motors. Moreover, new products are being under development in response to various customers' requirement.

Key words non-oriented electrical steel, traction motor, HEV, EV

无取向电工钢的{100}织构控制与磁性能改进

毛卫民 杨 平

（北京科技大学，北京 100083）

通常人们采用优化热轧工艺、热轧常化退火、冷轧中间退火、优化最终退火工艺、微合金化等技术手段来改善无取向电工钢工业产品的织构以提高其磁性能。这些改进措施往往多表现为压制对磁性能不利的{111}织构，或在一定程度上提高有利的{100}织构；其对磁性能的改进效果有一定局限性，且按照这类技术路线作进一步改进的潜力日渐枯竭。因此需要针对不同无取向电工钢品种的特点开发全新的工艺路线。

高牌号无取向电工钢含有较高的硅，不发生奥氏体转变。从连铸开始，利用铁素体取向在生产加工过程的继承和遗传关系可使柱状晶区的{100}织构保留至成品板。研究表明，采用中等压下量可造成{100}织构一定程度保留，且会限制{111}织构的生成。适当的最终退火使特定{100}织构再次增强，并促使磁性能明显改善。

低牌号无取向电工钢会发生完全的奥氏体转变。奥氏体轧制变形织构和动态再结晶织构在相变时可根据 K-S 关系转变成铁素体的{100}织构，相变的体积效应还会产生弹性应力，并影响铁素体新晶粒的取向选择。铁素体的弹性各向异性，促使相变在变体选择时更倾向于产生低弹性应力的{100}织构。采用高导热的纯氢气氛加热则可造成钢板较显著的温度梯度及由表及里的相变过程，促进{100}晶粒长大，并明显改进退火钢板的磁性能。

中牌号无取向电工钢的硅含量通常介于高牌号和底牌号无取向电工钢之间，会发生部分奥氏体转变。加热过程中钢板始终保有全部或部分铁素体组织。若在钢板成分中适当增加碳含量以增强奥氏体的稳定性，最终退火时引入纯氢脱碳气氛使表面率先脱碳并转变成铁素体，而心部高碳奥氏体尚不能转变。这样随脱碳而实现的由表及里的相变可获得较好的{100}织构，并改善了磁性能。

对高硅电工钢(6.5%Si)可借助控制热轧粗晶组织，进而使冷轧产生带有{100}或<100>取向亚结构的剪切带，最终退火生成{310}<001>及{100}<043>等有利织构，以明显改善钢板磁性能。

综上，根据不同类型无取向电工钢的成分特点，需利用不同促成{100}织构的机制，包括利用晶粒取向的遗传关系、相变取向关系、铁素体弹性各向异性、脱碳控制原理等。由此以传统工业装备为背景，设计出新的工艺流程，进而大幅度增强{100}织构、消除不利的{111}织构，促使产品磁性能呈突破性改善。

Gas Tungsten Arc/Tungsten Inert Gas Welding of Laminations

Gwynne Johnston, Erik Hilinski

(Tempel Steel Company, Chicago, Illinois USA)

Abstract Process conditions and equipment required for GTAW/TIG welding are described in terms of fundamental theory. This covers the basics of the electrode, shielding gas, the effects of gaps and clamping pressure, and combines this theory with practical experience. Quantitative data is presented to describe the effects of weld current, weld speed, electrode diameter, electrode stand-off distance, and different grades of materials on the width and the depth of welds on laminations using electrical steel.

Key words GTAW, TIG, welding, steel laminations

电动汽车驱动电机用无取向硅钢产品的开发

陈　晓[1]　谢世殊[2]　王　波[2]

（1. 宝山钢铁股份有限公司硅钢部，上海　200941；
2. 宝山钢铁股份有限公司技术中心，上海　200941）

摘　要　论文阐述了宝钢低铁损、高磁感和优异的机械性能的电动汽车驱动电机用无取向硅钢新产品的研发。产品包括低铁损 AV 系列、高磁感 AHV、系列、APV 系列及高机械强度 AHS 系列等四个系列，满足了电动汽车驱动电机对无取向硅钢产品的各种性能需求。

关键词　无取向硅钢，铁损，磁感强度，电动汽车，驱动电机

Development of Non-oriented Silicon Steels for Traction Motor Use in Electric Vehicles

Chen Xiao[1], Xie Shishu[2], Wang Bo[2]

(1. Silicon Steel Department, Baoshan Iron & Steel Co., Ltd., Shanghai 200941, China;
2. Technical Center, Baoshan Iron & Steel Co., Ltd., Shanghai 200941, China)

Abstract The Non-oriented silicon steels for Traction motor of Electric vehicles have been developed in Baosteel, including loss loss AV series, high induction AHV series, super induction APV series and high strength AHS series, which meet all needs for traction motor use in electric vehicles (EV).

Key words non-oriented silicon steel, iron loss, magnetic induction strength, EV, traction motor

卷铁心变压器在牵引轨道交通供电系统中的应用

高国凯

(常州太平洋电力设备（集团）有限公司，江苏常州　213033)

摘　要　卷铁心变压器铁心采用取向硅钢带连续卷制而成，磁路连续紧密无气隙，是一种变压器结构和制造工艺创新，在基本不增加制造成本的前提下，实现降低变压器空载损耗，提高节能效果的一种解决方案。因受制于材料、铁心结构、工艺和生产设备配套等因素，直至近十年，立体卷铁心变压器才逐渐进入快速发展的商业化生产阶段。虽然这类变压器主要仅应用在 10kV 级配电网中，但 220kV 卷铁心牵引变压器的成功研发证明了卷铁心变压器在牵

引供电系统内推广是完全可行的。论文对卷铁心的技术发展趋势、应用及市场前景进行了展望。

关键词 卷铁心，220kV 牵引变压器，设计与制造

Application of Wound Core Transformer in Traction Power Supply System of Rail Transit

Gao Guokai

(Changzhou Percific Electrical Power Unit (Group) Co., Ltd.,
Changzhou Jiangsu 213033, China)

Abstract Wound core transformer made by winding grain-oriented silicon steel strip continuously to obtain continuous gapless magnetic circuit, which is an innovated solution to reduce no-load loss for energy saving without cost rise. However, it just entered fast commercialization stage in recently 10 years due to difficulties in materials, core structure, processing facility, etc. Although this type of wound core mainly used in 10kV class distribution transformers grid, the successful development of 220kV see the feasibility of application of wound core transformer in traction power supply system of rail transit massively. Paper also discussed the technology development trend, application and market needs of wound core transformers.

Key words wound core, 220kV traction transformer，design and manufacture

基于宝钢新型高效硅钢的高效电机降成本应用

张舟 石宙 张峰

（宝钢集团技术中心硅钢研究所）

摘 要 针对电机行业对高性价比材料的需求，宝钢开发了新型高效硅钢 AE 系列。本文介绍了基于宝钢新研发硅钢产品 AE-1 的改进型性能，通过设计和优化，充分发挥了该产品高磁感的特性，设计出性能优良、效率达标且成本更低的高效电机系列产品。

关键词 设计优化，降成本，IE2 高效电机

Design a Series of IE2 Motor with the New Style High Efficiency Electrical Steel for Cost-reducing

Zhang Zhou, Shi Zhou, Zhang Feng

(Silicon Steel Research Institution, R & D Center, Baosteel Group)

Abstract Baosteel has developed new high efficiency electrical steels named AE series, in order to meet the requirement of high cost performance ratio the electric motor industry demands. This paper introduced the performances of the new product AE-1, and designed a series of IE2 motors by giving fully play to its high magnetic strength, which can meet the requirement of efficiency regulation of national standard with lower cost.

Key words design optimization, cost-reducing, IE2 motor

含Nb低温高磁感取向硅钢热轧组织及织构

刘 彪 宋新莉 练容彪 贾 涓 范丽霞 袁泽喜

（省部共建耐火材料与冶金国家重点实验室武汉科技大学，湖北武汉 430081）

摘 要 钢借助光学显微镜（OM）和扫描电镜的附件电子背散射衍射仪（EBSD）分析了含铌与不含铌两组取向硅钢热轧板不同厚度位置的组织及织构，结果表明，两组试验钢热轧板沿厚度方向存在组织与织构梯度。表层主要得到多边形铁素体晶粒，心部仍为沿轧制方向拉长的变形晶粒，其中含铌钢表层晶粒尺寸更细小。两组热轧板表层主要得到高斯织构{110}<001>及{110}<112>、{112}<111>织构，其中含铌实验钢表层及次表层高斯织构含量比不含铌实验钢高。在两组实验钢的心部以立方织构为主，高斯织构消失，其中含铌钢心部还出现了 γ 纤维织构{111}<112>。

关键词 取向电工钢，高磁感，织构，铌

The Microstructure and Texture of Nb-bearing Low Temperature High Magnetism Induction Grain Oriented Silicon Steel

Liu Biao, Song Xinli, Lian Rongbiao, Jia Juan, Fan Lixia, Yuan Zexi

(State Key Laboratory of Refractory Materials and Metallurgy,
Wuhan University of Science and Technology, Wuhan 430081, China)

Abstract The microstructure and texture of different thickness of the Nb-bearing and Nb-free orientated silicon steels hot rolled steel sheets was investigated by Optical microscope(OM) and Electron backing scattering diffraction(EBSD) installed on Scanning electron microscope. The results show that there are microstructure and texture gradient along the different thickness of the two hot rolled steel sheets. The polygonal ferrite grains are obtained on the surface and there are elongation grains along rolling direction in the center of the hot rolled steel sheets and the grain sizes in the surface of the Nb-bearing silicon steel are smaller than that for the Nb-free steel. There are Goss texture {110}<001> and {110}<112> and {112}<111> texture in surface and subsurface of the two experimental silicon steels. The content of Goss texture {110}<001> is higher in the surface and subsurface for the Nb-bearing silicon steel. There are mainly the cubic texture in the centre of the two hot rolled silicon steels and the Goss texture disappears and there is γ-fiber texture {111}<112> for the Nb-bearing silicon steel.

Key words oriented electrical steel, high magnetic induction, texture, niobium

CSP流程生产无取向电工钢成品研究

樊立峰[1,2] 董瑞峰[1] 陆 斌[2] 何建中[2] 郭 锋[1] 仇圣桃[3]

（1. 内蒙古工业大学材料科学与工程学院，内蒙古呼和浩特 010051；

2. 包钢钢联股份有限公司，内蒙古包头 014010；
3. 钢铁研究总院连铸技术国家工程研究中心，北京 100081）

摘 要 研究了 CSP 流程生产的无取向电工钢 50W800 组织、织构及夹杂物特点，结果表明：(1)成品板为再结晶组织，晶粒尺寸为 43.68μm；(2)成品织构以{111}面织构为主，{111}<112>含量达到 35.8%；(3)成品板夹杂物以氧化物和硫化物的复合夹杂为主。

关键词 无取向电工钢，组织，织构，夹杂物

Study on Non-Oriented Electrical Steel Produced by CSP

Fan Lifeng[1,2], Dong Ruifeng[1], Lu Bin[2], He Jianzhong[2], Guo Feng[1], Qiu Shengtao[3]

(1. School of Materials Science and Engineering, Inner Mongolia University of Technology, Hohhot 010051,China;
2. Baotou Iron and Steel Group Co., Baotou 014010,China;
3. National Engineering Research Center of Continuous Casting Technology, Central Iron and Steel Research Institute, Beijing 100081, China)

Abstract The microstructure，texture and inclusion of product of Non-Oriented Electrical Steel produced in CSP Process were Studied，The results show：(1) the average grain size of product is 43.68μm；(2) the {111} texture is the main texture，the content of {111}<112> reached to 35.8%；(3) sulfide and oxide are the main composite type inclusions.

Key words non-oriented electrical steel, microstructure, texture, inclusion

无取向硅钢 W600 冶炼试制研究

陈 郑 廖 明 李 斌 李永祥

（重庆钢铁股份公司钢铁研究所，重庆 401258）

摘 要 重钢通过 210tLD-RH-CC 试生产了 7 炉 W600 无取向硅钢，分析了该浇次 W600 过程[C]变化、两种脱氧方式的合金收得率以及 RH 最佳工艺时间，为下一步的批量生产提供了技术支持。

关键词 W600 无取向硅钢，RH 精炼，脱氧合金化，过程控制

The Trial Steelmaking of W600 Non-oriented Silicon Steel

Chen Zheng, Liao Ming, Li Bin, Li Yongxiang

(Chongqing Iron & Steel Co., Ltd., Iron and Steel Institute, Chongqing 401258, China)

Abstract The W600 non-oriented silicon steel was first produced by 210LD-RH-CC in CISI. The △[C] during processes of W600, the yield of alloy with different deoxidation operations and optimum RH process time have been analyzed, which provides the technological support for mass production.

Key words W600 non-oriented silicon steel, RH treatment, deoxidation alloying, process control

轧制法制备无取向 6.5wt.%Si 高硅钢薄带的织构和磁性能

柳金龙 沙玉辉 张 芳 邵光帅 李保玉 左 良

（东北大学材料各向异性与织构教育部重点实验室，辽宁沈阳 110819）

摘 要 针对具有优异软磁性能的 6.5wt.%无取向高硅钢的轧制法制造，本文采用实验室开发和工程化实践相结合的方式，研究了轧制成形、组织和织构控制以及工程化等关键问题。采用传统的轧制和退火方法在实验室中制备了厚 0.30mm 的无取向 6.5wt.%Si 高硅钢薄带，薄带板形良好、厚度均匀、无明显边裂和内裂纹。通过调整热轧工艺，获得了理想的强 λ 再结晶织构，显著提高了薄带磁性能，B_8 和 $P_{10/50}$ 分别达到 1.41T 和 0.57W/kg。尝试进行了熔炼、铸造和热轧流程的工程化实践，最终在工厂的生产线上成功实现了宽 280mm 高硅钢热轧卷制造。

关键词 高硅钢，Fe-6.5%Si，织构，磁性能

Texture and Properties of 6.5wt.%Si Non-oriented High Silicon Steels Prepared by Rolling Method

Liu Jinlong, Sha Yuhui, Zhang Fang, Shao Guangshuai, Li Baoyu, Zuo Liang

(Key Laboratory for Anisotropy and Texture of Materials (Ministry of Education), Northeastern University, Shenyang 110819, China)

Abstract For the manufacturing of 6.5wt.%Si non-oriented high silicon steels having excellent soft magnetic properties, some key issues, including preparation of rolling, control of microstructure and texture, were investigated by the method of combination of laboratory study and engineering practice. Non-oriented high silicon steel sheets with thickness of 0.30mm were successfully rolled in laboratory. The sheets had good flatness, uniform thickness and no significant edge cracks and internal cracks. Furthermore, the idea stong λ recrystallization texture was obtained by the control of hot rolling process, leading to the significant increase of magnetic properties. B_8 and $P_{10/50}$ were up to 1.41T and 0.57W/kg, respectively. The study also attempted the engineering practice melting, casting and hot rolling process, finally, high silicon steel hot rolled coils with width of 280mm were successful prepared in the production line of factory.

Key words high silicon steel, Fe-6.5wt.%Si, texture, magnetic properties

硅钢中正常坯与交接坯洁净度对比

张立峰[1] 罗 艳[1] 任 英[1] 陈凌峰[2,3] 程 林[3] 胡志远[3]

（1. 北京科技大学冶金与生态工程学院，北京 100083; 2. 北京科技大学材料科学与工程学院，北京 100083; 3. 首钢股份公司迁安钢铁公司，河北迁安 064404）

摘　要　本文通过实验和热力学计算的方法对硅钢中正常坯与交接坯洁净度进行了研究。采用化学成分、全自动扫描电镜方法来分析铸坯中主要化学成分、夹杂物行为等研究。实验表明：铸坯中[C]、[S]、[Ti]中心偏析严重；正常坯中的 T.O.分布均匀，而交接坯中 T.O.波动相对较大。正常坯中夹杂物主要以 AlN、氧化物以及氧化物与 AlN 复合夹杂物为主，占 85%以上，交接坯主要以氧化物为主，占 85%以上。正常坯的夹杂物尺寸分布均匀，主要集中在＞1μm，而交接坯夹杂物尺寸主要集中在＜2μm。热力学计算表明：该钢种中主要夹杂物为氧化物、氮化物和硫化物。Al_2O_3、镁铝尖晶石析出温度高于 1800℃，AlN、MnS 析出温度分别为 1300℃、1150℃。

关键词　硅钢，铸坯，热力学计算

Cleanliness of Slabs under Steady and Unsteady State Casting in Silicon Steels

Zhang Lifeng[1], Luo Yan[1], Ren Ying[1], Chen Lingfeng[2,3], Cheng Lin[3], Hu Zhiyuan[3]

(1. Metallurgical and Ecological Engineering, University of Science and Technology Beijing, Beijing 100083, China; 2. Materials Science and Engineering, University of Science and Technology Beijing, Beijing 100083, China; 3. Qian'an Iron & Steel Co., Ltd., Shougang Group, Qian'an 064404, China)

Abstract　The cleanliness of continuous casting slabs under steady state casting and during ladle change in silicon steel was investigated through experiments and thermodynamic calculation. Main chemical composition and behavior of inclusions were detected by means of chemical analysis and automatic scanning electron microscopy. Experimental results showed that the centerline segregation of [C], [S] and [Ti] was severe. The distribution of T.O. content of slabs under steady state casting is more homogenous than that of slabs during ladle change. Major inclusions in slabs under steady state casting were AlN, oxides, complex inclusions of AlN and oxides, while over 85% inclusions in slabs during ladle change were oxides. Size distributions of inclusions in slabs under steady state casting and during ladle change were above 1μm and below 2μm, respectively. Thermodynamic calculation of silicon steel was carried out that prime inclusions were oxides, nitrides and sulfides. The precipitation temperatures of alumina and magnesia alumina spinel were both above 1800°C, and those of AlN and MnS were 1300℃ and 1150℃, respectively.

Key words　silicon steel, slab, thermodynamic calculation

电工钢片磁致伸缩的测量研究

张志高　龚文杰　林安利　贺　建　范　雯　王京平

（中国计量科学研究院，北京　100029）

摘　要　本文对电工钢片磁致伸缩测量方法进行了系统研究，比较了电工钢片磁致伸缩不同测量方法的优缺点。在国家质检公益性行业科研专项项目资助下，本文基于激光多普勒测振仪和单片磁导计，建立了一套高精度的电工钢片磁致伸缩测量装置。实现了无应力和应力下电工钢片的磁致伸缩的准确测量，磁致伸缩系数的分辨率优于 1×10^{-8}，应力范围为±20MPa。整套装置由计算机控制，采用数字磁感波形反馈技术和同步采样技术，保证测量过程的准确。按不同应力区间，评定了磁致伸缩测量装置的不确定度，包括磁感、应力和磁致伸缩系数。结合电工钢片在变压器中的应用特点，对装置中谐波注入和添加直流偏磁的特色功能进行了介绍，大大扩展了装置的应用范围。还在文中与国外磁致伸缩测量装置进行了比较。

关键词　磁致伸缩，电工钢片，应力，激光多普勒测振，谐波

Research on Measurement of Magnetostriction of Electrical Steel

Zhang Zhigao, Gong Wenjie, Lin Anli, He Jian, Fan Wen, Wang Jingping

(National Institute of Metrology, Beijing 100029, China)

Abstract The measurement methods of magnetostriction of electrical steel have been investigated systematically, and the advantages and disadvantages of measurement methods are compared to each other. Sponsored by Special Fund for Quality Inspection Research in the Public Interest of China, we built up a magnetostriction measuring system with high accuracy based on laser Doppler vibrometer and single sheet tester. The magnetostriction of electrical steel can be accurately measured under applied stress or zero stress, the resolution of magnetostriction coefficient is better than 1×10^{-8} and the stress is within ±20MPa. The measurement system is controlled by computer. The digital feedback and simultaneous sampling are used to ensure the accuracy of the measurement. For different stress regions, the measurement uncertainties of the system including flux density, stress and magnetostriction are evaluated. Considering the application of electrical steel in transformers, the special functions in our system such as harmonics injection and measurement of magnetostriction under DC-bias field are introduced, which will expand the application of the measuring system. The comparisons between other magnetostriction measuring system and us are presented as well.

Key words magnetostriction, electrical steel, stress, laser Doppler vibrometer, harmonics

不同脱氧工艺钢水增碳量的比较及原因探讨

郭亚东

（中冶京诚工程技术有限公司炼钢所，北京　100176）

摘　要 取向硅钢和IF钢增碳量存在较大差异。取向硅钢因成分和脱氧工艺产生低碱度渣，该渣易于侵蚀MgO-C渣线，使之产生结构性剥落，而MgO-C渣线具有较好抗IF钢的高碱度氧化性渣侵蚀性能。RH处理过程钢液面变化增大了渣线侵蚀面，处理结束液面上升使结构性剥落的MgO-C渣线有机会溶于钢水，表现为RH处理结束至转台硅钢钢水大量增碳，并提出了精确控制取向硅钢碳质量分数的措施。

关键词 IF钢，取向硅钢，MgO-C渣线，增碳

Increasing Carbon of Oriented Silicon Steel Comparison with IF Steel and Discussing the Mechanism

Guo Yadong

(Capital Engineering & Research Incorporation Limited Steelmaking and Continuous-casting Division, Beijing 100176, China)

Abstract Oriented silicon steel and IF steel by carbon are quite different. Oriented silicon steel composition and deoxidation processes generate low basicity slag which easily erodes MgO-C slag line to make a structural spalling, while the MgO-C slag line have better resistance to IF steel slag which is high basicity and high oxidative. Steel surface changes in the process of RH

treatment to increase the slag line erosion surface, and steel surface rises at the end and peeling MgO-C brick have the opportunity to dissolve in molten steel slag line, so that oriented silicon steel increases carbon largely from the end of RH treatment to Turntable. Measures are put forward to precisely control the carbon mass fraction of oriented silicon steel.
Key words IF steel, oriented silicon steel, MgO-C slag line, increasing carbon

无抑制剂取向硅钢二次再结晶影响因素分析

李 军[1] 吕 科[1,2] 赵 宇[1] 李 波[2]

（1. 安泰科技股份有限公司功能材料事业部，北京 100081；
2. 中国钢研科技集团有限公司，北京 100081）

摘 要 本文对无抑制剂法制备取向硅钢的二次再结晶过程进行实验分析，研究二次再结晶退火过程中的保护气氛、退火温度和退火时间等参数的影响，为后续制定二次再结晶退火工序奠定试验基础。实验结果表明，二次再结晶退火期间的退火升温/保温气氛对二次再结晶退火过程有着重要的影响。50%N_2+50%H_2和25%N_2+75%H_2气氛下均可以发生二次再结晶。退火温度的提升和退火时间的延长均能够促进二次再结晶的发展，但延长退火时间能更有效地促进二次晶粒的长大。
关键词 无抑制剂取向硅钢，保护气氛，退火温度，退火时间

The Research of the Influence Factors of Secondary Recrystallization of Grain-oriented Silicon Steel without Inhibitors

Li Jun[1], Lv Ke[1,2], Zhao Yu[1], Li Bo[2]

(1. Functional Materials Branch, Advanced Technology and Materials Co., Ltd., Beijing 100081,China;
2. China Iron and Steel Research Institute Group, Beijing 100081, China)

Abstract The grain-oriented silicon steel without inhibitors was prepared in the laboratory.The effects of the protective atmosphere, annealing temperature and annealing time on the secondary recrystallization annealing process were studied.The results show that the heating and heat preservation atmosphere have important influence on the secondary recrystallization annealing process.The secondary recrystallization can be occurred in both atmospheres of 50%N_2+50%H_2 and 25%N_2+75%H_2. The increase of both the annealing time and annealing temperature can promote the development of the secondary recrystallization, while the annealing time is more effective.
Key words oriented silicon steel without inhibitors, protective atmosphere, annealing temperature, annealing time

CSP 流程生产无取向硅钢的技术发展

夏雪兰 王立涛 裴英豪 董 梅

（马鞍山钢铁股份有限公司技术中心，安徽马鞍山 243000）

摘　要　本文介绍了马钢CSP流程生产无取向硅钢的现状及典型牌号产品性能。通过高牌号产品铸坯组织、热轧、常化酸洗、退火组织的差异分析，探讨了CSP流程与常规流程产品性能的差异。

关键词　CSP，无取向硅钢，高牌号

Development of Non-oriented Electrical Steel Produced by CSP Process

Xia Xuelan, Wang Litao, Pei Yinghao, Dong Mei

(Ma'anshan Iron and Steel Co., Ltd., Ma'anshan Anhui 243000, China)

Abstract　Current status of non-oriented electrical steel produced by CSP process in Masteel and its typical products properties were introduced in this paper. Through the analysis on microstructure in CSP casting slab, hot rolled sheet, normalizing plate, annealing plate of high grade non-oriented silicon steel, the reasons of these differences between CSP and conventional process were studied.

Key words　thin slab casting and rolling, non-oriented silicon steel, high grade

Cr对高硅钢组织及性能的影响研究

程朝阳[1]　刘　静[1]　陈文思[1]　林希峰[1]　朱家晨[1]　向志东[1]

张凤泉[2]　曾　春[2]

（1. 武汉科技大学 钢铁冶金及资源利用省部共建教育部重点实验室，湖北武汉　430081；

2. 武汉钢铁研究院，湖北武汉　430080）

摘　要　高硅钢具有优异的软磁性能，但其室温脆性严重。本文主要研究了添加不同含量的Cr对高硅钢组织、有序相及性能影响。通过常规的热轧与温轧工艺制备了厚度约为0.3 mm的不同Cr含量的高硅钢薄板，且温轧温度低于600℃。利用金相显微镜和透射电镜分别观察了高硅钢的微观组织和有序相，并测量了高硅钢的显微硬度和磁性能。试验结果表明：适量Cr的添加能改善高硅钢的可轧性，制备出较宽且板型较好的薄板。Cr的添加能有效细化高硅钢的晶粒，Cr含量越高细化作用越明显，同时能降低高硅钢的长程有序度和显微硬度。含Cr高硅钢可轧性的改善归因于晶粒细化和长程有序度降低的综合作用。但是，Cr的添加使高硅钢的磁性能劣化，尤其是使铁损值增大。

关键词　高硅钢，显微组织，有序相，轧制性能

Effects of Cr on Microstructure and Properties of Fe-6.5wt.% Si Silicon Steel

Cheng Zhaoyang[1], Liu Jing[1], Chen Wensi[1], Lin Xifeng[1], Zhu Jiachen[1],
Xiang Zhidong[1], Zhang Fengquan[2], Zeng Chun[2]

(1. Key Laboratory for Ferrous Metallurgy and Resources Utilization of Ministry of Education, Wuhan University of Science and Technology, Wuhan Hubei 430081, China;

2. Research & Development Center of WISCO, Wuhan Hubei 430080, China)

Abstract High silicon steel has excellent magnetic properties, but near zero ductility at room temperature. In this paper, the effects of Cr on the microstructure, ordered phases and properties of high silicon steel were investigated. 0.3 mm thickness 6.5%Si silicon steel sheets with various Cr contents were produced with normal hot rolling and warm rolling process with temperature below 600 °C. Optical micrograph and TEM were used to observe the microstructure and ordered phases, respectively, and the Vickers hardness and magnetic properties of 6.5wt% Si silicon steels with various Cr content were measured. The results showed that the addition of a suitable amount of Cr can improve the processability of high silicon steel, therefore, thin sheets in good shape with large width of good properties could be produced successfully. The addition of Cr will refine the grains of high silicon steels, ie., the higher Cr content, the finer grains. The addition of Cr will reduce the Vickers hardness and the degree of long distance order of the high silicon steel, therefore improve its processability However, the addition of Cr may slightly deteriorate the magnetic properties of high silicon steel, especially result in the increase of iron losses.

Key words high silicon steel, microstructure, ordered phases, processability

高磁感电工钢在高效电机成本控制中的应用

石 宙 张 舟 王 波

（宝山钢铁股份有限公司硅钢部硅钢研究所）

摘 要 针对符合IE2标准的电动机设计方案，通过使用多种50W600牌号硅钢片进行成本优化对比研究，指出高磁感电工钢在节约成本上的优势。

关键词 硅钢，高效电机，高磁感

Application of High Magnetic Induction Electrical Steel in Material Cost Optimization of IE2 Motor

Shi Zhou, Zhang Zhou, Wang Bo

(Research Institution, Electrical Steel Department, Baoshan Iron & Steel Co., Ltd.)

Abstract Aim to IE2 motor design scheme, research material cost optimization of motor through comparing application of different 50W600 electrical steel. Analysis on the comparative advantage of material cost in using high magnetic induction electrical steel.

Key words electrical steel, IE2 motor, high magnetic induction

调控热轧温度优化高硅钢薄板再结晶织构的研究

邵光帅 柳金龙 沙玉辉 张 芳 姚勇创 左 良

（东北大学材料各向异性与织构教育部重点实验室，辽宁沈阳 110819）

摘 要 本文采用轧制和退火的方法制造了厚度为0.30mm的高硅钢薄带，研究了热轧温度对热轧、温轧、冷轧和退火过程中织构演变的影响。研究表明，与 1100℃开轧、800℃终轧的热轧制度相比，1030℃开轧、750℃终轧的热轧制度有利于促进最终退火时 λ 再结晶织构的发展。其原因是，再结晶过程中 λ 织构极难形成，提升 λ 再结晶织构的主要方法就是增强热轧和温轧板中 λ 织构强度，利用初始的强 λ 织构特征，并辅助适当的冷轧和退火制度，使 λ 织构在最终退火后得以保留和发展，而较高的热轧温度会导致部分动态再结晶的发生，弱化初始的热轧 λ 织构。

关键词 高硅钢，织构，再结晶，热轧

Optimization of Recrystallization Texture of High Silicon Steel Sheets by Controlling Hot Rolling Temperature

Shao Guangshuai, Liu Jinlong, Sha Yuhui,
Zhang Fang, Yao Yongchuang, Zuo Liang

(Key Laboratory for Anisotropy and Texture of Materials (Ministry of Education),
Northeastern University, Shenyang 110819, China)

Abstract High silicon steel sheets with thickness of 0.30mm were prepared by rolling and annealing method. Effect of hot rolling temperature on the development of texture during hot rolling, warm rolling, cold rolling and final annealing was investigated. It is found that, compared with the hot rolling process with starting temperature of 1100℃ and finishing temperature of 800℃, the hot rolling process with starting temperature of 1030℃ and finishing temperature of 750℃ is beneficial to the λ recrystallization texture developed in the final annealing. The λ texture is extremely difficult to form during recrystallization, and increase of λ recrystallization texture in hot and warm rolling state is an important way of improvement of λ recrystallization texture, while higher temperatures will result in some dynamic recrystallization during hot rolling, weakening hot rolling λ texture.

Key words high silicon steel, texture, recrystallization, hot rolling

Development of η-fibers Textures in Sandwich Structures in Cold Rolled Fe-6.5 wt.% Si Sheets

Fang Xianshi, Wang Bo, Ma Aihua

(Research Institute, Baoshan Iron & Steel Co., Ltd., Shanghai 201900, China)

Abstract Texture evolution of Fe-6.5 wt.% Si sheets during cold rolling and recrystallization was investigated using macro-/micro-texture analyses. 90 mm wide and 0.28 mm thick cold rolled Fe-6.5 wt.% Si sheets were successfully fabricated by controlled hot rolling, warm rolling and cold rolling process. The cold rolling reduction severely affects the texture components of the sheet surface. Surface textures of the final cold rolled sheet with a sandwich-like structure are {100}<110>, η- and γ-fiber textures. With annealing of 1000 ℃ for 1.5 h, abundance of abnormal grown grains exhibit η-fiber orientation, which can be attributed to the sandwich structure of sheets with large orientation gradient and large density gradient of dislocation across the thickness section.

Key words Fe-6.5 wt.% Si sheets, cold rolling, texture, sandwich structure, recrystallization

无取向硅钢基板氧化对厚涂层产品的影响研究

李登峰　金冰忠　肖　盼　谢世殊　王　波

（宝钢集团中央研究院硅钢研究所）

摘　要　本文就无取向硅钢表面氧化对厚涂层产品性能影响开展了研究：分析了不同退火露点下形成的表面氧化层对涂层成膜性能、硅钢铁损和磁感、涂料润湿性、钢片冲剪加工性能产生的系列影响，指出合适表面氧化层的存在对厚涂层的成膜性能起到促进作用。

关键词　无取向硅钢，表面氧化层，厚涂层产品

Effect of Surface Oxidation on Non Oriented Silicon Steel with Thick Coating

Li Dengfeng, Jin Bingzhong, Xiao Pan, Xie Shishu, Wang Bo

(Silicon Department of Research Institute, Baosteel Group)

Abstract　The effects of the NGO's surface oxidation layer on thick coating product have been studied in this paper, ie., effects of surface oxidation under annealing condition with various dew points on the coating's, film-forming ability, the magnetic property, the wettability of the coating, and the punching ability. The results indicated that the surface oxide layer with suitable thickness could facilitate the thick coating's forming ability.

Key words　non-oriented silicon steel, surface oxidation layer, thick coating product

环境对铁基非晶带材表面质量和磁性的影响

马长松　黄　杰　孙焕德

(宝山钢铁股份有限公司宝钢研究院，上海　201900)

摘　要　本文通过模拟外部环境(温度 50℃、湿度 80%、放置 15 天)对铁基非晶带材表面质量和磁性能的影响，发现铁基非晶带材表面出现红色氧化铁皮，经 N_2 气保护退火后变为黑色氧化铁皮。测定单片磁性能，退火后表面有氧化铁皮条带相对正常条带铁损上升 52%，B8、B10 下降约 5%，B0.8 下降超过 10%，进一步铁损分离结果表明，涡流损耗变化不大，磁滞损耗升高。这主要是表面氧化铁皮阻碍磁畴移动、转动所致。

关键词　环境条件，铁基非晶带材，表面质量，磁性能

Effects of Ambient Condition on Surface Quality andMagnetic Properties of Fe-based Amorphous Ribbons

Ma Changsong, Huang Jie, Sun Huande

(Research Institute, Baoshan Iron & Steel Co., Ltd., Shanghai 201900，China)

Abstract In this paper, effects of ambient condition (50℃、80%humidity、15days)on surface quality and magnetic properties of Fe-based amorphous ribbons were investigated. Red iron oxides appear on the surface of amorphous ribbons and after annealing the color of oxides changes to black. Single sheet magnetic propert test shows that the iron loss increased by 52% and the B8, B10 decreased by about 5%, but B0.8 decreased by more than 10%. The iron loss split analysis shows that eddy current loss(Pe) changed little but evident increase of hysteresis loss (Ph) due to surface oxide's impeding domain moving, rotating.

Key words ambient condition, Fe-based amorphous ribbons, surface quality, magnetic properties

磁屏蔽用高磁导率电磁纯铁研究与开发

苗　晓　王新宇　胡志强

（山西太钢不锈钢股份有限公司，山西太原　030003）

摘　要 开发磁屏蔽纯铁材料新品要满足更高磁性能要求，通过研究化学成分、磁性热处理工艺对晶粒大小和磁性能的影响，提出了合适的铝含量和磁性热处理温度，实验室成功开发了磁屏蔽用高磁导率电磁纯铁，为工业化生产开发提供了技术储备。

关键词 磁屏蔽，高磁导率电磁纯铁，研究开发

Research and Development of High Permeability Magnetic Iron for Magnetic Shielding

Miao Xiao, Wang Xinyu, Hu Zhiqiang

(Shanxi Taigang Stainless Steel Co., Ltd., Taiyuan 030003, China)

Abstract New magnetic iron for magnetic shielding needs to meet higher magnetic requirements. Research about effects of chemical composition and magnetic heat treatment on grain size and magnetic properties was carried out. The appropriate aluminum content and magnetic heat treatment temperature were proposed. A high permeability magnetic iron for magnetic shielding was successfully developed in laboratory. Technical reserve was made for mass production.

Key words magnetic shielding, high permeability magnetic, research and development

发蓝工艺对环保涂层附着性影响的研究

黄昌国 邹 亮 郭建国 王 波

（宝山钢铁股份有限公司技术中心硅钢研究所，上海 201999）

摘 要 无铬环保K涂层是一种新开发的无取向硅钢涂层，在用户发蓝的过程中，发现DX气氛下容易造成涂层的脱落。为了解决该问题，本文解析了整个发蓝过程，研究了发蓝气氛、发蓝温度和发蓝时间对K涂层附着性的影响。发现在480℃，发蓝氧化层开始形成，在N_2加湿气氛下最佳发蓝温度在510~570℃，最佳发蓝时间在105~120min，在该条件下，K涂层附着性良好。此时，氧化层的厚度在1~1.4μm。而用户在DX条件下，气氛条件不均匀，部分附着性较好的铁心，氧化层厚度在1.5μm左右，和实验室结果一致，而附着性不好的铁心，氧化层厚度高达2.0μm左右。说明用户气氛的氧化性过强，导致氧化层过厚，附着性降低。进一步，结合用户DX气氛的条件和铁氧化的相图，从机理上解释了DX气氛加湿的氧化性过强，使形成氧化物的区域已经进入了疏松的FeO区域，不再是致密的Fe_3O_4区域，因此附着性下降显著。用户的发蓝温度合适，可以通过降低露点或者缩短发蓝时间的方法来改善产品的附着性。

关键词 无取向硅钢，环保涂层，发蓝，附着性

Study on the Effects of Bluing Conditions on the Adhesion of the Chromium Free Coating

Huang Changguo, Zou Liang, Guo Jianguo, Wang Bo

(Research Institute, Baoshan Iron & Steel Co., Ltd., Shanghai 201999, China)

Abstract Coating K is a chromium free coating that Baosteel develops for non-oriented silicon steel. Some users reflects that the coating adhesion is poor after bluing in DX atmosphere. In order to solve the problem, the bluing process is analysed and the effects of atmosphere type, bluing temperature and bluing time on the adhesion of coating K are investigated. It is found that the bluing oxide layer started to form at 480℃, and the optimum temperature is 510~570℃, the optimum time is 105~120min, while the adhesion is good. The thickness at the optimum condition is 1.0~1.4μm. While at the user DX atmosphere, the atmosphere is uneven. In some cores which adhesion are good, the thickness of oxide layer is about 1.5μm, which is consistent with the laboratory optimum result. But in some bad adhesion cores, the thickness is even 2.0μm. These results illustrates that the oxidizability of the user atmosphere is too strong. Further by the phase diagram of iron oxide, we can see the high oxidizability of the user forms loose FeO, rather than compact Fe_3O_4, and results in bad adhesion. The user's bluing temperature is suitable. The coating adhesion could be improved by reducing the dew point and the bluing time.

Key words non-oriented silicon steel, chromium free coating, bluing, adhesion

剪切对无取向硅钢磁性的影响研究

邹 亮 王 波 谢世殊 郝允卫

（宝钢股份研究院硅钢研究所）

摘　要　为减少电机能耗，对无取向硅钢片的选用和电机铁心设计与制造工艺的优化显得尤为重要。在电机铁心的制造过程中，减小硅钢片的磁性恶化极其关键。本文研究了剪切和线切割对硅钢片磁性恶化的影响。 通过研究剪切对不同无取向硅钢产品磁性的恶化作用，实现在现有剪切设备基础上，采取合理的制样方式使剪切对磁性影响最小。并初步探讨了不同切割方式对产品切面的宏观金相、粗糙度及显微硬度的影响，提出采用显微硬度来表征硅钢产品剪切应力。

关键词　硅钢，剪切，典型磁性，显微硬度

The Influence of Shearing Stress on the Magnetic Properties of Manufactured Non-oriented Silicon Steel Sheets

Zou Liang, Wang Bo, Xie Shishu, Hao Yunwei

（Silicon Steel Research Institute, Research Academy, Baosteel）

Abstract　To reduce motor energy consumption, proper selection of non-oriented silicon steel core material and design optimization of the motor is very important.In the motor core manufacturing process, it is extremely critical to reduce the deterioration of magnetic property of silicon steel .This paper investigated the effect of cutting method and cutting method on the deterioration of magnetic property of silicon steels. The results could be used for better sample preparation to minimize the magnetic deterioration.In this paper the impact of cutting method on characteristics of sheared surface, roughness and hardness has been discussed. It has been put forward to use micro-hardness test method to characterize the shear stress introduced during core,manufacturing process.

Key words　silicon steel, shearing, magnetic properties, micro-hardness

6.3　管线钢和管线管/ Pipeline Steel and Linepipe

国产 X80 ϕ1219mm×22mm 大口径厚壁螺旋埋弧焊管开发

毕宗岳[1,2]　黄晓辉[1,2]　牛　辉[1,2]　赵红波[1,2]　牛爱军[1,2]

（1. 中国石油宝鸡钢管公司钢管研究院，陕西宝鸡　721008；
2. 国家石油天然气管材工程技术研究中心，陕西宝鸡　721008）

摘　要　采用低 C、低 Mn 和 Mo–Cr–Nb 合金设计,开发出了以针状铁素体为主的 X80 钢级 22mm 厚壁热轧卷板；通过钢管成型工艺、焊接工艺、水压工艺等制管工艺优化和控制，开发出了国产大直径、厚壁 X80 ϕ1219mm×22mm 螺旋缝埋弧焊管。产品检测结果表明，管体屈服强度 557～675MPa, 管体抗拉强度 649～766MPa，焊接接头拉伸强度≥691MPa；–10℃下管母横向冲击功 255～457J，焊缝冲击功 108～222J，HAZ 冲击功 111～433J，0℃落锤撕裂试验 DWTT 剪切面积平均 97.5%， $FATT_{85\%}$均低于–20℃，管体和焊缝都具有较高的韧性；钢管静水压爆破试验起爆点位

于母材，爆破口呈100%韧断；环切法测得环向弹复量范围–55～–220mm，盲孔法测得环向残余应力–179～264MPa，与同规格、同钢级直缝管相当，具有较低的残余应力。产品经国家油气管材质量监督检验中心检测，并经管道局环焊试验，符合西三线技术条件和API 5L 及相关标准要求。产品千吨级试制表明，国内具备工业化批量生产能力。

关键词 大口径，厚壁，X80，大输量，螺旋焊管

R&D of X80 ϕ1219 mm×22mm Domestic SSAW Linepipe of Large-sized and Thick-Walled

Bi Zongyue[1,2], Huang Xiaohui[1,2], Niu Hui[1,2], Zhao Hongbo[1,2], Niu Aijun[1,2]

（1. Research Institute of Baoji Steel Pipe Co., Ltd., Petro China, Baoji 721008, China；
2. National Petroleum and Gas Tubular Goods Engineering Technology Research Center，Baoji 721008，China）

Abstract Using the technology of low C、low Mn and Mo–Cr–Nb alloying design, the domestic X80 grade,22mm thickness hot roll steel coil with AF structure have been developed. Through the optimization of forming parameters and welding parameters, control of residual stress, the X80 grade SSAW welded pipe with 22mm thickness was developed. Product test results showed that the yield strength of steel pipe body is 557～675MPa, tensile strength is 649～766MPa, the tensile strength of welded joint is greater than 691MPa. Under –10°c, the steel pipe impact energy is 255～457J, weld seams impact energy is 108～222J, HAZ impact energy is 111～433J, Under 0°c, the average shear area value of DWTT of is 97.5%, FATT85% are lower than –20℃,the results show that the pipe has modest level of mechanical properties and excellent toughness. The blasting test of the pipe showed that the weld pipe all blong to normal failure ,and all have 100% ductile fracture.Using slit ring method and blind hole method, the rebound and the residual stress distribution of surface has been checked that the rebound is range in –55～–220mm,the residual stress distribution of surface is range in the –179～264MPa. The products was checked by china national quality supervision testing and inspection center of oil tubular goods, that the physical and chemical properties are all meeting the relevant standards requirements. The products was checked by china petroleum pipeline bureau, that the ring welding test are all meeting or exceeding the technical requirements of 3rd west-east natural gas pipeline and API 5L . Kiloton trials showed that domestic have the capacity of industrial mass production.

Keywords large diameter, thick wall, X80, high heat input, SSAW

Characteristics of Self-Shielded Flux Cored Arc Welding Consumables for High Strength Steel Grades

Kuzmikova L., Han J., Zhu Z., Barbaro F.J.

(University of Wollongong, Northfields Ave, Wollongong, Australia)

Abstract The most widely used processes for girth welding of pipelines constructed from API X80 is mechanised GMAW and SMAW in combination with cellulosic electrodes. However, both of these processes face certain limitations. Mechanised GMAW is not suitable for tie-ins, applications in difficult terrain and due to high mobilisation cost for use on short sections. Whereas SMAW with cellulosic electrodes presents significant challenges in terms of satisfying weld metal strength (over)matching requirements and high risk of HACC that is particularly critical for high strength large diameter thick wall pipelines. Low hydrogen basic electrodes have also been increasingly used especially those designed for vertical down application. However, the cost of these consumables is high and due to the unique operational characteristics require extensive welder training. Self-shielded FCAW (SS-FCAW) consumables offer a feasible alternative to these limitations.

However, in recent years an increasing number of cases have been reported where the use of the SS-FCAW process produced welds with extremely poor toughness particularly when applied to high strength line pipe. This paper describes some initial results of weld metal dilution effects on Charpy impact toughness, which is part of an ongoing investigation into the performance of SS-FCAW consumables.

Key words SS-FCAW, weld consumable, dilution, elemental analysis, API 5L X80, Charpy impact test

高钢级管线钢冷弯加工对性能的影响研究

冯 斌　刘 宇　范玉然

（中国石油天然气管道科学研究院，河北廊坊　065000）

摘　要　冷弯管制作是长输油气管道建设施工的重要环节之一，高钢级、大口径冷弯管的制作难度明显加大，其质量性能的稳定性，以及服役安全性也显得更为重要。本文针对高钢级管线钢冷弯加工变形对拉伸性能的影响进行了较系统的研究，揭示出了其影响趋势及影响程度。研究表明，冷弯变形对高钢级管线钢拉伸屈服强度、屈强比、均匀伸长率、应力应变曲线等产生显著影响，而对抗拉强度、断后伸长率的影响较小。在冷弯管制作时，应严格控制冷弯变形量，避免出现屈强比显著增大、塑性严重下降等问题。

关键词　冷弯管，变形量，力学性能，变化趋势

Research on Cold Bending Process and Performances for High-grade Pipeline Steels

Feng Bin，Liu Yu，Fan Yuran

(Pipeline Research Institute of China National Petroleum Corporation，Langfang 065000，China)

Abstract　Making cold bend is one of the important link for long distance oil and gas pipeline construction, the production difficulty of large diameter and high grade cold bend is obviously increased, the stability of quality, performance, service security is more important. for cold bend. This paper analyzed and studied systemically the influence of cold bending deformation on tensile properties, reveals its influence trends and influence degree. Studies have shown that cold bending deformation have very obvious influence on the yield strength ,the Y/T ratio, uniform elongation and stress strain curve for high grade pipeline steels, but has little effect on the tensile strength, break elongation. Therefore in the cold bending production, should control strictly the cold bending deformation, so that avoid yield ratio to increasing significantly, plasticity to declining seriously and other issues.

Key words　cold bend, deformation, mechanical properties, variation

超厚规格 X70 管线钢热轧卷板组织性能研究

牛　涛[1]　安成钢[1]　姜永文[1]　吴新朗[2]　武军宽[2]　代晓莉[1]

（1. 首钢技术研究院，北京　100043；2. 首钢股份公司迁安钢铁公司，河北迁安　064404）

摘　要　本文在2250mm热连轧产线上采用低碳高Nb+Mo/Ni合金化的低碳当量成分体系开发出超厚规格25.4mm X70管线钢热轧卷板，并对其力学性能、显微组织进行了分析。结果表明，试制钢卷具有良好成型性能(屈强比≤0.84、均匀延伸率≥6%)、高韧性(–60℃冲击≥250J、DWTT韧脆转变温度–30℃左右)的特点。利用扫描电镜、透射电镜对X70的显微组织与M/A岛形貌进行了系统分析，结果表明，钢卷组织主要由细小均匀的针状铁素体构成，高密度位错与析出粒子的交互作用保证了钢卷的强度；组织中M/A岛细小弥散，尺寸在3μm以内，分布于晶内与晶界上；衍射结果表明其主要由奥氏体及马氏体构成，马氏体具有典型的孪晶特征；热变形后超快冷有利于M/A岛的细化，保证了钢卷良好的低温韧性。

关键词　超厚规格，管线钢，力学性能，显微组织，M/A岛

Research on Microstructure and Property of Ultra-thick X70 Hot-rolled Pipeline Coil

Niu Tao[1], An Chenggang[1], Jiang Yongwen[1], Wu Xinlang[2], Wu Junkuan[2], Dai Xiaoli[1]

(1. Shougang Research Institute of Technology, Beijing 100043, China;
2. Shougang Qian'an Iron & Steel Co., Ltd., Qian'an Hebei 064404, China)

Abstract　In this study, ultra-thick X70 pipeline steel with thickness of 25.4mm has been developed and produced using 2250mm hot rolling line, which employs a low carbon equivalent composition system was, including low C, Mn, high Nb alloying and Mo/Ni. Mechanical property and microstructure were researched in this study. Results show the great formability (Y/T ratio ≤0.84, uniform elongation ≥6%) and high toughness (CVN ≥250J at –60℃, FATT of DWTT lower than –30℃). Microstructure and M/A island morphology of the steel has been investigated using scanning electron microscope (SEM) and transmission electron microscope (TEM). It shows that the microstructure mainly consists of refined homogenous acicular ferrite, in which the interaction of high-density dislocation and precipitates help to guarantee the strength of steel. Refined M/A islands disperse in the grain or on the grain boundaries with diameter less than 3μm. Diffraction results show that M/A island consists of austenite and martensite with typical twin charcateristics. Ultra-fast cooling after rolling deformation helps to refine M/A island, and so that to increase the toughness of coil with high gauges.

Key words　ultra-thick, pipeline steel, mechanical properties, microstructure, M/A island

基于应变设计地区用X80管线管的研制

王　旭[1]　陈小伟[2]　李国鹏[1]　张志明[2]　闵祥玲[1]

(1. 中国石油渤海装备研究院，天津　300457；2. 中国石油渤海装备巨龙钢管公司，河北　062658)

摘　要　介绍了基于应变设计用大应变管道要求及材料设计原理，分析了厚壁X80钢级大应变管线钢研制中存在的问题和主要难点，即高强度对双相组织中硬相比例增加的要求与硬相增加对均匀延伸率的不利影响。通过实验确定了采用铁素体+贝氏体双相组织思路、双相组织比例要求及板材的力学性能要求，并分析了制管前后、热时效前后的板-管主要性能变化规律，开发并批量生产了具有优良性能的X80大应变管线管。

关键词　基于应变设计，X80，双相组织，大变形，管线管，性能

Development of X 80 Linepipe for Strain-based Design Pipeline

Wang Xu[1], Chen Xiaowei[2], Li Guopeng[1], Zhang Zhiming[2], Min Xiangling[1]

(1.CNPC Bohai Equipment Research Institute, Tianjin 300457, China; 2. Julong Steel Pipe Company of CNPC Bohai Equipmen Manufacturing Co., Ltd., Hebei 062658, China)

Abstract This paper introduces the requirements of linepipes and concept of material design for strain-based design pipelines, analyzes the problems and main difficulties met during the development of thick wall X80 high strain linepipe steel, which mainly focus on that higher strength claims a fraction increase of hard phase in dual phase microstructure, while the increase of hard phase has a bad influence on uniform elongation. By means of testing, we determined the concept of ferrite+bainite dual phase microstructure, specified the fraction requirements of dual phase microstructure and the mechanical properties requirements of steel plates, and analyzed the change rule of main performance before and after pipe manufacturing, before and after pipe thermal aging.Finally, X80 high strain linepipes with excellent performance were developed and mass produced.

Key words strain-based design, X80, dual phase microstructure, high deformable, linpipes, performance

合金元素对 X80 管线钢焊接性能的影响

孔祥磊 黄国建 付魁军 刘芳芳 黄明浩 张英慧

（鞍钢集团钢铁研究院，辽宁鞍山 114009）

摘 要 本文利用热模拟试验和对接焊试验研究 X80 管线钢母材中合金元素 Mo、Ni、Cr 等对焊缝和热影响区冲击性能的影响。结果表明，母材合金元素对焊缝和热影响区的冲击性能有明显影响，随母材合金设计中 Mo、Ni 含量的提高以及 Cr 含量的降低，焊缝和热影响区的冲击韧性更为优异。利用金相显微镜观察焊接试样组织，焊缝以针状铁素体为主，并有较少量的先共析铁素体，沿原始奥氏体晶界有颗粒状析出物。两者比较，Mo、Ni 含量高、Cr 含量低的试样焊缝先共析铁素体更发达，晶界析出物更少。不过由于 Mo、Ni 属于贵重元素，实际生产需要综合评价经济因素和性能指标要求合理设计。

关键词 热模拟，对接焊，合金，冲击性能

Effect of Pipe Body Alloy on Weldability of X80 Steel

Kong Xianglei, Huang Guojian, Fu Kuijun, Liu Fangfang, Huang Minghao, Zhang Yinghui

(Iron & Steel Research Institute of Angang Group, Anshan Liaoning 114009, China)

Abstract Effect of pipe body alloy such as Mo, Ni, Cr on impact property of pipe weld and hot-affected zone(HAZ) of X80 steel was investigated by thermal simulation test and butt welding test. The result shows that, there is an obvious relationship between pipe body alloy and the toughness of weld and HAZ, the more content of Mo, Ni and less of Cr in the body, the better of impact toughness of weld and HAZ. Metallographic microscope was used to compare microstructures of welding samples, every welded seam microstructure is mainly acicular ferrite and a little proeutectoid ferrite, and with some granular precipitations on original austenite grain boundary, the difference is that there are more proeutectoid ferrite and less precipitations of the sample with more content of Mo, Ni and less of Cr in the body. Because of the high price of Mo and Ni, alloy design must be considered comprehensively with the cost and property requirements in the production.

Key words thermal simulation, butt welding, alloy, impact toughness

海底管线用厚规格钢板的研制与开发

李少坡[1,2] 丁文华[1,2] 李家鼎[1,2] 张 海[1] 李战军[1,2]

（1. 首钢技术研究院，北京 100043；
2. 北京市能源用钢工程技术研究中心，北京 100043）

摘 要 针对海底油气输送管道用高钢级 DNV 485FD 管线钢的技术要求，从成分设计、连续冷却相变规律、钢坯冶炼、连铸和钢板显微组织控制的角度，详细介绍了首钢海底管线用厚规格 DNV 485FD 钢板生产的关键技术。2011~2014 年期间，结合首钢 4300mm 宽厚板生产线特点，采用上述成分体系，通过控制钢板的开始冷却温度、冷却速度和终止冷却温度，获得以多边形铁素体、准多边形铁素体、粒状贝氏体和板条贝氏体混合的多相组织，首钢成功开发了 31.8mm 厚 DNV 485FD 钢板和 40.5mm 厚 DNV 485FD 钢板，钢板分别通过国内宝世顺管厂和番禺管厂的制管，产品获得了良好的拉伸强度和优异的低温落锤韧性。

关键词 厚规格，DNV 485FD，深海管线，落锤

Research and Production of Thick-wall Pipeline Plate for Deep Water

Li Shaopo[1,2], Ding Wenhua[1,2], Li Jiading[1,2], Zhang Hai[1], Li Zhanjun[1,2]

(1. Shougang Research Institute of Technology, Beijing 100043, China;
2. Beijing Energy Steel Engineering Technology Research Center, Beijing 100043, China)

Abstract According to the technical requirement of DNV 485FD pipeline steel used in deep water Project, key production technology of pipeline steel in Shougang was introduced in details from the angle of composition design, continuous cooling transformation, steel making, continuous casting and microstructure control technologies. During 2011~2014, the productions of 31.8 mm DNV 485FD and 40.5 mm DNV 485FD for deep water projects were developed in Shougang 4300mm rolling line. By controlling the cooling start temperature, cooling stop temperature, and cooling speed, the microstructure consist of polygonal ferrite, quasi-polygonal ferrite, granular bainite and lath bainite was abtained. The plates were piped in PCK pipe and BSS pipe respectively domestic, the products showed good tensile strength and excellent low temperature DWTT toughness.

Key words thick wall, DNV 485FD, deep water pipe, DWTT

冷却速度对超高强度 X90 钢相变行为的影响

崔 雷[1,2] 徐进桥[1] 孔君华[1] 郭 斌[1] 李利巍[1] 邹 航[1] 刘文艳[1]

（1. 武汉钢铁(集团)公司 研究院，湖北武汉 430080；
2. 武汉科技大学材料与冶金学院，湖北武汉 430080）

摘 要 本文采用 Formastor-F 全自动相变仪进行了不同冷却速度下超高强度 X90 钢的连续冷却试验，针对冷却速度对相变开始温度、维氏硬度和显微组织的影响进行了分析研究。结果表明，冷却速度的对数与相变开始温度以及维氏硬度之间均存在线性关系，且随着冷却速度的升高，相变开始温度逐渐降低而维氏硬度逐渐升高；随着冷却速度的升高，X90 钢中铁素体的形貌依次呈现由块状、针状、板条状；同时发现，冷却速度的增加促进钢中 M/A 组织的产生和弥散化，且随着冷却速度增加 M/A 的形貌由块状和粒状逐渐细化与片层化。

关键词 冷却速度，X90，相变行为，M/A

The Effect of Cooling Rate on the Transition Behaviors of Ultra Strength steel X90

Cui Lei[1,2], Xu Jinqiao[1], Kong Junhua[1], Guo Bin[1],
Li Liwei[1], Zou Hang[1], Liu Wenyan[1]

(1. Reseach and Development Center of WISCO, Wuhan Hubei 430080, China;
2. College of Material and Metallurgy of WUST, Wuhan Hubei 430080, China)

Abstract The continuous cooling experiment of ultra strength steel X90 on Formastor-F fully automatic phase change apparatus was introduced in the paper, in which the effect of cooling rate on phase transition temperature, Vivtorinox hardness and microstructure was researched. It was found that the relationship between logarithm of cooling rate and the value of phase transition temperature / Vivtorinox hardness was linearity. Furthermore, along with the increase of cooling rate, the phase transition temperature reduced while the Vivtorinox hardness rose gradually. The morphology of ferrite in X90 was transform into the shape of massive, acicular and slablike along with the raise of cooling rate. Ulteriorly, the increase of cooling rate promoted the produce and dispersion of M/A, meanwhile the shape of M/A changed from massive and granular into laminar.

Key words cooling rate, X90, transition behaviors, M/A

轧制工艺对厚规格管线钢组织和性能的影响

刘宏亮[1] 黄 健[1] 郑 中[1] 赵 迪[1] 刘清友[2] 姜茂发[3]

（1. 本钢集团有限公司，辽宁本溪 117000；2. 中国钢研科技有限公司，北京 100081；
3. 东北大学材料与冶金学院，辽宁沈阳 110819）

摘 要 通常认为增加精轧应变可以有效细化管线钢组织，从而提高管线钢的强度和韧性，然而，随着产品厚度的增加，该调整方法受设备的限制，成为生产技术瓶颈。本研究采用 Gleeble-2000 热模拟试验机、金相分析、扫描电镜技术等系统研究了轧制工艺对管线钢组织和性能的影响。研究结果表明：厚规格产品韧性不符合标准归因于边部和心部冷却梯度增加，边部以粒状贝氏体组织为主，而心部存在多边形铁素体，这种组织梯度最终引起韧性的恶化。热模拟分析结果表明，通过减小精轧应变增加铁素体孕育时间，可以有效避免这一问题。依据研究结果，结合理论分析制定了减小中间坯厚度的生产工艺，并进行了 X70 和 X80 工业试验，管线钢强度和韧性均有明显增加，韧性不稳定问题得到解决，最终形成了稳定、批量生产厚规格管线钢产品的新工艺。

关键词 轧制，管线钢，组织，性能

Effect of Rolling Process on Microstructure and Properties in Thickness Pipeline Steel

Liu Hongliang[1], Huang Jian[1], Zheng Zhong[1], Zhao Di[1], Liu Qingyou[2], Jiang Mao-fa[3]

(1. Benxi Steel Group Corporation, Benxi 117000, China;

2. China Iron & Steel Research Institute Group，Beijing 100081, China;

3. School of Materials and Metallurgy, Northeastern University, Shenyang 110819, China)

Abstract It is widely acknowledged that increasing fine rolling strain can effectively refine the microstructure, and contributing greatly to improve the strength and toughness of pipeline steel. However, with increasing the thickness of the product, this method restricted by the capacity of equipment, and become bottlenecks in the production. In this paper Gleeble-2000 thermal simulation test machine, metallographic analysis, SEM technology, et al. has been used to study the effect of rolling process on the microstructure and properties in pipeline steels. The results show that: because of the increased gradient of cooling rate between edge and core in thickness specification product, the microstructure of edge is mainly characteristic of granular bainite (GB), but polygonal ferrite (PF) increased in core. This property resulted in toughness deterioration. Thermal simulation analysis results show that, the PF transformation incubation time can be increased by reducing the rolling strain, which is useful to avoid the problem. According to the results and theoretical analysis, the production process of reducing the thickness of intermediate billet is formulated. And industrial experiments were carried out with X70 and X80 pipeline steels. Surprisingly, the strength and toughness were significantly increased, and the problem of toughness instability was solved as expected. At last, a new technology for producing thickness pipeline steel is developed.

Key words rolling, pipeline steel, microstructure, properties

管线钢管纤维素焊条高速环焊接头韧性分析

刘 硕

（宝山钢铁股份有限公司研究院焊接与腐蚀防护技术研究所）

摘 要 针对宝钢 X65 级 UOE 管线钢管，应用纤维素焊条进行高速环缝打底焊接和热焊，填充盖面焊接选用低氢焊条，并对环焊接头进行冲击韧性和 CTOD 断裂韧性分析。同时，将纤维素焊条高速环焊接头与全部低氢焊条环焊接头在冲击韧性和断裂韧性方面进行比较，结果表明：无论是焊态还是消氢处理态，针对 X65 这种较低强度钢管，纤维素焊条高速环焊接头均具有良好的冲击韧性和断裂韧性。纤维素焊条高速环缝焊接在高强度管线钢管中的应用有待于进一步研究。

关键词　X65，纤维素焊条，高速焊接，环焊缝，冲击韧性，断裂韧性

Toughness Analysis for Pipeline Cellulose Electrode High Speed Girth Welding Joint

Liu Shuo

(Institute for Welding and Surface Technology of R&D Center, Baoshan Iron & Steel Co., Ltd.)

Abstract　Cellulose electrodes high speed welding has been used in the root beads and hot beads for Baosteel's X65 pipeline, while low hydrogen electrodes in filler and cap beads. Impact toughness and fracture toughness based on CTOD have been analyzed for the above girth weld joint. Meanwhile, comparison has been conducted in impact toughness and CTOD fracture toughness for cellulose electrode high speed girth weld joint and all low hydrogen electrode girth weld joint. As a result, either the welding state or the hydrogen released state, the impact toughness and CTOD fracture toughness are fairly good for the low strength X65 pipeline cellulose electrode high speed welding joint. More researches are expected to carried out in the application of cellulose electrode high speed welding for high strength pipeline.

Key words　X65, cellulose electrode, high speed welding, girth weld, impact toughness, fracture toughness

Effect of Strain Aging on the Tensile Property of an X100 Pipeline Steel with Different Microstructures

Lu Chunjie,　Qu Jinbo,　Yang Han

(Institute of Research of Iron and Steel (IRIS), Sha-steel, Zhangjiagang 215625, China)

Abstract　Effect of strain aging on the tensile property of an X100 pipeline steel with different microstructures was investigated by optical microscope, scanning electron microscope, and so on. It is found that the strength increases, while elongation decreases as the increase of pre-strain, but both have virtually no change with the increase of either aging temperature or aging time. Granular bainite plus polygonal ferrite (dual-phase) microstructure exhibits lower yield ratio than granular bainite (single-phase) microstructure in the as-rolled state, but both microstructures show similar yield ratios after strain aging. The extent of increase in strength due to strain aging is higher for the plates with either dual-phase microstructure or low finish cooling temperature. The amount of martensite/austenite (M/A) constituents is also found to have a significant effect on the strain aging behavior. A higher amount of M/A constituents leads to a lower yield ratio, in both dual-phase and single-phase microstructures, before and after strain aging.

Key words　pipeline steel, strain aging, microstructure, martensite/austenite

Influence of Centreline Segregation on Charpy Impact Properties of Line Pipe Steel

Su Lihong[1], Li Huijun[1], Lu Cheng[1], Li Jintao[1], Leigh Fletcher[1], Ian Simpson[1], Frank Barbaro[1], Zheng Lei[2], Bai Mingzhuo[2], Shen Jianlan[2], Qu Xianyong[2]

(1. School of Mechanical, Materials and Mechatronics Engineering, University of Wollongong, Wollongong, Australia; 2. Baoshan Iron & Steel Co., Ltd., Shanghai, China)

Abstract Centreline segregation occurs as a positive concentration of alloying elements in the mid-thickness region of continuously cast slab. Depending upon the severity of elemental segregation, variations in microstructure and non-metallic inclusions through the final product thickness will affect mechanical properties and potentially downstream processing such as weldability, particularly for high strength line pipe. Charpy impact tests were conducted API 5L grade X65 line pipe steel hot rolled from slabs with different levels of segregation, rated as Mannesmann 2.0 and 1.4. The results showed that centreline segregation does have an influence on the Charpy impact properties of line pipe steel. The segregation fraction in hot rolled stripsamples was in accordance with that assessed in the cast slabs and the segregated regions in hot rolled strip samples were found to be discontinuous. Charpy specimens located in segregated regions resulted in lower Charpy impact energy and exhibited linear features associated with microstructural features in the mid-thickness region of the rolled strip. The difference in recorded Charpy toughness can be considered minor and associated with the observed differences in pearlite structure at the centreline and level of MnS inclusions.

Key words centreline segregation, charpy impact test, line pipe steel

高性能钢组织控制及ϕ1422×32mm K60-E2 低温管线管用钢板开发

聂文金[1] 张晓兵[1] 林涛铸[1] S.V.Subramanian[2] 马小平[2] 尚成嘉[3]

（1.江苏沙钢集团有限公司总工办，江苏张家港 215625；2. 麦克马斯特大学材料科学与工程系，加拿大；3.北京科技大学材料科学与工程学院，北京 100083）

摘 要 本文基于高性能钢组织调控技术，获得了 3 种中温转变组织类型的高强度宽厚钢板，采用金相及 EBSD 技术分析了这 3 种钢的组织结构与力学性能之间的关系，发现为获得高韧性、低屈强比、高性能厚规格管线钢板，组织结构中要有高位错密度组织和高比例晶体缺陷，同时，要抑制大尺寸 MA 岛组元的形成。沙钢基于该类组织结构的控制技术，在国内率先开发出难度极高的大口径厚壁ϕ1422×32mm K60-E2 低温管线管用热轧钢板。钢板在俄罗斯某管厂进行制管试验，钢板质量和所制直缝埋弧焊管性能均满足俄天然气公司技术要求，包括钢管高压水爆试验。

关键词 宽厚管线钢板，高性能，组织控制，EBSD

Microstructure Control of the High Performance Steel and Development of Heavy Pipeline Steel Plates with High Toughness at a Low Temperature for ϕ1422×32mm K60-E2 LSAW Pipes

Nie Wenjin[1], Zhang Xiaobing[1], Lin Taozhu[1], S.V.Subramanian[2], Ma Xiaoping[2], Shang Chengjia[3]

(1.Chief Engineer Office, Jiangsu Sha-Steel Group Co., Ltd., Zhangjiagang 215625, China; 2. Department of Material Science and Engineering, McMaster University, Canada; 3. Department of Material Science and Engineering, University of Science and Technology Beijing, Beijing 100083, China)

Abstract Based on the microstructure-controlling technology of a high performance pipeline steel, three kinds of high strength heavy pipeline steel plates with intermediate transaforamtion structures have been produced. The microstructures cooperated with the mechanical properties have been studied by optical microstructure and EBSD technology. It is found that a higger density of dislocation, a more crystal defects, and a smaller size MA islands in macrostructure are necessary to get the low-temperature high toughness and the lower of the yield ratio of high strength heavy pipeline steel plates. The high strenth and low-temperature high toughness heavy pipeling steel plates were developed in Shagang with the help of the microstructure-controlling technology for ϕ1422×32mm K60-E2 LSAW pipes. The trial of pipe manufacuring was carried out in a Russia pipe factory and all test results are met the requiremnets of Gazpron including the high water pressure blast test of weled pipes.

Key words　heavy pipeling steel plates, high-performance steel, microstructure control, EBSD

Line Pipe for the QCLNG Pipeline in Queensland, Australia

Fletcher Leigh [1], Lei Jin[3], Gui Guangzheng[2], Zheng Lei[2]

(1. Welding and Pipeline Integrity, Braidwood, Australia; 2. Baoshan Iron & Steel Co., Ltd., Shanghai 200941, China; 3. Bao Australia Pty Ltd.)

Abstract　Baosteel China was awarded a contract by QGC, a BG Group business for the manufacture of 550km of UOE 42-inch longitudinal seam SAW API 5L X70 steel pipe for its Queensland Curtis LNG Project. The pipe was manufactured and coated in Shanghai, with deliveries to Brisbane and Gladstone starting in September 2010 and continuing until June 2011. The pipe represents the collection header pipeline and 340km main export line that is transporting gas to Gladstone from QGC's coal seam gas fields in the Surat Basin in southern Queensland. The pipeline was the first of three large diameter gas transmission pipelines laid onshore in Australia for the export liquefied natural gas, and was the first major export of high strength UOE line pipe outside of China. The paper describes risk management processes and technical specification of pipe for the project and Baosteel's process capability and quality management that have enabled the stringent requirements to be met for this very important project.

Key words　UOE line pipe, quality control, gas transmission pipeline

宝钢 X70M 厚板管线钢质量控制

王　波

（宝钢股份公司制造管理部，上海　201900）

摘　要　为了应对管线钢用户对 X70M 厚板管线钢在夹杂物、板坯偏析、强度波动范围和 DWTT 性能等方面越来越严苛的要求，宝钢采用产品质量先期策划（APQP）方法和一贯制质量管理模式，优化生产工艺，识别关键工序控制点并加强过程稳定性控制，明显提高了 X70M 厚板管线钢的性能稳定性和内质水平,生产的钢板及加工后的钢管满足用户的高标准要求。

关键词　X70M 厚板管线钢，质量控制，产品质量先期策划，夹杂物，偏析，性能稳定性

Quality Control of X70M Pipeline Plate in Baosteel

Wang Bo

(Products & Technique Management Department of Baoshan Iron & Steel Co., Ltd., Shanghai 201900, China)

Abstract In order to deal with more stringent requirements for X70M pipeline plates in inclusions, slab segregation, strength fluctuation range and the DWTT performance, Advanced Product Quality Planning and Control Plan (APQP) and integrated quality management mode were adopted to optimize production process, to identify key control process and to strengthen the stability of process control. Mechanical property stability and internal quality performance level of these plates were correspondingly increased, plates and pipes made from these plates also meet the user's high standard requirements.

Key words X70M pipeline plates, quality control, APQP, inclusions, segregation, stability of mechanical property

具有优异低温韧性和抗应变性能的X80管线钢板/管显微组织识别

Andrea Di Schino[1] 郑 磊[2] 章传国[2] Giorgio Porcu[1]

（1. Centro Sviluppo Materiali SpA, Via di Castel Romano 100, 00128 Roma, Italy;
2. 宝山钢铁股份有限公司，上海 201900）

摘 要 随着全球对天然气能源需求的增长，长输管道建设向地震、北极等区域延伸，而由于存在永冻现象这些区域易产生大的地面位移,为确保永冻区域管道的安全运行，要求其所用管材需具有抗应变能力，即管体纵向具有低的屈强比、高均匀延伸率以及拱顶（Roundhouse）型应力应变曲线特征，同时管体横向具有高的强度，此外还需确保具有高的低温韧性。合适的钢板设计是获得合适钢管性能的基础，就板材设计而言，显微组织控制是关键因素，而这需通过优选的化学成分和正确的工艺设计来实现。宝钢和CSM公司合作开展了X80应变设计钢管开发合作项目，宝钢研究院首先进行了实验室轧制实验以获得不同的显微组织，结果表明除了化学成分、轧制工艺参数等影响显微组织的形成外，同时开冷温度、停冷温度等冷却参数也有显著影响。本文对不同材料的力学性能进行评估，同时识别出具有优异应变潜能的X80钢管用钢板的显微组织。

关键词 应变设计，管线钢，组织性能

Microstructure Identification for X80 Plates/pipes with Low Temperature Toughness and Enhanced Strain Capacity

Andrea Di Schino[1], Zheng Lei[2], Zhang Chuanguo[2], Giorgio Porcu[1]

(1. Centro Sviluppo Materiali SpA, Via di Castel Romano 100, 00128 Roma, Italy;
2. Baoshan Iron & Steel Co., Ltd., Shanghai 201900, China)

Abstract Due to the increasing demand for natural gas, the construction of long-distance pipelines through seismically active regions or arctic regions with ground movement caused by permafrost phenomena will become more and more necessary. To guarantee the safe operation of those pipelines, the pipe material has to fulfill strain-based design requirements. Hence in longitudinal direction low yield-to-tensile ratios, high uniform elongation values and a roundhouse shape of the stress-strain curve combined with sufficient strength values in transverse direction are essential. Moreover, satisfactory low temperature toughness has to be guaranteed. An adequate plate metallurgical design is fundamental for appropriate pipe properties achievement. As far as concerns the plate design the understanding and the control of microstructure are the key factors, achieved by an adequate steel chemical composition and proper process parameters. In the framework of a co-operation between Baosteel and Centro Sviluppo Materiali (CSM), a project has been started aimed at manufacturing X80 strain based designed pipes. As a starting point pilot trials have been carried out at Baosteel Research Center in order to produce different microstructures. Besides the steel chemical composition, the cooling process has the most significant influence on the formation of the microstructure: in order to assess the effect of the cooling process, the same rolling schedule was adopted for producing the different test materials, obtained varying the start cooling and finish cooling temperatures.

The microstructure and mechanical properties of the different test materials were assessed and the best microstructure for the plates for X80 pipes with enhanced strain capacity has been identified.

Key words strain-based design, pipeline steel, microstructure and mechanical properties

Characterization of a Low Manganese Niobium Microalloyed Pipeline Steel for Sour Service

Gonzales Mario[1], Hincapié Duberney[1], Goldenstein Helio[1],
Barbaro Frank [2], Gray Malcolm[3]

(1. Department of Metallurgical and Materials Engineering, Polytechnic School, University of Sao Paulo. Sao Paulo, SP Brazil;
2. Department of Materials, Mechanical & Mechatronics, University of Wollongong, Wollongong, Australia;
3. Microalloyed Steel Institute, Houston Texas, United States)

Abstract The safe transport of sour gas requires steels with enhanced hydrogen induced cracking (HIC) resistance. Centerline segregation bands of high hardness microconstituents and non-metallic inclusions are the heterogeneities usually linked to HIC failures in pipeline steels. Full-scale production heats of a low manganese, niobium microalloyed steel were produced by different steel mills for application in sour gas environments.

This new steel concept has been manufactured industrially in order to demonstrate the capability to achieve high strength and fracture toughness at low Pcm and with improved control of the centerline segregation and of the inclusion dispersion. Two different pipes with low manganese level (<0.29wt%) were tested to evaluate the manganese induced effect of sulphur solubility on inclusion dispersion.

The new alloy design was able to provide both effective strengthening and exceptional fracture toughness, achieving API 5L X65 and in some cases up to X80. Analysis of the pipe microstructure have been performed and consisted of fine grained ferrite with a small volume fraction of uniformly dispersed inclusions. Most importantly, the microstructure was uniform in both the longitudinal and transverse directions and exhibited exceptional consistency along the thickness. Samples were submitted to mechanical property and HIC tests according to API 5L and NACE standards TM0284-2003 without any sign of cracking. The steel concept was compared with the steel market to API X65 pipe and showed similar performance. The results, so far, confirm that this new alloy design provides the unique opportunity to economically produce a high strength pipe steel grade with improved sour service resistance.

Key words API X65, sour service, HIC, H_2S, low manganese, niobium

特殊需求的 UOE 管线管开发及应用

谢仕强 王 波 黄卫锋 郑 磊 张 备 章传国 徐国栋 吴扣根

（宝山钢铁股份有限公司，上海 201900）

摘　要　介绍了宝钢 UOE 管线管生产线的特点。分析了近年来针对国际管线工程对管线管在可焊接性判定、强度、韧性、外观几何尺寸及检验方法等方面的一些特殊需求及发展特点,并总结了宝钢在具有特殊需求的 UOE 管线管方面的开发进展和工程应用。

关键词　UOE，管线管，特殊需求，开发，应用

Development Experience of UOE Line Pipe with Special Requirements

Xie Shiqiang, Wang Bo, Huang Weifeng, Zheng Lei, Zhang Bei,
Zhang Chuanguo, Xu Guodong, Wu Kougen

(Baoshan Iron & Steel Co., Ltd., Shanghai 201900, China)

Abstract　Some characteristics of baosteel's UOE pipe processing line was introduced in this paper. Depending on pipeline projects over the world in recent years, some special requirements and the charaateristics of the linepipe, such as weldability determination, strength, toughness, geometry size and appearance inspection methods etc., were analysed. And the development and application of the UOE line pipe with the special requirements at Baosteel were also introduced in this paper.

Key words　UOE, linepipe, special requirements, development, application

退火工艺对高强度热轧带钢组织和性能的影响

孙磊磊　章传国　郑　磊

（宝山钢铁股份有限公司研究院，上海　201900）

摘　要　热轧高强度贝氏体钢的显微组织和力学性能对温度有较高的敏感性。本实验对一种热轧高强钢在 500℃、540℃、580℃、620℃进行了不同时间的退火处理，利用金相显微镜和扫描电子显微镜分析了显微组织的变化，并检测拉伸性能。结果表明：热轧态组织主要为贝氏体和 M/A 组元，在较低温度，即 500℃、540℃退火时，有少量 M/A 分解成碳化物，屈服强度有一定程度增加，抗拉强度变化不显著；在较高温度，即 580℃、620℃退火时，随着时间的增加，组织中的 M/A 组元基本全部分解成碳化物，贝氏体子板条界面逐渐消失，铁素体含量增加，材料的屈服强度受析出强化和贝氏体组织退化的综合作用，随着保温时间的增加先显著升高，再呈现下降趋势，抗拉强度则因 M/A 分解而随保温时间的增加逐渐下降。

关键词　热轧高强钢，退火，贝氏体，显微组织，力学性能

The Influences of Annealing Process on the Microstructure and Mechanical Properties of Hot-rolled High-strength Steel

Sun Leilei, Zhang Chuanguo, Zheng Lei

(Research and Development Institute, Baoshan Iron & Steel Co., Ltd., Shanghai 201900, China)

Abstract　The microstructures and mechanical properties of hot-rolled high-strength bainite steel are highly sensitive to processing temperature. In present work, the hot-rolled steel specimens were annealed at 500, 540, 580 and 620°C, the mechanical properties of which were meassured and the microstructures were investigated by metallographical microscope and SEM. The original microstructures before annealing were composed of bainite and M/A mainly. After annealed at 500 and 540°C, part of M/A decomposed. The yield strength increased while tensile strength changed slightly. However, the microstructures of the specimens annealed at 580 and 620°C changed notably. Almost all the M/A decomposed to carbides, and the laths interfaces of bainite degenerated and transformed to ferrite. The yield strength increased at first and then declined with the prolonged annealing time, owing to the combined effects of precipitation strengthening and degeneration of bainite, while the tensile strength decreased notably caused by the decomposition of M/A.

Key words　hot-rolled high-strength steel, annealing, bainite, microstructure, mechanical property

X100 管线钢板调质的组织与性能

陈定乾

（湘潭钢铁有限公司宽厚板厂，湖南湘潭　411101）

摘　要　选择 14.2mm X100 管线钢板作为代表，进行试验性研究。实验结果表明 910℃淬火后，随着回火温度的升高，X100 样板屈服强度先增加后降低，抗拉强度逐渐降低，605℃回火降低到最低值 834MPa，从而导致屈强比呈现升高趋势；605℃回火，落锤剪切面中脆性区增多，落锤性能变差；870℃两相区淬火组织中奥氏体晶粒大幅度细化，经过 530℃回火后，细化的马氏体或者贝氏体组织中出现亚结构的回复软化，板条边界钝化和 M/A 组元分解产生的析出强化机制联合作用结果使得该条件下 X100 管线钢板获得最佳的综合性能。

关键词　X100 管线钢板，淬火，回火，力学性能，微观组织

Research on Microstructure and Properties of Quenching X100 Pipeline

Chen Dingqian

(Heavy Plate Factory, Xiangtan Iron and Steel(XISC) Co., Ltd., Xiangtan Hunan 411101, China)

Abstract　14.2mm X100 pipeline with different quenching process were investigated by means of mechanical property tests and microstructure analysis. The results show that after quenching at 910℃, the yield strength the first increased and then declined as the tempering temperature increased, while tensile-strength decreased gradually and reached the lowest 834MPa at 605℃, those were result to the ratio of yield strength increased and the DWTT value became the worst. Two-phase quenching at 870℃ and the tempering at 530℃ X100 pipeline steel plate had the optimum comprehensive properties. This could be attributed to the fine-grand austenite microstructure after quenching at 870℃, the combined effect of recovery and softening of the dislocation sub-structure, the blunting mechanism along strip boundaries and the precipitation strengthening from decomposition of massive M-A after tempering at 530℃.

Key words　X100 pipeline steel plate, quenching, tempering, mechanical properties, microstructure

基于CRACKWISE的X65管线直焊缝焊接缺陷工程临界评估

曹 能 王怀龙

（宝山钢铁股份有限公司研究院焊接与腐蚀防护技术研究所，上海 201900）

摘 要 文章以ϕ762mm×28.6mm（X65）的深海管线为研究对象，对管线直焊缝焊接接头区域的断裂韧性进行了测定，同时利用CRACKWISE软件对管线服役过程中直焊缝的缺陷容忍度进行了评估。对管线服役条件下所允许的最大裂纹尺寸及设计寿命下最大初始裂纹尺寸进行了计算，为管线生产过程中缺陷超标焊管的判废与否提供判断依据。

关键词 工程临界评估ECA，断裂韧性，CTOD，管线

Engineering Critical Assessment on Welding flaw of X65 UOE Seam Weld

Cao Neng ,Wang Huailong

(Baoshan Iron & Steel Co., Ltd., R&D Center, Department for Welding and Corrosion Protection, Shanghai 201900, China)

Abstract In this article, we focused on ϕ762mm×28.6mm（X65）offshore UOE, fracture toughness of its seam weld had been tested, tolerance of welding flaws had been calculated. The maximum size of welding flaw below which the safe operation of gas transportation can be assured provides an criterion to qualify the quality of UOE pipe.

Key words engineering critical assessment, fracture toughness, CTOD, UOE

管线钢边部翘皮缺陷形成机理探析

焦晋沙 李振山 杨文静 夏碧峰 费书梅 崔全法

（首钢股份公司迁安钢铁公司，河北迁安 064400）

摘 要 本文针对管线钢热轧板出现批量边部翘皮缺陷事件进行了分析研究，利用扫描电子显微镜对边部翘皮缺陷进行了微观分析，发现在翘皮区域出现了铁、锰氧化物和含Fe、Mn、Si、Ti的高温氧化质点，同时对出现缺陷的回炉坯及同浇次库存剩余板坯质量进行了复核，对此次边部翘皮缺陷形成原因进行机理探析。得出板坯切角后的残余铁瘤、豁口缺陷是导致此次边部翘皮缺陷的主要因素，同时热轧过程除鳞不彻底等也是造成边部翘皮缺陷的可能因素，通过加强板坯质量管理，确保板坯无缺陷轧制能够有效降低边部翘皮发生率。

关键词 边部翘皮，扫描电镜，连铸板坯，切角

Formation Mechanism Analysis of the Edge Shell Defect on the Pipeline Steel

Jiao Jinsha, Li Zhenshan, Yang Wenjing,
Xia Bifeng, Fei Shumei, Cui Quanfa

(Shougang Qian'an Iron & Steel Company, Qian'an 064400, China)

Abstract The edge shell defect appeared on the surface of pipeline steel hot rolled plate were analyzed and studied in this paper. The Fe, Mn oxides and Fe, Mn, Si, Ti high temperature oxidation mass in edge shell area were found through microscopic analysis by scanning electron microscope. Reheating and rolling process of slabs and the same cast slabs were also checked to analysis the formation mechanism of edge shell. The analysis showed that the main reason of edge shell was the residual iron tumor and gap defects after cutting slab corner, and other possible reason was the incompletely scale removal process of the hot rolling and so on. Strengthening the slab quality management, in order to ensure slabs without defects could reduce incidence of the edge shell effectively.

Key words edge shell, scanning electron microscopy, continuous casting slab, cutting angle

DWTT in Thicker Gauge Hot Rolled Coils for Line Pipe Application

Pradeep Agarwal, Gunna Venkata Ramana, Devasish Mishra,
Ashish Chandra, Gajraj Singh Rathore

(JSW Steel Ltd., Vijayanagar Works. Toranagallu, Karnataka, India)

Abstract Ductile failure in DWTT (Drop Weight Tear Test) at subzero temperatures is one of the key requirements for line pipe steels. As the temperature drops or thickness increases, the ductile failure mode is difficult to achieve in DWTT testing. JSW Steel has successfully developed new generation Nb-microalloyed line pipe steels for heavy gauges (up to 20mm) for grades up to API X-80 grade for the first time in India through hot strip mill. It has been observed that prior austenite grain boundaries act as major crack arresters due to the high angle of misorientation at the grain boundaries. In order to achieve ductile DWTT failure, the crack propagation by brittle Griffith cracks needs to be arrested by high angle boundaries with their interspacing lesser than that of the critical Griffith crack length. Hence refinement of prior austenite grain size which requires optimization of multiple dynamic and static recrystallisation phenomena coupled with reduced grain growth at rolling temperatures, leads to improved DWTT performance.

This paper describes the various aspects of steel chemistry design and the hot rolling process optimization to achieve subzero temperature DWTT in thicker gauge line pipe steels. Multiple online trials have been conducted at JSW Steel, Vijayanagar works for achieving the same, such that the effects of chemistry and process variables on final microstructures and mechanical properties have been established. Further, a correlation table has been prepared for predicting DWTT transition temperature based on the largest prior austenite grain size in the transfer bar microstructure before entering into the finishing mill. This table has proven to be very useful in alloy design and in optimizing hot rolling process window to achieve superior DWTT performance for different end thicknesses in line pipe steels.

Key words DWTT (drop weight tear test), dynamic recrystallisation, static recrystallisation, prior austenite grain size

俄罗斯高硅管线钢 K52 的组织性能研究

安成钢[1] 牛 涛[1] 吴新朗[2] 姜永文[1] 代晓莉[1] 陈 斌[1]

（1. 首钢技术研究院，北京 100043；2. 首钢股份公司迁安钢铁公司，河北迁安 064404）

摘 要 本文对俄罗斯高硅管线钢 K52 的成分、性能及组织进行了研究，并将其与 API X56 管线钢进行了对比研究。研究结果表明，K52 在成分上采用高硅低锰的成分体系，在强度上与 API X56 管线钢相近。但 K52 对塑形和韧性要求更高，要求屈强比≤0.86，–20℃ V 型冲击功和–60℃的 U 型冲击功。对于俄罗斯高硅管线钢 K52，由于其高硅低锰的成分体系，热轧过程中其氧化铁皮较为严重，较低的出炉温度可以减少氧化铁皮厚度。K52 在焊接过程中易产生含硅氧化物夹杂，对焊缝冲击韧性造成不利影响。

关键词 K52，高硅含量，管线钢，氧化铁皮

Microstructure and Mechanical Properties of Russian High-silicon K52

An Chenggang[1], Niu Tao[1], Wu Xinlang[2], Jiang Yongwen[1], Dai Xiaoli[1], Chen Bin[1]

(1. Shougang Research Institute of Technology, Beijing 100043, China;
2. Shougang Qian'an Iron & Steel Co., Ltd., Qian'an 064404, China)

Abstract In this paper, the composition, mechanical properties and microsruture of Russian high-silicon pipeline steel K52 were studied and compared with API X56. The results showed that the strength of K52 with high-silicon and low-manganese was similar to the strength of API X56. The moulding and toughness standards of K52 were higher than API X56. The yield ratio should be lower than 0.86. The higher impact energy with V-notch specimen at –20℃ and U-notch specimen at –60℃ should be satisfied. The oxide scale of K52 was more thichness because of the high-silic. The thickness of oxide scale could be reduced by the lower tapping temperature during hot rolling. The inclusion of silicon oxide was bad fot the toughness of welding line, which was forming during the welding process.

Key words K52, high-silicon, pipeline steel, oxide scale

轧后冷却速度和 Mo 含量对 X80 管线钢组织和性能的影响研究

周 峰[1] 刘自立[2]

（1. 宝钢集团广东韶关钢铁有限公司，广东韶关 512199；
2. 宝钢集团中央研究院，上海 201999）

摘 要 为适应管线工程经济性和安全性的要求，管线钢必须具有优良的强度、韧性、焊接性、耐腐蚀能力，目前国内外主要采用微合金化及控轧控冷工艺生产 X80 管线钢。本文进行了热模拟实验及中试实验研究，结果表明不

同的 Mo 含量对 X80 管线钢的组织和性能有显著的影响，Mo 可以抑制先共析铁素体和珠光体的形成，降低针状铁素体形成的冷却速度，扩大针状铁素体的形成范围。采用传统加速冷却工艺的高 Mo 管线钢与采用新型超快速冷却工艺的低 Mo 管线钢微观组织类型相同。因此采用新型超快速冷却工艺降低 X80 管线钢中贵重金属 Mo 降低生产成本完全可行。

关键词 超快速冷却，X80 管线钢，Mo 含量，微观组织

Study on Effects of Cooling Speed after Rolling and Mo Content on the Microstructure and Mechanical Properties of X80 Pipeline Steel

Zhou Feng[1], Liu Zili[2]

(1. Baosteel Group Guangdong Shaoguan Iron and Steel Co., Ltd., Shaoguan 512199, China;
2. Central Research Institute of Baosteel Group, Shanghai 201999, China)

Abstract In order to adapt to the requirement of the economy and safety of pipeline engineering, pipeline steels require high strength, toughness, weldability and resistance corrosion , At present Microalloying and thermal-mechanical control processing (TMCP) are the main technique for the product of X80 pipeline steels.In this paper, through laboratory study and pilot production,Results showed that different levels of Mo addition had a remarkable effect on the microstructures and mechanical properties of the investigated pipeline steels. The pro-eutectoid ferrite and pearlite formation was inhibited in the high-Mo steel and acicular ferrite was obtained over a wide range of cooling rates, whereas the dominant acicular ferrite microstructure could only be obtained when the cooling rates. Very similar microstructures and mechanical properties were obtained in the low-Mo steel produced with ultra-fast cooling and in the high-Mo steel produced by the conventional accelerated continuous cooling. It was proved by simulation and pilot production that high strength low alloy steels such as pipeline steels, could be produced using the novel ultra fast cooling which also reduce alloy Mo cost.

Key words UFC, X80 pipeline steel, Mo content, microstructures

管线管 DWTT 试样方法对试验结果的影响

柏明卓　郑　磊

（宝钢集团有限公司中央研究院，上海　201900）

摘　要 DWTT 是测试管线钢韧性性能指标的重要试验，本文通过对比不同试样方法的 DWTT 试验，研究了缺口类型及试样减薄和减薄方式对 DWTT 性能结果的影响。试验结果表明，压制缺口和人字缺口两种缺口类型对剪切面积比的判定没有很大影响，但在高韧性钢中压制缺口容易引起异常断口，采用人字缺口可以有效减少异常断口的发生，可以增加检测数据的稳定性。对于大厚度管线钢，壁厚尺寸效应对 DWTT 脆性转变的影响十分显著，按 API 标准等效温度条件下减薄试样往往提高相应 SA%判定值。减薄试样更能反映材料本身的韧脆转变特性，内壁减薄、双面减薄、外壁减薄对 DWTT 性能结果的影响也有所不同。

关键词 DWTT，管线管，试样方法

Effect of Specimen Preparation on Result of Linepipe DWTT Test

Bai Mingzhuo, Zheng Lei

(Central Research Institute of Baosteel Group Co., Ltd., Shanghai 201900, China)

Abstract Drop-weight tear test is one of the most important property tests on the linepipe steels. In the paper, the effects of different specimen preparation by different notch and thickness reduction methods were studied. The results showed pressed notch and chevron notch had little effect on the shear area percent of DWTT specimens, but specimens with pressed notches increased the frequency of abnormal fractures. Chevron notched specimens helped to increase the reliability of test results. Thickness effect showed significant to the ductile-brittle transition temperature, the thickness-reduced specimens showed better shear area percent than the full-wall specimens on the temperature reduction conditions with the thickness reduction as per API specification. The thickness-reduced specimens showed the natural material ductile-brittle transition, and the way of the thickness reduction had the different effect on the DWTT by machining the inside surface, the outside surface or both surfaces.

Key words DWTT, linepipe, specimen preparation

X65 管线钢连续冷却转变行为研究

刘文艳　徐进桥　袁桂莲　黄治军　郑江鹏　缪　凯

（武钢研究院，湖北武汉　430080）

摘　要 用膨胀法结合金相-硬度法，在 Formastor-F 相变仪上测定了 X65 管线钢过冷奥氏体连续冷却转变曲线(即 CCT 图)，得到了不同冷却速度条件下，X65 管线钢的组织转变和硬度变化特征。结果表明：在极快的冷却速度条件下(100~150℃/s)，转变产物为贝氏体。在 20~50℃/s 冷速范围内，得到贝氏体和铁素体组织，随着冷却速度减慢(<50℃/s)，贝氏体数量开始减少，铁素体数量增加。冷却速度为 5℃/s 时开始形成珠光体组织，在 0.05~5℃/s 冷却速度范围内，得到的组织为铁素体和珠光体。实际生产中，在 20~50℃/s 冷却速度范围内，可以得到以贝氏体和铁素体为主的组织。奥氏体向铁素体转变温度随着冷速的增加而降低，在 0.05~50℃/s 冷却条件下，铁素体转变开始温度范围为 667~758℃。冷却速度为 10~50℃/s，维氏硬度平均值为 166~185 HV。150℃/s 极快的冷却速度条件下，硬度平均值为 207 HV，说明 X65 钢淬硬倾向小。

关键词 X65，CCT 图，贝氏体，铁素体，珠光体

Continuous Cooling Transformation of X65 Pipeline Steel

Liu Wenyan, Xu Jinqiao, Yuan Guilian, Huang Zhijun, Zheng Jiangpeng, Miao Kai

(Research and Development Center, Wuhan Iron & Steel (Group) Corp., Wuhan 430080, China)

Abstract Continuous cooling transformation (CCT) diagram was determined for X65 pipeline steel based on the dilatometric method by using Formastor-F dilatometer. Microstructures and properties of X65 were demonstrated under the condition of different cooling rates. The CCT diagram shows that the microstructure of steels cooled at very rapid cooling rates (100~150℃/s) exhibits bainite. The microstructures at the cooling rates arranging from 20 to 50 ℃/s are characterized by bainite and ferrite. The slower the cooling rates become, the more ferrite forms and the less bainite forms when the cooling rate is slower than 50 ℃/s. Pearlite begins to form at the cooling rate of 0.5 ℃/s. Ferrite and pearlite are obtained over a wide range of cooling rates (0.05~5 ℃/s). Bainite and ferrite are ideal microstructures in production line and can be obtained when the cooling rates are selected between 20~50℃/s. The transformation temperature of $\gamma \rightarrow \alpha$ that decreases with the cooling rates increasing is between 667~758℃. The average Vickers hardness value is 166~185 HV under the condition of 10~50℃/s. The Vickers hardness obtains 207 HV when the cooling rate is 150℃/s, which demonstrates that the steel has a low quench-hardening tendency.

Key words　X65, CCT diagram, bainite, ferrite, pearlite

石油套管 J55 边裂缺陷的分析与研究

李　强　王文录　孙　毅　张志强

（邯郸钢铁集团公司技术中心，河北邯郸　056015）

摘　要　在连铸轧制的钢板表面有沿轧制方向的裂纹。采用化学成分分析，宏、微观检验等方法对裂纹进行了分析。结果表明，裂纹中存在氧化物及其脱碳等缺陷，这说明连铸铸坯在轧制前已存在裂纹并在轧前加热炉内裂纹发生氧化和脱碳，导致轧制后的钢板表面出现裂纹，分析了连铸坯质量引起边裂缺陷的机理，并提出了相应的预防措施。

关键词　表面边裂，金相检验，连铸铸坯

Analysis and Research of J55 Casing Edge Crack Defects

Li Qiang, Wang Wenlu, Sun Yi, Zhang Zhiqiang

(Handan Iron and Steel Group Company Technology Center, Handan Hebei 056015, China)

Abstract　There are some longitudinal surface cracks appear on the continuous casting slab and consequently on the rolling steel plate.By means of the chemical composition analysis, macro-scopic observation and microstructure examination, the results showed that there is oxide in the crack, and the decarburizing and the pit-oxide are exsited in the continuous casting and rolling process. All these factors caused the longitudinal surface cracks appeared on the steel plate after rolling.

Key words　the longitudinal surface cracks, metallographical examination, continuous casting slab

精轧压下比对厚规格管线钢 DWTT 性能的影响

章传国[1,2]　郑　磊[2]　孙磊磊[2]　丁　晨[2]　翟启杰[1]

（1. 上海大学材料科学与工程学院，上海　200072；2. 宝山钢铁股份有限公司，上海　201900）

摘　要　采用实验室轧制试验和 EBSD 分析方法，研究了精轧压下比对 30mm 规格 X65 管线钢组织性能的影响规

律，重点分析了 DWTT 韧性的改善机理。试验结果表明，精轧压下比由 2.1 倍上升至 6.3 倍时，试验钢板的强度及夏比冲击韧性变化不明显，精轧压下比为 4~5 倍时获得最优的 DWTT 性能。显微组织分析表明，4.1 倍压下比时试验钢相变得到细小均匀的多边形铁素体显微组织，由于多边形铁素体主要以大角度晶界为主，且同位晶界比例较高，晶界能量较低，有利于提升钢的抗动态撕裂性能。

关键词 管线管，厚规格，组织性能，压下比，EBSD

Effect of Transfer Bar Ratio on DWTT Performance for Heavy Gauge Pipeline Steel

Zhang Chuanguo[1,2], Zheng Lei[2], Sun Leilei[2], Ding Chen[2], Zhai Qijie[1]

(School of Material Science & Technology, Shanghai University, Shanghai 200072, China;
2. Research Institute, Baoshan Iron & Steel Co., Ltd., Shanghai 201900, China)

Abstract By using the way of pilot rolling and EBSD, the effect of transfer bar ratio on microstructure and mechanical properties were studied for 30mm X65 pipeline steel, enphasized on the analysis of the mechanism for DWTT improvement. The results show that there are small change on strength and Charpy impact energy when the transfer bar ratio increased from 2.1 to 6.3, while the better DWTT performance were obtained with the transfer bar ratio of 4~5. The microstructure characterized by uniform and fine polygonal ferrit which mainly contain large angle boundary and CSL boundary. The low energy for large angel boundary and CSL boundary is helpful to advance the resistance to crack propagation.

Key words line pipe, heavy gauge, microstructure and properties, transfer bar ratio, EBSD

消应力退火对 X100 级管线管组织性能的影响

张清清　章传国　郑　磊

（宝钢集团中央研究院钢管技术中心，上海　201900）

摘　要 本文研究了 610℃消应力退火热处理对 X100 管线管的影响规律，通过金相显微镜表征了退火前后 X100 管线管的显微组织，对比分析了退火前后管线管的室温拉伸及低温冲击性能，并对断口形貌进行了观察。结果表明，退火热处理后 X100 管线管显微组织中贝氏体铁素体含量减少，多边形铁素体含量增多。X100 管线管的横向和纵向抗拉强度下降约 60~70MPa，延伸率提高 2%~3%，–20℃和–40℃低温冲击韧性有所下降，退火后出现断口分离。

关键词 X100 管线管，退火，显微组织，强度，冲击韧性

Effects of Stress Relief Annealing on the Microstructures and Properties of X100 Steel Pipe

Zhang Qingqing, Zhang Chuanguo, Zheng Lei

(Center Research Institute of Baosteel Group, Shanghai 201900, China)

Abstract The effects of stress relief annealing at 610℃ on the microstructures and mechanical properties of X100 steel pipe were studied. The microstructure were characterized by optical microscope, scanning electric microscope and the mechanical properties were tested before and after annealing. The results show that the volume fraction of lath bainite decreases and the volume fraction of ferrite increases after annealing. Both the transverse and longitudinal tensile strength decrease by 60~70 MPa, while the elongation increases by 2%~3%. The impact toughnesses at –20℃ and –40℃ decrease and notch separations occur in the material after annealing.

Key words X100 steel pipe, annealing, microstructure, strength, impact toughness

6.4 不锈钢/Stainless Steel

6.4.1 铁素体及马氏体不锈钢 Ferritic and Martensitic Stainless Steel

Oxidation and Sticking of Stainless Steels in Hot Rolling

Zhao Jingwei[1], Jiang Zhengyi[1], Cheng Xiawei[1], Hao Liang[1], Wei Dongbin[1], Ma Li[2], Luo Ming[2], Peng Jianguo[2], Du Wei[2], Luo Suzhen[2], Jiang Laizhu[2]

(1. School of Mechanical, Materials and Mechatronic Engineering, University of Wollongong, NSW 2522, Australia;
2. Stainless Steel Technical Centre, Baosteel Research Institute (R&D Centre), Baoshan Iron & Steel Co., Ltd., Shanghai 200431, China)

Abstract The continuous oxidation behaviour of a B445J1M ferritic stainless steel (FSS) was investigated in humid atmosphere containing 18% water vapour in the temperature range of 1000 to 1200℃. The isothermal oxidation behaviour of indefinite chill (IC) roll material was also studied in both dry and humid atmospheres at a temperature range of 550 to 700℃. The results indicate that the formation of Cr-rich oxide follows a parabolic law over time before breakaway oxidation occurs, and breakaway oxidation occurs at 1150℃ for the B445J1M steel. Temperature of 1150℃ is critical for the B445J1M steel that breakaway oxidation takes place and more oxides begin to form. In the dry air, the oxidation kinetics of the IC roll show that the mass gain raises gradually with the increase of temperature until 650℃, and the mass gain at 700℃ is lower than that at 650℃. In the humid atmosphere, however, the mass gain shows a linear trend and obviously goes up with the increase of oxidation temperature. The oxide scale formed in humid atmosphere adheres firmly to the matrix, whereas the oxide scale generated in dry air tends to peel off. In order to solve the sticking problem of FSS B445J1M, reheating temperature higher than 1150℃ before hot rolling and practical application of IC work rolls are suggested.

Key words ferritic stainless steel, hot rolling, sticking, oxidation

Current Status and Future Prospect of Ultra-pure Ferritic Stainless Steels in Baosteel

Jiang Laizhu[1], Hu Xuefa[2]

(1. Baosteel Research Institute, Shanghai, China; 2. Baosteel Stainless Co., Ltd., Baosteel

Group, Shanghai, China)

Abstract The ultra pure ferritic stainless steels have been extensively developed in Baosteel recently over the past 10 years. Such group of stainless steels is featured low carbon plus low nitrogen contents, titanium and /or niobium microalloyed, good corrosion resistance, formability and weldability. In addition, different surface states, such as 2D, 2B, BA, HL, etc., have been gained through different manufacturing processes. The integrated production technology from steelmaking, hot rolling to annealing and pickling, cold rolling, etc, has been developed and improved consistently aiming for better surface quality and application properties to meet the higher requirement of customers. The number of new products is increased from 1 to 28 covering 16%~28% Cr and 0~3% Mo alloying, while the production volume is unbelievably increased from 90 tonnes in the year 2004 to 210000 tonnes last year. Those products are applied in many advanced areas, such as home appliance, construction and building, auto exhausting pipe and even energy industry, etc.

Key words ferritic stainless steel, production technology, application

Effect of Aging Treatment on Bending Workability of Dual Phase High Strength Stainless Steel

Naoki Hirakawa

(Stainless Steel & High Alloy R&D Dept. Steel & Technology Dev. Lab., Nisshin Steel Co., Ltd., 4976 Nomura Minami-Machi, Syunan, Yamaguchi 746-8666)

Abstract A ferrite-martensite dual-phase stainless steel (0.07%C-2%Ni-16.3%Cr alloy) having good formability has been developed. However, when this steel is bent toward transverse direction, its bending workability is inferior to when bent toward longitudinal direction, because this steel has anisotropy of bending workability and ferrite-martensite lamellar structure formed by rolling one way.

It was confirmed that the anisotropy of bending workability was improved by aging treatment in this investigation. This study was carried out to examine the effect of the aging treatment at various temperatures ranging from 250℃ to 450 ℃ for 1~24h. As a result, bend tests showed that the bending workability had been improved by the aging treatment at temperatures between 300℃ for 9h and 400℃ for 9h while maintaining strength.

In the case of an aging treatment at temperature between 300℃ for 9h and 400℃ for 9h, the Cottrell atmosphere is formed by carbon in ferrite phase, therefore the hardness of ferrite phase increases. On the other hand, the hardness of martensitic phase decreases by tempering. This fact shows that an appropriate aging treatment decreases the strength difference between ferrite phase and martensitic phase. The decreasing of the strength difference reduces concentration of deformation in ferrite phase. The improvement of bending workability is attributed mainly to the decreasing of the strength difference between ferrite and martensite.

Key words dual phase stainless steel, high strength, workability, aging

高强度高成形性能铁素体不锈钢制备工艺研究

谢胜涛　刘振宇　王国栋

（东北大学轧制技术及连轧自动化国家重点实验室，辽宁沈阳　110819）

摘　要　铁素体不锈钢是一种的成本低廉的无镍型不锈钢，但其强韧性和成形性能也偏低，无法应用于承力结构件。针对这一问题，研究了410S和430不锈钢冷轧板的淬火–配分工艺，优化了其奥氏化温度、淬火温度和配分温度。在原有铁素体中引入了尽量多的马氏体和残余奥氏体，残余奥氏体主要以片状分布于马氏体板条之间。在同等淬火–配分工艺下，410S与430相比，后者的残余奥氏体平均宽度较大、总量较多。淬火–配分板较常规退火板，强度显著提高；较淬火–回火板，强度稍低，但屈服连续，延伸率、强塑积和n值均较高。在同等淬火–配分工艺下，410S与430相比，后者的强度较低，但延伸率、强塑积和n值均较高。经淬火-配分热处理后的铁素体不锈钢有望在汽车结构件等领域得到应用。

关键词　汽车用，高性能，节约型，不锈钢，淬火–配分

Study of Manufacturing Process for High Strength and High Formability Ferritic Stainless Steels

Xie Shengtao, Liu Zhenyu, Wang Guodong

(State Key Laboratory of Rolling and Automation, Northeastern University, Shenyang 110819, China)

Abstract　Ferritic stainless is a typical cost-saving stainless steel containing no nickel. However, because it possesses relatively low strength, low formability and low toughness, it cannot be used as load-bearing structure components. In the present work, quenching and partitioning (Q&P) treatment was applied to cold rolled sheets of 410S and 430 stainless steels. The austenitizing, quenching and partitioning temperatures were studied and optimized to obtain as much martensite and retained austenite as possible in original ferrite matrix. The retained austenites with plate-like morphologies are mainly distributed between martensitic laths. After equal Q&P treatments, the retained austenite was larger in average width and more in total amount in 430 stainless steel than in 410S one. Compared to normally annealed sheets, the Q&P treated ones exhibited obviously higher strength. Compared to the Q&T treated sheets, the Q&P treated ones obtained slightly lower strength, but continuous yielding and higher elongation, the product of strength and elongation (PSE), and the n-value. After equal Q&P treatments, 430 stainless steel sheet exhibited lower strength but higher elongation, PSE, and the n-value than those of the 410S sheets.

Key words　automotive, high-performance, cost-saving, stainless steel, quenching & partitioning

超纯铁素体不锈钢表面白色条纹缺陷形成机理研究

段豪剑[1]　张立峰[1]　任 英[1]　张 莹[1]　李 实[2]　王立江[2]

（1. 北京科技大学冶金与生态工程学院，北京　100083；2. 宝钢不锈钢有限公司，上海　200431）

摘　要　超纯铁素体不锈钢中的白色条纹缺陷是指在含Ti超纯铁素体不锈钢冷轧板表面出现的大量宽0.5~1.5mm、长0.2~1.0m的白亮色条纹。本文通过SEM-EDS检测手段对白色条纹缺陷进行了观测，发现白色条纹缺陷的产生与冷轧板表面形成的大量TiN夹杂物富集有关。通过自动扫描电镜Aspex观测与统计连铸坯、热轧板和冷轧板表面TiN夹杂物的分布，发现在连铸坯、热轧板和冷轧板表面均有TiN夹杂物沿拉坯方向或轧制方向的局部富集现象，并且在冷轧板表面白色条纹缺陷位置与TiN夹杂物的富集位置相一致。由此认为超纯铁素体不锈钢中的白色条纹缺陷的形成机理为：在连铸坯表面生成的大量TiN夹杂物在局部富集，在热轧过程中局部富集的TiN夹杂物将被轧制延伸，而冷轧过程中富集的TiN夹杂物将暴露于冷轧板表面，并且促使冷轧板表面形成凹坑和褶皱状的轧制痕迹，从而使得在含Ti超纯铁素体不锈钢冷轧板表面出现大量的白亮色条纹，即形成白色条纹缺陷。

关键词　超纯铁素体不锈钢，白色条纹缺陷，形成机理

Formation Mechanism of White Stripe Defectsin Ultra-pure Ferritic Stainless Steels

Duan Haojian[1], Zhang Lifeng[1], Ren Ying[1], Zhang Ying[1],
Li Shi[2], Wang Lijiang[2]

(1. School of Metallurgical and Ecological Engineering, University of Science and Technology Beijing, Beijing 100083, China; 2. Baosteel Stainless Steel Co. , Ltd. ,Shanghai 200431, China)

Abstract White stripe defectswere white stripes with 0.5~1.5 mm in width and 0.2~1.0 m in length on the surface of cold-rolled Ti-stabilized ultra-pure ferritic stainless steels sheets. In the current paper, white stripe defects were analyzed using SEM-EDS system. It was found that white stripe defects were related to the enrichment of numerous TiN inclusions on the surface of cold-rolled sheets. Distributions of TiN inclusions at the slab, hot-rolled plate and cold-rolled sheet surface layers were analyzed using automated SEM-EDS system Aspex. It was found that TiN inclusions were enriched along the casting direction or the rolling direction. Based on the experimental results, formation mechanism of white stripe defects in ultra-pure ferritic stainless steels was proposed. Numerous TiN inclusions were locally enriched at the surface layer of slab and then extended in the process of hot rolling. Further more, enriched TiN inclusions were exposed to the surface of cold-rolled sheets and formed pits and roll marks during the cold rolling process. So, white stripe defects were formed on the cold-rolled Ti-stabilized ultra-pure ferritic stainless steels sheets surface.

Key words ultra-pure ferritic stainless steels, white stripe defects, formation mechanism

Microalloying Effects on Microstructure and Mechanical Properties of 18Cr-2Mo Ferritic Stainless Steel Plate

Han Jian[1], Li Huijun[1], Barbaro Frank[1, 2], Jiang Laizhu[1], Xu Haigang[3]

(1. School of Mechanical, Materials and Mechatronic Engineering, University of Wollongong, Wollongong NSW 2522, Australia; 2. CBMM Technology Suisse, 14, Rue du Rhone, Geneve 1204, Switzerland; 3. Baoshan Iron & Steel Co., Ltd., Shanghai 200431, China)

Abstract The dramatic increase in use of ferritic stainless steel (FSS) products has been primarily driven by the economic benefits over traditional austenitic stainless steels (ASS) but also the now well-established corrosion resistance. FSS grades however possess a relatively high ductile to brittle transition temperature (DBTT) especially when the plate thickness increases beyond 5 mm. Further economic benefits therefore depend on enhancing the base level toughness of this classic body centered cubic grade.

Microalloying is the main factor that determines the final mechanical properties in most steel grades. For FSS grades, it is essential that the detrimental effects of interstitials (C, N and O) be controlled by the addition of stabilising elements, such as Nb, Ti, V. It is the Cr and Mo balance in 18Cr-2Mo FSS, together with stabilising additions, that provide the enhanced corrosion resistance. The benefits of Nb in terms of effective stabilization and control of grains size, essential for both strength/toughness, avoids the detrimental effect of Ti on surface quality and the limitation of V in terms of effective stabilization and so offers a unique solution to increasing the product range of FSS plate.

Based on commercial Thermo-Calc software, a range of potential alloy designs was produced to evaluate the mechanical properties and corrosion resistance for a new thick plate FSS grade. Four alloy designs based on a Nb an Nb+Ti stabilised 18Cr-2Mo design were evaluated in terms of microstructure, viz grain size and precipitate species, tensile strength and Charpy impact toughness. Optical microscopy (OM), scanning electron microscopy (SEM), electron backscatter diffraction (EBSD), and also transmission electron microscopy (TEM) was utilized to characterize and correlate the performance of each alloy design. It was demonstrated that Nb provided both effective stabilization and grain refinement, through the formation of a fine dispersion of Nb rich carbonitrides, to increase the DBTT. It was also revealed that the grain misorientation distribution and grain boundary character distribution (GBCD) are not the main factors related to ultimate mechanical properties and will be the subject of this complete paper submission.

Key words　ferritic stainless steel, thick plate, microalloying, microstructure, mechanical property

The Development and Application of Modern Nb Microalloyed Ferritic Stainless Steel

Zhang Wei[1], Mariana Oliveira[2], Guo Aimin[1]

(1. CITIC-CBMM Microalloying Technology Center XinyuanNanlu, Chaoyang, Beijing 100004, China;
2. Companhia Brasileira de Metalurgia e Mineração)

Abstract　Ferritic stainless steel is being paid attention by steel industry and downstream users relying on its good over-all properties and low cost. With improvement of metallurgical technology and researching on fundamental theory, further development is obtained in properties optimization and quality stabilization of modern ferritic stainless steel, its marked feature are super purification and Nb/Nb-Ti stabilization. Nowadays, modern ferritic stainless steel is used widely in auto industry, home appliances, solar water heater, construction and decoration fields. The paper focuses on the effect mechanism of Nb technology on modern ferritic stainless steel and summarizes its application progress in typical downstream industries in China. Nb has an important role in the development of the ferritic stainless steel family since it adds value by improving final properties such as formability, weldability and corrosion resistance. home appliance and solar water heater are more open to ferritic stainless steel applications since they are volume driven and more cost sensitive.Nb-bearing modern ferritic stainless steels will have a bright application future in home appliance, solar water heater and other fields.

Failure of Tertiary Oxide Scales on Ferritic Stainless Steels in Hot Rolling

Cheng Xiawei[1], Jiang Zhengyi[1], Wei Dongbin[1], Zhao Jingwei[1], Hao Liang[1], Peng Jianguo[2], Luo Ming[2], Ma Li[2], Du Wei[2], Luo Suzhen[2], Jiang Laizhu[2]

(1. School of Mechanical, Materials and Mechatronic Engineering, University of Wollongong, NSW 2522, Australia;
2. Baosteel Research Institute (R&D Centre), Baoshan Iron & Steel Co., Ltd., Shanghai 200431, China)

Abstract　Oxidation kinetics of two stainless steels was studied at 1000–1150°C from 0 to 600 s in humid air containing 18% water vapour by a Thermo Gravimetric Analysis (TGA). Incubation time decreases significantly with increasing temperature. Short time oxidation and tensile tests of the ferritic stainless steels SUS430 and B445J1M were carried out in the Gleeble 3500 thermo-mechanical simulator between 1000 and 1150 °C for 180 s in a humid atmosphere with 18% water vapour. Two strain rates 0.2 and 4 s^{-1} were used in the tensile tests. The results show that the temperature has significant effect on the thickness and compositions of the oxide scale during the short time oxidation. The tertiary scale failure pattern under tensile stress depends on the steel grades, temperature and strain rate. On the steel SUS 430, the compositions of the oxide scale begin to change from a thin chromia layer to multi-layers as the temperature is higher than 1150 °C in the short time oxidation. Two oxide scale failure modes occur on multi-layer oxide scale indicating that there is a weak interface between the iron oxides and Fe-Cr spinel and also the weak iron oxides. For B445J1M, the oxide scale formed at higher temperature exhibited less dense oxide scale cracks and a very thin oxide scale reappeared on the exposed steel substrate. This indicates that the sticking is more likely to occur at 1000 °C than that at or above 1050 °C.

Key words ferritic stainless steel, oxidation, tertiary oxide scale, failure

Study on W-alloyed Type-429 Ferritic Stainless Steel for Exhaust System

Li Xin[1,2], Bi Hongyun[2], Chen Liqing[1]

(1. State Key Laboratory of Rolling and Automation, Northeastern University, Shenyang 110819, China;
2. Research Institute of Baosteel, State Key Laboratory of Development and Application Technology of Automobile Steels (Baosteel Group), Shanghai 201900, China)

Abstract Type-429 Ferritic stainless steel is an low cost material to be used in hot end of automotive exhaust system, this grade was widely used for the fabrication of manifold, which usually work at 800℃ -900℃. For developing a new kind of steel which is available at 1000℃, effect of W addition on room-temperature mechanical properties, high temperature strength and oxidation resistance of type-429 ferritic stainless steel were discussed in present study. The test result shows that W can enhance the strength in both room-temperature and 1000℃, oxidation resistance is also improved significantly with the increase of W content. However, W addition tend to degrade the adhesion of oxide layer, cause a spallation of oxide layer during a cold and hot alternate condition.

Key words ferritic stainless steel, high temperature strength, oxidation resistance

退火温度对超纯铁素体不锈钢再结晶组织和织构的影响

王彧薇 蔡庆伍 张 聪

（北京科技大学冶金工程研究院，北京 100083）

摘 要 利用电子背散射衍射技术（EBSD）分析了冷轧后退火温度分别在 880℃、920℃、960℃、1000℃和 1040 ℃的条件下，00Cr17 钢的显微组织和织构。结果表明：随着退火温度的升高，再结晶晶粒生成并长大，α 纤维织构减弱，而 γ 纤维织构增强，且织构强点为{111}<112>不变；再结晶完成后晶粒长大的实质是 γ 取向的再结晶晶粒消耗（吞并）其他取向的晶粒。

关键词 00Cr17 铁素体不锈钢，退火温度，织构，再结晶

Effect of Annealing Temperatureon Recrystallization Microstructure and Texture of Ultra-pure Ferritic Stainless Steel

Wang Yuwei, Cai Qingwu, Zhang Cong

(National Engineering Research Center for Advanced Rolling Technology, University of Science and Technology Beijing, Beijing 100083, China)

Abstract In this paper, Electron Backscattered Diffraction (EBSD) wasmainly applied to analyze the recry stallization microstructure and texture of 00Cr17 cold rolling samples annealing at 880℃, 920℃, 960℃, 1000℃ and 1040℃. The results show that, with the increase of annealing temperature, recrystallization grains formed and grew up. At the same time, α fiber texture decreased in intensity and γ fiber texture increased in intensity with a constant strong point {111}<112>. The essence of growth of recrystallization grains after recrystallization finishing is the consumption of grains with other orientations by recrystallization grains with gamma orientation.

Key words 00Cr17 stainless steel, annealing temperature, texture, recrystallization

超纯铁素体不锈钢中 TiN 的析出热力学和动力学研究

张　帆[1]　李光强[1]　万响亮[1]　陈兆平[2]

（1. 武汉科技大学钢铁冶金及资源利用省部共建教育部重点实验室，湖北武汉　430081；
2. 宝山钢铁股份有限公司中央研究院，上海　201900）

摘　要　为了控制 TiN 夹杂对钢材性能的不利影响，对 TiN 在超纯铁素体不锈钢冶炼和凝固过程中析出的热力学和动力学过程进行了分析。结果表明，在 Cr 含量为 16%~23%的条件下，超纯铁素体不锈钢的最佳 Ti 含量为 0.1%~0.3%；当 w(Ti)=0.3%时，将 w(N)控制在 0.0046%以下才能避免在液态钢中生成 TiN 夹杂；超纯铁素体不锈钢凝固过程中 Ti 的富集程度稍低于 N；凝固末期残余液相中[Ti]和[N]的浓度分别提高到其初始浓度的 2.5 倍和 3.2 倍；随着钢液的逐渐凝固，凝固前沿温度和凝固前沿钛氮平衡浓度积下降，凝固前沿钛氮实际浓度积升高；降低钢中氮含量能推迟 TiN 析出时间，使 TiN 夹杂在凝固末期形成，可减小 TiN 夹杂尺寸；降低钢中 Ti 含量也能推迟 TiN 析出时间，TiN 夹杂最终尺寸减小较少；在凝固过程中增加冷却速度能明显细化 TiN 夹杂的粒径。

关键词　超纯铁素体不锈钢，TiN 夹杂，析出，热力学分析，动力学分析

Thermodynamics and Kinetic of TiN Precipitation in Ultra-Pure Ferritic Stainless Steel

Zhang Fan[1], Li Guangqiang[1], Wan Xiangliang[1], Chen Zhaoping[2]

(1.The Key Laboratory for Ferrous Metallurgy and Resources Utilization of Ministry of Education, Wuhan University of Science and Technology, Wuhan Hubei 430081, China;
2.The Central Research Institute, Baoshan Iron & Steel Co., Ltd., Shanghai 201900, China)

Abstract In order to control the adverse effect of TiN in ultra-pure ferritic stainless steel, the thermodynamic and kinetics precipitation conditions in smelting and solidification process were studied. The results show that, the best content of [Ti] in the ultra-pure ferritic stainless steel which contains [Cr] 16%~23% is 0.1% to 0.3%; and when [Ti]=0.3mass%, controlling [N] to less than 46×10^{-4} mass% can avoid TiN formation in liquid steel. The segregation degree of Ti is less than that of N, and at the end of solidification, the content of [Ti] and [N] in residual liquid increase to 2.5 times and 3.2 times of their initial contents, respectively. Along with gradual solidification of liquid steel, in the solidification frontier the temperature and equilibrium of [%Ti][%N] go down, and the real equilibrium of [%Ti][%N] rises. To reduce [%N] can limit TiN precipitation at the end of solidification and reduce the size of TiN inclusion effectively. [%Ti] reduction also postpone TiN precipitation at the end of solidification and reduce the size of TiN inclusion slightly. Moreover, the size of TiN inclusion decrease effectively with increasing the cooling rate during solidification.

Key words ultra-pure ferritic stainless steels, TiN inclusion, precipitation, thermodynamics, kinetics

新型含 Sn 高 N 经济型 Cr17 超级马氏体不锈钢

冉庆选 李 钧 张自兴 肖学山

（上海大学材料研究所，上海 200072）

摘 要 本文设计了全新含 Sn 高 N 经济型超级马氏体不锈钢 00Cr17Ni5N0.2Sn0.4。实验材料首先在 1050 ℃ 处理 5 min 水淬。然后在不同的温度条件下(520 ℃、560 ℃、600 ℃ 、640 ℃、680 ℃)进行回火处理 2 h 空冷。利用光学显微镜（OM）、扫描电子显微镜（SEM）、透射电子显微镜（TEM）以及 X 射线衍射分析仪（XRD）研究实验钢回火态微观组织。通过室温拉伸实验、室温冲击实验检测材料在回火状态下的力学性能。通过极化曲线测试来表征材料耐点蚀性能。实验结果表明，回火处理后基体组织为板条马氏体，并且存在少量的δ-铁素体。600 ℃ 回火处理后板条马氏体之间以及板条内部均有逆变奥氏体产生并残留至室温，材料表现为最佳的综合力学性能：屈服强度为 600 MPa，抗拉强度达到 950 MPa，断裂伸长率为 17 %，强韧搭配较好。该状态下实验钢在 3.5 wt. %氯化钠溶液中点蚀电位约为 275 mV。通过对比表明，所设计的超级马氏体不锈钢 00Cr17Ni5N0.2Sn0.4 的综合性能相当于或者优于标准钢 0Cr17Ni5Mo。

关键词 超级马氏体不锈钢，微观结构，力学性能，腐蚀性能

Novel Sn-bearing, High N Economical Cr17 Super Martensitic Stainless Steel

Ran Qingxuan, Li Jun, Zhang Zixing, Xiao Xueshan

(Institute of Materials, Shanghai University, Shanghai 200072, China)

Abstract Novel economical Sn-bearing, high N super martensitic stainless steel 00Cr17Ni5N0.2Sn0.4 has been designed and investigated in this paper. The material was first solution treated at 1050 ℃ for 5 min followed by water quenching, and then subjected to different tempering temperatures (520 ℃、560 ℃、600 ℃ 、640 ℃、680 ℃)for 2 h followed by cooling in air. The microstructures after tempering treatment were investigated by optical microscopy (OM), scanning electron microscopy (SEM), transmission electron microscope (TEM) and X-ray diffraction analysis (XRD). Tensile tests and impact tests were conducted to analyze mechanical properties at room temperature. Polarization curve tests were used to analyze the corrosion resistance of the temper-treated steel. The results showed that lath martensite together with some -ferrite generated after tempering heat treatment. Retained austenite appeared after tempering at 600 ℃ for 2 h, and the material showed the best comprehensive mechanical properties: the yield strength, tensile strength and elongation were about 600 MPa, 950 MPa and 17 % respectively, which showed good combination of strength and toughness. Pitting potential was about 275 mV in 3.5 wt. % NaCl solution. The comprehensive performance of super martensitic stainless steel 00Cr17Ni5N0.2Sn0.4 is as good as that of 0Cr17Ni5Mo stainless steel, and even better.

Key words super martensitic stainless steels, microstructure, mechanical property, corrosion property

淬火温度对 0Cr13 不锈钢铸坯组织的影响

段路昭[1,2] 崔 娟[1,2] 白李国[1,2] 王笑丹[1,2]

（1. 邢台钢铁有限责任公司，河北邢台 054027；
2. 河北省线材工程技术研究中心，河北邢台 054027）

摘 要 通过对 0Cr13 铁素体不锈钢铸坯原始组织及不同温度保温-淬火后的金相组织的分析，研究了不同保温温度对其高温奥氏体及淬火马氏体形态的影响。结果表明：试验温度范围内，随着保温温度的升高，淬火马氏体形貌由针叶状、块状向链状及碎块状分布，马氏体内部碳化物逐渐减少，且马氏体体积分数呈增加趋势。

关键词 0Cr13，奥氏体，马氏体

The Effect of Quenching Temperature on the Microstructure of 0Cr13 Stainless Steel Billet

Duan Luzhao[1,2], Cui Juan[1,2], Bai Liguo[1,2],Wang Xiaodan[1,2]

(1. Xingtai Iron & Steel Co., Ltd., Xingtai Hebei 054027, China;
2. Hebei Engineering Research Center for Wire Rod, Xingtai Hebei 054027, China)

Abstract Through the microstructure of the 0Cr13 billet and different soaking-quenching processes, discuss the effect of the soaking and queching temperature on the Austenite and Martensite morphology. The results shows that: during the range of experiment temperature, as the soaking temperature rises, the Martensite turns from needle and block to chain and pieces. In addition, the volume of carbide reduces and the volume of the Matensite enlarges as the soaking temperature rises.

Key words 0Cr13, Austenite, Martensite

6.4.2 奥氏体及双相不锈钢 Austenitic and Duplex Stainless Steel

Duplex Families and Applications: A Review

Charles Jacques

Abstract The paper is a summary of a review paper devoted on duplex stainless steels (DSS) key developments and main applications partially presented at DSS'14 Stresa conference[1]. The presentation takes advantage of my personal involvement in DSS developments

Key words duplex stainless steels, review paper, hyper-duplex, super-duplex, duplex, lean duplex grades

Development of High Strength and High Ductility Nanostructured Stainless Steels and Applications in Different Industries (Automotive, Energy, Bioimplant, Construction)

Lu Jian

(Centre for Advanced Structural Materials, Department of Mechanical and Biomedical Engineering, City University of Hong Kong)

The ability to create structural materials of high yield strength and yet high ductility has been a dream for materials scientists for a long time. The study of the mechanical behavior of the surface nanostructured materials using SMAT (Surface Mechanical Attrition Treatment) shows significant enhancements in mechanical properties of the nanostructured surface layer in different materials including stainless steels[1-5]. We summarize our recent works on the advanced nanostructured stainless steels with exceptional dual mechanical properties using multiscale metallurgical structure-driven design combined with advanced mechanical simulation. The effect of surface nanostructures on the mechanical behavior and on the failure mechanism of metallic material shows the possibility to develop a new strength gradient composite with a dual property of high strength and high ductility[6-8]. The results show three key mechanisms for the enhancement and the extraordinary properties of layered and nanostructured metallic stainless steel sheet. The computational models and experimental results successfully provide valuable information about the nanomaterials properties as a function nanostructure configuration (nanograins and nanotwins)[9-12]. The processing of nanomaterials[13-14] using mechanical processing and heat treatment have been studied at nanoscale and atomic scale. With a detailed knowledge of the processing using high speed camera, we were able to accurately estimate the strain rate at different depths by analytical modeling and to study the correlation between the resulting microstructures and the strain/strain rate history of the material[13]. The material studies[15-18] using nanomechanics based experimental investigations (nanoindentation and nano-pillar tests) can reveal the effects of the atomic structure and nanostructure gradient on the mechanical behaviors. The failure mechanisms studies at nano-, micro- and macroscopic scale can provide efficient ways to enhance the ductility of materials using the general approach of strain non localization. In addition of the basic property of tensile strength, the nanostructured stainless steel can also enhance the fatigue behaviors[19-21]. Due to the nanostructure generated, the SMAT can modify drastically the diffusion behavior nitriding of stainless steel. The maximum hardness and thickness of treated layer can be significantly improved[22] to enhance the wear resistance of stainless steel. The nanostructured stainless steel is also an interested family of new materials that can change the dynamic properties under ballistic loading[23-24] to provide additional design solution. Due to its extraordinary possible variety of nanostructures, the new family of nanostructured stainless steel is an important element of high strength and high ductility materials[25]. Combined with advanced mechanical design using optimization and lattice structures concept[26-27], many other potential applications in different industries such as automotive, energy, bioimplant, construction can be anticipated. The results of some recent feasibility demonstration studied will be presented.

Overview of the Research on Candidate Materials for Advanced Ultra-supercritical Boiler Tubes

Wang Yan, Xu Fanghong, Fang Xudong, Li Yang, Fan Guangwei, Li Jianmin

(State Key Laboratory of Advanced Stainless Steel, Taiyuan Iron and Steel Group Company Limited, Taiyuan, China)

Abstract The drive to increase the efficiency of pulverized coal-fired power plants has led to raised steam conditions with operating temperature and pressure of 700℃/35MPa. A stress rupture life of 100000 hours at 750℃/100MPa stress as well as the metal loss of less than 2mm in 200000 hours are required for these advanced ultra-supercritical boiler tubes. This paper summarized the research on the four candidate superalloys, i.e. Inconel740H、617B、HR6W and Sanicro25 at domestic and foreign.

Key words　ultra-supercritical, Inconel740H, 617B

宝钢含氮奥氏体不锈钢的开发和应用

潘世华　季灯平

（宝钢不锈钢有限公司，上海　200431）

摘　要　含氮不锈钢的开发一直是不锈钢产品技术领域的研究热点之一。本文介绍了近年来宝钢开发的含氮奥氏体不锈钢产品及其应用情况，主要包括控氮系列产品 304N 和 316LN，含氮节镍型产品 BN 系列，以及高氮不锈钢 HNS 系列。研究结果表明，N 能显著提高产品的强度和耐腐蚀性能，通过 N 元素的合金化，可以在大幅度减少 Ni 元素使用的前提下，得到耐腐蚀性能和 304 不锈钢相当的经济型奥氏体不锈钢。另外，采用科学的氮合金化手段，常压下还获得了含氮量 0.8%的高氮不锈钢，其点蚀电位达到 1.0V 以上。目前，宝钢开发的含氮奥氏体不锈钢已经在冷藏集装箱、储运罐箱和制品领域得到了广泛的应用。

关键词　N，奥氏体不锈钢，腐蚀性能

Development and Application of Nitrogen Bearing Austenitic Stainless Steels in Baosteel

Pan Shihua, Ji Dengping

(Baosteel Stainless Steel Co., Ltd., Shanghai 200431, China)

Abstract　Nitrogen bearing stainless steel is always a research hotpot in the field of stainless steel product and technology. In this paper, nitrogen bearing austenitic stainless steels developed by Baosteel and their applications are introduced, and these steels are nitrogen controlled products 304N and 316LN, nitrogen containing economical products BN series and high nitrogen stainless steel HNS series. The study results show that nitrogen can significantly improve the strength and corrosion resistance of the products. By the mean of nitrogen alloying, economical austenitic stainless steels with considerably less nickel than 304 can be obtained and their corrosion resistance are almost the same as 304. Furthermore, by a scientific approach of nitrogen alloying, high nitrogen steel of 0.8% nitrogen content is fabricated under the non-pressurised condition, and the pitting potential of high nitrogen steel is more than 1.0V. At present, nitrogen bearing steels developed by Baosteel are widely used in the fields of cryogenic storage container, transportation container and household wares.

Key words　Nitrogen, austenitic stainless steel, corrosion resistance

The Effect of Tensile Deformation on Strain Induced Martensite and Pitting Corrosion Resistance for a New Developed Lean Duplex Stainless Steel

Hu Jincheng, Song Hongmei

Abstract Different tensile deformation was imposed to the new lean duplex stainless steel with chemical composition of 20.5Cr-4Mn-0.16N, and then the strain induced martenstie (SIM) was studied by the combination of ferrite detector, X-ray diffraction, scanning electric microscope and transmission electric microscope. The effect of strain induced martensite (SIM) on pitting corrosion resistance was also investigated by tests for pitting potential and weight loss rate in $FeCl_3$ solution. The results showed that strain induced martensite transformed through path of $\gamma\rightarrow\varepsilon\rightarrow\alpha'$, and about 17.6% α' martensite formed when the tensile deformation increased to 20%. Meanwhile, the SIM had no evident effect on the pitting corrosion resistance.

Key words stain induced martensite, lean duplex stainless steel, pitting corrosion

晶粒细化对 301LN 奥氏体不锈钢形变机制和力学性能的影响

万响亮 许德明 李光强

（武汉科技大学钢铁冶金及资源利用省部共建教育部重点实验室，湖北武汉 430081）

摘 要 利用相逆转变原理采用冷轧使得亚稳态的奥氏体组织转变为形变马氏体，随后通过不同温度和时间退火处理分别获得平均晶粒尺寸为500nm的纳米晶/超细晶和22μm的粗晶301LN奥氏体不锈钢。通过拉伸实验得到301LN奥氏体不锈钢的力学性能；采用TEM观察了形变量为0.1时试样的组织结构并利用SEM观察了断口形貌特征。结果表明 301LN 奥氏体不锈钢中晶粒尺寸由粗晶细化到纳米晶/超细晶，屈服强度提高 2.3 倍。低屈服强度的粗晶301LN奥氏体不锈钢塑性变形时发生TRIP效应，得到良好的塑性；而高屈服强度的纳米晶/超细晶301LN奥氏体不锈钢通过TWIP效应获得优良塑性；不同形变机制下获得的拉伸试样断口均为韧性断裂。形变机制由TRIP效应变化为TWIP效应归因于晶粒细化导致奥氏体稳定性大幅度提高。

关键词 奥氏体不锈钢，晶粒细化，形变机制，力学性能，奥氏体稳定性

Effect of Grain Refinement on Deformation Mechanism and Mechanical Properties of 301LN Austenitic Stainless Steel

Wan Xiangliang, Xu Deming, Li Guangqiang

(The Key Laboratory for Ferrous Metallurgy and Resources Utilization of Ministry of Education, Wuhan University of Science and Technology, Wuhan 430081, China)

Abstract The concept of phase reversion involving cold deformation of metastable austenite to generate strain-induced martensite, followed by different temperature-time annealing sequence, was used to obtain grain size of nanograined/ultrafine-grained and coarse-grained 301LN austenitic stainless steels. The mechanical properties of 301LN austenitic stainless steels with different grain size was obtained by tensile testing. the deformation microstructure and fracture surface were analyzed by TEM and SEM observations, respectively. The results indicated that the yield strength improved 2.3 times, when the grain size of 301 LN austenitic stainless steel was refined from coarse-grained structure to nanograined/ultrafine-grained structures. The strain-induced martensite transformation contributed to excellent ductility in the low strength coarse-grained 301 LN austenitic stainless steel, while in the high strength nanograined/ultrafine-grained 301 LN austenitic stainless steel, the high ductility is due to deformation twinning. Interestingly, both of the fracture mechanisms of the two steel with different deformation behavior were ductile fracture. The change in deformation mechanism from strain-induced martensite in the coarse-grained structure to deformation winning in nanograined/ ultrafine-grained structures is related to the increased stability of austenite with decrease in grain size.

Key words austenitic stainless steels, grain refinement, deformation mechanism, mechanical property, austenite stability

超级奥氏体不锈钢654SMO组织与性能研究

张树才[1] 姜周华[1] 李花兵[1,2] 冯 浩[1] 张彬彬[1] 耿 鑫[1]
范光伟[2] 张 威[2] 林企曾[2]

（1. 东北大学材料与冶金学院，辽宁沈阳 110004；2. 山西太原钢铁集团，山西太原 030003）

摘 要 本文对超级奥氏体不锈钢654SMO的析出热力学及动力学、热加工性能、抗高温氧化性能、力学性能和耐腐蚀性能进行了研究。654SMO具有迅速析出动力学，鼻尖温度约为1000℃。钢中主要析出相为σ相、Laves相和氮化物，临界形核温度分别约为1170℃、800℃和1200℃。654SMO抗高温氧化性能极差，主要原因：MoO_3的形成与挥发破坏氧化层完整性，促使空气中氮渗入基体形成Cr_2N，致使保护性Cr氧化层难以形成。654SMO热变形激活能Q=530kJ/mol，最佳热加工窗口为变形温度1150~1250℃，应变速率0.001~0.01s^{-1}。654SMO室温力学性能优于C-276，断口为等轴形韧窝；低温冲击韧性优异，随实验温度降低，断裂机理：拉长或等轴形韧窝→浅坑形韧窝→类解理和浅平韧窝混合型。高Cr、Mo、N含量使超级奥氏体不锈钢654SMO具有优异的耐点腐蚀性能和耐均匀腐蚀性能。

关键词 超级奥氏体不锈钢，热加工，高温氧化，力学性能，腐蚀

Research on the Microstructure and Properties of Super Austenitic Stainless Steel 654SMO

Zhang Shucai[1], Jiang Zhouhua[1], Li Huabing[1,2], Feng Hao[1], Zhang Binbin[1], Geng Xin[1], Fan Guangwei[2], Zhang Wei[2], Lin Qizeng[2]

(1. School of Materials and Metallurgy，Northeastern University，Shenyang Liaoning 110004, China；
2. Shanxi Taigang Stainless Steel Co., Ltd., Taiyuan Shanxi 030003, China)

Abstract The thermodynamicsandkinetics of precipitation, hot workability, oxidation resistance at high temperature, mechanical properties and corrosion resistance of 654SMO were investigated. The kinetics of precipitation of 654SMO was very fast, and the "nose" temperature of precipitation was found to be about 1000℃. The main precipitate phases were σ phase, Laves phase and nitride, whose critical nucleation temperatures were about 1170℃, 800℃ and 1200℃, respectively. The oxidation resistance of 654SMO at high temperature was relatively poor, which was because the volatilization of MoO_3 compromised the integrity of the oxide layer and prompted the formation of Cr_2N, reducing the ability to form a protective layer of Cr_2O_3.Theactivation energy of deformation was determined as530kJ/mol, and the best hot working temperatures range from 1150°C to 1250°C and the strain rates range from 0.001s^{-1} to 0.01s^{-1}.654SMOhad good tensile mechanical properties than C-276 at room temperature, and the fracture shownequiaxed dimple. In addition, it had excellent toughness; the change of fracture patterns of the steel was prolonged/equiaxed dimple→ shallow dimple→ mixture of quasi-cleavage facet and dimple. The high Cr, Mo, N contents contributed to the excellent pitting and general corrosion resistance of the steel.

Key words superaustenitic stainless steel, hot workability, oxidation, mechanical properties, corrosion

9Cr18 不锈钢材料触变成形的组织和性能研究

王永金 宋仁伯 李亚萍

（北京科技大学材料科学与工程学院，北京 100083）

摘 要 通过触变成形工艺制备 9Cr18 不锈钢材料制件，与传统热处理工艺对比，探讨触变成形制件的强韧性及耐蚀性能。研究结果表明，材料在半固态触变成形过程中，固液两相表现出不同的流动特性。熔融液态金属流动至材料表层，形成一层细小枝晶的表层，覆盖保留在内部的固相奥氏体组织，显微组织呈梯度分布。由于内部的固相奥氏体组织，材料的塑性得到提高，加工硬化能力显著上升，使材料表现出内部强韧性，与外侧硬度较高的枝晶层共同配合，表现出功能梯度特性。触变成形制件的电化学行为不同于常规热处理材料，腐蚀电流密度有所降低，但过钝化区出现了化学反应阶段。

关键词 触变成形，功能梯度材料，力学性能，电化学行为

Research on the Microstructure and Properties of 9Cr18 Stainless Steel Through Thixoforming

Wang Yongjin, Song Renbo, Li Yaping

(School of Material Science and Engineering, University of Science and Technology Beijing, Beijing, 100083, China)

Abstract Specimen of 9Cr18 stainless steel was fabricated through thixoforming process. Mechanical and corrosion properties of thixoforned specimen were compared with conventional heat treatment specimen. The results showed that the solid and liquid phase exhibited different flow behavior during thixoforming. Melted liquid flew to the surface and then formed fine dendrite, which covered the retained solid inside. With the inner austensite, the plastic and work hardening properties increased and then the material inside showed good strength, which cooperated with the hard dendrite surface. The electrochemical behavior of thixoformed specimen was different from that of the conventional heat treatment specimen. Corrosion current density was reduced, while chemical reaction occurred after the passivation zone.

Key words thixoforming, functionally graded material (FGM), mechanical property, electrochemical behavior

奥氏体不锈钢热轧氧化铁皮研究

杨 超 任建斌 董 涛 范 方

（宝钢德盛不锈钢有限公司，福建福州 350601）

摘 要 在奥氏体不锈钢热轧板生产过程中，发现氧化铁皮缺陷持续偏高，对公司产品质量和效益产生不利的影响。

本文通过在轧制线上进行不同加热炉生产的氧化铁皮层厚度、结构研究等的试验分析，1号炉生产的黑皮表面氧化铁皮厚度比2号炉生产的薄，1号炉黑斑处氧化铁皮层厚度为12~15μm，无黑斑处的氧化铁皮层厚度在10~13μm，而2号炉的氧化铁皮厚度为16~23μm，1号炉氧化铁皮较2号炉均匀。蓄热式加热炉生产的不锈钢黑皮卷氧化铁皮层结构：靠近基体氧化物以FeO为主，夹杂少量Cr、Mn、Cu氧化物；靠近表层氧化物以Fe_2O_3为主。本研究为本厂两座加热炉产品的差异性提供了一定的依据，利于产品的集中攻关，有助于提高产品质量。通过我厂的集中攻关，目前热轧氧化铁皮缺陷已得到了有效的控制。

关键词　氧化铁皮，加热温度，热轧

Research on the Scale of Austenitic Stainless Steel Hot Rolled

Yang Chao, Ren Jianbin, Dong Tao, Fan Fang

(Baosteel Desheng Stainless Steel Ltd., Fuzhou 350601, China)

Abstract　In the process of austenitic stainless steel hot rolling plate, a warming trend of iron oxide skin defect was found, which generated adverse effects to the product quality and efficiency of company. Based on the rolling line for production of iron oxide layer thickness, different heating furnace structure study of test analysis. The thickness of black surface oxide produced by 1# furnace is thinner than what 2 # furnace product , dark spots in 1 # furnace iron oxide layer thickness is 12 ~ 15μm ,other's iron oxide layer thickness of 1 # furnace in 10 ~ 13 μm, while the thickness of iron oxide layer in 2 # furnace is 16~23μm ,compared with 1# furnace is more even.Regenerative heating furnace production of stainless steel black iron oxide layer structure: close to the matrix oxide with FeO is given priority to, with a small amount of Cr, Mn, Cu oxides; near the surface of the oxide, Fe_2O_3 is primary.This research for our products have provided the certain basis, the difference of the heating furnace to product of intensive research, help to improve the quality of the product. By focusing on research of our plant, currently hot-rolled iron oxide skin defect has been effectively controlled.

Key words　scale, heating temperature, hot rolling

节镍奥氏体不锈钢冷轧延迟开裂行为研究

任建斌　杨　超　董　涛　范　方　郭文静

（宝钢德盛不锈钢有限公司，福建福州　350600）

摘　要　节镍奥氏体不锈钢具有优异的力学性能、一定的耐蚀性能以及较低的成本，在面板、制品和构件等领域应用广泛。由于其奥氏体相为亚稳定，冷轧特别是冬季时容易出现延迟断裂问题。本文研究了节镍奥氏体不锈钢冷加工行为，分析了延迟开裂的影响因素。结果材料加工硬化和马氏体相变强化的复合作用是延迟开裂的主要原因。通过调整化学成分，降低形变马氏体应变温度$M_{d30/50}$，可以有效提高材料抑制延迟开裂的能力。均匀的板厚、优良的板形和合理的冷加工工艺，可以进一步降低延迟开裂风险。

关键词　节镍奥氏体不锈钢，马氏体相变，延迟开裂，冷轧

Research on the Delayed Cracking of Low Nickel Austenitic Stainless Steel in Cold Rolling Process

Ren Jianbin, Yang Chao, Dong Tao, Fan Fang, Guo Wenjing

(Baosteel Desheng Stainless Steel Ltd., Fuzhou Fujian 350600, China)

Abstract Low-nickel austenitic stainless steel was widely used in board, kitchen ware and structure due to its excellent mechanism properties, certain corrosion resistance and low cost. The delayed cracking was easily induced in cold rolling process because the austenitic phase was unstable, especially in winter. The paper investigated the deformation mechanism during cold rolling and analyzed the influence factors on delayed cracking. The results indicated that hardening and martensite transformation were the main reasons. The cracking could be suppressed by composition designing to raise the $M_{d30/50}$ temperature. Moreover, the even thickness, good shape of strip and a suitable cold rolling process were also beneficial for delayed cracking.

Key words low-nickel austenitic stainless steel, matensite transformation, delayed cracking, cold rolling

304、904L 和 TiAl 合金耐铝液腐蚀及机制研究

王　剑　李　越　孙向雷　韩培德

（太原理工大学材料科学与工程学院，太原　030024）

摘　要　铝等有色金属熔炼过程中的连续测温已成为制约铝及其产品的生产效率及质量的难题。开展延长热电偶保护套管寿命研究具有重要的意义。本文选取 304 不锈钢，904L 不锈钢和 TiAl 合金进行了耐 750℃铝液腐蚀实验。结果发现：304 不锈钢经高温铝液腐蚀后产生 $FeAl_3$ 和 Fe_2Al_5 金属间化合物，随着不锈钢套管在铝液腐蚀时间的增加，$FeAl_3$ 和 Fe_2Al_5 金属间化合物容易形成不连续破碎、不能稳定存在的腐蚀层组织，腐蚀层与液铝相接触，在较高温铝液热应力等的作用下会脱落，造成腐蚀。904L 不锈钢由于含有较高的镍和钼含量，与高温 Al 液反应生成了 $NiAl_3$ 等金属间化合物，其熔点高于铁铝金属间化合物，因而较 304 显示了较好的耐铝液腐蚀性能。TiAl 合金由于形成了更高熔点的 $TiAl_3$ 化合物，阻碍了 Al 原子的扩散，因而显示出最佳的耐铝液腐蚀性。

关键词　不锈钢，钛铝合金，铝液，腐蚀，热电偶

氮对 25Cr-2Ni-10Mn-3Mo-xN 双相不锈钢组织和性能的影响

赵钧良　郑　卫

（宝钢特钢有限公司，上海　200940）

摘　要　双相不锈钢由奥氏体和铁素体组成的，铁素体含量在 30%~70%，它是一种发展很快的钢类，已有 Cr18 型、Cr22 型、Cr25 型和 Cr29 型。25Cr-2Ni-10Mn-3Mo-xN 双相不锈钢是 S32750（Cr25 型）的改进型，以 Mn 和 N 来替代部分 Ni。本文是在实验室条件下，研究了不同 N 含量下，温度与组织和铁素体含量的关系；N 对腐蚀性能和

力学性能的影响；得出25Cr-2Ni-10Mn-3Mo-0.4N双相不锈钢随着N含量的增加，1000℃左右的σ相的溶解峰越来越弱；在1050℃热处理下，钢中的铁素体含量在50%左右，点蚀电位均在1050mV以上，抗拉强度大于900MPa。为此，钢中N含量在0.40%为较佳，这样有利于它的加工和使用。

关键词　双相钢，氮含量，组织

The Effect of Nitrogen on the Microstructure and Properties of 25Cr-2Ni-10Mn-3Mo-xN Duplex Stainless Steel

Zhao Junliang, Zheng Wei

(Baosteel Special Metals Co., Ltd., Shanghai 200940, China)

Abstract　The duplex stainless steel is normally composed of austenite and ferrite , and the content of ferrite is 30%~70%, and has developed Cr18, Cr22 and Cr25. The 25Cr-2Ni-10Mn-3Mo-xN duplex stainless steel is developed on based of S32750, N and Mn can be used to replace the expensive Ni. It is one of the best cost-benefit engineering materials. It is simply studied the effect of Nitrogen on its phase structure and ferrite phase in the different temperature state, its corrosion resistance and its mechanical properies in the different of N content. It has a better property of corrosion resistance, breakdown potential is higher then 1050mv in 3.5%NaCl solution at 25℃. At 1050℃ heat treatment temperature, its content of ferrite is about 50% and the tensile strength is higher then 900MPa. The better content of N is higher than 0.40%.

Key words　duplex steel, content of N, microstructure

镧铈混合稀土对2205双相不锈钢组织及力学性能的影响

刘　晓　马利飞　李运刚

（华北理工大学现代冶金技术教育部重点实验室，河北唐山　063009）

摘　要　采用扫描电镜（SEM）、X射线能谱仪（EDS）、金相显微镜及冲击性能测试等方法研究了镧铈混合稀土对2205双相不锈钢组织及力学性能的影响。结果表明：稀土元素细化了2205双相不锈钢的显微组织，稀土元素优先富集在相界及其附近地区，延缓了σ相的析出，减少了σ相的析出量；含量为0.04%的镧铈混合稀土通过细化晶粒、变质夹杂物和延缓σ相的析出提高了时效后2205双相不锈钢的冲击韧性，20℃时提高约40%，促使其断裂机制从解理断裂向韧窝断裂转变。

关键词　稀土，双相不锈钢，组织，冲击韧性

Effect of La and Ce on Microstructure and Mechanical Properties of 2205 Duplex Stainless Steel

Liu Xiao, Ma Lifei, Li Yungang

(Key Laboratory of Modern Metallurgy Technology, Ministry of Education, North China University of

Science & Technology, Tangshan 063009, China)

Abstract Effect of La and Ce on microstructure and mechanical properties of 2205 duplex stainless steel (DSS) were investigated by scanning electron microscopy (SEM), energy dispersive spectrometer (EDS), metallographic microscope and impact test. The results show that the microstructure of 2205 duplex stainless steel is refined by the addition of rare earth and RE is preferentially enriched on the boundary and nearby, and the precipitation of σ phase is delayed. The impact value with 0.04% RE is increased 40% at the temperature of 20℃ by RE refining grain, modifying inclusions and delaying the precipitation of σ phase. And the fracture morphology is transformed from cleavage to ductile fracture.

Key words rare earth, duplex stainless steel (DSS), microstructure, impact toughness

终轧温度对高氮不锈钢微观组织的影响

赵英利　李建新　张云飞　杨现亮　李　杰

（河北钢铁技术研究总院，河北石家庄　050021）

摘　要 采用光学电镜、扫描电镜对不同终轧温度下的高氮不锈钢微观组织进行了观察。实验结果表明：终轧温度较高时，高氮不锈钢热轧后微观组织为奥氏体组织和少量铁素体，在奥氏体和铁素体晶界处分布有少量片层状的σ析出相；终轧温度较低时，铁素体含量增多，在奥氏体和铁素体界面析出的σ析出相显著增多，且在铁素体晶内存在有大量的σ析出相，σ相呈块状；σ析出相是富含Cr、Mo的析出相，其优先在奥氏体和铁素体晶界形核析出，随终轧温度的降低，σ析出相开始在铁素体晶内析出长大。

关键词 高氮不锈钢，终轧温度，微观组织

Effect of Finishing Rolling Temperature on Microstructure of High Nitrogen Stainless Steel

Zhao Yingli, Li Jianxin, Zhang Yunfei, Yang Xianliang, Li Jie

(Hebei Iron and Steel Technology Research Institute, Shijiazhuang 050021, China)

Abstract The microstructure of high nitrogen stainless steel at different finishing rolling temperature was observed by optical microscopy and scanning electron microscopy. The results indicate that the lamellar sigma phase precipitated at austenite and ferrite grain boundaries, the microstructure of high nitrogen stainless steel after hot rolling are austenite and a little ferrite; while finish rolling temperature is low, ferrite content and sigma phase precipitated at austenite and ferrite grain boundaries are significantly increased, and internal ferrite exist a large number of precipitate phase, the sigma phase is massive; sigma phase is rich in Cr, Mo, and preferential nucleated at grain boundaries of austenite and ferrite, sigma phase began to grow in ferrite with the final rolling temperature decrease.

Key words high nitrogen stainless steel, finishing rolling temperature, microstructure

6.4.3　不锈钢的加工及应用　Processing and Application

A Review of the Markets for Stainless Steel

John Rowe

(Secretary-General International Stainless Steel Forum Brussels)

Abstract　The presentation will take the form of a PowerPoint presentation, covering macro-economic data for the principal economies of the world, including production data.

It will include details of stainless steel production, on a global and regional basis, with historical data from 2001. I will include the ISSF forecasts for production for the principle market regions - Europe and Africa; the Americas; Asia (without China); and China.

There will be an estimate of current stock levels; a comparative of the growth of stainless steel and its main competing materials; and a report on order intake.

An Overview on the Adoption of Stainless Steel for Drinking Water Distribution Systems

Gaetano Ronchi

(BAC-IMOA)

Abstract　The potable water cyclefrom collection to tap is long and complex. Treatment, storage, transportation, distribution imply a large number of parameters that affect pipe corrosion, including water quality and composition, flow conditions, biological activity, and corrosion inhibitors.

Facts concerning public health, in particular, should give a significant boost to the use of stainless steel.A deep knowledge of the sanitary systems market dynamics (regional traditions, weakness of the competitor materials, end-user feeling) will be in the future the key point to work on in order to step up its presence.

A general survey of the status of adoption of stainless steel in Europe, USA and Japan will be presented with specific attention to the European market.

Development of Stainless Steel Products in Eastern Special Steel

Liu Xiaoya

(Zhenshi Group Eastern Special Steel Co., Ltd.)

Eastern Special Steel Co., Ltd was established by Zhenshi Holding Group Co., Ltd after acquiring the 100% ownership of the former Jiaxing Eastern Iron & Steel Co., Ltd in July 2007. The company imported 80-ton energy-saving environment-friendly CONSTEEL electric arc furnaces from Italy, possess 90-ton AOD converters, 90-ton ladle furnaces, a 1600 slab continuous casters, 1800 Steckel mills，medium wide plate finishing equipment, annealing and picking equipment. The major products are wide and thick stainless steel hot rolled coil, marine grade stainless steel, watch grade stainless steel, heat-resistant stainless steel and duplex stainless steel, which were used in food and beverage equipment, pressure vessel, watch parts, kitchenware, heat exchangers, etc. The products were adopting the manufacture license of special equipment, certificate of pressure equipment directive (PED), certificate of ASME, certificate of CCS, DNV, ABS, LR, GL, BV.

不锈钢在汽车 SCR 模拟环境中的腐蚀行为研究

徐泽瀚[1] 张国利[1] 李谋成[1] 毕洪运[2]

（1. 上海大学材料研究所，上海 200072；
2. 宝钢集团中央研究院不锈钢技术中心，上海 200431）

摘 要 运用 600℃废气氧化与 80℃冷凝液浸泡方法模拟柴油机 SCR 内部服役环境，对比研究了铁素体不锈钢 441 和奥氏体不锈钢 304 的冷凝液腐蚀行为。结果表明：经有/无尿素的废气氧化后，441 不锈钢在冷凝液中的腐蚀均处于钝化状态，而 304 不锈钢的腐蚀均处于活化状态；阳极极化曲线测量后，441 不锈钢表面形成了点蚀坑，304 不锈钢部分表面区域有明显的晶间腐蚀特征，氧化环境中引入尿素对两种不锈钢的前述局部腐蚀都具有增强作用。

关键词 汽车排气系统，腐蚀，选择性催化还原(SCR)，不锈钢，尿素

Corrosion Behavior of Stainless Steels in Simulated Automotive SCR Environments

Xu Zehan[1], Zhang Guoli[1], Li Moucheng[1], Bi Hongyun[2]

(1. Institute of Materials, Shanghai University, Shanghai 200072, China;
2. Stainless Steel Technical Center, Research Institute, Baosteel Group Corporation, Shanghai 200431, China)

Abstract The corrosion behavior of type 304 and 441 stainless steels were investigated in the simulated environments of diesel exhaust SCR system by using tests of exhaust gas oxidation at 600°C and condensates immersion at 80°C. The results indicate that, after 600°C -oxidation in the exhaust gases with or without addition of urea, 441 stainless steel shows passive corrosion state in condensate solutions, whereas 304 stainless steel is in active corrosion state. After the polarization curve measurements, some pits formed on 441 stainless steel surfaces, and intergranular corrosion feature appeared on part of 304 stainless steel surfaces. As urea was added into the exhaust gas, it can accelerate the aforementioned types of localized corrosion on both oxidized 304 and 441 stainless steels.

Key words automotive exhaust system, corrosion, SCR, stainless steel, urea

Reduction of Cr during Smelting Treatment of Stainless Steel Dust

Zhang Yanling, Jia Xinlei, Guo Wenming

(State Key Laboratory of Advanced Metallurgy, University of Science and Technology, Beijing 100083, China)

Abstract Staniless steel dust contains a remarkable content of Cr oxides which is worthrecycling.As a practical method, Smelting treatment is chosen to recover Cr resource in stainless steel dust. Distribution of Cr between the recovered melt and the residual slag phase depends on the activity coefficient of Cr in the metal phase (γ_{Cr}) and the activity coefficient of CrO in the slag (γ_{CrO}). Both γ_{Cr} inFe-Cr-Ni-C (-Si/Al) melt and γ_{CrO} in CaO-SiO_2-MgO-Al_2O_3 slag under different situations were firstlypredicted through thermodynamic calculation. Experimental research on reduction of Cr during smelting treatment of stainless steel dust was performed in this study. Theresults suggested that the recovery ratios of valuable metals increased with increasing %CaO/%SiO_2 of the residual slag to 1.17 and then decreased.A good linear relationship betweenthe apparent distribution ratio of Cr ($L_{Cr}^{m/s}$) and γ_{CrO}/γ_{Cr} was observed.

Key words stainless steel dust, activity coefficient, partition ratios, chromium

红土镍矿固相还原过程热分析动力学

吕学明 邱杰 刘猛 吕学伟

（重庆大学材料科学与工程学院，重庆 400044）

摘 要 利用法国 Sataram 公司型号为 Setsys EvoTG-DTA1750 热分析仪，在升温速率分别为 10ºC/min、15ºC/min、20ºC/min 条件下对红土镍矿固相还原过程进行非等温多重扫描速率分析。针对不同升温速率下的热失重曲线曲线，分别采用 Flynn-Wall-Ozawa（FWO）法、Starink 法以及 Kissinger-Akahira-Sunose（KAS）法等不同的转化率法计算其活化能，利用 Malek 法确定机理函数，分析了红土镍矿固相还原机理。结果表明：红土镍矿失重率与加热温度密切相关，而升温速率对其影响很小；依照红土镍矿还原转化率，固相还原过程可分为 0~40%、40%~70%、70%~100% 三个阶段，平均反应活化能分别为 71.19 kJ/mol、140.38 kJ/mol 和 370.21 kJ/mol；实验曲线在转化率 0~0.4 范围内与 35 号标准曲线重合性较好，所对应的函数为反应级数方程，机理为 n=4，机理函数的微分形式 $f(\alpha)=\frac{1}{4}(1-\alpha)^{-3}$；实验曲线在转化率 0.4~0.7 范围内，并不与任一曲线重合性较好，说明在该阶段，红土镍矿还原过程的机理函数不止一种；实验曲线在转化率 0.7~1.0 范围内与 39 号标准曲线重合性较好，所对应的函数为指数法则，机理为 n=1，加速形 $\alpha-t$ 曲线，机理函数的微分形式 $f(\alpha)=\alpha$。

关键词 红土镍矿，固相还原，非等温，热分析动力学

Thermal Analysis Kinetics of the Solid-State Reduction for Nickel Laterite

Lv Xueming, Qiu Jie, Liu Meng, Lv Xuewei

(College of Materials Science and Engineering, Chongqing University, Chongqing 400044, China)

Abstract The Non-isothermal multiple scan rate analysis of solid-state reduction for nickel laterite were determined by Setsys EvoTG-DTA1750 of Sataram company. Experiments were carried out at four heating rates of 10ºC/min、15ºC/min、 20ºC/min, respectively. According to the TA lines of different heating rates, the activation energies were calculated through three iso-conversional methods named the Flynn-Wall-Ozawa（FWO）, Starink and Kissinger-Akahira-Sunose（KAS）methods. The most probable mechanism function were determined by the Malek method, then the reduction process dynamics of laterite nickel ore were studied. The results show that: the process of the solid-state reductionfor nickel laterite is divided into three steps, 0~40%、40%~70%、70%~100%, respectively, according to the conversion rate of the reduction of nickel laterite, the average activation energy are 71.19kJ/mol、140.38kJ/mol and 370.21kJ/mol, respectively. Within the range of 0~40% of reaction degree,the most probable mechanism function fits the reaction order equation well, n=4,the formula is $f(\alpha)=\frac{1}{4}(1-\alpha)^{-3}$; Within the range of 70%~100% of reaction degree, experiment cuvers do not meet any theoretical curves, this shows that more than one mechanism function match in this stage; Within the range of 70%~100% of reaction degree, the most probable mechanism function fits the law of exponent well, n=1, curves of $\alpha - t$, the formula is $f(\alpha)=\alpha$.

Key words nickel laterite, solid-statereduction, non-isothermal, thermal analysis kinetics

Accelerated Descaling and Polishing of Ferritic Stainless Steel during the HCl-based Pickling Process

Yue Yingying, Liu Chengjun, Shi Peiyang, Jiang Maofa

(Key Laboratory for Ecological Metallurgy of Multimetalllic Ores (Ministry of Education), Northeastern University, Shenyang, China)

Abstract The mixtures of nitric acid and hydrofluoric acids are typically utilized for the pickling process of stainless steel. However, with the increasing environmental and safety consciousness due to nitrates and nitrogen oxides (NO_x), the use of the mixed acids, especially the nitric acid, has been recently questioned. In this study, the mixtures of hydrochloric acid and oxidant were employed as the environmental friendly pickling solution for ferritic stainless steel. In hydrochloric acid solution, both E_{corr} and i_{corr} of stainless steel increases with the elevation of HCl contents which is beneficial for improving pickling efficiency. However, the oxide scales cannot be removed completely via increasing HCl concentration solely and the intergaranular corrosion of base metal aggravates accordingly. Addition of oxidant can boost the corrosion potential of stainless steel significantly and change the electrode action resulting in desired descaling efficiency. At high oxidant content (0.6mol/L), a corrosion product film accumulates onto the surface and the corrosion is then governed by the mass-transport at the film/stainless steel interface. The random dissolving of metal ions because of the film leads to polishing of stainless steel surface and the local corrosion is suppressed. Under the experimental condition, the designed HCl-based pickling solution achieved accelerated descaling and polishing of ferritic stainless steel.

Key words oxidant, pickling, corrosion rate, anodic polishing

搅拌摩擦焊接高氮奥氏体不锈钢的组织与性能研究

冯　浩　李花兵　姜周华　朱红春　张树才　李　磊

（东北大学材料与冶金学院，辽宁沈阳　110819）

摘　要　本文对 Cr-Mn-Mo-N 系高氮奥氏体不锈钢进行搅拌摩擦焊接，在同一焊接速度（100mm/min）、不同搅拌头转速（400 r/min、500 r/min 和 600r/min）下，焊缝均无明显的沟槽、孔洞等表面缺陷，焊核区氮含量与母材基本相同。转速为 400 r/min 时焊接接头无内部缺陷，而转速增加到 500 r/min 和 600 r/min 时，焊核区分别形成了裂纹及裂纹和孔洞复合缺陷。焊核区的晶粒得到明显细化，热影响区的晶粒未发生明显长大。母材区的位错密度较低，而焊核区的位错密度均较高，焊接过程中动态再结晶不完全。焊核区的平均硬度明显高于母材，且随搅拌头转速的增加，焊核区硬度略微降低。在转速为 400 r/min 和 500 r/min 时，焊接件的屈服强度和抗拉强度都高于母材；600 r/min 的焊接件拉伸强度和伸长率均较低，发生脆性断裂。焊核区的点蚀电位（$E_{b,10}$）略低于母材，晶间腐蚀敏感度（I_r/I_a 值）略高于母材，且随转速的增加，$E_{b,10}$ 逐渐降低，I_r/I_a 值逐渐增大，表明焊核区耐点腐蚀和晶间腐蚀性均低于母材，且随转速提高而降低。

关键词　搅拌摩擦焊，高氮奥氏体不锈钢，组织，力学性能，耐腐蚀性能

Microstructure and Properties of Friction Stir Welded High Nitrogen Austenitic Stainless Steel

Feng Hao, Li Huabing, Jiang Zhouhua, Zhu Hongchun,
Zhang Shucai, Li Lei

(School of Materials and Metallurgy, Northeastern University, Shenyang 110819, China)

Abstract　Friction stir welding was applied to high nitrogen austenitic stainless steel (HNAS) of Cr-Mn-Mo-N. No groove-like defects were observed along the weld line at fixed welding speed (100mm/min) and different rotational speed (400 r/min, 500 r/min and 600r/min). And the nitrogen content of the welds was almost identical to that of base metal (BM). The cross-section of joint at rotational speed of 400r/min had no internal defects; however when the rotational speed increased to 500r/min and 600r/min, cracks and holes were observed in the stir zone (SZ). The grains in SZ were significantly refined, and grains in heat affected zone (HAZ) did not experience significant growth. The dislocation density of BM was lower than that of SZ at different rotational speed, and the dynamic recrystallization was incomplete during welding process. The average hardness of SZ was significantly higher than that of BM, and slightly decreases with the increment of rotational speed. The yield strength and tensile strength of samples at 400 r/min and 500 r/min were higher than that of BM. Brittle fracture occurred at 600 r/min and tensile strength and elongation were obviously low. The pitting potential (E_b,10) of SZ were slightly lower than that of BM and degree of sensitization (Ir/Ia) were slightly higher, indicating lower resistance of SZ to pitting and intergranular corrosion. With the increasing of rotational speed, Eb,10 of SZ gradually reduced and Ir/Ia increased, indicating the reduction of corrosion resistance of SZ.

Key words　friction stir welding, high nitrogen austenitic stainless steel, microstructure, mechanical properties, corrosion resistance

Latest Improvements for Stainless Steel 20Hi Rolling Process: Ultimate Technical Breakthrough for Advanced Rolling Efficiency and Capex/Opex Optimization

Baudu Florent, Calcoen Olivier, Ernst De La Graete Conrad, Freliez Jérémie

Abstract　In the field of stainless steel flat product cold rolling, the 20-Hi mill has clearly proved its technological superiority compared to other rolling mill types such as 6-High rolling mill or split housing 20-High rolling mill. However, nowadays, steel producers are under considerable competitive pressure. They are asked for perfect management of Capital expenditure (CAPEX) and operating expenditure (OPEX) but addressing meanwhile the new humans and environmental challenges such as safety, human factor, environmental pollution and energy consumption.

Fives DMS has integrated in its rolling mill design number of new features addressing these new challenges:

(1) New design of strip and work roll spraying;

(2) New design of strip wipers SCP® 3.0;

(3) Improvement of fume exhaust;

(4) New concept of flatness actuators;

(5) New concept of mandrel greasing.

Key words　20-high, cluster mill, advanced spray boards, strip wipers, SCP, fume exhaust, flatness, roll shifting, push-push, mandrel greasing

森吉米尔轧机非稳态轧制力预设定模型优化研究

刘亚军　钱 华　李 实　包玉龙　李 明

(宁波宝新不锈钢有限公司，浙江宁波　315807)

摘　要　某钢厂森吉米尔轧机轧制生产某牌号不锈钢时，由于非稳态轧制力预设定偏差较大，加之AGC控制能力有限，使非稳态轧制力波动较大，带钢厚度偏差较大。针对这一现场问题，本文基于工业现场数据采集，通过理论研究和数据分析，修正了现有轧制力设定结构，建立了非稳态过程轧制力预设定模型，该轧制力预设定模型重点针对原模型中的摩擦系数模型进行了优化，找到了相邻道次非稳态轧制阶段摩擦系数的关系，用上一道次计算所得摩擦系数预报下一道次的摩擦系数，并且通过仿真分析和应用研究验证了该方法的可行性。修正后的轧制力预设定模型提高了非稳态轧制过程中轧辊与带钢接触摩擦系数的计算精度，提高了轧制力预设定精度，同时为建立非稳态AGC控制参数在线动态计算模型提供了依据。

关键词　森吉米尔轧机，非稳态，轧制力预设定，AGC

The Study of Optimizing Sendzimir Mills' Pre-set Rolling Force Model in Unsteady Rolling Process

Liu Yajun, Qian Hua, Li Shi, Bao Yulong, Li Ming

(Ningbo Baoxin Stainless Steel Co., Ningbo Zhejiang 315807, China)

Abstract　In the process of rolling stainless steels by Sendzimir Mills in one steel company, we find that because of the inaccurate Pre-set rolling force and the limited AGC capability in unsteady rolling process, the actual rolling force has larger fluctuations and the strip thickness after rolling has larger deviation. Against this problem, the paper fixed the existing rolling force setting structure and established the Pre-set rolling force model in unsteady rolling process based on the industrial data analysis and theoretical study. New Pre-set rolling force model focuses on optimizing the original friction coefficient model, finds the connection of friction coefficients between neighboring passes in unsteady rolling process and forecasts the friction coefficient of next pass based on the previous friction coefficient. We also have verified the feasibility of the method by simulation analysis and applied research.The fixed Pre-set rolling force model improves the precision of predicted friction coefficient and Pre-set rolling force in unsteady rolling process. What's more, the accurate Pre-set rolling force provides the basis for establishing the dynamic model of calculating AGC control parameters online.

Key words　Sendzimir Mills, unsteady rolling process, pre-set rolling force, AGC

探讨430钢种脏污缺陷控制技术

郑其勇　王咏波　林隆声　季灯平

(宝钢不锈钢有限公司，上海　200431)

摘　要　430是应用十分广泛的一种铁素体不锈钢。本文对比了国内外430钢种的生产工艺，分析了宝钢生产430带钢出现表面脏污的原因，并提出了改进对策。保证Na_2SO_4溶液的清洁，适当提高酸洗能力和增强清洗能力是避免430冷轧带钢脏污缺陷的有效措施。

关键词　430不锈钢，酸洗，脏污

To Investigate the Technology of Control 430 Stainless Steel Dirt Defect

Zheng Qiyong, Wang Yongbo, Lin Longsheng, Ji Dengping

(Baosteel Stainless Steel Co., Ltd., Shanghai 200431,China)

Abstract　430 is a widely used ferritic stainless steel. In this paper, the technique of production of 430 stainless steel between domestic and foreign manufacturers is compared, the reason for dirt on 430 stainless steel strip surface is analyzed, and approaches of improvement are presented. To make sure the cleanness of Na_2SO_4 solution, raise pickling ability and increase cleaning are all effective methods to avoid the defect of dirt on 430 stainless steel cold rolled strip.

Key words　430 stainless steel, pickling, dirt

铁素体不锈钢冷连轧热划伤成因与控制措施的探讨

曹　勇　沈继程　蔡冬敏　韩武强

（宝钢不锈钢有限公司冷轧厂，上海　200431）

摘　要　热划伤一般是指在带钢表面存在条状划痕，沿轧制方向呈不规则分布，宽度一般如发丝粗细，也有几个毫米宽，是冷连轧特别是铁素体不锈钢冷连轧生产中经常遇到的技术性难题，产生机理复杂，影响因素较多。本文基于科研项目“不锈钢冷连轧带钢表面热划伤预报模型的研究与开发”对铁素体不锈钢冷连轧过程热划伤现象出现规律和轧制润滑机理的分析研究，对铁素体不锈钢冷连轧过程中热划伤的成因、影响因素及预防措施进行了较为全面的论述，并根据对宝钢不锈钢冷轧厂五机架冷连轧机组生产铁素体不锈钢时的实际工艺参数分析，提出了预防轧制过程中热划伤产生的相关建议。

关键词　不锈钢，热划伤，乳化液

Ferritic Stainless Steel Cold Tandem Rolling

Cao Yong, Shen Jicheng, Cai Dongmin, Han Wuqiang

(Cold Rolling Plant. Baosteel Stainless Steel Co., Ltd., Shanghai 200431, China)

Abstract　Heat Scratch is defined as scratch on strip surface, which is distributed irregular along the rolling direction. The scratch is like hairline, some of it is several millimeter width. Heat scratch is a difficult technical problem we usually meet for cold tandem rolling, especially for ferritic stainless steel cold tandem rolling. The mechanism of heat scratch is complex and it is caused by several factors. The paper is based on research project, which named Research on Heat Scratch Prediction on Strip Surface for Stainless Steel Cold Tandem Rolling, in which fully discusses the causes and preventive measures of heat scratch and provides some advice for controlling heat scratch when rolling, according to the analysis of actual process parameters of ferritic stainless steel rolling on five-stand tandem rolling mill in Baosteel Stainless Steel Co., Ltd.

Key words stainless steel, heat scratch, emulsion

不锈钢精炼钢包流场优化研究

王承顺[1] 成国光[1] 李六一[1,2] 张建国[2] 徐昌松[2] 屈志东[1] 刘 扬[1]

（1. 北京科技大学钢铁冶金新技术国家重点实验室，北京 100083；
2. 四川西南不锈钢有限责任公司，四川 106083）

摘 要 本研究根据相似原理，按照修正的 Froude 准数相等的原则建立了一套 1:4 的钢包物理模型，通过对钢包底部透气砖的 18 个喷吹位置不同流量下进行混匀时间的测定，优化了钢包底部透气砖吹氩位置。结果表明：原型钢包底部透气砖位置不合理，导致钢包内钢液流动不合理，钢包中部区域出现较大的死区，混匀时间太长；通过综合考虑钢包底部透气砖在不同半径处以及同一半径的不同相对位置优化底部透气砖的位置，钢包内钢液的流动得到了改善，使得钢包中部死区体积明显减小，并且在不同透气量下钢包内钢液的混匀时间减小均超过 15%。最后进行工业实验，初步结果反映较好。

关键词 钢包吹氩，物理模拟，流场优化

The Optimization of Flow Field of Stainless Steel Refining Ladle

Wang Chengshun[1], Cheng Guoguang[1], Li Liuyi[1,2],
Zhang Jianguo[2], Xu Changsong[2], Qu Zhidong[1], Liu Yang[1]

(1.State Key Laboratory of Advanced Metallurgy, University of Science and Technology Beijing, Beijing 100083, China; 2. Southwest Stainless steel Co., Ltd., Sichuan 106083, China)

Abstract Based on the similar principles and Froude criterion, the 1:4 ladle physical model is established. The position of plug at the ladle bottom is optimized by testing 18 different position of plug. Results show that the position of prototype ladle bottom plug is not reasonable, leading to the reasonable flow of molten steel in refining ladle. Dead zone volume is too large and mixing time is too long. By comprehensive consideration of ladle bottom plug position in different radius and the relative position of the same radius, the position of plug was optimized. After optimizing the position of plug, the flow of molten steel in refining ladle is improved, the central dead zone volume is reduced, and the mixing time is decreased more than 15% under different gas blowing rate. It got very good effect in the industrial experiment.

Key words argon blowing of ladle, physical simulation, optimization of flow field

AOD 炉冶炼不锈钢时脱硫的动力学分析

杜晓建

（山西太钢不锈钢股份有限公司炼钢二厂，山西太原 030003）

摘　要　在 AOD 炉内的冶炼过程中，针对用不同的反应温度和供气强度进行取样分析钢中的硫含量。结果显示：脱硫过程的控制步骤为扩散过程[S]从钢液通过边界层到钢-渣界面，其活化能为 10.59kJ/mol。

关键词　不锈钢，AOD 炉，脱硫，控制步骤

Kinetic Analysis of Desulfurization during Stainless Steel Smelting in AOD

Du Xiaojian

(No.2 Steelmaking Plant of Shanxi Taigang Stainless Steel Co., Ltd., Taiyuan 030003, China)

Abstract　During stainless steel smelting in AOD, sulfur content were analyzed at different temperature and at different stirring intensity. The results showed that the controlling step was the diffusion of sulfur from molten steel to the steel-slag interface.

Key words　stainless steel, AOD convertor, desulfurization, controlling step

提高 60tAOD 炉炉龄的生产实践

冯文甫　叶凡新　曹红波　吴广海　郭志彬

（邢台钢铁有限责任公司炼钢厂，河北邢台　054027）

摘　要　通过学习 AOD 炉耐材侵蚀的机理以及对实际生产数据分析，采取降低渣料消耗、推行一次脱硫操作、优化萤石加入配比、缩短冶炼周期，提高作业率、使用合适 Si 含量的高碳铬铁等措施后，使 60tAOD 炉的炉龄由之前的平均 60 炉达到 90 炉，取得一定的经济效益。

关键词　AOD 炉，成本，不锈钢，生产实践

Process Practice for Increasing Lining Life of 60t AOD

Feng Wenfu，Ye Fanxin，Cao Hongbo，Wu Guanghai, Guo Zhibin

(Steelmaking Plant, Xingtai Iron & Steel Corp. Ltd., Xingtai 054027, China)

Abstract　Through the study of AOD furnace refractory erosion mechanism and the analysis of actual production data, take reducing slag consumption, the implementation of a desulfurization operation and optimization of adding fluorite ratio, shorten the smelting cycle, improve the operation rate and the use of appropriate Si content in high carbon ferrochrome and other measures, so that 60tAOD furnace lining life is preceded by an average of 60 furnace reached 90 furnace, to obtain certain economic benefits.

Key words　AOD, cost, stainless steel, production practice

抗菌不锈钢的研究进展

徐鸣悦　王　丛　胡　凯　武明雨　李运刚

（华北理工大学冶金与能源学院，河北唐山　063009）

摘　要　介绍了抗菌不锈钢的产生原因，综述了抗菌不锈钢的概念、发展过程、分类及其制备工艺，分析了抗菌不锈钢的抗菌机制，指出了抗菌不锈钢制备工艺仍存在的问题及展望。

关键词　抗菌不锈钢，分类，制备技术，抗菌机制

Research Progress in Antibacterial Stainless Steel

Xu Mingyue, Wang Cong, Hu Kai, Wu Mingyu, Li Yungang

(College of Metallurgy and Energy, North China University of Science and Technology,
Tangshan 063009,China)

Abstract　The reason of antibacterial stainless steel is introduced. Concept, development process, classification and preparation process of antibacterial stainless steel are all reviewed. Antibacterial mechanism of antibacterial stainless steel is analyzed. Problems and prospects of fabrication process of antibacterial stainless steel are discussed.

Key words　antibacterial stainless steel, classification, preparation technology, antibacterial mechanism

不锈钢冶炼的研究进展

武明雨　胡　凯　王　丛　徐鸣悦　李运刚

（华北理工大学冶金与能源学院，河北唐山　063009）

摘　要　不锈钢材料由于其独特的耐蚀性、抗腐性和耐酸性，广泛应用于建筑、厨房、电气、运输和工业机械等领域。本文介绍了国内外不锈钢冶炼的发展动态，综述了马氏体不锈钢，奥氏体不锈钢，铁素体不锈钢三种钢种的冶炼工艺，对比分析了各钢种冶炼的优缺点，并提出了相关的改进措施，最后对不锈钢冶炼的发展提出了展望。

关键词　奥氏体不锈钢，马氏体不锈钢，铁素体不锈钢，冶炼工艺

Present Research and Progress on Stainless Steels

Wu Mingyu, Hu Kai, Wang Cong, Xu Mingyue, Li Yungang

(College of Metallurgy and Energy, North China University of Science and Technology, Tangshan 063009，China)

Abstract Stainless steel is widely used in construction, kitchen, electrical, transportation and industrial machinery and other field because of its unique corrosion resistance, corrosion resistance and acid resistance. Development trends of stainless steel smelting at home and abroad are introduced in this paper. The smelting process of martensitic stainless steel, austenitic stainless steel, ferritic stainless steel three steel are reviewed. The advantages and disadvantages of the steel smelting process is comparatively analyzed, and the relevant improvement measures is put forward. Finally, the development of stainless steel smelting is prospected.

Key words austenitic stainless steels, martenstic stainless steel, ferritic stainless steel, production process

6.5 特殊钢/Special Steel

The Development of Boiler Pipes Used for 600+℃ A-USC-PP in China

Liu Zhengdong[1], Bao Hansheng[1], Chen Zhengzong[1], Xu Songqian[2], Yan Peng[2], Zhao Haiping[2], Wang Qijiang[2], Yang Yujun[3], Zhang Peng[3], Lei Bingwang[4]

(1. Central Iron and Steel Research Institute (CISRI), Beijing 100081, China; 2. Baosteel Co., Ltd., Shanghai 201900, China; 3. Fushun Special Steel Co., Ltd., Liaoning 113001, China; 4. Inner Mongolia North Heavy Industries Group Co., Ltd, Inner Mongolia 014033, China)

Abstract This paper introduces the progress of boiler pipes used for the manufacturing of 600+℃ advanced ultra-super-critical (A-USC) fossil fuel power plants (PP) in china, with the emphasis on the detailed advancements of G115 and CN617 pipes, including technical exploration, industrial production and microstructure-property investigation. G115 is a novel ferritic heat resistant steels developed by CISRI, which is an impressive candidate material to make pipes for the temperature up to 650℃. CN617 (or C-HRA-3) is a recent modification of Inconel617B and the CN617 pipe with the dimension of ϕ460 mm×80 mm was successfully manufactured in China. Some newly available data associated with above materials will be released. G115 and CN617 are imposing candidate materials for the manufacturing of 600+℃ advanced ultra-super-critical (A-USC) fossil fuel power plants (PP) in china.

Key words 600+℃ advanced ultra-super-critical (A-USC), G115 steel, CN617 alloy

GH2132 合金冷拉棒材热处理工艺研究

张月红　陆　勇　王　勇　周　进　吴　静

（宝钢特钢有限公司，上海　200940）

摘　要 GH2132 合金是一种 25Ni-15Cr-Fe 基沉淀硬化型变形高温合金，主要通过时效析出的 γ'相 $Ni_3(Ti, Al)$强化，变形后的热处理工艺可以控制γ'析出相，从而影响其组织和性能。本文采用不同的固溶温度和固溶时间对 GH2132 合金进行热处理，研究热处理工艺对 GH2132 合金晶粒度和性能的影响。试验结果表明，相同的热处理条件下，GH2132 合金棒材冷拉态晶粒的长大趋势比热轧态晶粒大。文中将宝钢生产的 GH2132 合金冷拉棒材的热处理工艺与国外相同牌号相近规格棒材的热处理工艺进行对比分析。试验结果表明，采用相同的热处理工艺，宝钢生产的 GH2132 合金冷拉棒材和国外公司的产品在晶粒度和力学性能上水平相当。

关键词 GH2132，热处理，晶粒度，性能

Research on Heat Treatment Technology of GH2132 Alloy Cold-drawing Bar

Zhang Yuehong, Lu Yong, Wang Yong, Zhou Jin, Wu Jing

(Baosteel Special Steel Co., Ltd., Shanghai 200940, China)

Abstract GH2132 alloy is a 25Ni-15Cr-Fe based precipitation hardening alloy during high temperature deformation, mainly through precipitation of gamma' phase Ni_3 (Ti, Al) strengthening. Heat treatment after deformation process can control γ' precipitate, thus affecting the microstructure and properties. In this paper, heat treatment ways of GH2132 alloy with different heat treatment temperature and holding time were studied. Effect of heat treatment process on grain size and properties of GH2132 alloy was researched. Under the same heat treatment, grain of GH2132 cold-drawing bar grown up much faster than hot-rolling bar. Analyzed with the foreign similar grade specifications alloy, grain size and properties of Baosteel GH2132 cold-drawing bar were as good as production produced by foreign corporation, after same heat treatment technology.

Key words GH2132, heat treatment, grain size, properties

700℃超超临界机组用 C-HRA-1 合金管的研制

王婷婷 徐松乾 赵海平

（宝钢集团中央研究院，上海 201900）

摘 要 700℃超超临界机组是目前全世界燃煤发电的最新技术，高温材料的研究与制造是关键。本文以工业化规模研制了 700℃超超临界机组用 C-HRA-1 合金管，研究了其长期时效期组织稳定性及高温拉伸、持久等力学性能。合金管试验分析结果表明，C-HRA-1 合金管 750℃和 800℃长期时效过程中未发现有害相析出，组织稳定性良好；高温拉伸强度与 Inconel740H 相当，750℃持久性能优于 Inconel740H，800℃持久性能与 Inconel740H 相当。

关键词 700℃超超临界，镍基合金管，组织稳定性，力学性能

Investigation on C-HRA-1 Suerpalloys Tube for 700℃ A-USC Plants

Wang Tingting, Xu Songqian, Zhao Haiping

(Research Institute, Baoshan Iron & Steel Co. , Ltd. , Shanghai 201900, China)

Abstract 700℃advanced ultra-supercritical (A-USC) plants with a higher efficency is the most latest technology of

coal-fired power generation worldwide, and the key is the manufacture of high temperature materials. The research progress of the Ni-based suerpalloys tube C-HRA-1 for 700℃ A-USC plants is introduced in this paper. Furthermore, microstructural stability during long-term thermal exposure at 750℃and 800℃are investigated by OM, SEM, TEM, and tensile properties and creep rupture properties of C-HRA-1 tube are also studied to provide data and theoretical support for the practical application of the tube. Experimental results show that no dentrimental phases are found during long-term thermal exposure at 750℃and 800℃and the microstructural stability of C-HRA-1 tube is excellent. Furthermore, 750℃creep rupture life of C-HRA-1 tube is better than that of Inconel740H, and 800℃creep rupture life and high temperature tensile strength of both alloy is comparable.
Key words 700℃ advanced ultra-supercritical, Ni based superalloy tube, microstructural stability, mechanical properties

700℃超超临界高压锅炉管的研发进展

徐松乾 赵海平 王婷婷

（宝钢集团中央研究院，上海 201900）

摘 要 随着节能减排要求的提高，我国的火电技术，正在有序开发更高温度更高压力的700℃超超临界火电机组，由此需要开发适用于这种工况的高压锅炉管。在研发成功600℃超超临界高压锅炉管的基础上，进行了700℃超超临界高压锅炉用合金管的研发，共研发成功C-HRA-1、C-HRA-3和GH984G三个合金牌号，各种规格的合金管，对各生产工序的制造技术进行了系统研究，掌握了控制成品管性能的方法，钢管性能达到设计单位的要求。
关键词 700℃超超临界，高压锅炉管，进展

Progress of 700℃ Advanced Ultra-supercritical (A-USC) Boiler Tubes

Xu Songqian, Zhao Haiping, Wang Tingting

(Research Institute, Baosteel Group Corporation, Shanghai 201900, China)

Abstract Along with higher demand of energy-saving and emission-reduction，coal-fired power technology is developing higher temperature and pressure 700℃ Advanced Ultra-supercritical(A-USC) technology in China, the boiler tube for 700℃ A-USC plants is also developing. On the basis of 600℃ USC boiler tubes successful development, , three alloys (C-HRA-1, C-HRA-3 and GH984G) and various dimensions tubes had been developed for 700℃ A-USC. Manufacture technologys of every process were studied systemic. Property of tubes can be controled by various methods to meet the requirement of design unit.
Key words 700℃ advanced ultra-supercritical, boiler tube, manufacture technology

高氮不锈钢开发和应用的最新进展

姜周华 朱红春 李花兵 冯浩 李阳 刘福斌

（东北大学材料与冶金学院，辽宁沈阳 110819）

摘 要 本文介绍了高氮不锈钢加压制备理论、冶炼工艺、焊接技术以及品种开发方面的最近研究进展。给出了具有广泛适用性的预测不锈钢合金体系在常压和加压下的氮溶解度计算公式以及算固相中氮含量的计算结果。研究结果表明，压力增加使得钢锭与锭模之间的传热效果显著提高，冷却速度加快，局部凝固时间缩短，从而显著细化了晶粒尺寸。指出气相渗氮和微弧增氮的加压电渣重熔工艺是制备高氮不锈钢最具发展前景的手段。我们在实验室研究基础上，成功地开发了常压下电渣重熔高氮不锈钢的工业化技术，并研制了氮含量 0.69%的 600~1000MW 大型发电机护环。开发了实验室加压电渣炉和加压感应炉以及相关工艺技术。利用氮气保护的 FSW 技术成功实现了高氮奥氏体不锈钢 18Cr-18Mn-2Mo-0.96N（2.4mm 厚）的焊接。最后重点介绍了高氮马氏体不锈钢以及高氮奥氏体不锈钢的性能特点和应用情况。

关键词 高氮不锈钢，加压冶金，加压电渣重熔，搅拌摩擦焊接

Latest Progress in Development and Application of High Nitrogen Stainless Steels

Jiang Zhouhua, Zhu Hongchun, Li Huabing, Feng Hao, Li Yang, Liu Fubin

（School of Materials & Metallurgy, Northeastern University, Shenyang Liaoning 110819, China）

Abstract In this paper, the recent research progress of high nitrogen stainless steel, the theory of the pressurized metallurgy, the melting process, the welding technology and the development of different high nitrogen steels are introduced. The calculation formula of nitrogen solubility and the calculation results of the nitrogen content in solid phase were given. The results show that pressurizing significantly improve the heat transfer effect between ingot and mold, cooling speed, shorten the local solidification time, thereby significantly refines the grain size. It is pointed out that the process of gas phase nitriding and micro arc increasing nitrogen is the most promising method for preparing high nitrogen stainless steel. On the basis of laboratory research, the industrial technology of high nitrogen stainless steel was successfully developed, and the retaining ring for 600-1000MW of the large generator with 0.69% nitrogen content was developed. The development of laboratory pressurized electroslag furnace and induction furnace pressure and related technology was realized. The welding of high nitrogen austenitic stainless steel 18Cr-18Mn-2Mo-0.96N (2.4mm thick) was successfully realized by FSW technology. The performance characteristics and application of high nitrogen martensitic stainless steel and high nitrogen austenitic stainless steel were introduced.

Key words high nitrogen steel, pressurized metallurgy, pressurized electroslag remelting, friction stir welding

A Three-Dimensional Comprehensive Model for the Prediction of Macrosegregation in Electroslag Remelting Ingot

Wang Qiang[1], Li Bao kuan[1], Tsukihashi Fumitaka[2]

(1. School of Materials and Metallurgy, Northeastern University, Shenyang, China;
2. Department of Advanced Materials Science, Graduate School of Frontier Sciences, The University of Tokyo, Chiba, Japan)

Abstract A transient three-dimensional (3D) comprehensive model is established to understand the macrosegregation in the electroslag remelting (ESR) process. The electromagnetism, two-phase flow and heat transfer are included. The volume of fluid (VOF) approach is employed to trace the metal droplet. The solidification is modeled by an enthalpy-based technique. The solute transport is analyzed by the continuum mixture model. A reasonable agreement is obtained between the experiment and simulation. The results indicate that the liquid composition in mushy zone becomes enriched in Ni. The interdendritic metal with a higher Ni is replaced by the Ni poor metal carried by the downward flow in the pool. The Ni composition accumulates at the pool bottom and the concentration increases with time. The species movement is dominated by the thermal buoyancy because of the forced cooling. A negative segregation in the lower part and a positive segregation in the upper part are formed in the ingot. Thanks to the rapid solidification, the Ni is immobilized before moving to the other place resulting in a lower segregation level. The segregation becomes severer with the increasing current. The maximal positive and negative segregation indexes increase from 0.02742 to 0.03226 and from -0.01346 to -0.01561 with the current ranging from 1000 A to 2000 A.

Key words electroslag remelting, macrosegregation, numerical simulation, electromagnetism, heat transfer

水韧温度对 Fe-26Mn-7Al-1.3C 奥氏体钢组织与性能的影响

谭志东[1] 宋仁伯[1] 彭世广[1] 郭 客[2,3] 王忠红[3]

（1. 北京科技大学材料科学与工程学院，北京 100083；2. 辽宁科技大学，辽宁鞍山 114051；3. 鞍钢集团矿业设计研究总院，辽宁鞍山 114004）

摘 要 利用力学性能检测、X 射线衍射分析、显微组织及析出物能谱分析，研究了不同的水韧温度对 Fe-26Mn-7Al-1.3C 奥氏体钢的组织与性能的影响。其结果表明：水韧处理工艺对实验钢的显微组织和力学性能有显著影响，适宜的水韧处理工艺（1050℃×1h）可使实验钢的综合力学性能达到最佳值，其冲击韧性（V 型缺口）为 263.9J/cm^2，硬度为 220.3HB，抗拉强度为 767.9MPa，屈服强度为 395.5MPa，断后伸长率为 58.8%。随着水韧温度的升高，粗大的碳化物逐渐溶于奥氏体基体中，使组织均匀化，但晶粒度也随之增大；通过 TEM 和 EDS 分析，这种粗大的碳化物为$(Fe, Mn)_3AlC_x$的 κ 系碳化物，会急剧增加钢的脆性断裂行为，对材料的性能不利。

关键词 Fe-26Mn-7Al-1.3C 奥氏体钢，水韧处理，显微组织，力学性能

Effect of Water Toughening Temperature on Microstructure and Properties of Fe-26Mn-7Al-1.3C Austenitic Steel

Tan Zhidong[1], Song Renbo[1], Peng Shiguang[1], Guo Ke[2,3], Wang Zhonghong[3]

(1. School of Material Science and Engineering, University of Science and Technology Beijing, Beijing 100083, China; 2. University of Science and Technology Liaoning, Anshan Liaoning 114051, China; 3. Angang Group Mining Engineering Corporation, Anshan Liaoning 114004, China)

Abstract Effects of different water toughening temperatures on microstructure and properties of Fe-26Mn-7Al-1.3C austenitic steel were investigated by mechanical properties testing, X-ray diffraction (XRD), optical microscopy (OM), scanning electron microscopy (SEM), transmission electron microscope (TEM) and energy dispersive spectrometer (EDS). Results showed that water toughening treatment had a great effect on the microstructure and properties of the experiment steel. With suitable technology of water toughening treatment (1050℃×1h), the experiment steel had the best comprehensive performance with the impact toughness values (V-notch) of 263.9J/cm^2, hardness of 220.3HB, tensile strength of 767.9MPa, yield strength of 395.5MPa, and elongation after fracture of 58.8%. With the increase of water toughening temperature, massive carbides gradually dissolved in austenitic matrix to make the microstructure more uniform, but grain size increased as well. Through TEM and EDS analysis, the massive carbide was $(Fe, Mn)_3AlC_x$, which made an exponential increase in the brittle fracture behavior and was adverse to the performance of the steel.

Key words Fe-26Mn-7Al-1.3C austenitic steel, water toughening, microstructure, mechanical property

含 Si/Al 低温贝氏体钢的研究与应用

张福成[1,2] 杨志南[1,2] 郑春雷[1] 龙晓燕[1]
王明明[1] 王艳辉[1] 赵佳丽[1]

（1. 燕山大学亚稳材料制备技术与科学国家重点实验室，河北秦皇岛 066004；
2. 燕山大学国家冷轧板带装备及工艺工程研究中心，河北秦皇岛 066004）

摘 要 低温贝氏体钢以其优异的综合性能而受到广泛关注，然而其存在制造工艺周期长、高的碳含量使其焊接性能差等不足限制了其工业化应用。我们通过合金元素调整、工艺优化设计，在低温贝氏体钢方面取得了如下研究成果：（1）通过连续冷却转变工艺在中碳钢中成功获得低温贝氏体组织，这种组织不仅具有高的强度、冲击韧性，还具有非常高的断裂韧性。（2）首次将低温贝氏体组织拓展到渗碳钢中，发展了表面为高碳硬贝氏体组织、心部为高韧性低碳马氏体的渗碳钢，其耐磨性和疲劳性能均要优于现在常用的马氏体渗碳钢，推动了渗碳硬化技术的发展；（3）将高碳高硅钢制备低温贝氏体组织的工艺周期缩短至几个小时；这三种低温贝氏体钢可用于辙叉、钢轨、轴承、齿轮、螺栓、弹簧、耐磨钢板、等零部件的制造，具有非常广阔的应用前景。

关键词 低温贝氏体，渗碳钢，中碳钢

Research and Application of Low Temperature Bainite Steel Containing Si/Al

Zhang Fucheng[1,2], Yang Zhinan[1,2], Zheng Chunlei[1], Long Xiaoyan[1],
Wang Mingming[1], Wang Yanhui[1], Zhao Jiali [1]

(1.State Key Laboratory of Metastable Materials Science and Technology,
Yanshan University, Qinhuangdao Hebei 066004, China;
2.National Engineering Research Center for Equipment and Technology of Cold Strip Rolling,
Yanshan University, Qinhuangdao Hebei 066004, China)

Abstract The low temperature bainite (LTB) steel has gotten a lot of attention for its excellent comprehensive mechanical properties. However, it still has some disadvantages, such as a long production period, poor welding property for high carbon content, which restrict its application. Via regulating alloying element and optimizing the manufacturing process, the following results has been obtained: (1); we obtained the LTB in medium carbon steel via continue cooling phase transformation process after cooling from high temperature, which not only exhibited a high strength, high impact toughness, but also a high fracture toughness, as compared with other bainite steels; (2) and we firstly expanded the hard bainite microstructure into the carburized steel and developed a carburized steel with a hard bainite in the surface and a low carbon martensite interior, which augers well for the development of case-hardening technology based on LTB; (3) and we also shortened the period for manufacturing LTB microstructure in high carbon and high silicon steel to a few hours. These three kinds of LTB steels can be used to manufacture many kinds of parts, such as crossing, rail, bearing, gear, bolt, spring, wear resistant steel plate, which have a broad application prospects.

Key words low temperature bainite, carburized steel, medium carbon steel

Study on Precipitates Evolution in 10Ni3MnCuAl Steel after Aging

Luo Yi[1,2]

(1. Research and Development Center of Wuhan Iron and Steel (Group) Company, 430080, China;
2. School of Materials Science and Engineering, Shanghai University, 200072, China)

Abstract The precipitation hardened steel 10Ni3MnCuAl after solution heat treatment and aging at 510 ℃ is investigated by transmission electron microscopy (TEM) and three dimensional atom probe (3DAP). The results show that the Ni, Mn, Al and Cu distribute uniformly in the steel after solution heat treatment, but they form precipitates which are NiAl observed by TEM and are multicomponent clusters reconstructed by 3DAP after aging and cause the hardness increase. The chemical compositions of NiAl in 10Ni3MnCuAl steel are far from that of stoichiometric binary NiAl, because Ni, Mn, Al and Cu coexist in the same position, which leading to NiAl precipitates containng a certain content of Fe, Cu and Mn. The amounts of Ni and Al in the precipitates increase during further aging at the expense of Fe. Their chemistry converges towards the stoichiometricm composition of NiAl, but a little Mn, Fe, and Cu are still incorporated even after aging for 100 h. The observed interfacial segregation becomes more pronounced with increasing aging time, thereby forming a distinct shell containing Ni, Al and Mn adjacent to the Cu-rich precipitate cores, and the distance between concentration peak of Cu and those of Ni and Al becomes wider with the aging time, which is related to the precipitation character of Cu in α-ferrite.

Key words aging hardened, precipitate, 3DAP, 10Ni3MnCuAl Steel

电渣重熔 GCr15 轴承钢化学成分变化及夹杂物特性研究

李世健[1] 杨 亮[1] 成国光[1] 陈 列[2] 严清忠[2] 赵海东[2]

（1. 北京科技大学钢铁冶金新技术国家重点实验室，北京 100083；
2. 西宁特殊钢股份有限公司，青海西宁 810005）

摘 要 本文针对国内某特钢厂所生产的电渣重熔 GCr15 轴承钢轧材化学成分及夹杂物特性进行研究，并对其化学成分变化的原因、夹杂物成分与化学成分的关系进行分析。结果表明：电渣重熔后轧材 Al、Si 含量相对母材出现一定程度的降低，轧材氧含量相对母材出现一定程度的增加；Al 烧损程度越大，钢中氧含量越高；钢中夹杂物主要以单一 Al_2O_3，$xMgO.yAl_2O_3$，$xCaO.yAl_2O_3$ 为主，并含有少量以 Al_2O_3 为核心，以 TiN 为外围的复合夹杂物；铝含量低的轧材以 Al_2O_3（Mg）夹杂物为主，铝含量高的轧材以 $xMgO.yAl_2O_3$，$xCaO.yAl_2O_3$ 夹杂物为主。

关键词 电渣重熔，GCr15 轴承钢，成分变化，夹杂物特性

Study on the Steel Composition Variation and Inclusions Characteristics of GCr15 Bearing Steel through ESR

Li Shijian[1], Yang Liang[1], Cheng Guoguang[1], Chen Lie[2], Yan Qingzhong[2], Zhao Haidong[2]

(1. State Key Lab. of Advanced Metallurgy, University of Science and Technology Beijing, Beijing 100083, China;

2. Xining Special Steel Co., Ltd., Xining 810005, China)

Abstract This paper focuses on the chemical composition and inclusion characteristics of rolled bars through ESR (Electro-slag Refining) process, which are produced in a special steel plant. The reason of chemical composition variation, relationship between inclusions and chemical composition are studied. The results show that the Al, Si contents of rolled bars are lower than that in the electrode while oxygen content of rolled bars is higher than that in the electrode. With the increasing oxidation extent of Al, the oxygen content increases obviously. The inclusions are mainly the single Al_2O_3, xMgO.yAl_2O_3, xCaO.yAl_2O_3, and few are multiphase inclusions with Al_2O_3 as the core and TiN as the periphery. Besides, the inclusions in the rolled bar with low Al content are mainly Al_2O_3(Mg), and inclusions in the rolled bar with high Al content are mainly xMgO.yAl_2O_3, xCaO.yAl_2O_3.

Key words ESR, GCr15 bearing steel, composition variation, inclusions characteristics

压力对高氮钢宏观组织的影响规律

朱红春 姜周华 李花兵 刘国海 张树才 冯 浩 李可斌

（东北大学材料与冶金学院，辽宁沈阳 110819）

摘 要 本文在凝固压力为 0.5MPa、0.85MPa、1.2MPa 的条件下，利用 25kg 加压感应炉，在尺寸不同的铸型中，制备出了高氮奥氏体不锈钢 P2000（氮含量约 0.9%），并结合 Thermo-Calc 热力计算软件和低倍组织检测手段，研究了压力对高氮不锈钢宏观组织的影响规律；结果表明，增大压力和冷却速率能够有效避免铸锭内气孔的形成，加压可增强液相的补缩能力，明显减少中心疏松深度和面积，提高铸锭致密度，此外，加压能够通过改善铸型-铸锭间换热条件，改变柱状晶生长方向(铸锭中部枝晶生长方向与水平方向的夹角均随之减小)，扩大柱状晶区，从而对铸锭宏观组织产生影响。

关键词 压力，高氮钢，宏观组织，气孔

Effect of Pressure on the Macrostructure of High Nitrogen Steel

Zhu Hongchun, Jiang Zhouhua, Li Huabing, Liu Guohai, Zhang Shucai, Feng Hao, Li Kebin

(School of Materials & Metallurgy, Northeastern University, Shenyang Liaoning 110819, China)

Abstract In this article, the high nitrogen austenitic stainless steel P2000 (nitrogen content is about 0.9wt%) has been prepared under the different solidification pressure (0.5MPa, 0.85MPa, 1.2MPa) in two kinds of molds. The 25kg pressurized induction furnace was used.Thermodynamic computing software Thermo-Calc and macroscopic observation were combined to analyze the influence of the pressure on the macrostructure of high nitrogen steel. It was shown that increasing pressure and cooling rate decreases a tendency to form nitrogen gas pores slightly.The pressurization canimprove the capacity of liquid feeding, decrease the depth and area of porosity and increase thecompactness of macrostructure in the ingots.In addition, Pressure has a significant impact on the other features of macrostructure of ingots, such as change the growth direction (the angle between columnar crystals and horizontal direction are decreased in the middle of ingot) and enlarge the zone of the dendrite by improving interfacial heat transfer from the ingot to mold.

Key words pressure, high nitrogen steel, macrostructure, pore

中锰耐磨钢的冲击滚动复合摩擦磨损性能研究

王庆良[1]　强颖怀[1]　张　恒[2]　赵　欣[2]　陈　辉[1]

（1. 中国矿业大学材料科学与工程学院，江苏徐州　221116；
2. 宝钢特钢有限公司，上海　200940）

摘　要　选用热轧中锰耐磨钢为实验材料，Hardox 马氏体耐磨钢为对比材料，在 M2000 多功能摩擦磨损试验机上研究了冲击和滚动滚复合摩擦磨损性能，利用 XRD、SEM 和 TEM 等分析了组织转变及磨损机理。实验结果表明，热轧中锰钢比 Hardox 马氏体耐磨钢表现出更好的抗冲滚磨料磨损性能。冲滚磨损表面存在厚度达 1000μm 的硬化层，最高显微硬度达 490HV，洛氏硬度达 53HRC，加工硬化十分显著。位错强化、形变孪晶和马氏体相变是中锰钢硬化和抗磨损性能改善的主要原因，其磨损机制以凿削破坏为主，伴随局部的疲劳剥落破坏。

关键词　热轧中锰钢，冲滚磨损，磨料磨损，磨损机理

Studies of Impact and Rolling Composite Wear Properties for Hot Rolled Medium Manganese Steel

Wang Qingliang[1], Qiang Yinghuai[1], Zhang Heng[2], Zhao Xin[2], Chen Hui[1]

(1. School of Material Science and Engineering, China University of Mining and Technology, Xuzhou 221116, China; 2. Baosteel Special Steel Co., Ltd., Shanghai 200940, China)

Abstract　Hot rolling medium manganese steel and Hardox martensite steel were selected as the testing material in this paper. The impact-rolling composite wear properties of these steels were studied in the experimental machine of M2000. XRD, SEM and TEM were used to investigate the structural transformation and wear mechanism of medium manganese steel. The results show that the hot rolled medium manganese steel reveals the better resistance to impact-rolling composite wear properties compared with the Hardox martensite wear resistant steel in these testing conditions. Approximately 1000μm hardened layer in thickness is formed in worn subsurface. The highest Vivtorinox hardness is about 490HV. And the highest Rockwell hardness is about 53HRC. The work hardening is very significant. Dislocation strengthening, deformation twinning and martensitic transformation are the main reasons to improve the work hardening and wear resistance for hot rolled medium manganese steel. The drilling wear and fatigue fracture were the main wear mechanisms.

Key words　hot rolling medium manganese steel, impact-rolling wear, abrasive wear, wear mechanism

Effect of Aging on Microstructures and Properties of the Mo-alloyed Fe-36Ni Invar Alloy

Sun Zhonghua[1], Li Jianxin[1], Chang Jinbao[1], Zhang Yunfei[1], Peng Huifen[2]

(1. Hebei Iron and Steel Technology Research Institute, Hebei Iron and Steel Group, Shijiazhuang 052165, China;

2. School of Materials Science and Engineering, Hebei University of Technology, Tianjin 300130, China)

Abstract The Mo-alloyed invar alloy is prepared by the vacuum induction melting furnace. The investigations of mechanical properties, phase constitutions, microstructures, thermal expansion features are performed. It was found that precipitation of the Mo_2C secondary phase could be tailored by adjusting the aging processes, and that the highest strength of 820 MPa and good elongation of 35%, together with a low CTE value of 3.37×10^{-6}/℃ were obtained when solution treating at 1050℃ for 1h and than aging at 525℃ for 3h. Increase in the aging time resulted in the Mo_2C phase to aggregate and grow at the grain or sub-grain boundaries of the austenite, concomitant with a deterioration in strength and thermal expansion of the materials.

Key words invar alloy, Mo alloying, strengthening, thermal expansion, aging

锻造用钢锭质量控制

赵俊学[1] 葛蓓蕾[1] 李小明[1] 仇圣桃[2] 唐雯聃[1]

（1. 西安建筑科技大学冶金工程学院，陕西西安 710055；
2. 钢铁研究总院国家连铸中心，北京 100081）

摘 要 钢锭质量问题直接影响到锻件合格率，从冶炼、铸锭环节进行质量控制意义重大。本文结合生产企业锻件质量统计和检测，对钢锭质量问题进行了梳理与分析，针对锻件裂纹、中心缺陷两个主要方面，提出了改进建议。经实施，锻件裂纹率由5.10%下降到3.08%，探伤合格率由80%提高为95%，同时钢锭利用率也明显提高。

关键词 钢锭质量，缺陷，裂纹，探伤合格率

Quality Control of Ingot for Forging

Zhao Junxue[1], Ge Beilei[1], Li Xiaoming[1], Qiu Shengtao[2], Tang Wendan[1]

(1. School of Metallurgical Engineering, Xi'an University of Architecture and Technology, Xi'an Shanxi 710055, China; 2. National Engineering Research Center of Continuous Casting Technology, Central Iron and Steel Research Institute, Beijing 100081, China)

Abstract the rate of forged parts up to grade can be determined by ingot quality. So quality control in metallurgical process as well as casting is of great significance. In this paper, the ingot quality problems is classified and analyzed based on the production. Pointed at the main quality problems, such as cracking, porosity and internal defects, some suggestion was given. The cracking ratio of the forged part decreased from 5.1% to 3.08%. the flaw detection past rate increased from 80% to 95%. And the ingot utilization ratio is promoted obviously at the same time.

Key words ingot quality, flaw, crack, flaw detection past rate

免退火型 ML33Cr 冷镦盘条的研发

李宝秀[1,2] 郭明仪[1,2] 车国庆[1,2] 李世琳[1,2]

（1. 邢台钢铁有限责任公司，河北邢台　054027；
2. 河北省线材工程技术研究中心，河北邢台　054027）

摘　要　通过优化成分设计、制定合理的工艺路线和控轧控冷工艺、结合下游制造业的生产工艺特点采用正公差孔型轧制开发了免退火型 ML33Cr 热轧盘条。经检验，金相组织为均匀的铁素体+珠光体，表面硬度为 85HRB，较好地满足了技术要求。计算了 ML33Cr 的淬透性曲线，推断其产品规格不大于ϕ31mm 时可以淬透。

关键词　免退火，ML33Cr，正公差孔型，表面硬度，淬透性

Study on Non-annealed Cold Heading Steel Wire Rod ML33Cr

Li Baoxiu[1,2], Guo Mingyi[1,2], Che Guoqing[1,2], Li Shilin[1,2]

(1. Xingtai Iron and Steel Corp., Ltd., Xingtai Hebei 054027, China;
2. Hebei Engineering Research Center for Wire Rod, Xingtai Hebei 054027, China)

Abstract　Through optimizing composition design, establishing rational technological route and controlled rolling and controlled cooling process, Adopting plus tolerance pass combined with production process characteristics of the tour industry, it developes non annealed type ML33Cr hot rolled wire rods. After inspection, the microstructure is uniform ferrite + pearlite, the surface hardness is 85HRB, which meets the technical requirements well. The permeability curves of ML33Cr quenching were calculated, deduce that the product of not greater than 31mm diameter can be quenched.

Key words　non annealed, ML33Cr, plus tolerance groove, surface hardness, hardenability

电渣重熔过程数学模型研究及过程仿真软件介绍

李　青[1]　杜　斌[1]　王资兴[2]　谢树元[1]　陈濛潇[2]　代鹏超[2]

（1. 宝钢集团中央研究院自动化所，上海　201900；
2. 宝钢集团中央研究院特钢技术中心，上海　200940）

摘　要　利用自主开发的多物理场计算程序，紧密结合电渣重熔（ESR）工艺过程特点，建立了适用于 ESR 过程数值模拟的数学模型，并开发了通用的数值模拟软件。目前，模型可以实现自电极熔化起始直至铸锭模冷结束的全过程的瞬态模拟计算，计算涵盖熔炼过程的电磁场、流动、传热、熔化及凝固多个物理过程，给出了熔炼过程温度及液相体积分数分布、熔池及两相区形状等工艺控制所关心的特征信息的变化过程，同时计算了铸锭质量密切关联的局部凝固时间、冷却速度、枝晶间距、黑斑判定修正瑞利数 Ra 等参数。模型计算能与实验分析的结果很好吻合。经过发展和完善，本模拟软件可以成为工艺质量分析及过程优化的得力工具，并为新产品和工艺研究提供重要的技术支撑。

关键词　电渣重熔，数学模型，过程仿真，软件

An Introduction of the Research on the Mathematical Model of Eletroslag Remelting Process and the Process Simulation Software

Li Qing[1], Du Bin[1], Wang Zixing[2], Xie Shuyuan[1], Chen Mengxiao[2], Dai Pengchao[2]

(1. Automation Research Center, Research Institute(R&D Center), Baoshan Iron & Steel Co., Ltd., Shanghai 201900, China; 2. Technical Center of Special Steel, Research Institute(R&D Center), Baoshan Iron & Steel Co., Ltd., Shanghai 200940, China)

Abstract Based on the self-developed multiple physics calculation program, in close consideration of characteristics of electroslag remelting (ESR) process, a mathematical model for numerical simulation of ESR process was established and a general simulation software was developed. Till now, the model could realize transient simulation of the whole process from the melting start point of electrode to the end of cooling stage of the ingot in the mould. The computation covers physics of electromagnetic, fluid flow, heat transfer, and melting and solidification during the remelting process giving the dynamic variations of distributions of temperature, liquid phase volume fraction, shapes of melt pool, and two-phase region, etc. concerned with the process control, and the values of parameters of local solidification time, cooling rate, dendrite spacing, and the criterion of the modified number of Ra for freckle formation, etc. closely related to the quality of ingot. The results of the model calculations agree with the experimental results fairly well. With the development and improvement of the model, the simulation software could be an useful tool for the analysis of process and quality and process optimization and provide important technical support for the research of new product and technology development.

Key words eletroslag remelting, mathematical model, process simulation, software

液析碳化物对 H13 热模具钢力学性能的影响机理

王 辉 徐 锟 卢守栋 胡现龙

（宝钢轧辊科技有限责任公司，江苏常州 213019）

摘 要 为研究液析碳化物对 H13 热模具钢力学性能的影响行为，通过拉伸试验与冲击试验研究不同液析碳化物等级对材料力学性能的影响规律，并利用金相显微镜及场发射电子探针观察冲击试样断口金相组织及碳化物分布、大小及类型；通过高温扩散退火处理，研究其对液析碳化物的影响。结果表明：液析碳化物级别越高，材料的冲击功越低，液析碳化物主要以 V_xC_y 化合物形式存在，呈点、链状，且硬而脆，严重割离基体，是材料内部的应力集中点及疲劳裂纹扩展源头。该类碳化物在高温下难熔，通过热处理很难消除；颗粒越多、越大，冲击功越低；其次，通过高温扩散退火处理，可以改善其形貌和大小，提高冲击性能。减少或消除液析碳化物缺陷应设计合理的电渣重熔冶炼工艺和锻造成型工艺。

关键词 H13，液析碳化物，场发射电子探针，力学性能

Effect Mechanism of Eutectoid Carbide on Mechanical Properties of H13 Hot Die Steel

Wang Hui, Xu Kun, Lu Shoudong, Hu Xianlong

(Baosteel Roll Technology Co., Ltd., Changzhou 213019, China)

Abstract In order to research the effect of eutectoid carbide on mechanical properties of H13 hot die steel, Different eutectoid carbide level on the properties of mechanics of materials are studied by tensile test and impact test research. The distribution, size and type of carbide and metallographic structure of impact fracture were observed and analysised by metallographic microscope and field emission electronic probe. And the influence of eutectoid carbide were studied by the high temperature diffusion annealing treatment. The results show that the higher level of liquid eutectoid carbide, the lower impact value of material. The eutectiod carbide are mainly exist in V_xC_y with shaping in spot and catenarian, serious isolated matrix for its hard and brittle, and it's also the internal stress and fatigue crack starting point of material. It's very refractory alloy under the high temperature, and it's very difficult to eliminate through the heat treatment. The A_{KV} will be lower with its size and number increase. The morphology and size can be improved by high temperature diffusion annealing processing and then the A_{KV} properties be improved. To reduce or eliminate eutectoid carbide defects should design reasonable electroslag remelting process and forging process.

Key words H13, eutectoid carbide, field emission electronic probe, mechanical properties

Investigation of Longitudinal Surface Cracks in a Continuous Casting Slab of High-carbon Steel

Guo Liangliang[1], Xu Zhengqi[1], Shi Jianbin[2]

(1. Baoshan Iron & Steel Co., Ltd., Shanghai 201900, China;
2. Baosteel Special Steel Co., Ltd., Shanghai 200940, China)

Abstract Based on a series of related investigation, a mechanism for the formation of longitudinal surface cracks on continuous casting slab of high carbon steel has been studied. The high-temperature tensile test of slab samples, metallographic and scanning electron microscopy studies, as well as mathematic model of prediction of heat flux and shell thickness in continuous casting, have been carried out. The results showed that high carbon steel was found to have much lower liquidus temperature and wider brittle temperature range immediately after solidification than that of low carbon steel. Concentrations of the element K and Na, which are contained in the mold fluxes, were not found in cracks of this study. The longitudinal cracks occur first close to the solidification front observed in situ using confocal laser scanning microscope, where the ductility is extremely low, and the shell growth was slower than others leading to the thinner shell and depressed shrinkage owing to the entrapment of molten flux slightly below the meniscus. Furthermore, pouring temperature of high carbon steel is about 100 °C lower than low carbon steel, so it is hard to form stable liquid flux near the meniscus in short time at the beginning of casting. There is not enough liquid slag film, as well as crystalline slag film, providing enough lubrication between the shell and mold resulting in higher friction force that induce or aggravate cracks. Therefore, the homogeneity of mold fluxes and initial solidification in mold should be improved to solve the slab surface defects.

Key words longitudinal cracks, high-carbon steel, continuous casting

高等级轿车用齿轮钢 20MnCr5H-1 的开发

丁忠　林俊

（宝钢特钢有限公司长材事业部，上海　200940）

摘　要　宝钢特钢有限公司采用电弧炉+炉外精炼+模铸的工艺成功开发了 20MnCr5H-1 高等级轿车用齿轮钢，通过控制电弧炉出钢[P]和出钢温度，LF 钢包炉实现成分精控，冶炼过程精确控制 C、Si 和 S 元素的范围，获得满足成分要求的钢水。试验结果表明：采用该工艺生产高等级轿车用 20MnCr5H-1 齿轮钢，成品材氧含量达到了 15ppm 以下，A 细夹杂物得到了有效控制，淬透性带宽仅 4HRC，其冶金实物质量达到了国外进口材的水平。

关键词 20MnCr5H-1，齿轮钢，精炼工艺，淬透性，夹杂物

The Development of 20MnCr5H-1 Used for the High Quality Automobile Gear Steel

Ding Zhong，Lin Jun

(Baosteel Special Metals Co., Ltd., Long Products Business Unit，Shanghai 200940, China)

Abstract The high quality gear steel of 20MnCr5H-1 was realized successfully by "EAF→ Secondary Refining→Mould cast" processing in route in BaoSteel Special Steel Co..With controlling arc furnace tapping [P] and tapping temperature, by precise control of composition with Ladle furnace and precise control of C，Si and S in the liquid steel in conformity to the chemical composition specification can be obtained. Experimental results show that: the total oxygen content was less than 15ppm; the grade of A thin inclusion in steel can be effectively controlled, the bandwidth of hardenability was only 4 HRC, the metallurgical quality of the 20MnCr5H-1 achieved the imported material level.

Key words 20MnCr5H-1, gear steel, refining technical, hardenability, inclusion

合工钢 S2M 脱碳特性的研究

朱 平[1] 赵 亮[2]

（1. 宝钢特钢有限公司，上海 200940；2. 宝山钢铁股份有限公司研究院，上海 200940）

摘 要 研究了不同加热温度、加热时间对合工钢 S2M 脱碳层深度和组织的影响。结果表明，700℃以下，S2M 钢表面不会产生脱碳层，超过 700℃以后，其脱碳层深度随着保温时间的延长逐渐增大，但增加速度变慢。与时间相比，脱碳对温度更敏感，在 700~900℃加热时，脱碳组织以全脱碳层为主，而且全脱碳层深度在 800℃达到最大值；在 950~1050℃加热时，脱碳组织以部分脱碳层为主，总脱碳层增加十分缓慢；但是超过 1050℃加热时，总脱碳层深度又开始明显增加。综合各方因素考虑，合工钢 S2M 线材轧制时的加热温度应为 950~1050℃。

关键词 合金工具钢，脱碳，线材，组织

Research of Decarburization Characteristics of S2M

Zhu Ping[1], Zhao Liang[2]

(1. Baosteel Special Material Co., Ltd., Shanghai 200940, China; 2.Research Institute, Center Institute, Baoshan Iron & Steel Co., Ltd., Shanghai 200940, China)

Abstract The effect of heating temperature and holding time on the decarburized layer depth and the microstructure during heating process of alloy tool steel S2M. The results show that, under 700 ℃, S2M steel surface decarburization layer does not produce. The decarburization layer depth increases gradually with increasing the temperature when S2M steel is heated over 700℃，then it slow to increase. Decarburization is more sensitive to temperature than to the time. The major decarburized microstructure is fully decarburized layer at 700~900℃.There is the largest depth of decarburization layer at 800℃. When heating temperature at 950~1050℃，the major decarburized microstructure is Semi-decarburized layer，the full decarburization layer increase slowly。The fully depth of decarburization layer began to increase significantly when heating temperature exceeds 1050℃. Considering the various factors, the rolling heating temperature of alloy steel S2M wire should be at 950~1050℃.

Key words　alloy tool steel, decarburization, wire, microstructure

轴承钢 GCr15 控温轧制探索

李学保

（宝钢集团广东韶关钢铁有限公司，广东韶关　512123）

摘　要　韶钢使用旧的控温轧制工艺生产 GCr15 轴承钢，由于对轴承钢网状碳化物形成的温度了解不深，穿水冷却工艺不完善，控制返红温度不合理，导致网状碳化物等级偏高，无法满足客户需求。2014 年底通过合理优化控温工艺，根据不同规格采用控轧控冷或者轧后控冷两种控温轧制方式，有效地控制网状碳化物析出，结合相关控温轧制理论对参数进行分析比较，寻找出适合韶钢生产 GCr15 轴承钢的工艺。目前韶钢 GCr15 轴承钢网状碳化物等级可以得到有效控制，采用轧后控冷工艺控制轧件最高返红温度在 600~720℃，可以保证网状碳化物级别达到 2.5 级以下；采用控轧控冷工艺控制轧件终轧温度在 780~800℃，轧后最高返红温度在 680~720℃，可以保证网状碳化物级别达到 2.0 级以下。韶钢改进控温工艺后轴承钢可以满足各种客户要求。

关键词　GCr15 轴承钢，网状碳化物，　控温轧制，工艺参数

The Exploration of Temperature Controlled Rolling of Bearing Steel GCr15

Li Xuebao

(Baosteel Group Shaoguan Iron & Steel Co., Ltd., Shaoguan 512123，China)

Abstract　Shaogang use the old temperature-controlled rolling parameter for production of GCr15 bearing steel,and Because of the less understand of the formation of the network carbide in bearing steel,water cooling process is not perfect and not suit shaogang ,and the temperature of the red back is unreasonable,the network carbide is high grade,and can not meet the needs of customers.According to the different specifications,choose the controlled rolling and cooling or the post-rolling cooling.By the end of 2014, through the rational optimization of temperature control technology to effectively control the precipitation of network carbide, in combination with the relevant of temperature-controlled rolling theory, analyze the parameter comparison to find a suitable process of Shaoguan Iron and steel to production of GCr15 bearing steel.At present,the network carbide grade of Shaogang bearing steel GCr15 can be effectively controlled. Using the controlled rolling and cooling the highest return temperature at 600 to 720 DEG C,and it can be ensure the carbide network level reached 2.5 below;Using post-rolling cooling the highest return temperature at 780 to 800 DEG C, and it can ensure the carbide network level reached 2.0 below; After the improve of temperature control process, shaogang bearing steel can meet all kinds of the needs of customers.

Key words　bearing steel GCr15, network carbide, temperature-controlled rolling, technological parameter

Analysis of the Rare Non-ferrous Metal Recycling Situation and Suggestions for Domestic Industry

Huang Lei, Tian Peiyu, Chen Hai, Xu Yimin

(Baosteel Special Metals Co., Ltd., Shanghai 200940, China)

Abstract Recycling of rare non-ferrous metals is an important method for reducing the cost of rare material, protecting the non renewable mineral resources, energy conservation and environmental protection in the non-ferrous metals melting industry. It is widely used in this field home and abroad. The processing and utilization of revert is the key for non-ferrous metals recycling. This paper firstly introduce the general situation of the on-ferrous metals recycling industry in domestic and foreign, and then analysis the relationship of the main participates and the benefit for using the revert. Thirdly, this paper analysis the patents relating to the processing and utilization of revert during 2004 to 2014 and the main revert processing technology. Finally, it is pointed that the rare non-ferrous metals recycling in domestic still has a great gap compared with the international counterparts, so it is needed to strengthen the overall coordination ability, revert processing and utilization capacity, industry standards , qualifications management and etc.

Key words nickel, titanium, recycling, revert processing technology

大型盾构机用轴承钢 ZWZ12 动态连续冷却相变规律研究

黄 超[1] 张朝磊[1] 赵 敏[2] 杨 勇[1] 刘雅政[1]

（1. 北京科技大学材料科学与工程学院，北京 100083；
2. 抚顺特殊钢股份有限公司，辽宁抚顺 113001）

摘 要 采用 Gleeble-1500 热模拟试验机，通过金相显微镜、扫描电镜观察及硬度测试，研究了 ZWZ12 轴承钢在 900 °C 变形 50%之后的连续冷却相变规律。结果表明，ZWZ12 钢连续冷却转变仅有高温转变区（相变组织为先共析碳化物和珠光体组织）和低温转变区（相变组织为马氏体），实验条件下没有发现中温区的贝氏体转变。ZWZ12 钢具有很好的淬透性，临界冷速在 1~3 °C/s 之间，马氏体转变温度约为 220 °C；冷速低于 1 °C/s 时，ZWZ12 钢的珠光体转变发生在 700～600 °C 之间。随着冷速的增加，ZWZ12 钢的显微硬度由 0.1 °C/s 的 393.1 HV 增加到 15 °C/s 的 849.2 HV，但显微硬度增加的速度逐渐减慢，冷速大于 3 °C/s 后硬度趋于平缓。

关键词 轴承钢，CCT 曲线，硬度，马氏体

Investigation of Continuous Cooling Phase Transformation on ZWZ12 Steel for Bearing of Large Shield Machine

Huang Chao[1], Zhang Chaolei[1], Zhao Min[2], Yang Yong[1], Liu Yazheng[1]

(1. School of Materials Science and Engineering; University of Science and Technology Beijing,

Beijing 100083, China; 2. Fushun Special Steel Co., Ltd., Fushun Liaoning 113001, China)

Abstract The continuous cooling process of ZWZ12 bearing steel rolled at 900 °C with reduction of 50% was simulated on a Gleeble-1500 themal simulator and investigated by means of optical microscope, scanning electron microscopy and hardness test. The results show that only pearlite and martensite but no bainite transformation are found in the continuous cooling transformation of ZWZ12 steel. With critical cooling rate of 1~3 °C/s and martensite start temperature about 220 °C, ZWZ12 steel have good hardenability and a narrow pearlite transition zone of 600~700 °C when cooled lower than 1 °C/s. The microhardness of ZWZ12 steel increases from 393.1 HV at 0.1 °C/s to 849.2 HV at 15 °C/s with increasing cooling rate, but the increase of microhardness gradually slows down and almost remains unchanged when cooling rate larger than 3 °C/s.

Key words bearing steel, CCT curve, hardness, martensite

超超临界机组叶片用 10Cr12Ni3Mo2VN 钢淬火工艺研究

李俊儒[1]　宋明强[1]　张朝磊[1]　陈 列[2]　佐 辉[2]　钱才让[2]　刘雅政[1]

（1. 北京科技大学材料科学与工程学院，北京　100083；
2. 西宁特殊钢股份有限公司，青海西宁　810005）

摘　要　通过热处理实验，研究了淬火温度和淬火加热时间对 10Cr12Ni3Mo2VN 钢组织和力学性能的影响规律。实验结果表明，980~1100℃，淬火温度对强度影响不大，但随淬火温度升高，塑性和韧性下降，且当淬火温度由1040℃升高至 1070℃时，韧性降低更为明显；1040℃淬火加热时间在 45~135min 之间时，淬火加热时间对力学性能影响较小，随淬火加热时间延长，强度变化不大，塑性和韧性小幅下降。随淬火温度升高，固溶效果增强，晶粒尺寸增大，10Cr12Ni3Mo2VN 钢晶粒粗化温度约为 1040~1050℃。淬火温度对 10Cr12Ni3Mo2VN 钢力学性能的影响因素中，随淬火温度变化晶粒尺寸的变化是主要影响因素，固溶效果的变化对力学性能作用较小，随淬火温度升高和淬火加热时间延长，塑性和韧性下降主要是由晶粒尺寸增大造成的。

关键词　10Cr12Ni3Mo2VN，热处理，力学性能，晶粒粗化

Investigation of Quenching Process on 10Cr12Ni3Mo2VN Blade Steel for Ultra Supercritical Unit

Li Junru[1], Song Mingqiang[1], Zhang Chaolei[1], Chen Lie[2], Zuo Hui[2], Qian Cairang[2], Liu Yazheng[1]

(1. School of Materials Science and Engineering；University of Science and Technology Beijing, Beijing 100083，China；2. Xining Special Steel Co., Ltd., Xining Qinghai 810005，China)

Abstract The influence of quenching process on microstructure and mechanical property of 10Cr12Ni3Mo2VN steel has been studied by heat treatment experiments. The experimental results show that quenching process has a little impact on strength at 980~1100℃. The plastic property and impact property reduces with the quenching temperature rises and the heating time prolongs. And the impact property has a significant reduce with the quenching temperature raising from 1040 ℃ to 1050℃. Most of the precipitated phase dissolved after quenching and the solution of precipitated phase becomes more efficiently with the quenching temperature rises. The grain size increases with the quenching temperature rises and the grain coarsening temperature is about 1040~1050℃. Grain size increasing is the main reason lead to plastic property and impact property reduces with the quenching temperature rises and the heating time prolongs.

Key words 10Cr12Ni3Mo2VN，heat treatment，mechanical property，grain coarsening

回火温度和 Ni 对盾构机用大尺寸轴承套圈用钢组织性能的影响

蒋 波 董正强 周乐育 张朝磊 刘雅政

（北京科技大学材料科学与工程学院，北京 100083）

摘 要 本文利用冲击实验、硬度测试实验、FESEM 扫描电镜以及 TEM 透射电镜，研究了回火温度和合金元素 Ni 对新开发的盾构机轴承套圈用钢调质后组织和性能的影响。研究结果表明，淬火温度不变，回火温度从 500℃升高到 650℃，冲击韧性逐渐升高，断口由准解理形貌向韧窝形貌过渡，回火硬度逐渐降低，回火温度为 650℃时，冲击韧性最高为 107.2J，回火硬度最低为 272.3HB。回火过程中，随着回火温度的升高，回火索氏体中碳化物逐渐由马氏体板条界和马氏体束边界的连续条状碳化物加马氏体内针状碳化物，向短棒状碳化物过渡；同时板条马氏体逐渐发生回复和再结晶，转变为等轴状 α 相。低温回火条件下，组织中的连续的条状碳化物对材料的冲击韧性十分不利。新开发的实验钢，由于 C 元素和 Ni 元素的加入，在一定程度上抑制了回火过程中铁素体的回复与再结晶，对冲击韧性不利，但 Ni 元素的加入大大促进了碳化物的弥散细小分布，有利于冲击韧性的提高。

关键词 回火温度，调质钢，碳化物，韧性，硬度

Effect of Tempering Temperature and Ni on the Microstructure and Properties of Large Size Bearing Ring Steel for Tunneling Boring Machine

Jiang Bo, Dong Zhengqiang, Zhou Leyu, Zhang Chaolei, Liu Yazheng

(School of Materials Science and Engineering, University of Science and Technology Beijing, Beijing 100083, China)

Abstract The effect of tempering temperature and alloying element Ni on the microstructures and properties of the newly designed quenched and tempered steel for bearing ring for Tunneling Boring Machine was analyzed by using toughness test, Brinell hardness test, field emission scanning electron microscope (FESEM) and transmission electron microscope (TEM). Results show that the toughness increases, the hardness decreases and the fracture surface changes from quasi-cleavage to dimple with the tempering temperature from 500 to 650 ºC. When the tempering temperature is 650 ºC, the toughness and the hardness are 107.2 J and 272.3 HB, respectively. Accompanying with the increase of tempering temperature, the carbides in the tempered microstructure gradually changes from the continuous banded carbide distributed along the boundary of martensite lath and block and the acicular carbide in the martensite lath to the short-rod like carbide. At the same time, the recovery and recrystallization of martensite lath occur and thus the equiaxed ferrite is formed. The continuous banded carbide is detrimental to the toughness of steel when the tempering temperature is relatively low. To some content, the recovery and recrystallization of martensite are suppressed by addition of C and Ni in the newly developed steel, which is disadvantageous to the toughness. However, the addition of Ni can largely promote the dispersedly distribution of carbide and improve the toughness.

Key words tempering temperature, quenched and tempered steel, carbide, toughness, hardness

超超临界机组叶片用 KT5331 钢低温冲击韧性研究

宋明强[1]　李俊儒[1]　张朝磊[1]　陈　列[2]　佐　辉[2]　钱才让[2]　刘雅政[1]

（1. 北京科技大学材料科学与工程学院，北京　100083；
2. 西宁特殊钢股份有限公司，青海西宁　810005）

摘　要　通过正交实验和固溶实验，研究了热处理工艺对 KT5331 钢组织和低温冲击韧性的影响规律。实验结果表明，热处理态试验钢显微组织均为回火马氏体组织，断口机理为穿晶解理断裂。1040~1120℃淬火温度对低温冲击韧性影响较大，并且随淬火温度升高，低温冲击功先上升后下降。当淬火温度低于 1080℃时，低温冲击值较低是由原始奥氏体晶粒出现明显晶粒异常长大现象导致，当淬火温度高于 1080℃时，低温冲击值下降是由原始奥氏体晶粒长大导致。淬回火参数中回火温度是影响低温冲击韧性的主要参数，适当提高回火温度可有效改善低温冲击韧性。1080 °C 保温 60 min 以上淬火，720 °C 以上保温 2 h 回火可有效地提高 KT5331 钢的低温冲击性能。并且调质处理前 1150 °C 固溶处理可小幅改善低温冲击性能。

关键词　KT5331，正交试验，低温冲击

Investigation of Low-temperature Impact on KT5331 Blade Steel for Ultra Supercritical Unit

Song Mingqiang[1], Li Junru[1], Zhang Chaolei[1], Chen Lie[2], Zuo Hui[2], Qian Cairang[2], Liu Yazheng[1]

（1. School of Materials Science and Engineering; University of Science and Technology Beijing, Beijing 100083, China, 2. Xining Special Steel Co., Ltd., Xining Qinghai 810005, China）

Abstract　The influence of heat treatment on microstructure and mechanical property of KT5331 steel has been studied by orthogonal and solution experiments. The experimental results show that microstructure of heat treated is tempered martensite and the fracture is transgranular cleavage.　quenching temperature has great impact on low-temperature impact at 1040~1120℃, and with the increase of quenching temperature, the low-temperature impact begins to rise then falls. When the quenching temperature is below 1080, the lower low-temperature impact was caused by abnormal grain growth of primary austenite. However, when the quenching temperature is above 1080, the lower low-temperature impact was caused by the increase of austenite grain size. The tempering temperature is the main parameter and the high tempering temperature increases the low-temperature impact. The best heat treatment process for low-temperature impact is that quenching at 1080 °C after holding for 1 hour, and tempering at 720 °C. And the low-temperature impact can be inproved by solution treatment before the quenching and tempering heat treatment.

Key words　KT5331, orthogonal experiment, low-temperature impact

热处理工艺对 NM500 耐磨钢力学性能的影响

张　新　宋仁伯　温二丁

（北京科技大学材料科学与工程学院，北京 100083）

摘 要 本文利用正交试验法，研究了淬火温度、淬火方式、回火温度、回火时间四个因素对力学性能的影响，运用极差分析方法分析了正交试验结果。结果表明，热处理工艺参数对布氏硬度的影响程度次序为：淬火方式>回火温度>回火时间>淬火温度；对冲击韧性的影响程度次序为：淬火方式>回火时间>淬火温度>回火时间。随着淬火温度的升高，布氏硬度和冲击功均呈先降低后升高的趋势。随着回火温度的升高，布氏硬度呈下降趋势，冲击功先升高后降低且波动较小。布氏硬度和冲击功均随着回火时间的延长而降低。确定了最佳热处理工艺制度为 880 ℃保温 30 min 水冷，200℃保温 30 min 回火。所得最佳力学性能为：硬度 531HB，抗拉强度 1724.8MPa，屈服强度为 1340.8MPa，延伸率 11.6%，–20℃冲击功 23.3J。

关键词 NM500，正交试验，耐磨钢，力学性能

Effect of Heat Treatment on the Mechanical Properties of NM500 Wear-resistant Steel

Zhang Xin, Song Renbo, Wen Erding

(School of Materials Science and Engineering, University of Science and Technology Beijing, Beijing 100083, China)

Abstract The effect of quenching temperature, quenching medium, tempering temperature and tempering time on mechanical properties were researched by orthogonal experiment. The best heat treatment process was obtained by means of range analysis. The results show the influence degree of heat treatment process parameters. Brinell hardness: quenching medium>tempering temperature>tempering time>quenching temperature; Impact energy: quenching medium>tempering time>quenching temperature>tempering temperature. With quenching temperature increases, either brinell hardness or impact energy decreased before increasing. With tempering temperature increases, the brinell hardness decreased and the impact energy increased before decreasing. As the extension of tempering time, the brinell hardness and impact energy decreased.The heat treatment process was optimized as water quenching after holding at 880 ℃ for 30 min and then tempering at 200 ℃ for 30 min. At these optimized heat treatment parameters, the best mechanical properties: the brinell hardness of 531 HB, the impact energy (–20 ℃) of 23.3 J, the tensile strength of 1724.8 MPa, the yield strength of 1340.8 MPa, the elongation of 11.6%.

Key words NM500, orthogonal experiment, wear-resistant steel, mechanical properties

稀土 La 在真空感应熔炼 300M 钢的脱氧和脱硫研究

王 承 龚 伟 姜周华 陈常勇

（东北大学材料与冶金学院，辽宁沈阳 110819）

摘 要 低合金超高强度钢因为较差的断裂韧性和抗应力腐蚀能力，其应用受到限制。超纯净冶炼能够把杂质元素的含量控制到很低，从而显著地提高超高强度钢的断裂韧性。稀土 La 具有很强的脱氧及脱硫能力，因此本文研究了在真空感应炉冶炼 300M 钢时添加 La 进行脱氧脱硫。结果表明，微量的 La 可以进行深脱氧，并且大幅度地降低

钢中硫含量，钢中 La 含量为 0.0084%和 0.015%时，钢中的氧含量均降到 0.0003%，硫含量分别降到 0.0013%和 0.0011%。但是，当钢中 La 含量过多时，反而会使钢中的氧含量、硫含量均增加，造成反作用，钢中 La 含量为 0.032%时，钢中氧含量和硫含量分别增加到 0.0008%和 0.0025%。另外，在保证钢中 Al 含量符合要求的前提下，在进行稀土处理之前，先利用适量的 Al 尽量降低氧含量，可以在得到相同效果下减少 La 的用量，降低成本。

关键词 低合金超高强度钢，脱氧，脱硫，稀土 La

Deoxidation and Desulfurization Investigation of Rare Earth La Addition during VIM 300M Steel

Wang Cheng, Gong Wei, Jiang Zhouhua, Chen Changyong

(School of Materials & Metallurgy, Northeastern University, Shenyang 110819, China)

Abstract The application of ultrahigh strength low alloy (UHSAL) steel is restricted because of the poor fracture toughness and resistance to stress-corrosion. The ultra pure melting is capable of observably improving the fracture toughness because by which the content of impurity elements can be controlled to very low. The La has a strong deoxidation and desulfurization capability, so in this paper deoxidation and desulfurization of rare earth La which added during VIM 300M steel were investigated. The results indicated that the La was capable of deep desulfurizing and reducing the sulfur content drastically. For the La contents of 0.0084% and 0.015% in steels, the oxygen contents both ware 0.0003% and the sulfur contents were respectively 0.0013% and 0.0011%. But the oxygen content and sulfur content both increased when the La content in steel was too much, the oxygen content and sulfur content respectively increased to 0.0008% and 0.0025% when the La content was 0.032%. In addition, under the premise of guaranteeing the Al content conformed to the requirements, the same effect could be obtained by suing less rare earth La resulting in cost reduction when an appropriate amount of Al was added to largely lower the oxygen content before adding La.

Key words ultrahigh strength low alloy steel, deoxidation, desulfurization, rare earth La

1Cr21Ni5Ti 不锈钢的高温氧化行为研究

侯 栋 董艳伍 姜周华 曹玉龙 曹海波 张新法

程中堂 王进鹏 汪 祥

（东北大学材料与冶金学院，辽宁沈阳 110819）

摘 要 研究了不同温度下 1Cr21Ni5Ti 不锈钢的高温氧化行为和氧化膜的组成及演变机制。结果表明：1Cr21Ni5Ti 钢的氧化增重符合抛物线规律，其激活能为 231.433 kJ/mol；分析结果表明目标钢种在 800℃和 1000℃下氧化 100 min 后表层氧化膜的主要物相为$(Fe, Cr)_2O_3$，1100℃和 1200℃下主要为 Fe_2O_3；SEM-EDS 以及热力学分析表明氧首先与活泼元素铬反应生成 Cr_2O_3 初始氧化膜，随后基体中的铁离子溶于 Cr_2O_3 膜并形成$(Fe, Cr)_2O_3$ 氧化膜，同时氧离子经过$(Fe, Cr)_2O_3$ 氧化膜将氧化膜/基体处的 Cr 氧化形成新的 Cr2O3，此时基体内呈富铁贫铬状态，最终基体中的铁离子扩散通过$(Fe, Cr)_2O_3$ 氧化膜，从而在试样表层形成 Fe_2O_3 的氧化层。

关键词 1Cr21Ni5Ti，高温氧化，氧化动力学，氧化膜

Study on Oxidation Behavior of 1Cr21Ni5Ti Stainless Steel

Hou Dong, Dong Yanwu, Jiang Zhouhua, Cao Yulong, Cao Haibo, Zhang Xinfa, Cheng Zhongtang, Wang Jinpeng, Wang Xiang

(School of Materials & Metallurgy, Northeastern University, Shenyang 110819, China)

Abstract The oxidation behavior of $1Cr_21Ni_5Ti$ steel at different temperature and formation mechanism of oxide film were studied. The results indicated that the curves of gained mass of 1Cr21Ni5Ti were liner to parabola and its activation energy is 231.433 kJ/mol. The analysis showed that the phase of oxide film at 800℃ and 1000℃ for 100 min were $(Fe, Cr)_2O_3$, 1100℃ and 1200℃ were Fe_2O_3, respectively. SEM-EDS analysis showed that the O would react with Cr at the beginning time, and then the $(Fe, Cr)_2O_3$ formed after Fe dissolved into the Cr_2O_3. At the same time, the matrix at oxide film with rich Fe and poor Cr would occur as the O passes through the $(Fe, Cr)_2O_3$ oxide film and reacts with Cr. As a result, the Fe dissolving into the $(Fe, Cr)_2O_3$ oxide film makes the sample have the Fe_2O_3 oxide film.

Key words 1Cr21Ni5Ti, oxidation at high temperature, oxidation kinetics, oxide film

极地船舶用低温钢发展

叶其斌 刘振宇 王国栋

（东北大学轧制技术及连轧自动化国家重点实验室，辽宁沈阳 110819）

摘 要 北极地区能源和贸易航线潜力受到越来越多关注，促进了大型高技术极地运输破冰船舶的需求与发展，对满足极地服役条件的极地船舶用低温钢提出了更高的要求，具有优异低温韧性和易焊接性的更高强度级别极地船舶用低温钢是发展趋势。然而目前缺乏专门的极地船舶材料国际规范，极地船舶建造只能采用现有规范中的钢级。实船调查显示目前低温断裂评价准则与极地服役条件存在着明显的差距。我国完全具备自主开发极地船舶用低温钢的装备条件与技术研发能力，但与俄罗斯、日本、韩国、芬兰等极地船舶用低温钢领先水平相比有着明显的差距。建议采用低碳当量成分设计的新 TMCP 工艺技术路线，应用纳米相强韧化理论和适合大线能量的冶金技术研究成果，重视配套焊材与工艺开发，建立极地服役环境下的低温钢安全断裂评价准则。

关键词 极地用钢，低温韧性，极地准则，北极开发

Abstract The great potential of energy storage and shipping route in Arctic Ocean draws more and more interest, which enhances the demand and development of large high-tech business ice breaker. This requires arctic steels with higher performance to servesafely in extreme harsh environment. Higher strength grade steels with excellent cryogenic toughness and weldability are required to construct polar ships in the future. However, the absence of international specification on arctic materials makes that the steel grades specified by the current rules of classification societies have to be used in construction. The investigation on ice-breakers shows that the current criterion of toughness at low temperature is not suitable to evaluate the performance in the real service conditions of arctic steels. Even though equipment and technology of steel manufacture have the capability to develop arctic steels, there is a clear gap between China and world leading countries, such as Russia, Japan, Korea, and Finland. It thus suggests the technology development of arctic steels employing new TMCP route based on low Ceq alloying design in combination with nano-phase strengthening mechanism and metallurgy progress for high heat input welding. Moreover, the suitable welding consumables and process are both important to be developed. Finally, the safety criterion should be improved to evaluate the steel performance in real arctic conditions.

Key words　arctic steel, cryogenic toughness, polar code, arctic development

UNS N06600 合金电渣孔洞缺陷分析与解决

李　博[1]　王　琛[1]　沈海军[2]

（1. 宝钢特钢有限公司特材事业部，上海　200940；2. 宝钢股份研究院，上海　201900）

摘　要　分析了电渣重熔 UNS N06600 尾部孔洞的宏观特征及显微特征，剖析了孔洞的形成机理，并针对此提出避免形成孔洞的措施，为实际生产提供帮助。

关键词　镍基合金，电渣重熔，尾部孔洞

Analysis and Solution of Porosity in Electro-slag Remelting Ingot of UNS N06600 Alloy

Li Bo[1], Wang Chen[1], Shen Haijun[2]

(1. Special Steel Business Division of Baoshan Iron & Steel Co., Ltd., Shanghai 200940, China;
2. R&D Center of Baoshan Iron & Steel Co., Ltd., Shanghai 201900, China)

Abstract　Macro and micro character of porosity in electro-slag remelting of UNS N06600 alloy is discussed, and the mechanism of its formation was obtained. Methods of avoiding porosity are proposed, which is helpful to the improvement of ingot quality.

Key words　UNS N06600 alloy, electro-slag remelting, porosity of foot section

GH738 合金 γ′相溶解规律研究

代朋超[1]　曹秀丽[2]　魏志刚[1]　陈国胜[1]

（1. 宝山钢铁股份有限公司研究院，上海　201900；
2. 宝钢特钢有限公司，上海　200940）

摘　要　分别在 1000℃、1020℃、1040℃、1060℃保温 1h、3h 处理，观察组织中 γ′相的数量、尺寸情况。实验结果表明：在 1000℃保温处理，合金中 γ′相发生溶解，随着保温时间延长到 3h，相尺寸发生粗化长大，达到 400nm 左右。随着保温温度的升高，合金组织中 γ′相数量逐渐减少。当温度达到 1040℃，合金组织中 γ′相出现不均匀溶解。当温度达到 1060℃时，γ′相完全溶解。在同一温度下，随着保温时间延长，γ′相的尺寸不断增大。硬度测试发现，随着处理温度的提高、保温时间的延长，合金硬度呈降低趋势。

关键词　GH738 合金，γ′相，溶解规律

The Investigation of γ′ Phase Disolution Mechanism in GH738 Alloy

Dai Pengchao[1], Cao Xiuli[2], Wei Zhigang[1], Chen Guosheng[1]

(1. Research Institute, Baoshan Iron & Steel Co. , Ltd., Shanghai 201900, China;
2. Baosteel Special Steel Co., Ltd., Shanghai 200940, China)

Abstract The samples were heated to 1000℃、1020℃、1040℃ and 1060℃ for 1 and 3h. The quantity and size of γ′ phase were observed. The results show that when heated at 1000℃, the γ′ phase dissolves, and as the duration time extends to 3h, the size of γ′ phase reaches to 400nm. With the temperature increasing, the quantity of γ′ phase decreases. When the temperature reaches 1040℃, the γ′ phase has a heterogeneous dissolution, and when it reaches to 1060℃, the γ′ phase disolves completely. When at the same temperature, the size of the phase increases as the duration time extended. The hardness test result illustrates that the hardness decreases as the temperature rises and the duration time extends.

Key words GH738 alloy, γ′ phase, dissolution mechanism

N08825 电渣扁锭轧制表面开裂原因分析

陈濛潇[1] 徐文亮[1] 彭予民[1] 李 博[2]

（1. 宝山钢铁股份有限公司研究院，上海 200940; 2. 宝钢特钢有限公司，上海 200940）

摘 要 通过对N08825扁锭轧制开坯裂纹处取样，进行低倍检测、金相观察以及成分分析发现裂纹附近存在疑似重熔残渣。因此推测由于电渣重熔初期工艺设定的不合理导致未形成流动性良好的熔池和渣池,造成渣沟、裹渣等缺陷从而使电渣扁锭尾部塑性较差在轧制开坯时容易开裂。在工艺调整中通过提高电渣重熔化渣期输入功率以及缩短化渣期时间，电渣锭尾部表面质量有大幅改善且基本解决了后道轧制开坯表面开裂的问题。

关键词 电渣重熔，开坯裂纹，表面质量

Analysis of Causes of Crack Formation in N08825 Slab

Chen Mengxiao[1], Xu Wenliang[1], Peng Yumin[1], Li Bo[2]

(1. Research Institute of Baoshan Iron & Steel Co., Ltd., Shanghai 200940, China; 2. Special Steel Branch of Baoshan Steel Corporation, Shanghai 200940, China)

Abstract Through sampling around the crackle on N08825 slab and after macrostructure inspection、microstructure inspection and energy spectrum analysis around the crackle, we find suspected ESR used slag around the crackle. So it is suspected that the abnormal of the start phase leads to the defects like surface slag runner、enclosed slag and so on，while these defects make the ductility of ingot bottom worse and is likely to crack during the cogging process. By improving the input power of start phase and decreasing the time of start phase during the electroslag remelting process, the surface quality of ingot bottom has improved greatly and the problem of cracking during the cogging process has also been solved.

Key words electroslag remelting, cogging crackle, surface quality

马氏体不锈钢X20Cr13中$FATT_{50}$的研究

苏瑞平

（宝钢特钢有限公司，上海 200940）

摘 要 X20Cr13钢与国内牌号2Cr13钢相近，该类钢在国内外是研究比较成熟，应用比较广泛的耐热型不锈钢，在核电、石油、能源等领域均得到广泛应用。但如果在汽轮机叶片和汽轮机转子等设备中应用，相对复杂的服役环境对其力学性能的要求则越来越高。韧脆转变温度$FATT_{50}$是汽轮机叶片应用要求中一项重要的技术指标。本文主要研究了X20Cr13马氏体耐热不锈钢化学成分及热处理工艺对其$FATT_{50}$技术参数的影响。结果发现，X20Cr13马氏体不锈钢$FATT_{50}$的参数控制明显取决于化学成分和回火温度：（1）材料中降低C含量和Mo含量对$FATT_{50}$影响较大， 降低C含量或者提高Mo含量，可以降低材料的韧脆转变温度$FATT_{50}$；（2）随着回火温度提高，$FATT_{50}$相应地降低。经970℃保温1h油冷+680℃保温2h空冷调质后，材料的综合力学性能达到良好匹配。

关键词 叶片钢，$FATT_{50}$，化学成分，热处理

Study on $FATT_{50}$ in a Martensitic Stainless Steel X20Cr13

Su Ruiping

(Baosteel Special Steel Co., Ltd., Shanghai 200940, China)

Abstract X20Cr13 stainless steel has the same chemical composition with the domestic grade 2Cr13 steel. And it has been relatively mature at home and abroad, which is widely applied in nuclear power plant, oil ,energy and other infrastructures. But once applied in turbine blades and turbine rotor moderate sophisticated equipments which serve in severe environment, its performance requirements are getting higher and higher. The transition temperature from ductile to brittle state is called FATT (Fracture Appearance Transition Temperature), when metallic material is in low-temperature conditions. $FATT_{50}$ means the real temperature when proportion of ductility area and brittleness area is 1:1, which is an extremely important technical indicator in stainless steel X20Cr13. This paper studied the effects of chemical composition and heat treatment process on $FATT_{50}$. The results show that chemical composition and the tempering temperature are decisive to control the parameter $FATT_{50}$ in material: (1) the observed results show that C content and Mo content in this material have great effects on $FATT_{50}$, and $FATT_{50}$ value declines with the decrease of C content or increase of Mo content; (2) keep the temper temperature higher during the heat treatment could keep $FATT_{50}$ value in a lower level. And comprehensive mechanical properties could be got if the material is heat treated as follow process: quenched at 970℃ for 1 hour and tempered at 680℃ for 2 hours.

Key words stainless steel for turbine blade, $FATT_{50}$, chemical composition, heat treatment

热处理对含Co型铁素体耐热钢组织和性能的影响

马煜林[1] 刘 越[1] 古金涛[1] 穆建华[2] 刘春明[1]

（1. 东北大学材料与冶金学院，沈阳 110819；2. 辽宁福鞍重工股份有限公司，鞍山 114016）

摘 要 (超)临界机组不断提高的蒸汽参数，使对耐热钢的性能要求也在不断提高，本课题在探索研究适合更高蒸汽参数的超(超)临界机组用铁素体耐热钢为背景下展开研究。本文以一种含 Co、B 分别为 1.0%和 0.022%的新型铁素体耐热钢为研究对象，通过对比讨论不同正火温度和冷却方式对微观组织和室温力学性能的影响规律，提出优化正火温度和冷却方式。结果发现：随正火温度升高，奥氏体化充分，板条状回火马氏体组织粗化，原奥氏体晶粒尺寸增大，抗拉强度和屈服强度增加，正火温度在 1050℃时性能最好；相同正火温度条件下，经过风冷的试样能够得到组织更加致密的板条状马氏体，但空冷试样的抗拉强度和屈服强度较高，断口处韧窝尺寸较大，说明韧性较好，冷却方式选择空冷为宜；经过回火后试样的晶界处出现大量析出相，经过分析确定为 $M_{23}C_6$ 型碳化物。

关键词 铁素体耐热钢，正火温度，冷却方式，组织与性能

Effect of Heat Treatment on Microstructure and Properties of a Co-bearing Ferritic Heat-resistant Steel

Ma Yulin[1], Liu Yue[1], Gu Jintao[1], Mu Jianhua[2], Liu Chunming[1]

(1. School of Material & Metallurgy, Northeastern University, Shenyang 110819;
2. Liaoning Fu-An Foundry Group Co., Ltd., Anshan 114016)

Abstract (Ultra) Supercritical Units increasing steam parameters, so that the performance requirements of heat-resistant steel is also rising, the subject of exploring study based on higher steam parameters for ultra (ultra) supercritical unit ferritic heat-resistant steel background research for the next start. In this paper, containing Co, B content of 1.0% and 0.022%, respectively, of the new ferritic heat-resistant steels for the study, discuss different normalizing temperature and cooling influence of temperature on the microstructure and mechanical properties by comparison, and optimization of the normalizing temperature and cooling. The results show that; with normalizing temperature, austenitizing sufficiently tempered lath martensite coarsening austenite grain size increases, increase the tensile strength and yield strength, when normalizing temperature at 1050℃ best performance; at the same normalizing temperature conditions can be obtained through the air-cooled sample tissue more dense lath martensite, but air-cooled high tensile strength and yield strength of the specimen, the fracture dimple size than large, indicating toughness, air-cooled cooling mode selection is appropriate; the grain boundaries after tempering precipitates a large number of samples, after analysis identified as $M_{23}C_6$ type carbide.

Key words ferritic heat resistant steel, normalizing temperature, cooling mode, organization and performance

不同 Nb 含量对 D2 钢力学性能和组织的影响

孙绍恒[1] 赵爱民[1,2] 尹鸿祥[2] 裴 伟[2] 曾尚武[2]

（1. 北京科技大学钢铁共性技术协同创新中心，北京 100083；
2. 北京科技大学冶金工程研究院，北京 100083）

摘 要 D2 钢是一种高碳高铬莱氏体型冷作模具钢，为了研究 Nb 对 D2 钢性能的影响，本文设计了当 Nb 含量分别为 0.23%、0.46%、0.52%时，对 D2 钢力学性能和组织的影响。试样经过锻造，淬火和回火后进行各项性能测试，采用洛氏硬度计测试硬度，摆锤冲击韧性试验机上测量试样断裂的冲击功，热场发射扫描电镜观察组织，XRD 测定残余奥氏体的量。试验结果表明，在不同热处理阶段硬度变化不大，当 Nb 含量为 0.23%时各阶段硬度均为最高，还发现在不同试样中硬度随着 Nb 含量的增加有降低的趋势，但是韧性却在提高，当 Nb 含量为 0.52%时为最高，

可达 73.85J/cm²。Nb 含量的提高还能使铸态组织中网状碳化物减少，晶粒细化，经过最终热处理后，组织中的碳化物更加细小弥散分布，并且组织中残余奥氏体含量减少，马氏体中含碳量降低。

关键词　Nb 含量，硬度，韧性，残余奥氏体

Impact of Different Nb Content on the Mechanical Properties and Microstructure of D2 Steel

Sun Shaoheng[1], Zhao Aimin[1,2], Yin Hongxiang[2], Pei Wei[2], Zeng Shangwu[2]

(1. Collaborative Innovation Center of Steel Technology, University of Science and Technology Beijing, Beijing 100083, China; 2. Engineering Research Institute, University of Science and Technology Beijing, Beijing 100083, China)

Abstract　D2 steel is a kind of ledeburite cold working die steel with high carbon and high chromium. In this paper, in order to study the impact of Nb on D2 steel, different Nb contents were designed——0.23%, 0.46% and 0.52%—on the mechanical properties and microstructure of D2 steel. Various performances were tested after forging, quenching and tempering. Hardness was tested by Rockwell hardness tester. The impact energy was tested by pendulum impact toughness testing machine. The microstructure was watched by scanning electron microscope (SEM). The volume of retained austenite was tested by X-ray diffraction (XRD). Experimental results show that the hardness changes little in different phases of heat treatment, and it is the highest when the Nb content is 0.23%, besides they also show that hardness of different sample has a tendency to reduce with the increase of Nb content, but the toughness is improving, and it is the highest, 73.85J/cm², when the Nb content is 0.52%. The raise of Nb content can also reduce carbide network in the casting microstructure and refine grain size. After the final heat treatment, carbides in the microstructure become smaller and disperse more uniformly. What's more, the volume of retained austenite in the microstructure and carbon content in martensite reduce.

Key words　content of Nb, hardness, toughness, retained austenite

热处理参数对 40CrNiMo 钢硬度和韧性的影响

梅　珍　蒋　波　董正强　周乐育　张朝磊　刘雅政

（北京科技大学材料科学与工程学院，北京　100083）

摘　要　本文通过热处理实验，结合光学显微镜、SEM 扫描电镜观察、硬度测试及冲击功测试，研究了不同淬火和回火温度下 40CrNiMo 钢布氏硬度和-20℃冲击功的变化特点。结果表明：回火温度为 625℃时，随着淬火温度由 800℃升高到 875℃，实验钢回火硬度由 273HB 逐渐增加到 289HB，淬火温度继续升高，硬度下降；冲击功在淬火温度为 800℃时最高，为 127J，淬火温度继续升高，冲击功逐渐下降但不明显。淬火温度一定，为 800℃时，随着回火温度由 500℃升高到 650℃，回火硬度逐渐降低，冲击功逐渐增加，回火温度为 650℃时，回火硬度最低，为 280HB，冲击功最高，为 137J，断口形貌由小且浅韧窝向大且深韧窝转变。随着回火温度的降低，马氏体发生回复和再结晶程度增加，马氏体原始位相逐渐消失，碳化物逐渐析出并长大，影响硬度和冲击功的主要因素是马氏体基体的软化作用。综合研究，回火温度是影响实验钢硬度和冲击功的主要参数，淬火温度影响不明显。

关键词　淬火温度，回火温度，硬度，冲击韧性

Effect of Heat Treatment on Hardness and Toughness of 40CrNiMo Steel

Mei Zhen, Jiang Bo, Dong Zhengqiang, Zhou Leyu, Zhang Chaolei, Liu Yazheng

(School of Materials Science and Engineering, University of Science and Technology Beijing, Beijing 100083, China)

Abstract Effect of quenching and tempering temperature on Brinell Hardness and -20 ℃ impact energy of steel 40CrNiMo was investigated by heat treatment experiment, optical microscope, scanning electron microscope, hardness test and impact test. Results show that when tempering temperature is 625 ℃, the tempered hardness of steel 40CrNiMo rises from 273HB to 289HB as the quenching temperature increases from 800 ℃ to 875 ℃, and decreases as the quenching temperature continues to increase. The impact energy decreases from 127J to 113J with the increasing quenching temperature incrementally. When quenching temperature is 800 ℃, hardness presents a downward trend, and impact energy shows a rise with the tempering temperature increasing from 500 ℃ to 650 ℃. Tempered sorbite and carbide morphology and distribution are the main factors affecting the hardness and impact energy. Comprehensively, effect of tempering temperature on hardness and impact energy of 40CrNiMo is more obvious than that of quenching temperature.

Key words quenching temperature, tempering temperature, hardness, impact energy

C 含量及热处理对 Fe-36Ni 系因瓦合金膨胀性能的影响

王雪听　陆建生　田玉新

(宝山钢铁股份有限公司中央研究院特钢技术中心，上海　200940)

摘　要　Fe-36Ni 合金由于具有极低的膨胀系数而得到广泛使用，C 作为一种间隙原子，其含量及其在组织中的存在状态直接影响合金膨胀性能的稳定性。本文研究了 C 含量对因瓦合金膨胀性能的影响，并分析了不同热处理工艺对常温膨胀性能的影响规律，发现 C 含量在 0.01~0.05%范围内，随着其含量的增加，膨胀系数随之增加，两者呈线性关系；试样经过固溶处理后，膨胀系数有明显的降低，固溶+时效处理后，膨胀系数略微升高；合理的热处理工艺有助于提高合金膨胀性能稳定性。

关键词　因瓦合金，C 含量，热处理，膨胀性能

Effects of C Element and Heat Treatment on Fe-36Ni Alloy Thermal Expansion Ability

Wang Xueting, Lu Jiansheng, TianYuxin

(Baoshan Iron & Steel Co., Ltd. Special Steel R&D Center, Shanghai 200940, China)

Abstract Fe-36Ni alloy is widely used for its extremely small thermal expansion. The amount of element C and its status in this alloy will directly affect the stability of thermal expansion. In this article, several different amount of C element Fe-36Ni alloy are smelted, the effects of C element and heat treatment on thermal expansion are studied. It shows that: in

the range of 0.01~0.05%，with the C content increase，coefficient of thermal expansion also increase. C content and CTE show a linear relationship. CTE decrease significantly after solution. After the aging treatment, CTE increases slightly.
Key words　Invar alloy, C content, heat treatment, coefficient of thermal expansion (CTE)

900MPa 超高强度钢层状撕裂原因分析及对策

杜　明　马玉喜　张友登　陈　浮　陶军晖　宋　畅

（武钢研究院，湖北武汉　430080）

摘　要　利用金相显微镜对发生层状撕裂的钢板进行了显微分析，发现产生抗层状撕裂的主要原因是钢板中心偏析的存在，由于切割时受热应力影响，并受沿着钢板厚度方向作用的拉应力影响下，使得裂纹起始于钢板切割面的热影响区贝氏体组织，扩展到焊接热影响区的中心带状组织。可通过改善焊接接头的设计、焊接工艺以及选用抗层状撕裂钢材的方式进行预防。

关键词　超高强度钢，层状撕裂，Z 向性能

Lamellar Tearing of 900MPa Ultra High Strength Steel Cause Analysis and Countermeasures

Du Ming, Ma Yuxi, Zhang Youdeng, Chen Fu, Tao Junhui, Song Chang

(Research and Development Center of WISCO, Wuhan 430080, China)

Abstract　By means of metallographic microscope for lamellar tearing of steel has carried on the microscopic analysis, found to produce lamellar tearing resistance is the main reason of the steel plate center segregation, as a result of when the cutting heat stress, and is affected by the tensile stress of the role in the direction of the steel plate thickness, makes the crack began cutting heat affected zone of the bainite steel group, extend to the center of welding heat affected zone, banded structure.By improving the design of welded joint, welding technology, and choose to prevent lamellar tearing resistant steel.
Key words　ultra high strength steel, lamellar tearing, properties of Zdirection

7 能源与环保

Energy Saving and Environmental Protection

炼铁与原料

炼钢与连铸

轧制与热处理

表面与涂镀

钢材深加工

钢铁材料

★ 能源与环保

冶金设备与工程

冶金自动化与智能管控

冶金分析

信息情报

Fundamental Approach to a Low-carbon Sintering Process of Iron Ores

Kazuya Fujino, Taichi Murakami, Eiki Kasai

(Graduate School of Environmental Studies, Tohoku University, Sendai, Japan)

Abstract Low-carbon sintering has been an important target as well as low-carbon operation of blast furnace (BF), since CO_2 emission from the iron ore sintering occupies about 14% of the total emissions of an integrated steelworks. It is notable that the backgrounds on the low-carbon technology are a little different between the iron ore sintering and BF processes. A decrease in reducing agent rate of BF may lead to decreases in the formed amount of BF gas and/or coke oven gas. Hence, a certain amount of external energy input will be necessary in order to compensate the energy shortage in the downstream processes. On the contrary, the sintering process usually does not supply a large amount of energy to other processes. It means that lowering the amount of carbonaceous materials used in the sintering process will efficiently lead to a reduction of CO_2 emission. Therefore, development of an innovative low-carbon sintering technology is strongly required, which can also produce quality sinter, adapting to the significant change in the iron ore property.

Here, a possible approach to a low-carbon sintering technology is discussed considering the characteristics of the sintering process. Dusts, scales and magnetite ores are taken up as potential materials to replace coke as alternative agglomeration agents, since they contain lower oxides of iron, *i.e.*, Fe, FeO and Fe_3O_4 and metallic iron, which give heat through oxidation reactions. Oxidation behaviors of Fe in the sintering bed is examined based on the experimental results.

Key words reduction of CO_2 emission, alternative agglomeration agent, metallic iron, oxidation reaction, coke combustion

Hot Slag Processing Aiming for Efficient Metal, Mineral and Energy Recovery – a Swerea MEFOS Perspective

Ye Guozhu

(Department of Metallurgy and Environment, Swerea MEFOS, Sweden)

Abstract The metallurgical industry generates every year huge amounts of slag. The annual amount of generated steelmaking slags alone is more than 100 million tons. The steel slag is normally used as construction materials, for instance, for road and river bank construction. Fully utilization of this huge amount of byproduct is limited by its leaching and volume stability properties. In addition to the useful minerals in the slag, there is also a considerable amount of feasible heat lost in the hot slag. Swerea MEFOS have been involved in a number of hot slag processing projects in the past 20 years. These projects included slag stabilization to improve the physical and leaching properties, slag reduction for metal recovery, dry slag granulation for heat recovery and for reduced leaching of heavy metals, etc. This paper will highlight some of the projects with focus on the wellknown IPBM and VILD projects. The extensive pilot experiences of Swerea MEFOS have shown that there is a wide range of options for economic and ecological recovery of metals, minerals and energy from hot metallurgical slag.

Key words hot slag processing, recovery of metal, mineral and energy

Assessment of Low Temperature Waste Heat Utilization in the Steel Industry through Process Integration Approach

Wang Chuan

(Swerea MEFOS, Box 812, SE-971 25 Luleå, Sweden)

Abstract This article presents a techno-economic evaluation of various options for recovery and reuse of low temperature heat (LTH) with temperatures below 350°C. A process integration approach has been applied to illustrate the investment strategies for a typical European integrated steel plant towards positive energy and environmental effects. The modelling results indicate a realistic CO_2 reduction potential of 0.32%-1.5% and an energy saving potential of 0.36%-2.5% by LTH reuse on a steel production site with the pay-back time below three years. The results are only valid for the simulated generic site. For the results implementation at real steel plants, local boundary conditions should be considered.

Key words low temperature heat, flue gas, steel industry, process integration

New Developments in Energy Efficiency and Energy Recovery for BOF Dry Dedusting Systems

Peter Puschitz, Philipp Aufreiter, Thomas Kurzmann, Florian Weineis, Thomas Steinparzer

(Primetals Technologies)

Abstract This paper describes the improvements and new developments made in the area of energy saving for dedusting system operations, in particular for dry dedusting system, and increased energy recovery. It provides integrated environmental solutions for steelmaking and it has always been a driving force of development and innovation to make the processes most efficient with regard to environmental performance as well as operating costs. These solutions are available both for new dedusting systems as well as for the upgrading and modernization of existing steelmaking facilities. It has set up a separate business division called "ECO Solution" to combine all environmental and energy saving technologies.

Key words BOF, DDS, converter, dry dedusting, ESP, energy efficiency, steam production, cooling stack extension, energy savings

Life Cycle Environmental Performance Assessment of Steel Products

Broadbent Clare, Yang Ping

(World Steel Association, Brussels, Belgium)

Abstract With anincreasing focus on environmental issues, steel as the world's mostproduced material (1.6 billion tonnes per year), is facing ahuge challenge. Therefore, it is important to understandthe environmental performance of the steel industry.

Life cycle assessment (LCA) is a tool that can be used to evaluate the environmental impact of products, which can be helpful to understand where improvements can be made throughout their full life cycle, from raw material extraction through to end of life recycling or disposal of a product. It can also be used to provide the data to support projects on energy conservation and emissions reduction as well as demonstrating the benefits of the recyclability of steel.

Through the introduction of life cycle assessment in the steel industry, this presentation explains how the steel industry is using LCA to calculate the environmental impact of steel products. The steel industry has been working on LCA for the past 20 years and has developed a database for 15 steel products, showing the environmental performance of steel. As the Environmental Product Declaration schemes require data to be less than 5 years old, the new steel LCI data collection started worldwide at the beginning of 2015 to keep the steel database updated and represent the recent technology and situation inthe steel industry. The steel industry LCI data that is released by worldsteel has been requested from all over the world for all different market sectors.

The presentation will also demonstrate case studies that highlight the importance of LCA for the steel industry to achieve sustainability to support decisions on material choices.The global transportation industry is an important source of air pollution (which accounts for around 23% of all man-made CO_2 emissions), and many regulations are requiring the reduction of emissions to air; thus the application of LCA in the automotive sector will be used as an example to explain how LCA can be implemented to support material decision making in product design from a life cycle perspective, and to give amore accurate picture of the environmental performance of the vehicles.

Key words life cycle assessment (LCA), steel products, steel industry, sustainability

Multimodule System for Wastewater Treatment by Photocatalysis

Sorrell Charles Christopher, Koshy Pramod, Gupta Sushil

(School of Materials Science and Engineering, UNSW Australia, Sydney, NSW 2052, Australia)

Abstract The present work presents the basic elements of materials and design for a modular reactor system for the removal of heavy metals, organic materials, and pathogens from wastewater. The primary module employs photocatalytic TiO_2 in packed helical tube form for the removal of organic materials and pathogens, which are decomposed to carbon dioxide and water. The secondary module employs similar materials but in packed bed form for the reduction of dissolved metal ions to the metallic state, which are removed from the photocatalysts by the reverse process. Additional tailored modules for alternative processes, such as sieving, filtration, ion exchange, and adsorption, can be added in series as required for the removal of site-specific water contaminants that require different or modified approaches. Both of the core modules are based on the UV irradiation of ceramic or polymeric beads decorated with TiO_2 particles. The present work summarises the critical design parameters for the modules and the overall benefits of the use of this technology for wastewater treatment in natural, domestic, and industrial settings, the latter of which include coal washing, tailings ponds, and cooling water reservoirs. The system is designed to be applicable in most environments as it is solar powered, portable, robust, simple to operate and maintain, and relatively inexpensive.

Key words wastewater, water purification, photocatalysis, TiO_2, reactor, heavy metals, organics, pathogens

我国钢铁工业“十三五”节能潜力分析

李新创 郜 学 姜晓东

（冶金工业规划研究院，北京 100711）

摘 要 系统总结了我国钢铁工业“十二五”期间节能工作所取得的巨大成绩，并剖析了目前钢铁工业节能工作存在的五大主要问题。文章进一步展望了未来“十三五”钢铁工业面临的节能形势和整个钢铁工业未来的发展趋势，在此基础上分析了未来我国钢铁工业主要节能潜力和方向。文章认为，“十三五”期间，钢铁工业技术节能的贡献平稳并呈下降趋势，结构节能贡献比例上升但同时存在较大变数，管理节能将逐步发挥更大潜力。

关键词 节能潜力，结构节能，管理节能，技术节能

Analysis on Potential of Energy Conservation of Chinese Steel Industry in the 13th Five-year Period

Li Xinchuang, Gao Xue, Jiang Xiaodong

(China Metallurgical Industry Planning and Research Institute, Beijing 100711, China)

Abstract This paper provides a comprehensive analysis on great achievements of energy conservation of Chinese Steel Industry in the 12th Five-year Period, and illustrates five main weaknesses existing in energy conservation. In addition, the paper makes a prospect for development trend of energy conservation and the whole industry in the 13th Five-year Period, as well as provides an analysis on potential and direction of energy conservation of Chinese Steel Industry over the coming years. According to the content, it is predicted that proportion of technical contribution will move steadily and show a downturn, proportion of contribution made by structure will move up and contribution made by management will show a greater potential step by step in the 13th Five-year period.

Key words potential of energy conservation, energy conservation by structural adjustment, energy conservation by management, energy conservation by technical improvement

Physical and Leaching Properties of Ceramic Tiles with Black Pigment Made from Stainless Steel Plant Dust

Zhu Renbo, Ma Guojun, Cai Yongsheng, Chen Yuxiang, Yang Tong, Duan Boyu, Xue Zhengliang

(Key Laboratory for Ferrous Metallurgy and Resources Utilization of Ministry of Education, Wuhan University of Science and Technology, Wuhan 430081, China)

Abstract Stainless steel plant (SSP) dust is a hazardous by-product of the stainless steelmaking industry. It contains large amounts of Fe, Ni and Cr, and can be potentially used as inorganic black pigment in the ceramic tile industry. In this paper, the physical and leaching properties of ceramic tiles with black pigment using stainless steel plant dust as a raw material were investigated by considering the process parameter of sintering temperature, sintering time, addition ratio of pigment, mineralizer addition and compress strength of sample preparation . The results show that the good physical properties of ceramic tiles can be obtained with 8% pigments addition and sample preparation pressure of 25 MPa sintering at 1200℃ for 30min. The controlling leaching mechanisms of Cr and Pb from the ceramic tiles are initial surface wash-off, while the leaching of Cd, Ni and Zn from the product are controlled by matrix diffusion.

Key words dust, heavy metals, physical properties, stainless steel plant, semi-dynamic leaching test

Control Methods and Operation Strategies for Improving Energy Saving in a Multi-cell Fan Based Cooling Tower System

Tsou Ying[1], Chang Chuncheng[1], Wu Chanwei[2], Jang Shishang[3]

(1. New Material Research & Development Dept., China Steel Corporation, Kaohsiung, Taiwan, R.O.C.,China;
2. Office of Energy and Environmental Affair, China Steel Corporation, Kaohsiung, Taiwan, R.O.C. ,China;
3. Department of Chemical Engineering, National Tsing Hua University, Hsinchu, Taiwan, R.O.C. ,China)

Abstract In cooling towers, the replacement of traditional single-speed fans with variable-frequency drive (VFD) fans or two-speed fans can achieve significant energy saving. However, concerns regarding frequent on/off switching and the lack of a well-devised controller discourage the widespread implementation. In this study, a temperature zone setting method is proposed to replace the traditional point setting method of fan control. And we also developed a multi-fans control strategy. Additionally, the highest allowed temperature of output water in the process is set as the upper limit of a zone in order to further conserve energy. Both strategies are comprehensively analyzed for a virtual cooling tower that uses operational data from an existing VFD-fan-based cooling tower system in CSC. The results of on-site implementation show energy savings of 38% for a 0.75 °C zone without increasing the on/off switching frequency. The proposed strategies were further verified via an on-line field experiment. The proposed methods can be universally and easily applied to any existing cooling tower with significant energy conservation.

Key words cooling tower, variable-frequency drive, zone setting control

钢铁全流程节能新技术研发的思考

向顺华

（宝钢中央研究院能源与环境研究所，上海 201999）

摘 要 本文就钢铁生产全流程中余热回收、工序变革及工业炉节能等方面提出了需要开发的新技术。介绍了能够大幅度降低能耗的电炉烟气余热回收技术、新型立式烧结技术、连铸与热轧低能耗连接技术和工业炉氧燃料烟气循环技术的新思路。

关键词 节能技术，工序节能，钢铁生产全流程，能源利用效率

The Thinking about Energy Saving New Technology for Whole Iron and Steel Manufacture Process

Xiang Shunhua

(Research Institute (R&D Center), Baosteel Group Corporation, Shanghai 201999, China)

Abstract It is suggested that the new technology would be developed for whole iron and steel manufacture process which relating to the waste heat recovery, the process improvement and the industrial furnace energy saving. The new ideas of EAF exhaust gas heat recovery , vertical sintering process , low-energy connecting of CCM and hot rolling , and oxyfuel flue gas recirculation are introduced.

Key words　energy saving technology, process energy saving, whole iron steel manufacture process , energy use efficiency

关于实现低碳绿色炼钢的若干设想

唐志永[1]　孙予罕[2]

（1. 中科院低碳转化科学与工程重点实验室副主任；

2. 中科院上海高等研究院书记、副院长，中科院低碳转化科学与工程重点实验室主任）

摘　要　钢铁工业是能源密集型的行业，消耗大量化石能源，排放大量CO_2。中国作为世界第一钢铁产量的大国，钢铁工业CO_2排放约占全国总排放的15%，同时炼钢过程中排放大量的污水。在世界性CO_2减排和环境保护的压力下，欧盟、日本、韩国等主要钢铁生产地区均制定了相应的规划和技术措施，中国的钢铁行业CO_2排放量现在仍维持在相对较高的水平，随着环境要求的日趋严格，有效减少钢铁生产CO_2和污水排放已成为中国钢铁工业亟待解决的问题，在目前产业转型升级和经济结构调整的背景下，中国的钢铁行业迫切需要低碳绿色的钢铁解决方案，寻找新的技术突破和模式创新来有效降低钢铁生产CO_2排放已成为亟待解决的问题。

基于钢铁企业CO_2过程排放模型，本文选取国内典型规模的钢铁生产系统进行计算分析，分析了各个工序对钢铁企业CO_2排放和能耗的不同影响。在分析传统炼钢工艺碳排放和能耗的基础上，结合目前中国在低碳技术方面的新突破，构建了低碳绿色钢铁解决方案（见图1）。一方面从耦合能源角度，通过二氧化碳与天然气的耦合，将二氧化碳重整转化为合成气，随后经膜分离装置将合成气分离为CO和H_2，CO直接用于高炉炼铁，替代高炉炼铁所需的焦炭，重整装置从而替代传统的焦化工序，可直接减少大量二氧化碳排放；另一方面重整副产的氢气与二氧化碳加氢生产甲醇，发展甲醇下游产业链；此外剩余的二氧化碳从生物循环角度通过生物质微藻系统，利用水热方式转化为活性炭/利用加氢方式转化为二元醇，从生命周期角度也大大降低了二氧化碳的排放。利用现场富含氮磷营养元素的废水作为微藻培养基，在实现微藻固碳的同时也实现了废水的资源化利用。

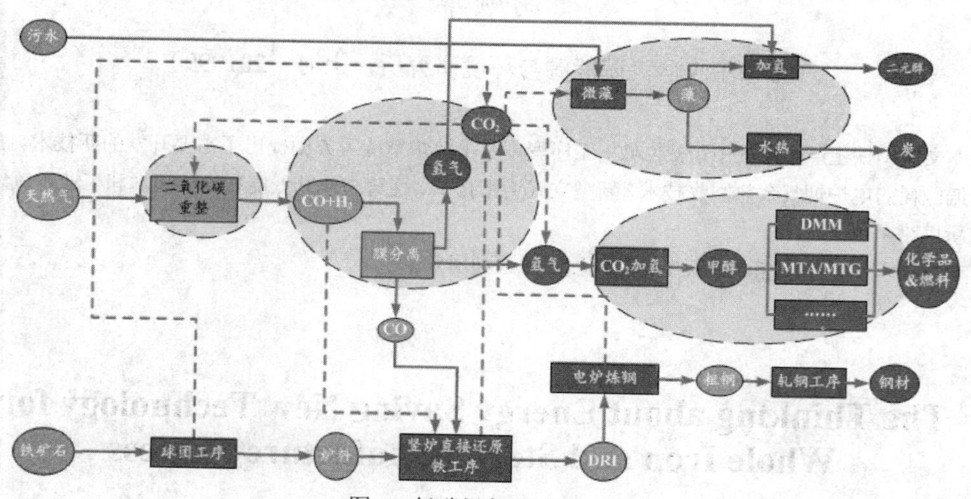

图1　低碳绿色钢铁解决方案

通过建立的技术经济模型，在不同的情景下对低碳绿色钢铁解决方案进行了技术经济评估，分析了低碳绿色钢铁解决方案与传统炼钢工艺的碳排放、能耗和经济性对比，低碳绿色钢铁解决方案具备了良好的社会效益和经济效益，在中国沿海地区有着很好的实施可行性，从核心技术支撑角度，相关核心技术在国内陆续进入或完成中试，为实施本方案提供了很好的核心技术支撑。

关键词 钢铁行业，二氧化碳排放，重整，微藻，低碳

High Efficiency Coking Wastewater Treatment Process

Cheng Chu-I, Chen Yanmin

(Engineer New Materials R&D Dept., R&D Division, China Steel Corporation, Taiwan, China)

Abstract Over 85% of COD in CSC coke-oven wastewater is removed by the conventional activated sludge system. In order to prevent the exceed quantity of incoming coke-oven wastewater that caused system upset, CSC has built a membrane bioreactor (MBR) to increase the capacity of the wastewater treatment system. The MBR is able to adapt unexpected high influent quantity and/or contaminant concentrations. The overall COD removal efficiency of coke oven wastewater treatment plant has been increased from 88% to 93%. In addition, CSC is planning to revamp the existing aeration tanks into a MBBR nitrogen removal process to meet the future ammonia standard of 20mg/L. The R&D department has established the start-up strategies and the upset recovery plans based on the pilot study in the field. Preliminary pilot results show that phenolics, thiocyanate and cyanate will certainly inhibit nitrification. Moreover, the overall ratio of ammonia to alkalinity (as $CaCO_3$) in the start-up period of shall be 1:9, which is higher than the theoretical value of 1:4. This suggests that autotrophic bacteria such as thiocyanate-degrading bacteria existing in the activated sludge would consume the alkalinity as a carbon source for growth. Furthermore, with the application of carriers in the nitrification tank, the biofilm on the carriers can increase 20% of ammonia conversion rate.

Key words coke-oven wastewater, membrane bioreactor, moving bed biofilm reactor (MBBR)

Low Temperature Recycling Process of EAF Dust with Plastic Wastes Containing Brominated Flame Retardants

Oleszek Sylwia[1,2], Grabda Mariusz[1,2], Shibata Etsuro[1], Nakamura Takashi[1]

(1. Institute of Multidisciplinary Research for Advanced Materials, Tohoku University, Sendai, Japan;
2. Institute of Environmental Engineering, Polish Academy of Sciences, Zabrze, Poland)

Abstract The fate of zinc, lead, and iron present in electric arc furnace dusts is investigated during thermal treatment of the dust with flame-retarded polycarbonate (PC-Br) resin containing tetrabromobisphenol A (TBBPA). Mixtures of these materials were heated under either oxidizing or inert conditions in a laboratory-scale furnace at 550 °C for 80 min. The resulting products were then characterized using X-ray diffraction, inductively coupled plasma, and ion chromatography. The results obtained indicate that inorganic bromine evolved from decomposing flame-retarded resin almost completely reacts with metal (zinc and lead) oxides present in the dust. The highest bromination rates of zinc (78% and 81%) and lead (90% and 93%) were obtained under oxidizing conditions during thermal treatment of mixture comprising PC-Br and EAF dust at ratios of 6:1 and 9:1 w/w, respectively. The metallic bromides formed were almost entirely vaporized from the solid residues. Iron remained in an iron-enriched solid residue. The results indicated that heating such mixtures under specific conditions enables a high separation of zinc and lead from iron-rich residues within an attractive temperature range. Heating under inert conditions had little effect on the effectiveness of the bromination-evaporation of zinc, but caused more abundant formation of char, decresed bromination of lead, and resulted in sporadic bromination of iron present in the dust.

Key words zinc recovery, EAF dust, waste BFRs-plastics, simultaneous recycling

高辐射覆层技术研究与应用的最新进展

周惠敏　刘常富　杨秀青　李　亮　张绍强　翟延飞

（山东慧敏科技开发有限公司，山东济南　250100）

摘　要　高辐射覆层技术自2005年应用以来，已在全国60余家钢铁企业的404座高炉热风炉和4座焦炉上应用，获得了显著的节能效果。近年来慧敏科技对高辐射覆层材料配方优化升级开发出绿色环保型无铬高辐射覆层材料；在覆层高温物理性能方面进行了深入研究，证明覆层材料具有优良的高温稳定性、高温抗折性能以及高温耐磨性能；通过对鞍钢、济钢、石横特钢等工业应用的长期跟踪，充分证实了高辐射覆层技术的长效性；起草的国家标准《高炉用高风温顶燃式热风炉节能技术规范》和行业标准《高辐射覆层蓄热量的测定与计算方法》、《轧钢加热炉节能运行技术要求》，已分别于2014年6月15日、2014年7月1日和2015年6月1日正式实施。高辐射覆层技术已列入国家发改委发布的《国家重点节能低碳技术推广目录》和《固定资产投资项目节能评估和审查工作指南（2014年本）》。

关键词　高辐射覆层技术，节能涂料，高温物理性能，国家标准，行业标准

The Application Progress of High Radiative Coating Technology in Ironmaking Industry

Zhou Huimin, Liu Changfu, Yang Xiuqing, Li Liang, Zhang Shaoqiang, Zhai Yanfei

(Shandong Huimin Science & Technology Co., Ltd., Jinan Shandong 250100, China)

Abstract　Since 2005, the high radiative coating technology has been applied to 408 hot blast stoves and coke ovens in more than 60 steel and iron enterprises, which has obtained obvious energy saving effects. In recent years, Shandong Huimin Science&Technology Co., Ltd. for high radiation coating material formulation optimization to develop green environmental protection without chromium high radiation coating material; In high temperature coating physical properties were studied and verified the cladding material excellent high temperature stability, high temperature flexural performance and high temperature wear resistance;by following up the industry application of Angang, Jigang and Shiheng steel plant, all the fact confirmed the long-term application of the high radiative coating technology. we draw up the national standard "Specifications of top combustion stove with high efficiency, high blast temperature in blast furnace", the industry standard "The Measuring and Calculation Method of Heat Storage Capacity on High Radiative Coating" and "Technical specification of energy-saving operation for reheating furnace in steel rolling" were implemented on June 15th in 2014, July 1st in 2014 and June 1st in 2015. This technology has been included in the national key energy-saving and low-carbon technology, "energy-saving assessment and review guidelines about fixed assets investment projects (2014 version)".

Key words　high radiative coating technology, energy saving coating,high-temperature physical property, national standard, industry standard

Resource Application and Development of Carbon Dioxide in the Ferrous Metallurgy Process

Zhu Rong[1,2], Wang Xueliang[1]

(1. School of Metallurgical and Ecological Engineering, University of Science and Technology Beijing, Beijing, China; 2. Beijing Key Laboratory of Special Melting and Preparation of High-End Metal Materials, Beijing, China)

Abstract A review of resource application and development of carbon dioxide in the ferrous metallurgy process is presented. Study on resource utilization of CO_2 in the steelmaking process is significant to the reduction of CO_2 emissions and coping with global warming. Based on the resource application of carbon dioxide in the whole process of ferrous metallurgy, the paper introduces the development and application of carbon dioxide in the sintering, BF, Converter, secondary refining, Continuous Casting and smelting process of stainless steel in recent years. According to the domestic and foreign research and application status, the paper analyzes the feasibility and metallurgical effects of the application of carbon dioxide in the ferrous metallurgy process.

The paper mainly introduces some new techniques: flue gas circulating sintering, blowing CO_2 through Blast Furnace tuyere and CO_2 as a pulverized coal carrier gas, top and bottom blowing CO_2 in the converter, Ladle Furnace and Electric arc Furnace bottom blowing CO_2, CO_2 as Continuous Casting shielding gas, CO_2 for stainless steel smelting, and CO_2 circulation combustion. CO_2 has a very broad application prospect in iron and steel metallurgy process, and the quantity of CO_2 utilization is expected to 100k/ton steel. It will effectively facilitate the progress of metallurgical technology, strongly promote energy conservation of metallurgical industry, and contribute to sustainable development.

Key words carbon dioxide, ferrous metallurgy process, resource application

Extraction of Phosphorous from Steelmaking Slag by Leaching

Kim Sun-Joong, Gao Xu, Uedaand Shigeru, Kitamura Shin-ya

(Institute of Multidisciplinary Research for Advanced Materials, Tohoku University, 2-1-1 Katahira, Aoba-ku, Sendai 980-8577, Japan)

Abstract In order to recyclephosphorusfrom steelmaking slag, a separation processby leaching has been proposed by current authors.The principle of this leaching process was to utilize the differencesin the water-solubility betweenthe C_2S-C_3 Psolid solution and the other phases. Experimentswere carried out to determinethe effect of pH on the dissolution behavior of various elements in the steelmaking slagwith and without free-CaO phase.When the pH was decreased, the dissolution behavior of Ca, Si, P and Fe from the steelmaking slag was increased. Furthermore, at a pH of 3, the dissolution ratio of P from steelmaking slag without the free-CaO was larger than that from steelmaking slag including free-CaO.

Key words steelmaking slag, C_2S-C_3P, P resources, leaching process

Recycling Use of Steel Dust Containing High Content of Potassium and Sodium

Zhong Yiwei, Guo Zhancheng, Gao Jintao

(State Key Laboratory of Advanced Metallurgy, University of Science and Technology Beijing, Beijing 100083, China)

Abstract As the production of crude steel, Iron and steel industry in China generates several million tons of fine dusts with a great deal of valuable metals such as, K, Na, Zn, Pb, which present major environmental problems. This paper is focused to analyze Zn/Pb removal and K/Na recovering from the metallurgical dust with high K/Na content. The physicochemical property, elements distribution characteristic of the typical metallurgical dusts are discussed. Consequently, the technical principles and methods for classification recycling and comprehensive utilization of metallurgical dust will be proposed, which will provide technical support for the industrial production. And then the industrial practice of metallurgical dust utilization and technique integration for KCl production are reviewed.

Key words metallurgical dusts; KCl, fine iron ore, recovering, recycling use

高炉冲渣水及冲渣乏汽余热回收系统的综合研究应用

刘 森[1] 赵忠东[2] 陈 健[3] 孟庆利[4] 贾兆杰[5] 李 静[6]

（1. 北京亿玮坤节能科技有限公司，北京 102205；
2. 河北钢铁邯郸分公司能源中心，河北邯郸 056015；3. 营口京华钢铁有限公司，辽宁 115005；
4. 天津市天重江天重工有限公司，天津 300402；5. 国能安融分布能源技术有限公司，
河北 050000；6. 北京科技大学，北京 100083）

摘 要 本文通过对高炉冲渣水及冲渣乏汽余热回收采暖和间接发电等综合利用成功案例的分析，为尚未开发和利用的高炉冲渣水及冲渣乏汽余热综合应用提供理论基础和技术支持。特别是针对 INBA 法和平流法冲渣方式下的渣水成功过滤及换热有创新性的运用，解决了高炉冲渣水余热回收应用过程中出现的堵塞、结垢和腐蚀问题，并能够高效提取渣水低温余热，换热温差小于 5℃，节能效果显著，投资回收期短，具有良好应用前景，为进一步拓宽高炉冲渣水全面高效回收应用奠定了技术和实践基础。节约能源，具有良好的应用前景。

关键词 高炉冲渣水，高炉冲渣乏汽，过滤，换热，采暖，间接发电

Research and Application of Waster Heat Recovery System of Slag Water and Slag Steam in Blast Furnace

Liu Sen[1], Zhao Zhongdong[2], Chen Jian[3], Meng Qingli[4], Jia Zhaojie[5], Li Jing[6]

(1. IVYQUEN Energy Saving Technologies Beijing, Beijing 102205, China; 2. Handan Iron and Steel Group Co., Ltd., Hebei 056015, China; 3. Minmetals Yingkou Medium Plate Co., Ltd., Liaoning 115005, China; 4. Tianjin Giant Heavy Industries Co., Ltd., Tianjin, 300402, China; 5. Guonenganrong Distribution Energy Technology Co., Ltd., 050000, China; 6. USTB, Beijing 100083, China)

Abstract This paper proposes an analysis of the successful engineering project of how to use the waste water including the slag water and slag steam in blast furnace to generate electricity power, and provides the theoretical basis and technical support for other same applications. The energy-saving effect is remarkable, especially the payback investment period is short, and saving energy has a good application prospect.

Key words slag water, slag steam, filtering, exchange heat, heating, indirect power generation

Effect of CaO/SiO_2 Mass Ratio on the Precipitation Behavior of Spinel Phase in Stainless Steel Slag

Cao Longhu, Liu Chengjun, Zhang Chi, Jiang Maofa

(Key Laboratory for Ecological Metallurgy of Multimetallic Mineral (Ministry of Education), Northeastern University, Liaoning Shenyang, China)

Abstract In the present work, synthetic samples of $CaO-SiO_2-MgO-Al_2O_3-FeO-Cr_2O_3$slagwere obtained to analyze the effect of CaO/SiO_2 mass ratioon the precipitation of spinel phase.The samples were heated to 1600°C for melting completely,and then slowly cooled to 1300°Cfor holding 30min beforequenching in water. The phases composition and precipitation behavior of spinel phasewere analyzed by X-ray powder diffraction (XRD) and scanning electron microscope-energydispersive spectroscopy (SEM-EDS). The law of conservation of mass and linear least square equation were used to calculate the mineral phase proportion in stainless steel slag. The leaching test was conducted for pointing out the chromium stability in different CaO/SiO_2 ratio slags. The results showed that Cr was mainly enriched in spinel phase and liquid phase. Slag composition with CaO/SiO_2=1.5 was beneficial to improve theprecipitation of spinel phase and reduce the chromium leaching amount.HigherCaO/SiO_2 ratioshould be avoided, which could lead to the formation of leachable chromium containing periclase phase and high chromium leaching. The average grain size of spinel phase was found to increase with decreasing CaO/SiO_2 ratio.

Key words stainless steel slag, spinel phase, thermodynamic, leaching, basicity

An Overview of Biomass Utilization in Ferrous Metallurgy Processes

Wei Rufei[1,2], Xu Chunbao (Charles)[1,3], Cang Daqiang[2]

(1. Department of Chemical and Biochemical Engineering, Western University, Ontario N6GA5B9, Canada; 2. School of Metallurgical and Ecological Engineering, University of Science and Technology Beijing, Beijing 100083, China; 3. School of Chemistry and Chemical Engineering, Anhui University of Technology, Ma'anshan, Anhui 243002, China)

Abstract Concerning over climate change and depletion of fossil energy, iron and steel industry has been facing enormous challenges and great pressure for seeking alternative and renewable energy such as biomass or bioenergy. This paper provides an overview of biomass utilization in ferrous metallurgy processes via the blast furnace (BF) - basic oxygen furnace (BOF) route, direct reduction (DR) - electric arc furnace (EAF) route, and scrap - EAF route. In the BF-BOF processes, biomass can be used as a fuel for iron ore sintering, or as a raw material for the production of bio-coke, and utilized for blast furnace injection. In the DR-EAF processes, direct reduced iron can be produced from iron ore and biomass pellets. In addition, biomass can be utilized in EAF through a cogeneration system. This paper also briefly reviews the environmental performance and biomass strategy of Canadian steel industry.

Key words biomass, iron making and steel making, greenhouse gases, environmental performance, canadian steel industry

高炉-转炉长流程钢铁企业能量流分析

马光宇 李卫东 张天赋 刘常鹏 赵爱华 张 宇

（鞍钢股份技术中心，辽宁鞍山 114009）

摘 要 本文通过建立能量流分析模型并以某公司西区500万吨钢生产基地为例，对各工序的能量流进行分析，结果表明能源损失占整个生产流程能源消耗的33%，界面损失占整个生产系统能源损失的15%，铁-钢界面和钢-轧界面的能量损失分别占整个生产流程中界面损失的41.5%、52%，故钢铁企业应从减少各工序能源损失和界面损失两个方面开展节能减排工作。

关键词　能量流，能源损失，界面损失

Integrated Steel Industry Energy Flow Analysis

Ma Guangyu, Li Weidong, Zhang Tianfu, Liu Changpeng, Zhao Aihua, Zhang Yu

(Technology Center of Ansteel Co., Ltd., Anshan Liaoning 114009, China)

Abstract　This paper presents a calculation model for integrated steel industry energy flow analysis and analyses the energy flow for a 5 million t steel production base in Ansteel. The results show that the process energy loss and interface energy loss accounts for 33% and 15% of the total energy consumption separately, energy loss at iron making/steelmaking and steelmaking/hot-rolling interface represent 41.5% and 52% of the total interface energy loss accordingly. Thus, the highlights of energy saving in steel industry lies in process and interface energy loss control.

Key words　energy flow, energy loss, interface energy loss

烧结矿余热罐式回收若干关键技术问题探讨

冯军胜[1]　董　辉[1]　蔡九菊[1]　张井凡[2]　张　凯[2]

（1. 东北大学材料与冶金学院，辽宁沈阳　110819；
2. 朝阳重型机械设备开发有限公司，辽宁朝阳　122000）

摘　要　针对传统烧结余热回收系统存在着漏风率高、余热回收效率偏低等难以克服的弊端，借鉴干熄炉的结构和工艺，提出了烧结矿余热竖罐式回收利用的结构和工艺流程，提出了罐式回收是否可行的两个关键问题，即罐体内烧结矿层阻力特性和气固传热问题。研究表明：影响料层阻力特性和气固传热特性的关键因素主要包括罐体内料层填充特性、冷却风量、冷却段高度；减小料层阻力损失、强化气固传热特性的主要技术手段是：增大罐体内烧结矿层空隙率，适当设置罐体内冷却段高度，适当设置热载体流量。

关键词　烧结，余热，回收利用，竖罐，干熄焦

Discussion on Some Key Technical Problems of Sinter Waste Heat Recovery and Utilization with Vertical Tank

Feng Junsheng[1], Dong Hui[1], Cai Jiuju[1], Zhang Jingfan[2], Zhang Kai[2]

(1. School of Materials and Metallurgy, Northeastern University, Shenyang 110819, China;
2. Chaoyang Heavy Machinery Equipment Development Co., Ltd., Chaoyang 122000, China)

Abstract　In view of the drawbacks of traditional sintering waste heat recovery system, such as higher air leakage rate, lower waste heat recovery rate, and so on, the structure and craft process of waste heat recovery and utilization for sinter in vertical tank are proposed by imitating the structure and process of coke dry quenching furnace. The two key problems affecting the feasibility of vertical tank are clearly showed, namely the problems of resistance characteristics and gas-solid heat transfer characteristics in sinter bed layer of vertical tank. The results shows that the main factors affecting resistance characteristics and gas-solid heat transfer characteristics in sinter bed layer are the bed layer packing characteristics, cooling air flow rate, height of cooling section. The main technical means of reducing bed layer resistance loss and strengthening gas-solid heat transfer characteristics are increasing the sinter bed layer voidage, setting up the height of cooling section and heat carrier flow appropriately.

Key words sinter, waste heat, recovery and utilization, vertical tank, CDQ

烧结烟气综合治理技术研发和实践

李咸伟 俞勇梅 石 磊 王如意 刘道清

（宝钢集团中央研究院，上海 201900）

摘 要 烧结排放烟气量大、温度偏低、污染物含量高且成分复杂，是钢铁行业烟气治理的难点和重点，已引起国内外的广泛关注。近年来，宝钢依托国内首套烧结废气循环中试装置和示范工程，针对烧结废气的特点，在循环烧结工艺系统设计和优化、二噁英类污染物源头抑制、循环烧结配矿优化、循环风机和管道耐磨设计、高效除尘设备开发、系统联动控制和稳定运行技术、烟气混合和切换、循环烟罩密封、风量氧量调节、热风循环烧结工艺控制和过程模型等方面进行了开创性研究，初步形成成套装备和技术。研究结果表明，烧结废气循环不但可以显著减少烧结工艺的废气排放总量及污染物排放量，还能回收烟气中的低温余热、节省烧结工序能耗，具有较大的节能减排和推广应用价值。

关键词 烧结，废气循环，烟气净化，二噁英，综合治理

Research and Industrial Application on Sintering Flue Gas Comprehensive Treatment

Li Xianwei, Yu Yongmei, Shi Lei, Wang Ruyi, Liu Daoqing

(R&D Center, Baoshan Iron & Steel Co., Ltd., Shanghai 201900, China)

Abstract For the sake of low temperature, large waste gas volume, high pollutants content and complicated compositions, sintering flue gas treatment has always been a tough problem in steel industry, which has attracted widespread attention at home and abroad. Recently, based on the first domestic industrial testing equipment and large-scale demonstration project of sintering flue gas recirculation(SFGR), Baosteel has been carried out a series of creative research and development on SFGR system design and optimization, dioxin-related contaminants source suppression, ore matching structure optimization, wear-resistant countermeasure on cycling fan and pipeline, development on high efficiency dust removal equipment, system control and stable operation strategy, flue gas mixing and switching control technology, circular hood sealing, oxygen content conditioning for recirculation gas, recirculation sintering system process control and model development, etc. Through Pilot Plant and Demonstration Project study, a complete set of equipments and technologies on SFGR process and advanced purification of emission gas has initially been formed. The results suggest that SFGR process can not only significantly reduce exhaust gas volume and pollutant emissions, but also can recovering low temperature waste heat and saving energy consumption of sintering, therefore the whole technologies possess great value in energy-saving, pollution emission reduction and sintering ore quality/yield improvement.

Key words sintering, sintering flue gas recirculation(SFGR), waste gas purification, dioxin, comprehensive treatment

Ammonia-containing Wastewater Reclamation with Membrane Absorption and Vacuum Membrane Distillation

Lu Jun, Li Baoan

(Chemical Engineering Research Center, School of Chemical Engineering and Technology, Tianjin University, Tianjin 300072, China)

Abstract Membrane absorption and vacuum membrane distillation were integrated to recover ammonia and water from saline ammonia-containing wastewater discharged by metallurgical plants. Ammonia was removed by membrane absorption to decrease the ammonia content to below 5mg/L. Then the wastewater was concentrated by vacuum membrane distillation to maximize fresh water recovery. Recovery of ammonia and fresh water reached 99.8% and 80% within 250 min and 160 min, respectively. Permeate flux was greatly restored through a cleaning technique. The experimental results demonstrate the potential for resource utilization of saline ammonia-containing wastewater by combining membrane absorption and vacuum membrane distillation.

Key words membrane absorption, membrane distillation, ammonia removal, water recovery

焦化废水微电解协同催化氧化预处理技术

田京雷[1] 李立业[1] 李彦光[2] 陈君安[2] 刘 欢[2]

（1. 河北钢铁技术研究总院，河北石家庄 052165；
2. 河北钢铁集团邯郸分公司，河北邯郸 056015）

摘 要 本文针对焦化废水可生化性差、难降解的特点，结合目前传统生化处理工艺特点，提出可显著提高废水可生化性的微电解协同催化氧化预处理技术。实验结果表明，微电解的最佳条件为：进水 pH 为 3.0，气水比为 3:1，反应时间为 2.5h；催化氧化试验最佳条件为：进水 pH 为 3.5，双氧水添加比例为 0.1%，反应时间为 0.5；在该条件下可实现对 COD、氨氮、挥发酚类物质的去除率分别达到 30%、20%、50%以上，并将废水的可生化由 0.26 提高至 0.45 以上，大幅度去除有毒有机物对生化系统的毒害作用。该技术作为焦化废水的预处理技术，在去除部分 COD、减少生化阶段污染负荷的同时，显著提高焦化废水的可生化性，实现好氧阶段处理效率提高。该技术作为废水预处理技术具有很大的技术和经济优势，是可实现、易操作的预处理技术，适合于高浓度、难降解有机废水。

关键词 焦化废水，预处理，微电解，催化氧化，可生化性

Coking Wastewater Pretreatment by Micro Electrolysis with Catalytic Oxidation

Tian Jinglei[1], Li Liye[1], Li Yanguang[2], Chen Jun'an[2], Liu Huan[2]

(1. Hebei Iron and Steel Technology Research Institute, Shijiazhuang Hebei 052165, China;
2. Hebei Iron and Steel Group Handan Company, Handan Hebei 056015, China)

Abstract Coking wastewater is one of the refractory wastewater with high concentration and low biodegradability. Based on the analysis of characteristics of coking wastewater, we put forward the technology which can significantly improve the biodegradation of wastewater, as micro electrolysis with catalytic oxidation pretreatment. We set up a medium test in Handan coking Plant. The experimental results show that the optimum condition of micro electrolysis are as follows: pH is 3.0, gas water ratio is 3:1, the reaction time is 2.5 h; the best conditions of catalytic oxidation test are as follows: pH is 3.5, and with a 0.1% hydrogen peroxide solution is added, the reaction time is 0.5 h. The removal rate of COD, ammonia nitrogen, volatile phenol material can reach more than 30%, 20%, 50% respectively, and the biodegradability of wastewater can be varied from 0.26 to above 0.45. Some pollutants can be removed by micro electrolysis with catalytic oxidation, and it can improved the biodegradability of wastewater significantly. Most of all, it can reduce the pollutant load and improve the processing efficiency. The technology have a great technical and economic advantages, and it is realizable, suitable for high concentration of refractory and low biodegradability organic wastewater as coking wastewater.

Key words coking wastewater, pretreatment, micro electrolysis, catalytic oxidation, biodegradability

TiO_2-氧化石墨烯负载 MnO_X 低温催化还原 NO_X 研究

卢熙宁

（冶金工业规划研究院，北京 100711）

摘 要 采用溶胶凝胶法制备不同质量比的 TiO_2-氧化石墨烯（TiO_2-GO）载体，通过超声波浸渍负载一定量的 MnO_X 活性组分，催化剂显示出良好的物理结构特性和优异的电子转移能力，有利于提升催化反应活性。通过场发射扫描电镜（FESEM）、X 射线衍射（XRD）、比表面积（BET）和 X 射线光电子能谱（XPS）对催化剂样品进行表征，并考察催化剂氨气低温催化还原 NO_X 的活性。结果表明，TiO_2-GO 载体中存在 GO 物相，且所有样品均为锐钛矿型 TiO_2 结构。MnO_X/TiO_2-GO 催化剂中存在多种不同价态的锰氧化物，特别是复合催化剂表面非化学计量的 MnO_X/Mn 有助于电子转移，从而提高催化剂的氧化还原能力。MnO_X/TiO_2-GO(125:1)催化剂的最佳 Mn 负载量为 9%，所有催化剂样品都表现出优良的 N_2 选择性。

关键词 氧化石墨烯，TiO_2-GO，选择性催化还原，低温，锰氧化物

Low-temperature Catalytic Reduction of NO_X Over TiO_2-graphene Oxide Supported with MnO_X

Lu Xining

(China Metallurgical Industry Planning & Research Institute, Beijing 100711, China)

Abstract TiO_2-graphene oxide(TiO_2-GO) nanocomposites were prepared by sol-gel method with different mass ratios of graphene. With MnO_X active component loaded by means of ultrasonic impregnation, the catalysts exhibited excellent physical structure and great electron transfer properties, which were in favor of the catalytic reaction activity. All the catalysts were characterized by FESEM, XRD, BET and XPS, meanwhile, the catalytic reduction activities of NO_X were studied under condition of low-temperature, with ammonia as the reductant. The results indicated the formation of graphene oxide in TiO_2-GO supports, revealed and TiO_2-GO were readily indexed to be anatase TiO_2 in all samples. Various valence state of manganese species coexisted in MnO_X/TiO_2-GO catalysts. Especially, non-stoichiometric (MnO_X/Mn) on the surface of composite catalyst was beneficial to the electron transfer, thereby the redox performance. The optimum mass

ratio of Mn in MnO_X/TiO_2-GO(125:1) was 9 wt.%, all the samples have showed excellent N_2 selectivity.
Key words　graphene oxide, TiO_2-GO, selective catalytic reduction, low temperature, manganese oxides

宝钢湛江钢铁炼铁厂环保新工艺技术的应用

梁利生　周　琦

（宝钢湛江钢铁有限公司炼铁厂，广东湛江　524072）

摘　要　国家“十二五”规划明确提出了将主要污染物排放总量显著减少作为社会经济发展的约束性指标，并强调了推进包括钢铁行业在内的二氧化硫和氮氧化物排放控制的治理，未来大气污染物排放的标准将会更为严格。宝钢湛江钢铁秉承宝钢“成为绿色产业链的驱动者”的绿色愿景，切实履行国家的环保新要求，全力打造一个“绿色钢铁基地”。本文首先简要分析了目前钢铁行业大气污染物排放和治理放情况，再结合宝钢湛江钢铁大气污染物排放控制的目标，介绍了炼铁厂采用的环保新工艺技术。
关键词　环境保护，烟气净化，烟粉尘，二氧化硫，氮氧化物

Application of New Environmental Protection Technolgies of Baosteel Zhanjiang Ironmaking Plant

Liang Lisheng, Zhou Qi

(Baosteel Zhanjiang Iron & Steel Co., Ltd., Ironmaking Division, Zhanjiang Guangdong 524072, China)

Abstract　China's twelfth Five-year plan claimed that obligation of reducing pollutants discharge should be going with economic growth, and emphasized the suppression of SO_2 and NO_x emissions in industry. The Baosteel zhanjiang Iron&Steel Co.,Ltd, which is a new iron&steel making works located in Guangdong province, sticks to Baosteel's green vision of becoming a green industrial chain driver and fulfills requirement of environmental protection by government. Target of zhanjiang works is going to be a green iron&steel making base. This paper generally analyzed the major air pollutants sourced and emitted in ironmaking process. Then, a comprehensive review of flue gas purification technologies employing for ironmaking was discussed. In terms of strict emission standards of air pollutants in China and company's target of emissions control, advanced environmental protection technologies were employed in zhanjiang works. This paper focused on the air pollutants in ironmaking process and introduced the technologies and facilities for flue gas purification in ironmaking plant.
Key words　environmental protection, flue gas purification, dust, SO_2, NO_x

太阳能光伏发电技术在宁钢的应用

李广军

（宁波钢铁有限公司能源环保部，浙江宁波　315807）

摘　要　随着现代工业的发展,全球能源危机和大气污染问题日益突出,新能源太阳能光伏发电系统其可持续性，清

洁无污染及安全性是未来能源的发展方向，受到许多国家的重视。随着太阳能光伏发电技术迅速发展,应用的规模和范围也在不断地扩大,已成为当今全球新能源发电领域的一个研究热点。本文简介了钢铁行业的用能情况，在介绍太阳能光伏发电基本原理的基础上,介绍了宁波钢铁光伏并网发电系统的项目基本情况、逆变器等主要设备情况，项目主要技术特点、投运后的测试数据、发电效益，系统投运后运行稳定，既节约了土地资源，又提供了清洁环保的能源，为缓解宁波市北仑区供电紧张、构建环境友好型社会具有较好的示范作用。论述了太阳能光伏发电技术的主要发展优势、发展历程,并分析了太阳能光伏发电技术的应用前景。

关键词 太阳能，光伏发电，可再生能源，最大功率点跟踪，孤岛效应

Application of Solar Photovoltaic Power Generation Technology in Ningbo Steel

Li Guangjun

(Energy and Environmental Department of Ningbo Iron & Steel Co., Ltd., Ningbo 315807， China)

Abstract With the development of modern industry, exhaustion of world energy and environment pollution is more serious.The new energy Solar photovoltaic power generation system and its sustainable , clean and without pollution and safety is the development direction of future energy. solar energy has attracted worldwide attention. With the development of Solar energy photovoltaic (PV) technology rapidly, it applied dimensions and area are continuously expanding, which has been one of the hot researches in the new energy generation area. This paper introduced the energy-using situation of iron and steel industry,on the basis of basic principle of solar photovoltaic power generation, Project of ningbo steel photovoltaic (PV) grid power generation system are introduced the basic situation, inverter and other main equipment, main technical characteristics, parameter test data, power generation benefits.After the system was put into operation stability, save land resource, and provides a clean and green energy, to alleviate the ningbo beilun district power supply tension, to build an environment-friendly society has good exemplary role.This paper discusses the main development of solar photovoltaic power generation technology advantage, development, and analyzes the application prospect of solar photovoltaic power generation technology.

Key words solar energy, photovoltaic generation, renewable energy, maximum power point tracking, islanding effect

Case Study of Greening Soil Remediation for Industrial Sites

Xi Lei

(Baosteel Comprehensive Industries Development Co., Ltd., Shanghai 201999, China)

Abstract This paper presented a case study of soil remediation which took place at an industrial site in shanghai. In this paper, a remediation method which was efficient, cheap, safe and green was introduced. It combined the process of “biopile-soil improvement-soil phytoremediation” to achieve a better restoration effect. It covered an introduction of the remediation mechanism and the implementation process of this method, where engineering implementation included the construction preparation, soil excavation, screening and rushing, operation process, planting greenery, acceptance check monitoring, etc. In order to assess the effectiveness of this method, the pH value, electrical conductivity and organic matter content of the soil were analyzed and evaluated. The results showed that after remediation, the average soil pH value decreased from 8.9 to 8.0, the EC value increased from 0.30mS/cm to 0.64mS/cm, and the organic matter content increased from 0.93% to 1.26%, which could meet the target for the third level of“Greening Soil” (CJ/T340—2011). The results indicated that this method would achieve “As Remediation As Production” and stood in the forefront of remedial options for future projects.

Key words industrial site, greening soil, engineering, remediation technology, amendment

炉渣成分对含铬钢渣中铬赋存状态的影响

李建立[1] 李孟雄[2] 吴鹏飞[3] 薛正良[1] 徐安军[4]

（1. 武汉科技大学钢铁冶金及资源利用省部共建教育部重点实验室，武汉 430081；
2. 江苏永钢集团有限公司炼钢二厂，张家港 215628；
3. 中国五矿集团公司，北京 100010；
4. 北京科技大学冶金与生态工程学院，北京 100083）

摘 要 不锈钢渣是不锈钢冶炼的副产品，因 Cr^{6+} 溶出风险而限制了不锈钢渣作为二次资源再利用。研究不锈钢渣中铬溶出的抑制措施对不锈钢渣的无害化处理和资源化利用具有重要意义。本文基于实验室实验，采用 FactSage 热力学计算模拟、SEM-eds 和 XRD 分析检测相结合的方法，研究 MnO 对不锈钢渣中铬赋存状态的影响。研究结果表明随着 MnO 含量的增加，炉渣中铬尖晶石的析出量和晶体粒径明显增加。当 MnO 含量由 0 增加至 6 wt%时，尖晶石的析出量由 3.86%增加至 14.97%，且晶体粒度均由~5μm 增加至~20μm。MnO 含量的增加明显改善铬的赋存状态，促使铬在铬尖晶石内富集，显著降低铬在非尖晶石相中的含量。

关键词 不锈钢渣，铬尖晶石，赋存状态，MnO

Effect of Slag Components on Chromium Occurrence in Stainless Steel Slag

Li Jianli[1], Li Mengxiong[2], Wu Pengfei[3], Xue Zhengliang[1], Xu Anjun[4]

(1. Key Laboratory for Ferrous Metallurgy and Resources Utilization of Ministry of Education, Wuhan University of Science and Technology, Wuhan 430081,China;
2. No.2 Steel-making Plant, Jiangsu Yonggang Group Co., Ltd., Zhangjiagang 215628, China;
3. China Minmetals Corporation, Beijing 100010, China;
4. School of Metallurgical and Ecological Engineering, University of Science and Technology Beijing, Beijing 100083, China)

Abstract The stainless steel slag is a by-product of the stainless steel making, while the elution risk of hexavalent chromium restrains its utilization as raw material in other fields. The work on the suppression measures of Cr leaching is significantly important to the harmless treatment and recycled utilization of the slag. On basic of the lab experiments, the effects of MnO on the occurrence of Cr were investigated, adopting some analysis methods, such as SEM-eds and XRD. The results show that MnO could increase the precipitated amount and size of spinel crystal. When the MnO content increases from 0 to 6 wt%, the spinel amount increases from 3.86% to 14.97%. The addition of MnO significantly promotes the enrichment of chromium in the Cr-Spinel phase, and reduces the content of chromium in other mineral phase, which is beneficial for the stabilization of chromium in the steel slag.

Key words stainless steel slag, Cr-spinel, occurrence state, MnO

转炉汽化冷却烟道破损因素分析及改进措施

樊 译 石 焱 胡长庆 张玉柱 冯英英

（华北理工大学冶金与能源学院，河北省现代冶金技术重点实验室，河北唐山 063000）

摘 要 本文分析了周期性交变热应力、水循环不均匀、水质、结渣、炉尘冲刷等因素对转炉汽化冷却烟道的影响，并提出通过保持自然循环最小流量、严格控制补给水水质、合理控制烟气流速、改善水冷管材质、完善烟道结构设计，以及开展汽化冷却烟道烟道破损机理研究，优化水冷管相关参数等有效改进措施，提高转炉汽化冷却烟道寿命。

关键词 转炉，汽化冷却烟道，破损，寿命

Damage Factor Analysis And Improvement Measures of Converter Vaporization Cooling Flue

Fan Yi, Shi Yan, Hu Changqing, Zhang Yuzhu, Feng Yingying

(North China University of Science and Technology, College of Metallurgy and Energy, Hebei Key Laboratory of Modern Metallurgy Technology, Tangshan Hebei 063000, China)

Abstract This article describes the cycle of alternating thermal stress, uneven water cycle, water quality, slagging, furnace dust flushing and other factors that affect the lifetime of converter vaporization cooling flue, and summarizes to keep the minimum natural cycle flow, control the water quality strictly, control of the flue gas flow rate reasonable, improving the water-cooled pipe material, improve the flue structure design and carry out research on vaporative cooling flue broken mechanism, optimize the relevant parameters of the cooling tube, improve the lifet of converter vaporization cooling flue.

Key words converter, vaporative cooling flue, damage, lifetime

Iron Resources Selective Beneficiation by Molten Oxidation from Copper Slag

Fan Yong[1], Shibata Etsuro[2], Iizuka Atsushi[2], Nakamura Takashi[2]

(1. Graduate School of Environmental Studies, Tohoku University
6-6-20 Aramaki-Aza-Aoba, Aoba-ku, Sendai, 980-8579, Japan;
2. Institute of Multidisciplinary Research for Advanced Materials (IMRAM), Tohoku University
2-1-1 Katahira, Aobaku, Sendai 980-8577, Japan)

Abstract Iron in copper slag is in the form of iron silicate. We proposed to oxidize the iron silicate to magnetite, allowing the selective recovery of the slag's iron content by magnetic separation. This proposal would effectively decrease the volume of dumped slag, and the recovered iron concentrate could be used for the future iron or steel making. The crystallization morphology of magnetite and hematite were studied as reaction products of the molten oxidation of copper slag. A gas atmosphere of 1 vol% of oxygen was found to effectively precipitate the iron in the molten slag as magnetite.

Key words molten oxidation, magnetite precipitation, copper slag

混合铁源制备磷酸铁锂材料

陈宇闻

（上海宝钢磁业有限公司，上海 201900）

摘 要 锂离子电池作为高性能的二次绿色电池，就具有高电压、高能量密度（包括体积比能力、质量比能量）、低的自放电率、使用温度范围宽、循环寿命场、无记忆效应等优点。目前主流的磷酸铁锂制备方法采用的铁源为磷酸铁或者是氧化铁。以磷酸铁为铁源制备的磷酸铁锂电性能优异，但是原料价格高昂。以氧化铁为铁源制备的磷酸铁锂成本低廉，但是材料性能不佳。采用磷酸铁和氧化特的混合铁源的制备方法，在保持磷酸铁工艺优异性能的同时减少原料成本。

关键词 铁源，磷酸铁锂，性能

The Preparation of Lithium Iron Phosphate Material Mixed Iron Sources

Chen Yuwen

(Shanghai Baosteel Magnetic Co., Ltd., Shanghai 201900, China)

Abstract Lithium ion batteries as a high performance second green battery, will have a high voltage, high energy density (including volume ratio, mass ratio, energy), low self-discharge rate, use temperature scope width, cycle life, no memory effect, etc. At present the mainstream of the preparation methods of lithium iron phosphate source for phosphoric acid iron or iron oxide of iron. Iron phosphate as the source of iron preparation of lithium iron phosphate performance is excellent, but the high price of raw materials. Preparation of iron oxide as the source of iron lithium iron phosphate as low cost, but poor material performance.

Using iron phosphate and oxidation, the preparation method of mixed iron source, while maintaining the excellent iron phosphate process performance to reduce raw material costs.

Key words the source of iron, lithium iron phosphate, performance

酸再生低锰氧化铁粉制备技术研究

魏 恒 赵忠民 赵生军 严 军

（上海宝钢磁业有限公司，上海 201900）

摘 要 钢板酸洗后产生的大量酸洗废液需要进行再生，利用酸再生工艺可生产高品质的氧化铁粉。目前国内大型钢铁公司大部分都引进了 Ruthner 公司的喷雾焙烧废酸处理技术，该方法由于废酸由喷嘴以喷雾的形式喷入焙烧炉进行反应，故生产出来的氧化铁粉具有较小的粒径，品质较高。该系统可以分为除硅和酸再生两部分，除硅是酸再生的前道工序，用于去除废酸中的二氧化硅等杂质，用来提高氧化铁粉的化学纯度。为了顺应市场对高锰钢种的旺盛需求，宝钢各冷轧机组中高锰钢的产量逐年增加，直接影响酸再生副产品氧化铁粉的质量等级。酸再生所生产的氧化铁粉主要用于制作磁性材料，因此氧化铁粉的化学纯度、物理性能及其一致性对软磁铁氧体的生产质量至关重要。本文试验得到最佳除硅工艺以及除锰试剂以降低锰含量制取高品质氧化铁粉。

关键词 酸再生, 氧化铁粉, 锰

Preparation of Iron Oxide with Low Mn in Acid Regeneration

Wei Heng, Zhao Zhongmin, Zhao Shengjun, Yan Jun

(Shanghai Baosteel Magnetics Co., Ltd., Shanghai 201900,China)

Abstract Pickling process of steel sheet will produce large amounts of pickling waste liquor which need to be regenerated. High quality of iron oxide is produced by acid regeneration process. Nowadays, Ruthner spray roasting technology is widely used to deal with pickling solutions. Because of the method of waste acid from nozzle to spray into the furnace for reaction, iron oxide is produced with smaller particle size and higher quality. The system can be divided into two parts of desilication and acid regeneration. Desilication which used to improve the chemical purity of iron oxide is the former procedure of acid regeneration for the removal of silica in the waste acid. In order to adapt the active demand of high manganese steel, the high manganese steel production of Baosteel cold rolling mills increases year by year. The increasing production directly affect the quality of iron oxide which is the by-products of acid regeneration. The iron oxide is mainly used for the production of magnetic materials, therefore the chemical purity, physical properties and consistency of iron oxide is quite important to the production of soft magnetic ferrite. In this paper, the process parameters of desilication will be optimized and high quality of iron oxide with low manganese will be produced.

Key words acid regeneration, iron oxide, manganese

改质高炉配渣制备烧结矿渣微晶玻璃：Na^+和K^+对烧结-结晶行为影响的对比分析

刘 峰[1] 张建良[1] 国宏伟[2]

（1. 北京科技大学冶金与生态工程学院，北京 100083；
2.苏州大学沙钢钢铁学院，江苏苏州 215021）

摘 要 本文以高炉配渣为原料，通过成分改质，采用烧结法进行了矿渣微晶玻璃的制备。在成分改质过程中，结合硅酸盐熔体的结构理论，文章创新地提出了“以 NBO/T 为基础”的改质规则，且在 NBO/T 相同的情况下，配制了三种成分体系以研究 Na^+和 K^+对烧结–结晶行为的不同影响。实验结果表明，改质操作所采用的 NBO/T，完全满足制备致密烧结矿渣微晶玻璃的要求，但由于 Na^+和 K^+的影响不同，所需要的最佳烧结温度以及烧结产品的残余孔隙度并不相同。其中，对于含 Na^+的体系而言，所需要的烧结温度较低，特别是当体系中同时含有一部分 K^+时，所需的烧结温度及最终烧结产品的孔隙度要更低一些，而当体系中的 Na^+全部被 K^+取代后，只能利用二次烧结在高温区内完成致密化过程。

关键词 高炉渣利用，烧结矿渣微晶玻璃，NBO/T，烧结-结晶行为，混合碱金属效应

Preparation of Slag Glass-ceramics Based on Compositionally Modified Synthetic Blast Furnace Slag: Comparison of the Effect of K^+ and Na^+ on the Sinter-crystallization Behavior

Liu Feng[1], Zhang Jianliang[1], Guo Hongwei[2]

(1. School of Metallurgical and Ecological Engineering, University of Science and Technology Beijing, Beijing 100083,China;
2. Shagang School of Iron and Steel, Soochow University, Suzhou 215021, China)

Abstract In this paper, a synthetic blast furnace slag was compositionally adjusted to prepare slag glass-ceramics via sintering route. In the course of composition modification, NBO/T-based formulation rule was proposed based on the structural theory of silicate melts. Besides, three different compositions with the same NBO/T value were used to investigate the difference between Na^+ and K^+ regarding their effect on the sinter-crystallization behavior. The experimental results show that, the NBO/T selected in this paper, regardless of the variation in Na^+ and K^+ content, is capable of producing dense sintered glass-ceramics. However, some differences in terms of the required optimum sintering temperature as well as the residual porosity of final products need to be clarified. The composition containing both Na^+ and K^+ is characterized by a lower sintering temperature and a higher densification degree. For the K^+-free composition, the maximum densification degree is slightly lower than the other two compositions. In contrast, even though Na^+-free composition is able to reach a densification comparable to the composition containing Na^+ and K^+, the secondary sintering process in high-temperature region is necessary to complete the densification.

Key words blast furnace slag utilization, slag glass-ceramics, NBO/T, sinter-crystallization behavior, mixed alkali effect

冶金焦活化、改性及在铁矿石烧结烟气中脱硝性能的研究

任 山 姚 璐 王小青 刘清才 孟 飞 杨 剑

（重庆大学材料科学与工程学院，重庆 400044）

摘 要 冶金行业排放的大气污染物中，NO_x是主要有害成分，对人类健康、人居和生态环境都产生了极其严重的影响。钢铁冶炼的整个生产流程中，烧结工序是 NO_x 的主要来源，因此，经济高效的烧结烟气脱硝技术是钢铁企业环境治理的重中之重。本研究针对传统活性炭脱硝效率低、成本高等缺点，采用冶金焦丁作为烟气脱硝的固相吸附剂。通过对冶金焦进行活化试验，发现在800℃条件下采用水蒸气法对冶金焦丁进行活化可以显著改变冶金焦的孔隙数量和尺寸；并且通过改性试验负载 V_2O_5 催化剂能够显著提高冶金焦的脱硝效率。冶金焦丁经过活化改性既能满足脱除 NO_x 的需求，还能充分利用钢铁企业的低价焦丁直接用于烟气治理，大幅度降低投资运营和环保维护成本。对冶金焦丁的高效利用和铁矿石烧结烟气治理意义重大。

关键词 冶金焦丁，活化，改性，脱硝

Study on the Activation, Modification and Sintering Flue Gas Denitration of Metallurgical Coke

Ren Shan, Yao Lu, Wang Xiaoqing, Liu Qingcai, Meng Fei, Yang Jian

(College of Materials Science and Engineering, Chongqing University, Chongqing 400044, China)

Abstract NO_x which is one of air pollution discharged by metallurgy industry has an important effect on people's health and environment. In the whole production process of iron and steel melting, NO_x comes mainly from sintering. Therefore, a cost-effective denitration technology of sintering flue gas is the priority of environmental management in steel companies. This study is aimed at the disadvantage of traditional activation carbon, such as low efficiency of denitration and high cost, the metallurgical coke is used as solid-phase absorbent of denitration. According to the activation experiment of metallurgical coke, the number and size of coke pore could increase obviously at 800℃. And the supported V_2O_5 catalyst could improve the efficiency of denitration obviously by modification experiment. Metallurgical coke which is activated and modified both satisfies to denitrate and uses the cheap coke of steel companies fully, then the cost of investment operation and environmental maintenance could decrease significantly. It could well make sense to use metallurgical coke efficiently and administer sintering flue gas of iron ores.

Key words metallurgical coke, activation, modification, denitration

石灰石泥饼复配矿物掺合料活性研究

徐　兵　朱煜东　汪　毅

（宝钢发展有限公司，上海　201900）

摘　要　宝钢石灰石泥饼与石灰石成分接近，CaO 含量高,S 含量低，其品质较高。本实验采用烘干后的石灰石泥饼为原料，分别同烘干后的宝钢高炉水淬渣进行混磨复配试验和单独粉磨复配 S95 矿粉试验。实验结果表明混磨未体现出石灰石泥饼的易磨性优势，随着泥饼掺量增加，混磨粉料比表面积呈下降趋势，复配矿物掺合料 7d 和 28d 活性明显下降。石灰石泥饼单独粉磨后复配 S95 矿粉效果较好；综合比较复配 3%~10%石灰石泥饼粉，S95 矿粉中掺加 5%一定细度石灰石泥饼粉时性能最佳，7d 活性提高 1.1%，28d 活性仅下降 3.5%，流动度提高 2%。试验证明宝钢石灰石泥饼复配矿物掺合料具有技术可行性。针对石灰石泥饼含水 20%的不利因素，提出了相应的资源化利用构想。

关键词　石灰石泥饼，掺合料，活性，混凝土

Activity Experimental Study of Mineral Admixtures Compound with Limestone Mud Cake

Xu Bing, Zhu Yudong, Wang Yi

(Baosteel Development Co., Ltd., Shanghai 201900, China)

Abstract The composition of limestone mud cake in Baosteel is similar to limestone,it has high calcium oxide content and low in sulphur content.The experiment used dried limestone mud cake as raw material,the first set adopted grinding compound with dried blast furnace slag ,the second set adopted separate grinding powder compound with S95 GGBS.The results of experiment show that mixing grinding can not reflect the advantage of grindability of limestone mud cake. With the increasing dosage of mud cake, the specific surface area of mixed grinding powder is on the decline, the 7d and 28d activity of compound mineral admixtures decrease remarkably. The effect of separate grinding limestone mud cake compound with S95 GGBS is good. After comprehensive comparison with 3%~10% powder of limestone mud cake, adding 5% certain fineness powder of limestone mud cake to S95 GGBS has best performance:7d activity increased by 1.1%, 28d activity fell by only 3.5%, fluidity increased by 2%. Trials show that compounding Baosteel limestone mud cake to mineral admixtures have technical feasibility. In view of the unfavorable factor that limestone mud cake has 20% moisture content, the corresponding conception of resource utilization has been put forward.

Key words limestone mud cake, admixture, activity, concrete

不锈钢酸洗废水生物脱氮工程设计及运行实践

李　勇

（宝钢工程技术集团有限公司，上海　201999）

摘　要　介绍了钢铁厂不锈钢酸洗废水的来源、危害，以及不锈钢酸洗废水的生物脱氮原理以及实际工程案例。实践证明，采用合理的工艺，在特定的碳源、合适的碳氮比以及反应条件下，不锈钢酸洗废水经过处理后总氮完全能满足最新的《钢铁工业水污染物排放标准》的标准要求。

关键词　不锈钢酸洗废水，生物反硝化脱氮，C/N，碳源，水温

Design and Operation Practice of Biological Nitrogen Removal in Stainless Steel Pickling Wastewater

Li Yong

(Baosteel Engineering & Technology Group Co., Ltd., Shanghai 201999, China)

Abstract This paper introduces the source and damage of stainless steel pickling waste water and the principle of biological nitrogen removal and the practical engineering case. Practice shows that the process is reasonable, in particular carbon source, suitable carbon and nitrogen ratio and reaction conditions, stainless steel pickling waste water after treatment after total nitrogen can fully meet the requirements of the new "iron and steel industrial water pollutant discharge standard" standard.

Key words stainless steel pickling wastewater, biological denitrification denitrification, C/N, carbon source, water temperature

铬泥回流技术在冷轧含铬废水处理中的应用研究

刘尚超[1]　薛改凤[1]　段建峰[2]　陶　灿[2]

（1. 武钢研究院，武汉　430080；　2. 武钢能源动力总厂供水厂，武汉　430083）

摘　要　含铬废水一直是冶金行业废水的处理难点之一。本文以钢铁行业较多使用的还原沉淀法处理含铬废水工艺作为研究对象，使用浊度为表征对象，通过实验来研究在还原沉淀法絮凝沉淀段中使用铬泥回流来强化絮凝沉淀效果。实验结果表明，在各回流比的实验结果中，铬泥回流技术对絮凝沉淀池中的铬沉淀具有较好的效果，废水中的浊度得到明显改善。

关键词　含铬废水，铬泥回流，浊度，总铬

Study on Chromium Sludge Recirculation Technology in Treatment for Wastewater Containing Chromium

Liu Shangchao[1], Xue Gaifeng[1], Duan Jianfeng[2], Tao Can[2]

(1. Research and Development Institute of WISCO, Wuhan 430080, China; 2.Water Supply Pant, Energy Sources and Power Plant, Wuhan Iron and Steel (Group) Corp., Wuhan 430083, China)

Abstract　Chromium containing wastewater has been one of the difficulties in the metallurgical industry wastewater. In this paper, reduction—precipitation method which is widely used in iron and steel industry is studied to how to improve removal of chromium. In the experiment turbidity was used as the characterization of the wastewater. Serial experiments in reducing precipitation flocculation section with chromium sludge recirculation are to strengthen flocculation effect. The experimental results show that, chromium sludge recirculation technology has a good effect on the flocculation sedimentation tank in chromium precipitation and turbidity of wastewater was obviously improved

Key words　wastewater containing chromium, chrome sludge reflux, turbidity, total chromium

低温烟气余热回收及应用

汪　毅　李晓东

（宝钢发展有限公司，上海　201900）

摘　要　宝钢股份厂区存在大量低温烟气余热无法作为生产辅助用能加以利用，同时宝钢集团内一些浴室使用燃油燃煤锅炉用于职工生活热水热源。然而传统余热回收问题存在容易发生腐蚀、经济性差等问题。宝钢发展工业环境保障部在股份公司能源环保部热力分厂采用复合相变换热器技术回收利用股份能环部 4 号低压锅炉排放的低温烟气余热生产生活热水，原锅炉排烟温度由 194℃降低到 130℃，每小时平均产生温升 65℃热水 21.4t，年节约标煤1350.4t,节能效果明显，与传统换热器比较有效克服了低温腐蚀，换热性能好，运行平稳。经过 3 个月试运行达到了取代燃油燃煤浴室锅炉为职工提供生活热水预期目标，推进了宝钢节能减排，提升了能源利用效率，为高效利用低温烟气提供了新路径。

关键词　相变复合换热器，低温烟气，酸露点，排烟温度

Utilizing Recycling Waste Heat of Low Temperature Fume

Wang Yi, Li Xiaodong

(Baosteel Development Co., Ltd., Shanghai 201900, China)

Abstract　There is a mass of waste heat of low temperature fume in the plants of Baosteel corporation which can't be reused as auxiliary heat in production.Meanwhile,some shower room in Baosteel group use oil fired boiler and coal fired boiler for supplying stuff and workers with heat of domestic hot water.However,it usually occurs corrosion and also has a poor economical efficiency in traditional method of waste heat recovery.Baosteel Comprehensive Industries Development Co., Ltd,Industrial Environment Security Division use the FXH for recycling waste heat of low temperature fume which is from Baosteel Corporation,Energy and Environment division's 4th low-pressure boiler in the Thermal plant,the temperature of discharging smoke reduced from 194℃ to 130℃,producing 21.4 ton 65 degree water average every hour, saving 1350.4 ton of standard coal every year,the effect of saving energy is obvious.Referring to traditional heat exchanger the FXH overcoming corrosion in low temperature and also being better in heat exchange performance during stable operation.Via 3 months of pilot working it has reached the objective that replacing the fired boiler and coal fired boiler with FXH to supply stuff and workers with domestic hot water.It could boost the energy saving and pollution reduction work in Baosteel,improving energy efficiency,and show a new approach in reusing low temperature fume efficiently.

Key words　FXH, low temperature fume, acid dew point, exhaust gas temperature

Thermodynamic Calculation of Equilibrium between Carbon, FeO-containing Slag and CO-CO_2-H_2O Gas

Wu Yan[1,2], Matsuura Hiroyuki[2], Yuan Zhangfu[1], Tsukihashi Fumitaka[2]

(1. College of Engineering, Peking University, Beijing, China;
2. Graduate School of Frontier Sciences, The University of Tokyo, Kashiwa, Japan)

Abstract　High temperature slags could act as not only heat carrier and catalyst but also reactants in some chemical reactions to produce syngas due to the individual components in slags. A new method is put forward to utilize the thermal energy of converter slag and generate CO and H_2 gases which could be reused as energy. Thermodynamics of the reaction between carbon, FeO-CaO-SiO_2 slag and CO-CO_2-H_2O gas was studied by thermodynamic calculation and the effects of slag composition, slag temperature and added carbon amount on the behavior of the production of H_2 and CO gases were clarified.

Key words　FeO-containing slag, steam, fuel gas, carbon, thermodynamic calculation

烧结工序中低温余热回收潜力分析

赵　斌[1]　晁双双[1]　屈婷婷[1]　杨　鹤[1]　王晓旭[2]

（1. 华北理工大学河北省现代冶金技术重点实验室，河北唐山　063009；
2. 南京凯盛开能环保能源有限公司，江苏南京　210036）

摘　要　随着钢铁工业节能减排的深入，烧结过程中低温余热资源的回收技术成为研究的热点。针对钢铁工业中烧结矿冷却和显热提取及再利用过程，依据热力学第一定律和第二定律，在冷却机、余热锅炉和汽轮发电机组耦合响应机制的基础上，建立余热资源回收率和吨矿余热发电量模型，选定三个工况计算了吨矿极限发电量和理论发电量，分析了影响烧结余热发电负荷的因素，给出了烧结余热发电负荷提升的方法，并基于场协同原理及梯级利用原则提出了烧结旋冷机的结构。研究成果可为烧结工序余热集成回收与梯级利用提供理论基础和技术支撑。

关键词　烧结，显热回收，单位发电指标，旋冷机

Potential Analysis of Middle and Low Temperature Waste Heat Recovery in Sintering Process

Zhao Bin[1], Chao Shuangshuang[1], Qu Tingting[1], Yang He[1], Wang Xiaoxu[2]

（1. Hebei Key Laboratory of Modern Metallurgy Technology, North China University of Science and Technology, Tangshan Hebei 063009, China；2. Nanjing Kesen Kenen Environment and Energy Co., Ltd., Nanjing Jiangsu 210036, China）

Abstract With the proceeding of energy saving and emission reduction in steel industry, the technology of middle and low temperature waste heat recovery in sintering process has become a hot research. In view of the process of cooling and sensible heat extraction and reuse of sinter in steel industry, according to the first and second law of thermodynamics, based on the coupling response mechanism of the cooler, the waste heat boiler and steam turbine generator unit, the waste heat recovery rate and tons of ore waste heat power generation model was established, the limited and theoretical power generation of tons sinter was calculated under three cases, the factor of influence of the sintering waste heat power generation load was analyzed, the method of ascension of sintering waste heat power generation load was given, and the structure of sintering helical cooler was put forward based on the field synergy and cascade utilization principle. The results can provide theoretical basis and technical support for the recovery and cascade utilization of waste heat in sintering process.

Key words sintering, sensible heat recovery, unit power generation indicator, helical cooler

65 t 转炉干法除尘系统汽化冷却烟道数值模拟

李海英　滕军华　张　滔

（华北理工大学冶金与能源学院，河北唐山　063009）

摘　要　转炉干法除尘技术因除尘效率高、节水效果好、能源消耗和运行费用低、系统阻力小、使用寿命长、维护维修少等优点，在国际上被认定为今后的发展方向。目前干法除尘技术已获得世界各国的普遍重视和采用。汽化冷却烟道是连接转炉和蒸发冷却器的设备，在整个除尘工艺中十分重要。采用计算流体力学（CFD）方法对国内某钢厂 65 t 转炉汽化冷却烟道内的烟气流动状况进行了数值模拟，分析了烟气速度场、温度场等在汽化冷却烟道内不同截面的分布，讨论了烟道内容易发生爆管的部位及在汽化冷却烟道末端速度分布的不均匀现象等问题，对实际运行时汽化冷却烟道的维护管理有较强的参考价值。

关键词　转炉，干法除尘，汽化冷却烟道，数值模拟

Numercial Simulation of the Vaporization Cooling Flue in the 65t Converter Dry Dedusting System

Li Haiying, Teng Junhua, Zhang Tao

(North China University of Science and Technology, College of Metallurgy and Energy, Tangshan 063009, China)

Abstract Converter dry dedusting technique is consider as the development direction in the word because of high dedusting efficiency, saving water, low energy consumption and operating expenses, low system drag, long using age, less maintenance and so on. Currently, converter dry dedusting technique has been widely used in the world. Evaporative cooling flue is a important device in the dedusting system which connects the converter and the evaporative cooler. The gas flow process in the evaporative cooling flue of 65t converter is numerically simulated by means of computational fluid dynamics(CFD).The gas temperature and velocity of different section in the evaporative cooling flue is analyzed. Phenomenon of burst pipe and velocity inhomogeneity in the evaporative cooling flue is also discussed. So this paper has strong reference value for maintenance and management of the evaporative cooling flue.

Key words converter, dry dedusting, evaporative cooling flue, numerical simulation

转炉 LT 干法除尘工艺应用存在问题及解决方法

李海英　张　滔　滕军华　贾永丽

（华北理工大学冶金与能源学院，河北唐山　063009）

摘　要　通过转炉干法除尘系统与湿法除尘系统对比，转炉干法除尘更具优势且更符合我国冶金工业可持续发展的要求。总结介绍了转炉 LT 干法除尘技术应用及发展现状。分析了转炉干法除尘系统在钢铁企业实际运行中系统泄爆、静电除尘器极线断裂、蒸发冷却器积灰、输灰系统故障问题的产生原因及解决方案；并对转炉干法除尘及煤气回收过程中易出现的安全问题提出了相应的解决措施。减少或避免设备故障的发生，对于转炉干法除尘系统稳定运行，提高工作效率有重要意义。

关键词　转炉 LT 干法除尘，安全，解决措施

Problems and Solving Methods of the Converter LT Dry Dedusting System

Li Haiying, Zhang Tao, Teng Junhua, Jia Yongli

(North China University of Science and Technology, College of Metallurgy and Energy, Tangshan Hebei 063009, China)

Abstract Comparison between dry method and wet method of flue de-dusting of converter, dry dedusting of Converter have more advantages in the sustainable development of the metallurgical industry requirement. introduce application and technology development status of LT dedust for converter. Analysis the Problem Causes and Solutions of the dry dust removal system of converter in the actual operation, such as explosion venting panel ,discharge electrode fracture of ESP , evaporative cooler fouling, conveying system failure and puts some corresponding measures of the security problem of dry dedusting and gas recovery system. To reduce or avoid the happening of equipment failure, has important significance for the dry dust removal system of converter and stable operation of converter, improve the work efficiency

Key words dry dedusting of converter，safety，countermeasure

转炉干法除尘灰及 OG 泥冷固球团工艺研究

武国平　周　宏　魏永义　秦友照　张　涛

（北京首钢国际工程技术有限公司，北京 100043）

摘 要 转炉煤气经干法除尘后产生的除尘灰或湿法除尘产生的OG泥具有粒度细、含铁量高的特点，是可回收的二次资源。介绍了采用转炉干法除尘灰和湿法OG泥生产冷固球团的工艺以及实际应用案例，并介绍了冷固球团工艺的主要设备及其特点。通过配料计算并采用冷固球团工艺，生产出符合转炉使用要求的冷固球团，冷固球团可代替烧结矿用作转炉冷却剂和造渣剂，具有很好的经济效益、环境效益和社会效益。冷固球团工艺是钢铁企业完成节能减排目标的重要支撑技术。

关键词 转炉，干法除尘灰，OG泥，冷固球团

Research on Process of Cold-hardended Pellet Using Converter Dry Dust and OG Sludge

Wu Guoping, Zhou Hong, Wei Yongyi, Qin Youzhao, Zhang Tao

(Beijing Shougang International Engineering Technology Co., Ltd., Beijing 100043, China)

Abstract The dust which is generated by dry de-dusting or OG sludge which is generated by wet de-dusting of converter gas，with the characteristics of fine particle size and high iron content，is a recyclable secondary resource. The process and the practical application cases of the cold-hardended pellets produced by converter dry dust and wet OG sludge are introduced, and the main equipments and their characteristics of the cold-hardended pellet process are introduced. Cold-hardened pellets that meet the requirements of converter can be produced by calculating the charging mixture and using cold-hardened technology. Replace sinter with cold-hardened pellets as the agents of coolants and converter slag is with good economic, environmental and social benefits. The process technology of cold-hardened pellets is an important support technology for steel companies to complete the new emission reduction targets.

Key words converter,dry dust, OG sludge, cold-hardened pellet

鞍钢高炉冲渣水余热供暖实践

曲 超 王罡世 黄显保 何 嵩

（鞍钢股份有限公司能源管控中心，辽宁鞍山 114001）

摘 要 本文描述了鞍钢高炉冲渣水余热利用的现状，分析了冲渣水余热供暖面临的技术难题，通过选取目前最先进的冲渣水换热设备，回收冲渣水余热资源的同时取得了巨大经济效益，达到节能降耗的目的。提高能源资源的利用效率、减少对环境的污染，为加速绿色转型、融入城市经济圈提供了重要支撑，为其他企业高炉冲渣水余热供暖提供借鉴。

关键词 高炉冲渣水，余热利用，节能降耗

Ansteel Blast Furnace Slag Flushing Water Heat Utilization for Central Heating Application

Qu Chao, Wang Gangshi, Huang Xianbao, He Song

(Energy Management and Control Center of Ansteel, Anshan 114001, China)

Abstract This paper introduces the waste heat utilization situation of Ansteel blast furnace slag flushing water. Several challenging technical problems are analyzed for implementing slag flushing water exhausted heat utilization. Through applying the latest special heat transfer devices for slag flushing water, the exhausted heat can be recycled to make great economic benefits and achieve energy conservation. The slag flushing water exhausted heat utilization can increase energy utilizing efficiency and reduce environmental pollution, which provides an important support for shifting towards a green economy and assimilating into city economic circle. This provides reference for waste heat utilization of other steel plants.

Key words blast furnace slag flushing water, waste heat utilization, energy conservation

燃煤锅炉脱硝技术在鞍钢的应用

何 嵩[1] 黄显保[1] 孙 亮[1] 李丛康[1] 曲 超[1] 贾 振[2]

（1. 鞍钢股份有限公司能源管控中心，辽宁鞍山 114002；
2. 鞍钢集团钢铁研究院，辽宁鞍山 114003）

摘 要 针对鞍钢股份能源管控中心中央电站 2 台 220t/h 燃煤锅炉存在氮氧化物（NO_x）排放浓度超标的问题，对其脱硝系统采用混合脱硝技术低氮燃烧（ROFA）+SNCR 系统+内置式 SCR 系统进行改造。改造后，NO_x 排放浓度低于国家标准，氨逃逸率大幅降低，取得较为显著的节能和环保效果。

关键词 燃煤锅炉，氮氧化物，脱硝技术

The Application of Denitrification Technologies on Coal-fired Boilers in Ansteel

He Song[1], Huang Xianbao[1], Sun Liang[1], Li Congkang[1], Qu Chao[1], Jia Zhen[2]

(1.Energy Management and Control Centre of Ansteel ,Anshan,114002, China;
2. Iron & Steel Research Institute of Ansteel, Anshan Liaoning 114003, China)

Abstract To resolve the problem of the NO_x concentration exceeding discharge standard, two 220t/h coal-fired boilers used in central power station of energy management & control of Ansteel have been retrofitted for removing NO_x through applying hybrid denitrification technology of Rotating Opposed Fired Air(ROFA), Selective Non-Catalytic Reduction(SNCR) and Selective Catalytic Reduction(SCR). After the renovation, the effluent NO_x concentration is lower than that of national standard, and ammonia escape rate is reduced greatly. Then remarkable energy saving and environment protection effects have been achieved.

Key words coal-fired boiler, NO_x, denitrification technology

脱硫废液预处理工艺及其改进

郑晓雷 李 志

（鞍钢股份有限公司鲅鱼圈钢铁分公司炼焦部，辽宁营口 115007）

摘　要　鞍钢股份鲅鱼圈钢铁分公司炼焦部脱硫废液预处理工艺为焦化废水生物脱氮脱酚系统改造工程的一部分，该工艺采用中科院过程工程研究所研发的脱氰技术（IPE-DCN），针对脱硫废液中含有高浓度 CN^-、S^{2-}毒害生化处理系统的问题，将脱硫废液先进行单独的脱氰脱硫预处理。在该工艺建设、调试及运行期间，我们针对工艺中存在的不足之处做了大量改进。本文重点对脱硫废液预处理工艺作简要的说明，并介绍该工艺的改进情况。

关键词　脱硫废液，预处理，工艺，改进

Coke Oven Gas Desulfurization Wastewater Pretreatment Process and Improvement

Zheng Xiaolei, Li Zhi

(Ansteel Bayuquan Iron & Steel Subsidiary Department of Coking, Yingkou Liaoning 115007, China)

Abstract　Coke Oven Gas Desulfurization Wastewater Pretreatment Process of AnSteel Bayuquan Iron & Steel Subsidiary Coking Department is a part of the Biological Nitrogen & Phenol Removal of Coking Wastewater System Renovation Project, the process uses the cyanide removal technology (IPE-DCN) developed by the Chinese Academy of Sciences Institute of Process Engineering to do decyanation & desulfurization pretreatment for Coke Oven Gas Desulfurization Wastewater in order to solve the problem of that the Coke Oven Gas Desulfurization Wastewater containing high concentrations of CN^- & S^{2-} which poisoned the biochemical treatment system. For existing inadequacies in the process we made a lot of improvements during constructing, commissioning and running the process. This article focuses on the pretreatment process of the Coke Oven Gas Desulfurization Wastewater with a brief description, and describes the improvement to the process.

Key words　desulfurization, pretreatment, process, improvement

精益化能源管理模式的实践与创新

高　军　白世宏　王　荣　李富强

（鞍钢股份有限公司冷轧厂，辽宁鞍山　114021）

摘　要　本文介绍了冷轧厂精益化能源管理模式是针对先进能源管理系统的能源介质在线监测、主要用能设备和生产过程的能效分析与优化控制、能源介质管理、分析与优化技术方面的新研究内容及其进展进行了详细的阐述。通过建立精益化能源管理模式，实现各工序能源实时消耗的全面监控，优化考核评价机制，实施管理及技术节能降耗，开展能效对标，改进生产组织、选择最佳工艺路径，实现能源成本最优，提升产品成本的竞争力，加快和推动废气余热回收利用，蒸汽冷凝水回收，工业废水回用项目等各种更加前沿的节能技术的应用。通过精益化能源管理应用和实施，实现冷轧厂节能降耗的可持续发展。在钢铁企业能源管理系统中推广应用这些新的研究成果，将有效提升钢铁企业能源管理的效果。

关键词　精益化，提升，能源管理，效果

The Practice and Innovation of Extractive Benefit Energy Management

Gao Jun, Bai Shihong, Wang Rong, Li Fuqiang

(Cold Strip Works of Angang Steel Co., Ltd., Anshan Liaoning 114021, China)

Abstract It brings forward extractive benefit energy management. It clarifies the online measure of energy medium for the advanced energy management system, the main equipment for the energy consumption, energy and effect analysis and optimization control, energy medium management and the progressing in the details. We can achieve the application of advanced technique of energy-efficient via extractive benefit energy management, such as realizing the complete energy management for working procedure, optimizing checking and valuing system, putting in practice the reduce energy consumption of management and technique, developing energy efficiency benchmarking, improving the organism of production and choosing the best process route, realizing optimization control of energy consumption ,expanding the competitive power of cost ,accelerating and driving the using of the recovery of waste heat of exhaust gas, the steam of condensate water, waste water etc. It can achieve the sustainable development of energy conservation vie the application of the extractive benefit energy management.The popularization and application of the advanced research for the energy management system in the steel enterprise will improve the effect of the energy management in the steel enterprise.

Key words extractive benefit, improve, energy management, effect

钢管酸洗废硫酸再生工艺对比分析与实践

张永亮 李 伟

（宝山钢铁股份有限公司钢管事业部，上海 201900）

摘 要 本文通过不同结晶法废硫酸再生工艺的对比分析，和蒸发浓缩和调酸两种冷冻结晶法工艺方案比较和经济性分析，确定了适用于宝钢精密钢管废硫酸再生工艺方法即调酸-冷冻结晶法；介绍了调酸-冷冻结晶法废硫酸工艺在精密钢管废硫酸再生处理的工业应用实践情况，实践表明此工艺可有效再生硫酸，节省废硫酸处置费用，具有推广应用价值。

关键词 废硫酸，再生，钢管

Comparison of Different Recovery Process for Waste Sulfuric Acid from Steel Tube Acid Washing Mill and Its Practice

Zhang Yongliang, Li Wei

(Tupe Pipe and Bar Business Unit Baosteel, Shanghai 201900, China)

Abstract This paper compares and analyzes the advantages and disadvantages of the waste sulfuric acid regeneration process based on different principles, and through the comparison of the economic analysis, the paper determined the waste sulfuric acid recycling process method, and the method of regulating the acid and freezing.After the economic comparison between the cost of disposal of waste sulfuric acid and the cost of the construction project, building a project to dispose the waste sulfuric acid has significant economic benefits.

Key words waste sulfuric acid，recovery, steel tube

炼钢散料灰作烧结配料试验研究

徐鹏飞 杨大正 于淑娟 张大奎 耿继双 侯洪宇 钱 峰 王向锋

（鞍钢股份有限公司技术中心，辽宁鞍山 114009）

摘　要　为实现炼钢散料灰的资源化利用，把其作为烧结配料进行试验研究。在固定原料配比条件下，用炼钢散料灰替代部分石灰石和菱镁石，改变炼钢散料灰的加入量，替代率分别为 0%、20%、40%、60%、80%、100%。通过烧结杯烧结试验，检测烧结矿成品率、转鼓强度、化学成分和冶金性能指标，得出炼钢散料灰可以代替石灰石和菱镁石作熔剂制备烧结矿，并且在代替 40%的时候为最佳。

关键词　炼钢散料灰，配比，烧结矿，冶金性能指标

Study on Using Granule Material Dust in Steel-making as Sinter Mixture

Xu Pengfei, Yang Dazheng, Yu Shujuan, Zhang Dakui, Geng Jishuang,
Hou Hongyu, Qian Feng, Wang Xiangfeng

(Technology Centre of Angang Steel Co., Ltd., Anshan Liaoning 114009, China)

Abstract　In order to reuse granule material dust in steel-making, the experiment using granule material dust in steel-making as sinter mixture was tested. In this paper, granule material dust in steel-making was used as a substitute for limestone and magnesite, the rate of substitution was 0%、20%、40%、60%、80% and 100% with constant other mixture proportion. By sintering pot test, the yield of sinter, the drum strength of sinter, chemical constituents and metallurgical properties were obtained. The results show that granule material dust in steel-making can be used as a substitute for limestone and magnesite, and the best rate of substitution is 40%.

Key words　granule material dust in steel-making, proportion, sinter, metallurgical properties

离子色谱法测定钢铁工业用水中的阴阳离子

王　飞　胡绍伟　陈　鹏　刘　芳　王　永　徐鹏飞　徐　伟

（鞍钢股份有限公司技术中心，辽宁鞍山　114009）

摘　要　研究了利用直接电导检测-离子色谱法快速测定钢铁工业用中常见的 10 种阴阳离子（F^-、Cl^-、NO_2^-、NO_3^-、SO_4^{2-}、Na^+、NH_4^+、K^+、Ca^{2+}、Mg^{2+}）的方法。分析结果表明：几种离子线性关系良好，检出限为 0.01~0.13 mg/L，相对标准偏差在 0.13% ~1.93 %之间。该方法的结果准确，操作简单、快捷，对水样适应性好。

关键词　离子色谱，阴阳离子，测定，钢铁工业

Determination of Common Ions in the Water from Iron and Steel Industry by Ion Chromatography

Wang Fei, Hu Shaowei, Chen Peng, Liu Fang, Wang Yong, Xu Pengfei, Xu Wei

(Ansteel Technology Center, Anshan 114009, China)

Abstract　A method for the determination of common ions (F^-、Cl^-、NO_2^-、NO_3^-、SO_4^{2-}、Na^+、NH_4^+、K^+、Ca^{2+}、Mg^{2+}) in the water from iron and steel industry by ion chromatography using direct conductivity detection was proposed. The analysis results showed that the common ions linear relationship was good, with detection limits 0.01~ 0.13mg/L and the relative standard deviation among 0.13 % ~ 1.93%. This method was accurate, simple, fast, and has good adaptility to various water samples.

Key words　ion chromatography, cations and anions, determination, iron and steel industry

热固红泥块强度影响因素研究

刘金刚[1]　刘　成[2]　郝　宁[1]　李战军[1]

（1. 首钢技术研究院，北京　100043；2. 首秦金属材料有限公司，山东秦皇岛　066326）

摘　要　通过对某厂生产排出的三种固废：转炉 OG 泥、氧化铁皮和除尘灰进行不同配比，并部分配比采用黏结剂（SiO_2-Al_2O_3）的情况下研究得出：含除尘灰时，转炉 OG 泥、氧化铁皮可全部回收利用，除尘灰可部分回收利用，其最佳配比为 60%转炉 OG 泥+25%氧化铁皮+10%除尘灰+5%黏结剂；不含除尘灰时，适合比例的转炉 OG 泥和氧化铁皮须配加 7%黏结剂才能满足抗压强度要求。实验结果证明：除尘灰对提高抗压强度有明显作用，可将黏结剂用量由 7%降至 2%~5%。

关键词　固废，OG 泥，氧化铁皮，除尘灰，黏结剂

Study on Influential Factors of Baked-Consolidation-Block Strength

Liu Jingang[1], Liu Cheng[2], Hao Ning[1], Li Zhanjun[1]

(1. Shougang Research Institute of Technical, Beijing 100043, China;
2. Shouqin Metal Material Company Ltd., Qinhuangdao 066326, China)

Abstract　Through the study of different proportions on OG dust, iron scale and precipitator dust , partially with the binder, it can be concluded that, OG dust and iron scale can be fully recycled, precipitator dust can be partially recycled, the best ratio is 60% OG dust + 25% iron scale + 10% precipitator dust + 5% binder; When free of precipitator dust, 7% binder should be added into the certain ratio of OG dust and iron scale to meet the compressive strength. Experimental results prove: precipitator dust plays a significant role in improving the compressive strength, which reduce the binder proportion from 7% to 2% ~5%.

Key words　solid waste, OG dust, iron scale, precipitator dust, binder

鞍钢鲅鱼圈分公司达标外排水处理工艺

李成江

（鞍钢股份有限公司鲅鱼圈钢铁分公司，辽宁营口　115007）

摘　要　介绍了鞍钢鲅鱼圈钢厂生产污水处理的整体工艺，该系统采用 CAF 气浮、机械加速澄清和纤维束过滤作为主体处理工艺。在 CAF 气浮池前设置污水调节池，调节池除了具备对污水均质和均量的作用外，还在一定程度上起到沉淀和澄清的作用。CAF 气浮技术在去除钢厂污水中的悬浮物和油类等指标中起到非常明显的效果，在去

除大量 SS 的同时，还能够有效降低废水中非溶解性 COD 和 BOD_5 的含量。气浮出水进入加装斜管的机械加速澄清池，澄清池加装斜管可以大幅提高沉淀效果，使 SS 和胶体等得到进一步去除，由于澄清池外排泥中会夹带部分油类，使得澄清池对油也具备一定的去除作用。介绍了氨氮的来源和简易处理工艺，通过控制含氨氮污水的排出，采用各点收集和集中处理的方法，使出水氨氮含量达到 5mg/L 以内，实现了达标排放。

关键词 钢厂生产污水，涡凹气浮，氨氮，达标排放

Treatment Technology of External Drainage of Bayuquan Branch of Anshan Iron and Steel Group Corporation

Li Chengjiang

(Anshan Iron & Steel Group, Bayuquan Iron and Steel Company Limited, Yingkou Liaoning 115007, China)

Abstract This paper introduces the whole process of the production of wastewater treatment in Bayuquan steel plant in. The system uses the technology of the CAF air floatation, the mechanical cleaning and the fiber bundle filtration as the main body. In the CAF air floatation tank, it is provided with the function of adjusting the pool, which has the function of the homogeneous and the average amount of the wastewater, and the function of the water tank is still in a certain degree. Caf flotation technology in the removal of steel mill effluent suspended solids and oil and other indicators to very obvious effect, the removal of a large number of SS and, also can effectively reduce the non soluble COD and BOD_5 in wastewater. Flotation effluent into installation of inclined tube mechanical accelerated clarification tank, clarifier installation of inclined tube can significantly improve the effect of precipitation, the SS and colloids were further removed due to clarification pool outside the row of mud will entrain some of oil, making the clarifier of oil also has certain removal effect. The sources of ammonia nitrogen and simple treatment process by controlling the content of ammonia nitrogen in waste water discharged by the collection and centralized processing method, the effluent ammonia nitrogen content of 5mg / L within, to achieve the discharge standards.

Key words wastewater of steel production, the cavitation air flotation, ammonia nitrogen, discharging standard

鞍钢高炉冲渣水溢流问题的分析及技术改进

韩淑峰 于成忠 赵正洪 孟凡双 田业军

（鞍钢股份有限公司炼铁总厂，辽宁鞍山 114021）

摘 要 鞍钢高炉冲渣工艺多样、复杂，其中 1 号、2 号、3 号和 4 号高炉为茵芭工艺，5 号高炉为嘉恒工艺；7 高炉和 11 高炉为轮法工艺，针对不同的高炉冲渣工艺特点，分析影响高炉冲渣水溢流的问题，并提出一系列的改进措施，通过对冲渣工艺的优化，冲渣系统外来水源的有效控制和岗位操作技能的提高，实现高炉冲渣水内部动态平衡，逐步减少高炉冲渣水溢流外排，达到节能减排的目的。

关键词 冲渣水，溢流，分析，控制

The Analysis and Improvement of BF Slag Granulation Water Overflow in Ansteel

Han Shufeng, Yu Chengzhong, Zhao Zhenghong, Meng Fanshuang, Tian Yejun

(Ironmaking Work of Ansteel Co., Ltd., Anshan Liaoning 114021, China)

Abstract There are many type of BF slag granulation systems in Ansteel, which include INBA process in 1#、2#、3# and 4# BF, Jiaheng process in 5# BF and Graining process in 7# and 11# BF. The reasons to slag granulation water overflow have been analyzed accordingly and improvements have been carried out, which include BF slag granulation optimum, the efficient control of slag granulation water resource and improvement of operating skill. The dynamic balance for slag granulation water is established in ironmaking work, the overflow of slag granulation water has been reduced gradually and the aim of energy saving and emission reduce is achieved.

Key words slag granulation water, overflow, analysis, control

鞍钢第二发电厂锅炉吸风机变频改造方案分析

李汉儒 李宝山 吴 猛

（鞍山钢铁集团公司第二发电厂，辽宁鞍山 114012）

摘 要 对以合同能源管理模式实施的变频改造项目的设计方案进行重点分析，提出了节能改造方案的制定对改造成本的影响，而节能量的多少又是能否以合同能源管理模式实施的关键。进而分析了在进行方案设计时应注意的问题，为节能项目合同能源管理模式的实施提供了技术参考。

关键词 合同能源管理, 变频改造, 方案设计, 节能量

Analysis on the Frequency Conversion Transformation for the Boiler Section Fan of Ansteel Second Power Plant

Li Hanru, Li Baoshan, Wu Meng

(Ansteel Second Power Plant, Anshan Liaoning 114012, China)

Abstract The design scheme of frequency conversion transformation project implemented basing on the contract energy management mode is emphatically analyzed, putting forward the influence towards the transformation cost made by the formulation of energy-saving transformation scheme. The energy-saving quantity is the key factor for the implementation of contract energy management mode. And the problems that should be paid attention when designing the scheme are further analyzed. It can provide technical reference to the implementation of contract energy management mode.

Key words contract energy management, frequency conversion transformation, scheme design, energy saving quantity

空分设备冷状态下的快速启动、避免氧气纯度波动的方法

徐作宇 于 泳 董昕宏

（鞍钢股份能源管控中心，辽宁鞍山 114021）

摘　要　简介了鞍钢引进的35000m³/h内压缩流程空分设备的流程特点和性能参数，简要分析了目前空分设备冷状态下启动时间长、投粗氩塔氧气纯度出现波动的原因，提出了缩短启动时间及避免氧气纯度波动的方法。

关键词　大型空分设备，内压缩流程，冷状态，快速启动，氧气纯度波动，冷量

Cold Air Sepearation Equipment of Quick Start to Avoid the Method of Oxygen Purity Fluctuations

Xu Zuoyu, Yu Yong, Dong Xinhong

(Ansteel Energy Control Center, Anshan Liaoning, 114021, China)

Abstract　Introduction to the 35000m³/h Angang introduction to the internal compression process of the air separation equipment of process characteristics and performance parameters, a brief analysis of the at present in the air separation equipment, cold state start for a long time, investment and crude argon column oxygen purity fluctuation reasons, proposed to shorten the startup time and avoid the oxygen purity fluctuation method.

Key words　large air separation equipment, internal compression process, cold condition, quick start, oxygen purity fluctuations, cooling capacity

鞍钢外排废水达标排放技术研究

白旭强[1]　龙海萍[1]　刘　芳[2]

（1. 鞍钢股份有限公司能源管控中心，辽宁鞍山　114021；
2. 鞍钢股份有限公司技术中心，辽宁鞍山　114009）

摘　要　钢铁工业在工业领域里是耗水和排污大户，如何加强污水治理，提高水的利用率是当前研究的重点。鞍钢在水系统优化的基础上，采用将高污染的焦化废水、煤气洗涤水等集中处理后分级使用，并将南大沟水回收至西大沟进一步处理的新方法，可明显提高外排水水质，使废水排放指标氨氮<5mg/L、COD<50mg/L，符合国家和辽宁省规定的排放标准，实现达标排放。

关键词　污水治理, 水利用率, 氨氮, COD

A Research for Ansteel Emissions Wastewater Discharging Standard Technology

Bai Xuqiang[1], Long Haiping[1], Liu Fang[2]

(1. Energy Management & Control Centre of Angang Steel Co., Ltd., Anshan Liaoning 114021, China;
2. Technology Center of Angang Steel Co., Ltd., Anshan Liaoning 114009, China)

Abstract　Iron and steel industry is the most water consumption and heavy emitters in the field of industrial, how to strengthening sewage disposal and improve the utilization rate of water is the focus of current research. Base on the water system optimization of ansteel, using a new method which classification after focus by high pollution of the coking waste water and coal gas washing water , then take south ditch water recycling to west ditch for further processing, and drainage water quality improved obviously. That conforms to the emission standards of countries and provisions, ammonia nitrogen<5mg/L、COD<50mg/L, and realized discharging standard.

Key words sewage disposal, water utilization, ammonia nitrogen, COD

烧结烟气脱硫灰 $CaSO_3$ 转化试验研究

耿继双 王东山 张大奎 徐鹏飞 钱 峰 侯洪宇 杨大正

（鞍钢股份有限公司技术中心，辽宁鞍山 114009）

摘 要 分析了鞍钢烧结烟气脱硫灰的化学成分及再利用的影响因素，按照其特性及资源化利用限制因素，分别进行随温度、搅拌时间、氧化剂量及催化剂等对脱硫灰 SO_3^{2-} 转化影响试验，确定适合工业化应用的技术方案及结论意见，实现 $CaSO_3$ 向 $CaSO_4$ 低成本及有效的转化，解决制约烧结烟气脱硫灰的资源化利用的障碍。

关键词 烧结烟气，脱硫灰，资源化利用

The Research on the $CaSO_3$ Translation in Sintering Flue Gas Desulfurization Ash

Geng Jishuang, Wang Dongshan, Zhang Dakui, Xu Pengfei, Qian Feng, Hou Hongyu, Yang Dazheng

(Technology Center of Angang Steel Co., Ltd., Anshan 114009, China)

Abstract We analyzed the chemical composition and utilization factors of the desulfurization ash in Ansteel. According to the infection factors of the ash utilization, we did several experiments such as the infection of temperature, stirring time, oxidizing agent amount, catalyzer amount on the SO_3^{2-}. The most suitable method was found. The translation from $CaSO_3$ to $CaSO_4$ was carried out with low cost through a effective way.

Key words sintering flue gas, desulfurization ash, utilization

热轧油泥清洗试验研究

杨大正[1] 齐殿威[1] 杨立军[2] 马光宇[1] 耿继双[1] 徐鹏飞[1] 张大奎[1]

（1. 鞍钢股份有限公司技术中心，辽宁鞍山 114009；2. 鞍钢股份热轧厂，辽宁鞍山 114009）

摘 要 本文介绍了热轧油泥中油的来源、种类和目前油泥的处理方法,采用常温清洗方法去除油泥中的油分，在实验室正交实验的基础上进一步进行了中试试验，试验结果表明：在常温、pH 值为 8~10、油泥与水重量比为 1：4 的情况下，按每吨油泥加入 12kg 6501、12kg 的 LAS 和 9kg 的 STPP，可使含油率为 16%的热轧油泥洗后含油率小于 5%，作为烧结原料利用，同时废油回收利用。清洗每吨油泥效益为 518 元。本试验采用的油泥湿法除油技术，不仅避免了传统油泥焙烧造成的烟气二次污染、而且处理成本较低，为大量堆存的含油率高达 16%的热轧油泥循环

利用找到了新路。

关键词 热轧，油泥，清洗

Experimental Study on Cleanout of Oil Sludge Bearing Iron Scales Resulted in Hot Rolling

Yang Dazheng[1], Qi Dianwei[1], Yang Lijun[2], Ma Guangyu[1], Geng Jishuang[1], Xu Pengfei[1], Zhang Dakui[1]

(1. Technology Center of Angang Steel Co., Ltd., Anshan 114009, China;
2. Hot Rolled Mill of Angang Steel Co., Ltd., Anshan, 114009, China)

Abstract The origins and kinds of oil in oil sludge bearing iron scales and the methods for dispose of oil sludge bearing iron scales were introduced. Based on the orthogonal experiment for removing the oil content in oil sludge by cleaning method at normal temperature in the laboratory the pilot testing for removing the content was carried out then. Experimental results show that the oil length of oil sludge bearing iron scales resulted in hot rolling with the oil length of 16% which was cleaned can be controlled up to 5% by adding 12kg of 6501, 12kg of LAS and 9kg of STPP into per ton of oil sludge at room temperature provided that the pH value is between 8 and 10 and the ratio of oil sludge and water is 1 to 4. And thus the obtained iron scales can be recycled as sintering raw materials and waste oil can also be recycled. According to the calculation the economic benefit of 518 yuan can be achieved by cleaning per ton of oil sludge. The wet process for oil removal was used in experiments introduced in the paper, which not only avoids the secondary pollution caused by the fume produced by application of the traditional technology for roasting oil sludge, but also the cost for dispose of oil sludge is relatively low. So by this the new way for recycling a large amount of oil sludge resulted in hot rolling with the oil length as high as 19% stockpiled was found.

Key words hot rolling, oil sludge bearing iron scales, cleanout

鞍钢火车受料槽扬尘治理

李小丽 孙兴鹤

（鞍钢股份有限公司炼铁总厂，辽宁鞍山 114021）

摘 要 为解决钢铁行业火车受料槽卸料时粉尘污染，本文介绍了火车受料槽卸料时的产尘机理，通过理论分析和计算，合理确定了受料槽卸料除尘的密闭方式、抽风量的大小、抽风点的位置及数量等。该技术用于鞍钢三烧原料区域受料槽扬尘治理取得了良好的除尘效果。

关键词 受料槽，扬尘，产尘机理，密闭，抽风量，效果

Train Receiving Groove by Dust Control in Anshan Steel Company

Li Xiaoli, Sun Xinghe

(Ironmaking Plant of Anshan Steel Co., Ltd., Anshan 114021, China)

Abstract　In the iron and steel industry, in order to solve the dust pollution when discharging goods through the chute, this paper introduces the mechanism of dust producing during discharging. According theoretical analysis and calculation, this paper figures out some reasonable data including sealed type of dedusting, the wind pressure of exhaust fan, the quantity and positions of air exhausters, etc. The technology used to solve the dust pollution during chute-discharging in AISC No.3 sintering workshop has obtained some efficient achievements in the dedusting aspect.

Key words　the chute, dust, the mechanism dust producing, air exhauster, volume, effectiveness

高炉-转炉钢铁生产流程碳排放强度分析

张天赋　马光宇　李卫东　贾　振　徐　伟　王东山

（鞍钢股份技术中心，辽宁鞍山　114009）

摘　要　本文根据高炉/转炉钢铁生产流程的特点，确定出碳排放强度的计算边界和计算方法，并以某 500 万吨钢生产基地的碳排放强度进行了计算。由计算结果可看出：高炉/转炉钢铁生产流程碳排放源主要为煤，约占总排放的 90%；通过含碳物质的外销和余热余能的回收利用可减少钢铁企业碳排放。

关键词　高炉-转炉工艺，碳排放，余能回收

Carbon Print Analysis in BF-BOF Route Steel Industry

Zhang Tianfu, Ma Guangyu, Li Weidong, Jia Zhen, Xu Wei, Wang Dongshan

(Technology Center of Ansteel Co., Ltd., Anshan Liaoning 114009, China)

Abstract　Based on the BF-BOF route process in steel industry, the CO_2 calculation boundary and method have been defined and the CO_2 emission intensity of a 500 million ton steel production base is calculated. The result shows that the main CO_2 emission comes from coal consumption in steel industry which accounts for about 90% of the total emission and the materials containing carbon and residual energy recovery and reuse could be the efficient way to cut down CO_2 emission in steel industy.

Key words　BF-BOF process, carbon emission, residual energy recovery

钢铁企业大气污染物产排源分析

王　珲

（中冶建筑研究总院有限公司，北京　100088）

摘　要　本文以一个年产 1000 万吨全流程钢铁企业为例，调研、计算了原料、烧结、球团、炼焦、炼铁、炼钢、轧钢、自备电站和石灰窑等工序的烟粉尘、SO_2 和 NO_x 的年产、排量。在此基础上分析了钢铁企业内烟粉尘、SO_2 和 NO_X 的重点产排工序及其占全厂的比例。同时，总结了钢铁企业大气污染治理设施在运行中存在的问题，提出了对烟气净化设施运行状态进行在线监控的建议。

关键词　钢铁企业，烟粉尘，SO_2 和 NO_x，产排源，防控现状

Analysis on Emission Sources of Air Pollutants in Iron and Steel Enterprise

Wang Hui

(Central Research Institute of Building and Construction Co., Ltd., MCC, Beijing 100088, China)

Abstract Dust, SO_2 and NO_X annual emissions of raw material field, sintering, pellet, coking, blast furnace, converter, rolling, captive power plant and lime kiln in a typical China iron and steel enterprise were researched and accounted. The outputs and emissions of three air pollutants in each process were calculated. The most improtant processes of dust, SO_2 and NO_X emissions were analyzed. And then, the some problems and suggestions of air pollutants control and monitoring in iron and steel industry were gived.

Key words iron and steel enterprise, dust, SO_2 and NO_X, emission sources, control situation

电凝并增效在烧结机头烟尘净化中的应用

何 剑[1] 刘道清[2] 周茂军[2] 徐国胜[1]

(1. 西安理工大学，陕西西安 710048；2. 宝山钢铁股份有限公司，上海 201900)

摘 要 本文提出了一种新型双极荷电凝并方法及装置，采用荷电与凝并一体化的思路，实现了全电场凝并，并在宝钢一烧机头烟尘净化增效改造中成功应用，测试结果显示：新增凝并器对主系统生产无影响，与电凝并器停运相比，电凝聚器正常工作时，电除尘器后 PM_{10} 排放量下降了 48.90%，$PM_{2.5}$ 下降了 51.94%，PM_1 下降了 91.17%，总尘排放浓度下降了 22.6%。表明该凝并器能够显著提高电除尘器的除尘效率，尤其是对微细粉尘的净化率。

关键词 电凝并，微细粉尘，增效，净化

Application of Electrostatic Agglomerator in Auxiliary Dust-collection for Sintering Machine Head

He Jian[1], Liu Daoqing[2], Zhou Maojun[2], Xu Guosheng[1]

(1. Xi'an University of Technology, Xi'an 710048, China;
2. Baoshan Iron and Steel Co., Ltd., Shanghai 201900, China)

Abstract An new method and device of fine particle bipolar charging and agglomeration was presented, in which particles charged and agglomerated simultaneously, and agglomeration happened throughout the electrostatic field. The application in the renovation of dust-collection system for 1# sintering machine of Baoshan Iron and Steel Co., Ltd has achieved ideal effect. The measurements showed addition of electrostatic agglomerator has little effect on sinter production, PM_{10} reduced by 48.9% and $PM_{2.5}$ by 51.94, PM_1 by 91.17%, dust emission by 22.6% at the outlet of electrostatic precipitator after electrostatic agglomerator put into operation, which proved that electrostatic agglomerator can markedly improve the collection efficiency of not only electrostatic precipitator but fine particles especially.

Key words electrostatic agglomeration, fine particle, efficiency enhancing, purification

烧结烟气综合治理技术现状与展望

石　磊　李咸伟

（宝山钢铁股份有限公司研究院，上海　201900）

摘　要　烧结废气温度偏低、废气量大、污染物含量高且成分复杂，是钢铁行业低温余热利用和废气治理的难点和重点。通过几十家典型钢企、十余种代表性治理工艺的现场调研，比较了国内烧结烟气脱硫设施建设、运行的基本情况及技术经济性，提出了烧结烟气综合治理技术的发展方向，最后介绍了宝钢近年来在烧结烟气治理领域的进展。

关键词　烧结烟气，综合治理，脱硫，脱硝

Comprehensive Treatment Technlogy Status and Prospect on Sintering Flue Gas

Shi Lei, Li Xianwei

(R&D Center, Baoshan Iron & Steel Co., Ltd., Shanghai 201900, China)

Abstract　For the sake of low temperature, large waste gas volume, high pollutants content and complicated compositions, waste heat recovery and emission gas treatment has always been a tough problem in steel industry, which has attracted widespread attention at home and abroad. Through investigations on dozens of typical steel enterprises and representative treatment processes, construction, operation and technical economy indicators of domestic flue gas desulphurization facilities are compared, then sintering flue gas treatment technology trend is proposed, finally, related research and industrial applications of Baosteel sintering flue gas treatment is introduced.

Key words　sintering flue gas, comprehensive treatment, desulphurization, denitration

链算机-回转窑生产永磁铁氧体预烧料的探索与实践

卜二军　刘红艳　孙胜英

（河北钢铁集团邯钢公司技术中心，河北邯郸　056015）

摘　要　本文介绍了邯钢采用链算机-回转窑生产铁氧体预烧料的主要工艺技术。从铁鳞的预处理、配方优化、添加剂量、预氧化温度、预烧温度、预烧料质量等方面进行控制，并讨论了铁鳞成分和粒度对永磁铁氧体预烧料性能的影响。锶永磁铁氧体磁性能稳定达到 Br：400mT，Hcb：240kA/m，Hcj：250 kA/m，（BH）max：31.5kJ/m^3以上。

关键词　铁鳞，铁氧体预烧料，链算机-回转窑，磁性能

Exploration and Practice of Permanent Magnetic Ferrite Pre-sintered Materials by Chain Grate - Rotary Kiln

Bu Erjun, Liu Hongyan, Sun Shengying

(Technology Center of Hansteel Group of HBIS，Handan Hebei, 056015, China)

Abstract This paper introduces the main technology of permanent magnetic ferrite pre-sintered materials by chain grate-rotary kiln in Hansteel Group.The key techniques including pre-treatment of iron scale，ingredient optimization，additive usage amount，the temperature of pre-oxidation and pre-sintered， product quality are controled. And the effects of composition and particle size on the iron scale for pre-sintered materials magnetic properties are discussed. In the production the final magnetic performance stably achieve Br：400 mT， Hcb：240 kA/m，Hcj：250 kA/m，（BH）m：31.5 kJ/m^3.

Key words iron scale, permanent magnetic ferrite pre-sintered materials, chain grate-rotary kiln, magnetic properties

烧结烟气二噁英减排综合控制技术研究

俞勇梅 李咸伟 王跃飞

（宝山钢铁股份有限公司，上海 201900）

摘 要 新的《钢铁烧结、球团工业大气污染物排放标准》要求2015年1月1日起，烧结机头烟气二噁英排放要小于 0.5ng-TEQ/m^3。本文分别从源头抑制、过程控制以及末端治理的角度，对一些减排控制技术以及相关试验结果进行了整理和归纳。试验结果表明，通过控制抑制剂的添加量，烧结过程中二噁英源头抑制技术可减少 50%以上的二噁英生成量；烧结废气循环技术也具有较好的二噁英减排效果，当废气循环量为20%左右时，可减少35%左右的二噁英排放量；携流式吸附脱除技术是一种行之有效的末端处理技术，通过控制吸附剂的喷入量，二噁英的去除效率可达70%以上。

关键词 PCDD/Fs，二噁英，循环烧结，抑制剂

The Study of Reduction in Dioxin Emissions of Iron Ore Sintering

Yu Yongmei, Li Xianwei, Wang Yuefei

（Baoshan Iron & Steel Co., Ltd., Shanghai 201900, China）

Abstract According to the Emission Standard of Air Pollutants For Sintering And Pelletizing of Iron And Steel Industry, the new limit of Dioxin emissions of sintering is 0.5ng-TEQ/m^3. It has been implemented on Jan.1st 2015. The technologies of Dioxin reduction and the results of industrial test are shown in this article. The suppression technologies from the source of sintering showed over than 50% of the formation of Dioxin were successfully suppressed during the sintering process. Selected waste recirculation is beneficial to reduce the emission of Dioxin. Almost 35% Dioxin was reduced when the volume of waste gas was around 20%. The technology of Absorbent Injection is an end-of-pipe treatment and effective to control the Dioxin emission. The maximum removal rate 50% is reached depends on the volume of absorbent injection.

Key words　PCDD/Fs, dioxin, sintering, inhibitors

活性炭法烧结烟气净化技术研究及应用

叶恒棣　魏进超　刘昌齐

（中冶长天国际工程有限责任公司，湖南长沙　410007）

摘　要　活性炭法烟气净化技术由于其技术的先进性已成为钢铁企业实现绿色及可持续发展的一个重要技术选择。活性炭法具有高效脱硫脱硝脱二噁英及重金属性能，其工艺简单，占地面积小，系统无二次污染物产生且副产物可回收利用。相比于“半干法+SCR”，总投资更具有优越性。本文研究了新型低温炭基催化剂，研制了具有多床层模块化结构的脱除塔、集加热与冷却于一体的活性炭再生塔及具有多点卸料功能的链式斗提机，并对系统进行了仿真研究，结果表明该系统流场均匀，达到设计要求，能实现稳定运行。目前中冶长天已与宝钢湛江钢铁有限公司签订了2台550m^{2}烧结机烟气净化系统工程承包合同，随着该技术成本的降低和产业化的进一步推广，将会产生巨大的社会及经济效益。

关键词　活性炭，活性焦，烟气净化，烧结

Research and Application of Activated Carbon Method for Sintering Flue Gas Purification

Ye Hengdi, Wei Jinchao, Liu Changqi

(Zhongye Changtian International Engineering Co., Ltd., Changsha Hunan 410007, China)

Abstract　Activated carbon (AC) method sintering flue gas purification, as an advanced technology, is considered as an important option for green and sustainable development of steel and iron enterprises. AC method flue gas purification technology could remove SO_2, NOx, dioxin and heavy metals from flue gas with high efficiency. Its process is simple and requires small area. What's more, the system does not generate secondary pollutants and the by-products could be recycled. Compared with “semidry desulfurization + SCR”, AC method is more advantageous in total investment. In this article, a new type carbon based catalyst is studied; key equipments such as modular structured removal tower with multi bed layer, AC regeneration tower with heating and cooling function, chain bucket elevator with multi discharge points are studied and manufactured; simulation study of the whole system is carried out. Simulation study result shows that the system has a uniform flow field and that it can meet the design requirements and achieves stable operation. At present, Zhongye Changtian International Engineering Co., Ltd and Baosteel Zhanjiang Engineering Co., Ltd has signed contract for 2 550m^2 sintering machine flue gas purification project. Huge social and economic benefits would be generated as this technology's cost reduces and its industrialization is further promoted.

Key words　activated carbon, activated coke, gas purification, sintering

邯钢高炉煤气干法除尘创新及实践应用

胡雷周　刘铁岭　邱　耕

（河北钢铁股份有限公司邯郸分公司设备动力部，河北邯郸　056015）

摘 要 邯钢8号高炉建设时，配套新建了煤气干法布袋除尘：布袋系统均为双排布置，选择低压长袋氮气脉冲布袋除尘器，低压氮气脉冲反吹形式，使用气力输灰工艺，并联新建一套调压阀组和消音器，新建除盐脱盐塔及配套小泵房。经过运行经验摸索累计、吸收消化，经过改革、技术创新，设备工艺较为成熟，具备推广条件；邯钢还有4号、5号、7号高炉使用的湿法煤气除尘，对其他高炉除尘系统进行了改干法除尘技术升级节能改造，避免了常规干法除尘系统存在的缺陷，收到了良好效果。同时为提高发电量，对湿法TRT进行干法改造。干法除尘优点：能耗低、占地小、污染小，配套TRT发电量比湿法提高25%~35%，经济效益好。

关键词 高炉煤气，除尘，脱盐塔

Hansteel Blast Furnace Dry-type Dedusting System Innovation and Practice

Hu Leizhou, Liu Tieling, Qiu Geng

(Handan Steel of Hebei Iron & Steel Group, Handan 056015, China)

Abstract Hansteel adopted the dry-type bag dedusting system in 8# blast furnace. This bag system, arranged in two rows, had one low-pressure long bag nitrogen pulse dust filter, which blew ash with low-pressure nitrogen pulse in opposite direction. the dry-type bag dedusting systemalso had a pressure regulating valve, a silencer, a desaltingtower and a small pumping station. The equipment was improved to be mature by experience accumulation and technical innovation in its test running and could be promoted. The upgrade of 4#5#7#blast furnace withwet-type dedusting system could avoid the defect of conventional dedusting technology. Meanwhile technological remould by replacing wet-type system with dry-type system could increase the generating capacity by 25%~35%. This dry-type bag dedusting method has advantages of low energy consumption, small footprint, less pollution, and good economic benefits.

Key words BFG, dust practical, desalting column

从智能电网到能源互联网及对宝钢电能使用的启示

陈阿平

（宝钢金属有限公司，上海 200940）

摘 要 能源互联网将代表未来信息与能源-电力技术深度融合的必然趋势，是新一代工业革命大潮的重要标志，也是智能电网的扩展和延伸。能源互联网借鉴互联网思维和理念，构建新型信息-能源融合“广域网”，它以大电网为“主干网”，以微电网、分布式能源等能量自治单元为“局域网”，以开放对等的信息-能源一体化架构，真正实现能量的双向按需传输和动态平衡使用，因此可以最大限度的适应新能源的接入。本文在介绍能源互联网的基本概念、架构与组成的基础上，汇总分析了国内外能源互联网的研究现状和发展趋势，最后，探讨了能源互联网对宝钢电能使用的启示。

关键词 能源互联网，智能电网，能源路由器，新能源，微电网

Abstract Energy Internet will represent the inevitable trend of deep integration about information and energy and electricity technology, It is an important symbol of new generation of industrial revolution, and also is the expansion and extension of the smart grid. Energy Internet build new WAN(Wide Area Network) of integration of information and energy, It truly realize the two-way on-demand transmission and the use of dynamic balance of energy, with power grid as the "backbone", and with energy autonomous unit of micro-grid and distributed energy as LAN(Local Area Network),and with open peer integration framework of information and energy . therefore, it can uttermost join up the new energy. This paper introduced the basic concept, framework and composition of energy Internet, summarized the research status and development trend of domestic and overseas energy internet, Finally, this paper discussed energy internet on baosteel electricity use.

Key words energy internet, smart grid, energy router, new energy, micro-grid

热轧主轧机冷却风机节能改造的研究和应用

陈 枫

（上海宝菱电气控制设备有限公司，上海 201900）

摘 要 随着全球钢铁行业需求增长放缓，钢铁行业面临产能过剩、环境保护等发展和调整的共同难题。需要我们通过节能减排、能源管理、流程优化等措施推进钢铁工业的清洁化生产。本文以宝钢 1880 热轧产线节能为实际案例，浅析热轧产线上节能改造的应用。热轧产线上用电和节电要从技术和管理两方面着手节约用电，提高用电效率。以主电机冷却散热为切入口，可采用优化流程、变频调速改造等方法，减少冷却风机满负荷长时间运转造成的电能损耗，使其在多数情况下运行在经济运行点上，合理优化供电系统，合理分配与平衡负荷，优化运行。项目 2014 年下半年投运后，性能稳定可靠，节能效果良好，节能率大于 60%，年节电 320 万度，达到合同要求，满足性能考核指标，具备标准化广泛推广应用的条件。

关键词 节能改造，变频调速自动化控制

浅谈黑体技术在棒线加热炉上的应用案例分析

杨文滨 黄 立 孔令斌

（上海宝钢节能环保技术有限公司，上海 201999）

摘 要 针对某钢厂棒线加热炉炉膛内采用黑体元件节能技术，即在炉膛内布置一定数量的黑体元件，增大炉内传热面积，提高炉膛发射率，更能对炉膛内的热射线进行有效的调控，使之从漫射的无序状态调控到有序，直接射向钢坯，提高对钢坯的辐照度和热射线的到位率，强化辐射传热，使炉温均匀性增强，炉衬的使用寿命延长，最终达到节约能源、减少排放的目的。根据该加热炉膛的内部结构，共计布置黑体元件 15432 只，对改造前后的实际生产数据进行比较，可知改造前后平均单耗下降 12%，全年创效在 250 万元以上，取得了很好的节能效果，具有很好的推广价值。

关键词 黑体，加热炉，节能率

The Application Result of Black-body Technology in the Wire Rod Heating Furnace

Yang Wenbin, Huang Li, Kong Lingbin

(Shanghai Baosteel Energy Service Co., Ltd., Shanghai 201999, China)

Abstract Black body technology could be used to manage the heat rays in the firing space of furnace. The black body installed in the firing space could enlarge the heat transfer area and increase the emissivity rate of the firing space. In the meantime, black body could help effectively control the heat rays from randomly diffusion into orderly focusing onto the steel billet. This will help increase the irradiation of steel billet and the arrival rate of heat rays, therefore strengthen the heat transfer by radiation and save energy significantly. The actual production data after using this technology in the wire rod heating furnace show that the energy saving rate reached 12% with a good energy saving effect compared with that without implement of this technology.

Key words black body, heating furnace, energy saving rate

轧钢生产能源介质供应的节能降耗技术现状

徐言东[1] 张战波[2] 王利伟[3] 程知松[1] 詹智敏[1]

（1. 北京科技大学高效轧制国家工程研究中心，北京 100083；
2. 北京中冶设备研究设计总院有限公司，北京 100029；
3. 鞍山灵山工业经济区管理委员会，辽宁鞍山 114021）

摘 要 本文对轧钢生产能源介质供应的节能降耗技术进行了描述，指出了各类技术的优缺点，提出了“解决轧钢这个最后一道工序的节能降耗，需因地制宜采用不同的技术”这一观点，使用户以较少的投资获得最佳的节能效果。

关键词 轧钢生产，能源介质，节能降耗

Actuality of Energy-saving and Consumption-reducing Technology in Steel Rolling Energy Sources and Medium Supply

Xu Yandong[1], Zhang Zhanbo[2], Wang Liwei[3], Cheng Zhisong[1], Zhan Zhimin[1]

(1. NERCAR of University of Science and Technology Beijing, Beijing 100083, China;
2. China-Metallurgy Equipment Design and Research Institute Co., Ltd., Beijing 100029, China;
3. Anshan Lingshan Industrial Economy Management Committee, Anshan Liaoning 114021, China)

Abstract This paper describes energy-saving and consumption-reducing technology in steel rolling energy sources and medium supply system, points out various technology advantage and disadvantage, and puts forward a viewpoint that to implement energy-saving and consumption-reducing in steel rolling, various technology will be used according to local conditions. So the enterprises can use less investment to achieve better energy-saving result.

Key words steel rolling, energy sources and medium, energy-saving and consumption-reducing

钢渣水洗尘泥制陶粒的试验研究

李灿华 刘 思 焦立新

（武汉钢铁集团金属资源有限责任公司，湖北武汉 430082）

摘 要 以钢渣水洗尘泥为原料，配以 30%粉煤灰、30%黏土为辅料，经搅拌、成球、干燥、500℃预热 10min、1080℃焙烧并保温 15 min，成功烧制的陶粒堆积密度 539kg/m^3，表观密度 1063kg/m^3，吸水率 2.4%，筒压强度 4.5MPa，满足 GB/T 17431—2010 规定的普通轻集料 600 级的技术要求。而且，利用钢渣尘泥烧制陶粒浸出液中的重金属浓度远远低于《危险废物鉴别标准 浸出毒性鉴别》GB 5085.3—2007 所规定的最高允许浓度，说明经过高温焙烧后的陶粒对重金属产生了很好的固化效果。

关键词 钢渣水洗尘泥，制备，陶粒，焙烧

Experimental Research on Preparation of Ceramsite Using Steel-slag Sludge

Li Canhua, Liu Si, Jiao Lixin

(Metallurgical Resource Co., Ltd., of WISCO, Wuhan 430082, China)

Abstract The high-strength light weight ceramsite was successfully produced by using steel-slag sludgemainly, together with fly-ash (30%) and clay(30%) for the accessories. It was prepared by a series ways: mixed and squeezed into a ball, dried, preheated (500℃,warnl-up 10min), calcined (calcination temperature of 1080℃and thermal insulation 15 min). The results show that it can meet the state standard GB/T 17431—2010 level 600 with its packing density of 539kg/m^3, apparent density of 1063kg/m^3, coefficient of water absorption of 2.4%, and cylinder compressive strength of 4.5MPa. Further, the heavy metal ions in the ceramsite leaching concentration were far below the maximum allowable concentration of the standard GB 5085.3—2007 and it shows that ceramsite roasted by high temperature has very good curing effect on heavy metal ions.

Key words steel-slag sludge, preparation, ceramsite, calcination

高炉熔渣直接纤维化制备矿渣棉实验研究

龙 跃[1,2] 杜培培[1,2] 李智慧[1,2,3] 张良进[1,2] 张玉柱[1,2]

（1. 华北理工大学冶金与能源学院，河北唐山 063009；
2. 华北理工大学现代冶金技术教育部重点实验室，河北唐山 063009；
3.东北大学材料与冶金学院，辽宁沈阳 110004）

摘 要 通过利用喷吹法和四辊离心法制备矿渣棉，研究在不同制备方法条件下，矿渣棉其化学稳定性、表观特征、含水率及渣球含量与酸度系数之间的关系。结果表明：在满足成纤要求的前提下，高炉渣酸度系数应控制在 1.2~1.4；利用喷吹工艺制备的矿渣棉直径较细，含水率较低，成纤率较低，渣球含量较高纤维质量较差；相对于喷吹工艺离心法制备的矿渣棉直径相对较粗，含水率大，渣球含量较少，成纤率较高，纤维质量较好。

关键词 高炉渣，矿渣棉，喷吹，离心

Experimental Study of Blast Furnace Slag Slag Wool Preparation Directly Fibrosis

Long Yue[1,2], Du Peipei[1,2], Li Zhihui[1,2,3], Zhang Liangjin[1,2], Zhang Yuzhu[1,2]

(1. School of Metallurgy and Energy, North China University of Science and Technology, Tangshan, 063009, China; 2. Modern Metallurgical Technique Key Laboratory of Ministry of Education,North China University of Science and Technology, Tangshan, 063009, China; 3. College of Material and Metallurgy, Northeastern University, Shenyang, 110004, China)

Abstract The slag cotton was prepared by spray blowing and four centrifugal centrifugal preparation. Under different methods, the relationship between slag wool, chemical stability, surface characteristic, water content and slag content and the acidity coefficient is studied. The results showed that under the premise of meet the requirements of fiber forming, the acidity coefficient of blast furnace slag should be controlled at 1.2~1.4, the diameter of slag cotton of blowing process is smaller, the water content is low, but the content of slag ball is high, the diameter of slag cotton of centrifugation is relatively coarse, the water content is high, the content of slag ball is low.

Key words blast furnace slag, slag wool, injection, centrifugal spinning

加速溶剂萃取-气相色谱/质谱法测定土壤中多环芳烃

凌 冰 黄 晓 周宏明

（宝钢工程集团宝钢工业技术服务有限公司，上海 201900）

摘 要 本文建立了加速溶剂萃取-气相色谱 / 质谱联用检测土壤中多环芳烃的方法，优化了试验条件，该方法的回收率在 80% ~105% 之间，多环芳烃的检出限在 1.6~4.8μg/kg 之间，通过对实际样品的测定表明，本方法能够满足土壤分析的要求。

关键词 加速溶剂萃取，多环芳烃，气质联用

Determination of Polycyclic Aromatic Hydrocarbons in Soil by Accelerated Solvent Extraction Eoupled with GC-MS

Ling Bing, Huang Xiao, Zhou Hongming

(Baosteel Environment Monitoring Center, Shanghai 201900,China)

Abstract 16 kinds of Polycyclic Aromatic Hydrocarbons in soil were determined by GC-MS was established. The detective condition was optimized. The matrix spike recovery ranged from 80％ and 105％ . The Method detection limit was between 1.6~4.8μg/kg. The results of actual samples determination shows the method can meet the demand of soil analysis.

Key words acceleration solvent extraction, polycyclic aromatic hydrocarbons, GC-MS

强化烧结烟气 SO_2 富集的生产实验

何 峰[1] 杜 力[1] 富田武[2]

（1. 华菱湘潭钢铁有限公司，湖南湘潭 411101；
2. 中日合作“大气氮氧化物总量控制项目”，北京 100012）

摘 要 湘钢新二烧在维持生产参数基本不变的条件下，利用烧结机尾部可切换风箱改变烟气导向，在烧结终点前 SO_2 浓度仍较高的风箱段增加导入脱硫烟道的风量，实现了烧结烟气 SO_2 进一步向脱硫烟道富集，从而有效降低了非脱硫烟道中 SO_2 浓度，得到了非脱硫烟道烟气 SO_2 浓度低于新排放限值的有益结果，对于目前在役的主流大型烧结机为应对新排放标准而对非脱硫烟道追加脱硫设施提供了新的低成本的技术思路。

关键词 烧结，选择性脱硫，强化富集，新排放标准

Productive Experiment to Intensify the Beneficiation of SO_2 in Sinteing Gas

He Feng[1], Du Li[1], Takeshi Tomita[2]

(1. Valin Xiangtan Iron & Steel Co., Xiangtan Hunan 411101, China;
2. China-Japan Cooperated “The Project for Total Emission Control of Nitrogen Oxide in Atmosphere”, Beijing, 100012, China)

Abstract By utilizing the switchable wind boxes on tail of sintering machine to change the flow direction of the exhaust gas, and with keeping main manufacturing parameters as usual, Xiangtan Iron & Steel Co., on their New No.2 sintering machine, tested to increase the ratio of the exhaust gas leading in desulfurization duct just before the end of sintering, which gas contained with still relatively high SO_2 concentration, so, they intensified the beneficiation of SO_2 in sintering gas into desulfurization duct, and then the SO_2 concentration of the sintering gas in non-desulfurization duct was correspondingly reduced effectively, and obtained an useful result that the SO_2 concentration of non-desulfurization duct was decreased to below the new emission limit of SO_2 concentration according to national standard. However, for all the main large sintering machines in-service tending to be added new desulfurization devices for its non-desulfurization duct to meet the national new emission standard, this work showed a new and low-cast technological choice.

Key words sintering, selective desulphurization, intensified beneficiation, new emission standard

Application of Hydrothermal Treatment and HHP for Recycling of Steel Industry by-Products

Jung Woo-Gwang[1], Gu Bong-Ju[1], Kang Ki Seong[2]

(1. Department of Advanced Materials Engineering, Graduate School of Kookmin University

77 Jeongneung-ro, Seongbuk-gu, Seoul, 136-702, Korea;
2. Dong Do Basalt Ind. Co., Gyeongju-si, 780-802, Korea)

Abstract In order to investigate the possibility on the developing higher value-added products by recycling steel industry slags, the hydrothermal treatment and hydrothermal hot pressing were madeusing BF slag and quartz or flyash.The characterization of fabricated pellets was made on the mechanical strength and micro hardness, as well as XRD analysis and SEM observation.In the hydrothermal treatment, the tobermorite ($5CaO•6SiO_2•5H_2O$) phase was found in XRD patterns for the samples with 20%, 30% and 40% quartz addition and with 10% flyash addition. The tobermorite peaks were shown for the sample with 10% flyash addition. Maximum strength was found on the samples with 20% quartz addition and 10% flyash addition. However, the hydrothermal reaction in the mixed powder was not active in the hydrothermal hot pressing in the present experimental condition. Improvement in the experimental parameters will be necessary to get hydrothermal reaction in the process of hydrothermal hot pressing.

Key words slag, recycling, hydrothermal, hydrothermal hot pressing, high strength material

上海宝钢工业园区大气 $PM_{2.5}$ 元素污染特征及溯源分析

张 锋[1] 陈正勇[2]

（1. 宝钢工程集团上海宝钢工业技术服务有限公司，上海 201900；
2. 江苏天瑞仪器股份有限公司，江苏昆山 215300）

摘 要 上海宝钢工业园区是大型钢铁冶炼工业区，通过对该地区 2014 年 3 月 19 日到 2014 年 5 月 19 日大气颗粒物浓度及 $PM_{2.5}$ 中元素浓度进行连续在线监测，以重金属元素为污染物示踪因子进行污染源溯源分析，通过主成分分析及各元素主成分与污染玫瑰图结合的研究方法，解析污染贡献的方位。结果表明：研究时间段内上海宝钢工业园区 $PM_{2.5}$ 与 PM_{10} 平均浓度分别为 53.49 $\mu g/m^3$ 和 74.89 $\mu g/m^3$；将化学元素分为钢铁冶炼污染、冶金堆场排放和钢铁冶炼伴生矿污染三个污染源；钢铁冶炼污染是引起当地 $PM_{2.5}$ 浓度较高的主要因素；钢铁冶炼污染来源于工业园区的东南方向，冶金堆场排放来源于工业园区的北方，钢铁冶炼伴生矿污染来源于工业园区的东方。

关键词 $PM_{2.5}$，大气元素，主成分分析，溯源

Study on the Pollution Characteristics and Source Apportionment of Particulate Matter Based on the Elements in Baosteel Factory District

Zhang Feng[1], Chen Zhengyong[2]

（1. Baosteel Engineering & Technology Group Co., Ltd., Shanghai Baosteel Industry Technological Service Co., Ltd., Shanghai 201900, China; 2. Jiangsu Skyray Instrument Co., Ltd., Kunshan 215300, China）

Abstract An atmospheric particulate matter and chemical elements of PM_{10} and $PM_{2.5}$ in Baosteel factory district from March 19, 2014 to May 19, 2014 were studied in this paper. The pollution source of atmospheric particulates and its elements was studied by principal component analysis and correlation analysis. The results revealed that the mean mass concentration of $PM_{2.5}$ and PM_{10} was 53.49 $\mu g/m^3$ and 74.89$\mu g/m^3$, respectively. By principal component analysis and correlation analysis, there were three sources of heavy metal element including steel smelting pollution, metallurgical yard emissions and steel smelting associated ore pollution. Iron and steel smelting pollution is a major cause of high local concentrations of $PM_{2.5}$. Steel smelting pollution comes from the southeast direction of monitoring sites, metallurgy yard emissions from monitoring sites north, steel smelting associated ore pollution derived from the east of monitoring sites.

Key words $PM_{2.5}$, chemical elements, principal component analysis, source apportionment

提高高炉煤气计量准确率的方法浅析

冯桂红

（上海梅山钢铁股份有限公司，江苏南京 210039）

摘 要 文章介绍了梅钢高炉煤气能源计量检测信息系统组成、设计原则、关键技术、公司内部厂与厂之间结算及外销计量检测点分布情况，对目前公司高炉煤气计量存在的问题进行了分析，为提高高炉煤气计量准确率提出了具体的改进措施，从而减小了高炉煤气计量产销差。该系统的投用，为公司节能降本增效发挥了强有力的支撑作用。

关键词 高炉煤气，能源计量，检测，管理效能，信息化

Analysis to Improve the Metering Accuracy of Blast Furnace Gas

Feng Guihong

(Shanghai Meishan Iron & Steel Co., Ltd., Nanjing 210039, China)

Abstract Energy metering information system framework, design principle, key technology, the inter plant settlement and the metering detection spots for export of blast furnace gas in Meigang are introduced. Metering problems of blast furnace gas are analyzed. the specific measures to improve metering accuracy of blast furnace gas are put forward, thereby metering difference between production and marketing of blast furnace gas is reduced. This system plays an important role in energy-saving, cost-reduction and benefit-enhancing.

Key words blast furnace gas, energy metering, detection, management efficiency, information

冷轧浓碱废水系统设计及运行优化

王崇武 陈 琦

（宝钢工程技术集团有限公司，上海 201999）

摘 要 冷轧废水是冶金行业水质最复杂、种类最多、最难处理的废水之一，本文详细介绍了某钢铁企业冷轧浓碱废水系统的设计和运行优化过程。设计内容包括工艺流程的选择、工艺参数的计算、主要设备的选型等。系统以气浮设备作为预处理部分核心设备，以“生物接触氧化+斜板沉淀”作为生化部分核心处理工艺，出水与冷轧其他废水混合经深度处理后外排。调试过程中对影响系统稳定运行之处进行了优化改进或提出了改进措施，使系统更完善。系统设置了 DCS 控制中心，配备了多个在线仪表，自动化水平高，运行维护方便。系统处理效果达到或优于设计

水平，证明利用生化法处理冷轧浓碱废水是切实可行的，可作为相似工程的借鉴和参考。

关键词 冷轧浓碱废水，设计，优化，气浮设备，生物接触氧化法，斜板沉淀，DCS控制中心

Design and Run Optimization of Cold Mill Strong Alkaline Waste Water System

Wang Chongwu, Chen Qi

(Baosteel Engineering & Technology Group Co., Ltd., Shanghai 201999, China)

Abstract Cold mill industry waste water is one of the the most complex, most species, most difficult treatment of Metallurgical industry .This paper describes the design and run optimization progress of cold mill strong alkaline waste water system servicing for a iron and steel enterprise. Design content includes the secletion of process, the calculation of process parameters and the selection of major equipments,etc. The system uses float equipment as the nuclear equipment and uses "bio-contact oxidation process+ inclined plate sedimentation" as the nuclear process. The effluent mixed with other cold mill wastewater efflux after deep treatment. The system sets the DCS system control center, equipped multiple line instrument, its automation level is high and the operation and maintenance is convenient.System processing effect to be or better than the design level. Which proofed that using biochemical process treatment ccentrated alkali waste water is feasible. It can be used for reference of similar project.

Key words cold mill strong alkaline waste water, design, optimization, float equipment, bio-contact oxidation process, nclined plate sedimentation, DCS system control center

烧结料层对SO_2的吸收

于 恒[1,2] 张春霞[1] 王海风[1] 王志花[3]

（1. 钢铁研究总院先进钢铁流程及材料国家重点实验室，北京 100081；

2. 东北大学材料与冶金学院，沈阳 110819；

3. 中国钢研科技集团有限公司，北京 100081）

摘 要 为研究烧结烟气循环过程中烧结料层对SO_2的吸收，分别进行烧结矿和烧结混合料对SO_2的吸收试验，并进行热力学分析。结果可知，烧结矿对SO_2无吸收作用，烧结混合料对SO_2有吸收作用，主要是由于CaO和$CaCO_3$与SO_2的反应。混合料中S的吸收率主要受$CaCO_3$的分解温度和CaO吸收SO_2的反应平衡温度影响，高于平衡温度后，混合料中的S开始脱除，在1000℃时S含量与烧结矿差不多。

关键词 烧结料层，SO_2吸收，烧结混合料，热力学分析

Absorption of SO_2 by the Sinter Layers

Yu Heng[1,2], Zhang Chunxia[1], Wang Haifeng[1], Wang Zhihua[3]

(1. State Key Laboratory of Advanced Steel Process and Products, Central Iron and Steel Research Institute, Beijing 100081, China;

2. School of Materials and Metallurgy, Northeastern University, Shenyang 110819, China;
3. China Iron & Steel Research Institute Group, Beijing 100081, China)

Abstract In order to research the SO_2 absorption by sinter layers during the flue gas recirculation, the experiments of SO_2 absorption with sinter ore and sinter mixture are proceeded, as well as thermodynamic analysis. The results show that, sinter ore couldn't absorb SO_2, while sinter mixture could absorb SO_2, mainly because of the reactions of CaO, $CaCO_3$ and SO_2. The S absorptivity of sinter mixture is influenced by the decomposition temperature of $CaCO_3$ and the reaction equilibrium temperature of CaO and SO_2, when it's higher than the equilibrium temperature, the S content in sinter mixture is removed, and the S content nearly equal to that of sinter ore in 1000℃.

Key words sinter layers, SO_2 absorption, sinter mixture, thermodynamic analysis

钢铁企业低温热能的回收利用

墙新奇

（宝钢集团新疆八钢有限公司，乌鲁木齐 830022）

摘 要 钢铁企业的低温热能种类较多，总量很大，并具有再生性，但这部分热能回收利用的很少，存在节能的机会和潜力。本文对钢铁企业的低温热能按其载体的不同，分为热水低温热能和废气低温热能，对钢铁企业各工序低温热能存在的主要形式进行了统计，对目前钢铁企业低温热能利用的技术和用途进行了论述。本文重点介绍了冷电循环技术的工作原理，并对冷电循环利用低温热能的机理进行了研究，以钢铁企业的热水低温热能的回收利用为例对冷电循环的应用前景进行了展示，从而进一步揭示钢铁企业低温热能的潜力，以引起对低温热能回收利用的关注。

关键词 钢铁企业，低温热能，冷电循环

Low Temperature Thermal Energy Recycling of the Iron and Steel Enterprises

Qiang Xinqi

(Baosteel Group Xinjiang Baiyi Iron & Steel Co., Ltd., Urumqi 830022, China)

Abstract Iron and steel enterprises of low temperature thermal energy sort is more, total amount is large, and have a reproducible, but this part of the thermal energy recycling rarely, exist opportunities and potential of energy saving. In this paper, the irzon and steel enterprises of low temperature thermal energy carrier according to the different, divided into hot water low temperature thermal energy and waste gas low temperature thermal energy. The main form of existing in the process of low temperature thermal energy for statistics. The current technology of low temperature thermal energy utilization in iron and steel enterprises and uses are discussed. This paper mainly introduces the working principle of the cold electricity circulation technology. Cold electricity circulation mechanism of low temperature heat were studied.In iron and steel enterprises of low temperature hot water thermal energy recycling, for example on the cold electricity circulation applications, to further reveal the potential of low temperature heat energy, to the attention of low temperature thermal energy to be recycled.

Key words the iron and steel enterprise, low temperature thermal energy, cold electricity circulation

蓄热式中间包烘烤器在宝钢炼钢厂的应用

薛立秋　田正宏　李冬梅

（宝钢股份公司炼钢厂，上海　201900）

摘　要　本文简述了蓄热式中间包烘烤器的工作原理、性能特点及烘烤器热效率的计算方法，通过计算蓄热式烘烤器的热效率，可以看出蓄热式烘烤器的热效率较直燃式烘烤器有大幅提高。文章还介绍了宝钢炼钢厂三号连铸机中间包蓄热式烘烤器改造后的应用实绩，改造后的蓄热式烘烤器预热中间包效果良好，中间包平均温度达到或超过1150℃，浸入式水口预热温度达到650℃以上，完全满足中间包烘烤的工艺要求。通过计算蓄热式中间包烘烤器的节能率，得出了蓄热式中间包烘烤器比普通烘烤器具有节能优势的结论，节能率达到35%以上，在日常生产中能够大大降低中间包烘烤的燃气消耗量，采用蓄热式烘烤器预热中间包有显著的经济效益，具有广泛的推广应用价值。

关键词　连铸，中间包烘烤，蓄热式

Application of Regenerative Roaster for Preheating Tundish in Steelmaking Plant of Baosteel

Xue Liqiu, Tian Zhenghong, Li Dongmei

(Steelmaking Plant, Baoshan Iron & Steel Co.,Ltd.,Shanghai 201900,China)

Abstract　The article briefly describes the working principle and characteristics of the regenerative roaster as well as the formula of thermal efficiency calculation has been introduced,it can be seen that the regenerative roaster is much more effective than the normal preheating device by calculation of the thermal efficiency of the two kind of preheating device.The practical application of preheating device revamping for No.3 continuous caster in steelmaking plant of baosteel are also been introduced in this article.The achievements of revamping on the preheating device is positive, the preheating temperature of tundish are reaching or higher than 1150℃ , the submerged nozzle preheating temperature is above 650 ℃, fully match the requirement of the preheating tundish process. The paper draws a conclusion that the regenerative roaster is much better than the normal preheater on energy saving by calculation of the energy saving rate of the regenerative roaster, its energy saving rate is more than 35%, meanwhile the gas consumption is extremely reduced during the preheating of tundish. The remarkable economic benefits has been achieved by adapting the regenerative roaster,it is really worth for the application value.

Key words　continuous casting, preheating tundish, regenerative roaster

首钢京唐烟气治理技术研究与应用

王代军[1,2,3]　吴胜利[1,2]

（1. 北京科技大学高效钢铁冶金国家重点实验室，北京　100083；

2. 北京科技大学冶金与生态工程学院，北京　100083；

3. 北京首钢国际工程技术有限公司，北京　100043）

摘　要　针对首钢京唐烧结脱硫的教训及太钢烧结烟气治理的成功经验，首钢京唐球团烟气治理采取活性焦干法技术，副产品将实现 1046 万~1784 万元/年收益。Fluent 软件研究表明，6 个吸附塔单元烟气流量差异最大值仅为 4.939%，整个塔内烟气均匀。双级吸附塔可适应 SO_2 浓度的波动，在保证较高的脱硫效率情况下，同时，脱硝效率达到 50%以上。与日本住友技术比较，研发的双级吸附塔，不仅脱除效率高，而且一次性投资、运行费用低。与 MgO 湿法、SDA 半干法相比，活性焦干法具有广阔的发展和应用前景。

关键词　烧结，球团，脱硫，脱硝，活性焦

Study and Application of Flue Gas Treatment Technology in Shougang Jingtang

Wang Daijun[1,2,3], Wu Shengli[1,2]

(1.State Key Laboratory of Advanced Metallurgy, University of Science and Technology Beijing, Beijing 100083, China;

2. School of Metallurgical and Ecological Engineering, University of Science and Technology Beijing, Beijing 100083, China;

3.Beijing Shougang International Engineering Technology Co., Ltd., Beijing 100043, China)

Abstract　According to the lessons of Shougang Jingtang sintering and the successful experience of Tisco sintering flue gas desulfurization, the active coke flue gas treatment technology is adopted in Shougang Jingtang pellet plant. The revenue of 1046~1784mpa will be achieved by by-product. Fluent software research showed that, the difference of the flue gas flow rate of the 6 adsorption towers unit was only 4.939%, the flue gas in the tower was well-distributed. The double stage adsorption tower could withstand the fluctuation of SO_2 concentration. In the case of high efficiency of desulfurization, denitrification efficiency was reached to more than 50%. Compared with Japan Sumitomo technology, the double stage adsorption tower, not only removal efficiency is high, but also one-time investment and operating costs are low. Compared with MgO wet and SDA semi dry process, the active coke dry process has much broad development and application prospect.

Key words　sintering, pelletizing, desulphurization, denitration, active coke

喷淋式钢管淬火装置水循环节能设计

姚发宏　余　伟　程知松

（北京科技大学设计研究院有限公司，北京　100083）

摘　要　针对喷淋式钢管淬火装置用水特点，对水循环系统进行优化，将钢管淬火时间及间歇时间内的排水进行独立设计，节省了一半的重复循环处理水量，不但达到了节能降耗的效果，而且在一定程度上降低了工程投资以及生产运行成本。同时，根据工程设计经验，介绍了优化后的内喷回水系统、外淋回水系统、铁皮坑及泵站的设计方案及思路。

关键词　钢管，热处理，喷淋式淬火装置，水循环，节能设计

Energy Saving Design of Water Circulating System in the Spray Steel Pipe Quenching Device

Yao Fahong, Yu Wei, Cheng Zhisong

(USTB Design & Research Institute Co., Ltd., Beijing 100083, China)

Abstract In allusion to the characteristics of the water spray type steel quenching device, optimize the water cycle system, the steel pipe quenching time and interval time of drainage, an independent design, save half of the water recirculation treatment, not only achieved the effect of saving energy and reducing consumption, and to a certain extent, reduce the engineering investment and operation cost. At the same time, according to the engineering design experience, the paper introduces the optimized in spray water system, rain water system, scale pit and pump station design and train of thought.

Key words steel pipe, heat treatment, spray quenching device, water circulation, energy saving design

钢铁联合企业焦化废水处理控制措施探讨

张 垒[1] 冷 婷[2] 龚晓萍[3] 薛改凤[1] 常红兵[1] 吴高明[1]

（1. 武汉钢铁（集团）公司研究院，湖北武汉 430080；2. 武汉钢铁股份公司制造部，湖北武汉 430080；3. 武汉钢铁（集团）公司安环部，湖北武汉 430080）

摘 要 针对焦化废水水质特点和目前环保要求，介绍了国内外焦化废水处理技术研究现状，探讨了钢铁联合企业焦化废水深度处理及回用的几种方法和控制措施，提出采用清洁生产技术，从源头减少焦化酚氰废水产生量，然后利用钢铁企业用水特点，在企业内部实施分质供水，寻找焦化废水消纳途径，减少废水外排，为钢铁联合企业焦化废水处理及回用方案选择和研究提供参考。

关键词 钢铁联合企业，焦化废水，控制措施

Discussion on Coking Wastewater Treatment and Control Measures in Iron and Steel Enterprises

Zhang Lei[1], Leng Ting[2], Gong Xiaoping[3], Xue Gaifeng[1], Chang Hongbing[1]， Wu Gaoming[1]

(1. R&D Center of WISCO, Wuhan Hubei 430080, China; 2. Manufacture Section of WISCO, Wuhan Hubei 430080, China; 3. Safety and Environment Section of WISCO, Wuhan Hubei 430080, China)

Abstract According to the water quality characteristics of coking wastewater and the environmental protection requirements, several methods and control measures of coking wastewater treatment were introduced in the effluent from iron and steel enterprises, which can provide a reference for process selection and research on treatment of coking wastewater in iron and steel enterprise.

Key words iron and steel enterprise, coking wastewater, control measures

浅谈提高武钢中水回用量措施

明金阳　张汉华　胡爱群　周爱军　周　文　周　非

（武汉钢铁股份有限公司能源动力总厂供水厂，湖北武汉　430083）

摘　要　介绍了武钢回用水现状，对武钢废水综合利用过程中依然存在的一些问题进行了简要探讨。并结合武钢在实际运行过程中的经验，提出了相应的辅助措施，使武钢回用水量实现稳步提升，降低了武钢新水消耗，为武钢节能减排工作奠定了一定了基础。

关键词　回用水，再利用，节能

The Measures on Improving the Amount of Reuse Water in WISCO

Ming Jinyang, Zhang Hanhua, Hu Aiqun, Zhou Aijun, Zhou Wen, Zhou Fei

(Water Supply Plant, Wuhan Iron Steel Co., Wuhan 430083, China)

Abstract　This paper introduced the reuse water situation of WISCO, and the issues that still exist in the process of utilization of the wastewater treatment in WISCO are discussed briefly. Combined with experience in the actual operation in WISCO, and proposed supplementary measures accordingly, so that improve the reuse water steadily in WISCO , and reducing new water consumption, laid the foundation of the energy conservation in WISCO.

Key words　reuse water, recycling, energy saving

浅谈稳定武钢 CSP 二冷水系统供水压力的措施

蔡　健　汪　颖　陈　涛　陈　彪

（武钢能源动力总厂供水厂，湖北武汉　430080）

摘　要　针对武钢 CSP36#水站连铸浊环水外供水泵 PC2 系统压力波动大，设计工艺存在缺陷，能耗高的现状；通过对该系统设备实际运行工况参数的长期监测，计算出水泵与流体输送相匹配的最佳工况点，采用更换高效节能的三元流叶轮、用电动调节阀替换持压泄压阀, 并对供水压力控制工艺进行技术改造的一系列措施，实现系统的恒压供水控制。

关键词　三元流叶轮，持压泄压阀，电动调节阀，节能改造

Introduction of Measure of Stabilizing the Water Supply Pressure of Secondary Cooling Water System of CSP of WISCO

Cai Jian, Wang Ying, Chen Tao, Chen Biao

(Water Supply Plant of Energy Power Plant of WISCO, Wuhan Hubei 430080, China)

Abstract The big pressure fluctuations of turbid circulating water of Continuous Casting in No.36 water station of CSP of WISCO is introduced in this paper. Defect of design and technology and high energy consuming were found in this system. According to the long-term monitoring of the actual condition parameter of equipment of this system, the best condition point of water pump was calculated. Electric-control valve was instead of pressure relief valve, and energy-efficient three-dimensional flow impeller was used in this system. Technological transformation made the water supply pressure constant.

Keywords three-dimensional flow impeller, pressure relief valve, electric control valve, energy conservation transformation

生物接触氧化法+MBR在含油废水处理中的应用及运行维护

刘海英 彭 斌 陶 灿 梁 刚 章茂晨

（武汉钢铁公司能源动力总厂供水厂，武汉 430083）

摘 要 随着工业废水处理技术的不断进步，生物接触氧化法+ MBR在武钢冷轧含油废水处理工艺中的应用越来越广，随之在日常运行维护过程中出现了各类异常情况。本文主要针对生物接触氧化法+ MBR在生产运行中逐渐出现的生物单元恶化造成的污泥膨胀，频繁的冲击负荷造成污泥活性下降，大量的菌胶团老化脱落，进而使MBR膜污堵情况加剧，通量陡然下降，膜抽吸压力不断升高的异常现象利用科学手段进行诊断。通过不断的研究和反复试验，逐步摸索出消除膜污堵的方案，并组织进行实施，较好地解决了上述问题，同时，我们也在此过程中积累了许多“双膜法”运行的宝贵经验，并且，我们对上述运行中的出现的异常情况提出我们的日常预防和整改措施，以期供大家参考。

关键词 膜法，冷轧含油废水，诊断，处置

Abstract With the continuous progress of membrane technology, membrane application in the field of industrial wastewater treatment is more and more mature. Biological contact oxidation+MBR is finding widely application in the cold rolling oil-bearing wastewater of wisco.In this paper scientific means were used to diagnosis all kind of anomalies during the application of contact oxidation +MBR process，such as sludge bulking, caused by frequent impact load, which causing a lot of bacteria group increased aging and fall off. The falling bacteria group deposit block up of the MBR membrane, which caused the membrane flux declined precipitously. Solution had been found during continuous research and repeated experiment, which solves the questions for eliminate membrane fouling and biological membrane fast film forming simultaneously. During the process of research, we also accumulated a lot of valuable experience about double-film method. At last, the preventive and corrective measures for the abnormal situation occurred in the use of the above routine were proposed for your reference.

Key words biological contact oxidation/MBR, cold-rolling oily wastewater, diagnosis, disposal method

高炉煤气洗涤水脱除氰化物技术研究

段建峰 俞 琴 贺 琨

（武钢能源动力总厂供水厂，湖北武汉　430080）

摘　要　氰化物污染物一直是冶金行业高炉煤气洗涤水排污水的处理难点之一。本文针对武钢高炉煤洗涤水排污水氰化物污染物含量较高的特点，在实验室采用 NaCl-NaOH 法处理高炉煤气洗涤水排污水的氰化物，实验结果表明，NaCl-NaOH 法处理高炉煤气洗涤水排污水中的氰化物是可行的，且具有较高的处理效率。最后文章介绍了武钢高炉煤气洗涤水排污水的脱氰工程.

关键词　煤气洗涤水排污水，NaCl-NaOH 法，氰污染物，脱氰工程

Study on Cyanide Removal of BF Gas Scouring Water

Duan Jianfeng, Yu Qin, He Kun

(Water Supply Pant, Energy Sources and Power Plant, Wuhan Iron and Steel (Group) Corp., Wuhan Hubei 430080, China)

Abstract　Cyanide pollutant is one of the difficult points in the treatment of blast furnace gas washing wastewater in metallurgical industry. Based on the characteristics of cyanide in the sewage from pollutant concentration .Gas scouring water was with the method of NaCl-NaOH treatment of blast furnace gas washing wastewater cyanide. Experimental results show that cyanide in BF gas washing water of sewage water treatment with NaCl-NaOH method is feasible, and has higher efficiency. Finally, the cyanide removal of the washing water discharged from the blast furnace gas of Wuhan Iron and steel is introduced.

Key words　gas scouring water, method of NaCl-NaOH, cyanide, removal of cyanide

一种改进的动态电压恢复器电压补偿策略

蔡惠红[1]　邱　军[1]　林新春[2]　谭　俊[2]

（1. 武钢股份能源动力总厂，湖北武汉　430080；2. 华中科技大学，湖北武汉　430074）

摘　要　动态电压恢复器是一种串联型电压补偿装置，该装置能够在很短的时间内检测到电网电压跌落，然后输出一个补偿电压使负荷电压恢复到跌落前的状态。其中储能单元为动态电压恢复器提供补偿期间所需的能量，其成本也一定程度上限制了 DVR 的推广。首先，本文分析了现有 DVR 补偿策略对储能单元电压及容量的要求，并指出这三种补偿策略的优势与不足。在此基础上，本文提出了一种新的电压补偿策略。该策略综合了常见补偿策略的优点，通过在补偿过程中适时地切换补偿策略，保证在负荷电压不出现相位跳变的前提下，延长 DVR 补偿时间。然后，本文通过实验验证了该补偿策略的可行性与有效性。因此，在满足一定技术指标的前提下，该策略能更有效地利用直流储能，从而减小储能容量、降低整个装置的成本。

关键词　动态电压恢复器，储能，电压补偿策略

An Improved Voltage Compensation Method for Dynamic Voltage Restorer

Cai Huihong[1], Qiu Jun[1], Lin Xinchun[2], Tan Jun[2]

（1. The Energy and Power Company of WISCO, Wuhan Hubei 430080, China;
2. Huazhong University of Science and Technology, Wuhan Hubei 430074, China）

Abstract Dynamic voltage restorer is a series connected voltage compensator. This device could quickly detect voltage sags and inject compensation voltages to restore the load voltage. The energy storage device, which provides the energy during the compensation, is an important part of the whole system, but its cost impedes the popularization of dynamic voltage restorer. Firstly, this paper analyzed the existing voltage compensation methods and found the advantages of each methods. Based on the comparisons, this paper proposed a improved compensation method, which synthesized the advantages of existing methods. The voltage compensation time of this method can be extended by translating the voltage compensation methods during the whole compensation process. Then, the proposed voltage compensation method is verified by experimental results. So this paper proposed a new voltage compensation method to utilize the energy more efficiently, and then reduces the capacity and cost of the dynamic voltage restorer.

Keywords dynamic voltage restorer, energy storage, voltage compensation method

旋流式火星捕集器内流场分析

葛玉华

摘　要 本文介绍通过利用计算机仿真模拟技术，对宝钢四烧结机尾除尘器的旋流式火星捕集器内含尘气体进行流场分析，从而了解和判定该火星捕集器原设计的问题及产生这些问题的原因，为其优化改造及日常维护提供依据。

关键词 火星捕集器，宝钢四烧结，含尘气体，流场，计算机仿真

富氧助燃技术在活性石灰回转窑的应用实践

张洪雷[1]　饶发明[2]　季佳善[2]　王　悦[1]　刘　黎[1]　徐国涛[1]

（1. 武汉钢铁集团公司研究院，湖北武汉　430080；
2. 武钢矿业有限责任公司乌龙泉矿，湖北武汉　430213）

摘　要 将膜法制氧、富氧助燃工艺设备在武钢多座活性石灰回转窑进行了应用，本文介绍了其工艺流程及主要设备，并在实际工况下综合评测了其使用效果，该工艺经适当改造可适用于回转窑等动态窑炉。采用该工艺有效提高了回转窑的燃烧效率及各项运行指标，显著降低了产品的煤耗及温室气体排放。

关键词 富氧助燃，石灰，回转窑

The Application of Oxygen-enriched Combustion Process in Active Lime Rotary Kiln

Zhang Honglei[1], Rao Faming[2], Ji Jiashan[2], Wang Yue[1], Liu Li[1], Xu Guotao[1]

（1. Research and Development Center of WISCO, Wuhan 430080, China;

2．Wulongquan Mine, MINERAL Co., of WISCO, Wuhan 430213, China)

Abstract The membrane technology for oxygen enrichment combustion-assistance was applied in active lime rotary kiln. This paper introduces the process technique and main equipments, evaluates its application result in practical production, shows that this technology can effectively improve the operation index of rotary kiln，reduce the coal consumption and greenhouse gas emissions.

Key words oxygen-enriched combustion, lime, rotary kiln

8　冶金设备与工程

Metallurgical Equipment and Engineering Technology

大型液压传动布料器无料钟的性能测试与分析

曾 攀[1] 雷丽萍[1] 李聪聪[1] 宋江腾[1] 郑 军[2]

(1. 北京市海淀区清华大学机械工程系，北京 100084；
2. 重庆市中冶赛迪工程技术股份有限公司，重庆 401122)

摘 要 本文针对大型液压传动布料器无料钟的主体结构，在冷态及热态下运行状态下，具体就布料器的曲柄、万向框架、溜槽花键和顶盖等主要承力部件进行应变测试，考虑不同溜槽倾角下的周期性条件为：40°→35°→30°→25°→20°→15°→10°→5°→40°，实时获得测试点应变的变化规律。对于高炉布料器关键部位的耳轴及轴套进行磨损预测，首先保障真实磨程的前提下，进行加速实验，并找到加速条件下的磨损关系，同时对高炉布料器中的耳轴与轴套的受力状态进行分析，基于模拟仿真，最后对高炉布料器中的耳轴与轴套的磨损状况进行评价。

关键词 大型布料器，无料钟，性能测试

Performance Test and Analysis of Large Hydraulic Charging Equipment without Bell in BF Top

Zeng Pan[1], Lei Liping[1], Li Congcong[1], Song Jiangteng[1], Zheng Jun[2]

(1. Tsinghua University, Beijing 100084, China; 2. Chongqing CISDI Engineering Co., Ltd., Chongqing 401122, China)

Abstract For the large hydraulic charging equipment without bell in BF top, under cold state and hot state running state, the paper tested the specific strain varying of crank, universal framework, chute spline, and the top main cover parts, considering different chute angle under the periodic condition: 40°→35°→30°→25°→20°→15°→10°→5°→40°, to obtain the real-time strain varying rule of test points. For blast furnace charging equipment, the wear of the key positions in contact couple of ear shaft and shaft sleeve was predicted. Under the premise of guaranting real wearing distance, the accelerated wear was tested, and the rule was found. Based on the accelerated wear rule, the numerical simulation, the wear state of large hydraulic charging equipment was evaluated.

Key words large charging equipment, BF top without Bell, performance test

Application of Low Voltage Pulsed Magnetic Field in Solidification of Metals

Yang Yuansheng, Li Yingju, Luo Tianjiao, Ji Huanming, Feng Xiaohui

(Institute of Metal Research, Chinese Academy of Sciences, Shenyang, China)

Abstract The grain refinement of static castings and DC-casting ingots was achieved with a low-voltage pulsed magnetic field setup. Through the study of the interaction between melt and the low-voltage pulsed magnetic field (LVPMF), a 3-D model was established for simulation of the electromagnetic force, melt flow, heat and mass transfers. The effects of the

LVPMF on the solidified structure of silicon steel and Ni-based superalloy as well as magnesium alloy were investigated, and the optimized processing parameters were obtained for grain refinement of the alloys. Based on the dynamic theories for crystal growth under melt flow and the LMK marginally stable theory, the mechanics of the interface morphology changing from typical dendrite to globular morphologies with the LVPMF was explained. Combined with the simulation results on electromagnetic vibration and convection caused by the LVPMF and experimental results, the grain refinement mechanism was discussed in terms of nucleation and growth theories.

Key words pulsed magnetic field, casting, grain refinement, electromagnetic vibration, melt convection

上海大学在电磁连铸方面的研究工作

任忠鸣 雷作胜 钟云波 邓 康

（上海大学，上海市现代冶金与材料制备重点实验室）

摘 要 电磁冶金是20世纪90年代以来涌现出来的一个新的分支，电磁连铸是其中的一个重要组成部分。本文简单综述了上海大学在电磁连铸方面的一些研究工作，包括软接触结晶器电磁连铸技术，小方坯电磁旋流水口技术，针对小方坯二冷段的间歇变向电磁搅拌参数优化以及针对板坯连铸电磁搅拌和电磁制动研究而建立的，基于低熔点液态金属的结晶器电磁控流技术研究实验平台等。

Some Research Work on Electromagnetic Continuous Casting of Shanghai University

Ren Zhongming, Lei Zuosheng, Zhong Yunbo, Deng Kang

(Shanghai Key Laboratory of Modern Metallurgy and Material Processing, Shanghai University)

Abstract Electromagnetic metallurgy is a new branch emerged from 1990s, among which electromagnetic continuous casting charactering by several technologies applying different kinds of magnetic field is an important part in both scientific research and industrial application. This paper gives a general review on some research work fulfilled by Shanghai key laboratory of modern metallurgy and material processing on electromagnetic continuous casting including soft-contact mold electromagnetic continuous casting, electromagnetic swirling nozzle for billets CC, intermittently reversing direction electromagnetic stirring for billets CC, electromagnetic stirring and electromagnetic brake for slab CC.

宝钢连铸电磁搅拌装置的系统集成与应用

李华刚 连井涛

（上海宝信软件股份有限公司机电一体化事业部，上海 201900）

摘 要 过连铸电磁搅拌装置是满足工艺，生产高质量、高附加值钢铁产品的特殊冶金装备之一。在2009年以前宝钢连铸电磁搅拌一直引进外方技术，但从2009依靠宝钢研究院完成6CC铸流辊式搅拌科研项目，科研成功后由宝信软件进行 6CC 技改，从此宝钢连铸电磁搅拌走出了一条自主集成的成功之路。通过近几年科研研究及工程应用实践中，目前宝钢已全面掌握连铸电磁搅拌装置系统内的电磁搅拌本体、冷却水系统、低频专用电源、控制系统及对外抗干扰系统的五大核心技术。本文介绍的宝钢连铸电磁搅拌装置的系统集成已成功在宝钢 8CC 末端电磁搅拌装置系统、宝钢 5CC 结晶器电磁搅拌装置系统、宝钢 1CC 结晶器电磁搅拌装置系统、宝钢 5CC 铸流电磁搅拌装置系统中得到应用，运行至今，系统工作情况稳定、运行良好。

关键词 电磁搅拌，连铸，系统集成

The System Integration and Application for the Electromagnetic Stirring Device of Continuous Casting in Baosteel

Li Huagang, Lian Jingtao

(Mechanical Department, Shanghai Baosight Software Ltd., Shanghai 201900, China)

Abstract The electromagnetic continuous casting device is a kind of special metallurgy equipment which is used to meet the process, improve the production of high quality, and produce high value-added steel products. Before 2009, the electromagnetic stirring of Baosteel continuous casting has been introduced into the foreign technology, but the continuous casting electromagnetic stirring roller research has been successful in 2009, at the same time, Baosteel produced their own electromagnetic stirring rollers which were used in Baosteel's 6CC, since then, the electromagnetic stirring of the Baosteel's continuous casting has been on a successful way of independent integration. Through R&D and engineering investment in recent years , now Baosteel has fully mastered the five core technologies of the system of electromagnetic stirring device which are composed of the electromagnetic stirring body part, the cooling water system, the low frequency special power supply, the control system, and the anti-Interference system. The system integration and application for the electromagnetic stirring device of continuous casting which is introduced in this paper, has been successfully applied in the 8CC F-EMS system, the 5CC M-MES system, 1CC M-MES system,5CC S-MES system. The integration system works stably, running well since it is online.

Key words electromagnetic stirring, continuous casting, system integration

电磁搅拌对 Incoloy800HT 连铸坯凝固组织的影响

王 菲 徐 宇 王恩刚 邓安元 张兴武

（东北大学材料电磁过程教育部重点实验室，辽宁沈阳 110819）

摘 要 本文以镍基耐蚀高温合金 Incoloy800HT 为研究对象，采用连铸方法制备横断面为 100mm×225 mm 的连铸坯，并研究了线性电磁搅拌和连铸工艺参数对 Incoloy800HT 高温合金连铸坯凝固组织的影响。研究结果表明，施加线性电磁搅拌可以细化 Incoloy800HT 连铸坯的等轴晶晶粒组织，提高铸坯等轴晶比例，并显著改善常规连铸坯凝固组织中的穿晶现象。当二冷区冷却强度较高时，拉坯速率对未施加电磁搅拌的连铸坯的晶粒尺寸和等轴晶比例影响不大；施加电磁搅拌后，当拉坯速率由 200mm/min 增加至 600mm/min 时，电磁搅拌对铸坯晶粒组织的细化作用更加明显，铸坯断面等轴晶比例明显扩大。当降低二冷区冷却强度时，铸坯内部所受电磁搅拌作用的液相区增加，随着拉坯速率增加至 600mm/min 时，等轴晶晶粒尺寸减小到 1.05mm，截面等轴晶比例增大至 49.75%。

关键词 高温合金，电磁搅拌，凝固，连铸

Effect of Electromagnetic Stirring on Solidification Structure of Incoloy800HT Superalloy Strand

Wang Fei, Xu Yu, Wang Engang, Deng Anyuan, Zhang Xingwu

(Key Laboratory of Electromagnetic Processing of Materials (Ministry of Education) Northeastern University, Shenyang 110819, China)

Abstract The Ni-based superalloy Incoloy800HT slabs were produced by continuous casting process, and the effect of linear electromagnetic stirring (L-EMS) and the parameters of continuous casting on the solidification structure of superalloy Incoloy800HT slab were investigated. The results show that the equiaxed grains were obviously refined and the equiaxed ratio is increased with L-EMS. The influence of casting speed on the solidification structure is small under high secondary cooling rate and without of L-EMS; When L-EMS is applied, as the casting speed increases from 200mm/min to 600mm/min, the grains of strand are refined and the ratio of equiaxed grains is increased under the high secondary cooling condition. The area of the liquid zone of strands that can be affected by EMS is increased, which is attributed to the reduction of the secondary cooling rate and the increasing of casting speed, the gains size is further refined to 1.05mm, and the ratio of equiaxed grains is increased to 49.75%.

Key words superalloy, linear electromagnetic stirring, solidification, continuous casting

通道式电磁感应加热中间包电磁场、流场和温度场耦合的数值模拟研究

岳 强[1,2,3] 张 炯[3] 陆 娟[3] 张立峰[2] 肖 红[1] 李爱武[1]

（1. 湖南中科电气股份有限公司，湖南岳阳 414000；
2. 北京科技大学冶金与生态学院，北京 100083；
3. 安徽工业大学冶金工程学院，安徽马鞍山 243002）

摘 要 中间包电磁感应加热技术，可以有效地补偿钢液的热损失，利于实现低过热度恒温浇注，进而增加等轴晶的比率，改善铸坯质量。中间包感应加热技术越来越受到人们的广泛重视。本研究建立了感应加热电磁、流动和传热的三维数学模型，并对模型的可靠性进行了验证。研究表明，感应电流在注流室、分配室以及两个通道间形成了电流回路，通道内的电流密度远大于注流室与浇注室的电流密度。由于集肤效应与邻近效应，通道靠近线圈一侧的磁场比其他区域要大。感应电流与磁场相互作用产生指向通道中心的电磁力，通道内的钢液产生箍缩力，感应电流与磁场的不均匀导致电磁力偏心分布，钢液在偏心电磁力的作用下，会产生旋转。焦耳热主要集中分布于通道内，与电流密度分布一致。频率从 50 Hz 增加到 500 Hz 时，通道内焦耳热从 3.98×10^7 W/m^3 增加到 2.29×10^8 W/m^3，电磁力由 1.42×10^6 N/m^3 增加到 2.59×10^6 N/m^3。

关键词 中间包，感应加热，电磁场，数值模拟

Numerical Simulation of Electromagnetic, Flow and Heat Transfer for Tundish with Channel Induction Heating

Yue Qiang[1, 2, 3], Zhang Jiong[3], Lu Juan[3], Zhang Lifeng[2], Xiao Hong[1], Li Aiwu[1]

(1. Hunan Zhongke Electric Co., Ltd., Yueyang 414000, China;
2. School of Metallurgical and Ecological Engineering, University of Science and Technology Beijing, Beijing 100083, China;
3. School of Metallurgy Engineering, Anhui University of Technology, Ma' anshan 243002, China)

Abstract Induction heating technology in the tundish, can effectively compensate the heat loss of liquid steel, which achieves low superheat constant temperature casting, thus increasing the ratio of equiaxed grains, and improve the quality of billet. A 3D mathematical model of induction heating electromagnetic field,which is validated with the accuracy of the model based on the previous experimental data and the feasibility of the calculation method. The results show that induced current is formed current circuit in the pouring chamber, discharging chamber and between two channels where current density is greater than other regions. Magnetic field is mainly distributed in the channel, because of the skin effect and proximity effect,on one side of the channel near the coil magnetic field is bigger than other areas. The interaction of induced current and magnetic field produces electromagnetic force pointing to the center of the channel, which has pinch effect on flows through the channel of the liquid steel. The uneven induced current and the uneven eccentric field lead to eccentric distribution of the electromagneticforce, which the liquid steel in eccentric under the action of electromagnetic force will forward rotating flow. Joule heating mainly distributed in the channel that can quickly heat flows through the channel of the liquid steel in order to increase the temperature of the liquid steel rapidly in line with current density distribution. When current frequency from 50 Hz increases to 500 Hz, joule heating density from 3.98×10^7 W/m^3 increases to 2.29×10^8 W/m^3, and the electromagnetic force from 1.42×10^6 N/m^3 increases to 2.59×10^6 N/m^3.

Key words tundish, induction heating, electromagnetic field, numerical simulation

电磁搅拌作用下板坯连铸结晶器内流场优化数值模拟研究

金小礼[1] 周月明[1] 雷作胜[2]

（1. 宝钢集团中央研究院，上海 201999；
2. 上海大学现代冶金与材料制备重点实验室，上海 200072）

摘 要 采用数值模拟方法，对板坯连铸结晶器内的流场进行计算，分析了不同连铸工艺参数和电磁搅拌参数对流场分布的影响，利用液面卷渣指数（MFEI）和流场均匀性指数（VUI）对结晶器内的流场进行评估，为流场结构优化提供理论依据，并为电磁搅拌参数的优化提供指导。研究成果在生产实践中得到了验证，取得了较好的应用效果。

关键词 板坯连铸，流场分布，电磁搅拌，结晶器，参数优化

Study on Numerical Simulation of Fluid Flow in Slab Continuous Casting Mold under Electromagnetic Stirring

Jin Xiaoli[1], Zhou Yueming[1], Lei Zuosheng[2]

(1. Center Research Institute of Baosteel Group, Shanghai 201999, China;
2. Shanghai Key Laboratory of Modern Metallurgy & Materials Processing, Shanghai 200072, China)

Abstract A three-dimensional mathematical model was established in present study, it was used to calculate the fluid flow in slab continuous casting mold under electromagnetic stirring force. The flow structure and distribution in different continuous casting parameters and stirring current were studied. Based on the calculation results, the Mold Flux Entrapment Index (MFEI) in free surface and Velocity Oniform Index (VUI) were used to evaluate the flow field in mold, with which the theoretical basis for the optimization of the flow field structure was provided, and also the optimization method of electromagnetic stirring parameters was suggested.

Key words slab continuous casting, fluid flow, electromagnetic stirring, mold, parameters optimization

State of the Art of New Technologies for Electromagnetic Metallurgy Based on Clean Steel Production

Liu Xingan, Li Dewei, Wang Qiang, Su Zhijian,
Zhu Xiaowei, He Jicheng

(Key Laboratory of Electromagnetic Processing of Materials (Ministry of Education),
Northeastern University, Shenyang 110004 China)

Abstract The development and promotion of clean steel technology is one of the hot issues and it has important applications in steel industry, although there are some difficulties should be resolved. New generation of electromagnetic metallurgical technologies could dramatically improve steel cleanliness, so as to achieve the purpose of improving the quality of steel and increasing productivity. In this paper, two representative technologies are introduced. One is an electromagnetic induction-controlled automated steel-teeming, the other one is a technology of electromagnetic swirling flow in the submerged entry nozzle (SEN). The basic idea of the first one is using an induction coil located in the nozzle brick to melt part or all of the new well-packing material (i.e., Fe-C alloy with a similar composition of molten steel), which replaces the conventional nozzle sand, and to obtain smooth automated steel teeming. In the latter technology, a rotating electromagnetic field is set up around the SEN to induce swirling flow in it by the Lorentz force without contact with the steel. This technology could avoid nozzle clogging, and has great effect on improving the uniformity and stability of the outlet flow. It could also reduce the meniscus fluctuation, homogenize the distributions of flow and temperature in the mold. The development and promotion of these kind of electromagnetic metallurgical technologies has great significance for the production of clean steel in future.

Key words electromagnetic metallurgy, clean steel, electromagnetic steel-teeming, electromagnetic swirling flow, continuous casting, electromagnetic processing of materials

热镀锌液的电磁连续净化

董安平[1]　疏　达[1]　冒飞飞[1]　王　俊[1,2]　孙宝德[1,2]

（1. 上海市先进高温材料及精密成形重点实验室（上海交通大学），上海　200240；
2. 上海交通大学金属基复合材料国家重点实验室，上海　200240）

摘　要　锌渣缺陷是影响热镀锌钢板表面质量的最重要因素之一。本文通过数值分析计算了电磁净化过程中锌渣的运动轨迹，阐明了锌渣电磁分离的机理；设计并搭建了包括加热保温系统、锌液循环系统、电磁净化系统、移动平台和控制系统组成的热镀锌液净化中试试验平台装置，开展了GA镀锌液的连续净化中试试验，考察了中试条件下锌渣的去除效率以及锌锅内锌渣含量和尺寸分布的变化。结果表明，锌锅中锌渣的去除效率可达80%以上。

关键词　热镀锌液，电磁净化，锌渣

Continues Electromagnetic Purification of Hot Dip Galvanized Zinc Melt

Dong Anping[1], Shu Da[1], Mao Feifei[1], Wang Jun[1,2], Sun Baode[1,2]

(1.Shanghai Key Laboratory of Advanced High-temperature Materials and Precision Forming, Shanghai Jiaotong University, Shanghai 200240, China; 2.The State Key Laboratory of Metal Matrix Composites, Shanghai Jiaotong University, Shanghai 200240, China)

Abstract　Zinc dross defect is one of the most important factors affect the surface quality of hot dip galvanized steel plate. The trajectory of zinc dross during electromagnetic purification process was calculated by using numerical method; A large flow of hot-dip galvanizing zinc melt electromagnetic purification equipment was designed and built up, which grouped by heat insulation system, zinc liquid circulation system, electromagnetic purification system, mobile platform and control system. GA zinc melts purification experiments were carried out on this equipment, the results shows that the purification efficiency can reach above 80%.

Key words　hot dip galvanizing zinc melt, electromagnetic separation, zinc dross

铜合金水平电磁连续铸造的研究

李廷举[1]　郭丽娟[2]

（1. 大连理工大学材料科学与工程学院，辽宁大连　116024;
2. TDK大连电子有限公司，辽宁大连　116600）

摘　要　研究和实施了电磁场在铜合金棒、铜及铜合金管和铜合金板带坯水平连续铸造中的应用，获得了高质量的铸坯。例如，采用非真空水平电磁连续铸造的高强高导Cu-Cr-Zr铸坯制备的高铁用接触导线的平均拉强度610 MPa，导电率79.85%IACS，批量应用于京沪高铁，2010年10月3日在京沪高铁线通过了最高运营时速486.1公里的实验。

关键词 电磁场，连续铸造，铜合金

Study on the Horizontal Electromagnetic Continuous Casting of Copper Alloy

Li Tingju[1], Guo Lijuan[2]

(1. School of Material Science and Engineering, Dalian University of Technology, Dalian 116024, China;
2. Tdk Dalian Electronic Co., Ltd., Dalian 116600, China)

Abstract The in this paper, the application of electromagnetic field on the horizontal electromagnetic continuous casting of Cu and Cu alloys was studied, which were prepared in different shapes, e.g. the rods, pipes, plates and strips, with a result of obtaining the high-quality ingots. The high strength and high conductivity CuCrZr contact wires of high-speed railway, made by horizontal electromagnetic continuous casting under non vacuum atmosphere, possessed average tensile strength o f 610 MPa and conductivity of 79.85 %IACS. They had been widely used in Beijing-Shanghai high-speed railway. The cont act wires were successfully passed the experiment of the highest operating speed of 486.1 km/h in the Beijing-Shanghai hig h-speed railway on October 3, 2010.

Key words electromagnetic field, electromagnetic continuous casting, Cu alloys

Induction and Steel Finishing Lines: Major Trends End of 2015

Lovens Jean

(Inductotherm Coating Equipment, Herstal Belgium)

Abstract Cold steel strip processing lines are increasingly becoming more and more sophisticated. They include critical heat treatment sections to produce high-strength grade steel strips particularly for automotive applications (Multi-Phase & Trip). After fast cooling, annealed materials are rapidly and efficiently reheated under protective atmosphere (some of which containing a high H_2 concentration) in overaging or tempering zones to control precisely the steel microstructure through an improved temperature control. Very often, new steel strip production lines include also in line sophisticated post treatments sections to improve further more strip surface quality.

Inductive solutions also open new ways to improve the furnace transition behavior which increases indirectly the capacity of the processing line and results in savings in the production costs.

Inductotherm Coating Equipment, a major partner of the steel industry, developed various inductive technologies answering the needs of the metallurgy to continuously improve the performance of existing or new production lines in the interests of their customers (high quality, high yield, high economical efficiency, high flexibility ...). From the induction heating point of view, the first concern is the required high power density level of these strip heaters. The second concern is the modification of the electromagnetic characteristics of these new high strength steels (martensic / austenitic content). This is not only a problem for the heaters under protective atmosphere but it is also a problem for the galvannealing furnaces. We adapted our standard designs to improve their load matching capability to be compatible with normal and high strength steels.

Key words strip production line, inductive heater, protective atmosphere

电磁搅拌技术及其应用

侯亚雄　彭立新　沈长华

（湖南科美达电气股份有限公司，湖南岳阳　414000）

摘　要　本文主要介绍了电磁搅拌技术的工作原理和理论基础，介绍了电磁搅拌技术在冶金行业的典型应用：方圆坯连铸电磁搅拌，板坯连铸电磁搅拌，有色金属熔炼电磁搅拌，坩埚电磁搅拌等。文中分别介绍了目前国际国内电磁搅拌技术在上述领域中的应用特点，并介绍了科美达公司开发的电磁搅拌设备在上述领域中的应用实践，取得的冶金效果与成果。

关键词　电磁搅拌，连续铸钢，冶金效果，螺旋磁场

Electromagnetic Stirrer Technology & Application

Hou Yaxiong, Peng Lixin, Shen Changhua

(Hunan Kemeida Electric Co., Ltd., Yueyang Hunan 414000, China)

Abstract　Electromagnetic Stirrer (EMS) working principle & theoretical basis are introduced as well as its typical applications in metallurgical industry, such as EMS for billet/round continuous casting machine, for slab caster, for non-ferrous melting & for crucible, etc. EMS application features in above industries in China & abroad is also introduced with practical operations & metallurgical effects achieved by Hunan Kemeida Company.

Key words　electromagnetic stirrer, continuous casting, metallurgical effects, spiral magnetic field

宝钢感应加热技术的研究进展

吴存有　周月明　金小礼

（宝钢研究院，上海　201900）

摘　要　感应加热技术可以显著改善材料的组织和力学性能，由于新产品开发和对产品质量提出的更高要求，宝钢原有感应加热技术已难以满足生产提出的新要求。感应加热技术的关键在于如何精确地将加热温度控制在一个合理的范围之内，为此研究过程中着重开展了加热设备能力及加热工艺参数等系统参数的研究。研究过程中，利用计算机仿真模拟加热过程，并通过物理实验修正了计算模型参数，优化加热工艺参数。利用仿真计算大大减少了实验的次数，利用仿真计算模型给出的参数可以直接指导工艺实验和生产实践，提高了研究效率。本文着重介绍了宝钢在支承辊整体感应加热技术、TMCP感应加热技术等方面的研究进展。

关键词　感应加热，支承辊，钢管，TMCP

Research of Induction Heating in Baosteel

Wu Cunyou, Zhou Yueming, Jin Xiaoli

(Baoshan Iron & Steel Co., Ltd., Research Institute (R&D Center), Shanghai 201900, China)

Abstract The structure and physical properties of metal materials can be improved by induction heating treatment, due to the development of new product and higher requirement of product properties, the current existing induction heating technology of BaoSteel can not satisfy the new requirement raised by the product line anymore. How to control the temperature to an appropriate range is the key of induction heating technology, thus researches focused on the parameters of induction heating system, including those of induction heating equipment and heating process. During the researches, computer emulation was used to simulate the induction heating process, physical experiment was then carried out to verify and improve the simulation modal, the optimized induction heating parameters were finally proposed. By using computer simulation, the physical experiment times were dramatically decreased, and the proposed heating paraments can be used as a guidance to physical experiment and product line, which increased the efficiency of researches. This paper focuses on the development of induction heating technology in BaoSteel, which includes integer induction heating technology of back-up roller, the inner induction technology of steel pipe, the induction heating technology of TMCP and so on.

Key words induction heating, back-up roller, steel pipe, TMCP

感应加热在钢铁生产中的应用与新进展

季凌云　李守智

（西安美泰电气科技有限公司，陕西西安　710119）

摘　要　本文对感应加热装备技术在钢铁生产中的应用进行了分析，阐述了国内外感应加热技术的进展及感应加热装置在宝钢的应用。并从新技术的应用、电力电子器件和检测技术等方面，说明提高电源性能的措施。另从感应加热电源角度，就一些技术思考及应用谈几点想法，以供商榷。

关键词　电力电子，感应加热，电源装置，检测系统

Development Trend and Thinking about the Technology of the Induction Heating Power Supply

Ji Lingyun, Li Shouzhi

(Xi'an MATTEL Electric Technology Co., Ltd., Xi'an 710119, China)

Abstract The induction heating technology is used in the steel production. In this paper discuss the domestic induction heating equipment used in baosteel and focus on the application new technology and the test instrument for the power electronic devices. Some thinking about the technological advance of the induction heating power supply is also introduced for discussion.

Key words power electronics, induction heating, power supply, technological thinking

热挤压替代压余坯料工艺感应加热数值模拟研究

雷作胜[1] 朱宏达[2] 郭加宏[2] 周月明[3] 金小礼[3] 高 齐[3]

（1. 上海大学省部共建特殊钢冶金与制备国家重点实验室，上海 200072；
2. 上海大学应用数学与力学研究所，上海 200072；
3. 宝钢中央研究院设备所，上海 200019）

摘 要 针对压余坯料感应加热过程建立了电磁场-温度场耦合的二维轴对称数学模型，通过将数值模拟的结果与物理模拟实验结果进行了比较，验证了数学模型的正确性。为实现在最短时间内达到“内外温差较小，整体温度在目标温度”这一控制目标，在对大量计算结果进行分析的基础上，认为锯齿形加热功率曲线是较优的加热方式。这一方法为工程上寻求优化的加热工艺提供了定量的参考。

关键词 感应加热，替代压余，热挤压，数值模拟

Mathematical Modeling of Induction Heating of Discard Substitution Block for Billet Hot Extrusion Process

Lei Zuosheng[1], Zhu Hongda[2], Guo Jiahong[2], Zhou Yueming[3], Jin Xiaoli[3], Gao Qi[3]

(1. State Key Laboratory of Advanced Special Steel, Shanghai University, Shanghai 200072, China;
2. Shanghai Institute of Applied Mathematics and Mechanics, Shanghai University, Shanghai 200072, China;
3. Baosteel Central Research Institute, Shanghai 200019, China)

Abstract A magnetic field and temperature field coupled mathematical model was proposed to calculated the induction heating process of discard substitution block for billet hot extrusion process. The mathematical model is validated by compared the simulation results with the temperature measurement of a physical modeling. Base on systematical analysis of calculation results, a quantitative sawtooth induction power curve was proposed in order to realize the aim of most well distributed temperature field in the block in the least induction time.

Key words induction heating, discard substitution, hot extrusion, mathematical modeling

Development of Transverse Flux Induction Heating for Wide Rang Size Strip

Hou Xiaoguang[1,2], Li Jun[2], Zhou Yueming[2]

(1. Key Laboratory of Electromagnetic Processing of Materials (Ministry of Education), Northeastern University, Shenyang 110016, China; 2. Baoshan Iron & Steel Co., Ltd., Shanghai 201900, China)

Abstract The concept and its characteristics of transverse flux induction heating (TFIH) were introduced, and several typical cases of TFIH development for wide rang size strip were mainly reviewed. Then, according to the latest research progress of TFIH in Baosteel, a few key problems of TFIH for industrial applications were discussed. Finally, the development trend and application prospect of TFIH are prospected positively through analyzing the change of numbers of the related TFIH patents and papers.

Key words induction heating, transverse flux induction heating, longitudinal flux induction heating

Z-Mill Roll Shop Auto Roll Identification and Storage Yard Management

Yeh Ho-Tien

(Electrical & Control Department, China Steel Corporation, Kaohsiung 81233, China Taiwan)

Abstract Generally, in Hot Strip Mill or Cold Mill, the roll ID is etched on head/tail end side face of the roll for visual identification. But in Z-Mill the work roll is too small (65~100mm) to etch roll ID and because of the thrust bearing for work roll, etching roll ID on side face is not allowed. The only way to identify the rolls is to etch the ID on the head side neck which is a small area. Besides, the Z-Mill has to change work rolls at least one time for each coil rolled, which means larger amount of work rolls compare with conventional mills. The rolls have to be stored in the pallet which can store up to 6 rolls and the pallet can be piled up to 5 layers, it is difficult for visual identification in the storage yard. For roll ID identification, CSC R&D department designed the ring shape RFID tags (2mm ring width and thickness, 20/30mm diameter) which fits and glued inside the center hole of the roll head/tail end side. CSC Electrical & Control department configured and integrated the interfaces between the fixed RFID readers and the Roll Shop Management System(RSMS) for automatic roll ID identification; and also design the Android tablet computer application which use Blue Tooth to connect to the portable RFID reader for roll ID、pallet ID and location ID readings, and use wireless network (Wi-Fi) to update the RSMS yard database to facilitate the roll storage yard management.

Key words Z-Mill, RFID, blue tooth, android tablet, Wi-Fi

宝钢五号连铸机综合改造简介

谭希华　沈轶奇

（宝钢工程技术集团有限公司，上海　201999）

摘　要　本文介绍了宝钢五号连铸机综合改造的背景、改造内容与目标，说明了连铸机改造方案设计采用的关键技术，同时介绍了连铸机改造后的使用效果。

关键词　连铸机，改造，设计，关键技术

Brief Introduction of Baosteel No.5 Continuous Caster Revamping Project

Tan Xihua, Shen Yiqi

(Baosteel Engineering & Technology Group Co., Ltd., Shanghai 201999, China)

Abstract The article introduces the background, scope and objectives of Baosteel No.5 continuous slab caster revamping project. It describes the key technologies which are used in the revamping designing of the project. Finally the article introduces the results of the revamping project.

Key words continuous caster, revamping, design, key technology

Liquid Metal Modelling of Continuous Steel Casting

Gerbeth Gunter, Wondrak Thomas, Stefani Frank,
Shevchenko Natalia, Eckert Sven, Timmel Klaus
(Helmholtz-Zentrum Dresden-Rossendorf, Institute of Fluid Dynamics, MHD Department,
Bautzner Landstr. 400, D-01328 Dresden, Germany)

Abstract Model experiments with low melting point liquid metals are an important tool to investigate the flow structure and related transport processes in melt flows relevant for metallurgical applications. We present recent results from the three LIMMCAST facilities working either with room-temperature alloy GaInSn or with the alloy SnBi at temperatures of 200~350°C. The main value of cold metal laboratory experiments consists in the capabilities to obtain quantitative flow measurements with a reasonable spatial and temporal resolution, which is essential for code validation. Experimental results are presented covering the following phenomena: contactless electromagnetic tomography of the flow in the mold, flow monitoring by ultrasonic sensors, mold flow under the influence of an electromagnetic brake, injection of argon bubbles through the stopper rod, X-ray visualization of gas bubble two-phase flow in the nozzle and in the mold.

Key words continuous casting, physical modeling, flow measurements, magnetic field, flow control, electromagnetic brake

Principle of Thermal Spray Nanotechnology and Applications of Nanostructured Coatings in Steel Industry

Marth Charlie, Tian Qingfen
(Shanghai Inframat Nanotechnology Company, Ltd.)

Abstract Nanostructure science and technology is a broad and interdisciplinary area of research and development activity that has been growing explosively worldwide in the past decade years. Nanotechnology has the ability to synthesize, characterize and manipulate matter at the less than 100nm length scale. It has the great potential for revolutionizing the ways in which materials and products are manufactured and the range and nature of functionalities that can be accessed and managed in nanoscale. Nanomaterials as the cornerstones of nanoscience and nanotechnology have demonstrated uniqueness in physics, chemistry, structure array and function. Therefore, nanomaterials and their processing are already having a significant commercial impact on industrial applications, which will assuredly increase in the future.

Thermal spray nanotechnology is to produce a nanostructured deposit or coating containing nano-particles and or nanograins mostly from nanometric or submicrometric material feedstock by thermal spray process. The thermal spray processes for producing a nanostructured coating typically include thermal spray using nano-particle agglomerate feedstock, liquid solution precursor or suspension, or amorphous material feedstock. The main method for thermal spraying of nanocoating includes the sequence of nanoparticle synthesis, agglomeration, particle reconstitution and air plasma spray (APS) or high velocity oxy-fuel spray (HVOF) of nano-feedstock. In comparison with conventional microstructured coatings, the nanostructured coatings have demonstrated dramatic improvement in coating durability, reliability, adaptability and function. Those attractive properties associated with a nanocoating may include, but not limited to superior resistance to corrosion, wear, erosion, impact damage, cracking and spallation. In addition, a nanocoating can provide enhanced function and unique property of catalytic conversion, chemistry homogeneity, optical emission, anti-bacteria and biology. Therefore, there are extensive applications and broad markets for the use of nanostructured thin films, membranes and coatings.

Thermal spray nanotechnology has been developed and applied to enhance the protection and performance of critical components and equipment used in steel industry after near decade of our dedicated and extensive research and development of thermal spray nanotechnology. These nanocoatings are used for steel production lines, including continuous galvanizing line (CGL), continuous annealing line (CAL) and continuous casting line (CCL). The custom-designed nanocoatings are verified to significantly extend the longevity and reliability of the components coated and be cost-effective. A group of thermal spray nanocoatings, either alloy, ceramic or composite, have been developed and implement to resolve those issues and challenge associated with steel production. Typical cases of nanocoatings used for critical components in CGL, CAL, CCL and blast furnace gas recovery systems will be reviewed and discussed in details.

纳米喷涂技术的技术原理和在钢铁行业中的应用

Marth Charlie Tian Qingfen

(英佛曼纳米科技有限公司(上海))

摘 要 纳米结构科学和技术是一门涉及广阔和学科交叉的研究与开发活动，在过去的十年内世界地呈现爆炸式成长。纳米科技可以在小于 100nm 的长度尺度上开展材料的合成，表征和操作。其提供了巨大的机会以革命性的方式进行产品的开发和制造，进而实现了在纳米尺度上进行功能化的序列存取和管理。 纳米材料作为纳米科学和技术的基石已经显示了其在物理学，化学组分，结构矩阵和功能性等多方面的独特性。因此，纳米材料和其工艺处理方法对于工业化应用已经产生了巨大的商业冲击力，并十分确定在未来将进一步成长。

纳米热喷涂技术是采用纳米级或亚微米级粉料通过热喷涂工艺方法制备出包含纳米颗粒以及纳米晶粒的纳米结构的涂层。制备纳米涂层的热喷涂工艺方法通常可采用纳米颗粒聚合的固态粉料，液态溶液前驱液或纳米颗粒的悬浮体，或非晶态材料作为制备涂层的粉料。纳米热喷涂的主要工艺方法可以概括为几个主要次序和步骤，包括纳米材料合成，颗粒聚合，颗粒构造成型和等离子热喷涂（APS）或高速氧焰热喷涂（HVOF）制备沉积涂层。与通常的微米级结构的涂层加以对比，纳米结构的涂层在涂层的耐久性，可靠性，适用性和功能性方面均有显著的改善。 纳米涂层所具有吸引力的性能不仅限制于卓越的抵抗腐蚀，磨损，磨蚀，冲击性破坏，开裂和剥落的能力。此外，纳米涂层可以提供增强的功能和独特的性能，例如催化转化、化学均一性、光学发射、抗细菌和生物学等。因此，纳米结构涂层在薄层，薄膜和涂层的应用方面有广阔的市场前景。

热喷涂纳米技术经过我们仅十年的致力研究和开发，已经应用于钢铁行业的关键设备和仪器的保护和性能的增强。研发的纳米涂层应用于多条钢铁生产线，包括连续热镀金属生产线（CGL）,连续退火生产线（CAL）和连续铸造生产线（CCL）。客制化设计的纳米涂层已被证实可以大幅度延长施加涂层部件的使用寿命和可靠性，并可节省费用。包括合金，陶瓷和复合材料为基体的纳米涂层的家族已经被研发成功并加以应用，可以应对钢铁制造中存在的问题和挑战。纳米涂层在关键设备上的典型应用案例，包括 CGL、CAL、CCL 和高炉煤气回收和利用系统等，将加以详细的介绍和讨论。

Development Trends of Mold Flow Control in Slab Casting

Zhong Alex-Yuntao[2], Pan Hanyu [2], Jacobson Nils [1], Sedén Martin [1]

(1.ABB AB, Metallurgy Products, Terminalvägen 24, 721 59 Västerås, SWEDEN;

2. ABB Metallurgy, ABB (China) Limited, Beijing, China)

Abstract In view of its recognized potential in the improvement of as cast quality of slab, the interest for electromagnetic devices in slab casting has been steadily increasing over the years. By the application of electromagnetic devices for stirring and/or braking in the mold, the surface and subsurface quality of the slab is greatly improved. Various casting conditions with respect to cast qualities, casting velocities and cast dimensions require, however, different electromagnetic approaches with regard to technical solution to achieve best possible result. Casters with low throughput normally require a stirring (AC) function while casters with high output require a braking (DC) function. The challenge has been to design an electromagnetic device which is versatile and flexible in achievingthe best quality improvements within a broad range of throughput.

Key words FC Mold, cleanliness, surface quality, slab

RFID Tags and Auto-Labeling Machine for Heavy Plate Products Identification and Tracing

Pan Shengbo, Wu Ruimin

(Baoshan Iron & Steel Co., Ltd., Shanghai 201900, China)

Abstract RFID (radio frequency identification) is an emerging technology used for identifying objects via radio waves. Its many advantages make it very suitable for steel industry, but because of metal shielding, it is a challenge to deploy the radio frequency in a metal environment. This paper briefly introduces the existing RFID applications in steel industry. By improving the existing RFID tags, we have designed two kinds of RFID tags for heavy plate product identification and tracking. The first one is a paper-based tag, whose RFID inlay hangs out from the plate, makes it has a good readability. Another is a metal tag with long metal strip, which can be mounted on the side surface of the plates. Their reading performances are tested in laboratory environments. We have also developed a corresponding automatic labelling machine. The machine has an automatic hopper for feeding usable tags, and uses an industrial robot as the carrier, to carry a labelling head to complete labelling jobs including tag picking, object detection and tag labeling. The entire working process of the machine on the field is described, can adapt the rapid and continuous steel production. This research of RFID technology for heavy plates lays a foundation for further practical application.

Key words RFID, heavy plate, tag, auto-labeling machine, steel industry

用于连铸机的高可靠性的步进控制液压缸

陈剑波[1]　村上志郎[2]　内野雄幸[2]　村上弘记[3]　星野修二[4]

（1．IHI（集团）石川岛（上海）管理有限公司，上海　200120；2．IHI（集团）IHI 机械系统株式会社，日本东京　135-8710；3．IHI（集团）技术开发本部，日本东京　135-8710；4．IHI（集团）产业系统本部，日本东京　135-8710）

摘　要　日本 IHI（集团）公司（原石川岛播磨重工业株式会社）作为大量使用伺服液压缸的重工业设备开发制造企业，为解决常用电磁伺服液压缸在恶劣的工况下的可靠性问题，在 1971 年开始改建电磁伺服液压缸，开发出步进控制液压缸（简称步进控制缸或步进缸，也被称为数字控制缸）。使用大扭矩的步进电机代替小推力的电磁铁，可以很容易克服因铸件砂粒以及焊渣和加工毛刺等引起的电磁伺服阀卡死故障；使用可靠的精密丝杆机构代替易受干扰的不稳定的电气位移传感器，解决反馈信号的稳定性和可靠性；特别是，根据步进电机的强脉冲电流在恶劣环境下的传输中不易衰减和不受干扰的高可靠性特点，使用其脉冲电流的脉冲数代替弱小的电磁控制电流来进行数字控制，容易确保宝钢连续铸造结晶器的控制精度和可靠性。

关键词　连续铸造，结晶器，液压缸，伺服控制，步进控制

Hydraulic Cylinder with Stepping Control of High Reliability Used in Continuous Caster

Chen Jianbo[1]，MURAKAMI Sirou[2], UCHINO Noriyuki[2],
MURAKAMI hiroki[3], HOSHINO shuuji[4]

(1. Ishikawajima (Shanghai) Management Co., Ltd., IHI Corporation, Shanghai 200120, China;
2. IHI Machinery and Furnace Co., Ltd., IHI Corporation, Tokyo 135-8710, Japan;
3. Technology Development Headquarters, IHI Corporation, Tokyo 135-8710 Japan;
4. Machinery Operations, IHI Corporation, Tokyo 135-8710 Japan)

Abstract　In 1971, IHI Corporation, as a heavy industries corporation used a large number of hydraulic servo cylinder, began to be engaged in solving the reliability problems of the common hydraulic cylinder with electromagnetic servo valve under the adverse conditions and developed this hydraulic cylinder with stepping control. It uses the stepping motor with a large torque instead of the electromagnet with a small thrust to be easily overcome the valve failure caused by casting sand, welding slag and machining burrs, and uses the reliable precision screw rod mechanism instead of the unstable electric displacement sensor to solve the feedback signal stability and reliability. In particular, according to the high reliability characteristics of strong pulse current of stepping motor which is not easy to decay and affected by the interference in signal transmission, the pulse number of pulse current is used to make the digital control instead of the weak electromagnetic current to be easy to ensure the control accuracy and reliability of the continuous casting mould in Baosteel Corporation.

Key words　continuous casting, mould, hydraulic cylinder, servo control, stepping control

高强度钢板残余应力消减方法研究

苏愿晓　张清东　曾杰伟

（北京科技大学机械工程学院，北京　100083）

摘　要　高强度钢板作为板带钢中的一种新型产品，具有显著的减重节材和节能减排的效果。高强度钢板在轧制过程中会产生残余应力，而在影响构件寿命的因素中，残余应力是很关键的一项。通过外加交变磁场和通过机械振动（高频、低频）改善材料的力学性能是两种新型的处理方法。在这两种方法的基础上，自行设计适用于高强钢这种柔性体装卡的振动时效和磁场处理设备，探索研究了高强度钢板在振动时效和磁场处理的不同工况试验下，其残余应力的消减和均化情况。本试验材料选取 DP590 高强度钢板试样，利用 X 射线应力测定仪测量振动时效和磁场处理前后残余应力大小，并进行比较。通过研究表明，振动时效和磁场处理对于消减高强度钢板的残余应力均有一定作用，为科学研究和生产实际提供了一定的参考价值。

关键词　残余应力，高强钢，振动时效，磁处理

The Study on the Way of Residual Stress Reduction in High Strength Steel

Su Yuanxiao, Zhang Qingdong, Zeng Jiewei

(School of Mechanical Engineering, University of Science and Technology Beijing, Beijing 100083, China)

Abstract As a new product in the steel strip, high strength steel has a significant effect on material and energy saving. Residual stress will be produced in the rolling process, and in the factors affecting the service life of components, the residual stress is a key point. There are some new processing method to improve the material mechanical properties, one is adding the alternating magnetic field, and another is through the mechanical vibration treatment (high frequency and low frequency). On the basis of the two methods, the vibration stress relief (VSR) and the magnetic field treatment equipment which is suitable for the flexible body such as high strength steel which is designed all by myself, in order to study whether the two methods in different working conditions have reduction and homogenization effect on the residual stress of high strength steel. The material of the experiment is DP590 high strength steel, the residual stress of the sample before and after the treatment is measured by X-ray stress measurement instrument. Studies shows that VSR and magnetic field treatment both have a certain effect on cutting the residual stress of high strength steel, which provide some reference value for scientific study and production practice.

Key words residual stress, high strength steel, vibration stress relief, magnetic treatment

Weakly Supervised CNNs based Surface Defect Localization and Recognition

Zhang Yun[1], Li Wei[1], Song Yonghong[1], He Yonghui[2]

(1. Institute of Artificial Intelligence and Robotics, Xi'an Jiaotong University, Xi'an Shanxi 710049, China; 2. Equipment Research Institute, R&D Center of Baoshan Iron & Steel Co., Ltd., Shanghai 200941, China)

Abstract Automatic quality monitoring for steel plate is a typical challenge in industries. Although there have been several research results at home and abroad, the requirements of defect detection and recognition for low-contrast complex steel plate, are still unable to satisfy. Currently, computer vision technology based surface defect detection and recognition approaches for steel plate are increasingly concerned and taken. An effective approach can not only reduce working intensity and improve working efficiency, but also retrieve financial lost for the enterprise, which has important theoretical significance and practical application prospects. For the fact that there is less annotated data and more un-annotated data in dataset, a weakly supervised approach of surface defect location and recognition for steel plate, was presented with convolutional neural networks. First, the annotated data of defects was used to learn and extract the features by supervised convolutional neural networks. Second, the features was transferred to another net as mid-level defect representations, and then, the un-annotated data of defects was used to train the new multi-scale net. Third, by composite analysis of the multi-scale results, the final defects of the steel plate were located and recognized. Furthermore, the proposed approach has a robust and wonderful performance, which was verified by lots of experiments.

Key words defect detection, convolutional neural networks, weakly supervised

大型高炉热风管系开裂原因的调查

陈　辉　马泽军　孙　健　武建龙　梁海龙　刘文运

（首钢技术研究院，北京　100043）

摘　要　某大型高炉运行3年后，热风管系陆续发现粘稠液相流出物、砂眼、开裂，是失效烧出，利用金相显微镜、扫描电镜和能谱分析的手段，调查表明，液相粘稠流出物含有大量的$(NO_3)^-$，钢材开裂形式为沿晶开裂，说明存在典型的硝酸根腐蚀，失效的波纹管存在高温疲劳断裂倾向。因此，热风管道的长期安全运行需要从腐蚀防护与防止管壁接触高温空气两方面努力。

关键词　大高炉，热风管系，腐蚀

Finite Element Analysis of a Vertical Edger Longevity

Zhou Chenn[1], Sun Yuanbang[1], Wu Bin[1], Nolar Mitchell[2], Cox Jeffery[2], Klootwyk Jason[2], Chatman Cliff[2]

(1. Center for Innovation through Visualization and Simulation (CIVS) Purdue University Calumet, 2200 169th Street, Hammond, IN, USA 46323; 2. ArcelorMittal 3210 Watling Street 2-975, East Chicago, IN, USA 46312)

Abstract　A vertical edger is a critical component of the roughing mill, used to reduce the width of steel slabs. When a vertical edgerexperiences a failure, it directly affects the production of the roughing mill. One such failure that has occurred took place in the R2 housing, which is not a typical location for failure to occur.The housing is designed to withstand significant force, but the maximum threshold for duration and intensity before failure occur are unknown. In addition to the force, many other factors affect theremaining life of a component, and life prediction is difficult predict. Furthermore, new steels seem to be pushing the roughing mills harder than before, and it is crucial to know how far they can be pushed before components break. The purpose of this research is to simulate the edging process using Finite Element Analysis (FEA) model.By performing a numerical analysis of the components in a vertical edger, an improved understanding of the mechanical behavior and potential failures can be identified to improve safety.

Key words　roughing mill, vertical edger, FEA, longevity extrapolation, fatigue analysis

高效环保冷镦钢丝表面酸洗处理机组开发

马世峰　杨和来　范应奇

（宝钢工程技术集团有限公司，上海　201999）

摘　要　本文介绍了冷镦钢丝表面酸洗处理的工艺及装备配置，对机组布置进行分析，重点介绍了开发机组的主要技术特点，以及实现高效、节能、环保生产的解决方案。

关键词 冷镦钢丝，酸洗，表面处理，高效、节能、环保

Development of ECO- Cold Heading Steel Pickling & Coating Process Line

Ma Shifeng, Yang Helai, Fan Yingqi

(Baosteel Engineering & Technology Group Co., Ltd., Shanghai 201999,China)

Abstract This paper introduces the process & related equipment for cold heading steel pickling & coating line, puts forward analysis for layout, the main technical feature , and the solution for ECO

Key words cold heading steel, pickling, coating, ECO

An Investigation into the Wear Behaviour of High Speed Steel Work Roll Material under Service Conditions

Tieu Kiet[1], Zhu Hongtao[1], Deng Guanyu [1], Zhu Qiang[1], Zhang Jie [1], Wu Qiong[2], Fan Qun[2], Sun Dale[2]

(1. School of Mechanical, Materials and Mechatronic Engineering, University of Wollongong, New South Wales 2522, Australia;
2. Baosteel Central Research Institute, Baosteel, Shanghai 200941, China)

Abstract HSS work rolls have been utilized for both early and last finishing stands in hot strip mills due to their high hardness, good wear resistance and high temperature properties. During hot rolling, the work rolls are subjected to rapid changes of mechanical and thermal loads on their surface layers due to the alternated contact with the hot strip and the cooling water in every revolution. It is important to understand the thermal and wear behaviours of HSS work rolls under industrial service conditions. In the present study, HSS roll samples from damaged rolls during hot rolling service have been characterised to explore the role of oxide scale in the deterioration process of hot roll surface during rolling. From nano-indentation/nano-scratch tests, the modulus of elasticity, hardness of the oxide scale and the scale thickness were found to affect the friction of the oxide surface. A series of wear tests on HSS was carried out on a high temperature oxidation/wear roller-on-disc wear test rig which can emulate the contact conditions of hot rolling. Further investigation on the wear behaviour of HSS material was then carried out according to the temperature guidance provided by a FEM modelling analysis of the temperature evolution within a work roll based on Baosteel hot strip rolling conditions.

Key words high speed steel, work roll, oxidation, wear, hot strip rolling

物流仿真技术在多品种混流生产机组设计中的应用

谢小成

（宝钢工程技术集团有限公司，上海 201999）

摘　要　针对多品种混流生产线存在的问题，对混流生产线进行了分析研究，然后介绍物流仿真技术的特点和作用。最后，结合实践应用案例，详细描述物流仿真技术在混流生产线设计中的应用过程。

关键词　物流仿真技术，多品种混流生产

The Application of Logistics Simulation Technology in the Design of Multi-species Mixed Production Unit

Xie Xiaocheng

(Baosteel Engineering & Technology Group Co., Ltd., Shanghai 201999, China)

Abstract　According to the problem of mixed flow production line, the research on mixed flow production line was done, and then described the characteristics and role of logistics simulation technology. Finally, with an application cases, the application process of the logistics simulation technology in mixed production lines design was detailed.

Key words　logistics simulation technology, multi-species mixed production

湛江钢铁转炉二次除尘系统的技术创新

张挺峰　孙　洁

（宝钢工程技术集团有限公司，上海　201999）

摘　要　对湛江转炉炼钢二次除尘系统的方案确定过程进行分析总结，论述了除尘系统的设计与工艺生产密切结合的重要性，通过优化烟气捕集方式和系统划分等技术创新，合理确定系统设计风量。在保证除尘系统使用效果的基础上有效控制总风量，实现除尘系统高效性和运行经济性的目标。

关键词　转炉炼钢，二次除尘，烟气捕集，系统划分

The Technical Innovation of the Zhanjiang BOF-Steelmaking Secondary Dedusting System

Zhang Tingfeng, Sun Jie

(Baosteel Engineering & Technology Group Co., Ltd., Shanghai 201999,China)

Abstract　This paper summaries the plan determination process of the Zhanjiang BOF-steelmaking secondary dedusting system, discusses the close relationship between the design of th dedusting system and the production process.According to features of the BOF-steelmaking secondary dedusting system, by optimizing the flue gas capture mode and the technology innovation such as system division ,the air flow of the dedusting system is reasonably determined. On the basis of the guarantee of the dedusting effect, the total air flow of the dedusting system is effectively controlled, the high efficiency and economic operation of the dedusting system is achieved.

Key words　BOF steelmaking, secondary dedusting, flue gas collecting, system division

高性能大功率中压交直交轧机变频系统开发与应用

张勇军[1] 尚 敬[2] 胡家喜[2] 郝春辉[1]

（1. 北京科技大学高效轧制国家工程研究中心，北京 100083；
2. 南车株洲电力机车研究所有限公司，湖南株洲 412001）

摘 要 基于双 PWM 控制的高性能三电平中压交直交系统已成为轧机主传动的主流控制设备。本文主要介绍国产化高性能中压交直交三电平变频调速装置的开发与研制关键技术及构成，分别对系统网侧与电机侧数学模型，以及整流侧的双闭环控制和逆变侧的气隙磁链控制进行了介绍。该系统已成功应用于某热轧线粗轧机的大功率电励磁同步电机的驱动，现场实验结果证明，该系统性能良好、运行可靠，完全满足高性能轧制工艺需求，为推动我国自主高性能交直交调速装备步入国际先进行列迈出坚实一步。

关键词 轧机主传动，交直交变频，电励磁同步电机，气隙磁链控制

Development and Application of High-performance AC-DC-AC Middle-voltage Frequency-variable Inverter in Main Drive System for Rolling Mill

Zhang Yongjun[1], Shang Jing[2], Hu Jiaxi[2], Hao Chunhui[1]

(1. National Engineering Research Center for Advanced Rolling Technology of USTB, Beijing 100083, China;
2. CSR Zhuzhou Institute Co., Ltd., Zhuzhou 412001, China)

Abstract As the focal point of the present energy-saving emission reduction, high-performance AC-DC-ACmiddle-voltage frequency-variable inverter in mill main drive system is the mainstream of large mechanical drive equipment. The core technology based on independent innovation about double three-level frequency conversion system of electrically excited synchronous machine is presented in this article. The double closed loop control with space vector modulation for PWM rectifier and the air gap flux control for PWM inverter are adopted. This system is successfully used in a main drive system for a hot rough mill. Testing results show that the system has reliable performance, and the system can completely meet technological requirements of high-performance rolling.It is an important step to promote our country's own high performance AC speed control equipment toward the international advanced level.

Key words mill main drive, AC-DC-AC frequency-variable, electrically excited synchronous machine, air gap flux control

防止开卷机带尾损伤的研究

孙长津

（鞍钢股份有限公司冷轧厂，辽宁鞍山 114021）

摘　要　本文通过对鞍钢股份冷轧厂二号线清洗机组开卷机卷筒上的剩余带尾部分进行了研究。从带材在卷筒上的几种损伤形式入手，详细分析了带尾涨碎、带尾擦伤两种带尾损伤方式形成原理，并推算出带尾损伤的边界条件与带尾剩余量之间的函数关系，并结合传统卷筒的设计方式——钢卷在卷筒上与卷筒之间不发生滑移条件推算出卷筒涨径压力的可用区间。在可用区间内，使用安全裕度等值方式得出涨径压力与剩余卷径关系的最优曲线。

关键词　冷轧，带尾，成材率，涨径力

The Study on the Prevention of Band Tail Injury on Uncoiler

Sun Changjin

(The Cold Rolling Plant of Ansteel Co.,Ltd., Anshan 114021, China)

Abstract　In this article, based on the two line of Ansteel Cold Rolling Plant cleaning residual tail part processing uncoiler reelis studied. Starting from the several forms of damage in the strip, a detailed analysis of the tail, tail up brokenscratch two tail damage form principle, and calculated the function relationship between the band tail damageboundary condition and the remaining amount of the tail, and -- steel coil on the drum and the drum between theno slip condition calculate the available interval drum up diameter pressure combined with the traditional design method of drum. In the available range, the use of safety margin equivalent way optimal curve expanding pressure and the residual volume relationship.

Key words　cold rolling, band tail, the yield, expanding force

非稳态连续退火过程带钢张应力分布仿真分析

张晓华[1]　徐　刚[2]　王英杰[2]　张清东[1]

（1. 北京科技大学机械工程学院，北京　100083；
2. 宝山钢铁股份有限公司冷轧厂，上海　200941）

摘　要　在连续退火过程中，横向局部张力集中导致的不均匀拉伸变形是带钢产生瓢曲的主要原因。工业生产过程中，在变规格或加减速阶段连续退火炉内更易发生此类缺陷，因此需重点研究非稳态连续退火过程中带钢的横向张力分布，探究炉内带钢横向张力的影响因素及分布规律。针对带钢变规格或加减速等非稳态过程，通过建立导向辊-带钢一体化有限元模型，分析了不同的导向辊辊形对炉内带钢横向张力分布的影响以及减速、加速过程中带钢与导向辊间出现的相对滑动状态对带钢横向张力分布的影响。结果表明，导向辊辊形可改善带钢横向张力分布；而导向辊与带钢之间的相对滑动状态会引起炉内靠近炉辊的带钢横向张力发生较大幅度的波动，导致带钢产生不均匀的纵向拉伸变形而引起板形缺陷。

关键词　连续退火，张力，有限元，瓢曲

The Simulation Analysis about the Strip Tension Stess Distribution in the Process of Unsteady Continuous Annealing

Zhang Xiaohua[1], Xu Gang[2], Wang Yingjie[2], Zhang Qingdong[1]

（1. School of Mechanical Engineering, University of Science and Technology Beijing, Beijing 100083, China;
2. Cold Rolling Plant,Baoshan Iron & Steel Co., Ltd., Baosteel Branch, Shanghai 200941, China）

Abstract In the process of the continuous annealing, the uneven tensile deformation which is caused by the lateral local stresses concentration is a leading reason of the strip buckling phenomenon. In the process of industrial production, this kind of defect is more likely to happen in the continuous annealing furnace when switching the strip specification or changing the speed of the strip. It is important to study not only the transverse distribution of tension but also its influence factors and the regularities of distribution in the process of continuous annealing. The influence of the profiles of guide roller and the relative sliding state that happens when changing the speed of the strip or switching the strip specification will be studied with the established integrated simulation model of the guide roller and strip. Different relative sliding state is described with different friction coefficient. The simulation results show that the profiles of guide roller will contribute to improving the transverse distribution of tension. In the unsteady continuous annealing process, the relative sliding between the roller and the strip may cause the significant fluctuations on the transverse distribution of tension,which is more likely to cause the uneven longitudinal tensile deformation on the strip. The non-uniform deformation will cause the sheet shape defect.

Key words continuous annealing, tension, finite element method,buckling

不锈钢/铜/钛复合板复合界面元素扩散行为研究

冯 哲 张清东 张勃洋

（北京科技大学机械工程学院，北京 100083）

摘 要 金属层状复合材料的结合层宽度是影响金属层状复合材料物理、力学性能的根本因素，而轧制后复合材料结合层宽度的变化主要取决于热处理工艺。以不锈钢/铜/钛复合板为研究对象，对其进行不同加热温度及不同保温时间的热处理，获得了不同热处理条件下复合板复合界面微观组织及元素扩散行为规律，揭示了热处理工艺与复合界面元素扩散的内在联系，为不锈钢/铜/钛金属复合板的热处理工艺提供了相应指导。实验结果表明：热处理工艺可明显增加复合板结合层宽度提高复合板力学性能，热处理工艺的保温时间及加热温度对结合层宽度的影响基本呈线性规律；且成品复合板的不锈钢/铜和铜/钛初始结合层宽度存在较大差距，但复合板经过热处理后这种差距明显减小。

关键词 热处理，元素扩散，复合板，结合层

Research on Elements Diffusion Behavior of Composite Interface of Stainless Steel/Cu/Ti Clad Plate

Feng Zhe, Zhang Qingdong, Zhang Boyang

（School of Mechanical Engineering, University of Science and Technology Beijing, Beijing 100083,China）

Abstract The width of composite interface is a fundamental factor to influence the physical and mechanical properties of the clad plate, and the change of width of composite interface of the clad plate which has been rolled depends mainly on heat treatment process. Based on the research on elements diffusion and microstructure of Stainless steel/Cu/Ti clad plate with the different heating temperature and holding time of heat treatment process, the internal relation between elements diffusion of composite interface and the heat treatment is revealed, and the test and theoretical foundation for the subsequent heat treatment of Stainless steel/Cu/Ti clad plate is provided. The results indicate that, for the Stainless steel/Cu/Ti clad plate which has been rolled, the roughness and cleanliness of stainless steel and titanium surface before rolling has a great influence on width of composite interface of Stainless steel/Cu/Ti clad plate which is the final product, but this effect can be diminished by heat treatment process after rolling, the mechanical properties of the clad plate is improved by increasing the width of composite interface. And the influence of the holding time and heating temperature of the heat treatment process for the width of composite interface is basically a linear change.

Key words　heat treatment, elements diffusion, clad plate, composite interface

基于内聚力模型的金属复合板再轧制减薄有限元模拟

冯　哲　张清东　张立元

（北京科技大学机械工程学院，北京　100083）

摘　要　再轧制减薄工艺是将已完成复合工艺的金属复合板通过再轧制的方法使其厚度变薄、表面质量变好、各方面性能也有所增强的工艺。本文利用二维弹塑性有限元法，对不锈钢-碳钢-不锈钢复合板轧制减薄模型进行研究，为了更加真实还原不锈钢-碳钢复合板结合界面的实际情况，采用内聚力模型对结合界面进行模拟，可以为复合板再轧制减薄工艺提供更加准确的指导。分析结果表明：在复合板再轧制减薄过程中，随着压下量的提高，稳定波动阶段平均轧制力会随之线性增加，而轧后厚度会随之线性减小，且压下量是改变轧后厚度的主要原因；随着不锈钢层厚度与碳钢层厚度比的增加，所需的轧制力会线性升高，而界面层受到的应力会相应减小，对界面的损伤也会相应降低。

关键词　内聚力模型，再轧制减薄，金属复合板

The Finite Element Analysis on Thickness-reduction by Rolling of Metal Clad Plate with Cohesive Zone Model

Feng Zhe, Zhang Qingdong, Zhang Liyuan

（School of Mechanical Engineering, University of Science and Technology Beijing, Beijing 100083,China）

Abstract　The thickness reduction by rolling is a process which can make the thickness of a composite metal clad plate thinning, the surface quality of it better and the various aspects performance of it enhancing. The model of the thickness-reduction by rolling of Stainless steel/Carbon steel/Stainless steel clad plate is studied by the method of 2d elastic-plastic finite element. In order to restore the actual situation of bonding interface of Stainless steel/Carbon steel clad plate, the cohesive unit is adopted to simulate the bonding interface. So the more accurate guidance of the thickness-reduction by rolling of clad plate is provided. The analysis indicate that, in the process of the thickness-reduction by rolling, with the improvement of rolling reduction, the average rolling force in the stable fluctuations phase has linear increase, but the thickness after rolling has linear decrease, and the rolling reduction is the main reason for the change thickness after rolling. The change of the rolling reduction has a little impact on the maximum SDEG value. With the increase of the thickness ratio of stainless steel and carbon steel, the rolling force has linear increase, but the stress in bonding interface has decrease, so the damage of bonding interface reduces.

Key words　cohesive zone model, thickness-reduction by rolling, metal clad plate

带钢冷轧机工作辊表面微观形貌耐磨性研究

李 根[1] 叶学卫[2] 张清东[1] 马 磊[1]

（1. 北京科技大学机械工程学院，北京 100083；
2. 宝山钢铁股份有限公司冷轧厂，上海 200941）

摘 要 钢板表面微观形貌由轧辊表面微观形貌（磨削及毛化制备初始形貌或磨损后形貌）通过有/无润滑介质界面轧制塑性变形转印到带钢上而形成。其中工作辊表面磨损量为微米或亚微米级的表面微观形貌磨损，直接影响成品带钢表面质量和生产成本。针对高强/超高强钢生产过程中轧制力大引起的工作辊表面微观形貌磨损较快这一生产实际，在原有的电火花毛化加工工作辊表面基础上，增加工作辊表面超精磨工艺，以期提高工作辊表面形貌的耐磨性。在此基础上，利用元胞自动机方法建立轧辊表面微观形貌磨损仿真模型，分别模拟电火花毛化工作辊和先电火花毛化再超精磨工作辊表面微观形貌的磨损演变过程，并通过现场工业试验验证仿真结果的正确性。理论和实验都证明了先电火花毛化再超精磨加工工艺可明显提高工作辊表面微观形貌的耐磨性。

关键词 高强钢，磨损仿真，元胞自动机，超精磨

Study on the Wear Resistance of the Cold Rolling Mill

Li Gen[1], Ye Xuewei[2], Zhang Qingdong[1], Ma Lei[1]

(1. School of Mechanical Engineering, University of Science and Technology Beijing, Beijing 100083, China;
2. Cold Rolling Plant, Baoshan Iron & Steel Co., Ltd., Baosteel Branch, Shanghai 200941, China)

Abstract The morphology of the surface of the steel plate(preparation of initial morphology or the morphology of the worn surface by grinding and texturing) is formed by the plastic deformation which through the interface of a / no lubrication medium.The surface wear of the work roll is a micron.andsurface morphology wear directly affects the surface quality and production cost of finished products.In the production process of high strength and ultra high strength steel, the surface morphology of the work roll surface is much faster.On the basis of the surface of the work roll surface, the superfinishingultra precision grinding process is increased, which is to improve the wear resistance of the surface morphology of the work roll.On this basis, wear simulation model of roll surface micro topography is established by using the theory of cellular automata were simulated EDT of work roll and SF ultra precision grinding surface of work roll wear evolution process, and the correctness of the on-site industrial test to verify the simulation results.Theoretical and experimental results show that the wear resistance of the surface morphology of the working rolls can be improved obviously by the first EDT of the SF.

Key words shigh strength steel, wear simulation, cellular automata, superfininshing

多辊矫直机板形调控性能仿真研究

焦宗寒[1] 卢兴福[1] 冯 哲[1] 孙大乐[2] 王学敏[2] 张清东[1]

（1. 北京科技大学机械工程学院，北京　100083；2. 宝山钢铁股份有限公司，上海　200941）

摘　要　热轧高强钢板在热处理过程中，会在板带内部引起不均匀内应力，进而出现各种各样的板形不良情况，其后增设的矫直机就是为了改善板形，消除内部残余应力。矫直机的作用在于通过合理的辊缝设定使板带发生较大的弹塑性变形，实现多次反复弯曲以达到矫正板带板形平直性缺陷、消除钢板瓢曲和翘曲、降低残余内应力的目的。矫直机矫正作用的发挥对板带最终产品质量的控制起到重要的作用。矫直模型是矫直机及矫直技术的核心关键，其计算精度直接决定生产节奏和产品质量的稳定性。因此本文研究了及辊缝设定和弯辊作用对板形的改善效果，深入研究多辊矫直机的板形矫正行为与规律、参数化矫直过程仿真模型、合理的矫直工艺策略与规程和矫直过程建模方法与手段。

关键词　多辊矫直，板形缺陷，压下量，弯辊量

Simulation and Analysis on Shape Control Behavior of Multi-roller Straightener

Jiao Zonghan[1], Lu Xingfu[1], Feng Zhe[1], Sun Dale[2], Wang Xuemin[2], Zhang Qingdong[1]

(1. School of Mechanical Engineering, University of Science and Technology Beijing, Beijing 100083, China;
2. Baoshan Iron & Steel Co., Ltd., Shanghai 200941, China)

Abstract　High strength hot-rolled plate could have residual stress distribution and shape defects in the process of heat treatment. Roller levelling is the last straightening process in the cut-to-length line, eliminating these shape defects by an alternating elastic-plastic bending of the sheet with a decline of bending intensity. The roller levelling technology would eliminate waves, curl and residual stress. The model is the key to the straightening technology. The calculation accuracy directly determines the stability of the product quality. Therefore, this project is intended to research the roll gap setting and roll bending effect on the residual stress distribution and shape defects. Study on the behavior of the profile correction of multi roll straightening machine and parametric straightening process simulation model. The simulation and analysis research have laid the groundwork for the future work.

Key words　roller levelling, flatness defects, intermesh, convexity of the rollers

棒材近表面旋转超声横波检测校准方法及工艺分析优化

胡柏上

（宝钢集团韶关钢铁有限公司，广东韶关　512123）

摘　要　棒材近表面自动旋转超声横波扫查的斜探头调节难度大、耗时长、调节效果差、每次调节重复一致性差，没有规范一致的调节方法、检测结果存在近表面区漏检和表面误报等不确定问题，以上问题在设备投用后的较长时间未能得到有效解决。本文首先针对以上问题进行试验分析，确定斜探头调节最佳方法，使斜探头调节操作规范、准确、高效、且调节效果一致性好，然后对棒材近表面自动旋转超声横波扫查工艺进行推导计算，由此分析现有工

艺的不足，并给出更换规格时的斜探头调节工艺优化及实施方法。最后介绍棒材近表面横波扫查关系式在定量评价设备运行精度要求和材料外形要求方面的应用，为确保设备运行稳定性和检测效果分析提供理论依据，为检测结果准确性提供了有效保障。

关键词 棒材近表面区，自动旋转超声检测，横波扫查校准，扫查工艺优化，工艺评价

Calibration Method and Process Analysis for the Ultrasonic Shear Wave Inspection of the Near Surface of Bar

Hu Baishang

(Baosteel Group Shaoguan Iron and Steel Co., Ltd., Shaoguan Guangdong 512123, China)

Abstract Bar near surface automatic rotary ultrasonic shear wave scanning oblique probe regulation is difficult, time-consuming and poor adjustment effect, each adjustment repetitive consistency is poor, not of the same specification adjustment method, test results are uncertain problem near surface detection and surface false positives, the above problems in equipment put into use for a long time failed to get effectively resolved. Firstly in view of the above problems are tested and analyzed. Determine the oblique probe adjusting method is the best, the oblique probe adjustment operation standardized, accurate, efficient, and moderating effect of consistent good, then the bar near surface automatic rotary ultrasonic shear wave sweeping check process was deduced, thus to analyze the deficiency of the existing technology is given, and when replacement specification oblique probe regulating process optimization and implementation method. Finally, the bar near surface shear wave sweeping check in the quantitative evaluation of equipment running accuracy and shape requirements of application, to ensure the stability of equipment operation and test results analysis provide theoretical basis, to provide effective guarantee for the accuracy of the detection results.

Key words near surface area, automatic rotation ultrasonic testing, transverse wave scanning, scanning process optimization, process evaluation

极薄带钢非稳态轧制过程板形仿真研究

张亚文[1] 刘亚军[2] 张清东[1]

（1. 北京科技大学机械工程学院，北京 100083；
2. 宁波宝新不锈钢有限公司，浙江宁波 315807）

摘　要 非稳态轧制阶段一般伴随着张力的连续变化或轧制速度的连续变化，过焊缝轧制阶段是一种典型的非稳态轧制阶段，表现为轧制速度的连续变化。某钢铁厂某条高速冷连轧机组在过焊缝轧制时存在较为严重的板形缺陷，针对这一问题，结合现场实际情况，选取过焊缝轧制过程中几个不同的轧制速度时刻进行分析，通过有限元软件仿真分析了轧制速度改变对板形的影响，弯辊力改变对板形的影响，并在此基础之上提出了非稳态轧制过程中轧制速度连续变化时板形的弯辊力补偿控制策略，为现场非稳态轧制过程中板形控制提供了依据。

关键词 非稳态，板形，仿真分析，弯辊力补偿

The Simulation Analysis about the Shape of Extremely Thin Strip in the Unsteady Rolling Stage

Zhang Yawen[1], Liu Yajun[2], Zhang Qingdong[1]

(1. School of Mechanical Engineering, University of Science and Technology Beijing, Beijing 100083, China;
2. Ningbo Baoxin Stainless Steel Co., Ningbo Zhejiang 315807, China)

Abstract Generally, rolling tension or speed changesduring theunsteady rolling stage.The over weld rolling stage is one kindof unsteady rolling stages and the rolling speed is continuously changed in this stage. In the high-speed cold rollingprocess of one steel company,there were some seriousdefects of strip shape during the over weld rolling stage.To solve this problem,the paper analyzes some typical processes by selecting several different rolling speeds in the over weld rolling stage considering the actual situation of the scene,and finds the impact of rolling speed and bending force on the strip shape.What's more, the paper puts forward the bending force compensation control strategy for improvingthe quality of strip shape in theunsteady rolling stage,which provides the basis for controllingstrip shape in theunsteady rolling stage.

Key words unsteady rolling stage, strip shape, simulation analysis, bending force compensation control

烧结主抽风机电气系统构成与维护策略

于小光[1,2] 李维亚[2] 常文兴[2]

（1. 华北理工大学电气工程学院，河北唐山 063000；
2. 河北钢铁集团唐钢公司，河北唐山 063016）

摘 要 简述了主抽风机在烧结系统生产中的地位及对于烧结区配电网的重要影响，明确了设备的重要性。提出了将主抽风机电气系统列入状态维修的维修模式。简述主抽风机电气系统的构成，逐项分析了同步电机、水电阻降压启动系统、高压及保护柜、直流励磁装置构成特点。论述了主抽风机的启动过程及同步运行要求。着重分析电气系统组成部分设备的点检维护方法，提出依托设备改进、日常点检、定修手动试车检测方法保设备的可靠性，并就励磁装置常见故障总结归纳了应急处理方向。通过正确的维护策略，实现了主抽风机的长周期稳定运行，为电气设备维护积累了宝贵经验。

关键词 主抽风机，电气系统，构成，维护策略

Composition of Electric System for Main Drawing Fan in Sintering Machine and Its Maintenance Strategy

Yu Xiaoguang[1,2], Li Weiya[2], Chang Wenxing[2]

(1. College of Electrical Engineering, Huabei University of Science and Technology, Tangshan 063000, China; 2. Tangshan Iron and Steel Company, Hebei Iron and Steel Group, Tangshan 063016, China)

Abstract Briefly described the status in the sintering system production of the main drawing fan, and it's important influence in power distribution network of sintering area, describe the importance of equipment. The electric system of main drawing fan is repaired accord to it's status. Briefly described composition of electric system, such as synchronous motor, water resistance, high-voltage board and excitation cabinet. Start-up procedure and synchronous operation is detailed description. Electrical equipment maintenance methods are analyzed, also talk about transformation equipment, routine inspection, manual test to keep the equipment reliable. Emergency management measures is listed in order to eliminate the fault of the excitation cabinet. The main drawing fan operate normal as long as 8 years via the suit maintenance strategy, meanwhile the technical level of electrical engineer is improved.

Key words main drawing fan, electric system, composition, maintenance strategy

Development of Reappearance and Analysis System of Section Characteristics in Hot Continuous Mill Unit

Zhou Lianlian[1], Qian Cheng[2], Bai Xiaoye[2], Liu Yaxing[2], Bai Zhenhua[2]

(1. Electronic Experiment Center, Yanshan University, Qinhuangdao Hebei 066004, China;
2. National Engineering Research Center for Equipment and Technology of C.S.R.,
Yanshan University, Qinhuangdao Hebei 066004, China)

Abstract The measurement data about section characteristic and the crown meter only serve the current process in hot continuous mill unit production processes, and can not be applied in quality objection analysis and guidance of parameters optimization of production process in the down-stream product, which results in low data utilization ratio. In order to solve this problem ,the ordinary four-roll hot continuous mill unit was taken as the object, fully considering the hot continuous mill unit equipment and process characteristics. Making use of relevant function of crown meter and data acquisition system, a set of reappearance and analysis technology of section characteristics for hot continuous mill unit has been put forward. The corresponding reappearance and analysis system of section characteristics was developed, and was applied to a certain 2050 hot continuous mill unit, which has achieved good results. Not only was the different moment of rolling strip thickness distribution dynamic reappearance, but also the 3D distribution of the section characteristics of the whole volume strip and the distribution characteristic of the crown and the wedge are given. This technology effectively improved the quality of the hot rolling process of the product and reduced the quality complaints from the down-stream user and created the larger economic benefits.

Key words hot continuous mill, section characteristic, crown, wedge

宁波钢铁热轧厂 1780mm 打捆机常见故障分析及处理

马 克

（宁波钢铁有限公司设备部，浙江宁波 315800）

摘 要 打捆机是用来对成品钢卷打捆包装的设备，宁波钢铁打捆机为日本钢板工业生产的 FA-32NH 型气动打捆机，操作模式分为手动、半自动和全自动三种，控制采用西门子 S7-300PLC。本文通过对打捆机常见故障的总结，分析故障产生的原因，提出相应的解决措施。

关键词 打捆机，故障，处理方法

Failure Analysis of Bundler and Its Solution in Ningbo Steel 1780mm Unit

Ma Ke

(Equipment Department of Ningbo Steel Co., Ltd., Ningbo 315800,China)

Abstract　Bundler is used for packaging of finished coil devices,NingBo Steel bundlers were made from Japan.There are FA-32NH type pneumatic bundlers. They operating mode divided into three kinds,there are manual, semi-automatic and fully automatic.There were controlled by Siemens S7-300 system.Based on the bundlers common fault summarize, analyze the cause, propose solutions.

Key words　bundler, fault, solution

高综合性能输送带——芳纶芯带的研制与应用

向　何　姚学功　曾宗义　孙业斌

（宁波钢铁有限公司，浙江宁波　315800）

摘　要　抗拉强度、黏合强度、抗冲击性及节能性等综合性能优异的输送带是厂家研发的方向，也是用户的追求，而输送带的性能取决于其面胶和芯层，常规的输送带为达到使用要求的抗拉强度：强度要求偏低点的采用多层芯的织物芯输送带，强度要求高的采用绳径较大的钢丝带，这样输送带的总厚度较厚、重量偏重，而面胶又较薄，造成输送带使用较耗能、寿命偏短。本文以耐高温带为研究对象，对其芯层和面胶深入研究和试验，在不降低芯层抗拉强度前提下，将芯层由多层芯改为单层芯，减薄减轻芯层厚度及重量，同时改善面胶的耐高温性能，并适当增厚上盖胶，延长输送带的使用寿命，通过上述研究及各项性能试验，研制出综合性能优异的新型输送带，并进行了成功的应用，取得了较大的经济效益。

关键词　输送带，综合性能，芳纶带，芯层

The Development and Application of the High Integrated Performance Conveyor Belt- Aramid Fiber Belt

Xiang He, Yao Xuegong, Zeng Zongyi, Sun Yebin

(Ningbo Steel Co., Ltd., Ningbo 315800, China)

Abstract　The conveyor belt with tensile strength, bonded strength, anti-impact and energy conservation and so on , excellent integrated performance,is the factory research and development direction, the factory research and development direction, also is user's pursue, but the conveyor belt performance is decided by its rubber and the core layer.Conventional conveyor belt for use to achieve the requirements of tensile strength: strength requirements of the low points of the core multilayer fabric core conveyor belt, high strength the rope diameter larger steel ribbon, such delivery with the total thickness of the thicker, emphasis on weight, and surface glue and thinner, resulting in conveyer belt is energy consumption, life is too short. This article take the thermostable belt as the object of study, and improve the surface adhesive resistant high temperature performance and properly thickened rubber cover, developed with excellent comprehensive performance new conveyor belt, and the successful application, has made great economic benefits.

Key words conveyor belt,integrated performance, aramid fiber belt, the core layer

换热器管束腐蚀减薄的准确检测方法

张雪峰

（上海金艺检测技术有限公司，上海 201900）

摘 要 以往我们一般采用便携式超声波探伤仪或测厚仪进行检测。然而由于便携式超声波仪器需要探头直接接触到工件表面，因此当被测工件处于封闭空间或者探头与工件被测表面耦合不良的时候，仪器容易出现误判或者无法判断的情况。而换热器管束一般都是密集排列的管箱，每次仅能检测局部，无法准确评估换热器的整体情况。本文介绍一种基于脉冲回波检测法，专门针对换热器管束检测的技术旋转超声检测，经过对锅炉换热器管束样件进行实验，在管束样件中发现缺陷已图像显示，深度、厚度、位置、减薄量等信息可以直观显示在图像中。能准确评估设备的状况，可以为后续修复工作能提供精确的数据保障。

关键词 换热器，腐蚀，旋转超声

An Accurate Detection Method of the Heat Exchanger Tube Bundle Corrosion Thinning

Zhang Xuefeng

(Shanghai Jinyi Inspection Technology Co., Ltd., Shanghai 201900, China)

Abstract In the past, we usually use portable ultrasonic flaw detector or line thickness gauge to emasure the thickness. However, because of the portable ultrasonic instrument need probe directly contacts with the workpiece surface, when the measured workpiece is located in a confined space or the coupling between probe and the workpiece surface to be tested is bad, the instrument is prone to misjudge or is unable to judge. Heat exchanger tube bundle are generally of densely packed tube box, only foreign circle tube bundle canbe tested, the whole situation of heat exchanger is unable to be accurately assessed. A test technique for boiler tube bundle is used in this paper, after testing, the found defect of boiler tube bundle samples is displayed, about its image, depth, thickness, and the informations such as location, amount of thinning are visual displayed in the image. The test results can provide accurate data for subsequent repair work.

Key words heat exchanger, corrosion, rotary ultrasonic

宝钢新品热轧工作辊缺陷的超声检测

黄 玲

（上海金艺检测技术有限公司，上海 201900）

摘 要 为了防止宝钢常用新品热轧工作辊在上机后就立即产生失效等问题，使宝钢轧线生产产生不必要的损失，超声检测是对宝钢常用新品热轧工作辊质量的有力保证。经过系列的现场检测，经分类研究发现，宝钢常用新品热轧工作辊的主要缺陷为工作层缺陷、外层厚度偏薄、弱结合、辊身辊颈及卡环槽局部底波衰减等几大类。在发现以上问题时，现场就要引起警觉，防止其在生产过程中造成轧辊的剥落掉块、裂纹、爆裂、断辊，以至最终成为轧机卡钢、停产的祸端。

本文主要简单阐述了宝钢三条热轧线常用新品热轧工作辊超声检测中的缺陷，通过整套的超声检测总结分析其形成的主要原因和发展方向，为控制宝钢常用新品热轧工作的合格率做保证。同时为厂家的制造改进给予一定的帮助和方向。

关键词　新品热轧工作辊，轧辊失效，超声波检测

Ultrasonic Inspection of Defects in New Hot Rolls of Baosteel

Huang Ling

(Shanghai Jinyi Inspection Technology Co., Ltd., Shanghai 201900, China)

Abstract　In order to prevent the immediate failure of hot working roll newly used and the corresponding losses, ultrasonic testing is a strong quality guarantee for new hot roll products, which is commonly used in Baosteel quality control. After a series of field detection and the classification study, we found that the main defects of the new hot rolling work roll are faults in the defect layer, the thickness thinning,of the outer layer weak combination, and the bottom wave attenuation of body neck and card ring groove partial of several major categories. The scene should be on alert, if above problems were found, and prepare to do some prevention work for latent accidents might be caused by roll flaking off block, crack, crack broken,,and so on

This paper simply expounds the ultrasonic defect detection of new hot rolling work rolls,in three hot strip rolling lines of Baosteel, by a set of ultrasonic testing summary and analysis of main reason, and suggests the defect's formation and development direction for the control of new products commonly used in Baosteel hot rolling qualified rate guarantee and also for the manufacturer.

Key words　new hot rolling work roll, roll failure, ultrasonic testing.

Innovation of Backup Rolls for Skin Pass Mills

Ji Weiguo, Weyand Gerd, Habitzki Klaus

(Gontermann-Peipers GmbH, Hauptstrasse 20, 57074 Siegen, Germany)

Abstract　Gontermann-Peipers GmbH (GP) Germany developed a new, high alloyed steel grade (Hot working steel) for the manufacturing of back-up rolls in Skin Pass Mills. The advanced GP double pour casting processes provides highest flexibility in being quite open in rolls work layer chemistry and hence microstructure with no influence on core respectively neck properties. GP took the process advantage to provide a brand new back-up roll grade with remarkably reduced roll wear while operating with textured work rolls. In the meantime several successful trials have been made with this new roll grade on different Skin Pass Mills. As a result the performance of back-up roll and thus availability of mill will be improved huge. Besides the work rolls get benefits to extent the campaign time. The surface quality has been improved due to reduced phenomenon of chattering in stand. So that not only specific costs for mill will drop. Also the products have increased their value due to high quality. This quite new innovative back-up roll grade has been launched already in 2007. With enormously positive feedback by using the field of applications is since 2014.

Key words　back-up roll, skin pass mill, wear, chattering, DCI core

高炉无料钟炉顶称量料罐封头的分析与研究

蒋治浩

(北京首钢国际工程技术有限公司设备开发成套部，北京 100043)

摘 要 采用ANSYS分析软件，针对多种形式的高炉串罐无料钟炉顶称量料罐封头，进行受力仿真计算，并对计算结果进行分析、比较，其计算结果可为料罐施工图设计提供更精准的计算依据。

关键词 料罐封头，ANSYS有限元法，分析与研究

Analysis and Research of Weighing Tank Sealed Head for Blast Furnace Bell-less Top

Jiang Zhihao

(BSIET Mechanical Dept., Beijing 100043, China)

Abstract In view of the various of blast furnace series tank bell-less top weighing material tank sealed head, ANSYS analysis software is used for force simulation calculation, analyze and compare the calculated results, the calculation results can provide more accurate calculation basis for material tank construction drawing design.

Key words tank sealed head, ANSYS finite element method, analysis and research

盲板力作用下高炉炉壳应力及变形分析

刘 奇 程树森

（北京科技大学冶金与生态工程学院，北京 100083）

摘 要 采用弹性力学分析方法研究了盲板力作用下高炉炉壳的应力分布及变形行为。结果显示，高炉炉壳受盲板力作用下最大等效应力达到 160MPa，位于炉缸底对应炉壳位置，低于炉壳材料的屈服强度。盲板力作用下，高炉炉壳不易发生开裂破损问题。风口开孔变形后呈近似椭圆形，风口开孔将对风口套产生过大的约束作用。炉壳设计中应在风口开孔位置增加抗变形元件，减小风口开孔变形，以削弱风口开孔对风口套的约束作用。

关键词 高炉，盲板力，炉壳，应力，变形

Stress and Deformation Analysis of Blast Furnace Shell Subjected to Blind Flange Force

Liu Qi, Cheng Shusen

(Metallurgy and Ecology Engineering School, University of Science and Technology of Beijing, Beijing 100083, China)

Abstract　The stress distribution and deformation behavior of blast furnace shell subjected to blind flange force were investigated by elastic mechanics method. The results show that the maximum Von Mises equivalent stress of the blast furnace shell is about 160MPa, which nears to the bottom of the hearth, lower than the yield strength of blast furnace shell steel. The blast furnace shell subjected to blind flange force will be less likely to crack. The tuyere hole deforms into approximate ellipse, which can lead to excessive constraint at the tuyere cooler. The anti-deformation component should be installed at the tuyere hole used to decrease the deformation degree of the tuyere hole, and weaken the constraint at the tuyere cooler.

Key words　blast furnace, blind flange force, shell, stress, deformation

不定形耐火材料新技术在高炉维修方面的应用

章荣会　徐吉龙　刘贯重　于运祥　邓乐锐

（北京联合荣大工程材料有限责任公司，北京　101400）

摘　要　本文介绍了几种炼铁系统相关的耐火材料新技术，包括：1. 采用金属陶瓷材料制作的一种预挂渣皮的新型冷却壁镶砖，其强度可达 150MPa 以上，代替传统镶砖，能够避免从燕尾部分断裂，延长使用寿命。导热率较高，利于形成稳定渣皮层。2. 高炉内衬湿法喷注造衬技术替代传统半干法喷注技术在国内外超过 160 座高炉上得以应用。喷注内衬强度高，反弹低，寿命长，具有显著的社会效益和经济效益。3. 开发的铁口修补用纳米 SiO_2 结合浇注料能够解决铁口热态修复问题，该材料可实现快速烘烤，不爆裂，施工工期短，使用寿命长，一次修复可使用半年以上。4. 炉缸整体泵送浇注维修技术能够对风口组合砖及炉缸陶瓷杯及陶瓷垫进行快速修复，工期一般不超过一周，整体浇注陶瓷杯可安全使用 2 年以上。

关键词　高炉，耐火材料，金属陶瓷，喷注，炉缸浇注

The Applications of Unshaped Refractories's New Technology in Blast Furnace Maintenance

Zhang Ronghui, Xu Jilong, Liu Guanzhong, Yu Yunxiang, Deng Lerui

(Beijing Allied Rongda Engineering Material Co., Ltd., Beijing 101400, China)

Abstract　This paper has introduced several new refractory material technology relating iron-making-system, which included: 1. using metal ceramic materials to make a new cooling wall brick lines as a pre-slag skin, and its strength value can reach 150MPa above, which would replace traditional brick lines to avoid the crack from dovetail groove and prolong service life. Furthermore, High thermal conductivity is conducive to the formation of stable slag. 2. wet jet technology has replaced the traditional semi-dry spray technology, which applied on above 160 domestic and foreign blast furnaces. The former has brought significant social benefits and economic benefits because of its high strength, low rebound, and long service life. 3. the technology of the tap-hole-repair has adopted nano-SiO_2 combined with castable materials, which has solved hot-fix problems, and the material can realize fast firing with no burst and short construction period, besides, the service life is expected for more than half a year. 4. the technology of pump sending casting-repair on tuyere combined bricks and hearth ceramic-cup, which construction time is generally less than seven days, and integral casting ceramic cup can be safely used for more than 2 years.

Key words blast furnace, refractory, metal ceramic materials, shotcrete, hearth ceramic-cup

板坯连铸结晶器电磁搅拌物理模拟和数值模拟研究

雷作胜[1] 韦如军[1] 李 彬[1] 钟云波[1] 任忠鸣[1]
周月明[2] 吴存有[2] 金小礼[2]

（1. 上海大学现代冶金与材料制备重点实验室，上海 200072；
2. 宝钢中央研究院设备所，上海 200019）

摘 要 本文采用物理模拟和数值模拟的方法，对电磁搅拌下板坯结晶器内的流场进行了测量和计算。考察了不同连铸工艺和搅拌电流作用下的流场结构和分布特征，提出了自由液面的卷渣指数（MFEI）和结晶器内流场均匀性指数（VUI），介绍了上述指数对板坯结晶器-内流场电磁搅拌效果的判定方法，进而提出了搅拌参数的综合优化方法。

关键词 板坯连铸，结晶器，电磁搅拌，参数优化

Physical Modeling and Numerical Simulation of Electromagnetic Stirring in Slab Continuous Casting Mold

Lei Zuosheng[1], Wei Rujun[1], Li Bin[1], Zhong Yunbo[1], Ren Zhongming[1],
Zhou Yueming[2], Wu Cunyou[2], Jin Xiaoli[2]

(1. Shanghai Key Laboratory of Modern Metallurgy & Materials Processing, Shanghai 200072,China;
2. Baosteel Central Research Institute, Shanghai 200019,China)

Abstract The flow field in slab continuous casting mold with electromagnetic stirring was measured by physical modeling and calculated by numerical simulation under different casting parameters and stirring currents. Mold Flux Entrapment Index and Velocity Uniform Index were proposed in order to evaluate flow field in the mold. Based on these two indexes, optimization stirring parameters were proposed under different casting situation.

Key words slab continuous casting, mold, electromagnetic stirring, optimization parameters

Cracking During Welding Pipeline Steel

Aucott Lee[1], Wen Shuwen[2], Dong Hongbiao[1]

(1. Department of Engineering, University of Leicester, University Road, Leicester, LE1 7RH, UK;
2. Tata Steel Europe Ltd., Swinden Technology Centre, Rotherham, UK)

Abstract Weld solidification cracking, also known as hot cracking, is an important issue in welding of pipeline steel. If undetected, the cracking defects can act as stress concentration sites which lead to premature failure via fatigue, as well as offer favourable sites for hydrogen assisted cracking and stress corrosion cracking. Because of this, solidification cracks are considered the most deleterious of all welding defects. In this study weld solidification cracking in pipeline steel has been

successfully observed in-situ for the first time using a high speed, high energy, synchrotron X-ray radioscopy approach. Analysis of the in-situ radiography sequence revealed the solidification cracking initiates in the weld sub-surface trailing the welding electrode. Pore like features are then observed in ex-situ 3D tomographic reconstructions of the cracking network, exclusive to the region of solidification crack initiation observed through in-situ radiography. These observations provide new insights into cracking initiation and growth during welding of pipeline steel.
Key words　cracking, welding, solidification, pipeline steel

板坯连铸机横移台车辊道控制系统改进

余潭慧[1]　蔡光富[2]　李　欣[1]　李文杰[1]　周　伟[1]

（1. 鞍钢集团鞍钢股份炼钢总厂；2. 鞍钢集团鞍钢股份炼铁总厂，辽宁鞍山　114021）

摘　要　原设计横移台车上辊道控制系统采用传统的接触器控制方式，电机采用双速电机，通过 Y-△变换实现高低速切换，制动系统采用机械抱闸。该系统已经严重老化，电机频繁掉电，给生产造成极大的影响，将接触器控制方式替换为交流变频调速系统，取消双速电机，使用单套绕组的笼型三相异步电动机，保证调速精度，实现原有的控制、操作和显示功能。大大降低横移台车的设备故障率，缩短设备的检修时间，保证了横移台车辊道的稳定运行。
关键词　横移台车，辊道，控制系统，改进

Improvement of Slab Continuous Caster Traversing Trolley Roller Control System

Yu Tanhui[1], Cai Guangfu[2], Li Xin[1], Li Wenjie[1], Zhou Wei[1]

(1. Anshan Iron & Steel Corporation, Ansteel Co., Ltd., Steel Making Plant;
2 . Anshan Iron & Steel Corporation, Ansteel Co., Ltd., Iron Making Plant, Anshan 114021, China)

Abstract　The original design of　traversing trolley roller control system adopts traditional contactor control mode, the motor adopts double speed motor, high speed switching by Y- Δ transform, the braking system adopts the mechanical brake.The system has serious aging, motor frequency power down, caused great impact to production.Through the contactor control mode to change AC variable frequency speed regulation system, cancel the double speed motor, squirrel cage three-phase induction motor using a single set of winding, ensure accuracy of speed, To achieve the original control, operations and display function. Greatly reduce the equipment failure rate of traverse trolley, shorten the time of equipment maintenance, to ensure stable operation of the traversing trolley roller.
Key words　the traversing trolley, roller, control system, improve

干熄焦锅炉爆管问题研究

孔　弢　张允东　陈艳伟

（鞍钢股份有限公司，辽宁鞍山　114000）

摘 要 由于国家新环保法的颁布、实施，炼焦厂对于干熄焦工序要求日益严格，也就迫使干熄焦系统的故障率必须降低，而作为主要部分的余热锅炉吊挂管结构周期性损坏甚至出现过爆管等问题直接影响干熄焦工序的秩序，本文主要通过分析爆管原因，钢结构受力分析，提出解决方案。

关键词 吊挂管，爆管，断口形貌

Study on Tube Explosion of Boiler

Kong Tao, Zhang Yundong, Chen Yanwei

(Angang Steel Co., Anshan, Liaoning, 114000, China)

Abstract Because of the promulgation of the new national environmental law, implementation, coking plant the CDQ process of increasingly stringent requirements, will force the CDQ system failure rate must be reduced, as the main part of the waste heat boiler hanging pipe periodic structure damage and even burst pipe directly affect the CDQ process order, this paper mainly through the analysis of pipe explosion reason, the stress analysis of steel structure, put forward solutions.

Key words hanging of pipe, burst pipe, fracture morphology

No1 纠偏辊摆动架剪切应力和拉伸应力的分析

王云良[1] 王 新[2] 李红雨[1] 杨军荣[1]

（1. 鞍钢股份有限公司冷轧厂，辽宁鞍山 114021；
2. 鞍钢股份有限公司线材厂，辽宁鞍山 114021）

摘 要 针对酸洗机组No1纠偏辊摆动架在运行过程中的摆动座与横梁之间焊缝开焊现象，通过计算验证焊接的设计强度是否符合要求，分析了焊接的理论强度在实际应用中遇到的具体问题，指出了纠偏辊装配图在实际应用中设计缺陷的问题，并通过改造摆动架侧面结构杜绝摆动架与固定架横梁相互干涉的方法，使问题得到最终解决。

关键词 纠偏框架，焊接强度，联接

Analysis on Shear Stress and Tensile Stress of The Swivel Frame of the No1 Steering Roll

Wang Yunliang[1], Wang Xin[2], Li Hongyu[1], Yang Junrong[1]

(1. Angang Iron & Steel Group Cold Strip Works; 2. Wire Rod Mill, Anshan Liaoning 114021, China)

Abstract The paper mainly discusses the welding seam destruction of No1 steering roll swivel frame and connection beam while working. Through calculation, it verifies whether the designed strength of the weld joint fulfills the specification and gives the solutions to the design faults in the assembly drawings. It finally solves the interference problems through reconstruction of the base frame and swivel frame.

Key words steering frame, weld strength, joint

厚板轧机主传动轴轴套拆卸

姜世伟　尚春姝　赵立军　肖争光　魏长杰　贾丹江

（鞍钢股份鲅鱼圈钢铁分公司设备保障部，辽宁营口　115007）

摘　要　很多大型传动部件轴和套之间的连接采用了大过盈量无键连接，在装配作业过程中涉及到轴和套的拆卸，而大型圆柱面大过盈连接无损拆卸一直是机械拆装作业中的难点。对于一般轴套类装配件，现场采用的传统拆卸方法可分为两种，即利用气割或机械加工的方法直接破坏包容件拆卸和热膨胀法拆卸。这两种方法虽然可行，但都属于破坏性拆卸，拆卸后零件在拆卸作业过程中极易受到破坏导致不能再利用，且工程量大，费用昂贵，生产周期长。而采用液压膨胀后拉出的方法则很好地解决了这一难题，既减少了工作量，又节约了成本，很好地做到了无损伤拆卸。某钢厂在主轴改造过程中涉及到了电机侧安全联轴器内套的拆卸，属于典型的大型圆柱面大过盈连接无损拆卸问题。文中涉及大过盈轴套保护性拆卸方式及计算过程，对同类作业具有一定的借鉴意义和参考价值。

关键词　厚板，轧机，传动轴，轴套，过盈，拆卸

The Main Drive Spindle Safety Coupling Inner Sleeve Disassembling of Plate Mill

Jiang Shiwei, Shang Chunshu, Zhao Lijun, Xiao Zhengguang, Wei Changjie, Jia Danjiang

(The Equipment Supporting of Ansteel Bayuquan Iron & Steel Subsidiary, Yingkou Liaoning, 115007, China)

Abstract　Many heavy duty driving transmission shaft and coupling uses high interference fit-none key connection. During some maintenance operation, disassemble sleeve from shaft is needed. However, none damage disassembling of large dimension column connection with high interference fit is always a difficult task in such mechanical assembling and disassembling operation. To such kind of shaft-sleeve connection, there are two traditional disassembling methods, method of flame cutting or mechanical cutting which directly destroy the containing parts, and method of thermal expanding to disassemble. Though applicable, they are all destructive disassembling. Parts is easily to be damaged during disassembling and no longer useable after disassembling, meanwhile, the disassembling process is high work load, high cost, and long time work. A method of hydraulic expansion and pull out can be a good solution to this difficulty, not only reducing work load, but also reducing working cost, none destructive disassembling achieved well. During a spindle revamping project in a steel plant, the safety coupling motor side inner sleeve disassemble is involved, such disassembling is a large dimension column high interference fit connection disassembling. This article discusses the large dimension column high interference fit shaft-sleeve disassembling methods and calculation process, it has signification of guideline and reference to the same kind operation.

Key words　plate, rolling mill, driving spindle,sleeve, interference fit, disassembling

激光对中找正的原理及应用

李俊峰　王　越　杨忠杰　李富强

（鞍钢股份有限公司冷轧厂，辽宁鞍山　114021）

摘　要　本文讲述了轴不对中对设备的危害，介绍了激光对中仪的结构，系统地阐述了激光对中的测量原理，通过使用激光对中仪对鞍钢冷轧厂张力辊的故障进行现场检测及调整，讲解激光对中找正在实际中的应用。

关键词　激光对中，偏差，联轴器

The Principle and Application of Laser Centring

Li Junfeng, Wang Yue, Yang Zhongjie, Li Fuqiang

(The Cold Rolling Plant of Ansteel Co., Ltd., Anshan Liaoning 114021, China)

Abstract　The article introduces the harm to the mechanical equipment when the centering offset of the couple is not good. It introduces the structure of the Laser centering Device , it expatiates on the principle of measure,we measured and adjusted the Tension Roll of the cold rolling plant of Ansteel co.,Ltd by using the Laser centering Device ,it explains the application of the Laser centering Device in the workshop.

Key words　laser centering, offset, couple

焦炉放散点火系统的应用与改进

朱庆庙　谭　啸　庞克亮　陈立哲　魏　威

（鞍山钢铁股份有限公司炼焦总厂，辽宁鞍山　114001）

摘　要　本文针对焦炉生产过程中能够产生重大环境污染的环保装置进行了系统实践应用研究试验，提出了焦炉集气管自动放散点火装置的系统恢复及使用办法，并针对不同情况提出了现场生产岗位所需采取的不同的处理措施。同时针对焦炉自动放散点火装置日常的维护及保养提出了要求，以保证此套装置在应急条件下能够发挥作用，避免出现影响恶劣的大面积环境污染现象。

关键词　荒煤气，自动放散点火系统，应用与改进

Application and Improvement of Ignition Bleeding System on Coke Oven

Zhu Qingmiao, Tan Xiao, Pang Keliang, Chen Lizhe, Wei Wei

(General Coking Works of Ansteel Company Limited，Anshan Liaoning 114001, China)

Abstract　Based on coke oven production process to avoid major environmental pollution devices for practical application of the system of environmental protection and research experiments presented coke oven collector auto emission ignition systems restoration and the procedures to use and production posts is proposed for different circumstances required different treatment measures taken. While coke oven automatic relief requested ignition device for routine maintenance and repairs, to ensure that this unit can play a role in emergency conditions to avoid bad influence of large-area pollution.

Key words　crude oil gas, auto emission ignition system, application and improvement

干熄焦一次除尘膨胀节在线修复技术

朱庆庙　谭　啸　叶　亮　陈立哲　魏　威

（鞍山钢铁股份有限公司炼焦总厂，辽宁鞍山　114001）

摘　要　本文针对干熄焦系统生产中的薄弱部位：一次除尘出口膨胀节。膨胀节用以吸收干熄炉砌体与一次除尘砌体及锅炉炉管间的热膨胀，消除装焦过程中三个主体间的相对位移。由于热胀冷缩容易出现损坏，造成膨胀节内部浇筑料脱落，外部高温的问题。经过多次年修并在线进行修复此膨胀节，两年的生产运行未发现任何问题，延长了干熄焦的年修周期，并降低了年修费用。

关键词　一次除尘，膨胀节，在线修复，措施

Online Repair Technology of CDQ Primary Dedusting Expansion Joints

Zhu Qingmiao, Tan Xiao, Ye Liang, Chen Lizhe, Wei Wei

(General Coking Works of Ansteel Company Limited, Anshan Liaoning 114001, China)

Abstract　This paper aims at the weak parts in production of CDQ system that export expansion section of primary dedusting expansion joints. Expansion joints to absorb the CDQ dust masonry and primary dedusting masonry and thermal expansion of boiler tube, eliminate the loading process of relative displacement between the three subjects. Due to thermal expansion and contraction of easily damaged, resulting in expansion joint placement within the material loss and external temperature problem. After repeated annual repair and online fix for this expansion section, production operation runs without any problem for two years. It extends years of repair cycle of CDQ system and reduces the costs in annual repair.

Key words　primary dedusting, expansion joints, online fix, measure

提高5500轧机牌坊轧制精度、耐久性和稳定性的技术研究

潘凯华　罗　军　乔　馨　姜世伟　李靖年　李　丰　于金洲　韩　旭

（鞍钢股份有限公司鲅鱼圈钢铁分公司，辽宁营口　115007）

摘　要　利用激光熔覆技术具有减少基体热影响区域、减少基体温升及材料耐磨耐腐蚀等优点，采用该技术修复鲅鱼圈厚板部5500轧机牌坊可以减少对牌坊的损伤，激光熔覆后耐磨能力提高10倍以上，显著提高牌坊的使用寿命，同时可以改善轧制稳定性，改善后的牌坊精度为产品质量的提升和设备本体的稳定均提供了有力保证。

关键词　轧机牌坊精度，激光熔覆，稳定性，牌坊间隙

Abstract　Using laser cladding technology is to reduce substrate heat affected zone、decrease substrate temperature and material wear-resisting corrosion resistance, etc.Using the technology to repair the 5500 rolling mill Bayuquan plate arch can reduce the damage to the memorial arch.After laser cladding wear-resistant ability increased by more than 10 times, improve the service life of the memorial arch.At the same time it can improve the stability of rolling.Improved accuracy of memorial arch for the ascension of the quality of the product and the stability of the equipment ontology provides a powerful guarantee.

Key words　Mill arch precision, laser cladding, stability, arch clearance

桥式起重机车轮偏斜的激光检测技术

齐鹏程　严　力　李江宁

（鞍钢集团公司，辽宁鞍山　114021）

摘　要　本文根据冶金行业桥式起重机普遍存在的啃轨现象，结合现场检修实际情况与经验，详细介绍了桥式起重机车轮水平偏斜与垂直偏斜的激光检测技术，推导出相应的偏斜量和调整量的计算公式，实现对车轮偏斜的准确调整，避免因车轮偏斜而造成啃轨、车轮损坏等后果，保证起重机的良好运行状态。

关键词　激光检测，桥式起重机，车轮偏斜，调整

The Laser Detection Technology of Bridge Crane Wheel Deflection

Qi Pengcheng, Yan Li, Li Jiangning

(Ansteel Group, Anshan Liaoning 114021,China)

Abstract　Based on widespread metallurgical gnaw rail phenomena of the bridge crane in combination with the practical situation of on-site inspection and experience summary, the paper introduces in detail the bridge crane wheels horizontal deflection and vertical deflection of the laser detection technology, deduces the corresponding formula for calculation of deviation and adjust the volume wheel deflection for accurate adjustment, avoid gnaw rail, wheel damage caused by wheel deflection consequences, such as to ensure good running status of the crane.

Key words　laser detection, bridge crane, wheel deflection, adjustment

极薄板带钢S型翘曲变形行为研究

李　宇[1]　李秀军[2]　张清东[1]

（1. 北京科技大学机械工程学院，北京　100083；

2. 宝山钢铁股份有限公司冷轧薄板厂，上海　200941）

摘　要　针对极薄板带钢生产过程中产生的S型翘曲缺陷，提出了此类缺陷的产生机理，即轧制平整后带钢纵向延伸沿厚度方向存在不均匀分布，同时在靠近带钢宽度方向两边区域存在延伸的正负变化，而中间区域延伸均匀，导致了带钢两侧发生弯曲变形从而出现S型翘曲缺陷。在上述分析的基础上应用ABAQUS有限元软件建立了S型翘曲的数值仿真模型。模型计算结果表明，带钢翘高与带钢上下表面的延伸差、带钢宽度成正比，而与带钢厚度、带钢中间的均匀延伸宽度成反比；同时发现纵向延伸宽向假设为阶跃函数分布时对应的翘曲值最高，二次函数分布时对应的翘曲值最小。

关键词　薄带钢，S型翘曲，有限元，纵向延伸

S Warping Deformation of Thin Steel Strips

Li Yu[1], Li Xiujun[2], Zhang Qingdong[1]

(1. School of Mechanical Engineering, University of Science and Technology Beijing, Beijing 100083, China;
2. Baoshan Iron & Steel Co., Ltd., Shanghai 200941, China)

Abstract　In the light of S warping defects produced in ultra thin steel strips during productive process, the generation mechanism of this warping are put forward.It is indicated that the S warping deformation is caused by the uneven distribution of plastic deformation along the thickness direction during temper rolling and there is an extension of the positive and negative in the direction of the width of the strip when the middle region extends uniformly resulting in bending deformation of strip on both sides.Simulation analysis of S warping behavior is established using ABAQUS finite element software.According to the analysis results,the height of the strip is proportional to the extension of the upper and lower surface or width of the strip,which is in inverse proportion to the width of the strip thickness and the average width of the strip.The warping value corresponding to the longitudinal extension of the step function is the highest, and the two time function distribution is the lowest.

Key words　thin steel strips, S warping, finite element, longitudinal extension

干熄焦焦罐旋转影响因素分析及改进

郑晓雷　赵　华

（鞍钢股份鲅鱼圈钢铁分公司炼焦部，辽宁营口　115007）

摘　要　伴随着国家生态化发展的趋势，因在同样焦炭质量下干熄焦技术能降低约5%配煤成本、能有效回收焦炉输出热量的30%左右、吨焦炭节水0.4～0.6m^{3}以及不产生含酚、氰等有害气体及粉尘量较少等节能环保的优势，干熄焦技术在国内各焦化企业广泛应用。但干熄焦在生产运行过程中暴露出了许多问题，特别在红焦装入干熄炉前故障率较高，严重制约了干熄焦运行得平稳性，给安全生产带来了较大隐患。本文通过对干熄焦旋转焦罐传动、制动过程进行阐述，围绕制约干熄焦旋转焦罐正常传动、制动的各种因素，通过设置干熄焦传动装置挡焦板、回转台下辊轮槽排灰排水孔及使干熄焦焦罐传动、制动系统等故障率大幅度降低，安全可靠性得到了大大提高，有效地保证了干熄焦系统的稳定运行。

关键词　焦罐，旋转，传动，制动

CDQ Coke Cans Rotation Factors Analysis and Improvement

Zheng Xiaolei, Zhao Hua

(Ansteel Bayuquan Iron & Steel Subsidiary Department of Coking, Yingkou Liaoning 115007, China)

Abstract Along with the trend of the development of the national ecological, coke dry quenching technology can reduce about 5% of coal blending cost, about 30% of the effective recovery coke oven heat output, the advantage of energy conservation and environmental protection such as tons coke saving water 0.4 ~ 0.6 m^3, and don't produce harmful gases such as phenol cyanide and less dust dry , coke dry quenching technology is widely used in domestic coking enterprises, but dry quenching has exposed many problems in the process of production, especially in front of the red coke into dry quenching furnace failure rate is higher, severely restricted the dry quenching operation stability, brought large hidden trouble for safety production. This paper describes the CDQ coke cans transmission, braking, rotation around the constraints CDQ Coke cans normal transmission, brake a variety of factors, by setting the transmission gear CDQ Coke board, turning the audience roller slot row gray drainage holes and parallel increase slowdown stop limit and other measures to effectively reduce the CDQ coke cans rotation frequency is not in place appear to meet the production needs of CDQ.

Key words coke cans, rotate, drive, brake

电液动煤塔放料装置控制系统改进

于庆泉

（鞍钢股份鲅鱼圈钢铁分公司炼焦部，辽宁营口 115007）

摘 要 煤塔是炼焦生产中用于存放煤原料的筒仓，筒仓中的煤原料通过电液动煤塔放料装置输送至装煤车。每个煤塔配置有若干套电液动放料装置，每套电液动放料装置由双扇形仓门、电液推杆、仓门位置检测限位、开关信号传输开关等设备组成。电液动煤塔放料装置控制系统普遍采用继电器控制系统。该控制系统功能单一、故障率高，导致煤塔漏煤事故频发。本文提出了一套适应焦化现场工况的，高可靠性、高智能性、可实现人机交互的基于西门子 PLC 的电液动煤塔放料装置控制系统。此技术已应用于鞍钢股份鲅鱼圈钢铁分公司炼焦部 1 号煤塔及 2 号煤塔，提高了煤塔放料装置控制系统的可靠性、稳定性、可维护性，提高了生产效率，降低了生产成本。

关键词 煤塔，PLC，控制系统

The Improvement for Electro-hydraulic Coal Feeding Device Control System

Yu Qingquan

(Ansteel Bayuquan Iron & Steel Subsidiary Department of Coking, Yingkou Liaoning 115007, China)

Abstract Coal tower stores coal in coke production, coal is transported to the coal charging car by electro-hydraulic feeding device of the coal tower. Each coal tower configurations some set of electro-hydraulic feeding device, each electro-hydraulic feeding device haves double fan bin gate, electro-hydraulic putt, warehouse door position detection limit position, switch signal transmission switch equipment. The control system for electro-hydraulic coal tower feeding device always adopts the relay control system in the past.That control system's function is poor, fault rate is high, always leading to the coal leakage accidents. This paper presents a control system based on Siemens PLC that adapt coking field condition, high reliability, high intelligence, can achieve human-computer interaction. This technology has been used in BaYuQuan branch company of Angang Steel Company Limited. It has improved the reliability, stability and maintainability of the coal tower feeding device control system,improved the production efficiency, reduced the production cost.

Key words　coal tower, PLC, control system

X80管线钢快速感应回火组织性能变化研究

范宇静　麻永林　张云龙　邢淑清

（内蒙古科技大学材料与冶金学院，内蒙古包头　014010）

摘　要　为研究快速感应回火条件下钢材组织和性能变化规律，以 X80 管线钢为研究对象，利用自制电磁感应加热装置及箱式电阻炉分别对其进行快速感应回火和传统回火实验，采用金相显微镜、扫描电镜观察不同回火条件下X80 管线钢的组织形貌及析出相尺寸、分布，并测定其低温冲击韧性。结果表明：快速感应回火工艺参数下的试样性能均优于同参数下的传统回火试样。其中回火温度为 590℃，保温时间为 90s 时，回火后的贝氏体（针状铁素体）组织明显细化，析出的碳氮化合物更弥散细小、均匀地分布于回火基体组织上，其–40℃和–60℃温度下的低温冲击功均最高，分别为 430.5J 和 351.3J，而传统回火试样的最高值分别为 323.2J 和 312.1J，均低于快速感应回火试样。

关键词　X80 管线钢，快速感应回火，回火组织，析出相，低温韧性

Research on the Change of Microstructure and Property of Rapid Induction Tempering X80 Pipeline Steel

Fan Yujing, Ma Yonglin, Zhang Yunlong, Xing Shuqing

(School of Material and Metallurgy, Inner Mongolia University of Science and Technology, Baotou 014010, China)

Abstract　In order to study the change rule of microstructure and property of rapid induction tempering X80 pipeline steel, rapid induction tempering and conventional tempering experiments are carried out by using the self-made electromagnetic induction heating device and the box type resistance furnace, respectively. Microstructure morphology, precipitates size and distribution of X80 pipeline steel under the condition of different tempering were observed by means of Metallographic microscope and SEM, and its low temperature impact toughness was measured. The results show that all the performances of the rapid induction tempering samples are better than those of traditional tempering samples which have the same tempering parameters. When heating temperature is 590℃ and holding time is 90s, acicular ferrite obviously refined. The characteristic of carbonitride precipitation have more small dispersion, which evenly distributed on the tempering base microstructure. Low temperature impact work of –40℃ and –60℃ samples were the highest, they are 430.5J and 351.3J respectively, while the best values of the traditional tempering are 323.2J and 312.1J.

Key words　X80 pipeline steel, rapid induction tempering, tempering microstucture, precipitation phase, low temperature toughness

Numerical Investigation on Turbulent Multiple Jets Merging and Acoustic Characteristic in the Slab Scarfing Process

Li Yiming, Qi Fengsheng, Wang Xichun, Li Baokuan

(School of Materials and Metallurgy, Northeastern University, Shenyang 110819, China)

Abstract In order to predict the multiple jets screech phenomenon in the scarfing process of rolling slab, a three-dimensional transient mathematical model has been developed to investigate the turbulent multiple jets merging. Large eddy simulation (LES) with dynamic Smagorinsky sub-grid scale (SGS) approach was applied to simulate the turbulent multi-jet flow field. The acoustic field was calculated by the Ffowcs Williams-Hawkings (FW-H) integral equation. In consideration of the compressibility of high Mach number gas jets, density-based explicit formulation was adopted to solve the coupled system of governing equations, and the viscosity was approximated by Sutherland kinetic theory. The multiple jets merging acoustic characteristic was demonstrated, as well as the noise intensity were predicted with a good accuracy of practical data. The results indicated that multiple jets merging phenomenon is induced by the Coanda effect, and adding to the complexity and instability of flow field. The main factors affecting multiple jets merging are injection orifice diameter and distances between orifices. The overall sound pressure level (OASPL) profiles have a directive property, which is also influenced by the multiple jets merging. Controlling multiple jets turbulent intensity and taking advantage of the noise directionality are both good measures for noise reduction. The conclusion obtained in this study can provide valuable data to guide the development of manufacturing-green technology in the scarfing process.

Key words multiple jets, acoustic characteristic, noise predicted, coanda effect, scarfing process

厚板预矫直机支撑辊辊型优化

姚利松 孙大乐 范 群

（宝钢集团有限公司中央研究院，上海 201900）

摘 要 宝钢厚板预矫直机在生产过程中，工作辊和支撑辊辊面出现不均匀磨损和擦伤，严重影响了矫直后的钢板表面质量。由于厚板预矫直机的换辊周期较长，一般在半年以上，为满足生产要求，必须不定期停机对辊面进行在线人工打磨处理。针对厚板预矫直机辊面不均匀磨损和擦伤问题，利用有限元仿真技术开展了预矫直机支撑辊辊型优化。仿真计算结果表明，预矫直机支撑辊辊型优化可以改进辊面接触应力分布均匀性，显著降低辊间接触峰值应力，从而改善辊间的接触状态，减轻辊面不均匀磨损和擦伤现象。宝钢厚板预矫直机有长、短两种支撑辊。对于短支撑辊，仅优化支撑辊倒角形式及其参数即可；对于长支撑辊，需同时优化支撑辊辊面中部辊型和边部倒角形式及其参数。根据仿真结果，给出了预矫直机短支撑辊和长支撑辊的优化辊型。

关键词 矫直机，支撑辊，辊型优化，宽厚板

Optimization of Backup Roll Profile for Heavy Plate Pre-leveler

Yao Lisong, Sun Dale, Fan Qun

(Research Institute，Baosteel Group Corporation，Shanghai 201900, China)

Abstract In the production process of Baosteel Heavy Plate pre-leveler, the work rolls and backup rolls surface appear uneven wear and scratch, which seriously affecting the quality of the plate surface after straightening. Due to the roll changing period of the heavy plate pre-leveler is long, usually six months or more, in order to meet production requirements, unscheduled downtime for roll surface online artificial grinding is needed. Aiming at surface uneven wear and scratch of pre-leveler rolls, using the finite element simulation technology to carry out pre-leveler backup roll profile optimization. The simulation results show that the optimization of pre-leveler backup roll profile can improve the uniformity of contact stress distribution of the rolls, significantly reduces the contact peak stress, thereby improving the contact state of the rolls, reducing roll surface uneven wear and scratch. Baosteel Heavy Plate pre-leveler has two kinds of backup rolls, the long backup roll and the short backup roll. For short backup rolls, only need to optimize backup roll chamfer and its parameter. For long backup rolls, simultaneously optimization of backup roll middle part profile, edge chamfer shape and its parameters are needed. According to the simulation results, the optimal roll profile of short and long pre-leveler backup rolls were recommended.

Key words leveler, backup roll, roll profile optimization, wide and heavy plate

电磁搅拌对水口堵塞情况下结晶器内流场的影响

高 齐[1] 周月明[1] 雷作胜[2]

（1. 宝钢集团中央研究院，上海 201999;
2. 上海大学现代冶金与材料制备重点实验室，上海 200072）

摘 要 本文采用物理模拟的方法，以水银为实验介质，在板坯连铸过程中，在水口不同程度堵塞以及施加电磁搅拌情况下，采用超声波多普勒测速仪测量了结晶器模型内金属液的流速分布。结果表明：当水口无堵塞时，结晶器内流场为典型的双环流结构；水口堵塞会在结晶器内水口两侧造成严重的非对称流，即堵塞侧射流减小，非堵塞侧射流增大；当水口堵塞达到50%时，堵塞侧上环流消失，在弯月面靠近结晶器窄边处出现流动死区；电磁搅拌能修正由水口堵塞造成的偏流现象,提高结晶器内流场的对称性。

关键词 水口堵塞，电磁搅拌，结晶器，流场

Effect of Electromagnetic Stirring on Flow Field in Mold under Submerged Entry Nozzle Clogging Condition

Gao Qi[1], Zhou Yueming[1], Lei Zuosheng[2]

（1. Center Research Institute of Baosteel Group, Shanghai 201999, China;
2. Shanghai Key Laboratory of Modern Metallurgy & Materials Processing, Shanghai 200072, China)

Abstract A mercury physical model was developed to investigate the influences of electromagnetic stirring (EMS) on flow field in slab continuous casting mold while the submerged entry nozzle (SEN) was clogged. The velocity distribution in the mold with different SEN clogged rate (0%, 10%, 25%, 50%) and different EMS current (0A, 40A, 60A) was measured by ultrasonic Doppler velocimeter (UDV). The results show that when the SEN clogging is 0%, the flow field in the mold is typically a double roll structure. As the SEN clogging rate increasing, the flow distribution is transforming to more and more asymmetric. When the clogging rate is up to 50%, the up circulation disappears on the clogging side. In the mold, EMS can make the symmetry of the flow field better, by correcting the bias flow caused by SEN clogging.

Key words clogging of submerged entry nozzle, electromagnetic stirring, mold, flow field

几种无缝钢管缺陷的检测分析

曾海滨 王超峰

（宝山钢铁股份有限公司钢管事业部，上海 201900）

摘 要 本文介绍了一个无缝钢管缺陷的试验情况，记录了管坯上的人工缺陷在轧制前后产生变化的过程，通过人工测量、超声波检测、金相分析，剖析了缺陷的形貌特征。管坯上的人工纵向刻槽缺陷，长度方向平行于管坯的轴线，深度方向垂直于管坯的外表。斜轧穿孔、连轧、张力减径轧制后，缺陷在管料长度方向上产生了旋转和延伸，在管料横断面上产生了扭转和压缩。本文对缺陷变形进行分析，提出斜轧穿孔时的基本变形导致缺陷沿轴向延伸，沿径向压缩，附加变形引起缺陷沿长度方向旋转和横向扭转。本文针对几种无缝钢管自然纵向缺陷，结合轧管工艺，无损检测，金相试验，分析了缺陷产生的原因。结论是管坯带来的纵向缺陷在轧管过程中产生周向旋转和轴向延伸，连轧和张力减径产生的纵向缺陷沿轴向分布。超声波检测，可以初步判断出缺陷的大致方向。制定有针对性的措施，选用适当的无损检测方法，可以有效检出缺陷。

关键词 钢管，缺陷，检测，分析

Inspection and Analysis of Several Kinds of Seamless Steel Tube Defects

Zeng Haibin, Wang Chaofeng

(Steel Tubing Plant, Baosteel Iron & Steel Co., Ltd., Shanghai 201900, China)

Abstract This paper introduces the test of a seamless steel tube defects, and records the process of the artificial defects on the tube blank before and after rolling. The morphology characteristics of the defects were analyzed by means of manual measurement, ultrasonic testing and metallographic analysis. The longitudinal groove defect of the tube blank is parallel to the axis of the tube blank, and the depth direction is perpendicular to the surface of the tube. After rolling and tension reduction, the defects in the direction of the length of the pipe to produce a rotation and extension, in the pipe material on the cross section of the torsion and compression. In this paper the defect deformation analysis, proposed skew rolling piercing the basic deformation leads to defects extending along the axial direction, along the radial compression and additional deformation caused by defects along the length direction of rotation and lateral torsional. In this paper, we analyze the causes of the defects in the natural longitudinal defects of the seamless steel tube, the process of the rolling pipe, the nondestructive testing, the metallographic test. It is concluded that the longitudinal defects caused by tube bloom in the process of rolling pipe produce circumferential rotation and axial extension, and the longitudinal defects of continuous rolling diameter distribution along the axial direction. Ultrasonic testing can be a preliminary judgment of the general direction of the defect. To develop targeted measures, the use of appropriate non-destructive testing method, can effectively detect defects.

Key words steel tube, defects, inspection, analysis

螺杆式煤气压缩机密封的设计改进

施建军　韩凌俊

（宝钢股份设备部，上海　200941）

摘　要　用于从焦炉煤气制取氢气的制氢装置，其中螺杆式煤气压缩机在投入运行后不久即出现密封不良而导致内泄漏，工艺煤气进入润滑系统的故障。该密封的结构为充氮的碳环密封辅以迷宫密封，依据充氮口实际压力对泄漏量进行理论计算，分析判断密封泄漏的主要原因是设计不当，迷宫密封间隙与气体回流孔管径不匹配，从迷宫密封中逃逸出的工艺煤气气体流量大，超出回流管的实际回流能力，管道阻损加上背压，造成回流口的实际压力大于设计压力，超过充氮口压力形成“逆压差”，以至于向轴承室泄漏。经中外技术专家分析、研究，采取加大回流排气孔孔径、将密封氮气的回流管总管分开、回流管管径加大以及运行时提高轴封氮气压力等整改措施，煤气压缩机修复后运行正常，实测密封氮气压力与理论计算值相符，密封工作可靠无泄漏。

关键词　螺杆压缩机，煤气压缩机，碳环密封，迷宫密封

Design Improvement of Screw Coke Gas Compressor Sealing

Shi Jianjun, Han Lingjun

(Equipment Department, Baoshan Iron & Steel Co., Ltd., Shanghai 200941, China)

Abstract　For coke oven gas to produce hydrogen from hydrogen production device, the screw type coke gas compressor after put into operation shortly bad sealing and lead to leakage, process gas into the lubrication system fault. The sealing structure is filled with nitrogen and carbon ring sealing supplemented with labyrinth seal, according to the nitrogen export actual pressure on leakage was calculated, analysis to determine the main reason for the leakage of the seal is poorly designed, the labyrinth seal gap and the backflow of gas hole diameter mismatch, from the labyrinth seal escape escape process gas flow, beyond the actual ability to backflow backflow pipe, and the resistance loss of pipeline and pressure, resulting in a reflux inlet pressure is greater than the design pressure, over nitrogen pressure reverse pressure difference "that to the bearing chamber leakage. Through technical experts from home and abroad, analysis, take increasing reflux vent aperture, sealed nitrogen reflux pipe is separated and reflux pipe diameter to increase operation and improve the seal nitrogen pressure and the corrective measures, normal operation after the coke gas compressor repair measured sealing nitrogen pressure and the theory agrees with the value calculated, sealed reliable leak free.

Key words　screw compressor, coke gas compressor, carbon ring, labyrinth seal

Numerical Simulation of Milling Processes of UOE Edge Milling Machine and Influence Analysis on Accuracy

Wang Xuemin, Sun Fenglong

(Baoshan Iron & Steel Co., Ltd., Shanghai 201900, China)

Abstract To simulate edge milling process of the milling machine of a UOE welding pipe line, the geometric models of milling cutters and steel plate were built by Pro/E software, and the virtual milling process was simulated based on Open Inventor software. The material model of plate was built according to the physical properties of X80 API. In 10.3mm thick steel Case, the milling process was simulated respectively both for blunt edge milling cutter and for bevel milling cutter. The average value of milling force was obtained. The deformation and displacement contour of plate were achieved under milling force by ABAQUS software, and then the maximum deformation of plate was gotten after milling. The effect on the milling quality from cutter deviation was analyzed from the point of view of foundation offset. These studies provides theoretical reference and technical support for learning milling process, rational formulating of process parameters, the guarantee of plate milling accuracy and the normal running of the production line.

Key words UOE, milling machine, milling processes, numerical simulation, accuracy

Water Jet Quenching Machine Slit Flow Field Simulation and Structure Optimization

Wang Xuemin, He Xiaoming, Fan Qun

(Baoshan Iron ＆ Steel Co., Ltd., Shanghai 201900，China)

Abstract Heat treatment quenching machine is one of the key equipments for steel quenching and tempering treatment, whereas water spraying system is an important part of the quenching machine. High pressure outlet is designed to be a slit nozzle, also known as water jet. The water flow distribution uniformity for plate flatness is very critical. In this paper, by simulating analysis based on FLUENT software, the internal flow field distribution was obtained inside the quenching machine. By changing the internal structure of the slit jet, including the exit slit gap, outlet spray angle, flow hole quantity, nozzle length and web, the impact on outlet flow distribution was obtained. Results shows that slit gap, outlet spray angle and nozzle length don't affect the outlet water flow distribution uniformity significantly, however, flow hole quantity have some degree of impact , whereas web has significant effect. The analysis of the result provides the theoretical reference for the design and optimization of the structure of the water jet.

Key words quenching machine，water jet，structure design，plate shape

On Wear Morphology and Mechanism of Backup Roll

Wen Hongquan, Wu Qiong, Yao Lisong, Sun Dale

(Research Institute of Baosteel, Shanghai 201900, China)

Abstract The backup roll(BUR) is an important part in continuous rolling. The main failure form is wear of its surface . In this paper, to clarify the wear mechanism and find out the effective countermeasures to reduce the wear loss of BUR, taking samples from the normally off-line backup rolls in 2050 hot strip mill and 1550 cold strip mill of Baosteel, the surface wear morphology is investigated by means of the stereomicroscope, confocal laser scanning microscope, and scanning electron microscope, et al. The results show that the wear mechanism of backup roll is a combination of abrasive and corrosive wear. The hot rolling BUR is mainly of abrasive wear and corrosive wear, while the cold rolling BUR is mainly of abrasive wear. The main countermeasures to decrease the backup roll wear include: (1) strengthened the matrix strength (hot/cold rolling BUR); (2) fine matrix microstructure(hot/cold rolling BUR); (3) minimization of micro segregation (hot rolling BUR); (4) solution of undissolved carbides(hot rolling BUR).

Key words backup roll, wear morphology, wear mechanism, countermeasure

Research of an Internal Defects Detection Equipment for Steel Sheet Based on Magnetic Flux Leakage

Shi Guifen[1], He Yonghui[1], Zhang Qing[2]

(1. Engineering Institute of Baoshan Iron & Steel Co.,Ltd.,Shanghai 201900, China;
2. East China Normal University, Shanghai 200062, China)

Abstract An Equipment based on magnetic flux leakage method to inspect internal defects with diameters that is no more than 0.13mm is introduced. and it is able to The magnetic flux leakage method to inspect internal defects in steel sheet is based on ferro-magnetic material's magnetic phenomena in magnetic fields. It works by magnetic resistance sensors, which detect defects by closing to the steel strip with a small lift-off distance. The steel strip is magnetized int saturated status by a DC magnetizer. If there's any defects in a ferro-magnetic material, magnetic flux leakage field will appear in the surface of the defects and then it will be inspected by the sensors. After the signal is captured to process system, the defects' information can be displayed by filtering and processing the signal. After the signal is captured, the wavelet analysis method pre-processes the signal and then the size of defects can be calculated by statistical method or neural network method. In the end, the difficulties and technical trend for applications of magnetic flux leakage detecting technology in future has been discussed.

Key words magnetic flux leakage detection, internal defect, magnetizing

Contact Fatigue Damage Distribution form Assessment of Backup Rolls

Wu Qiong[1], Qin Xiaofeng[2], Sun Dale[1]

(1. Research Institute, Baoshan Iron & Steel Co., Ltd., Shanghai, China;
2. College of Mechanical Engineering, Taiyuan University of Technology, Taiyuan, China)

Abstract The contact stress distribution of backup rolls in a strip production line was analyzed by finite element method. Based on the contact stress, the fatigue damage in the axial section of backup roll was theoretically assessed and characterized by hardness test.The result revealed that the damage distribution of backup roll are consistent well with that obtained by theoretical method.

Key words contact stress, backup roll, fatigue damage, hardness

Local Template-based Segmentation Algorithm for Strip Steel Images

Peng Tiegen, He Yonghui

(Baoshan Iron & Steel Co., Ltd., Research Institute, Shanghai, China)

Abstract Inspired by the background subtraction method in motion detection algorithm, we proposed a new local template-based segmentation algorithm to detect the defects in steel strip images. The differences between the horizontal and the vertical templates were used to determine the location of the defect pixels. First, a block in original image is taken as template and moved it along the horizontal and vertical directions respectively to generate a relative motion between the template and the defects. Secondly, the defect pixel location is determined by the gravity center of the template differences. Experiments results show effectiveness of this segmentation algorithm.

Key words local template, image segmentation, strip steel

连铸塞棒紧固失效原因分析及措施

沈 康

（宝钢特钢有限公司，上海 200940）

摘 要 连铸机中间包塞棒机构一般多采用螺栓固定并通过钳臂上直通槽的结构进行固定和位置调整，主要依靠摩擦力来实现塞棒的紧固，由于受现场高温环境影响，钳臂机构易变形，止动弹簧易实效，从而导致传统安装结构易发生螺栓松脱，导致塞棒滑脱事故。通过对塞棒的安装方式进行改进，将钳臂的槽口改为L型，同时在塞棒上加装凸台垫圈，可确保及时在螺栓松脱的情况下，塞棒不会从钳臂中滑脱，确保使用安全。采用此结构的要点在于，通过钳臂的槽型改变，塞棒在脱离钳臂时需要转弯动作才能脱离，而由于加装了凸台垫圈，使得塞棒转弯半径增加。同时，钳臂的L型转弯半角远小于塞棒的转弯半径，利用两者转弯半径的不同来实现对塞棒行进路线进行限制固定，确保塞棒不会发生脱落。通过本文介绍为同型连铸机塞棒安装方式提供可靠的实施依据，可有效减少使用中塞棒脱离的事故发生。

关键词 连铸机，塞棒，安装，改进

Failure Analysis and Measures of Continuous Casting Stopper Rod Fastening

Shen Kang

(Baosteel Special Steel Co., Ltd., Shanghai 200940，China)

Abstract Caster tundish stopper rod mechanism generally use more bolts and fixed by clamp arm through groove structure is fixed and position adjustment, mainly rely on friction to achieve the fastening stopper, due to the high temperature environment impact of the scene, clamp arm mechanism easy deformation, stop spring easy effect, resulting in traditional mounting structure prone to loose bolts, lead to stopper slip accident. Through the stopper of the installation is improved, and the notch of the clamp arm is changed into a L type, at the same time in the stopper rod installing lug washer, ensure timely in the bolt loose situation, the stopper will not slipping from the clamp arm to ensure safe use. The point of this structure is that by changing the groove of the clamp arm, plug in the left forceps arm need turning action to detachment, and due to the installation of a lug boss washer, the stopper rod turning radius increased. At the same time, forceps arms of L-turn angle far less than stopper rod of the turning radius, using both turning radius to achieve different fixed limit on the stopper rod travel route, to ensure that the stopper will not happen to fall off. This paper introduces the installation method of the same type of continuous casting machine and provides a reliable basis for the effective reduction of the occurrence of the accident.

Key words continuous casting machine, the stopper rod, installation, improve

高分子复合波纹膨胀节在高炉煤气系统上的研发与应用

李 敏[1] 宫福利[2]

（1. 滕州市绿原机械制造有限责任公司，山东滕州 277500；
2. 山东莱钢永锋钢铁有限公司，山东齐河 251100）

摘 要 介绍了高分子复合波纹膨胀节的耐煤气腐蚀机理、优异性能及在 TRT 及 1080m3 高炉煤气（BFG）系统上的成功应用,从根本上解决了目前 316L、Incoloy800 镍基合金、超级奥氏体不锈钢 254SMo 等金属补偿器无法解决的晶界腐蚀问题，避免了因膨胀节泄漏造成的煤气泄漏事故，保证了高炉、TRT 及下游分厂生产的稳定性，为钢铁企业创造效益提供了可靠保障。

关键词 高分子，高炉煤气，TRT，波纹管膨胀节，煤气腐蚀

Development and Application of Macromolecule Polymer Complex Bellows Expansion Joints for the System of Blast Furnace Gas

Li Min[1], Gong Fuli[2]

(1. Tengzhou Lvyuan Machinery Manufacturing Co., Ltd., Tengzhou Shandong 277500, China;
2. Shandong Yongfeng Power Plant/ Ironworks,Qihe Shandong 251100, China)

Abstract Introduced macromolecule polymer complex bellows expansion joints corrosion mechanism, excellent performance and the successful application on the system of TRT and 1080m^3 blast furnace gas (BFG). The corrosion in grain boundary of metal bellows expansion joints made of 316L & Incoloy800 nickel alloy & super austenitic stainless steel 254SMo is solved fundamentally.

Key words macromolecule polymer, blast furnace gas(BFG), TRT, bellows expansion joints, gas corrosion

The Evaluation of Inclusions in Slab by the Focused Ultrasonic Method

Zhang Guoxing[1], Shen Yanwen[2]

(1. R&D Center of Baosteel company, No.655, Fujin Road, Baoshan District, Shanghai 201900, China;
2. Shanghai Jinyi Testing Company, Shanghai, 201900, China)

Abstract With modern steel manufacture capacity increasing, the demand for techniques of quick and accurate inclusion evaluation of the slab by large scale becomes urgent for the metallurgical technicians, especially the test carried out on the initial steel products. By these test results, the quality of further processed steel product can be forecasted. These techniques involve the investigation on inclusion's size, distribution, kind, and quantity. At present, the ultrasonic method is a suitable means to meet this demand. In this paper, inclusions detection sensitivity by focused ultrasonic method on the slab specimen is discussed, and an ultrasonic method by which inclusion as small as 100μm in the slab specimen of 30mm thick can be efficiently tested is proposed by the help of the focused ultrasonic transducer whose working frequency is 7.5MHz. The test result is verified by mechanical dissection and microscope observation at last, the experiment result demonstrates that not only it is feasible to carry out inclusion detection directly on the slab specimen by ultrasonic method, but also the ultrasonic detection sensitivity can be much higher than commonly estimated degree.

Key words ultrasonic, inclusions, slab, sensitivity, dissection

基于目前宝钢大型桥式起重机啃轨原因分析及处理

朱列昂 杨建华 颜 涛 贺 俊

（上海宝山钢铁股份有限公司炼钢厂运转车间，上海 201900）

摘 要 我国工业发展潜力大，大型桥式起重机作为应用范围最广的起重运输机械，作用于不同领域的各类车间、仓库，在室内或露天固定跨间作装卸、搬运及起重工作。它是所有大型工厂室内物流系统必不可少的运输机械设备。本文叙述桥式起重机的大车在日常作业运行中的啃轨现象及造成后果，从轨道缺陷、车轮缺陷、桥架变形等多方面分析了桥式起重机的啃轨原因，对各个方面的问题提出了处理措施，并对目前宝钢二炼钢 450t 行车进行了整改，现今运行正常。

关键词 桥式起重机，车轮，啃轨

Based on the Baosteel Large Bridge Crane Gnaw Rail Cause Analysis and Processing

Zhu Lieang, Yang Jianhua, Yan Tao, He Jun

(Shanghai Baosteel, Ltd., Steel Workshop Operation, Shanghai 201900, China)

Abstract China's industrial development potential is big, big bridge crane as the most widely application scope lifting transportation machinery, applied to different areas of the various workshops, warehouses, in indoor or outdoor fixed across intercropping loading and unloading, handling and lifting work. It is an essential part of all large factory indoor logistics system transport machinery equipment. This paper bridge crane cart gnaw rail in the daily operation run phenomenon and cause consequences, and wheels from orbit defects, bridge deformation analysis the causes of bridge crane gnaw rail, puts forward treatment measures on every aspect of the problem, and the no.2 steel-making plant of baosteel 450 t train operation for rectification, now operating normally.

Key words bridge crane, the wheel, gnaw rail

电石炉尾气的处理和综合利用

顾丽萍

（宝钢工程技术集团有限公司，上海 201999）

摘 要 电石乙炔是基本的化工原料，在化学工业的发展史上起过极为重要的作用。由于近年来世界石油化工高速发展，在发达国家已由乙烯、丙烯取代了电石乙炔的地位。根据我国化学工业的生产状况以及能源资源现状，电石作为化工原料还会在我国存在相当长的一段时间，而电石炉是高能耗、高污染设备，电石炉在生产过程中会产生大量高温的含尘尾气（烟气），这些尾气处理不当，会影响操作人员的健康，排入大气就会对环境产生污染。而电石炉的尾气中含有大量CO，又是一种可利用的能源介质。本文主要针对电石炉尾气的特性，对电石炉尾气进行收集、净化处理，使电石炉尾气满足清洁能源的要求，这样既改善了电石炉的操作环境，又变废为宝，节约了能源，提高了企业的经济效益，同时符合《电石行业准入条件》的规定。

关键词　电石，电石炉尾气，净化，综合利用

Manage and Integrate Utilize the Exhaust Gas of Calcium Carbide Furnace

Gu Liping

(Baosteel Engineering & Technology Group Co., Ltd., Shanghai 201999, China)

Abstract　Acetylene derived from calcium carbide is the elementary raw chemical material which had played a significant role in chemical industry developing history. As the result of the high speed expansion of world petroleum chemical in recent years, it has been replaced by ethane and propylene in developed countries. Based on the status of production and energy sources of chemical industry in our nation, calcium carbide will have been used as raw chemical material in a quite long time. As calcium carbide furnace is high expand energy and high pollution facility, it will produce a large quantity of high temperature exhausted gas during production process. Those exhausted gas will impact operators' health and cause environmental pollution if it was mishandled and been emitted into the atmosphere. There is a great amount of carbon monoxide in exhausted gas of calcium carbide furnace which can be used as a kind of energy medium. Basing on the characteristic of exhausted gas of calcium carbide furnace, this thesis mainly introduced how to collect, purify and manage those exhausted gas in order to meet the standard of clean energy. Thus it not only improved the operation environment of calcium carbide furnace but also make it possible to recycle and save energy which will increase the economic benefit of enterprise and finally correspond the regulation of *Access conditions of calcium carbide industry*.

Key words　calcium carbide, cxhausted gas of calcium carbide furnace, purify,integrate utilization

基于PATTERN表的转炉底吹控制

姚晓伟

（宝钢工程技术集团有限公司，上海　201999）

摘　要　将底吹仪表和阀门集成在一个阀站内，改善了仪表设备的使用性能，缩短了现场施工的周期。同时介绍了一种基于PATTERN表的转炉底吹自动控制方法，使得底吹的气体种类和流量能根据转炉的作业状态和顶吹氧量的比例自动进行调节。

关键词　转炉底吹，仪表阀站，pattern，氮气，氩气

Bottom Blow Control of BOF Based On PATTERN Table

Yao Xiaowei
(Baosteel Engineering & Technology Group Co., Ltd., Shanghai 201999, China)

Abstract It puts instruments and valves of BOF bottom blowing into a valve station to improve their function and reduce construction period. It introduces an auto control method of bottom blowing based on PATTERN table. The kinds and flow rate of bottom blowing gas can be auto tuned according to BOF's state and oxygen's percentage of top blowing.
Key words bof bottom blowing, instrument valve station, pattern, nitrogen, argon

一起冶金起重机失控故障分析

郭卫忠
（宝钢集团新疆八一钢铁有限公司检修中心，新疆乌鲁木齐 830022）

摘 要 起重机在工业企业中处于非常重要的地位，尤其是在炼钢厂里它是必不可少的主要生产设备，因此掌握起重机的复杂故障的分析方法将是工程技术人员必须认真学习的。本文就将通过对某公司炼钢分厂一台 220t 起重机的故障为例，从维护工排查处理方式调查、现场电气元件故障判断、PLC 程序监测、模拟故障发生等方面对本次故障进行了全面分析，总结了经验和教训。
关键词 起重机，PLC，有条件调用

A Fault Analysis with Metallurgical Crane Control

Guo Weizhong
(Baosteel Group Xinjiang Bayi Steel Co., Ltd., Maintenance Center, Urumchi 830022, China)

Abstract Crane plays a very important role in industrial enterprises. Especially in the steel mill. It is the main production equipment essential. So the engineering and technical personnel will be the need to seriously study and master the complex fault analysis method of crane. This paper will through the fault of Second Steelmaking Plant of a 220t crane as an example，From the maintenance treatment and investigation style questionnaire、Electric field component fault judgment、Monitor the PLC program、Simulate the fault occurrence etc，To conduct a comprehensive analysis，Summed up the experience and lessons.
Key words crane, PLC, conditional call

真空感应熔炼炉上的高效节能浇注系统

钱红兵

（应达工业（上海）有限公司，上海 201203）

摘 要 浇注对钢锭质量有非常大的影响，真空浇注工艺能消除钢液的二次氧化和对气体卷吸。VIM 配备性能优良的真空闸板阀，在真空熔炼室不被破真空的条件下，完成浇注。大体积的熔炼室便于生产操作，合理的浇注系统设计，使得 VIM 能够实现半连续生产，即使是大体积的熔炼室，生产过程中的抽真空时间也非常短。Consarc 的浇注系统采用中间包而不是流槽，缩短了整个浇注系统的长度；中间包的布置方向与感应炉出钢的液流方向一致，中间包可以被设计成窄条状，浇注系统的宽度变窄，在满足工艺要求的条件下尽能小。小而精密的浇注系统，减少抽真空的时间，带来生产效率的提高和能耗的降低。真空浇注系统中配备了工业电视与检测仪表，所采集的信号被传送到 VIM 的 SCADA 系统，由系统中的 PLC 统一处理。

关键词 真空感应熔炼，浇注系统，生产率，节能，中间包

High Efficiency and Energy Saving Pouring System of VIM

Qian Hongbing

(Inductotherm（Shanghai）Industrial Company, Shanghai 201203, China)

Abstract The casting has great influence on the quality of steel ingot. Vacuum casting process can eliminate either re-oxidation of molten steel or the gas entrainment. VIM equipped with excellent vacuum gate valve can complete casting ingot, in the condition of the vacuum melting chamber not being broken vacuum. As the VIM is semi continuous production, a reasonable design of pouring system makes sure that the time of large volume of the smelting chamber being vacuumized is very short during the production process. A large volume of the smelting chamber helps to operate process and improve production efficiency. The pouring system designed by Consarc equips with tundish rather than launder. It has shorter the length of the pouring system. The direction of tundish is in line with pour stream of induction furnace, so the width of tundish is in meet the process conditions as more narrow as possible. A small poring system reduces the vacuum pump operation time, and improves productivity and reduces energy consumption. The vacuum pouring system is equipped with industrial TV and detection instruments. The collected information is transferred to the SCADA system of VIM, and is handled by the PLC of the system.

Key words VIM, pouring system, productivity, energy saving, tundish

模块化数控程序在多品种小批量零件中的运用

谢红军 章 意

（江苏省常州宝菱重工机械有限公司，江苏常州 213019）

摘 要 随着市场日益国际化，企业生存环境发生了巨大的变化，产品生命周期缩短，客户需求多样化和个性化，交货期更加严格，多品种、小批量生产比例增大，传统的加工生产已满足不了需求，迫切需要采用一种通用合理化方法以提高效率、产品质量和档次，缩短生产周期，以提高企业的市场竞争能力，在这种环境所趋下数控模块化加工标准库孕育而生，数控加工模块化标准库是由产品系列化、组合化、通用化和标准化组成的。系列化的目的在于用有限品种和规格的产品来最大限度、且较经济合理地满足需求方对产品的要求。组合化是采用一些通用系列部件与较少数量的专用部件、零件组合而成的专用产品。通用化是借用原有产品的成熟零部件，不但能缩短制造周期，

降低成本，而且还增加了产品的质量可靠性。目前数控加工技术正从 ONC 向 CNC 转变，数控编程技术显得尤为重要，提高数控机床的连续运行时间，降低辅助生产时间才能保证高效加工制造，而数控模块化程序能保证机床的连续运行，通过编程人员编制的模块化加工程序，降低操作人员因为编制、调试程序而占用的等待时间，从而提高生产效率。因此数控模块化制造技术是提高生产效率、缩短生产周期的重要手段之一，也是制造技术发展的趋势。

关键词 模块化，参数编程，宏程序，插补，SIEMENS，FANUC

Application of Modularized Programs for Multi-specification and Small-batch Production

Xie Hongjun, Zhang Yi

(Baoling Heavy Industrial Machinery Co., Ltd., Changzhou, 213019, China)

Abstract With the market growing more internationalized, survival environment for enterprises has undergone tremendous changes. Short leading time products, customers' diversified and personalized requirements, more stringent delivery condition, increasing of multi-specification & small-batch, the traditional processing can not satisfy these demands, therefore it is very urgent for us to find a rational way to improve efficiency, quality& grades and shorten the production cycle as well to enhance the capacity of competition. Under this circumstances, the NC modular processing standard library has set up. NC processing modular standard library is consist of series, combination, universalization and standardized of products. The purpose of the series is to maximum, and more cost-effectively meet the requirements of the demand with limited varieties and specifications of products. The combination is a combined specific products with some common components and a small number of specialized components together. Universalization borrows from the original mature parts, it can not only shorten the manufacturing cycle, reduce costs, but also increase the quality and reliability of the product.At present, NC machining technology is changing from ONC to CNC, CNC programming is essential to improve the continuous operation time of CNC machines and reduce auxiliary production time in order to ensure efficient processing, while NC modular program can ensure continuous operation of the machine. It can improve productivity due to modular machining programs which programmed by programmer and decreased the waiting program time. So NC modular manufacturing technology is one of the important means to improve productivity, reduce production cycle, as well as manufacturing technology trends.

Key words modularized, parameter programming, macro program, interpolation, SIEMENS, FANUC

热轧卡罗塞尔卷取机结构分析及设计

梅如敏

（常州宝菱重工机械有限公司技术中心，江苏常州 213019）

摘 要 卡罗塞尔卷取机因其高速、连续、高可靠性等优点，已在现代化冷连轧生产线中广泛应用，但在热轧生产中很少应用，国内几乎没有对其结构及设计、工业应用的研究。本文根据国产化的第一台热轧卡罗塞尔卷取机，在分析、对比冷轧行星式卡罗塞尔卷取机的基础上，研究了热轧卡罗塞尔卷取机在总体设备布置、结构形式、主要结构设计特点，以满足热轧生产工艺的要求。着重介绍了热轧卡罗塞尔卷取机主体结构、支撑装置、卷筒、辅助设备等结构及设计。文中还总结了热轧卡罗塞尔卷取机设计方法及要点、主要参数的设计计算及选取经验，简要介绍了国产热轧卡罗塞尔卷取机现场使用情况及存在的问题。热轧卡罗塞尔卷取机的成功国产化及实际应用为其推广引用及深入研究奠定了基础，本文为热轧卡罗塞尔卷取机的自主设计及应用提供一些经验和方法。

关键词　热轧　卡罗塞尔卷取机　行星齿轮　结构分析　设计

The Analysis of Structure and Design of Carrousel Coiler for Hot Mill

Mei Rumin

(Technical Center of Changzhou Baoling Heavy & Industrial Machinery Co., Ltd.,
Changzhou 213019, China)

Abstract　Because of it's high speed ,continuity and high reliablity, carrousel tension reel had been applied to the cold tandem mill production line widely,but it rarely had been appiled to the hot mill production line. Even the research on the structure, design and industry application of carrousel coiler for hot mill is too few in China. Based on the first localization carrousel coiler and the analysis and comparison of planetary gear type carrousel tension reel for cold mill, this article studies the carrousel coiler's general plant layout ,structure type, main structure design characteristic for hot mill production technology in detail, especially it introduces the struture and design of carrousel coiler's main struture ,support mechanism , mandrel and accessory equipments etc.. It summarizes some correlative methods and experience for carrousel coiler design, supplies the method of main parameter calculation for motor power and reducer ratio. And it also introduces the service condition and problems at user's shop of the first localization carrousel coiler in brief. Carrousel coiler's successful localization laids the foundation for it's popularization and the study and in-depth reference.So this article can supply some methods and experience for carrousel coiler independent design and industry application.

Key words　hot mill, carrousel coiler, planetary gear, structure analysis, design

热镀锌机组炉鼻子及沉没辊在线监测方法研究

蔡正国

（宝钢工程集团宝钢技术金艺检测有限公司，上海　201900）

摘　要　冷轧热镀锌生产线中的沉没辊轴套和轴瓦沉浸于 450～550℃的锌液中，经受着非常严重的锌液腐蚀和磨损，工作寿命一般都较短。沉没辊的轴头断裂事故会造成较长时间停机，严重影响热镀锌机组的生产效率和镀锌板的生产成本。本文通过在锌锅沉没辊辊架上安装振动传感器，研究采用分形维数指标预报沉没辊运行状态劣化趋势的可行性；在炉鼻子内稳定辊轴承座上安装振动和温度传感器，采用振动速度和轴承冲击参数监测其振动和温度变化，在线预警热镀锌机组稳定辊的状态异常。通过跟踪研究沉没辊从上线到换辊使用周期的实际数据表明，沉没辊在不同运行状态下分形维数表现出不同的特征，沉没辊轴承磨损和轴头断裂等典型故障时分形维数较小，而运行正常时分形维数较大。该技术可实现对冷轧镀锌线炉鼻子及沉没辊的运行状态管理，避免因沉没辊断裂导致的设备非计划停机，支撑冷轧镀锌生产线的正常生产。

关键词　冷轧，炉鼻子，沉没辊，振动，预报

Research of On-line Monitoring for Furnace Nose and Sink Roll in Hot-dip Galvanizing Line

Cai Zhengguo

(Shanghai Jinyi Inspection Co., Ltd., Shanghai 201900,China)

Abstract In hot-dip galvanizing line the sink roll sleeve and bearing are immersed in molten zinc of 450 to 550℃ suffering from very severe liquid zinc corrosion and wear, life span is generally shorter.The breakage accidents will result in longer downtime and seriously affect the production efficiency of hot galvanizing and galvanized sheet production costs.The on-line monitoring and diagnosis method is declared for sink roll of galvanizing line in cold mill. The vibration acceleration sensors are mounted on frame of sink roll and stable roll with temperature sensors at the same time. It aims at strategy of forecasting the running trend on sink roll using fractal dimension of vibration waveform. The parameters such as vibration velocity and bearing pulse (gSe) as well as temperature are monitored on furnace nose and stable roll. By tracking the actual data from start to end fractal dimension of sink roll shows very clearly effective to evaluate the fault characteristics. That is definitely small when the running state is abnormal such as bearing wear and equipment fault before rupture of axis neck however the counterpart running at normal time is larger. The practical application shows that it is significant to condition monitoring and forecasting for sink roll and stable roll to avoid abnormal shutdown of hot-dip galvanizing line.

Key words cold mill, furnace nose, sink roll, vibration, forecasting

S 型转炉导渣技术在转炉炼钢上的应用

田志恒[1] 谢俊华[1] 杜开发[1] 何 航[2] 史志凌[2] 龙友锋[2]

（1. 衡阳镭目科技有限责任公司，湖南衡阳 421000；
2. 华菱集团湘潭钢铁有限责任公司五米板厂，湖南湘潭 411100）

摘 要 本文介绍 S 型转炉导渣技术在转炉炼钢上的应用，S 型转炉导渣法是变堵挡为疏导，在出钢快要结束时启动导渣装置，驱动导渣槽至准备位置，结合镭目转炉下渣检测技术，在下渣时刻导渣槽立即由准备位置伸向出钢口，把原先流向大包的渣通过导渣槽导流至渣盆或其他适当的地方，待完成导渣后迅速地回到初始位置。导渣法可有效防止渣进入大包，实现大包渣层薄、提高钢水洁净度的目的。

关键词 转炉炼钢，导渣技术，导渣槽，渣层薄，钢水洁净度

The Application of S Type Converter Slag Diversion System in Converter Steelmaking

Tian Zhiheng[1], Xie Junhua[1], Du Kaifa[1], He Hang[2], Shi Zhiling[2], Long Youfeng[2]

(1. Ramon Science & Technology Co.,Ltd.,Hengyang 421000, China;

2. Xiangtan Iron & Steel Co.,Ltd., Valin Group Xiangtan, Xingtan 411100, China)

Abstract The paper introduces the application case of S type converter slag diversion system in steel plant. S type converter slag diversion system is changed slag blocking to slag diversion, which starts the slag diversion device and slag diversion trough to their position place when the tapping is about to be finished, making the slag flowing to the trough or other proper places but the ladle, and later returning to their initial position rapidly after finishing slag diversion. The slag diversion method can prevent slag enter into ladle efficiently, making thin layer of ladle slag and improving the cleanness of molten steel.

Key words converter steelmaking, slag diversion technology, slag diversion trough, thin slag layer, improving molten steel cleanness

焦炉烟道吸力调节装置及其使用方法

王　宁

（宝钢工程技术集团有限公司，上海　201999）

摘　要　本文主要介绍了一种新型的焦炉烟道吸力调节装置，其具备操作简便，维护方便和经济适用等特点，能很好地满足焦炉生产过程中烟道吸力调节的需要。

关键词　焦炉，烟道，吸力调节，闸板

The Suction Adjustment Device of Coke Oven Flue and Its Application

Wang Ning

(Baosteel Engineering & Technology Group Co., Ltd., Shanghai 201999, China)

Abstract This paper mainly introduced a new kind of coke oven flue suction control device, it has simple operation, convenient maintenance and economical characteristics of well satisfy the need of regulation in the production of coke oven flue suction.

Key words coke oven, the flue, suction adjustment, flashboard

高速钢轧辊在梅钢 1422 生产中的应用

马叶红[1]　李欣波[2]

（1. 宝钢工程集团宝钢工业技术服务有限公司南京分公司；
2. 宝钢股份梅山钢铁公司热轧板厂，江苏南京　210039）

摘　要　为降低轧辊消耗、提高轧机作业率、改善产品质量，梅钢热轧 1422 轧机精轧机组开始试用高速钢轧辊。对高速钢轧辊使用过程中的表面氧化膜、耐磨性能、抗热裂纹和防剥落性、加工性能进行跟踪。实践证明，随着高

速钢轧辊的使用，换辊次数、磨削次数明显减少，轧制公里数大大提高，产品表面质量有了明显改善。

关键词 热轧宽带钢轧机，精轧机组，高速钢轧辊

Application of High-speed Steel Roll to Finishing Stands of Meishan 1422 Hot Strip Mill

Ma Yehong, Li Xinbo

Abstract In order to increase the operation rate of mill and decrease rollwaste. the high—speed steel roll is applied to finishing stands of hot wide strip mill in Baosteel of Meishan Iron and Steel Co., Ltd.. Its some properties are measured, such as surface oxsigen,resistance to wearing and surface roughness, resistance to hot crack and spalling, machining etc. meanwhile the methords of how to use high—speed steel roll are followed by the properties of the HSS rolls.The practical application results show that the times of roll changing and grounding are obviously decreased,as the same time,the surface qulity of hot slab is improved very well.

Key words hot wide strip mill, finishing mill, high-speed steel roll

利用振动分析技术提高引风机长周期运行

吴小树

（南京钢铁股份有限公司中板厂，南京 210035）

摘 要 通过对现代离线巡检监测系统的使用，提出利用振动分析技术的重要性，并就加热炉煤气引风机的振动现象，找出振动的原因，解决振动问题，并提出了预防措施。

关键词 引风机，振动，原因分析，长周期，运行

Improve Blower Life through Vibration Analysis Technology

Wu Xiaoshu

（Nanjing Iron and Steel Co., for Midium Plate Plant, Nanjing 210035, China）

Abstract Through use of mordern offline monitoring system, the importance of vibration analysis technology is come out. And in this article, according to vibration of Heating furnace gas blower, vibration reason is found and resolved methods are introduced.

Key words blower, vibration, analyses, long cycle, run

轧机牌坊现场防腐激光熔覆的应用

吴小树

（南京钢铁股份有限公司中板厂，南京 210035）

摘 要 轧机牌坊在恶劣工况条件下会出现不同程度的腐蚀，针对这种情况，通过堆焊恢复原尺寸后在表面采用耐腐蚀性能和抗氧化性能最好的S13（Co1）材料进行现场牌坊激光熔覆，通过一年半的现场使用验证，未发现开裂、脱落等异常现象，磨损量在设计范围内0.03mm/年。

关键词 轧机牌坊，激光熔覆，耐腐蚀，抗氧化性能

Application of Laser Melting Coating for Rolling Mill House Corrosion

Wu Xiaoshu

（Nanjing Iron and Steel Co., for Midium Plate Plant, Nanjing 210035, China）

Abstract The rolling mill house would be corroded in bad working condition. To avoid this bad situation, the rolling mill house was recovered to original size through overlaying welding, and then laser melting coating process was performed. The material S13(Co1) was used which has good anti-corrosion and anti-oxidation quality. After one and half year, there is no cracking and falling off, and the wearing capacity is 0.03mm per year within design scope.

Key words rolling mill house, laser melting coating, anti-corrosion, anti-oxidation

同步电动机自耦变降压启动动态仿真及分析

沈国芳[1] 贡兆良[2] 杨左勇[1]

（1. 上海金艺检测技术有限公司，上海 201900；2. 宝钢股份有限公司，上海 201900）

摘 要 论文介绍了宝钢如何基于 ETAP 软件的专用电动机 MA 启动模块，实现通过自耦变降压启动的一台27000kW 制氧空压机同步电动机的动态启动过程仿真和分析。文章介绍了仿真数据文件的建立，主要包括电动机等值模型数据文件、启动过程空压机负荷模型数据文件，以及自耦变模型数据文件；介绍了所研究的自耦变自身压降的等值方法，用于提高仿真精度；介绍了在基于ETAP软件的宝钢等效电网中，模拟各种运行方式下对27000kW制氧空压机同步电动机进行启动动态仿真，并对启动电流-时间-压降等进行计算。文章对仿真结果进行了分析，提出了相应的启动策略，结合早期设备调试阶段所录制的启动波形，及本文研究过程中所录制的启动波形完成了空压机电动机的继电保护整定优化，解决了现有的空压机电动机启动失败难题，有效提升了现场技术管理水平。

关键词 同步电动机，自耦变，动态启动仿真，保护整定

Dynamic Simulation and Analysis of syn. Motor Starting via Autotansf

Shen Guofang[1], Gong Zhaoliang[2], Yang Zuoyong[1]

(1. Shanghai Jinyi Inspection Technology Co., Ltd., Shanghai 201900, China;
2. Baoshan Iron & Steel Co.,Ltd., Shanghai 201900, China)

Abstract Based on ETAP thinking power software's MA module,Baosteel has built a synchronous motor starting simulaiton system and has made dynamic simulation and analysis of a 27000kW syn. motor.The paper describes three simulation module data files,including motor、compressor，and autotransformer;a smart equivalent method to solve the voltage drop of autotransformer,and improve simulation accuracy;different simulation results of the motor starting current,busbar and motor terminal voltage ,and total starting time,ect;analysis of the results , and further measures must be implemented .

Key words synchronous motor, autotransformer, dynamic starting simulation, relay setting

浅谈 PDCA 循环在板坯电磁搅拌系统维护管理中的应用

成建峰

（首钢迁安钢铁公司，河北迁安 064404）

摘 要 PDCA 循环是全面质量管理所应遵循的科学程序，是企业质量管理提升的重要工具。它的应用不仅体现在质量管理工作中，也被越来越多地发展和应用在其他各项管理活动中。本文简单介绍了 PDCA 循环的概念，并以迁钢公司板坯铸机电磁搅拌系统为背景重点介绍了 PDCA 循环管理模式在电磁搅拌系统的维护管理中的实践应用和取得的效果。

关键词 PDCA，电磁搅拌，维护管理

Introduce the Application of PDCA Cycling in Electromagnetic Stirring System's Maintenance and Management of Slab Caster

Cheng Jianfeng

(Shougang Qian'an Steel Corp, Qian'an Hebei 064404, China)

Abstract PDCA Cycling is a scientific law and program in total quality management ,and it's a important tool to promote management of enterprises. The application is not only embodied in quality management work, also been developed and applied to others management activities more and more. This paper simply introduces the concept of PDCA cycling, mainly introduces the practical application and the good effects of PDCA cycling management in electromagnetic stirring system's maintenance and management in Qian'an steel corporation.

Key words PDCA, electromagnetic stirring system, maintenance and management

宝钢三烧结主排变频起动装置的应用及典型故障案例

黄志刚[1] 潘世华[2] 刘 珧[1] 沈国芳[2]

（1. 宝山钢铁股份有限公司设备部，上海 200941；
2. 上海宝钢工业技术服务有限公司，上海 201900）

摘　要　宝钢三烧结主排风机采用了交直交电流型变频起动装置，由一套变频起动装置分别起动两台主排风机。变频起动装置从静止开始加速主排风机，当加速到同步速后，通过同步并网调整后，同步机几乎无冲击切换到工频运行。本文以宝钢三烧结主排风机变频起动装置为例，介绍了三烧结主排风机系统结构与主回路参数配置，介绍了交直交电流型变频起动装置的可控硅换流原理与同步机系统控制原理。并以三烧结主排为例，详细介绍变频起动过程中系统各部分之间的协调控制；在从变频向工频切换过程中，整步并网装置的控制方法，以及并网三要素的具体整定要求。并结合几个典型故障案例，介绍了同步电动机故障处理的思路、方法。最后对三烧结主排变频起动装置进行简单总结。

关键词　同步电动机，变频起动装置，负载换相逆变器

The Use of VVVF Startup of Synchronous Motor and Some Typical Faults

Huang Zhigang[1], Pan Shihua[2], Liu Yao[1], Shen Guofang[2]

(1. Equipment Department, Baoshan Iron & Steel Co., Ltd., Shanghai 200941, China;
2. Shanghai Baosteel Industry Technology Service Co., Ltd., Shanghai 201900, China)

Abstract　In combination with the start device of No.3 sinter main blower, this article mainly introduces the principle structure and starting process of the AC-DC-AC VVVF startup device.with the example of some typical faults ,it introduces some thinking ways and methods to solve the faults.

Key words　synchronous motor, VVVF startup, load-commution-inverter

承压类管状特种设备短期超载超温韧性失效分析浅谈

廖礼宝　邓　聪

（宝钢股份设备部，上海　201900）

摘　要　锅炉压力容器和压力管道等特种设备主要由管状承压类构件组成。由于该类设备始终处于高温、高压、冲刷及磨蚀等低周高应运行工况，其失效形式多种多样。再加上设计、制造、安装、使用、维护、修理与改造等各环节中难以避免地存在各种问题，因此，其失效原因分析往往变得更为复杂。为了分类总结 30 年来该类管状特种设备的技术与管理经验，文章针对短期超载超温失效，从断口金相与显微金相、宏观断口与微观断口形貌、SEM/TEM、EDAX/EDS、力学性能等方面进行了专题分析，结果发现：失效特征取决于工况，即该类设备分别在短期超载或超温、同时超载并超温、超载超温失效后伴有过热或过烧工况等条件下，均与相应的宏观和微观失效形貌特征一一对应。反之，依据特定的宏观与微观失效形貌特征可快速推测出相应的失效工况和失效原因。该研究提供的新颖短平快因果分析法，对设备技术与管理工作大有裨益。

关键词　承压类管状特种设备，短期，超载超温，韧性失效分析

A Probe into Toughness Failure with Short Period Overload and Super Temperature on Pressure-bearing Tubing Special Equipment

Liao Libao, Deng Cong

(Equipment Department, Baoshan Iron & Steel Co., Ltd., Shanghai 201900, China)

Abstract Special type equipment such as boiler, pressure vessel and pressure pipe are mainly consist of tube-shaped pressure parts. Such equipment are always in the case of high temperature, high pressure, washing, grinding and corrosion, low frequency and high stress and so forth that the failure form and mode are diversified. Because unavoidable differentia problem may emerge from design, manufacture, installation, use, maintenance, repair and reform etc. Thus the cause of failure may more often than not become sophisticated.To solve the problem and summarize 30 years experience in technology and management , fractography and metallography, macro and micro fracture characteristic, mechanical property,SEM analysis and failure mechanism were specially analyzed and tested.The result showed that failure typification was closely related to use condition. Under the condition of short period overload, super temperature,both overload and super temperature, short period overload and super temperature accompanying over heat and over burn, one to one correspondence macro and micro failure characteristic were found. Vice versa,according to the macro and micro failure characteristic, the working condition and the cause were deduced and elicited accordingly.Therefore, the research provided with a novel,rapid and high efficacious shortcut cause and effect analysis method which will benefit the technical and manageable affairs in the field of tube-shaped pressure type equipment.

Key words pressure-bearing tubing special type equipment, short period, overload and super temperature, toughness failure analysis

大型锻钢支撑辊残余应力评估技术

贺 强[1] 徐济进[2]

（1. 上海金艺检测技术有限公司，上海 201900；
2. 上海交通大学材料科学与工程学院，上海 200240）

摘 要 Cr4、Cr5 支撑辊具有高的耐磨性、抗断裂性、抗剥落性，其制作工艺复杂，最终热处理工艺导致的残余应力大小及分布将直接决定其工作性能和使用寿命，因此评估支撑辊表层及内部残余应力，为热处理工艺的优化和支撑辊的失效分析提供参考。本文采用残余应力测试及数值模拟技术，评估大型锻钢支撑辊残余应力大小及分布，同时分析最终热处理工艺参数（保温时间、回火温度）对残余应力的影响。结果表明，支撑辊热处理数值模拟的残余应力与两种应力测试方法的测量结果比较吻合，支撑辊表层的轴向应力和环向应力主要表现为压应力，距离表层一定深度显示为拉应力，而心部的应力基本接近于零。保温时间不仅对表层残余应力而且对内部残余应力都有显著的影响；回火温度只对表层压应力有显著影响，而对内部应力没有影响。

关键词 支撑辊，数值模拟，残余应力，盲孔法，X 射线衍射法

Residual Stress Assessment of the Large Forged Steel Bearing Roller

He Qiang[1], Xu Jijin[2]

(1. Shanghai Jinyi Inspection Technology Co., Ltd., Shanghai 201900, China;
2. School of Materials Science and Engineering, Shanghai Jiaotong University, Shanghai 200240, China)

Abstract Bearing roller of Cr4 or Cr5 has high wear resistance and fracture resistance and spalling resistance. Its production process is very complex. Final heat treatment process leads to residual stresses, which will directly determine the performance and service life. Therefore, it will provide a reference for heat treatment process optimization and failure analysis by assessing the residual stresses of the bearing roller on the surface and interior. The residual stress measurement and numerical simulation technology were employed to assess the residual stresses of the large forged steel bearing roller. The effects of the final heat treatment process parameters (holding time and tempering temperature) on residual stress were analyzed. The results show that the simulated residual stresses and the measured results are in good agreement. The axial stresses and hoop stresses on the bearing roller surface are compressive. The residual stresses from the surface to a certain depth are tensile. The residual stresses at the core are close to zero. Holding time has significant influence not only on the surface residual stress, but also on the internal residual stress. Tempering temperature only has a significant effect on the surface residual stress, but has no effect on the internal stress.

Key words　bearing roller, numerical simulation, residual stress, blind hole drilling method, X-ray diffraction

转向辊升速过程中带钢打滑的数学模型及其应用

袁文振　张宝平　杜国强

（宝钢股份公司冷轧厂，上海　201900）

摘　要　带钢与转向辊之间出现速度差，容易产生打滑，在带钢表面形成打滑伤，影响带钢表面质量。通过建立转向辊升降速过程中与带钢之间的运动学模型和力学模型，系统分析了转向辊升降速过程中带钢打滑的极限条件，结合现场实际，指出了在升降速过程中带钢打滑不唯一是由于转向辊粗糙度下降引起的，转向辊前后张力的波动也会导致带钢产生打滑，前后张力差的增大会导致带钢和转向辊的加速度增大，当增大到一定程度摩擦力矩不能提供转向辊转动所需的惯性力矩时即会出现打滑事故；前后张力差的减小会导致带钢和转向辊的加速度减小，当减小到一定程度，摩擦力矩与转向辊的阻动力矩相平衡，带钢将会和转向辊不再加速，并分析了前后张力的波动对处于“包裹段”带钢的摩擦力、支持力、加速度的影响。在此基础之上，指出了张力波动控制优化的目标值，并应用于现场，避免了打滑事故的再次发生。

关键词　转向辊，打滑，升速，数学模型

The Mathematical Model of Strip Sliding on Steering Roll during Acceleration Stage and Its Application

Yuan Wenzhen, Zhang Baoping, Du Guoqiang

(Baosteel Cold Rolling Plant, Shanghai 201900, China)

Abstract　The speed difference between the strip and the steering roll is easy to produce strip's sliding, and the surface quality of the strip is affected.By building up the kinematics model and mechanics model of the strip and the steering roll during its acceleration and deceleration stage,the limiting conditions of strip's sliding in the process of the steering roll during its acceleration and deceleration stage is analyzed, combine with the actual,point out that in the process of acceleration and deceleration stage strip's sliding is not only caused by the decline of the steering roll's roughness,but also by the fluctuation of the forward tension and the backward tension,the increase of the difference between the forward tension and the backward tension will lead to the increase of the acceleration of the strip and the steering roll,when increasing to a certain degree that friction torque can not provide the torque required for the rotation of the steering roll,it will be strip's sliding accident.The decrease of the difference between the forward tension and the backward tension will lead to the decrease of the acceleration of the strip and the steering roll,when decreasing to a certain degree, the friction torque and the torque of the steering roll will balance, the strip and the steering roll will no longer accelerate, and analyze the influence of the fluctuation on the friction force,the support force, the acceleration of the strip in the "package",and on this base,point out the target of the tension fluctuation control optimization ,and put it into use,then avoid the slipping of accident happening again successfully.

Key words steering roll, sliding, acceleration, mechanics model

基于经验模式分解的轧机齿轮箱故障诊断

张建新 刘 晗

（上海宝钢工业技术服务有限公司，上海 201900）

摘 要 轧机齿轮箱故障直接影响着轧线的生产，对轧机齿轮箱进行状态管理是现代冶金设备管理的发展方向。本文通过对经验模式分解方法进行介绍并通过经验模式分解(EMD)方法对现场齿轮箱振动信号进行处理，对处理前后的振动信号进行比较并对处理后的振动信号进行分析,对齿轮箱故障进行诊断，对现场轧机齿轮箱状态进行把握，最终实现设备的状态管理。

关键词 轧机齿轮箱，经验模式分解，故障诊断

Based on the Empirical Mode Decomposition Mill Gearbox Fault Diagnosis

Zhang Jianxin, Liu Han

(Shanghai Baosteel Industry Technological Service Co., Ltd., Shanghai 201900, China)

Abstract Mill gear box failure directly affects the rolling line of production, mill gearbox state management is the development direction of modern metallurgical equipment management.In this paper, Based on the empirical mode decomposition method introduced and through empirical mode decomposition (EMD) method for field gearbox vibration signal processing, vibration signals were compared before and after treatment and to analyze the vibration signal processing for gearbox fault diagnosis, on-site mill gearbox state grasp, and ultimately the device state management.

Key words mill gearbox, empirical mode decomposition, fault diagnosis

万吨高炉大修运输车-DCMC 型自行式液压模块车

陈永昌 曹新杰 职山杰 梁 勇

（苏州大方特种车股份有限公司，江苏苏州 215151）

摘 要 本文对 DCMC 型自行式液压模块车的研发背景、技术性能、创新点进行了介绍。宝钢工程技术集团苏州大方公司以宝钢股份 3、4 号高炉原地快速大修炉体运输工程为契机，研制了载重量达 10000t 级的 DCMC 型自行

式液压模块车（简称：C 型模块车），该模块车具备超大载重、模块化、智能化、多模式转向等特点，大幅缩短了高炉停炉时间（3 号高炉停炉时间仅为 76 天），节约了高炉大修成本。该模块车成功替代了进口产品，打破了国外厂商对万吨级运载装备的技术垄断。同时作为超大型海工平台、大型船体的核心运载装备，必将助力我国冶金、海洋工程等行业的快速发展。

C 型模块车由多个 6 轴线模块、4 轴线模块及动力模块（PPU）组合而成。模块车采用 CAN 总线微电控制，液压驱动行走，液压独立转向方式，最大组拼规模可达到 1000 轴线，最大载重量可达 2 万吨。

关键词　高炉大修，模块运输车，大件运输，动力分散，分布式控制

Transportation of 10000 tons Blast Furnace-DCMC Type Self-propelled Modular Transporter

Chen Yongchang, Cao Xinjie, Zhi Shanjie, Liang Yong

(Suzhou Dafang Special Vehicle Co.,Ltd., Suzhou 215151, China)

Abstract　The paper introduce the Research background, Characteristic and Innovative points of DCMC type Self-propelled modular transporter. Suzhou Dafang Special Vehicle Co.,Ltd (member of Baosteel Group)developed the DCMC type self-propelled modular transporter（for short:C type modular transporter） which load capacity exceed 10000t,and the transporter had been successfully applied on the Baosteel No.3 and No.4 Blast furnace reconstruction project.This transporter has many advantages,such as large load capacity,modular assembled,intelligent control,multiple steering etc,the transporter can reduce the off production time of the blast furnace(for the Baosteel No.3 blast furnace reconstruction project ,the off production time is only 76 days),it can also save the cost of furnace reconstruction project.The modular transporter break the technique monopoly of imported products,and can be a good alternative to the imported products.As the key transport equipment of Oceanographic engineering platform and large ship hull,this modular transporter should be contribute to the rapid development of metallurgical and Oceanographic engineering industry. DCMC type self-propelled modular transporter is composed of one or more 6-axle modules , 4-axle modules and power pack units(PPU).The transporter is controlled by micro-electrical system based on Controller Area Network (CAN) ,the modular transporter is electric-hydraulic drive,multiple electric-hydraulic steering,it can be assembled with 1000 axle lines,and Max. load capacity can be 20000 tones.

Key words　blast furnace reconstruction, modular transporter, transportation of large components, distributed power, distributed control

冷轧厂滚筒式飞剪刀片精度调整方法

陈 尧　孟 宇　王 翔　金仁超

（宝钢工程集团上海宝钢工业技术服务有限公司，上海　201900）

摘　要　本文介绍了冷轧厂滚筒式飞剪的典型结构，分析了飞剪工作原理，并总结了提高飞剪刀片安装精度的调整方法。实践证明，适当的刀片间隙及重叠量对于飞剪的剪切质量及刀片的使用寿命非常重要。该方法对相同结构的滚筒式飞剪刀片更换安装有较大的参考价值。

关键词　滚筒式飞剪，间隙调整，重叠量，剪刃侧隙

Precision Adjustment Method of Roller Type Flying Shears in Cold Rolling Mill

Chen Yao, Meng Yu, Wang Xiang, Jin Renchao

(Baosteel Engineering & Technology Group Shanghai Baosteel Industy Technological Service Co., Ltd., Shanghai 201900, China)

Abstract This paper introduces the typical structure of roller type flying shear in cold rolling mill.The working principle of flying shear is analyzed.And the adjustment method of improving the installation accuracy of flying shears is summarized.The practice shows that the appropriate blade gap and overlap amount are very important to the shear quality and the service life of the blade.This method has great reference value for the replacement of the same structure of the roller type flying shears.

Key words drum type flying shear, gap adjustment, overlap, cutting edge gap

攀钢钒热轧板厂精轧 F1~F3 十字万向接轴寿命分析

胡学忠[1] 孙 良[1] 高 原[2] 腊国辉[2]

（1. 攀钢钒有限公司，四川攀枝花 617062；2. 西钢钒有限公司，四川西昌 615000）

摘 要 攀钢热轧板厂经过第三期技术改造和 R2 轧机更新后，R2 出口的中间坯厚度上限增加至 40mm，钢种范围扩大，生产薄规格产品的比重相应增大使得精轧机的负荷显著增大，而此负荷主要由 F1~F3 轧机分担。在如此高强度的工况下，F1~F3 的主传动十字万向接轴寿命明显缩短，其破坏形式为接轴叉股根部位置开裂。本文通过对十字包受力分析，校核叉股的强度，说明寿命缩短的原因，并提出延缓其劣化速度的改进措施。

关键词 十字万向接轴，叉股，疲劳强度，切应力，拉应力，负载率

Analysis of Cardan Joint Shaft Service Life for L1~F3 Mill Stands at HSM of Pangang

Hu Xuezhong[1], Sun Liang[1], Gao Yuan[2], La Guohui[2]

(1. Panzhihua Steel & Vanadium Co., Ltd., Panzhihua Sichuan 617062, China;
2. Xichang Steel & Vanadium Co., Ltd., Xichang Sichuan 615000, China)

Abstract After the third-phase technical revamping at HSM and renewal of R2 mill stand, the transfer bar thickness out of R2 stand has increased up to 40mm. With production of more and more steel grades, thin-gauge strips are accounting for a bigger proportion of the quantity, causing a drastic addition of work loads on finishing mills, mostly borne by F1~F3 mill stands. Under such heavy-load conditions, the service life of cardan joint shafts for F1~F3 millstands is obviously shortened with frequently-occurred cracks at the bottom of shaft forks. In this article, forces at the cardan joint shaft and strength of shaft forks are examined for analyzing the reason of a shorter life and improvement measures are illustrated in order to slow down its deterioration.

Key words cardan joint shaft, shaft fork, fatigue strength, shear stress, tensile, strength load rate

提高承压容器焊接合格率

何小鹏 汤小践 马保华

（武汉钢铁集团公司，湖北武汉 430080）

摘 要 承压容器是在冶金行业、能源工业等国民经济的各个部门都起着重要作用的设备。而由于密封、承压及介质等原因，容易发生爆炸、燃烧起火而危及人员、设备和财产的安全及污染环境的事故。焊接作为承压容器生产和维护的主要环节，当承压容器出现泄漏或者损坏需要更换和修复时，如何保证其焊接质量十分重要。本文通过分析武钢某厂承压设备出现的焊接缺陷及其原因，并提出切实有效的改进措施，以提高承压容器焊缝的一次合格率。

关键词 承压容器，缺陷，焊接工艺，焊接质量

Increase the Pressure Vessel Welding Qualified Rate

He Xiaopeng, Tang Xiaojian, Ma Baohua

(Wuhan Iron and Steel Group Corp, Wuhan 430080, China)

Abstract Pressure vessel is metallurgical industry, energy industry and so on the various sectors of the national economy plays an important role in equipment. And because the reasons, such as sealing, pressure and medium easy explosion, burning a fire and endanger the safety of the personnel, equipment and property and environmental pollution accident. Pressure vessel welding as a main link in production and maintenance, when there is leakage or damage to the pressure vessel in need of replacement and repair, how to ensure the welding quality is very important. In this article, through analysis of the pressure equipment energy power plant of wisco. The welding defects and the reasons, and put forward concrete and effective measures, in order to improve the qualified rate of pressure vessel weld.

Key words pressure vessel, the defect, welding process, the welding quality

Research on the Strip Running Deviation and Automatic Shape Control in the Cold Rolling Unit

He Ruying[1], Gu Tingquan[2], Xu Feng[1]

(1. Engineering Technology Center, Ningbo Baoxin Stainless Steel Co., Ltd., Ningbo 315000, China;
2. Research Institute, Baoshan Iron & Steel Co., Ltd., Shanghai 200940, China)

Abstract In allusion to the strip deviation problem of strip cold rolling mill in the rolling production process, due to the strip asymmetric wave shape, cross wedge, equipment installation deviation and unreasonable process, a set of automatic control model of cold rolled strip running deviation and shape is put forward for the first time, combined with the characteristics of the equipment and technology of cold rolling mill.Strip shapemeter are respectively arranged in the entrance, exit of mill,to real-time detect the shape. Minimize the amount of strip running deviation and the deviation between the rolled strip shape and the target shape after the shape adjusting mechanism is automatically adjusted, aiming at the rolled strip shape and the amount of strip running deviation. The application of the model, not only can reduce the incidence rate of strip cold rolling process deviation, improve unit productivity; but also can ensure the accuracy of strip shape control, improve the strip product quality, create greater economic benefits to the enterprise.

Key words cold rolling unit, shape, the amount of strip running deviation

Strip Shape Feed-forward Control Based on Mill Entrance Strip Shape

Qiao Aimin[1], Gu Tingquan[2], Qian Hua[1], Xu Feng[1]

(1. Engineering Technology Center, Ningbo Baoxin Stainless Steel Co., Ltd., Ningbo 315807, China;
2. Research Institute, Baoshan Iron & Steel Co., Ltd., Shanghai 200431, China)

Abstract Considering the problems caused by entrance strip shape of instability during the rolling process and quality of exit strip shape ,this paper proposed a strip shape feed-forward control method based on the change of mill entrance strip shape and constructed a feed-forward deviation calculation model and a strip shape feed-forward control model for tandem cold rolling mills, the correctness of which was verified by comparing with actual values .This method is applicable to the rolling mill equipped with strip shapemeter at the entrance and puts the stability of the rolling process and obtaining good exit strip shape as the control target .On the base of studies on the strip shape feed-forward control strategy and execution cycle of strip shape feed-forward control , a complete automatic flatness control system was established for a 20-roller stainless steel cold rolling mill. Production results showed that when the strip shape feed-forward control method was used, the stability and production capacity of cold rolling mill unit improved obviously, leading to further improvement in strip shape control accuracy and quality.

Key words tandem cold rolling mill, feed-forward control, entrance strip, strip shape

The Automatic Recognition Method of Cold Rolling Strip Running Deviation Based on Flatness Detection

Xu Feng[1], Gu Tingquan[2], Qiao Aimin[1]

(1. Engineering Technology Center, Ningbo Baoxin Stainless Steel Co., Ltd., Ningbo 315807, China;
2. Research Institute, Baoshan Iron & Steel Co., Ltd., Shanghai 200431, China)

Abstract With reference to the problem of strip running deviation caused by non symmetrical wave of incoming strip, cross wedge, equipment installation accuracy deviation, unreasonable process and etc in the cold rolling process, this study fully considered mechanical and technological characteristics of the cold rolling unit and proposed a calculation method for the strip running deviation direction and the amount of running deviation by putting the shape feedback value obtained from contact type shapemeter into the shape calculation model. This method is applied to the production practice, mainly to solve a situation that most manufacturers do not equip with edge detection instrument , with only the shape meter, to detect the strip deviation, considering the cost factor. The method has achieved good results, obtaining the accurate strip running deviation direction and the amount of running deviation, with further popularization and application values.

Key words the cold rolling unit, strip running deviation direction, amount of running deviation

宁钢高炉振动筛的改造研究

刘维勤[1] 王士彬[1] 王 耀[2]

（1. 宁波钢铁有限公司，浙江宁波 315800； 2. 上海西重所重型机械成套有限公司，上海 201900）

摘 要 宁钢高炉原设计高炉槽下焦炭筛分系统是筛分湿熄焦，高炉使用自产和外购焦炭是湿熄焦。由于宁钢焦炉搬迁后，增设干法熄焦系统，焦炭生产工艺发生变化，加之近年来环保要求水平提高，原设计槽下焦炭筛分设备不能满足生产需求，筛板寿命短，筛下物焦丁跑粗、现场环保问题比较突出。本文重点分析和研究了宁钢高炉振动筛改造的问题，在改造的过程中，通过焦炭振动筛形式对比分析，选择采用现有的筛体形式，将改造的重点集中到筛板的设计和制造上；重新选择振动筛筛板形式，选择合适的除尘参数，优化振动筛的关键部件，合理制定改造的方案。方案实施后，筛板寿命显著提高，现场环境改善，焦丁跑粗大幅度减少，提高冶金焦利用率，降低了生产成本，效果明显，可在国内高炉槽下筛系统推广应用。

关键词 焦炭，振动筛，高炉，EVA

The Reform and Analysis of the Vibrating Screen in Blast Furnace

Liu Weiqin[1], Wang Shibin[1], Wang Yao[2]

(1. Ningbo Iron & Steel Co., Ltd., Ningbo 315800, China;
2. Shanghai XZS Heavy Machinery Integration Co., Ltd., Shanghai 201900, China)

Abstract Before Ningbo Iron & Steel Co., Ltd. move to the new address, wet coke quenching is screened by the process system. Current coke process is producting dry quenching, and outsourcing is also wet coke quenching. The process of coke product is changed. In other way, the demand on environmental protection becomes stricter and stricter. This article introduced the vibrating screen problem of Ningbo steel blast furnace; put forward the scheme for reform the existing process and equipment. Through the transformation of vibrating screen, resetting the vibration sieve mesh size, optimization the partial component of screen, and management strengthen are used, improving the utilization ratio of coking, to eliminate nut coke loss, improve the environment of the site ,Reduce the cost of production. The equipment can be spread out and applied to many blast furnace.

Key words coke, vibrating screen, Blast furnace, EVA

钢刷密封在首钢京唐 580m^2 烧结环冷机中的应用实践

曹刚永

（首钢京唐钢铁联合有限责任公司，河北唐山 063200）

摘 要 环冷机作为烧结系统冷却成品烧结矿的重要设备，传统橡胶密封的环冷机漏风率过高是众多钢铁企业共同面临的难题。本文对原环冷机进行系统分析，有针对性地论证各个部位的漏风原因，最终确定治理方案。某钢铁厂

580m^{2}烧结环冷机通过引进新型钢刷密封和进行结构形式的改型升级，冷却效果明显改善，余热回收系统产蒸汽效率提高了25%，通过减少环冷鼓风机的运行时间，可实现电耗降低1189万/年，合计年创效益2099多万元。

关键词 钢刷密封，环冷机，余热回收

The Application and Practice of Steel Brush Seal in Modification of the 580m^2 Sinter Circular Cooler at Shougang Jingtang

Cao Gangyong

(Shougang Jingtang United Iron and Steel Co., Ltd., Tangshan 063200, China)

Abstract Circular cooler ,as an important equipment of sintering system cooling the finished sinter . Traditional rubber seal circular cooler air leakage rate is too high is a common problem faced by many steel companies . This article provides a comprehensive analysis of the original ring cooler,demonstrates specifically all parts of air leakage， has been finally confirmed the treatment scheme.A steel plant of the 580 m^2 sinter circular cooler effect improve obviously by introducing new steel brush seal and Upgrading the structure form of modification, Waste heat recovery system to produce the steam efficiency increases by 25%, By reducing the running time of cold ring blower，Power consumption reduced 11.89million/year , total years benefit is more than 20.99 million yuan.

Key words steel brush seal, circular cooler, waste heat recovery

BaoVision-WSIS-Online Surface Inspection System for High-speed Hot Wire

Peng Tiegen[1], He Yonghui[1], Tang Jingsong[2]

(1. Baoshan Iron & Steel Co., Ltd., Research Institute, Shanghai, China;
2. Steel Tube and Wire Department, Shanghai, China)

Abstract BaoVision-WSIS (Wire Surface Inspection System)-Online Surface inspection system (SIS) for high-speed hot wire which is developed by Baosteel is applied in wire factory. High illumination LED lighting and high speed imaging camera are used in this system for capturing the cylindrical surface images of high-speed hot wire. For difference diameter of wire, the system can realize auto-focus by adjusting the camera position and get sharply focus images. Surface images are transferred through the 10-Gigabit network and processed by the dedicated application computers. As from October 2014, BaoVision-WSIS has been used for surface defect detection in Baosteel and played an important role for promoting the quality of wire in the factory.

Key words surface inspection system (SIS), hot wire, BaoVision-WSIS

VOLVO液压挖掘机液压系统维修与维护

孙利民　付守利

（鞍山钢铁集团公司铁路运输公司，辽宁鞍山 114000）

摘　要　液压系统中液压油在元件和管路中的流动情况，外界是很难了解到的，所以给维护、维修带来了较多的困难，因此要求人们具备较强分析判断故障的能力。在机械、液压、电气诸多复杂的关系中找出关键部位并及时、准确加以维护和维修。

关键词　故障，维护，维修 System

9　冶金自动化与智能管控

Metallurgy Intelligent Management and Control

炼铁与原料

炼钢与连铸

轧制与热处理

表面与涂镀

钢材深加工

钢铁材料

能源与环保

冶金设备与工程

★　冶金自动化与智能管控

冶金分析

信息情报

智能制造的使能技术和中国制造的发展路径

彭　瑜

（上海工业自动化仪表研究院，上海　200233）

摘　要　工业 4.0 的架构包括工业物联网、服务互联网、智慧工厂和 CPS 系统等基本组成部分，由此分析智能制造的使能技术包括两类：目前可用的使能技术和需要进一步开发的使能技术。重点介绍工业物联网、标准化问题和构建新型工业网络等问题。并尝试讨论中国企业在此大转型过程中的路径和阶段划分。

关键词　工业 4.0，智慧工厂，智能制造，工业物联网，智能制造的标准化，软件定义联网 SDN，IEC 61131-3 的 OPC UA 信息模型

基于物质流能量流协同的钢铁企业能源系统管控技术

孙彦广

（中国钢研集团冶金自动化研究设计院，北京　100071）

摘　要　钢铁企业能源系统与钢材生产过程耦合紧密，能源管控需要考虑物质流与能量流协同。报告从钢铁企业系统节能角度，讨论了能量流网络模型与仿真、多介质多场景能源计划、反馈前馈结合能源调控、多层次多视角能效分析等关键技术问题，介绍了相关应用实例。

关键词　能量流网络模型，能效分析，能源调控

钢铁物流与智慧决策

霍佳震

（同济大学，上海　200092）

摘　要　运用现代优化技术、数据分析技术、智能专家决策技术等，针对大型钢铁企业（如宝钢）的入厂（inboound）物流、厂内物流和出厂（outbound）物流中的关键决策问题，进行理论和技术上的分析，并以典型的成功案例阐明其可行性、有效性，以及带来的巨大经济效益，为钢铁企业的物流管理和运营提供一种智慧决策的方向。

关键词　优化技术，决策，钢铁物流

宝钢供应链智能优化技术的研发

杜　斌

（宝钢集团中央研究院自动化所，上海　201900）

摘　要　在宝钢股份信息化基础系统建成的前提下，宝钢研究院自动化所从 2002 年开始进行钢铁业智能优化生产管理领域的控制与决策技术研究。研究分为三层架构：基础算法软件包（如线性规划、进化智能、离散事件仿真、自动建模等）、问题层（如计划排程、库存问题、物流仿真、合同优化等）、应用层（如热轧计划智能排程、厚板堆垛优化推荐、板坯余材智能充当、合金采购决策支持等），该结构使得日后的研究效率得到了提高，研究组织得以明晰。

十多年来，本领域研究人员，涉足了从马迹山港的铁矿石装卸智能优化、铸坯热卷冷卷的库处理与优化、碳钢不锈钢的混合排程、船板智能垛位推荐、智能配车、剪切加工中心的排产与优化套裁等跨供应链的智能优化技术研发和应用。在众多的项目中，由于各种原因，不成功率比我们在过程控制模型领域的项目高，而后者基本都成功。

在工业 4.0、大数据等新技术热潮下的今天，本报告在汇报宝钢成功案例的同时，也交流我们工作的体会以及相关技术决策的一些原则和建议。

关键词　柔性制造，智能优化，钢铁供应链，技术选择

工业 4.0 时代的工业软件

丛力群

（上海宝信软件股份有限公司，上海　201203）

摘　要　中国制造 2025 将智能制造作为两化融合的主攻方向，推进工业 4.0 战略，实现智能制造，需要将工业制造技术与新一代信息技术深度融合，如：物联网、云计算、三个集成、大数据、CPS……；每一项技术都要通过软件能力来实现。

工业 4.0 时代对软件提出了特殊的要求，以 IT 技术承载工业知识，形成工业软件，将工艺制造技术转化为 IT 配置规则，通过 CPS 来描述流程，模块化 CPS 组件实现可配置，意味着流程可以更加灵活，制造工艺可以通过配置而变得敏捷。

这集中反映了企业核心软件资产有效配置的重要理念，即：随着系统规模的扩大和需求的不断出现，在持续优化制造管理系统的同时，迫切需要建立有效的机制，来适应不同时期投入、不同供应商开发的工业软件，如何将这些存量的软件资产功能整合在同一个运营管理平台上，避免信息孤岛或系统孤岛，别让企业成为软件垃圾桶。

所以，推进工业 4.0，要研究相关技术系统的构建、开发、集成和运行的一个框架，构建了软件的应用程序和服务架构，最终形成企业级的 CPS，并以此为基础，建设智慧型运营中心。

开发具有自主知识产权的各类工业控制软件、研制集成开发平台是势在必行，有利于技术传承和保护，降低成本，对于国家的产业安全和工业信息安全至关重要；目前，工业软件的现状与制造业处于类似的境况，低端产能过剩，高端产品和品牌缺乏，整体竞争力不足，且面临着技术和品牌两个方面的挑战，要发挥行业知识和贴身服务的优势，综合系统设计、工程应用、产品服务提供一体化解决方案，从一开始就将智能设计到系统中。

关键词 智能制造，工业软件，IT 配置规则

钢铁企业工业 4.0 的终极目标

郭朝晖

（宝钢集团中央研究院自动化所，上海 201900）

摘　要 钢铁业是自动化、信息化水平最高的工业行业之一，是工业 3.0 时代的骄子，推进工业 4.0 的基础较好。设定终极目标，对技术战略布局和发展策略是有必要的。在工业 4.0 时代，现有许多工作岗位可能会实现无人值守，并以此为抓手实现更高的质量、更好的服务、更低的成本和排放。与此同时，许多“规则制定者”的技术岗位将应运而生并大力发展。

关键词 工业 4.0，智能制造，信息物理系统

六西格玛管理在钢铁产品质量检验方面的应用与实践

夏碧峰　崔全法　费书梅

（首钢股份公司迁安钢铁公司，河北迁安 064400）

摘　要 介绍了六西格玛管理系统，探讨了钢铁产品质量检验重要性及现状，应用六西格玛管理方法对产品质量检验变差进行改革探索，按照定义-D、测量-M、分析-A、改进-I、控制-C 五个阶段进行开展实施，大幅度地提升了质量检验的管理能力，提高了质检员的判定准确率，降低质量异议发生几率和失误后的经济风险，构建了具体特色的质量检验管控保障体系。

关键词 六西格玛管理，测量系统分析，改进，控制，标准

The Application and Practice of Six Sigma Management System on the Quality Inspection of Steel Products

Xia Bifeng, Cui Quanfa, Fei Shumei

(Shougang Qian'an Iron & Steel Company, Qian'an 064400, China)

Abstract This paper introduced the Six Sigma management system, discussed the importance and present situation of the quality inspection of steel products, and explored the variation of product quality inspection with Six Sigma management method, what put into effect by the five stages of definition (D), measurement (M), analysis (A), improvement (I) and control (C). The application and practice of six sigma management system greatly promoted the capacity of quality inspection management, put up the accuracy rate of quality decision, reduced quality objection probability and economic risk of fault, structured the quality inspection management control guarantee system with specific characteristic.

Key words Six Sigma management, measurement system analysis, improvement, control, standard

多系统信息关联与协同在铁水罐跟踪中的实践

刘毅斌

（新余钢铁集团有限公司技术中心，江西新余　338001）

摘　要　铁水罐是钢铁企业铁钢界面的重要载体，它的运行效率与效果直接关系到铁钢单位的生产运行组织、关系到企业的成本。本文介绍了作者在少投入的情况下，综合应用多种信息化技术实现铁水罐信息集成的经验与所进行的尝试。本公司铁水罐系统原有的管理比较分散，大部分依靠纸质记录、部分信息有单个应用系统进行数据储存，由于没有整合各方面的信息，在日常铁水罐的运行管理中存在诸多问题。本文通过对现状的全面分析，梳理了需要整合的信息内容，并界定了整合的范围，针对 SQL 数据库、现场检测原件、纸质材料等不同的信息对象提出了不同的数据采集与整合方案，并予以实施。最终完全达到了项目设计目标，提高了铁水罐的管理效率和使用效率，降低了铁水罐的耐材消耗，减少了运行使用成本。

关键词　铁水罐，多系统，关联，整合

Multi-system Information and Association Practice on Iron Ladle Tracking

Liu Yibin

(Xinyu Iron and Steel Group Co., Ltd., Xinyu Jiangxi 338001, China)

Abstract　The iron ladle is the important carrier in iron and steel enterprise interface. Its running efficiency and the effect is directly related to production run and the cost of the iron steel plant. in this paper introduces the author's attempt and experience on the integrated application of a variety of information technology to realize iron POTS of information integration in the case of less investment. The original management of hot iron system is more dispersed, most rely on paper records, some information stores in single application system. Because of lacking integration of all information, there exists many problems in operation management of the daily iron ladle. Through the comprehensive analysis on the present situation, collating the information which need to be integrated, and define the scope. Different data acquisition and integration scheme are proposed and implemented according to different information like: SQL database, original copy of field tests, the paper file and etc. Finally the target of the project design is achieved completely to improve the management efficiency and use efficiency of iron POTS, and reduce the consumption of refractory material for the iron ladle and the operation cost.

Key words　hot metal ladle, multi-system, relevance, integration

1580mm 热连轧机精轧区域带钢宽度缺陷分析及改进

陈志荣　王　喆

（宝山钢铁股份有限公司热轧厂，上海　201900）

摘 要 本文研究分析了1580mm热连轧精轧区域带钢的宽度缺陷及原因分析，针对带钢“水印点”原因造成的宽度拉窄、带钢头部穿带金属秒流量不匹配原因造成的宽度拉窄及带钢全长宽度超差情况提出了控制改进方案。这些针对控制改进方案在宝钢1580mm热连轧机应用后，带钢在精轧区域的宽度拉窄问题得到明显改善。

关键词 宽度缺陷，热连轧，控制改进

Analysis of the Width Defects in FM Area and the Corresponding Improvement Solutions for 1580 mm Hot Continuous Rolling Mill

Chen Zhirong, Wang Zhe

(Hot Rolling Plant of Baosteel Co., Ltd., Shanghai 201900,China)

Abstract In this article, the cause of width defects in FM area is analyzed. The corresponding control methods to solve the width defects by “SKI”, by un-matching metal flow and by the whole length out-of-tolerance have been proposed. These methods have been developed in the 1580 HSM and improved the strip width quality quite a lot.

Key words width defect, continuous rolling mill, control method

基于厂级控制的钢铁企业电网控制问题研究

郝 飞[1] 沈 军[2] 陈根军[1] 燕 飞[2]

（1. 南京南瑞继保电气有限公司系统软件研究所，江苏南京 211102；
2. 首钢京唐钢铁联合有限责任公司能环部，河北唐山 063210）

摘 要 按照厂级控制的设计方法对钢铁企业电网控制问题进行分析，并采用分层控制、逐级细化的思想对整个控制系统的进行重新设计，将原来散布在电力调度中的控制子系统融合起来，重新制定控制策略。通过在上层建立对未来电网的潮流变化趋势进行分析，在下层的协调优化控制环节，改进超前AGC和PDC的控制策略，提高了有功调节的响应速度；同时采用交叉迭代的思想对AGC与AVC的控制指令进行校核修正，得到综合协调控制指令，最大限度地降低了两类指令的相互影响，实现了系统经济性和安全性的综合优化协调控制。

关键词 自动电压控制，自动发电控制，最大需量控制，厂级控制，协调控制

Control Problem Analysis and Strategy Research for Iron and Steel Enterprise Based on Plantwide Control

Hao Fei[1], Shen Jun[2], Chen Genjun[1], Yan Fei[2]

(1. NR Electric Co., Ltd., System Software Institute, Nanjing 211102, China;
2. Shougang Jingtang United Iron & Steel Co., Ltd., Tangshan 063210, China)

Abstract According to the design method of plantwide control, the power grid control problems in the iron and steel enterprise are analyzed, hierarchical control, step by step to refine the idea of the control system is re-designed, the control system had spread in the power dispatching in together, to develop control strategies. Through the establishment of the future trend of power flow analysis in the upper layer, the coordination and optimization control on the bottom layer, improvement of control strategy of AGC and PDC in advance, these designs can improve the response speed of adjustment of active power control command at the same time. At the same time, the idea of the cross iteration can check and revise the AGC and AVC control orders, and obtain the comprehensive coordination control instruction, which can reduce the influence of two kinds of instructions, and achieve a comprehensive optimization of coordinated control system of economy and safety.

Key words automatic voltage control, automatic generation control, maximum power demand control, plantwide control, coordinated control

BIM 技术助力工厂设施全生命周期数字化

华　跃

（宝钢工程技术集团有限公司，上海　201900）

摘　要　本文分析了 BIM 技术在建筑领域的发展，论述了利用最新的 BIM 技术实现工厂设施全生命期数字化的好处。本文介绍了在工厂建设的规划方案阶段、建设阶段、设施维护管理阶段、设施改造阶段、以及报废搬迁阶段的应用。并通过 BIM 技术与传统方式的对比，分析了 BIM 为工业企业带来的效益。最后，本文分析了 BIM 技术在工业领域的推广应用所需要进一步研究解决的五大问题。

关键词　BIM，数字工厂，工业建筑

BIM Technology Facilitates Factory Facilities Digital In Plant Life Cycle

Hua Yue

(Baosteel Engineering & Technology Group Co., Ltd., Shanghai 201900, China)

Abstract This article summarized the development of BIM technology in construction and the profits it brings to industrial enterprise through realizing digital factory in all stages of production. It elaborated on the application of BIM technology in the stages of planning and constructing of factories, maintaining and reforming of facilities and scraping and relocating of factories. It also analyzed the profits BIM brings to enterprises by comparison to traditional methods. Finally, five major problems were pointed out that might be encountered during the promotion of BIM technology in industry.

Key words BIM，digital factory, industry building

钢铁板带生产库存结构与计划排程协同优化

乐　洋

（马鞍山钢铁股份有限公司制造部，安徽马鞍山　243003）

摘　要　当前钢铁企业板带生产普遍面临着订单结构复杂，多品种、小批次订单对计划排程影响较大的问题。传统

的生产计划排程对于库存管理偏向于物料状态的管理，多数情况下是针对现有库存来调整计划排程的过程。若需要进一步优化和改善生产过程环境，则需要从主动优化库存结构入手。本文以冷轧板带生产为例，根据历史订单数据的分析对比，根据常用过渡材需求分析，结合生产工艺需求主动建立灵活有效的过渡材库存，积极解决订单状态和生产过程的矛盾。以更优化的库存结构为基础，进一步支持计划排程过程，提高订单兑现率。

关键词 库存结构，优化排程，订单兑现率

Collaborative Optimization of Production Inventory Structure and Scheduling of Steel Sheet

Yue Yang

（Manufacturing Department of MA-Steel, Ma'anshan Anhui 243003, China）

Abstract At present, steel strip production is generally faced with some problems.Orders, for example, complex structure, many varieties, small batch order for greater influence on the project schedule and so on.Traditional production plan scheduling for inventory management to material status management, mostly in view of the existing inventory to adjust the plan scheduling process.If need to further optimize and improve the production environment, you need to, from the perspective of the initiative to optimize the inventory structure.As an example, this paper manufacture with production order according to the historical data analysis of the contrast, according to the commonly used transitional material demand analysis, combined with production process needs initiative to establish a flexible and effective transition of material inventory, solved the problem of order status and production process actively. On the basis of better inventory structure, further support plan scheduling process, improve the order completion.

Key words inventory structure, optimized scheduling, order fill rate

Research and Application of the Intelligent Control Systems for Iron Ore Agglomeration Process

Fan Xiaohui, Yang Guiming, Huang Xiaoxian, Chen Xuling, Gan Min

(School of Minerals Processing and Bioengineering, Central South University, Changsha 410083, China)

Abstract Iron ore agglomeration is an important step in iron & steel making process. As production scale enlarges and market competition increases, stabilizing or optimizing the sintering & pelletizing induration is a key approach to enhance production criteria and reduce energy consumption. Artificial intelligence and mathematical model are promising tools to deal with these complicated control problems, and recently, techniques such as fuzzy control, expert system, process model visualization have been applied in level-2 control of agglomeration process more and more widely. Based on the thorough understanding of the process and the above advanced control techniques, this paper mainly details two intelligent control systems for iron ore agglomeration developed by the authors, one is for typical sintering process and the other is for typical grate-kiln process. The systems adopt fuzzy control & fuzzy model, expert system and mathematical simulation. The corresponding control strategies for sintering process and grate-kiln process are firstly introduced, and then the developed systems are briefly described, system application and the running performance are finally highlighted. These two intelligent control systems can bring significant economic benefits, the promotion or further application will enhance the competitiveness of domestic iron & steel industry.

Key words iron ore agglomeration, mathematical model, fuzzy control, expert system

Weka 在带钢力学性能预测中的应用初探

万文骏　郭　强　荆丰伟　凌　智

（北京科技大学高效轧制国家工程研究中心，北京　100083）

摘　要　屈服强度、抗拉强度、延伸率等带钢力学性能指标是带钢质量的重要指标。带钢力学性能的预测一方面能够减少手动检验力学性能的工作量，另一方面是实现带钢力学性能在线调整的前提条件，所以具有重大意义。本文初步探究了使用 Weka 平台进行数据挖掘实现带钢力学性能预测的可行性。文中以某热连轧生产线的 SPA-H 钢种的实测数据为样本，使用了 C4.5 决策树算法的分类功能实现数据挖掘，最终得到决策树形式的性能预测模型，对样本数据的准确分类率达到 79.8%。

关键词　带钢力学性能预测，数据挖掘，决策树，Weka

Preliminary Study on Steel Strip's Mechanical Properties Prediction via Weka

Wan Wenjun，Guo Qiang，Jing Fengwei，Ling Zhi

（National Engineering Research Center of Advanced Rolling Technology of USTB，Beijing 100083,China）

Abstract　Yield strength,tensile strength,elongation are important indicators of strip's quality.On one hand,the mechanical properties prediction of strip can reduce the workload of manual testing;on the other hand,it is the precondition to achieve the online adjustment of mechanical properties of the steel strip. Therefore, realizing the offline mechanical properties prediction of strip is of great significance. This paper did some preliminary study on the feasibility of using Weka to realize offline mechanical properties prediction of the steel strip. In this paper, the real-time data set of SPA-H is from a hot strip rolling production line, and the decision tree algorithm C4.5 is used for data mining, finally a decision tree model is generated to realize offline mechanical properties prediction of striped steel. The model's correct classification rate is 79.8144%.

Key words　mechanical properties prediction of steel strip，data mining，decision tree，Weka

面向铸轧一体化生产的热轧批量计划编制方法与应用

郑　忠[1]　呼万哲[1]　高小强[2]　黄世鹏[1]　徐兆俊[1]

（1. 重庆大学材料科学与工程学院，重庆　400045；
2. 重庆大学经济与工商管理学院，重庆　400044）

摘　要　针对铸轧一体化生产条件下的热轧批量计划编制问题进行分析，重点讨论了带材与中厚板材两类不同产品轧制规程对计划编制的影响；从综合考虑连铸、加热和轧制三大工序的生产目标及相应工艺约束的角度，基于车辆路径问题建立了兼顾前后工序计划衔接的热轧批量计划编制数学模型，并设计了改进的元胞遗传算法进行模型求解。以某钢铁企业轧钢厂的带钢生产计划编制的实例数据，进行模型的仿真实验验证，检验了模型和算法的可行性和有效性。

关键词 热轧，生产批量计划，铸轧一体化，带材，中厚板

Batch Planning of Hot Rolling for Integrated Production of Continuous Casting and Rolling and Its Application

Zheng Zhong[1], Hu Wanzhe[1], Gao Xiaoqiang[2], Huang Shipeng[1], Xu Zhaojun[1]

(1. College of Material Science and Engineering, Chongqing University, Chongqing 400045, China;
2. College of Economics and Business Administration, Chongqing University, Chongqing 400044, China)

Abstract Batch planning of hot rolling for integrated production of continuous casting and rolling is analyzed and the impact of different rolling schedule between strip steel and medium plate on making production plan is discussed emphatically. A mathematical model based on vehicle routing is proposed, which has given consideration to the connection between continuous casting and rolling by synthetically considering various objectives and constraints of continuous casting, heating and rolling. A modified cellular genetic algorithm is proposed to solve it. The simulation experiment with the data of hot strip rolling production in a rolling mill of a domestic iron and steel enterprise is conducted. The feasibility and effectiveness of the model has been verified.

Key words hot rolling, production batch planning, integrated continuous casting and rolling, strip steel, medium plate

工业 4.0 技术对钢铁工业的推动作用

李立勋 费 鹏 赵 雷 金百刚

（鞍钢股份鲅鱼圈钢铁分公司，辽宁营口 115007）

摘 要 钢铁工业发展至今对于工艺与装备的集成越来越重视，装备与工艺是相辅相成的关系，一方面冶金工艺的进步和发展是冶金装备技术进步的主要动力，另一方面冶金装备的进步可以引发和促进冶金工艺的实现与进步。将现代冶金工艺原理及前沿研发理论应用于生产且适合冶金工作者的工程平台就是各类过程控制系统及系统的 APC 模型（Advanced Process Control）接口。工业 4.0 技术与冶金 APC 模型相结合共同为生产高品质钢材提供机理、装备、工艺、质量的保证。提高企业的核心竞争力。

关键词 钢铁行业，发展，冶金工艺，先进过程控制，工程平台，核心竞争力

The Driver Effect of Industry 4.0 Technology for Iron and Steel Industry

Li Lixun, Fei Peng, Zhao Lei, Jin Baigang

(Bayuquan Subsidiary of Ansteel Co., Ltd., Yingkou Liaoning 115007, China)

Abstract Iron and Steel Industry developed to now, put more and more important on embedding Metallurgical–Process and equipment. Metallurgical Process and device has closed relationship. Firstly, improvement of Metallurgical–Process drives equipment improvement. Secondary, Improvement of Metallurgical equipment supports Metallurgical–Process come true. The suitable Engineering-workbench for Process worker is Advanced Process Control system and its interface. Industry 4.0 Technology and Metallurgical APC model provide assurance for producing high quality steel. Enhance enterprise kernel competition.

Key words iron and steel industry, develop, metallurgical -process, advanced process control, engineering-workbench, kernel competition

鞍钢蒸汽管网管理系统和监测系统的开发和应用

贯 振[1] 林 科[2] 黄玉彬[3] 何 嵩[4] 占 炜[3] 张天赋[1]

（1. 鞍钢集团钢铁研究院，辽宁鞍山 114009；
2. 鞍山钢铁集团公司第二发电厂，辽宁鞍山 114003；
3. 鞍钢股份有限公司计量厂，辽宁鞍山 114002；
4. 鞍钢股份有限公司能源管控中心，辽宁鞍山 114002）

摘 要 针对鞍钢本部蒸汽管网系统复杂，计量仪表分布不合理，管理操作人员对管网内蒸汽运行状况和设备运行状态不清晰等问题。根据鞍钢蒸汽管网管理流程开发了蒸汽管网管理系统，以及根据蒸汽在管网内流动的物理过程开发了蒸汽管网在线监测系统。这两个系统的应用极大地提升鞍钢蒸汽管网管理水平，大幅度降低鞍钢蒸汽资源的消耗，对于鞍钢节能降耗具有重要意义。

关键词 蒸汽管网，管理系统，监测系统

Steam Network Management and Monitoring System Deveopment and Application in Ansteel

Jia Zhen[1], Lin Ke[2], Huang Yubin[3], He Song[4], Zhan Wei[3], Zhang Tianfu[1]

(1. Iron & Steel Research Institute of Angang Group, Anshan 114009, China;
2. No. 2 Power Plant of Ansteel,Anshan 114003, China;
3. Measuring Plant of Angang Company Limited ,Anshan 114002, China;
4. Energy Management Centre of Angang Company Limited ,Anshan 114002, China)

Abstract Based on the problems such as the steam network having a complex structure, unreasonable measurement meters distribution, operators not familiar with steam pipe and related equipment situation, this paper presents the steam network management system which is based on steam network management process in Ansteel and the steam network monitoring system which is accoriding to steam flow physical process in pipe. The application of the two system improves the management of steam network greatly, hence reduces the steam consumption and is significant to energy saving and reduction in Ansteel.

Key words steam network, management system, monitoring system

钢铁企业信息化之质量检查与分析

杨吉星

（中冶京诚工程技术有限公司，北京 100176）

摘 要 本文首先介绍了我国钢铁企业发展的现状及发展过程中遇到的一些问题，而后介绍了利用信息化的技术手段实现的钢铁企业产品质量管理中的质量检查与分析的环节，意图为国内众多钢铁企业的生产管理人员提供一个快速、直观、高效的质量事故检查与分析的方法，提高其工作效率，提前完成生产任务。

关键词 钢铁企业，信息化，质量检查，质量分析

Quality Check and Quality Analysis of Iron and Steel Enterprise Information Technology

Yang Jixing

(Capital Engineering & Research Incorporation Limited, Beijing 100176, China)

Abstract In this paper, first it introduces us the development of China's iron and steel enterprises in the current situation and some problems encountered in the development process, next, the quality check and quality analysis methods based on information technology of the production quality management is introduced in this paper too, it intention to provide a fast, intuitive, efficient quality accident investigation and analysis methods for the number of manager, so as to improve the work efficiency, complete the production tasks as soon as possible.

Key words iron and steel enterprise, promotion of information technology, quality check, quality analysis

厚板精整剪切线物流优化系统应用开发

黄可为[1] 易 剑[1] 林 云[1] 许中华[2] 巩荣剑[2] 杨晓军[2]

（1. 宝钢研究院，上海 201900；
2. 宝钢股份厚板部，上海 201900）

摘 要 针对宝钢厚板精整剪切线生产物流优化，通过详细分析研究厚板剪切过程，以及上、下线剪切物流的工艺特点，建立了兼顾合同优先级别的厚板部精整剪切线物流优化模型；在此基础上，设计、开发了旨在提升生产物流效率的在线模型应用系统，应用模型优化方法对瓶颈物流工位进行作业改善、下线钢板的垛位优化，以及统筹上下线物料调度，达到提升物料库存周转率，促进合同完成率的提升，确保剪切线最大效率地生产的目的。

关键词 厚板，精整，剪切，物流，优化模型

Application and Development of the Logistics Optimization System for Thick Plate Shear Line

Huang Kewei[1], Yi Jian[1], Lin Yun[1], Xu Zhonghua[2], Gong Rongjian[2], Yang Xiaojun[2]

(1. Research Institute, Baoshan Iron & Steel Co.,Ltd., Shanghai 201900, China;

2. Plate Plant , Baoshan Iron & Steel Co.,Ltd., Shanghai 201900, China)

Abstract　Aiming at the optimization of production logistics of Baosteel Heavy Plate finishing shear line, through detailed analysis of thick steel plate cutting process, and, off the assembly line process characteristics of shear logistics, the establishment of the contract priority of heavy plate finishing　shear line logistics optimization model considering; based on to design and develop the aims to enhance the efficiency of the production logistics model online application system, optimization method of logistics bottleneck station operations to improve application model, offline steel plate stacking optimization, and co-ordinate offline scheduling of material, to improve inventory turnover rate, promote the completion of the contract rate of improvement, to ensure that the production of the shear line at maximum efficiency.

Key words　plate plant, finishing, shear, logistic, optimization model

核主元分析在特征提取中的优化应用

李　静[1]　张智密[1]　陈　健[2]　孟庆利[3]　盛靖芳[4]

（1. 北京科技大学冶金工程研究院，北京　100083；

2. 营口京华钢铁有限公司，辽宁营口　115005；

3. 天津市天重江天重工有限公司，天津　300402；

4. 西门子（中国）有限公司，北京　100102）

摘　要　本文通过对核主元分析方法的理论研究，提出了一种在核函数参数选取问题上进行优化的方法，通过相似度函数求解最优核参数值。这种方法提出了一种参数选取规则，很大程度上减小了计算量，使其在工业过程的状态监测中的实用性得到进一步提高。通过对采样实验数据的仿真研究，利用不同工作状态下的数据特征较准确地实现了正常与故障工况数据的分离，取得了较好的分类识别效果，证明了这种方法的有效性。

关键词　主元分析，核函数，特征提取，状态分类

Optimization and Application of KPCA Method in Feature Extraction

Li Jing[1], Zhang Zhimi[1], Chen Jian[2], Meng Qingli[3], Sheng Jingfang[4]

(1. University of Science & Technology Beijing, Beijing 100083, China;

2. Minmetals Yingkou Medium Plate Co., Ltd., Yingkou 115005, China;

3. Tianjin Giant Heavy Industries Co., Ltd., Tianjin 300402, China;

4. Siemens Ltd., China, Beijing 100102, China)

Abstract This paper proposes an optimized method of how to select the parameters of kernel function based on the study of kernel principal component analysis. The method selects the optimal parameter by solving a similarity function and presents a new rule of parameter selection that greatly reduces the complexity of calculation so we can further improve the utility in the condition monitoring of different industrial processes. Through simulation study we accurately separate the sampled experiment data under normal and fault conditions using theextracted data features and get a better classification result thus proves the effectiveness of this method.

Key words principal component analysis, kernel function, feature extraction, state classification

Energy Optimization of Iron and Steel Enterprises

Lin Yu, Li Bing, Niu Honghai, Chen Jun

(Research Department of Nari-Relays Electric Co., Ltd., Jiangsu, China)

Abstract The cooperated optimization model of energy in iron and steel enterprises is built based on the research of coupling relationship of gas, steam and power. And by using the hierarchical decomposition method, the model is broken down into two parts: optimization of gas system and optimization of thermal and power system. The case analysis indicates that: the model can dispatch the energy of gas, steam and power reasonably, safely and efficiently when the production condition is changed, and improve the energy utilization efficiency.

Key words iron and steel enterprises, energy system, cooperated optimization, energy-saving

连续热镀锌机组锌液温度的精确控制

张 军 钱洪卫 杨建国 林传华 孟宪陆

（宝山钢铁股份有限公司冷轧厂，上海 200941）

摘 要 锌液温度的控制直接影响到连续热镀锌机组锌锅内锌渣的产生量和锌层中抑制层的成分及结构,进而影响到热镀锌产品的表面质量。对影响连续热镀锌机组锌液温度精确控制的因素进行了系统分析,通过改进带钢入锌锅温度的控制,锌液温度的检控精度和锌锅加锌的精确控制，优化锌锅辊预热温度控制等方法，实现锌液温度的精确控制。

关键词 热镀锌，锌液温度，锌锅

Accurate Control of Liquid Zinc Temperature in Zinc Pot of Hot-dip Galvanizing Line

Zhang Jun, Qian Hongwei, Yang Jianguo, Lin Chuanhua, Meng Xianlu

(Cold Rolling Plant, Baoshan Iron & Steel Co., Ltd., Shanghai 200941, China)

Abstract The liquid zinc temperature control will affect the formation of zinc dross in zinc pot and the composition and phase structure in hibiting layer of the product in CGL line. It will further a ffect the product s surface quality. Finally it will affect the surface qulity of the product. The ralated factors affected the presicion control for liquid zinc temperature were analyzed. The control method for strip temperature at the entrance of zinc pot, the detection accuracy and control method for liquid zinc temperature, the presicion control method for adding zinc ingots and the management mothod for zinc pot roll temperature and so on were improved and optimized, which will be lead to the precision control of liquid zinc temperature.

Key words　hot dip galvanizing, liquid zinc temperature, zinc pot

小方坯定重切割行为控制体系研究

王福斌[1]　陈至坤[1]　陈世超[2]　王　一[1]

（1. 华北理工大学电气工程学院，河北唐山　063009；
2. 唐山钢铁集团有限公司二钢轧厂，河北唐山　063016）

摘　要　钢坯定重切割是提高轧材成品率、降低能耗的关键。分析了钢坯图像尺寸精度及钢坯拉速对钢坯定重的影响，构建了钢坯定重切割行为控制体系结构。基于有限状态机原理将切割目标分解为定重切割的任务及动作，以钢坯理论重量、积分重量及预报重量作为行为控制模型各个状态之间转换驱动的条件，自主实现定重切割目标。所构建的模型是对钢坯定重切割自主行为控制进行的有益探索。

关键词　连铸钢坯，定重切割，行为控制体系，状态流

Research for Behavior Control System of Billet Weight Cutting

Wang Fubin[1], Chen Zhikun[1], Chen Shichao[2], Wang Yi[1]

(1. Electrical Engineering College, North China University of Science and Technology, Tangshan 063009, China; 2. Third company, Tangshan Hebei 063016, China)

Abstract　It is the key to improve the yield of rolled products and reduce energy consumption based on billet weight cutting. Analyzes the effect on billet weight by dimensional accuracy of billet image and drawing speed of billet, establishes behavior control architecture for billet weight cutting. Break the target of billet weight cutting into tasks and actions based on finite state machine, Use theoretical weight, integral weight and forecast weight of billet as conversion drive condition between different states of behavior control model, achieves the goal autonomously. The built model is beneficial exploration for autonomous behavior control of billet weight cutting.

Key words　casting billet, weight cutting, behavior control system, stateflow

转炉炉衬激光测厚自动定位方法的设计研究

罗辉林　田　陆　易定明

（湖南镭目科技有限公司，湖南长沙　410000）

摘　要　在利用激光对转炉炉衬进行测量之前，需要确定激光器与转炉的相对位置关系，即需要进行定位。对于以三维激光扫描仪作为激光器部分的转炉炉衬激光测厚仪来说，其可以实现自动定位。本文详细设计了以三维激光扫描仪作为激光器部分的转炉炉衬激光测厚仪的自动定位方法，并探讨了软件算法实现，经过分析得该方法具有定位速度快、准确性高、方便快捷的特点，能够满足炉衬厚度测量时对于准确性、速度和操作便利性的需求。。

关键词　三维激光扫描，标靶球，定位，罗德里格七参数

Design and Research of Automatic Positioning Method of Converter Lining Thickness Measurement with Laser

Luo Huilin, Tian Lu, Yi Dingming

Abstract　Before the measurement of converter lining in the use of laser, it is necessary to determine the relative position between the laser and the converter, namely, the laser needs to positioning. The 3D laser scanner as part of converter lining laser thickness gauge can realize the automatic positioning. In this paper, we have detailed designed the automatic positioning method for converter lining laser thickness gauge which using 3D laser scanner as component, and discussed the realization of software algorithm. Through the analysis of this automatic positioning method, we can see that this method has the features of fast speed, high positioning accuracy, fast and convenient, which can meet the demand of accuracy, speed and convenience for lining thickness measurement.

Key words　3D laser scanning, target ball, positioning, the Rodrigo Perion parameters

转炉副枪模型自动化控制系统研究与应用

左文瑞　张小兵　刘火红　余传铭　黄国洪　陈志贤

（宝钢集团韶关钢铁有限公司炼钢厂）

摘　要　本文主要介绍了转炉副枪模型自动化控制系统的结构、特点以及主要的控制功能。该系统的顺利投产，该模型系统通过仪表自动化模型、电气自动化模型和计算机模型的构建，实现了“一键式”自动炼钢，同时，通过各种冶炼数据的反馈使得炼钢达到科学化、数据化、全自动化，为公司特钢生产创造良好的理论基础。

关键词　转炉，自动化控制系统，副枪

Research and Application of EIC Control System for Converter Sublance Model

Abstract　This paper mainly introduces the structure, characteristics and main control functions of the automatic control system for the model of the converter. The smooth operation of the system, the model system through construction of instrument automation model, electrical automation model and computer model, the realization of the “one click” automatic steelmaking, at the same time, through all kinds of smelting data feedback makes the steel-making to achieve scientific, data, full automatic, special steel and production for the company to create a good theoretical basis.

Key words　converter, automation control system, sublance

板带轧机垂直振动系统的自抗扰控制研究

张瑞成 马寅洲

(华北理工大学电气工程学院，河北唐山 063009)

摘 要 建立了板带轧机两自由度垂直振动系统的数学模型，将自抗扰控制器应用到垂直振动系统中，设计了轧机垂直振动自抗扰控制系统，对系统中出现的振动现象进行控制。通过仿真实验证明了自抗扰控制器具有良好的快速性、精确性，且抗干扰能力强和鲁棒性能好，可以方便地控制整个系统的稳定运行，具有较高的应用价值，为轧机垂直系统的振动控制提供了一种新的有效途径。

关键词 轧机，垂直振动，自抗扰控制，混沌

Abstract The two degrees vertical vibration system model of the rolling mill was established. The ADRC technique is applied in vertical vibration system, the vertical vibration ADRC system is designed. The vibration phenomenons of rolling mill are controlled. The simulation shows that ADRC technique has a good rapidity, accuracy, stronger robustness and anti-jamming ability, which can easily control the stable operation of the whole system. It has high application value and provides a new way for vibration control of the vertical system for rolling mill.

Key words rolling mill, vertical vibration, active disturbance rejection control, chaos

带积分环节的 GM-AGC 的分析与应用

王云波 岳 淳

（中冶京诚工程技术有限公司电气与自动化工程技术所，北京 100176）

摘 要 本文在压力 AGC 通用表达形式的基础上，在考虑辊缝位置环的条件下，采用解析方式推导出了带积分环节的 GM-AGC 出口厚差表达式，并着重对系统的稳定性进行了定量分析和仿真验证，最终给出了积分时间常数的取值范围。所得结论可以作为工程上算法选择与参数设定的参考。

关键词 压力 AGC，模型，差分方程，稳定性，积分环节

Analysis and Application of GM-AGC with Integration Control

Wang Yunbo, Yue Chun

(Electrical & Automation Division，MCC Capital Engineering & Research Institute，Beijing 100176,China)

Abstract Based on the general equation for pressure AGC algorithm, the exit thickness equations for GM-AGC with integration control considering gap position loop are presented in this paper, and the quantitative analysis for the stability are derived and proved by simulation. The range for integration time are also given. The conclusion can be served as reference for selecting AGC parameter in engineering.

Key words pressure AGC，variable stiffness control，rapidity, stability, integration control

多台烧结余热锅炉并联运行的数学模型与调控方法

闫龙格　胡长庆　赵　凯

（华北理工大学冶金与能源学院，河北唐山　063009）

摘　要　针对烧结余热发电系统运行稳定性差、停机率较高的问题，本文基于概率理论，建立了多台烧结余热锅炉并联运行发电系统数学模型，分析了余热锅炉工作状态对余热发电系统稳定运行的影响，提出了烧结余热回收发电系统稳定运行的调控方案。采用现场蒸汽流量监测数据，计算结果表明：多台余热锅炉并联运行，发电机组连续稳定运行改善；4 台锅炉并联运行时，发电稳定运行概率在 0.95 以上，且随状态系数变化而变化的幅度最小。蒸汽量不足时采用优化计算调控后，状态系数 0.65 时，发电稳定运行的概率超过 0.962。

关键词　烧结余热回收，优化控制，概率论数学模型

Mathematical Model and Control Method of Multiple Parallel Sintering Waste Heat Boiler Operation

Yan Longge, Hu Changqing, Zhao Kai

(College of Metallurgy and Energy, North China University of Science and Technology, Tangshan Hebei 063009, China)

Abstract　Focusing on the issues of poor operation stability and higher outage rate for sintering waste heat power generation system, mathematical model of multi waste heat boilers parallel operation power system was established based on the theory of probability. The influence of the working status of waste heat boilers on the operation stability of the waste heat power generation system was analyzed, and the control scheme of the stable operation of the sintering waste heat recovery power generation system was put forward. Adopted the in situ vapor flow monitoring data, the calculation results show that, the operation continuity and stability of power generation system would be improved through parallel operation of multiple waste heat boiler. For the 4 sets oilers parallel operation system, the probability of generator stable operation is 0.95, and the change range with the change of state coefficient is smaller. When the amount of steam is insufficient, the probability of generating stable operation is over 0.962 by optimization calculation and regulation at the case of the state coefficient 0.65.

Key words　sintering waste heat recovery, optimal control, mathematical model

连铸钢包最少残留钢优化控制技术的研究

申屠理锋　胡继康　奚嘉奇

（宝钢集团中央研究院自动化所，上海　201900）

摘　要　本文根据连铸钢包最少残留钢优化控制技术的研发工作，介绍了一种应用于连铸浇注生产中能有效减少残留钢的优化控制技术，分析了钢包浇注过程中钢水流场变化形态及漩涡吸附卷渣的机理，叙述了整个系统的结构组成及功能特点，阐述了该技术应用后效益的体现，同时介绍了技术的效果及应用情况。

关键词　连铸，钢包，残留钢，优化控制

Research on Minimum Residual Steel Optimization Control Technology in Continuous Casting Ladle

Shentu Lifeng, Hu Jikang, Xi Jiaqi

(Automation Research Dept, Baosteel Co., Ltd., Research Institute, Shanghai 201900, China)

Abstract　This article introduces the minimum residual steel optimization control technology in Continuous Casting ladle, analyzes the transformation of steel flow field distribution and principle of slag entrapment phenomenon, describes the components, function, and features of the system, discusses the benefit of the technology creating, and presents the application and effects of the technology .

Key words　continuous casting, ladle, residual steel, optimization control

优化控制在退火炉温度控制中的实现

蔡　新

（宝钢特钢有限公司条钢厂，上海　200940）

摘　要　针对宝钢特钢有限公司条钢厂棒一精整5号退火炉温度控制精度要求很高，由双交叉限幅串级燃烧控制搭建起来的传统温度控制很难达到工艺控制要求。本文介绍了2009年条钢厂利用炉体改造机会，利用原有的控制系统资源，对控制功能和控制策略进行了优化，引入史密斯预估控制等优化控制算法，使温度回路控制精度由原来的±15℃提高到±3℃，既提升了产品退火质量，又节约了天然气消耗20%以上。

关键词　优化控制，温度控制，控制精度

Abstract　It is introduced that the temperature control precision of 5# finishing annealing furnace is high in Baosteel's special steel co., ltd bar iron factory, a traditional temperature control,which is built up the pairs of cross-limiting combustion control cascade, is difficult to achieve process control requirements. This article describes the opportunity of the furnace transforming in 2009, as the original control system resources is used,the control function and the control strategy are both optimized, and optimal control algorithm such as the Smith Predictor control is introducd so that the temperature control precision is enhanced, which is from the original loop ±15℃ to ±3℃.And now the product quality of annealing is enhanced,the consumption of natural gas is saved more than 20%.

Key words　optimal control, temperature control, control precision

10 冶金分析

Metallurgical Analysis

Investigation on Interface Inclusion and the Bond Behaviour of Hot-Rolled Stainless Steel Clad Plates

Yin Fuxing[1], Li Long[2], Zhang Xin[1], Wang Gongkai[1]

(1. Research Institute for Energy Equipment Materials Hebei University of Technology, Tianjin, China;
2. Technology Institute for Clad Materials, Yinbang Clad Material Co., Ltd., Wuxi, China)

Abstract The impurities on the bonded interface were analyzed using optical microscopy and scanning electron microscopy (SEM), and the formation mechanism of oxides were investigated and discussed. The results show that the impurities are mainly made of oxides. The chemical composition of oxides can also be related to vacuum inside the clad metal slab, and the oxide of silicon dominates at the vacuum degree of 100Pa or more and the oxide of aluminum at the vacuum degree of 0.01Pa. The experimental results also indicated that oxidation also become more evident with the vacuum degree reduction, and the proportion of inclusion increase to 50% from 10% with increase in vacuum degree from 0.1Pa to near 20Pa, which lead to the decrease of bonding strength from 440MPa to 350MPa. Interface bond behavior between stainless steel and carbon steel was suggested based on element distribution and microstructure evolution.

Key words hot rolled clad steel, vacuum degree, inclusion, bond behavior

Determination of Insoluble Aluminum Content in Steels by Laser-induced Breakdown Spectroscopy Method

Li Dongling[1], Zhang Yong[2], Liu Jia[1], Jia Yunhai[1]

(1. Central Iron & Steel Research Institute Group, Beijing 100081, China;
2. Sichuan University, Chengdu 100088, China)

Abstract The insoluble aluminum content in steels has an important influence on the quality of the steel. In this paper, laser-induced breakdown spectroscopy (LIBS) was used to analyze the insoluble aluminum content in steel samples with scanning mode. The threshold value which was used to distinguish soluble and insoluble aluminum was calculated by an iteration method. In this method the value of summation of the average intensity and 2.5 times of standard deviations was iterated. Total and soluble aluminum content calibration curves were generated and can be used for the determination of the total and soluble aluminum contents in steel samples. The insoluble aluminum content could be calculated by subtracting the soluble aluminum content from the total aluminum content. The insoluble aluminum contents in standard spectrum steel samples and real steel samples were rapidly determined using this mathematical model. The results agreed well with the value obtained by chemical methods.

Key words laser-induced breakdown spectroscopy, insoluble aluminum content, threshold value

The Application of Automated EBSD & EDS on SEM to the Complete Characterisation of Steel Samples

Goulden J., Lang C., Larsen K., Hiscock M.

(Oxford Instruments, UK)

X射线荧光光谱法快速测定煤灰中常量和微量元素

张 杰 王一凌

（鞍钢集团钢铁研究院，辽宁鞍山 114009）

摘 要 本文利用X射线荧光光谱仪进行了煤灰中常量和微量元素的测定；针对煤灰的基体结构进行了校准样品的人工配制；对试样处理方法、熔剂体系和稀释比例的选择进行了探讨；讨论了样品的烧灼量对分析结果准确度的影响和采取的校正手段；采用大比例稀释、熔融制样和基体校正，消除了试样的矿物效应、颗粒效应和基体的吸收-增强效应。实验表明，该法简便快速，精密度和准确度都较好，可以满足冶金行业对煤灰、煤矸石灰和焦炭灰的日常检测工作的需要。

关键词 X射线荧光光谱，煤灰，熔融法，基体效应，烧灼量

Rapid Determination of Major and Minor Elements in Coal Ash by X-ray Fluorescence Spectrometric Method

Zhang Jie, Wang Yiling

(Iron & Steel Research Institute of Angang Group, Anshan 114009, China)

Abstract In this paper a method for determination of the major and minor elements in coal ash by X-Ray Fluorescence spectrometry was researched ,and some man-made reference materials which is in the same matrix with coal ash were developed. The disposal method of sample, the system of flux, and the dilution rate by reagent were discussed. The paper also discussed the effect of sample's loss rate on the accuracy of the analysis results and the adopted correction method. The mineral effects and the absorption and enhancement effects of sample were removed by the large dilution rate and the fusion disposal and the matrix correction .The experiments prove that this method is convenient and speedy with very good precision and accuracy and could meet the demand of on-line analysis and everyday determination of the major and minor elements in coal ash.

Key words X-ray fluorescence spectrometry, coal ash, fused glass bead, matrix effects, loss rate

电感耦合等离子体质谱法在钢中痕量铌元素检测精准度影响的研究

杜士毅 顾红琴 孙 娟 费书梅 庞振兴 王明利

（首钢股份公司迁安钢铁公司，河北迁安 064404）

摘 要 本文对ICP-MS法分析钢中Nb元素的影响因素进行了分析。采用条件实验的方式确定了实验器皿为聚四氟乙烯烧杯和塑料容量瓶，加酸溶解方式为20mL水+5mL浓硝酸+1mL氢氟酸，数据稳定性更好。经测定，浓硝酸、

氢氟酸、纯铁Nb空白较高，对于Nb含量小于0.001%时影响较大。为了消除空白影响，检测须采用标准加入法绘制分析曲线，并对硝酸、氢氟酸进行纯化。利用回归分析、DOE分析等方式，确定了最佳溶解方式为低温溶解20min、确定了103Ph为实验内标元素、样品冲洗时间12s、等离子体功率1300W、雾化气流量0.8L/min、采样深度140×10^2μm、冷却气流量13 L/min。并对检测的关键环节，制定了控制方法。经过实测验证，ICP-MS法钢中痕量Nb元素分析准确度在0.0001%以下；精密度RSD均小于4%；测定下限为0.00002%，能够满足特殊品种钢冶炼的检测要求。

关键词 铌，电感耦合等离子质谱，钢，痕量

Study on the Test of Inductively Coupled Plasma Mass Spectrometry of Trace Elements in Niobium in Steel Precision Influence

Du Shiyi, Gu Hongqin, Sun Juan, Fei Shumei, Pang Zhenxing, Wang Mingli

(Shougang Co., Qian'an Steel Corp, Qian'an 064404, China)

Abstract In this article, influencing factors of analyzing of Nb in steel have been researched. on the ICP-MS method. The condition experiments to determine the experimental vessel for PTFE beaker and plastic volumetric flask, dissolved by 20mL H_2O , 5mL HNO_3 and 1mL HF, the stability of data is the best. The determination of nitric acid, hydrofluoric acid, iron Nb blank is higher,and the content of Nb is less than 0.001% is affected remarkably. In order to eliminate the influence of blank, test shall be the standard addition method to draw curves, and the nitric acid, hydrofluoric acid purification. By regression analysis,DOE analysis and other methods, to determine the best solution for low temperaturesoluble 20min, 103Ph was confirmed as the internal standard element, the samples offlushing time 12s, plasma power 1300W, nebulizer gas flow of 0.8L/min, sampling depth of 140 × 10^2 μm,cooling gas flow of 13 L/min. And it make the control method of test on the key piont. After experimental verification, the accuracy of the analysis of trace elements in the 0.0001% Nb ICP-MS steel; the RSD of precision was less than 4%; the lower limit of test is 0.00002%, which can meet the test requirements of special varieties of steel smelting.

Key words niobium, inductively coupled plasma mass spectrometry, steel, trace

Mechanical Properties of Micro-phases along <001> Direction of Directionally Solidified Nickel-base Super-alloy by In-Situ Nano-indentation

Ma Yaxin[1,2], Zou Yuming[1], Gao Yifei[1],
Zhang Yuefei[3], Du Yuanming[3]

(1. Central Iron and Steel Research Institute, National Analysis Center for Iron and Steel, Beijing, China;
2. Chengdu Aeronautic Polytechnic, Chengdu, China;
3. Beijing University of Technology, Beijing, China)

Abstract In this paper, using the in-situ nano-indentation device (SPM/SEM hybrid system) to test the micro-mechanical properties of γ′ phase and eutectic γ′ phase which along <001> direction of directionally solidified nickel-base super-alloy. The hardness and modulus of the eutectic γ′ is higher than γ′ phase. For the specimen ST (without aging treatment), the elastic modulus and hardness of the γ′ phase respectively is 110.7GPa and 1.45GPa, the elastic modulus and hardness of the eutectic γ′ phase respectively is 171.2GPa and 3.98GPa; for the specimen W-1000(aging 1000hours), the elastic modulus and hardness of the γ′ phase respectively is 105.4GPa and 1.35GPa, the elastic modulus and hardness of the eutectic γ′ phase respectively is 161.5GPa and 3.85GPa. After long-term high-temperature aging without loading, the change of the micro-mechanical properties of γ′ phase and eutectic γ′ phase which along <001> direction of directionally solidified nickel-base super-alloy is not apparent. Meanwhile the results of the FEM are in good agreement with the results of the experiments.

Key words super-alloy, in-situ nano-indentation, hardness, modulus

H13 模具钢中纳米析出物提取方法的探究

郭 闯 郭汉杰

（北京科技大学冶金与生态工程学院，北京 100083）

摘 要 为研究 H13 模具钢在不同的淬火、回火温度下以及不同位置析出物的类型、微观结构和粒度分布，进而通过对纳米析出物的控制实现钢铁材料的组织性能强化。实验中采用非水溶液电解的方式提取分离钢中的析出物，并针对 H13 钢的特点对电解液的配方进行了研究与优化，设计了一种以甲醇、乙酰丙酮、氯化锂为主，并辅以多种添加剂的电解液，严格控制了常规非水溶液电解中的结晶和胶体干扰等现象，并将电解液的 pH 值稳定在 7~8 的范围内。使用高分辨率透射电子显微镜对分离物进行观察，获得了拥有完整形貌和微观结构的电子图像，特别是尺寸在 10nm 以下范围内的对钢材起到明显析出强化作用的纳米析出物，并初步验证碳化析出物的形成机理。

关键词 H13，纳米析出物，非水溶液电解，析出强化，形成机理

Exploration of the Extraction Method on Nano-depositionin H13 Die Steel

Guo Chuang, Guo Hanjie

(University of Science and Technology Beijing, School of Metallurgical and Ecological Engineering, Beijing 100083, China)

Abstract Through the study of the type, microstructure and size distribution of the deposition under different quenching and temperingtemperature or different position, people can strengthen the performance of the steel by controlling the precipitation of the nano-deposition. Non-aqueous solution electrolysis method is used to extract the deposition in the steel, a study and optimum proposal on the electrolyte formulais raised according to the characteristics of H13 steel in this paper, designing a type of electrolyte mixed with methyl alcohol, diacetylmethane, LiCl and some other additives, strictlycontrolling the crystallization and the interference of colloid in traditionalelectrolysis with non-aqueous solution, and stabilizing the pH fluctuation between 7~8. Observing the isolates with HRTEM, researchers get the electronic image with integral morphology and microstructure, especially those deposition whose size is less than 10nm and can make obvious impact on precipitation strength,the observed results also verify the formation mechanism of these deposition.

Key words H13, Nano-deposition, Non-aqueous solution electrolysis, precipitation strength, formation mechanism

ICP-MS 法测定管线钢中痕量砷、铅、锡、锑、铋

孙 娟 庞振兴 顾红琴 王贵玉 金 伟 杜士毅

（首钢股份公司迁安钢铁公司， 河北迁安 064400）

摘　要　随着管线钢牌号的不断升级，纯净度越来越高，钢中有害元素（砷铅锡锑铋）含量控制日趋严格，化学湿法及 ICP-AES 法无法满足现场对痕量检测的需求。电感耦合等离子体质谱（ICP-MS）作为一种超痕量元素分析的强有力技术，具有极低的检出限和较宽的线性动态范围，可以满足现场痕量检验的要求。钢中砷、铅、锡、锑、铋的存在会对钢铁的性能产生重要的影响，因此测定其准确含量具有重要的意义。本文研究了 ICP-MS 法测定管线钢中痕量砷、铅、锡、锑、铋的分析方法、对样品溶解、仪器参数、干扰情况等进行了优化，并计算了方法检出限和方法检测下限。

关键词　ICP-MS，内标，检测下限，功率

Determination of Trace As, Pb, As, Sb, Bi in Pipe Line Steel by Inductively Coupled Plasma Mass Spectrometry

Sun Juan, Pang Zhenxing, Gu Hongqin, Wang Guiyu, Jin Wei, Du Shiyi

(Shougang Qian'an Iron & Steel Company, Qian'an 064400, China)

Abstract　With the continuous upgrading of pipeline steel grades, steel becomes more and more purer and the content of harmful elements(As,Pb,Sn,Sb,Bi) in steel is more and more stricted control.Chemical and ICP-AES method can not meet the demand for analysis in time. Inductively coupled plasma Mass spectrometry (ICP-MS) technology as a powerful ultra-trace elemental analysis,it has very low detection limits and wide linear dynamic range,can meet the requirements of ultra-trace elemental analysis. The element(As,Pb,Sn,Sb,Bi) will have a major impact to performance steel, and therefore accurate determination has important significance. This paper studies the ICP-MS method for the determination of trace analysis methods As,Pb,Sn,Sb,Bi in the steel.The dissolution of the sample, instrument parameters and the interference were optimized and we calculate the detection limit of the method and the method detection low limit.

Key words　ICP-MS, the internal standard, the detection limit, the power

Advancement in "Direct Solid Sampling" Glow Discharge Mass Spectroscopy (GDMS) for Trace Analysis of Advanced Alloys, Engineering Coatings, Thin Films and Coarse Surfaces

Wang Xinwei, Su Kenghsien, Liu Yan, PUTYERA Karol

(EAG, Inc., 103 Commerce Blvd., Liverpool, New York 13088, U.S.A.)

Abstract　Glow discharge mass spectroscopy (GDMS) is a direct solid sampling technique with growing importance for high sensitivity trace elemental analysis of advanced materials used in aerospace, semiconductor, energy and medical device industries. In this article, several case studies were given, covering the applications in quality control, production support and failure analysis. GDMS is the technique of choice for full survey analysis, in particular monitoring the critical bulk trace impurity in advanced alloys such as S, Se, Pb, etc., due to its exceptional long-term reproducibility (yearly measurement variation RSD < 10%). The GDMS equipped with a fast flow GD source is also a highly sensitive technique for depth profiling of trace impurities in engineering coatings (up to 100s mm thick) in the direct current mode, and in thin metal oxide film (100s nm thick) in the modulated GD mode. We also demonstrated that this GDMS configuration is also an effective tool for near surface chemical analysis of coarse surfaces.

Key words　glow discharge mass spectroscopy, trace analysis, depth profile, alloys, engineering coatings

Advanced in Situ (S)TEM Studies for Metallic Materials

Xu Qiang[1,2], Sairam K Malladi[1], Liu Chunhui[1,3], Chen Jianghua[3], Henny W Zandbergen[2]

(1. NCHREM, Kavli Institute of nanoscience, TU Delft, the Netherlands; 2. DENS Solutions B.V. the Netherlands; 3. Department of Materials science & Engineering, Hunan University, China)

Abstract The development of metallic materials fabrication and processing technology will face great challenges, such as energy saving or environment protection. To facilitate the R&D efficiency, materials computation and design has become an important technology of enhancing the controllability of the processing and the predictability of the properties. On the other hand, in situ transmission electron microscopy provides another high efficient solution for experimentally characterizing structure-property relations and processing-structure links at atomic-scale.

Key words blast furnace, energy-saving equipment, hot stove waste-gas heat recovery system, top-combustion hot stove, metallic burner, multi-vessel electrostatic precipitator, BF charging equipment

低合金耐磨钢的组织亚结构对性能的影响

关　云　余　立　吴立新　张彦文　欧阳珉路

（武汉钢铁集团公司研究院，湖北武汉　430081）

摘　要　利用扫描电镜、透射电镜及能谱仪对 3 种低合金耐磨钢的组织亚结构进行了观察分析，结合其力学性能进行了研究探讨。结果表明：当组织以位错型板条马氏体为主，且板条上弥散分布着大量细小针状 ε 碳化物及(Nb,Ti)(CN)析出相时,高密度位错与析出相缠结在一起，起到主要的硬化和强化作用，同时具备较好的韧性；当回火马氏体板条尺寸不均匀或存在淬火马氏体时，组织应力增大，塑韧性降低；当存在粗大板条贝氏体时，基体的连续性被破坏，同时降低其强度和韧性；对于回火马氏体组织而言，细小针状 ε 碳化物因为共格强化及数量优势，其强化作用大于弥散分布的(Nb,Ti)(CN)析出相。当 ε 碳化物尺寸增大或转变成杆状 M3C 型碳化物后，强化作用大幅减弱。

关键词　耐磨钢，马氏体，硬度，强度，塑韧性，析出相

Effect of the Microstructure Substructure of Low Alloy Wear-Resistant Steel on Property

Guan Yun, Yu Li, Wu Lixin, Zhang Yanwen, Ouyang Minlu

(Research and Development Center of Wuhan Iron and Steel (Group) Co., Wuhan Hubei 430081,China)

Abstract The microstructure substructures of three kinds of low alloy wear-resistant steel were observed by scanning electron microscope, transmission electron microscope and energy dispersive spectrometer, and its mechanical properties were discussed. The results showed that: when the microstructure was mainly compose of dislocation type lath martensite on which a large number of small needle-like ε carbide and (Nb, Ti) (CN) precipitates dispersed, high density of dislocations and precipitation intertwined together playing a major role in hardening and strengthening, therefore, wear-resistant steel had high strength and good toughness; when the tempered martensite lath size is uneven or quenched martensite appears, organizational stress increases and the plasticity and toughness decrease; when there is coarse lath bainite, continuity of the matrix is destroyed, and its strength and toughness decrease; for tempered martensite, the invigoration effect of fine needle ε carbide as coherent strengthen and volume advantages, was larger than dispersed (Nb,Ti) (CN) precipitates. When the ε carbide size increases or changes into a rod M3C type carbide, strengthening effect weakens significantly.

Key words wear-resistant steel, martensite, hardness, strength, plasticity and toughness, precipitates

Research on Low Temperature Embrittlement of Advanced High Strength Steel

Zhou Yedong[1], Ding Chen[1], Zhang Jianwei[1], Li Wei[2], Zhou Shu[2], Fang Jian[1]

(1. Testing Center, Baosteel Research Institute, Shanghai 201900, China;
2. Automotive Steel Research Institute, Baosteel Research Institute, Shanghai 201900, China)

Abstract Brittleness of advanced high strength steel (AHSS) induced by low temperature is one of automakers' greatest concerns, considering the wide application of AHSS in structural components of car body. The toughness and brittleness of several kinds of advanced high strength steel at low temperature have been investigated, by using instrumented Charpy impact method. The results show that traditional impact test is useful for evaluating sheet metal's low temperature toughness ans brittleness. The absorbed energy and fracture surfaces that change with temperature are compared. The effect of prestrain on the change of low temperature embrittlement has also been investigated. The results indicate that forming or prestrain has negative effect on low temperature toughness.

Key words low temperature embrittlement, charpy impact, prestrain

Failure Analysis of the Crack on the Beam of the Heavy Truck

Ding Chen, Liu Junliang, Hu Xiaoping

(Test Center, Baosteel Research Institute, Shanghai 201900, China)

Abstract In this article, the crack found on the beam of the heavy truck produced by a specific company is investigated. Fracture topography indicates the crack origins from the center of shearing plane and the failure mode should be attributed to low-circle fatigue crack. The tensile-compressive stress on the beam induced by acceleration and braking during driving is the main alternating stress source. The analysis of metallographic structure of crack origin region proves the correspondence between the M/A constituents of the segregation band and the micro-cracks on the shearing plane due to poor shearing quality. On the other hand, TiN inclusions accelerate the process of fatigue crack concurrently. Inspection and polishing on the shearing plane after plate slitting should be the effective method to eliminate the fatigue core.

Key words failure analysis, beam, crack, fatigue

Study and Comparison on the Determination Methods of Trace Bismuth in Low-alloy Steel

Zhao Junwei, Zhu Li

(Research Institute of Baoshan Iron & Steel Co., Ltd., Shanghai 201900, China)

Abstract Inductively coupled plasma atomic emission spectrometry (ICP-AES), hydride generation atomic fluorescence spectrometry (HG-AFS) and inductively coupled plasma mass spectrometry (ICP-MS) were studied for determination of trace bismuth in low-alloy steel. Spectral interferences from coexisting elements were investigated in the method of ICP-AES. Interference coefficient correction method was implemented to eliminate the interference from chromium. The detection limit is 19ng/mL and the detection range is from 0.002% to 0.2% in the method of ICP-AES. In the method of HG-AFS, the influences of acids, matrix and coexisting elements were investigated. With the purpose of eliminating the interference from coexisting elements, thiosemicarbazide-ascorbic acid was selected as masking agent. The detection limit is 0.26ng/mL and the detection range is from 0.00005% to 0.01% in the method of HG-AFS. In the method of ICP-MS, the non-spectral interferences arising from matrix effect and interface effect were calibrated efficiently through matrix matching and adding Rh as an internal standard, hence precision of long time determination was improved. The detection limit is 0.0005ng/mL and the detection range is from 0.00001% to 0.01% in the method of ICP-MS. When these three methods combined, the detection range of bismuth in low-alloy steel is from 0.00001% to 0.2%. The results of standard samples were basically consistent with the certified values, with high degree of precision. At last, the comparison on advantages and disadvantages of the three methods was discussed. With accurate and reliable results, methods of HG-AFS and ICP-MS were applied to the determination of trace bismuth in high-alloy steel.

Key words inductively coupled plasma atomic emission spectrometry, hydride generation atomic fluorescence spectrometry, inductively coupled plasma mass spectrometry, low-alloy steel, bismuth

Modern Creep Test for Future Material

Peter Ruchti

(Zwick GmbH Co., KG, Germany)

Key words strain model creep test method, fatigue creep test, crack/fracture mechanics creep test, and their application in industry

电感耦合等离子体原子发射光谱法测定钢中钨、钼、铌方法标准的研究

于媛君 亢德华 王 铁 杨丽荣 顾继红

（鞍钢集团钢铁研究院，辽宁鞍山 114009）

摘 要 研究建立了应用电感耦合等离子体原子发射光谱法（ICP-AES）测定钢中钨、钼、铌含量的分析方法标准。通过条件试验，研究确定了试样分解方法；确定了仪器最佳分析参数；通过基体匹配法和经验系数法校正光谱干扰，建立W、Mo、Nb元素工作曲线，消除基体干扰；待测元素分析谱线线性相关系数大于0.998，钨、钼、铌的检出限分别为0.037mg/L、0.0379mg/L 和 0.0629 mg/L。本方法适用于质量分数为0.005%~8.0%的钼、质量分数为0.005%~5.0%的铌和质量分数为0.005%~19.0%的钨含量的测定，多家实验室的方法精密度实验结果表明具有良好的重复性和再现性。

关键词 电感耦合等离子体，原子发射光谱法，钢，钨，钼，铌，方法标准

Research on Standard Method of Determination of W、Mo、Nb in Steels by Inductively Coupled Plasma Atomic Emission Spectrometry

Yu Yuanjun, Kang Dehua, Wang Tie, Yang Lirong, Gu Jihong

(Iron & Steel Research Institute of Ansteel Group, Anshan 114009, China)

Abstract Study to establish application of inductively coupled plasma atomic emission spectrometry (icp-aes) was developed for the determination of content of tungsten, molybdenum, niobium in steel. Through condition test, study to determine the sample decomposition method; The best analysis instrument parameters; By matrix matching method and experience of spectral interference correction coefficient method, establish W, Mo, Nb elements work curve, eliminate matrix interference; For the linear correlation coefficient is greater than 0.998, spectral lines, the detection limit of tungsten, molybdenum, niobium were 0.037mg/L, 0.0379mg/L and 0.0629mg/L. This method is suitable for the mass fraction of 0.005% ~ 8.0% molybdenum content, mass fraction of 0.005%~5.0% of niobium content and mass fraction is 0.005%~19.0% of the tungsten content determination. Several laboratory experimental results show that the precision of the method has good repeatability and reproducibility.

Key words inductively coupled plasma, atomic emission spectrometry, steel, tungsten, molybdenum, niobium, method standard

Determination of Vanadium in High-speed Tool Steel by Microwave Digestion-inductively Coupled Plasma Atomic Emission Spectrometry

Xia Peipei, Xu Yuancai

(Research Institute of Baoshan Iron & Steel Co., Ltd., Shanghai 201900, China)

Abstract After pretreatment by microwave digestion and selection of spectral line at the wavelength of 309.311 nm as the analytical line of vanadium, the content of vanadium in high-speed tool steel was determined by inductively coupled plasma atomic emission spectrometry. In our experiment, the parameters of microwave digestion was determined,the working parameters of the instrument was optimized and the suitable analytical line was chosen.Our experimental results illustrated that the determination method was not affected by the iron matrix,which could be eliminated by matrix matching. Combining the matrix matching and precision interpolation method, the precision and accuracy of the test could be improved. When the method was applied to determine certified reference materials and actual samples, the determination results were in keeping with the certified values.Furthermore,the method could be applied for the determination of high content of vanadium in high-speed tool steel.

Key words microwave digestion, inductively coupled plasma atomic emission spectrometry, high-speed tool steel, vanadium

Qualitative and Quantitative Analysis of Precipitate Phases in Nickel-based Corrosion Resistant Alloys with Different Isothermal Situation

Miao Lede, Zhang Yi, Yang Jianqiang,

Zhang Chunxia, Zhang Zhonghua

(Science and Technology, Shanghai 200237, China;
Central Research Institute, Baoshan Iron and Steel Co., Ltd., Shanghai 201900, China)

Abstract The capability of nickel-based corrosion resistant alloy is mostly influenced by the quantity, composition, size, distribution of precipitate phase in alloy. After isothermal treatment, nickel-based corrosion resistant alloy undergoes various decomposition processes which could form precipitates of inter-metallic phase and carbides. It is well known that these precipitates lead to a reduction in creep ductility and adversely affect the toughness and corrosion property. In present paper, qualitative and quantitative analyses of precipitate phases were carried out and the analytical procedure has been established. After choosing electrolyte and electrolytic system, electrolytic isolation was performed to extract precipitates from matrix.,The residues were collected by ultrasonic cleaning and filtration after galvanostatic electrolysis. Scanning electron microscope (SEM) and X-ray diffraction (XRD) were used to examine their structure, modality and size. The contents and elemental compositions of different precipitate phase were calculated by Rietveld method. Furthermore, the relation between percent of precipitate phase and capability of corrosion resistant was discussed. Our experiments demonstrated that the most of precipitate phase was formed at 900℃. For the precipitate phases at 900℃ with different aging time, the content and size showed significant increasing trend with the prolonging of aging time. The results of corrosion evaluation also showed the strong relativity between percent of precipitate phase and capability of corrosion resistant: with the increase of percent of precipitate phase, the average corrosion rate of material was also raised and the corrosion resistant capability of material was declined.

Key words nickel-based corrosion resistant alloys, inter-metallic phases, isothermal treatment, electrolysis, qualitative and quantitative analysis

Q345B 热轧钢卷力学性能与化学成分的关系研究

任　艳

（宁波钢铁有限公司，浙江宁波　315807）

摘　要　本文选取 248 个轧制条件相同的 Q345B 热轧钢卷，利用 Minitab 软件对选取的钢卷做力学性能和化学成分之间关系的统计回归分析，结果表明：随着碳、锰、铌含量的增加，下屈服强度是增大的；随着钛含量的增加，屈服强度是下降的。随着碳、硅、锰、铌含量的增加，抗拉强度增大；随着钛含量的增大，抗拉强度减小。随着锰含量的增加，断后伸长率减小；随着钛含量的增加，断后伸长率增加。为生产过程中化学成分的控制、降低质量成本提供指导。

关键词　力学性能，化学成分，Q345B

Study on Relationships between Mechanical Properties and Chemical Compositions of Q345B Hot Rolled Steel Coil

Ren Yan

(Ningbo Iron & Steel Co., Ltd., Ningbo 315807, China)

Abstract Two hundred and forty- eight hot roll coils of Q345B were selected in this article, These coils had the same production process and rolling conditions. The relationships between mechanical properties and chemical compositions were analyzed by Minitab software. The results showed that: The influence of C, Mn, Ti, and Nb contents on lower yield strength is important, Steel lower yield strength increases with the increases of C, Mn and Nb contents, and decreases with the increase of Ti content. The influence of C, Si, Mn, Ti, and Nb contents on tensile strength is important, Steel tensile strength increases with the increases of C, Si, Mn, and Nb contents, and decreases with the increase of Ti content. The influence of Mn and Ti contents on percentage elongation after fracture is important, Steel percentage elongation after fracture decreases with the increase of Mn content, and increases with the increase of Ti content. It is helpful to control chemical composition and reduce quality costs.

Key words mechanical property, chemical composition, Q345B

X 射线光电子能谱及扫描电镜对钢板表面半有机涂层价态结构的研究

张薛菲 马爱华 刘俊亮

（宝山钢铁集团中央研究院，上海 201900）

摘 要 应用 X 射线光电子能谱（XPS）及场发射扫描电镜（FE-SEM）对商用冷轧钢板表面半有机涂层进行元素深度价态分布及涂层分层的微观结构对比观察研究。综合对涂层内部不同深度不同元素的光电子谱线的分析和高分辨涂层截面组织形态的对比分析得到涂层表面上存在着 Mg、Fe、Cr 混合氧化物；在表面以下，涂层分为两层，第一层是以 Cr_2O_3、FeO 为主，混有少量 Cr_2C_3、Fe_3C、Mg_xC_y 化合物的多孔结构，厚 800nm；第二层是以 Cr_2O_3 为主的 200nm 连续薄膜；涂层和钢基体之间是厚 5000nm 的内氧化层，主要含有 SiO_2、Fe_3Si 和纯铁，层中不明显的界面或标志 SiO_2 到 Fe_3Si 的过渡。

关键词 化学价态，涂层，内氧化层

Chemical and Morphology Characterizations of Semi-organic Coated Steel using XPS and FE-SEM

Zhang Xuefei, Ma Aihua, Liu Junliang

(Baosteel Group Corporation, Research Institute, Shanghai 201900, China)

Abstract The coating layer structure of semi-organic coated steel and element chemical states is investigated by X-ray photoelectron spectroscopy(XPS) depth profiling and field emission scanning electron microscopy(FE-SEM). Through the analysis of photoelectron lines and Auger lines of different elements in different depth planes of the coating layer, it can be found that on the surface of the coating consists Mg, Fe, Cr mixed oxide. Under the surface, the coating is divided into two layers, the first layer is porous structure with thickness of 800nm constituted mainly by Cr_2O_3 and FeO compounds, accompanied by a small amount of Cr_2C_3, Fe_3C and Mg_xC_y mixture. The second coating layer is a 200nm thick continuous film composed by Cr_2O_3. Between the coating and the steel substrate is the 3rd layer, about 5000nm thick internal scale whose components are SiO_2, Fe_3Si and pure Fe. The indistinct interface in the 3rd layer imply the transition from silicon oxidation to iron silicide.

Key words chemical state, coating , inner scale

钢中珠光体片间距测量方法研究

孙宜强

（武钢研究院，湖北武汉 430080）

摘 要 珠光体片间距是珠光体组织的重要表征参数，为了准确表征珠光体片间距，对常见珠光体测量方法进行了分析和比较。首先用扫描电镜对某高碳热轧钢板的片状珠光体进行观察和拍照，然后分别采用最小片间距法、平均片间距法和真实片间距法三种测量方法对片间距进行测量，并对测量结果进行分析讨论，检测结果表明：最小片间距法的测量方法可操作性强，测量结果符合正态分布，平均值能准确代表测量值的中心，可以用来表征珠光体片间距；平均片间距法测量数据量大，测量结果不符合正态分布，平均值不能代表测量值的中心，建议使用众数作为平均片间距的表征值；由于真实片间距计算所需的平均片间距测量误差大，造成真实片间距的计算值与实际值偏差很大，真实片间距计算不适用于珠光体片间距的表征。

关键词 珠光体，片间距，测量方法

Study on Measurement Method of Pearlite Interlamellar Spacing

Sun Yiqiang

(Research and Development Center of WISCO, Wuhan 430080, China)

Abstract Pearlite interlamellar spacing is an important parameter of pearlite. Several measurement methods of pearlite interlamellar spacing were analyzed and compared to characterize the pearlite interlamellar spacing accurately. First SEM was used to observe and photo the pearlite in a high carbon hot rolled plate, and then the minimum, average and true pearlite interlamellar spacing were measured. The measure results were analyzed and discussed. The results showed: the minimum pearlite interlamellar spacing measurement method feasibility is strong, and the measure results accord to the normal distribution. The mean value can represent the center of the measure values. The average pearlite interlamellar spacing measurement method need more measure values, and the measure results do not accord to the normal distribution. The mean value cannot represent the center of the measure values, so the mode of the measure result was suggested to be the characterization value of the average pearlite interlamellar spacing measurement method. True pearlite interlamellar spacing measurement method was not suitable to characterize the pearlite interlamellar spacing.

Key words pearlite, interlamellar spacing, measurement method

两种强度级别冷轧双相钢的显微组织与力学性能研究

马家艳 关 云 邓照军 林承江

（武汉钢铁（集团）公司研究院，湖北武汉 430080）

摘　要　采用金相显微镜、扫描电镜和透射电镜观察了两种强度级别冷轧双相钢的微观组织特征并分析了组织与性能的关系。结果显示两个强度级别的双相钢组织均由铁素体和马氏体岛组成，且马氏体岛分布于铁素体晶界；随强度级别增加，双相钢的马氏体岛体积百分比增加，铁素体晶粒及马氏体岛尺寸减小，而伸长率下降。第二相强化、细晶强化是导致高强度级别双相钢强度明显增加的主要原因，而较小比例的铁素体软相是其塑性降低的主要因素。

关键词　冷轧双相钢，组织，性能

Microstructure and Mechanical Property Research of Two Kinds of Strength Level Cold-rolled Dual Phase Steel

Ma Jiayan, Guan Yun, Deng Zhaojun, Lin Chengjiang

(Research and Development Center of Wuhan Iron and Steel (Group) Co., Wuhan 430080, China)

Abstract　The microstructures of two kinds of strength level cold-rolled dual-phase steels were observed by optical microscope, scanning electron microscope and transmission electron microscope, and the relations between microstructure and properties were analyzed. The results show that microstructures of dual phase steels are all composed of ferrites and martensite islands which distribute on the ferrite grain boundary. With the increase of strength level of dual phase steel, the martensite content increases, ferrite grain size and martensite size decrease, and the elongation ratio decreases. The second phase strengthening and fine grain strengthening is the main reason which leads to the increase of the strength of the high strength grade dual phase steel, while the smaller proportion of ferrite soft phase is the main factor of the decrease of the plasticity.

Key words　two kinds of strength level dual-phase steel, microstructure, property

基于错误试验结果的室温拉伸试验国家标准推荐的方法 A 相关问题解析

李和平　徐惟诚　周　星

（上海宝钢工业技术服务有限公司检化验中心，上海　200941）

摘　要　近几年对于 GB/T 228.1(ISO 6892)推荐方法 A 的理由的真实性存在很多争议，各方都发表了所依据的数据图表。而根据科学常识同样的材料同样的试验方法不可能得出如此矛盾的结果。《理化检验-物理分册》组织争论的相关方及关注争议的 21 个实验室共同试验得出了一致的数据。根据 21 个实验室的数据分析，发现支撑推荐应变控制方法的数据和图表存在一个共同的问题：实际的试验条件不是所报告的试验条件；本文还分析了方法 A 推荐传统方法的下限速率的理由是错误的，错在误用了胡克定律。

关键词　拉伸试验, 横梁位移控制, 应变控制, 刚度修正

Analysis on the Reasons to Recommend Method A in GB/T 228.1(ISO 6892) are not true

Li Heping, Xu Weicheng, Zhou Xing

(Testing and Inspection Center, Shanghai Baosteel Industry Technological Service Co., Ltd., Shanghai 200941, China)

Abstract There are a lot of debate on the realness of the reasons to recommend method A in GB/T 228.1 (ISO 6892) in recent years. Both sides publish their test data and figures. As a common sense, same material same test condition can not get contradictory results. Not all of these test data and figures can be truth. The 21 laboratories (included different opinion companies) organized by the editorial department of "Physical Testing and chemical analysis part A: physical testing" did confirm tests and got coincident results. According to the test data, it is found out that there is a similar mistake in all figures and data supporting to recommend strain control method: the real test condition is not as reported. It is also explained in the paper that the reason to use low limit of tradition speed as required speed in method A is also based on a mistake. The mistake is misusing Hooke's law.

Key words tensile test, crosshead control, strain control, consider stiffness

Ni 含量对超低温钢性能和组织的影响

牟 丹 武会宾 唐 荻

（北京科技大学冶金工程研究院，北京 100083）

摘 要 本文研究了三种超低温钢的 Ni 含量对其力学性能和组织的影响。通过 thermo-calc 软件模拟计算了 Fe-Ni 相图，并用扫描电镜（SEM）、电子背散射（EBSD）、X 射线衍射仪（XRD）对试验钢的显微组织形貌，晶界分布和逆转变奥氏体的稳定性进行了研究。研究结果表明：超低温钢的强度随 Ni 含量的降低而降低，但低温冲击韧性并没有随着 Ni 含量的降低而降低，7Ni 钢完全满足 9Ni 钢的性能要求，且更经济，经 QLT 处理后，7Ni 钢的屈服强度 R_{eL} 为 603MPa，抗拉强度 R_m 为 686MPa，–192℃时的横向冲击吸收功 A_{kv} 为 244J；经 QLT 处理后生成的逆转变奥氏体稳定性非常好，经深冷处理后，奥氏体含量变化不超过 1%，而且在室温时存在 6%以上稳定的逆转变奥氏体时，即可获得良好的低温韧性。

关键词 超低温钢，7Ni 钢逆转变奥氏体

Influence of Ni Content on Microstructure and Properties of Super Low Temperature Steel

Mou Dan, Wu Huibin, Tang Di

(Research Institute of Metallurgical Engineering, USTB, Beijing 100083, China)

Abstract This paper researchesthe influence of Ni contentof three differentsuperlow temperature steelto the mechanical properties and microstructure and calculating the Fe-Ni Phase diagram by thermo-calc software and researching the microstructure morphology, distribution and the stability of the reversed austenite by SEM,EBSD and XRD.The research show that: with the decrease of the Ni content,the mechanical properties of superlow temperature steel reduce,but not the same with the charpy impact toughness, the mechanical properties of the intercritically quenched 7%Ni steel are equivalent to those of 9%Ni steel and more economically. Excellent mechanical properties can be achieved in the 7%Ni steel processed by the QLT heat treatment, showing yield strength(R_{eL}) greater than 603MPa, tensile strength (R_m) greater than 686MPa, charpy impact toughness (A_{kv}) is 244J even at –192℃. The stability of reversed austenite is great after QLT treatment. The volume fraction of reversed austenite change does not exceed 1% even in the super low temperature. Excellent charpy impact toughness can be achieved when there is more than 6% stable reversed austenite exist.

Key words super low temperature steel, 7%Ni reversed austenite

重轨钢中超低含量气体元素的快速测定

郑连杰　张　敏　秦晓峰

（河北钢铁集团邯钢公司技术中心，河北邯郸　056015）

摘　要 本方法利用 LECO TCH600 型氧氮氢分析仪测定重轨钢中超低含量气体元素。用线切割机从经轧制的钢材上取样，用碳化硅磨头和毡头对试样表面进行打磨与抛光，试样经剪切和超声波洗涤后进行分析。为保证基线稳定，设置了充分的脱气条件和分析延迟。通过实验，发现在较高的分析功率下，空白及试样中氢含量分析结果的精密度与低功率时没有显著区别；仪器经单标样校准后，工作曲线在用含量段准确度能够满足要求。由此确定了氧、氮、氢三元素同时测定和单标样校准的可行性，缩短了实验周期。用该方法分析标准物质，测定值与参考值一致，试样中氧、氮、氢分析的精密度（*n*=6）分别为 2.92%、0.86%和 3.68%。该方法快速简便，能够为重轨钢的生产与判定工作提供及时、可靠的数据。

关键词 重轨钢，氧，氮，氢，快速测定

Rapid Analysis of Ultra Low Gaseous Elements in Heavy Rail Steel

Zheng Lianjie, Zhang Min, Qin Xiaofeng

(Technique Center, Handan Iron and Steel Company, Handan 056015, China)

Abstract Concentration of ultra low gaseous elements in heavy rail steel were determined by LECO TCH600 hydrogen/nitrogen/oxygen determinator. The samples were obtained in rolled steel by wire cutting machine, burnished by carborundum drill head and felt drill head, cut and ultrasonic washed, and then analysed. Sufficient outgas conditions and analysis delay were configured to obtain stable baselines. The tests proved that the precision of hydrogen under high analysis power made little difference to that obtained under low analysis power. And the accuracy of the required sections of the working curves were satisfying. So the feasibility of simultaneous determination of oxygen, nitrogen and hydrogen and single standard calibration were proved, which reduced the time cost of the experiment. The analysis results of standard sample were accordant with the referenced value. The RSD (*n*=6%) oxygen, nitrogen and hydrogen in production samples were 2.92%, 0.86% and 3.68%. This method is fast and simple, and it can provide timely and accurate results for the production and quality estimation of heavy rail steel.

Key words heavy rail steel, oxygen, nitrogen, hydrogen, rapid analysis

一种高效炼钢炉前样品的全自动分析方法

杜士毅　王贵玉　王明利　金　伟　孙　娟

（首钢股份公司迁安钢铁公司，河北迁安 064404）

摘 要 本文介绍了一种炼钢圆柱样品的全自动分析方法。该方法通过将样品全自动切割为三部分，一部分小圆柱样进行自动铣样加工，进行自动直读光谱分析；一部分圆片样冲粒加工，进行全自动气体氧氮分析及自动红外碳硫分析；一部分样品尾部区域弃去。通过DOE分析、回归分析等方式确定了最佳样品切割进给速度16%，转速700r/min，铣刀进刀速度1000mm/min，深度0.1mm，样品水冷时间为20s，样品风冷时间为20s，使该方法直读样分析时间为240~250s（激发两点），氧氮样510~530s（分析两粒），碳硫样440~460s(分析两粒)。另外，本文对该方法常见设备故障的进行了研究和控制，使自动化运行率大于99.5%。综上，该方法具有分析速度快、稳定性高、自动化率高、一个样品能够完成炉前全元素的检验等优点，对于钢铁冶金企业的实施与推广具有较好的借鉴意义。

关键词 炉前检验，圆柱样，全自动分析

A Highly Efficient Steelmaking the Automatic Analysis Method of the Sample

Du Shiyi, Wang Guiyu, Wang Mingli, Jin Wei, Sun Juan

(Shougang Co., Qian'an Steel Corp, Qian'an 064404,China)

Abstract The article introduces a kind of Automatic analysis method for steelmaking samples.The automatic sample cutting into three parts, part of a small cylindrical samplesautomatic milling processing, automatic analysis of spark spectrum; part of a wafer like punching processing, analysis of automatic Oxygen and nitrogen gas analysis and automatic infrared carbon sulfur analysis; part of the tail region were abandoned. Through the DOE analysis, regression analysis and other methods to determine that the optimal sample cutting feed rate is 16%, rotate speed is 1000 mm/min, cutter feeding speed is 700r/min, milling sample depth is 0.1mm, cooling time is 20s, the sample cooling time is 20s, Using this method, spark spectrum analysis time is 240~250s (excitation at two points), oxygen and nitrogen gas analysis time is 510~530s (analysis of two) , infrared carbon sulfur analysis time is 440~460s (analysis of two). In addition, in the article, the method of common equipment failure was studied and control, automatic operating ratio greater than 99.5%. The method has fast analysis speed, high stability, high automation rate, a sample can be analyzed for all elements,for the implementation of the iron and steel metallurgy enterprises and promotion has good reference value.

Key words analysis for steelmaking, cylindrical sample, automatic analysis

光电直读光谱法测定铁基合金钢中铜元素

杨 琳 王 鹏 刘步婷

（湖北武汉钢铁股份有限公司质量检验中心，湖北武汉 430080）

摘 要 采用ARL-4460火花光电直读光谱仪中所提供的钢样分析程序测定高铜（含量约为0.8%～1.35%）铁基合金钢样品时，所得测定值已超出仪器原有工作曲线可测量范围。为此，利用ICP分析得出的内控标样，重新对高铜工作曲线进行校准。通过分析实验，对ARL-4460直读光谱仪中铜元素分析结果与内控定值进行了比较，从而对仪器内部参数予以调整，改善分析效果，为高铜铁基合金钢的分析提供准确、快速的分析数据。

关键词 火花直读光谱仪，Cu含量，分析，铁基合金钢

The Determination of Copper of the Iron-base Alloy Steels by Photoelectric Direct-reading Atomic Emission Spectrometry

Yang Lin, Wang Peng, Liu Buting

(Quality Inspection Center of WISCO, Wuhan 430080, China)

Abstract It was shown that in the determination of copper in sample of the iron-base alloy steels with high copper contents（0.8%～1.35%） using the steel analytical program supplied by the ARL-4460 photoelectric direct-reading atomicemission spectrometer, the results of out of rage were obtained. As a modification, the regression analytical curve for determination of copper in the steels were re-established by using control standard samples with high copper contents that it was determined by ICP instrument. According to analysis test , the analysis result data of copper in steel sample were compared, and the analysis effect was improved through slight adjustments to the internal parameter of instruments. The result data was send for ensuring the accurately and timely in the iron-base alloy steel.with high-copper contents.

Key words photoelectric emission spectrometry, copper, analysis, the iron-base alloy steels

镀镍板镀层的辉光放电光谱法解析

刘　洁　侯环宇　安　晖　蔡　啸　杨慧贤

（河北钢铁技术研究总院，河北石家庄　050000）

摘　要　本方法采用辉光放电光谱法对镀镍板样品进行逐层剥离，根据样品随着深度的辉光放电积分谱图，可以对镀镍板结构总体进行判定，还可以对镀镍板表面进行放大分析，了解镀镍板的基本结构。利用射频辉光放电光谱仪逐层分析的特点，对两种镀镍板进行了研究探讨，分析了镀镍板的镀层厚度及质量，利用辉光光谱仪分析镀镍板镀层质量相对稳定，精密度试验中其 RSD 值均小于 3%，通过 XRF 荧光光谱仪对其结果进行比对，比对结果基本一致。同时还分析了镀镍板基板成分，通过 ICP 光谱仪对其结果进行比对，比对结果基本一致。

关键词　辉光放电光谱法，镀镍板，镀层厚度，镀层质量，基板成分

Analysis of the Nickel Plate by Glow Discharge Optical Atomic Emission Spectrometry

Liu Jie, Hou Huanyu, An Hui, Cai Xiao, Yang Huixian

(Hebei General Research Institute of Iron and Steel Technology, Shijiazhuang 050000, China)

Abstract When nickel plate sample was ablated layer by layer by glow discharge optical atomic emission spectrometry, formula integral was set according to glow discharge integral spectrum from sample surface to its center for the purpose of judging the overall of the nickel plate,and analysising the amplification surface, and also understanding the basic structure of nickel-plated sheet.Two cases in the analysis of surface were studied and discussed by meand of radio frequency glow discharge optical emission spectrometry,which can analyze the sample surface layer by layer.The thickness and mass of coating in nickel plate were analyzed, the nickel plate coating mass is relatively stable, the precision test the RSD values were less than 3%, To compare its results by XRF fluorescence spectrometer, the results are basically the same. It also analyzes base-plate composition of the nickel plate, to compare its results by ICP spectroscopy, comparison results are basically the same.

Key words glow discharge spectrometry, nickel plate, coating thickness, coating mass, base-plate composition

钢中氧含量分析探讨

张希静　费书梅　崔全法

（首钢股份公司迁安钢铁公司，河北迁安　064404）

摘　要　本文利用 TC-600 氧氮分析仪，通过对测定钢中氧含量过程中可能存在的干扰因素进行分析，包括取样器的选择、制样工具的选择、清洗液和清洗时间的选择、试剂有效性的检查、设置合适的分析时间参数、称样量的选择、空白值的控制等。通过对上述因素进行严格控制，得到了较低的检出限和测定下限，保证了钢中氧含量分析结果的准确度和精密度。

关键词　钢，氧，分析

Analysis of Oxygen Content in Steel

Zhang Xijing, Fei Shumei, Cui Quanfa

(Shougang Qian’an Iron & Steel Compang, Qian’an 064404, China)

Abstract　Through the analysis of the steel oxygen content in the process of possible interfering factors on the determination, including the samplers, sample preparation tools, cleaning liquid and cleaning time, reagent validity check, set appropriate analysis parameters, sample selection, blank value control by TC-600 oxygen nitrogen analyzer in this paper. Through the strict control of the above factors, the detection limit and the lower limit are obtained, and the accuracy and precision of the oxygen content in the steel are ensured.

Key words　steel, oxygen, analysis

一种快速稳定的炉前铁样全自动分析方法

杜士毅　沈　涛　王贵玉　王明利　王　程　闫　丽

（首钢股份公司迁安钢铁公司，河北迁安　064404）

摘　要　本文介绍了一种炉前铁样的全自动分析方法。该方法通过将样品进行全自动打磨出符合条件的分析面，再进行 X 荧光光谱分析。本文将样品加工流程拆分为样品粗磨、粗粗磨后冷却、样品细磨、细磨后冷却四个过程，并对四个过程的关键参数进行了实验研究。确定最佳粗磨深度为 0.8mm,粗磨后冷却时间为 100s,细磨压力为 2.0bar,细磨冷却时间为 40s。本文确定了炉前铁样的仪器分析参数及分析方法，经过重现性测试，该分析方法精密、准确。

经过生产样品全流程的再现性测试，该检验流程的C、Si、Mn、P、S的相对极差均小于5%。此外，本文还对影响铁样全自动检测的接收站故障、分析仪器故障、软件系统故障、耗材更换、样品不合格问题进行了分析和控制。最终该炉前铁样自动分析方法的分析时间在350~400s之间，自动运行有效率达到99.8%。

关键词 炉前检验，铁样，全自动分析

A Fast and Stable Automatic Analysis Method for Iron Samples for Steelmaking

Du Shiyi, Shen Tao, Wang Guiyu, Wang Mingli, Wang Cheng, Yan Li

(Shougang Co., Qian'an Steel Corp, Qian'an 064404, China)

Abstract The article introduces a kind of Automatic analysis method for the iron sample for steelmaking. The sample is automatically polished, and it is analyzed by X-ray fluorescence spectrometer. In the article, the sample processing flow split the four processes, rough grinding, cooling of rough grinding, fine grinding and cool of fine grinding, and experimental study on the key parameters of four process. It is determined that the rough grinding depth is 0.8mm, the cooling time of rough grinding is 100s, fine grinding pressure is 2.0 bar, the cooling time of fine grinding is 40s. In the article, the analysis parameters and analysis methods of instrument is determined for iron samples for steelmaking, the analysis method is precise and accurate by repeatability test. The results of relative range for C, Si, Mn, P, S are less than 5% by the sample production process reproducibility test. In addition, the receiving station fault, analysis instrument failure, software system fault, supplies replacement, sample unqualified problem is analyzed and controlled for increasing the rate of automation. Finally, the analysis of time is between the 350~400s, and automatically run efficiency reach 99.8% by the automatic analysis methods for the iron sample for steelmaking.

Key words analysis for steelmaking, iron sample, automatic analysis

不锈钢阀片断裂失效分析

许竹桃　关　云　黄海娥　周元贵
欧阳珉路　刘　敏

（武汉钢铁（集团）公司研究院，湖北武汉　430080）

摘　要 采用化学分析、扫描电镜（SEM）、金相显微镜等对冰箱压缩机不锈钢吸气阀片断裂失效进行了分析。结果表明，阀片的化学成分为含钼元素的不锈钢，符合阀片制造工艺要求。该阀片断裂失效主要以疲劳断裂为主，疲劳裂纹起始于阀片的边缘区域，该区域长时间处于交变应力作用下反复受力，引起磨损严重，表面非常光滑；根据断面纹路扩展方向，可以观察到裂纹沿边部区域向中间区域扩展，两侧疲劳裂纹扩展相遇在中部形成了台阶形貌，具有典型的疲劳裂纹特征。扫描电镜下观察，失效阀片断裂区域附近的边缘存在缺陷，有明显的加工刀痕，深度达到20μm。由于不锈钢阀片组织均匀，均为回火马氏体+屈氏体+颗粒状碳化物，属于高强度钢，其裂纹萌生阶段在失效过程中占主要地位，因此，不锈钢阀片失效断裂区域的边缘加工质量差是阀片在边缘区域最早发生疲劳失效的主要原因。

关键词 不锈钢阀片，疲劳断裂，失效分析

The Fracture Failure Analysis on the Stainless Steel Sheet of the Compressor Valve

Xu Zhutao, Guan Yun, Huang Haie, Zhou Yuangui, Ouyang Minlu, Liu Min

(Research and Development Center of WISCO, Wuhan 430080,China)

Abstract The fracture failure of the stainless steel sheet of the compressor valve was analyzed by the scanning electron microscopy(SEM) and metallographic microscope. The results showed that the fatigue fracture was the main reason for the fracture failure of the valve sheet. The initiation area of the fatigue crack was at the edge of the valve sheet, and the crack area expanded to both side of the valve sheet and formed the step topography. The knife mark in the processing procedure at the edge of the fracture area was about 20μm, which caused the earliest fatigue failure at this area.

Key words stainless steel valve, fatigue fracture, failure analysis

80MnCr 控冷工艺对相变的影响

刘 毅 李 杰 张 鹏

（河北钢铁技术研究总院，河北石家庄 050000）

摘 要 通条性能不稳定、组织不均匀是 80MnCr 盘条在拉拔或扭转加工过程中发生断裂的主要原因之一，本文利用淬火膨胀仪绘制了 80MnCr 盘条的静态 CCT 曲线，采用切线法得到了相变过程中的临界转变温度点，通过金相显微镜观察了不同冷却速度条件下 80MnCr 盘条的析出组织，通过扫描电镜分析了不同冷却速度条件下 80MnCr 盘条的珠光体片层间距，研究了不同冷却参数对 80MnCr 盘条相变组织转变的影响。结果表明，80MnCr 盘条产生马氏体的临界冷却速度为 10℃/s，为避免产生马氏体组织，应控制冷速小于 10℃/s；同一吐丝温度下，随着冷速的提高，珠光体片层间距不断变细，当冷速为 7℃/s 时，珠光体片层间距为 60~104.8nm，生成屈氏体组织；冷速为 3~5℃/s 时，组织均匀，索氏体率为 90%以上，片层间距在 95~114.3nm，机械性能最佳。

关键词 80MnCr，CCT 曲线，索氏体，片层间距

Influence of Controlled Cooling Process on Phase Transition of High Carbon Steel 80MnCr

Liu Yi, Li Jie, Zhang Peng

(Hebei Iron & Steel Technology Research Institute, Shijiazhuang 050000, China)

Abstract Primary causes of drawing cleavage brittle fracture of high carbon steel 80MnCr are unstableentire length of performance and inhomogeneous microstructure. CCT curves with different cooling speedsof high carbon steel 80MnCr were measured by Quenching dilatometer. The phase critical transition temperatures were determined by tangent method.

The Precipitated phases with different cooling speeds were researched by metallurgical microscope. The Pearlite interlaminar spacing with different cooling speeds was researched by electron microscope. According to these results, the influences of controlled cooling process on phase transition of high carbon steel 80MnCr were researched. The experiment results show that martensite critical cooling rate of high carbon steel 80MnCr is 10℃/s. The cooling speed should be controlled less than 10℃/s to avoid martensitic transformation. At the same laying temperature, the pearlite interlaminar spacing gradually narrowed with the increasing of cooling speed. When the cooling temperature was 7℃/s, the pearlite interlaminar spacing was 60~104.8nm. It meant that troostite organization was generated at the cooling temperature of 7℃/s. When the cooling temperature was 3~5℃/s, the interlaminar spacing was 95~114.3nm, with uniform sorbitestructure and of 90% sorbite, the best mechanical properties were obtained.

Key words 80MnCr, CCT curve, sorbite, lamellar spacing

连续包覆机挤压轮开裂失效原因分析

欧阳珉路 张友登 韩荣东 张彦文

（武汉钢铁(集团)公司研究院，湖北武汉 430080）

摘 要 连续挤压包覆技术是在连续挤压技术的基础上发展起来的一种高质、高效、经济的新型塑性加工技术，已广泛应用于双金属复合导线、电缆护套、光缆护套等的生产上。挤压轮是连续包覆机上的最主要工作部件之一，其工作寿命直接影响到连续挤压的生产成本和生产效率。某厂家的一台连续包覆机在正常使用过程中，其挤压轮出现了早期开裂现象，通过高倍观察和微区成分分析等方法对开裂缺陷形成原因进行了检验和研究。结果表明该处裂纹为疲劳开裂，挤压轮表面的腐蚀凹坑为疲劳源；分析认为挤压轮存在的异常组织导致强度降低，致使在高温的工作环境下从挤压轮表面的热腐蚀凹坑开始迅速形成疲劳开裂，是造成该挤压轮早期失效的主要原因。为了避免此类缺陷的发生，在制造挤压轮的过程中需要控制好奥氏体化加热温度，避免过高的加热温度导致粗大的奥氏体晶粒产生，严重降低钢的强度和硬度。同时在使用挤压轮的过程中，应采取措施尽量减少挤压轮表面产生热力腐蚀缺陷，延长挤压轮的使用寿命。

关键词 挤压轮， H13R，疲劳开裂， 缺陷分析

The Crack Defect Analysis of Extrusion Wheel of Continuous Extrusion Cladding Machine

Ouyang Minlu, Zhang Youdeng, Han Rongdong, Zhang Yanwen

(Research and Development Center, Wuhan Iron and Steel (Group) Corporation, Wuhan 430080, China)

Abstract Continuous Extrusion Cladding, which is developed on the basis of Continuous Extrusion Forming, is a new economical technology of plastic forming with high-quality and high-efficiency. It has been applied to the manufacturing of metal wire, cable sheath and fiber-optic cable sheath. The Extrusion Wheel is one of the main parts of Continuous Extrusion Cladding Machine, of which the working life expectancy directly affect the cost and productivity. In the process of normal service, an early crack appeared in Extrusion Wheel of Continuous Extrusion Cladding Machine. The research shows this is a fatigue crack by examining and micro domain analyzing with a microscope and SEM. The source of the fatigue is an eroded pit on the surface of extrusion wheel. The analysis shows that due to an abnormal metallurgical structure of extrusion wheel that lowers the strength, thus leading to the fatigue crack in a short-term from the heat erosion pit on the surface of extrusion wheel under the high temperature. This is the main reason of the early crack appearing in the extrusion wheel. In order to avoid the defect, it is necessary to keep austenitizing temperature under control, thus avoiding the austenite grain coarsening created by overheated temperature, which lowers the strength and hardness of steel. In addition, measures should be taken to reduce the eroded defect on the surface to extend the life expectancy of the Extrusion Wheel during the operation of the Extrusion Wheel.

Key words　extrusion wheel, H13R, fatigue crack, defect analysis

线材表面横纹原因分析

张志明　刘金源　农之江　刘春林

（宝钢集团广东韶关钢铁有限公司，广东韶关　512123）

摘　要　对线材拉拔半成品钢丝目视观察到的表面疑似横裂纹进行了夹杂物、金相、电镜能谱和冷镦钢检测分析，在金相显微、背散射探头和二次电子探头下观察未发现表面裂纹；缺陷试样表面局部轻微打磨处理后在二次电子探头下观察表面光滑，颜色与钢基体一样呈均匀灰白色，确定钢丝表面目视疑似横裂纹应为没有向基体延伸、没有深度的横纹。横纹在线材和拉拔半成品钢丝上周期性出现，拉拔加工过程不能消除。钢材上的横纹对应在轧钢风冷辊道的搭接点处，因搭接点处冷却速度慢，集卷后搭接点比非搭接点处温度高，盘卷在 PF 运输线冷却过程搭接点处氧化铁皮开裂，再生成一层氧化铁皮，有新生成氧化铁皮部位去氧化铁皮后钢材表面出现横纹。通过适当降低轧制吐丝温度，调整风冷线辊道速度等措施可以防止钢材表面产生横纹。

关键词　线材，表面缺陷，横纹，氧化铁皮

Analyses in the Reason of Transverse Crack on the Surface of Wire Rod

Zhang Zhiming, Liu Jinyuan, Nong Zhijiang, Liu Chunlin

(Shaoguan Iron and Steel Co., Ltd., of Baosteel Group, Shaoguan 512123, China)

Abstract　The suspected surface transverse cracks observed in semi-finished drawing wire rod were studied by inclusions analysis,metallographic observation, electron microscopy observation, energy spectrum analysis and cold pier steel detection.The surface transverse cracks were not discovered through metalloscope,back scattering probe and secondary electron detector. The surface of defective samples were smooth and uniform gray after mild polishing under secondary electron detector,which was the same with steel matrix, and these suspected surface cracks were confirmed to be transverse cracks which were shallow and did not extend to substrate.The transverse cracks were generated in wire rod and semi-finished drawing wire rod periodically, which could not be removed in the drawing process.Transverse cracks existed in the overlapping point of air cooling roller, which was due to the slow cooling rate in overlapping point, and the temperature of overlapping point was higher than other place after gathering coil. Therefore, oxide scales crazed during cooling process in PF artery, and new oxide scales formed, which caused transverse cracks in steel surface after removing oxide scales.The transverse cracks can be prevented by reducing the temperature of spinning during rolling, adjusting the speed of air-cooled roller and so on.

Key words　wire rod, surface defect, transverse cracks, oxide scale

二安替吡啉甲烷光度法测定含钛冶金物料中二氧化钛含量

邓军华[1] 王一凌[1] 亢德华[1] 曹新全[2] 李 化[2]

（1. 鞍钢集团钢铁研究院，辽宁鞍山 114009；
2. 鞍钢集团股份有限公司质检中心，辽宁鞍山 114021）

摘 要 通过合理地优化分析条件，采用高灵敏性、高稳定性的二安替吡啉甲烷分光光度法，配合高量程分光光度计测定钛精矿、钒钛磁铁矿和高钛渣中二氧化钛含量。二氧化钛含量在1～15 μg/mL范围内符合比尔定律，波长420 nm，表观摩尔吸光系数为1.24×10^4 L/(mol·cm)，回归曲线方程：$A=1.5684\times w+0.0141$，线性相关系数为0.99995。样品结果与认定值、硫酸高铁铵滴定法的测定值相吻合，相对标准偏差(RSD，n=8)为0.15%～0.47%，方法的精密度优于现有行业标准，分析速度快，操作简单，值得推广。

关键词 二安替吡啉甲烷，分光光度法，二氧化钛，含钛冶金物料，高吸光度

Determination of Titanium Dioxide Content in Titanium Metallurgy Materials by Diantipyryl Methane Spectrophotometer

Deng Junhua[1], Wang Yiling[1], Kang Dehua[1], Cao Xinquan[2], Li Hua[2]

(1. Iron & Steel Research Institute of Angang Group, Anshan 114009, China;
2. Qualification Test Centre of Ansteel Co., Ltd., Anshan 114021, China)

Abstract Utilization diantipyryl methane spectrophotometer high sensitivity、stability and the high range of absorbance spectrophotometer. reasonably optimization analysis parameters, determination of titanium dioxide content in titanium ore, vanadium-titanium magnetite and high titanium slag by diantipyryl methane spectrophotometer. Beer's law was obeyed for titanium dioxide content in the range of 1～15 μg/mL , the apparent molar absorptivity was 1.24×10^4 L/(mol·cm) at 420 nm, the regression curve equation: $A=1.5684\times w+0.0141$, the correlation coefficient of 0.99995. The sample results were consistent with the certified values and those obtained by high ferric ammonium sulfate titration, the relative standard deviation(RSD, n=8)of 0.15% to 0.47%. The precision is better than existing industrial standards. The method of analysis speed, simple operation, worthy of promotion.

Key words diantipyryl methane, spectrophotometer, titanium dioxide, titanium metallurgy materials, high absorbance

含钛冶金物料中磷的分光光度法研究

王一凌 邓军华 亢德华

（鞍钢钢铁研究院，辽宁鞍山 114009）

摘　要　实验采用氢氟酸分离含钛冶金物料中硅，硼酸碳酸钠混合熔剂熔融的方法分解试样，选用硫酸浸取。解决了磷元素测定由于基体多样导致的酸溶消解难，碱熔过滤沉淀速度慢的问题，分离硅后硫酸的浸出速度加快，有效地防止钛的水解，同时避免了磷元素容易被沉淀物吸附损失导致结果偏低的问题，选择高锰酸钾为氧化剂保证了磷的完全氧化，确保检测结果的准确性。确定了显色酸度为1.1mol/L，吸收波长为700nm，钼酸铵溶液、抗坏血酸溶液和硫代硫酸钠-无水亚硫酸钠溶液的用量均为 5mL、硝酸铋溶液的用量为 15mL。优化实验条件后，磷的检测的范围在0.005 %～0.5%，方法的相对标准偏差小于8%，回收率在90%～110%之间，结果令人满意。

关键词　磷铋钼蓝，分光光度法，磷，含钛物料

Study of Spectrophotometer Method for Phosphorus Content in Titanium Materials

Wang Yiling, Deng Junhua, Kang Dehua

(Iron & Steel Research Institute of Angang Group, Anshan 114009, China)

Abstract　Experiment using hydrofluoric acid separating containing titanium metallurgy material of silicon, boric acid sodium carbonate mixed flux fusion method of decomposition of sample, selection of sulfuric acid leaching. To solve the phosphorus determination due to the matrix leading to a variety of acid dissolution difficult, alkali fusion filtering sedimentation speed is slow, the separation of silicon sulfuric acid leaching speed accelerate, effectively prevent the titanium hydrolysis, while avoiding the phosphorus element is easy to precipitate adsorption results in a loss results the problem of low and potassium permanganate as oxidizing agent to ensure the complete oxidation of phosphorus, to ensure the accuracy of test results. Determine the pH of color for 1.1mol/L, absorption wavelength to 700 nm, amount of ammonium molybdate solution, ascorbic acid solution and sodium thiosulfate anhydrous sodium sulfite solution was 5mL, bismuth nitrate solution dosage was 15ml. The range of the detection of phosphorus was 0.5% ~ 0.005%, the relative standard deviation was less than 8%, the recovery rate was between 90% and 110%. The results were satisfactory.

Key words　phosphor bismuth molybdenum blue, spectrophotometer, phosphorus, titanium materials

电感耦合等离子体原子发射光谱法测定炼钢增碳剂中杂质元素

亢德华　王　铁　于媛君　邓军华　王一凌

（鞍钢集团钢铁研究院，辽宁鞍山　114009）

摘　要　建立了利用电感耦合等离子体原子发射光谱法（ICP-AES 法）同时测定炼钢增碳剂中 5 种杂质元素的分析方法。研究了增碳剂试样的前处理方法，优选了适宜的仪器测定参数和分析谱线，采用基体匹配法进行基体效应的校正。实验结果表明，各元素分析谱线线性相关系数大于 0.9990，方法检出限在 0.012~0.83μg/mL 之间。该方法应用于类似基体国家标准物质分析，测定值与认定值相符，测定结果的相对标准偏差在 0.8% (Si)~5.8% (Ca) 范围。

关键词　电感耦合等离子体原子发射光谱法（ICP-AES 法），炼钢用增碳剂，杂质元素，基体匹配法

Determination of Impurity Elements in Carburetants for Steel-making by Inductively Coupled Plasma Atomic Emission Spectrometry

Kang Dehua, Wang Tie, Yu Yuanjun, Deng Junhua, Wang Yiling

(Iron & Steel Research Institute of Ansteel Group, Anshan 114009, China)

Abstract A method for simultaneous determination of 5 impurity elements in carburetants for steel-making by inductively coupled plasma atomic emission spectrometry was established. The sample pretreatment and dissolution method was studied, the optimum instrument working conditions and the analytical spectrum lines were optimized. Besides, the matrix match method was used to eliminate the matrix effect. The results showed that the linear correlation coefficient of all the elements were lager than 0.9990, the detection limits were between 0.012 ~0.83 μg/mL. The determination of certified reference materials with similar matrix showed that the measured results were in good agreement with the certified values, and the relative standard deviations were between 0.8% (Si)~5.8% (Ca).

Key words inductively coupled plasma atomic emission spectrometry (ICP-AES method), carburetants for steel-making, impurity element, matrix match method

一种高铝缓释脱氧剂中 Ti 元素的分析方法

闫　丽　王贵玉　庞振兴　杜士毅　刘飞宇　孙　娟

（首钢股份公司迁安钢铁公司，河北迁安　064400）

摘　要　本文叙述了用电感耦合等离子体发射光谱法测定高铝缓释脱氧剂中 Ti 元素的方法。试样采用酸溶后将残渣碱融合并再定容分析，通过系列实验，在酸溶解部分，对酸种类、酸溶解时间、酸加入量分析条件进行优化；对仪器的工作条件包括等离子体功率、雾化气流量、辅助气流量和冷却气流量进行优化。最终确定了一套最优的分析方法，方法回收率在 95%～105%，相对标准偏差小于 0.05（即 5%）。

关键词　高铝缓释脱氧剂，电感耦合等离子体发射光谱法，Ti

A Kind of High Aluminium Slow-release Deoxidizer in Ti Element Analysis Method

Yan Li, Wang Guiyu, Pang Zhenxing, Du Shiyi, Liu Feiyu, Sun Juan

(Qian'an Shougang Shares in the Company in Hebei Province Qian'an Iron and Steel Company, Qian'an 064400, China)

Abstract This paper describes the method of determining the Ti element in the high-Al sustained release oxygen release agent by inductively coupled plasma emission spectrometry.Sample with acid solution after the residue alkali fusion and constant volume analysis, through a series of experiments, in the acid soluble fraction of kinds of acid, acid dissolving time and acid amount analysis conditions were optimized. The working conditions of the instrument including plasma power, nebulizer gas flow, auxiliary air flow and cooling gas flow were optimized. Finally, a set of optimal analysis method was determined, the recovery rate of the method was 95% to 105%, and the relative standard deviation was less than 0.05 (i.e. 5%).

Key words　high-Al sustained release oxygen-release, inductively coupled plasma-emission spectrometry, Ti

Ti 微合金化 Q345B 的夹杂物分析

孙　强　米振莉　李志超　党　宁

（北京科技大学冶金工程研究院，北京　100083）

摘　要　使用扫描电镜技术和能谱分析技术对 Ti 微合金化 Q345B 钢种夹杂物的形貌和种类进行了分析，并解释了不同夹杂物产生的主要原因。实验结果表明，Q345B-Ti 钢板中的夹杂物主要包括 MgO 与钙硅酸盐复合夹杂、SiO_2 夹杂物、MnO 和 SiO_2 的复合夹杂物以及氧化物和硫化物的复合夹杂物，夹杂大多呈现块状且分布均匀，尺寸在 1~5μm 间。MgO 与钙硅酸盐复合夹杂来源于耐火材料脱落和二次氧化。$MnO·SiO_2$ 复合夹杂物尺寸较大，对钢的力学性能不利，可以通过增加吹氩时间和调节碱度减少其含量。元素组成复杂的复合夹杂物来源于结晶器卷渣，提高对结晶器的精确控制可以有效减少元素组成复杂的夹杂物。

关键词　Q345B-Ti，微合金化，夹杂物

Inclusion Analysis of Ti Microalloy Q345B

Sun Qiang, Mi Zhenli, Li Zhichao, Dang Ning

(Engineering Research Institute, University of Science and Technology Beijing, Beijing 100083, China)

Abstract　SEM and EDS techniques were used to analyze inclusions in Ti microalloy Q345B and dominant reasons caused different inclusions were explained. Experimental results indicate that, main types of inclusions in Q345B-Ti were composite inclusions of MgO and calcium silicate, SiO_2, $MnO·SiO_2$, composite inclusions of sulfide and oxide and complex composite inclusions. Inclusions were evenly distributed and their average size was between 1 μm and 5 μm. Composite inclusions of MgO and calcium silicate originated from falling of refractory materials and secondary oxidation. Average size of $MnO·SiO_2$ inclusions was large and could damage mechanical properties of Q345B-Ti and $MnO·SiO_2$ inclusions could be reduced by increasing time of argon blowing and adjusting basicity.Inclusions with complex compositions stated in this study originated from slag entrapment in mold, which could be escaped by increasing accuracy of controlling the crystallizer.

Key words　Q345B-Ti, microalloying, inclusion

电感耦合等离子体原子发射光谱法分析铝质浇注料中三氧化二铝

吕　琦　郭　芳　崔　隽　沈　克　贾丽晖　李小杰　古兵平　刘丽荣

（武钢股份质检中心，湖北武汉　430080）

摘 要 建立了用电感耦合等离子体发射光谱法（ICP-AES）测定浇注料中 Al_2O_3 含量的分析方法。提出了采用碳酸钠-硼酸混合熔剂熔融、稀盐酸浸取，以电感耦合等离子体-发射光谱法可以快速准确测定铝质浇注料中的 Al_2O_3 含量。在优化的分析条件及仪器工作参数下，Al_2O_3 测定结果与化学分析值或标准值一致，标准偏差小于 0.40 %，该方法快速、简便，结果准确可靠。

关键词 电感耦合等离子体-发射光谱，铝质浇注料，碱熔，基体匹配

Determination of Magnesian Refractories by ICP-AES

Lv Qi, Guo Fang, Cui Jun, Shen Ke, Jia Lihui, Li Xiaojie,
Gu Bingping, Liu Lirong

(Quality Inspect Centre of WISCO, Wuhan 430080, China)

Abstract Establish the method of determination of magnesian refractories by ICP-AES. The optimal testing parameters have been determined after optimization operation procedures of instrument, methods for sample solution. The results of the Al_2O_3 content between determination value and standard value of the standard samples and test samples are identical. The SD is less than 0.40%.

Key words ICP-AES, aluminum castable, alkali fusion, matrix matching

配对 T 检验在轻烧白云石中 SiO_2、CaO、MgO 成分分析的应用

闫 丽 费书梅 徐 佳 张希静 王 涛

（河北省迁安市首钢股份公司迁安钢铁公司，河北迁安 064400）

摘 要 利用六西格玛工具配对 T 检验对我公司自产轻烧白云石中 SiO_2、CaO、MgO 成分分析方法进行研究，通过对化学法和 X 射线荧光光谱法两种方法进行比较，最终确定了化学法分析自产轻烧白云石变更到 X 射线荧光光谱仪分析轻烧白云石中 SiO_2、CaO、MgO 的含量，使轻烧白云石检验满足生产检验数据准确性的同时使分析过程更加快捷、方便，满足我公司生产的要求。

关键词 轻烧白云石，分析，六西格玛，X 射线荧光光谱仪

The Paired T-test in Light-burned Dolomite SiO_2, CaO, MgO Component Analysis of the Application

Yan Li, Fei Shumei, Xu Jia, Zhang Xijing, Wang Tao

(Qian'an Shougang Shares in the Company in Hebei Province Qian'an Iron and Steel Company, Qian'an 064400, China)

Abstract Using six-sigma tools the paired t-test of our company produces light-burned dolomite of SiO_2, CaO, MgO component analysis method, through the chemical method and X-ray fluorescence spectrometry comparing two methods, finally determined the chemical analysis produced light-burned dolomite changes to the X-ray fluorescence spectrometer analysis of SiO_2, CaO, MgO component in light-burned dolomite, make the light-burned dolomite test meet the production test of the accuracy of the data at the same time make the analysis process more quickly Convenient, meeting the requirements of our company.

Key words light-burned dolomite, analysis, six sigma, X-ray fluorescence spectrometer

全自动直读光谱法测定生铁中 8 种元素

刘步婷 杨 琳 王 鹏 伍智娟

（湖北武汉钢铁集团公司质量检验中心，湖北武汉 430083）

摘 要 应用全自动直读光谱法测定了生铁中 8 种元素。提出了采用 HB3000 型磨样机制样，并探讨磨样参数的优化以及白口化程度对光谱测量的影响。优化了氩气纯度及流量、分析条件、光谱仪的清理等影响因素，给出了全自动分析的过程质量控制手段。在优化的试验条件下，8 种元素的相对标准偏差（n=11）均小于 5%。方法用于生铁试样分析，测定结果与 X 射线荧光光谱法的测定结果基本一致。

关键词 直读光谱法，生铁，硅，磷

Determination of 8 Elements in Pig Iron by Full Automatic Direct-Reading Spectrography

Liu Buting, Yang Lin, Wang Peng, Wu Zhijuan

(Quality Inspection Center of Wuhan Iron and Steel Group Company, Wuhan 430083, China)

Abstract Full automatic direct-reading spectrography was applied to determination of 8 elements in pig iron. The samples was prepared with HB3000 sample grinder, and optimizing the grinding sample parameters was discussed, and spectral measurement was impacted by the level white in pig iron. Influential factors, including argon gas purity and flow rate, analytical conditions and cleaning of spectrography were optimized. The method of process quality control of automatic analysis was provided . Under the optimized experimental condition, values of RSD's (n=11) of 8 elements found were less than 5%.The proposed method was applied to analysis of sample of pig iron, and results obtained were in conformity with by XRFS.

Key words direct-reading spectrography, pig iron, silicon, phosphorus

钢轨核伤产生机理分析及安全预测模型

费俊杰[1] 齐江华[1] 周剑华[1] 朱 敏[1] 徐 进[2]

（1. 武汉钢铁(集团)公司研究院，湖北武汉 430080；

2. 武汉铁路局工务处，湖北武汉 430071）

摘　要　通过线路无损探伤发现，在小半径曲线部分区域，钢轨使用一段时间后，在轨角内部出现核伤。通过对核伤钢轨的断口形貌进行观察、金相组织分析等一系列理化性能检验，认为钢轨出现核伤的原因是轮轨接触应力较大，而且应力作用点未发生移动，导致钢中未超标的细小夹杂物形成疲劳源并扩展成核伤，同时根据断裂力学模型对钢轨核伤安全临界尺寸进行了预测计算。

关键词　钢轨，核伤，疲劳裂纹，临界尺寸

Analysis of the Mechanism and Safety Appraisement Model of Rail Nucleus Flaw

Fei Junjie[1], Qi Jianghua[1], Zhou Jianhua[1], Zhu Min[1], Xu Jin[2]

(1. Research and Development Center of Wuhan Iron and Steel (Group) Co., Wuhan 430080, China;
2. Permanent Way Department of Wuhan Railway Administration, Wuhan 430071, China)

Abstract　The rail in individual small-radius curve had nucleus flaw after used for a period of time. The fracture morphology and metallographic structure of nucleus flaw were researched by SEM and optical microscope. It can be find that there has large wheel-rail contact stress in small-radius curve, and the stress point does not move, all of which lead the tiny inclusion (under the standard) near the stress point to form the fatigue source and then become the nucleus flaw. The safety appraisement of nucleus flaw has been done and the critical crack size of nucleus flaw has been estimated based on the fracture mechanics model.

Key words　rail, nucleus flaw, fatigue crack, critical dimension

一种高效全自动炉渣分析技术

王贵玉　杜士毅　王明利　金　伟　费书梅

（首钢股份公司迁安钢铁公司，河北迁安　064404）

摘　要　本文介绍了一种高效全自动炼钢、炼铁炉渣分析技术，该技术实现了一条生产线四种炉渣自动无相互干扰的分析。通过将炉渣样品的全自动分析过程拆分为样品传递、样品破碎、样品研磨、样品压片等8个环节，找到了可能影响全自动压片成功率和炉渣交互污染的关键点位。利用实验设计的方法，确定了转炉渣黏结剂加入量0粒、研磨时间120 s、压片压力120 kN、保压时间10 s；确定了保护渣、精炼渣、高炉渣黏结剂加入量12粒、压片压力为60 kN、研磨时间120 s、保压时间10 s。利用双样本t检验的方法确定了的清洁压头时间为25s、干洗时间为40s、物料清洗时间为10s。经过实测验证，在新参数条件下，四种炉渣的全自动压片合格率达到99%以上；四种炉渣CaO、MgO、SiO_2、Al_2O_3、TFe成分的检验准确度均满足要求。

关键词　炉渣，全自动，成分检验，样品加工

A high Efficient Automatic Slag Analysis Technique

Wang Guiyu, Du Shiyi, Wang Mingli, Jin Wei, Fei Shumei

(Shougang Co., Qian'an Steel Corp, Qian'an 064404,China)

Abstract The article introduces a kind of high-efficiency automatic analysis method for ironmaking and steelmaking slag,and the technology achieves the four kinds of slag automatically without mutual interference analysis on only one production line. Through the slag sample automatic analysis process having been divided into eight ,such as crushing, grinding, tabletting ,etc, The key points were found for affecting the success rate of automatic tablet and being slag interaction. By using the method of DOE, we confirmed that the the amount of binder is zero, milling time is 120s, tabletting pressure is 120KN, pressure maintaining time is 10s for converter slag. And we confirmed that the the amount of binder is twelve, milling time is 120s, tabletting pressure is 60KN, pressure maintaining time is 10s for Protecting slag, refining slag, blast furnace slag. Using two sample test method to determine that the cleaning head is 25s, dry cleaning time is 40 s, material cleaning time is 10s. Through actual measurement verification, under the condition of new parameters, four automatic slag tabletting qualified rate of 99% or more; Four kinds of slag, CaO, MgO, SiO_2, Al_2O_3, TFe composition test accuracy are meet the requirements.

Key words slag, automatic, composition analysis, the sample processing

11 信息情报

Information and Intelligence

冶金科技文献资源分析、推荐与利用

付　静　李春萌

（冶金工业信息标准研究院，北京　100730）

摘　要　文章紧扣钢铁行业发展的严峻形势，分析了信息情报工作在科技创新中的积极作用，介绍了5个国内核心的科技文献保障机构，并重点分析了国家科技图书文献中心冶金分中心的期刊、非刊等资源状况，对国外15项优质电子信息资源进行了分析和推荐，并结合行业发展需要，对文献资源的开发和利用工作进行了阐述，包括信息资源服务平台的打造，领域情报研究报告和信息情报跟踪等。对行业人士更多地了解业内的文献资源、更快捷地发现信息、更好地利用资源有很重要的参考意义。

关键词　冶金，文献资源，科技文献，数字资源，科技创新，利用

Analysis, Recommendation and Utilization of Metallurgical Science and Technology Literature Resources

Fu Jing, Li Chunmeng

(China Metallurgical Information and Standardization Institute, Beijing 100730, China)

Abstract　Closely linked with the development of iron and steel industry, the article analyzes the positive role of information work in scientific and technological innovation, and introduces the five core science and technology literature guarantee mechanism. Focusedon literature resource of the NSTL metallurgical center, 15 foreign high-quality electronic information resources are analyzed and recommended.Combined with the demand of industry, literature resource development and utilization are described, including the information resource service website, intelligence research report and information tracking. The article has a very important reference value to more understanding literature resources, more quickly finding information, better utilizing resources.

Key words　metallurgy, literature resources, science and technology literature, digital resources, scientific and technological innovation, utilization

按排名加权前后的专利权人影响力比较研究——以全球页岩气勘探开发技术为例

周群芳　吴　婕

（宝钢中央研究院情报中心，上海　201900）

摘　要　近几年，从专利引用的角度对专利权人的影响力进行评价在专利分析中得到了较好的应用。而随着合作研

发情况的普遍出现，专利权人申请专利时的排名顺序与其在领域中的影响力是否存在关联？为此，本文以全球页岩气勘探开发技术为例，对专利权人按排名顺序进行加权处理后，再计算其各项引用指数值，并与不考虑专利权人排序情况下的结果进行比对分析，从而得出相关结论。

关键词 专利权人，加权处理，H 指数，G 指数，W 指数

Comparative Study of the Influence of Patentee before and after Weighted Processing ——In the Field of Global Shale Gas Exploration and Development Technologies

Zhou Qunfang, Wu Jie

（Information Center of Research Institute ,Baoshan Iron & Steel Co., Ltd., Shanghai 201900, China）

Abstract In recent years, the method of evaluate the influence of the patentee through patents' citation analysis has widely used in patent analysis.With the widespread availability of collaborative R&D,is the patentee's ranking related to it's influence in the field?In order to draw some relevant conclusions, based on global shale gas exploration and development technologies, after the weighted processing of the patentee,the paper calculates the citation index, then makes comparison with the result of without patentee's weighted processing.

Key words patentee, weighted processing, H index, G index, W index

钢铁行业实施电子商务的关键问题分析

李新创 施灿涛 赵 峰

（冶金工业规划研究院，北京 100711）

摘 要 首先从国家政策环境、钢铁企业态度、电商平台的现状三个角度介绍了钢铁行业电子商务的发展概况，在对其进行剖析的基础上，总结了钢铁行业实施电子商务的优势所在：低成本与快捷性、商业模式开放性、信息资源共享性、优化交易模式、整合平台资源等，同时分析并阐述了钢铁行业电子商务发展的标准化、规模化、产业化的趋势。在积极进行转型升级的过程中，钢铁电商正面临着政策环境薄弱、平台规划不合理、钢铁企业应对不足、信用体系不完善和核心技术缺失等问题。为了更好地实施“工业 4.0”战略，结合钢铁行业的特征，提出了不断完善基础法规和标准化体系、从国家层面推进产品规范和质量规范的标准化、科学规划平台、创新商业模式、构建全产业链的生态系统等建议。

关键词 钢铁行业，电子商务，转型升级，工业 4.0

Key Issues of Iron and Steel Industry E-commerce Implementation

Li Xinchuang, Shi Cantao, Zhao Feng

(China Metallurgical Industry Planning and Research Institute, Beijing 100711, China)

Abstract It describes the current situation of iron and steel industry e-commerce from three aspects of national policy environment, the attitude of iron and steel enterprises. Then the advantages of low cost and quickness, the openness of business model, information sharing, optimize trading patterns, resource integration platform are summarized. On the basis of analysis, trends of standardization, scale, industrialization for iron and steel industry are proposed. In the process of transformation and upgrading, iron and steel industry e-commerce is facing such problems: weak policy environment, insufficient response of iron and steel enterprises, unreasonable platform planning, incomplete credit system and lack of core technology. A city or an industrial sector should form its own e-commerce patter featuring full dimensions and full integration to win the initiative to gain more economic benefits. We should pay close attention on transformation and upgrading and industry 4.0 and build an integrated e-commerce pattern of different levels. To better implement the "Industry 4.0" strategy, under the consideration of iron and steel industry characteristics, improving the basic regulations and standardization system, standardized specifications from the national level to promote the product specifications and quality, planning platform with science, innovating business model, constructing the ecosystem of the whole industry chain and other advice are proposed.

Key words iron and steel industry, electronic commerce, transformation and upgrading, industry 4.0

大数据时代钢铁企业的战略级信息应用

刘斓冰

（冶金工业信息标准研究院，北京 100730）

摘 要 本文介绍了大数据时代众多概念中的核心技术，将其与钢铁企业的信息化建设现状相结合，论述钢铁企业如何构建高性能的BI体系。并以成本管理系统为例，举例说明决策支持系统在企业运营管理过程中的核心价值。

关键词 钢铁企业，大数据，数据挖掘，商业智能，数据仓库

Strategic Level Information Application of Iron and Steel Enterprises in the Big Data Era

Liu Lanbing

(China Metallurgical Information and Standards Institute, Beijing 100730, China)

Abstract This paper introduces the core technology of many concepts in big data era, with the combination of iron and steel enterprise's informatization construction situation, discusses how to build high-performance BI system. And take the cost management system as an example, illustrates the core value of decision support system in the process of enterprise operation management.

Key words iron and steel enterprise, big data, data mining, business intelligence, data warehouse

基于本体的竞争对手产品分析研究

谷 俊 周群芳

（宝山钢铁股份有限公司，上海 201900）

摘　要　传统的竞争对手产品分析会花费情报部门大量的时间，在分析现有企业竞争对手产品分析过程的基础上，利用 Jena 语义工具包和 Lucene 搜索引擎工具包，通过已有的本体模型对检索词进行语义扩展，利用互联网信息抓取方法，在确保抓取结果全面的基础上，结合检索词语义扩展，特别是对产品名称和所在区域进行扩展与合并处理，从而得到较为准确而全面的关于竞争对手产品分析结果。

关键词　本体，竞争对手分析，语义扩展，统计分析

Study on Competitor Products Analysis Based on Ontology

Gu Jun, Zhou Qunfang

(Baoshan Iron and Steel Co., Ltd., Shanghai 201900, China)

Abstract　When the traditional competitor products analysis spends intelligence departments a lot of time, based on the analysis of the enterprise's competitor products analysis process, this paper extends the retrieval word by sematic extension. With the using the method of　internet information crawl, we extend and merge the products' names and products' marketing areas by sematic extension. At last, we can get accurate and comprehensive analysis result on competitors products.

Key words　ontology, competitor analysis, sematic extension, statistical analysis

钢铁本体的构建方法研究

印　康[1]　谷　俊[2]

(1. 华东师范大学商学院，上海　201143；
2. 上海宝山钢铁股份有限公司，上海　201900)

摘　要　领域本体在信息共享、数据集成和语义表达方面都具备强大的功能，近年来，越来越多的研究人员致力于领域本体的开发和应用中。目前在生物、医学、数字图书馆等研究领域都建立了标准的本体体系和许多可重用的本体资源库，使得领域专家能够使用它们来共享和利用领域中的信息。但是，本体在钢铁行业的应用成果却非常少，为了实现钢铁行业竞争情报的知识管理，本文采用领域本体来表示和组织钢铁领域知识，并针对钢铁行业资本密集的特点，探索切实可行的钢铁行业本体构建方法。首先介绍了领域本体的概念及常用构建方法，针对钢铁领域的特点建立适当的本体框架，通过延伸框架分支来完善本体的概念体系，并利用本体描述语言 OWL 和本体编辑工具 Protégé 5.0 构建了用于钢铁企业竞争情报分析的本体。

关键词　本体，钢铁领域，本体构建

Research of Iron and Steel Ontology Constructing Method

Yin Kang[1], Gu Jun[2]

(1.　School of Business, East China Normal University, Shanghai 201143, China;
2. Baosteel Co., Ltd., Shanghai 201900, China)

Abstract　Domain ontology has powerful function and semantic expression ways in the information sharing and data integration, in recent years, more and more researchers are committed to the development and application of the domain ontology. Currently in the research fields such as biology, medicine, digital library is the establishment of a standard system of ontology and many reusable ontology repository, the domain experts can use them to share and use the information in the field. However, the application of ontology in the steel industry results is very little, in order to realize the knowledge management of steel industry competitive intelligence, this paper USES the domain ontology to represent and organize steel domain knowledge, and according to the characteristics of the iron and steel industry capital intensive, explore the feasible ontology construction method of steel industry. First this paper introduces the concept of domain ontology and common building methods, according to the characteristics of the steel in the field of establishing appropriate ontology framework, by extending the framework branches to perfect the system of the concept of ontology, and by using ontology description language OWL ontology and editing tools protégé 5.0 build the ontology used in the iron and steel enterprise competitive intelligence analysis.

Key words ontology, steel sector, ontology constructing

新日铁住金在华钢管专利布局现状研究

王 刚 韩晓杰 朱婷婷

（上海宝山钢铁股份公司研究院，上海 201900）

摘 要 本文对新日铁住金公司在钢管产品与技术在华专利申请概况进行了研究，对其在钢管技术领域的专利布局情况进行了分类和统计，初步分析和阐述了新日铁住金公司在该领域的在华专利积累情况。根据其近年的专利技术方案，分析其技术布局和动向，挖掘出其技术领军人，概要性地整理了其技术发展路线，为我国国内钢管生产企业在该领域的研究策略提供参考。

关键词 新日铁住金，在华专利，无缝钢管，螺纹扣

The Research Status in China Steel Patent Layout NSSC

Wang Gang, Han Xiaojie, Zhu Tingting

(Shanghai Baoshan Iron & Steel Co., Ltd., Shanghai 201900, China)

Abstract This paper is a overview to patents in China of NSSMC about steel products and technology application on the steel pipe technology，and the patents layout were classified and statistical, preliminary analysis and elaboration. According to NSSMC patent technology plan in recent years, this paper analyzes the technical layout and trends, and digs out the technical development route and technology leaders, at last it provides a reference for the research of steel pipe enterprises in China.

Key words NSSMC, patent in China, seamless steel pipe, threaded joint

浅谈中央钢铁企业专利水平提升措施

魏建新

（武汉钢铁（集团）公司经济管理研究院，湖北武汉 430080）

摘　要　分析了专利技术对提升企业核心竞争力的意义，以及中央钢铁企业专利现状，提出了提升中央钢铁企业专利水平的对策措施：制定和实施专利战略与规划，进一步提升专利申请质量，加快国际化专利战略布局，有选择地进行对外专利实施许可与专利权转让，积极争创中国专利奖，建立健全专利的评价激励机制与管理体系，加强专利等知识产权文化建设。

关键词　专利，中央钢铁企业，对策措施

Discussion on the Countermeasures in Central State-owned Iron & Steel Enterprises Patent Level

Wei Jianxin

(Economic & Management Research Institute of WISCO, Wuhan 430080, China)

Abstract　This paper analyze the significance of patent technology to improve enterprise's core competitiveness, and the situation of patents in central state-owned iron and steel corporation, puts forward the measures to enhance the level of the central iron and steel enterprise patent: formulating and implementing the patent strategy and planning, further improving the quality of patent applications, speed up the internationalization of patent strategy, selectively to patent licensing and the patent right transfer, strive for China patent award, establishing the patent evaluation incentive mechanism and management system, strengthen the cultural construction of intellectual property.

Key words　paten, central state-owned iron & steel enterprises, countermeasures

汽车用高强塑积 TRIP 钢专利技术研发现状分析

陈 妍　李 侠　董 刚

（鞍钢股份有限公司技术中心，辽宁鞍山　114009）

摘　要　以全世界申请公开的强塑积在 24GPa%以上、强度在 700MPa 以上的汽车用 TRIP 钢专利为研究对象，对涉及的专利技术进行定量和定性分析，揭示其专利申请趋势、专利分布和专利技术研发趋势，研究不同申请人的技术开发路线、主要研究方向、专利布局以及研发技术空白点，为相关领域研发和战略布局提供支撑。

关键词　高强塑积，TRIP 钢，专利，研发现状

R&D Progress on High Strength-ductility Balance TRIP Steel for Automobile

Chen Yan, Li Xia, Dong Gang

(Technology Center, Angang Steel Co., Ltd., Anshan Liaoning 114009, China)

Abstract　The present paper studies on the patents applied for publicity worldwide of TRIP steel for the automobile. The strength-ductility balance of TRIP steel is above 24GPa%, while its strength is above 700MPa. Trough the quantitative and qualitative analysis of the related patent technology, the paper reveals the patents application trend, the patent distribution and the patent technology R&D trend. Meanwhile, the technology development route of various applicants, main research direction, patent layout and R&D technology void are researched in the paper. And it also provides support for the research and strategic layout of the related areas.

Key words high strength-ductility balance, TRIP steel, patent, research and development status

全球 Zn-Al-Mg 镀层钢板专利分析

周谊军 代云红 郑 瑞

（首钢技术研究院，北京 100043）

摘 要 本文采用 Orbit 专利分析平台，对 Zn-Al-Mg 镀层钢板领域全球专利发展趋势、主要国家及重点创新机构的专利布局和技术研发领域进行了分析。分析显示，日本是世界 Zn-Al-Mg 镀层钢板技术的引领者和主导者，我国近年来发展迅猛，但核心技术还很缺乏。本文以期帮助企业增强知识产权意识，了解 Zn-Al-Mg 镀层钢板产业发展的趋势和主要竞争对手专利申请情况，为技术合作及人才交流提供依据和参考。建议我国尽快制定 Zn-Al-Mg 镀层钢板领域知识产权发展战略，提高企业核心竞争力。

关键词 Zn-Al-Mg 镀层, 专利分析, 专利布局, Orbit

The Global Patents Analysis on Zn-Al-Mg Coated Steel Sheets

Zhou Yijun, Dai Yunhong, Zheng Rui

(Shougang Research Institute of Technology, Beijing 100043, China)

Abstract The profile of international patents of Zn-Al-Mg Coated Steel Sheets is analyzed from aspects of overall development trends, geographic distributions of priority and publication countries，key technology fields of top assignees and top inventors，based on Orbit patent analysis system database. It is found that Japan is technological leader in this field. China has been developing faster and faster in recent years，but lacks of key patents. This article intends to help Zn-Al-Mg Coated Steel Sheets enterprises to enhance their awareness of intellectual property right and understand the development trends and global patent application of Zn-Al-Mg Coated Steel Sheets, which is conducive to technical cooperation and personnel exchange. It is suggested that China should make plans on intellectual property development strategies for Zn-Al-Mg coated steel sheet to enhance the core competitiveness of enterprises.

Key words Zn-Al-Mg coating, patent analysis, geographic distributions of patents, Orbit

国内外特殊钢棒线材发展趋势分析

刘栋栋[1] 查春和[1] 杜婷婷[2]

（1. 冶金工业信息标准研究院，北京 100730；2. 冶金工业出版社，北京 100009）

摘 要 主要介绍了日本特殊钢棒线材的生产发展状况和日本神户制钢、JFE、新日铁等企业的新工艺，以及日本企业生产的一些新产品。此外，分品种详细论述了我国特殊钢棒线材的发展现状，并分析了我国特殊钢棒线材发展存在的问题，最后结合日本特殊钢棒线材发展经验，提出我国棒线材企业今后的发展方向。

关键词　特殊钢，棒线材，新技术，发展趋势

Production Status and Development of Special Steels Wire Rod and Bar in Domestic and Foreign Industry

Liu Dongdong[1], Zha Chunhe[1], Du Tingting[2]

(1. China Metallurgical Information and Standardization Institute, Beijing 100730, China;
2. Metallurgical Industry Press, Beijing 100009, China)

Abstract　The development of special steel wire rod and bar in Japan, the new technology and Some new products of Japanese Kobe Steel, JFE, Nippon Steel and other enterprises were introduced in this paper. In addition, this paper discussed and analyzed the development situation and problems of special steel wire rod and bar in China. At last, combined with the development experience of the special steel wire rod and bar in Japan, the future development direction of Chinese companies is put forward.

Key words　special steel, wire rod and bar, new technology, development

我国钢铁行业上市公司创新能力评价研究

王　诺[1]　谷　俊[2]

（1. 华东师范大学商学院，上海　200241；
2. 宝山钢铁股份有限公司，上海　201900）

摘　要　创新是企业发展进步的源泉，也是企业不断提高自身竞争力的重要因素，创新对于处于转型期间的钢铁企业更加重要。鉴于目前我国对钢铁企业创新能力评价方面的研究较少，为了找到提高钢铁企业创新能力的对策，本文研究了部分上市钢铁企业创新能力现状。通过本文的研究可以了解我国主要钢铁企业的创新能力现状，为钢铁企业今后的发展提供借鉴意义。本文根据创新相关理论构建了钢铁企业创新能力评价指标体系，并选取了我国钢铁行业上市企业进行了实证分析，运用因子分析法找到影响钢铁企业创新能力的关键因素，横向比较了钢铁企业之间的差距，并验证了评价模型的有效性，发现研发能力是钢铁企业创新能力的重要影响因子。最后根据评估结果针对不同的钢铁企业提高创新能力提出了相应的建议。

关键词　创新，钢铁企业，评价

Iron and Steel Industry Listed Company Innovation Ability Evaluation Research in China

Wang Nuo[1], Gu Jun[2]

（1. Business School, East China Normal University, Shanghai 200241, China；
2.　Baoshan Iron and Steel Co., Ltd., Shanghai 201900，China）

Abstract Innovation is the source of enterprise development and progress, but also an important factor for enterprises to continuously improve their own competitiveness, innovation is more important for iron and steel enterprises in the transition period. In view of the present research on the evaluation of the iron and steel enterprise innovation ability, In order to find the countermeasures to raise the capacity of iron and steel enterprise innovation, this paper studies the part listed steel companies innovation present situation. In this paper, we can understand the current situation of the innovation ability of the main iron and steel enterprises in China, and provide reference for the development of iron and steel enterprises in the future. According to the innovation theory, the evaluation index system of innovation capability of iron and steel enterprises is constructed. The key factors that influence the innovation capability of iron and steel enterprises are analyzed by means of factor analysis. Horizontal comparison of the gap between the iron and steel enterprises, verify the effectiveness of the evaluation model, found that R & D capability is an important factor in the innovation ability of iron and steel enterprises.

Key words innovation，iron and steel enterprises，evaluation

海工装备制造行业发展及海工平台用钢需求分析

郑 瑞 封娇洁 狄国标

（首钢技术研究院，北京 100043）

摘 要 全球海工装备市场近年保持活跃态势，我国政府对于海工装备产业的持续健康发展也非常重视，带动了海工装备产业的快速发展，也促进了海工平台用钢的需求和产品升级换代。文章概述了全球海工装备制造市场的竞争格局及发展趋势，着重介绍了国内海工装备的龙头企业和近期海工平台接单情况，指出目前国内海工装备新接订单主要集中在以中船重工、中远集团、中集来福士、招商局重工、中国船舶工业集团和振华重工6大船企为代表的企业手中。文章同时对国内外海工平台用钢的研发特点及发展趋势进行了分析，在此基础上对国内钢铁企业海工平台用钢的发展提出了建议。建议企业加强与行业联盟及下游优秀海工企业的合作，先期介入到下游行业，联合研发并推荐适用钢材，将满足用户的个性化需求视为企业未来发展新动力。

关键词 海工装备，钢板，海工平台

Analysis on the Development of Marine Equipment Manufacturing Industry and Offshore Platform Steel Demand

Zheng Rui, Feng Jiaojie, Di Guobiao

(Shougang Research Institute of Technology, Beijing 100043, China)

Abstract Global marine equipment market remained active in recent years, Chinese government also attached great importance to the sustained and healthy development of marine equipment industry，which led to the rapid development of marine equipment, also contributed to the demand for steel plate for offshore platform and product upgrades. The article summarized the overview of the competition pattern and development trend of global marine equipment manufacturing market, emphatically introduced the domestic leading enterprises and recent offshore platform orders，pointed out that new orders mainly concentrated in the CSIC, COSCO, Raffles, China Merchants Heavy Industry, China Shipbuilding Industry Corporation and Shanghai Zhenhua Heavy Industries. The article also analyzed characteristics and development trend of the steel plate for offshore platform. On this basis, several recommendations were made on the steel plate for offshore platform of China's steel industry. It was suggested that steel enterprises should cooperate with industry alliance and excellent down-stream enterprises, early intervene to downstream industries, and make meeting individual requirements as a new momentum in the future development of business.

Key words marine equipment, steel plate, offshore platform

大数据时代企业竞争情报系统构建
——以中国化信竞争情报平台为例

何洪优　石立杰　鲁　瑛

（中国化工信息中心，北京　100029）

摘　要　大数据时代，企业竞争情报工作面临新的机遇与挑战。竞争情报系统被越来越多地应用于企业竞争情报实践，为提升企业情报研究水平提供了有力的工具支持。本文以中国化信竞争情报平台为例，介绍了企业竞争情报系统的整体架构，并从信息采集、加工处理、分析挖掘和情报服务等方面提出了企业竞争情报系统的服务模式及构建建议，希望对我国企业开展竞争情报系统建设及管理工作有一定的借鉴意义。

关键词　企业情报服务，大数据，竞争情报系统，情报分析

Construction of Competitive Intelligence System for Corporation in Big Data Era: a Case Study of CNCIC CI System

He Hongyou, Shi Lijie, Lu Ying

(China National Chemical Information Centre, Beijing 100029, China)

Abstract　Under the big data circumstance, the competitive intelligence work for corporation faces a great deal of challenges and opportunities. This paper take CNCIC CI system as an example to give an introduction of the system architecture of the competitive intelligence system(CI system) for corporation, and give some suggestions for the service mode and the construction of the corporation CI system, which is a pioneer in the system construction and management of competitive intelligence work in Chinese enterprises.

Key words　corporation intelligence service, big data, competitive intelligence system, intelligence analysis

冶金信息网高级检索技巧介绍

杨宏章　付　静

（冶金工业信息标准研究院冶金信息研究所，北京　100730）

摘　要　本文从运算符、检索函数、特殊符号、保留字等几个方面的应用介绍了关于冶金信息网电子资源的高级检索技巧，这些技巧同样适用于其他以TRS(拓尔思)全文数据库存储的电子资源的检索。

关键词 冶金信息网，高级检索，运算符，检索函数，特殊符号，保留字

Advanced Retrieval Skills of Metalinfo.cn

Yang Hongzhang, Fu Jing

(China Metallurgical Information & Standardization Institute, Metallurgical Information Research Institute, Beijing 100730 ,China)

Abstract This paper presented the advanced retrieval skills about metalinfo.cn, in whichthe application of such as operator, search function, special symboland reserved wordis introduced.These skills are equally applicable to other electronic resources which are stored in TRS full text database.

Key words metalinfo.cn, advanced retrieval, operator, search function, special symbols, reserved word

油轮货油舱耐腐蚀钢专利分析

王俊海

（冶金工业信息标准研究院，北京 100730）

摘 要 介绍了油轮货油舱耐腐蚀钢的基本情况及使用背景，重点对新日铁住金、JFE、浦项以及中国的南钢、首钢、中国钢研集团、鞍钢、攀钢的油轮货油舱耐腐蚀钢的专利情况进行了分析。新日铁住金、神户制钢以及JFE拥有货油舱耐腐蚀钢领域的核心技术。浦项制铁、南钢和中国钢研集团在货油舱耐腐蚀钢领域的技术处于借鉴、突破攻关的研发阶段。通过专利申请情况分析看出，目前中国对货油舱耐腐蚀钢研发热度高，国内在该领域投入研发力量的企业较多。我国相关部门和钢铁生产企业关注国外竞争对手对中国的专利布局动向和其对抢占供中国市场的意图。

关键词 油轮货油舱用钢，耐腐蚀钢，专利分析，同族专利，IPC

Patent Analysis of the Corrosion Resistant Steel in the Cargo Oil Tank

Wang Junhai

(China Metallurgical Information and Standardization Institute, Beijing 100730, China)

Abstract This paper introduces the situation and application background of the corrosion resisting steel in the cargo oil tank. Patent analysis The corrosion resistant steel in the cargo oil tank about NSSC, JFE, POSCO and Nanjing Iron and Steel Co., Shougang steel Co., China Iron & Steel Research Institute Group,Anshan Iron and steel company,Panzhihua steel Co.Through the analysis of the patent application, the current domestic the corrosion resistant steel R & D attention, the domestic investment in the field more. The relevant departments of our country and the iron and steel production enterprises pay attention to foreign competitors to chinese patent distribution and its intention to seize the chinese market.

Key words steel for cargo oil tank, corrosion resistant steel, patent analysis, patent family, IPC

酒钢科技信息资源整合及平台建设构想

何成善

（酒泉钢铁（集团）有限责任公司，嘉峪关 735100）

摘　要　文章针对酒钢科技信息资源整合问题，阐述了科技信息资源整合的目的和必要性；分析了科技信息资源整合的目标；在预设科技信息资源整合实施原则的基础上，提出了科技信息资源整合网络平台要实现的基本功能；鉴于科技信息资源整合是一个企业全局性的考虑，文章结尾提出了酒钢实施科技信息资源整合的意见。

关键词　信息资源整合，数据库管理，信息规划，统一检索

The Conception of Sci-Tech Information Resources Integration and Platform Construction in JISCO

He Chengshan

(Jiuquan Iron & Steel (Group) Co., Ltd., Jiayuguan 735100, China)

Abstract　Aiming at sci-tech information resources integration in JISCO, the paper discussed the purpose and necessity of sci-tech information resources integration, analyzed the target of information resources integration, put forward the basic functions of integration platform on the basis of implementation principles. At last, the paper proposed the suggestions for information resources integration in view of the integrity of information resources integration for JISCO.

Key words　information resource integration, data-base management, information planning, unified index

浦项制铁汽车用钢专利分析

罗　晔[1]　王德义[2]　聂　闻[2]　季正明[3]

（1. 武汉钢铁(集团)公司研究院，湖北武汉　430080；

2. 中国冶金报社，北京　100029；

3. 无锡中检信安物联网检测技术有限公司，江苏无锡　214135）

摘　要　运用 Orbit 专利检索分析平台对浦项制铁汽车用钢的相关专利进行分析研究。结果显示：该项专利技术从 20 世纪 90 年代初才开始起步，2000 年以来迅速发展，但近几年专利成果申请数量明显减少。研发工作由浦项制铁主导，浦项产业科技研究院、韩国科学技术院及浦项科技大学共同参与，主要侧重于高锰镀锌薄钢板以及烘烤硬化冷轧钢板制造及轧制工艺方面的研究，该项专利技术主要在浦项制铁公司内部消化吸收，并为韩国科研部门和下游客户所应用，目前尚无同行竞争对手模仿，也未产生一定的国际影响力。

关键词　Orbit 专利平台，浦项制铁，汽车用钢，专利分析

Patent Analysis on POSCO's Steel for Automobile

Luo Ye[1], Wang Deyi[2], Nie Wen[2], Ji Zhengming[3]

(1. Research and Development Center of WISCO, Wuhan 430080, China;
2. China Metallurgical News, Beijing 100029, China;
3. CCIC-ISCCC IOT Testing Technology Co., Ltd., Wuxi 214135, China)

Abstract The statistical analysis of related patents about POSCO's steel sheets for automobile was based on the Orbit patent search and analysis platform. The results showed that this patent technology had barely started since the early 1990s, and developed rapidly since the year 2000,but recent years have witnessed a sharp decrease in the quantity of patents applied. The relevant R&D work was led by POSCO, with RIST , KAIST and POSTECH participating together, which mainly focused on the manufacturing method and rolling process for high manganese galvanied steel sheet and bake hardening cold-rolled steel sheet. The technology assimilation was largely achieved within the intra-company scope of POSCO, and the technology was only applied by some scientific research institutions and downstream vendors in Korea, at present, there is neither imitation by peer-competitors nor international influence to some extent.

Key words Orbit patent platform，POSCO，steel for automobile，patent analysis

移动“互联网＋”钢铁企业实时管理实践

贾生晖

（武汉钢铁股份有限公司冷轧薄板总厂，湖北武汉　430083）

摘　要　过去钢铁企业的成功主要靠的是外部机会与廉价资源。近几年随着互联网的兴起，许多企业患上了“互联网焦虑症”，拼命寻找所谓的“互联网思维”和“互联网转型模式”。其实，互联网精神的核心就是“平等、尊重、分享”加上“自主、掌控、意义”，这些均与人有关，尤其是与人的创造力有关。“群落”的核心就是以发挥人的创造力为本进行组织设计，应该成为新时代下组织变革的新方向。但能否变革成功，关键还在于企业领导者是否真正理解了互联网精神。本文以微信在企业管理中的应用实践，介绍了实现企业实时管理的一种途径。

关键词　互联网，群落，微信，企业管理

Mobile 'Internet +'：Real Time Management Practices in Iron & Steel Enterprise

Jia Shenghui

(Wuhan Iron & Steel Co., Ltd., Wuhan Hubei 430083, China)

Abstract In the past,the success of the iron and steel enterprises was mainly done by external opportunity and cheap resources. In recent years, along with the rise of the Internet, many enterprises suffer from anxiety neurosis, looking for

‘ Internet thought’ and ’Internet transformation model’ desperately . In fact, the core of Internet spirit is ‘equality, respect, sharing’ and ‘autonomy, control, meaning’, all these are related to people, especially related to creativity of people. The core of the ‘community’ is based on creativity of people to organize design, it should become to the new direction for organizational change in new era. But the key to success lies in wheather leaders understand the spirit of the Internet or not. Taking WeChat application in enterprise management practice, this paper introduces a way on enterprise management in real-time.

Key words Internet, community, WeChat, enterprise management

国内外企业在中国大陆地区薄带连铸领域的专利布局分析

王 强

(冶金工业信息标准研究院，北京 100730)

摘 要 对中国大陆地区薄带连铸专利信息进行了研究，重点分析了宝钢、蒂森克虏伯、浦项等国内外企业在中国大陆地区薄带连铸领域的专利保护情况。其中宝钢薄带连铸专利国内申请量为 138 件，保护领域涉及不锈钢、电工钢、侧封板、布流装置、铸造机、检测系统等，连同技术秘密一起构成了对薄带连铸技术的保护；东北大学薄带连铸专利国内申请量为 28 件，保护领域仅为电工钢薄带连铸技术及相关设备；蒂森克虏伯薄带连铸专利国内申请量为 43 件，保护领域涉及电工钢、不锈钢、侧封板、喷嘴、铸辊、铸造控制方法等，并且和于西纳有着良好的专利合作；浦项公司薄带连铸专利国内申请量为 18 件，保护领域涉及不锈钢、铸辊、控制领域，并且和浦项产业科学研究院有着良好的合作。

关键词 薄带连铸，专利申请，电工钢，不锈钢

Patent Analysis on Thin Strip Continuous Casting by the Enterprises Home and Abroad in Maimland China

Wang Qiang

(China Metallurgical Information and Standardization Institute, Beijing 100730, China)

Abstract The paper studied the patent information of thin strip continuous casting in mainland China, and analyzed emphatically the patent protection of thin strip continuous casting in mainland China for domestic and foreign enterprises such as Baosteel, ThyssenKrupp, POSCO. The number of patents for Baosteel reach 138, in the area of stainless steel、electrical steel、side dam、flow distribution device、cast machine、detection system.The patents and technical secret together protect the thin strip continuous casting technology for Baosteel. The number of patents for Northeastern University reach 138, in the area of thin strip continuous casting for electrical steel only; The number of patents for ThyssenKrupp reach 43, in the area of stainless steel、electrical steel、side dam、nozzle、casting roll、controlling of cast, and cooperation with USINOR in patent applications; The number of patents for POSCO reach 18, in the area of stainless steel、casting roll、controlling of cast, and cooperation with RES INST IND SCI&TECHNOLOGY at thin strip continuous casting technology.

Key words thin strip continuous casting, patent application，electrical steel，stainless steel

对中国企业“走出去”投资铁矿的思考

王晓波

（冶金工业规划研究院，北京 100711）

摘 要 本文通过分析未来全球铁矿石供求关系等影响铁矿石价格的因素，预测未来铁矿石价格走势，并分析了中国企业“走出去”开发铁矿的必要性，指出中国企业应抓住目前矿价处于低位的历史机遇期，走出去投资铁矿。同时，文章总结了近几年中国企业“走出去”的主要成功经验和失败教训，提出了中国企业未来在“走出去”时的相关建议。

关键词 中国企业铁矿，“走出去”，建议

Thinking on Chinese Enterprises Invest Overseasin Iron Ore Projects

Wang Xiaobo

(China Metallurgical Planning & Research Institute, Beijing 100711, China)

Abstract In this paper, by analyzing the factors which affect the price of iron ore in the future, such as the global iron ore supply and demand, to predict future prices of iron ore, and analyzing the necessity of Chinese enterprises go abroad to develop iron ore projects.It is pointed out that the Chinese enterprises should seize the opportunity that ironore price is currently at a low level of historical and go out investing iron ore projects. At the same time, this paper summarized the main successful experience and failure lessons of Chinese enterprises “going out” in recent years, and put forward some suggestions for Chinese enterprises to invest overseas in the future.

Key words Chinese enterprises, invest overseas, suggestions

“大数据”时代软科学研究单位对策研究

吴 瑾 文青英

（武钢经济管理研究院，湖北武汉 430080）

摘 要 文章首先对“大数据”的基本概念做了简要介绍，指出大数据具有数据量大、速度快时效高、数据类型多和价值密度低的特点。然后对软科学研究单位的数据分析目前所面临的问题进行了分析，揭示出6个方面的问题。接下来，针对这6个问题，文章从技术层面提出了4点应对策略：(1) 建立分类清晰的专业数据库，实现数据资源的有效共享；(2) 加强对多源数据的清洗与融合，提高数据的质量；(3) 注重对新的数据分析方法、技术和工具的应用，提高数据挖掘的效率；(4) 加强交流和调研，提高数据分析的针对性与应用价值。同时，文章从管理层面提出了 2 点应对策略：(1) 组建专业化的大数据研究团队；(2) 加强信息安全制度的建设。最后，文章指出，建立在数据分析和信息处理基础上的软科学研究单位应该抓住“大数据”带来的机遇、面对其赋予的挑战，以达到为企业

的经营决策提供更加科学依据的目的，从而不断提高企业在大数据时代的核心竞争力。

关键词　大数据，数据分析，信息处理，软科学研究单位

The Strategies of Soft Science Research Institution in "Big Data" Era

Wu Jin, Wen Qingying

(Economic & Management Research Institute of WISCO, Wuhan 430080, China)

Abstract　The paper briefly introduced the basic conception of "big data" at first, and pointed out the features of "big data": large quantity, high speed, diverse types, low density of value. Then, the paper analyzed the current problems of data analysis of soft science research institution, and revealed 6 questions. Next, the paper put forward 4 points of strategies at the technical level: (1) Establish specialized databases with clear classification to achieve effective sharing of data resources; (2) Strengthen cleaning and integration of multi-source data, to improve the quality of data; (3) Focus on the application of new methods, techniques and tools of data analysis, to improve the efficiency of data mining; (4) Strengthen the communication and investigation, to make data analysis more targeted and with applied value . At the same time, the paper put forward 2 points of strategies at the management level: (1) construct specialized research team of data analysis; (2) Strengthen the construction of information security system. At last, the paper pointed out, the soft science research institution should seize the opportunities and face the challenges, in order to provide a more scientific basis for the management decision of corporation, so as to continuously improve the core competitiveness of enterprises in the era of "big data".

Key words　big data, data analysis, information processing, soft science research institution

借力"一带一路"节点国家建设，中国钢铁走进哈萨克斯坦

鹿　宁　高　升

（冶金工业规划研究院，北京　100711）

摘　要　2014年5月，亚信峰会提出：中国将同各国一道，加快推进"丝绸之路经济带"和"21世纪海上丝绸之路"建设，尽早启动亚洲基础设施投资银行，更加深入参与区域合作进程，推动亚洲发展和安全相互促进、相得益彰。建设"一带一路"，是我国应对全球形势深刻变化、统筹国内国际两个大局作出的重大战略决策。与沿线新兴经济体和发展中国家以"和平、交流、理解、包容、合作、共赢"的精神开展互利合作，对外，可实现加快海外资源开发、转移优势产业、解决钢材贸易争端、加大技术和装备输出；对内，可统筹我国区域发展，实现东部经济发展能力持续增强，西部经济实力不断扩大的战略格局。"一带一路"鼓励中国钢铁过剩产能走出去，而哈萨克斯坦正需要大量钢材进口，同时引进外资在本国新建钢铁产能，以满足哈发展战略规划实施带来的钢材需求。中哈两国"一出、一进"的政策导向，为中国钢铁以海外建厂和钢材出口为途径，走进哈萨克斯坦创造了良好的机遇。同时，哈基础设施和工业发展滞后，钢材国际贸易争端加剧，因此，中国钢铁走进哈萨克斯坦也将面临不小的挑战。尽管我国钢铁走入哈萨克斯坦面临诸多威胁和不利环境，但哈国内巨大的钢材需求空间，对我国钢铁企业仍具诱惑。我国钢铁企业应借"一带一路"建设，通过海外建厂和钢材出口，将过剩产能输入哈萨克斯坦，在化解产能的同时，满足了哈基础设施建设和装备制造业钢材需求，同时带动哈国内钢铁企业装备、技术和

研发等整体水平的提升。从出口钢材品种来看，建议提高对哈棒线材和石油管出口比例；针对海外建厂，建议以合资方式为主，有利于快速熟悉环境、有限降低风险。

关键词 哈萨克斯坦，一带一路，化解产能

Chinese Steel Entered into Kazakhstan in Virtue of "One Belt, One Road"

Lu Ning, Gao Sheng

(China Metallurgical Industry Planning and Research Institute, Beijing 100711, China)

Abstract "One Belt, One Road" is a major strategic decision made by Chinese government to respond profound changes of global situation and to manage domestic and overseas status coordinately. Overcapacity of Chinese steel industry will be encouraged to move outside in virtue of implementation of "One Belt, One Road". And Kazakhstan needs to import large volume of steel, as well as to attract foreign investment to build new capacity satisfying steel demand increase caused by implementation of Kazakhstan Development Plan. Chinese steel companies should move overcapacity to Kazakhstan by means of overseas plants construction and steel export. This activity on the one hand will eliminate overcapacity of Chinese steel industry, on the other hand meet steel demand growth required by development of infrastructure and machinery industry in Kazakhstan as well as improve overall level of equipments, technologies and R & D of Kazakhstan steel industry. It is recommended to export more steel bar and oil pipe, and to establish steel plants through joint adventure to comply with market situation in Kazakhstan and to avoid any risk effectively.

Key words Kazakhstan, "One Belt, One Road", overcapacity elimination

基于光以太网铁路物流信息传输系统研制与应用

吴　非　韩庆宇　张红波

（鞍山钢铁集团公司铁路运输公司，辽宁鞍山　114021）

摘　要 随着时代的发展和铁路运输对信息的多样性、安全性、高速率的需求增长，对通信传输也提出了更高的要求。鞍钢铁路运输传输网络就存在着接口类型单一、传输速率低、保护方式不完善等问题。为适应铁运的发展，解决当前问题，基于光以太网铁路物流信息传输系统应运而生。新的系统使原有的多个独立系统能够更加紧密的融合在一起，结合网管统一管理，硬件设备更加简单、系统维护更加容易。

关键词 光以太网，信息传输，光环网

Based on Optical Ethernet Railway Logistics Information Transmission System Development and Application

Wu Fei, Han Qingyu, Zhang Hongbo

(Anshan Iron and Steel Group's Railway Transportation Company, Anshan Liaoning 114021, China)

Abstract With the development of era and railway transport to the diversity of information, security, high rate of demand growth, the ability of the transmission platform is also put forward higher requirements. Anshan iron and steel rail transport that existed in early railway transport network transmission equipment aging, lack of spare parts, model interface type single, low transmission rate, such problems as incomplete protection way. In order to adapt to the development of railway transport, and solve the current problems in transmission network, based on optical Ethernet railway logistics information transmission system arises at the historic moment. New system makes the original multiple independent system can more closely together, achieve complementary advantages, combined with the unified management of network, hardware device is more simple, system maintenance easier.

Key words optical Ethernet, information transmission, optical fiber ring network

国家科技图书文献中心会议文献服务现状分析

李春萌　王　梅

（冶金工业信息标准研究院，北京　100730）

摘　要　国家科技图书文献中心，简称 NSTL，目前在订文献 17000 多种，其中外文会议文献 8000 余种。NSTL 目前在订西文会议文献均在 NSTL 三期服务系统即 www.nstl.gov.cn 中有文摘级加工揭示。通过对 NSTL 三期服务系统中的尚未浏览和尚未原文传递服务等资源进行分析，查找尚未利用原因，通过语种分布分析、学科主题分布分析、馆藏分析等，找出尚未利用的原因。通过对 NSTL 三期系统中外文会议文献揭示程度、和功能的梳理，从而提出提高会议文献利用率的意见和建议，通过采取有针对性的措施，既发挥了国家科技图书文献中心的文献资源支撑和保障作用，又让用户更多地了解并使用国家科技图书文献中心订购的外文会议文献资源，从而提高 NSTL 经费的使用效率。建议外文会议文献利用率较低的成员的单位，加大资源的宣传力度；NSTL 三期服务系统进一步完善会议文献的题录、书目等揭示功能；加大尚未利用学科行业领域宣传力度，通过一系列措施提高外文会议文献的利用率。

关键词　NSTL，会议文献，利用，服务

The Present Situation Service Analysis of the NSTL Conference Proceedings

Li Chunmeng，Wang Mei

(China Metallurgical Information & Standardization Institute, Beijing 100730, Chian)

Abstract The National Science and Technology Library (NSTL) had more than 17000 kinds of documents such as serials, conference proceedings, reports, and etc. And there are over 8000 foreign language conference literature at present. All of the foreign conference proceedings has been processed. Users can see the titles and abstracts in the NSTL website (www.nstl.gov.cn) for free. The whole paper can be given to the users by email in 24 hours only if users use for personal research purpose. By analysis of the conference of the proceedings which are not be browsed and full-text demanded, we can know the characteristic of the unused papers. All of the data are just draw-out from the NSTL website system. The characteristic of the unused papers through analyzing the language, the industry subjects, the organization were found. The Research has yet to make use of the results to improve the rates of the papers, which indirectly reveal the NSTL resources supporting function, and let much more users know the conference proceedings NSTL has. Opinions and Suggestions for improving the utilization ratio of conference literature are launched. The members of the NSTL whose conference proceedings are used more worse, more effort they should take and resource of propaganda. The conference proceedings should be further processed on website of the NSTL service system such as catalogue and bibliography. Industries which conference proceedings unused much should be advertised and informed relevantly. By these means, the utilization rates of the foreign language conference proceedings can be improved.

Key words The National Science and Technology Library (NSTL), conference proceedings, use, service

专业图书馆利用微信开展信息服务的思考

陈 琦

（冶金工业信息标准研究院，北京 100730）

摘 要 微信是一种建立在通讯网络平台上，以手机为终端用户的新兴的信息传播方式。随着社会的发展，人们对信息获取的快捷性和便利性提出了更高的要求，微信在满足大众普通信息交流需求的同时，也为图书馆开展个性化的信息服务提供了一个全新的服务渠道。专业图书馆利用微信开展信息服务，是图书馆个性化服务模式的一种发展趋势。本文介绍了微信及其特点；分析了目前图书馆应用微信开展信息服务的现状，即微信作为一种信息聚合平台，主要可为图书馆用户提供实时在线咨询、FAQ 虚拟参考咨询、信息推送、信息共享等服务；提出了专业图书馆利用微信深化信息服务的策略，包括通过宣传推广扩大用户群、通过需求调研严选发布内容、开展信息咨询服务、开展主动信息推送服务等。

关键词 微信，专业图书馆，信息服务

Thinking of Professional Library Use WeChat to Develop the Information Services

Chen Qi

(China Metallurgical Information & Standardization Institute, Beijing 100730, China)

Abstract WeChat is a kind of based on communication network platform, to end users of mobile phones for emerging way of information dissemination. With the development of society, fast speed and convenience of access to information put forward higher request, WeChat in meet the demand of mass general information exchanges at the same time, also for the library to carry out the individualized information service provides a brand new service channel.Professional library information service conducted by WeChat is a development trend of library personalized service model.This paper introduces WeChat and its characteristics;analyses the current application of WeChat in library information service, the WeChat as an information aggregation platform, the main can provide real-time online consulting, faqs for library users virtual reference, information pushing, information sharing and other services;puts forward the professional library using WeChat deepening information service strategies, including through promotional expand user base, through the demand research strictly selected content, information consulting services, to carry out active information push service, etc. Application of WeChat public platform to carry out the professional library information services is still in its infancy, many aspects of development is not mature. With the continuous development of Internet technology, the WeChat will be constantly improve, professional library should make full use of WeChat public platform powerful features, to carry out the service marketing, to provide users with more convenient high quality service.

Key words WeChat，professional library，information services

连铸结晶器流动控制专利分析研究

韩晓杰　王　刚

（上海宝山钢铁股份公司研究院，上海　201900）

摘　要　在钢水的连铸过程中，结晶器内钢液流动不仅关系到结晶器的传热、夹杂物的上浮而且还影响铸坯的表面和内部质量。本文针对连铸结晶器流动控制相关专利文献进行分析，从专利分析角度说明相关技术研究动态和发展趋势。

关键词　结晶器，流动控制，专利分析

Patent Analysis of Flow Control in Continuous Casting Mold

Han Xiaojie, Wang Gang

(Baoshan Iron & Steel Co., Ltd., Research Institute, Shanghai 201900, China)

Abstract　In the continuous casting process of molten steel, molten steel flow in mold is related to heat transfer of mold, flotation of inclusions,the surface quality and internal quality of slab.We analyze the patent papers related to flow control in continuous casting mold,and the related technology research trends and development trends are explained from the perspective of patent analysis.

Key words　mold, flow control, patent analysis

企业信息资源网络保障平台的构建

赖碧波　姚昌国

（武汉钢铁（集团）公司研究院，武汉　430080）

摘　要　企业信息资源网络保障平台的构建，是企业信息化的重要组成部分，也是企业科研、经营决策的基础保障。本文从信息资源网络保障平台构建背景、平台构建、平台实施应用效果等三个方面进行了讨论。

关键词　信息资源保障，信息资源构建，科研，经营决策

On Construction of Information Resource Platform for Enterprise

Lai Bibo, Yao Changguo

(R&D Center of WISCO, Wuhan 430080, China)

Abstract The construction of enterprise information resource platform is not only one key parts of the informatization, but also abasis for the R&D, management and decision-making.The background, construction and inplimentary of the information resouirce platform, these three parts were discussed in this paper.

Key words information resource,information resource platform, research and development, management and decision-making

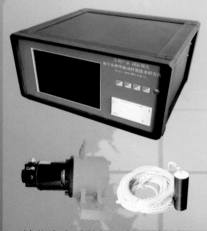

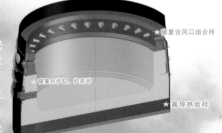